Water Reclamation Technologies for Safe Managed Aquifer Recharge

Water Reclamation Technologies for Safe Managed Aquifer Recharge

Christian Kazner, Thomas Wintgens and Peter Dillon

Publishing
London • New York

Published by

IWA Publishing
Alliance House
12 Caxton Street
London SW1H 0QS, UK
Telephone: +44 (0)20 7654 5500
Fax: +44 (0)20 7654 5555
Email: publications@iwap.co.uk
Web: www.iwapublishing.com

First published 2012

British Library Cataloguing in Publication Data
A CIP catalogue record for this book is available from the British Library

Library of Congress Cataloging-in-Publication Data
A catalog record for this book is available from the Library of Congress

ISBN 9781843393443 (Paperback)
ISBN 9781780400648 (eBook)

Contents

List of contributors

Christian ABEGGLEN, EAWAG Swiss Federal Institute for Environmental Science and Technology, Überlandstrasse 133 – PO Box 611, 8600 Dübendorf, Switzerland

Avi AHARONI, Mekorot Water Company Ltd., 9 Lincoln Street, P.O. Box 2012, Tel Aviv 61201, Israel

Gary AMY, Unesco IHE Institute for Water Education, Department of Urban Water and Sanitation, Westvest 7, 2611 AX Delft, The Netherlands
King Abdullah University of Science and Technology, KAUST, Water Desalination and Reuse Center, Al-Jazri Bldg Office 4231, Thuwal 23955-6900, Saudi Arabia

Josef ASMIN, EAWAG Swiss Federal Institute for Environmental Science and Technology, Überlandstrasse 133 – PO Box 611, 8600 Dübendorf, Switzerland

M. Neus AYUSO-GABELLA, Faculty of Pharmacy, University of Barcelona, Av. Joan XXIII, s/n, 08028 Barcelona, Spain

Susan BAGGETT, Cranfield Water Science Institute, Cranfield University, Cranfield, Beds MK43 0AL, United Kingdom

Jacob BEAR, Faculty of Civil & Environmental Engineering, Technion - Israel Institute of Technology, Haifa 32000, Israel

Jessica BENNER, Federal Institute of Hydrology BFG, Am Mainzer Tor 1, D-56068 Koblenz, Germany

Jacob BENSABAT, Environmental & Water Resources Engineering Ltd., P.O. Box 6770, 31067 Haifa, Israel

Axelle BESANCON, Cranfield Water Science Institute, Cranfield University, Cranfield, Beds MK43 0AL, United Kingdom

Davide BIXIO, European Commission, DG ELARG

Uta BÖCKELMANN, Technische Universität Berlin, Strasse des 17. Juni 135, 10623 Berlin, Germany

Marc BÖHLER, EAWAG Swiss Federal Institute for Environmental Science and Technology, Überlandstrasse 133 – PO Box 611, 8600 Dübendorf, Switzerland

Serge BROUYÈRE, Université de Liège, Chemin des Chevreuils 1, B-4000 Liège 1, Belgium

Johan CAUWENBERGHS, Aquafin NV, Dijkstraat 8, B-2630 Aartselaar, Belgium

Elise CARTMELL, Cranfield Water Science Institute, Cranfield University, Cranfield, Beds MK43 0AL, United Kingdom

Alma CHÁVEZ MEJÍA, Institute of Engineering, UNAM, Ciudad Universitaria, Coyoacan, México D.F., C.P. 04510, D.F.Mexico

Haim CHIKUREL, Mekorot Water Company Ltd., 9 Lincoln Street, P.O. Box 2012, Tel Aviv 61201, Israel

Bart DE HEYDER, Aquafin NV, Dijkstraat 8, B-2630 Aartselaar, Belgium

Wouter DE WILDE, Aquafin NV, Dijkstraat 8, B-2630 Aartselaar, Belgium

Peter DILLON, CSIRO Land and Water - Waite Campus, Waite Road - Gate 4, Glen Osmond SA 5064, Australia

Mathias ERNST, Centre for Water in Urban Areas, Technische Universität Berlin, Strasse des 17. Juni 135, 10623 Berlin, Germany

Guido FINK, Federal Institute of Hydrology BFG, Am Mainzer Tor 1, D-56068 Koblenz, Germany

Irina GAUS, Unité Traceurs Isotopiques et Datations, BRGM Service MMA, Bureau de Recherches Geologiques et Minieres, 3 av. C. Guillemin B.P. 36009, F-45060 Orléans cedex 2, France

Bettina GENTHE, CSIR Natural Resources and the Environment, P O Box 320, ZA 7599 Stellenbosch, South Africa

Richard GIBSON, Institute of Engineering, UNAM, Ciudad Universitaria, Coyoacan, México D.F., C.P. 04510, D.F. Mexico

Radu Constantin GOGU, GHG GeoHidroConsult S.R.L., Bd. 1 Decembrie 1918 nr 47, Sector 3, Bucuresti, Romania

Elisabeth GROHMANN, Technische Universität Berlin, Strasse des 17. Juni 135, 10623 Berlin, Germany

Yossi GUTMANN, Mekorot Water Company Ltd., 9 Lincoln Street, P.O. Box 2012, Tel Aviv 61201, Israel

Arne HEIN, Chair of Water Quality Control, Technische Universität Berlin, Strasse des 17. Juni 135, 10623 Berlin, Germany

Rita HOCHSTRAT, RWTH Aachen University, Department of Chemical Engineering, Turmstrasse 46, 52056 Aachen, Germany

Juliane HOLLENDER, EAWAG Swiss Federal Institute for Environmental Science and Technology, Überlandstrasse 133 – PO Box 611, 8600 Dübendorf, Switzerland

Bruce JEFFERSON, Cranfield Water Science Institute, Cranfield University, Cranfield, Beds MK43 0AL, United Kingdom

Paul JEFFREY, Cranfield Water Science Institute, Cranfield University, Cranfield, Beds MK43 0AL, United Kingdom

Martin JEKEL, Chair of Water Quality Control, Technische Universität Berlin, Strasse des 17. Juni 135, 10623 Berlin, Germany

Blanca JIMÉNEZ CISNEROS, Institute of Engineering, UNAM, Ciudad Universitaria, Coyoacan, México D.F., C.P. 04510, D.F.Mexico

Adriano JOSS, EAWAG Swiss Federal Institute for Environmental Science and Technology, Überlandstrasse 133 – PO Box 611, 8600 Dübendorf, Switzerland

Claus JØRGENSEN, DHI, Department of Environmental Riskassessment, Agern Allé 5, 2970 Hørsholm, Denmark

Christian KAZNER, Centre for Technology in Water and Wastewater, University of Technology Sydney (UTS), PO Box 123, Broadway, NSW 2007, Australia (former affiliation: RWTH Aachen University)

Wolfram KLOPPMANN, Unité Traceurs Isotopiques et Datations, BRGM Service MMA, Bureau de Recherches Geologiques et Minieres, 3 av. C. Guillemin B.P. 36009, F-45060 Orléans cedex 2, France

Irena KOPAC, Institute for Ecological Engineering IEI, Ljubljanska ulica 9, 2000 Maribor, Slovenia

Martin KRAUSS, EAWAG Swiss Federal Institute for Environmental Science and Technology, Überlandstrasse 133 – PO Box 611, 8600 Dübendorf, Switzerland

Sarah KREMER, CSIRO Land and Water - Waite Campus, Waite Road - Gate 4, Glen Osmond SA 5064, Australia

Guiseppe LAERA, CNR (National Research Council), Istituto di Ricerca Sulle Acque - Sezione di Bari, Via F. De Blasio, 5, 70123 Bari, Italy

Winson LAY, PUB - Public Utilities Board, 40 Scotts Road, Environment Building, Singapore 228231

Kristell LECORRE, Advanced Water Management Centre, University of Queensland, Research Road, Brisbane QLD 4072 (former affiliation: Cranfield University)

Lai Yoke LEE, Environmental Science & Engineering, Faculty of Engineering, National University of Singapore, 1 Engineering Drive 2, E1A-02-19, Singapore 117576

Caterina LEVANTESI, Consiglio Nazionale delle Ricerche, Istituto di Ricerca Sulle Acque, CNR - Water Research Institute, Area della Ricerca Roma 1, Montelibretti, 00015 Monterotondo (RM), Italy

João Paulo LOBO-FERREIRA, Laboratório Nacional de Engenharia Civil, Hydraulics and Environment Department, Groundwater Division, Av. do Brasil 101, PT-1700-066 Lisboa, Portugal

Catalina MAYA, Institute of Engineering, UNAM, Ciudad Universitaria, Coyoacan, México D.F., C.P. 04510, D.F. Mexico

Costantino MASCIOPINTO, CNR (National Research Council), Istituto di Ricerca Sulle Acque - Sezione di Bari, Via F. De Blasio, 5, 70123 Bari, Italy

Christa MCARDELL, EAWAG Swiss Federal Institute for Environmental Science and Technology, Überlandstrasse 133 – PO Box 611, 8600 Dübendorf, Switzerland

Thomas MELIN, RWTH Aachen University, Department of Chemical Engineering, Turmstrasse 46, 52056 Aachen, Germany

Konrad MIOTLINSKI, CSIRO Land and Water - Waite Campus, Waite Road - Gate 4, Glen Osmond SA 5064, Australia

Melanie MURO, Cranfield Water Science Institute, Cranfield University, Cranfield, Beds MK43 0AL, United Kingdom

How Yong NG, Environmental Science & Engineering, Faculty of Engineering, National University of Singapore, 1 Engineering Drive 2, E1A-02-19

Declan PAGE, CSIRO Land and Water - Waite Campus, Waite Road - Gate 4, Glen Osmond SA 5064, Australia

Paul PAVELIC, International Water Management Institute, Regional Office for South Asia Hyderabad Office, c/o International Crops Research Institute for the Semi-Arid Tropics (ICRISAT), 401/5, Patancheru 502324, Andhra Pradesh, India

Marie PETTENATI, Unité Traceurs Isotopiques et Datations, BRGM Service MMA, Bureau de Recherches Geologiques et Minieres, 3 av. C. Guillemin B.P. 36009, F-45060 Orléans cedex 2, France

Géraldine PICOT-COLBEAUX, Unité Traceurs Isotopiques et Datations, BRGM Service MMA, Bureau de Recherches Geologiques et Minieres, 3 av. C. Guillemin B.P. 36009, F-45060 Orléans cedex 2, France

Alfieri POLLICE, CNR (National Research Council), Istituto di Ricerca Sulle Acque - Sezione di Bari, Via F. De Blasio, 5, 70123 Bari, Italy

Mohammed Aziz RAHMAN, Geoscience Centre of the University of Göttingen, Dept. Applied Geology, Goldschmidtstr. 3, 37077 Göttingen, Germany

Rudi REGEL, United Water, 180 Greenhill Rd, Parkside, SA, 5063, Australia

Georg REIFFERSCHEID, Federal Institute of Hydrology BFG, Am Mainzer Tor 1, D-56068 Koblenz, Germany

Thorsten REEMTSMA, Chair of Water Quality Control, Technische Universität Berlin, Strasse des 17. Juni 135, 10623 Berlin, Germany

Stephanie RINCK-PFEIFFER, United Water, 180 Greenhill Rd, Parkside, SA, 5063, Australia

Bernd RUSTEBERG, Geoscience Centre of the University of Göttingen, Dept. Applied Geology, Goldschmidtstr. 3, 37077 Göttingen, Germany

Muath Abu SADAH, Palestinian Hydrology Group, PO Box 565, Al-Nahdah Bld., Emile Habibi St, Al-Masyoun, Ramallah

Miquel SALGOT, Faculty of Pharmacy, University of Barcelona, Av. Joan XXIII, s/n, 08028 Barcelona, Spain

Martin SAUTER, Geoscience Centre of the University of Göttingen, Dept. Applied Geology, Goldschmidtstr. 3, 37077 Göttingen, Germany

Michael SCHLÜSENER, Federal Institute of Hydrology BFG, Am Mainzer Tor 1, D-56068 Koblenz, Germany

Saroj K. SHARMA, UNESCO IHE Institute for Water Education, Department of Urban Water and Sanitation, Westvest 7, 2611 AX Delft, The Netherlands

Harry SEAH, PUB - Public Utilities Board, 40 Scotts Road, Environment Building, Singapore 228231

Hansruedi SIEGRIST, EAWAG Swiss Federal Institute for Environmental Science and Technology, Überlandstrasse 133 – PO Box 611, 8600 Dübendorf, Switzerland

Valter TANDOI, Consiglio Nazionale delle Ricerche, Istituto di Ricerca Sulle Acque, CNR - Water Research Institute, Area della Ricerca Roma 1, Montelibretti, 00015 Monterotondo (RM), Italy

Guihe TAO, PUB - Public Utilities Board, 40 Scotts Road, Environment Building, Singapore 228231

Thomas TERNES, Federal Institute of Hydrology BFG, Am Mainzer Tor 1, D-56068 Koblenz, Germany

Chris THOEYE, Aquafin NV, Dijkstraat 8, B-2630 Aartselaar, Belgium

Simon TOZE, CSIRO EcoSciences Precinct - Dutton Park, 41 Boggo Road, Dutton Park QLD 4102, Australia

Gideon TREDOUX, CSIR Natural Resources and the Environment, P O Box 320, ZA 7599 Stellenbosch, South Africa

Urs VAN GUNTEN, EAWAG Swiss Federal Institute for Environmental Science and Technology, Überlandstrasse 133 – PO Box 611, 8600 Dübendorf, Switzerland

Emmanuel VAN HOUTTE, I.W.V.A., Doornpannestraat 1, B-8670 Koksijde, Belgium

Marjoleine WEEMAES, Aquafin NV, Dijkstraat 8, B-2630 Aartselaar, Belgium

Thomas WINTGENS, Institute for Ecopreneurship, School of Life Sciences, University of Applied Sciences and Arts Northwestern Switzerland, Gründenstrasse 40, CH-4132 Muttenz, Switzerland
RWTH Aachen University, Department of Chemical Engineering, Turmstrasse 46, 52056 Aachen, Germany

Piotr WOJDA, Université de Liège, Chemin des Chevreuils 1, B-4000 Liège 1, Belgium

Liang YU, RWTH Aachen University, Department of Chemical Engineering, Turmstrasse 46, 52056 Aachen, Germany

Xuan ZHAO, Division of Environmental Technology, INET Institute of Nuclear and New Energy Technology, Tsinghua University, Beijing 100084, China

Cheng Xu ZHOU, Division of Environmental Technology, INET Institute of Nuclear and New Energy Technology, Tsinghua University, Beijing 100084, China

Acknowledgements

This book is based on European research activities in the field of Managed Aquifer Recharge funded by the European Commission during the 6th framework programme. It presents most of the project outcomes of the EU project RECLAIM WATER (www.reclaim-water.org, Contract-No. 018309) and selected highlights of the work from the EU project GABARDINE (www.gabardine-fp6.org). The projects were Specific Targeted Research Projects supported by the European Commission under the Thematic Priority 'Global Change and Ecosystems' of the 6th Framework Programme. The RECLAIM WATER project started in October 2005 and ran till December 2008.

Partners of the RECLAIM WATER project:

RWTH Aachen University – Department of Chemical Engineering, Germany, Christian Kazner, Diana Marquardt, Thomas Wintgens, Liang Yu, Rita Hochstrat, Gerhard Spalding, Thomas Melin

Consiglio Nazionale delle Ricerche - Istituto di Ricerca Sulle Acque, Italy, Guiseppe Laera, Rosanna La Mantia, Caterina Levantesi, Costantino Masciopinto, Alfieri Pollice, Valter Tandoi, Antonio Lopez

Technische Universität Berlin, Germany, Uta Böckelmann, Mathias Ernst, Elisabeth Grohmann, Arne Hein, Thorsten Reemtsma, Ulrich Szewzyk, Martin Jekel

Swiss Federal Institute for Environmental Science and Technology EAWAG, Switzerland, Josef Asmin, Christian Abegglen, Marc Böhler, Urs van Gunten, Adriano Joss, Martin Krauss, Christa McArdell, Juliane Hollender, Hansruedi Siegrist

Cranfield University – School of Water Science, United Kingdom, Axelle Besancon, Elise Cartmell, Kristell LeCorre, Susan Baggett, Melanie Muro, Bruce Jefferson, Paul Jeffrey

University of Barcelona, Spain, M. Neus Ayuso-Gabella, Miquel Salgot

DHI Water & Environment, Denmark, Bjørn Kaare Jensen, Claus Jørgensen

Institute for Ecological Engineering IEI, Slovenia, Irena Kopac

Ribo Technologies BV, The Netherlands, Michel van der Rest

Mekorot Water Company, Israel, Avi Aharoni, Haim Chikurel, Tom Kraitzer, Yossi Gutmann

UNESCO IHE, The Netherlands, Saroj K. Sharma, Gary Amy

Federal Institute of Hydrology BFG, Germany, Jessica Benner, Guido Fink, Georg Reifferscheid, Michael Schlüsener, Thomas Ternes

Tsinghua University INET, China, Cheng Xuzhou, Wang Jianlong, Zhao Xuan

Bureau de Recherches Geologiques et Minieres, France, Irina Gaus, Marie Pettenati, Géraldine Picot-Colbeaux, Wolfram Kloppmann

Aquafin NV, Belgium, Davide Bixio, Johan Cauwenberghs, Bart De heyder, Marjoleine Weemaes, Wouter de Wilde, Chris Thoeye

United Water, Australia, Rudi Regel, Stephanie Rinck-Pfeiffer

Commonwealth Scientific and Industrial Research Organisation - CSIRO Land and Water, Australia, Sarah Kremer, Konrad Miotlinski, Declan Page, Paul Pavelic, Simon Toze, Peter Dillon

National Autonomous University of Mexico UNAM – Institute of Engineering, Mexico, Richard Gibson, Catalina Maya, Alma Chávez Mejía, José Antonio Barrios Pérez, Blanca Jiménez Cisneros

CSIR – The Council for Scientific and Industrial Research, South Africa, Bettina Genthe, Gideon Tredoux

PUB – Public Utilities Board, Singapore, Kiran Kekre, Winson Lay, Puah Aik Num, Guihe Tao, Balakrishnan Viswanath, Harry Seah

National University of Singapore - Centre for Water Research, Singapore, Jiang Yong Hu, Lai Yoke Lee, Say Leong Ong, How Yong Ng

The support of the project through the utilities of the case study sites is gratefully acknowledged. Emmanuel van Houtte (IWVA/Belgium) is acknowledged for the very fruitful collaboration. The project partners are particularly thankful to the European Commission DG RTD project officers Avelino Gonzalez-Gonzalez and Panagiotis Balabanis for guidance and support throughout the project.

Foreword

Dr. Panagiotis Balabanis

Promoting research on innovative water technologies and developing affordable solutions for protecting our water resources and reverse the trend of their deterioration in Europe and worldwide is necessary for providing safe drinking water to the people, for sustaining our environment and for achieving Europe's sustainable growth and competiveness.

The complexity of current water resources problems, the need to help people affected by water shortages and lack access to safe drinking water, the need to cope with the deterioration of current water assets in urban areas, the design of appropriate waste water treatment systems, water reclamation and reuse schemes, the improvement of water efficiency in the industrial and agricultural sectors and the need to cope with mitigation and adaptation of climate change, are some examples, where technological research is needed, together with appropriate strategies and policies for sustainable water resources management and environmental protection.

Investing in technology and innovative solutions and strengthening public-private research funding in water would be essential for the development and competiveness of the European water sector, which is a major economic player.

Research on water technologies has been supported since the beginning of the European Union's environmental research programmes. However, this area of research has progressively gained further importance since the late 90ies and has been strengthened at the end of 6th EU Framework Programme for Research and Technological Development, with the launching of the European Environment Technology Action Plan (ETAP) and the establishment of the Water Supply and Sanitation Technology Platform (WssTP). With the emphasis on research, innovation and knowledge society for achieving a more dynamic, innovative and attractive Europe given in FP7, water technology research becomes a priority of the Environment Theme of the FP7 Cooperation programme. Activities in the field of water technology are relevant to the Water Framework Directive (WFD) and other water related policies (e.g. Nitrates Directive, Drinking Water Directive, Urban Waste Water Directives, Groundwater Directive, Flood Directive, EU action plan on Water Scarcity and Drought, etc). They also support other EU policies, for example, integrated pollution prevention and control (IPPC Directive), EU action plan on sustainable consumption and production and on sustainable industrial policy, as well as on policies related to climate change mitigation and adaptation. They also help to boost innovation in the water sector and overcome the various barriers that hinder their development and use of water technologies, thus contributing to the implementation of the Environmental Technologies Action Plan. With the additional importance given recently in the Europe 2020 strategy for the development of European Innovation Partnerships, it is evident that research on water technologies will have additional policy relevance in the near future.

I am very pleased to introduce this book which provides a synthesis of the main results of the RECLAIM WATER EU funded research project for many reasons. RECLAIM WATER addressed an important research concept, that is, managed aquifer recharge, which attracts more and more interest, not only from water resources managers but also from policy-makers who are seeking for a more knowledge based, effective and efficient implementation and further development of water policies. It gathered together key leading European universities and research centers working in water resources research and key scientists who paid equal attention not only to the scientific and technical aspects of their work, but made also a particular effort in highlighting the policy implication of their results, thus contributing in strengthening the link between, science, research and policy. In fact, from the results of the projects, it becomes evident that solutions to water stress problems are available, sustainable, economical and safe. Finally, RECLAIM WATER, helped to strengthen the international dimension of EU research by bringing together case studies and researchers between Europe and international cooperation partner countries.

Dr. Panagiotis Balabanis
Deputy Head of Unit
Directorate General for Research and Innovation
European Commission

Foreword

Dr. Valentina Lazarova

Water is often viewed as the symbol of life and was essential for the development of the great civilisations such as Crete, Egypt, Greece, Rome, and Angkor. As in the past, during the last three centuries, water had a crucial role for the industrialisation and the advance in science. Despite the progress in science and engineering, water plays and will play an even more important role for the human development in this millennium.

Water not only allows life to exist, helping people to produce food, goods and communicate, but also is very important for human wellbeing as part of our culture, art, music and architectural design. Probably the most fascinating interpretation of the variety and beauty of water is provided by Claude Monet in his painting, in particular the variations of light and colours on the water-lily pond in Giverny. Other artists were more fascinated by the destructive capacity of water with great storms, terrifying swirling waters and impressive waves. One of the most famous of them, Leonardo da Vinci, devoted also great efforts on innovative engineering solutions to control water.

Aquifer recharge is one of the most promising engineering solutions, enabling both to control the destructive capacity of water and preserve its benefits. Artificial recharge of aquifers is a quite recent technique, developed between 1950 and 1960 in Europe and the USA, which importance is expected to increase with climate change. It is important to stress that the most important impacts of climate change are on water, in terms of extreme events of floods and droughts, which will dramatically affect the population, environment and economic development.

During the last decade, aquifer recharge with recycled water is becoming one of the most rapidly growing water reuse applications due to the unique features of polishing, storage and transportation, as well as the lost of the "waste" identity of water. Various other fields of application are also developed, such as riverbank filtration, surface water storage, storage and polishing of storm water, control of saltwater intrusion into coastal aquifers. Despite the growing interest, aquifer recharge is not yet a common practice and remains a complex solution because the strong influence of local conditions, potential health risks and lack of regulations.

This book is the result of the tremendous effort of a large multi-disciplinary team to analyse the lessons learned from the existing would experience and latest advance in science and practice on managed aquifer recharge (MAR). A comprehensive analysis and guidance are provided, including:

- Recommendations for design and operation of treatment trains and aquifer systems,
- Assessment of methods and tools for water quality analysis, decision support tools and risk assessment and management,
- Examples of best practices.

In few words, this handbook demonstrates how a well designed and operated MAR project can be used as a water stress mitigation option regardless of the context. This outstanding contribution was made possible thanks to the holistic research approach and the substantial collaborative effort. A standard analysis protocol was used and applied for various recharge systems, including direct injection or infiltration using different alternative water sources and a number of pre-treatment options. It was demonstrated that in-depth knowledge of hydrogeological characteristics and the specific site conditions are essential for the success of a planned recharge scheme, while demonstration studies and water quality control strategy play an important role for health safety and public acceptance.

Even though this ambitious research program was funded by the European Commission, the research consortium included several research teams and field studies outside Europe, from Australia, Asia, Africa and America. The scientific relevance of the recommendations was ensured by the significant research efforts from several leading universities and technology institutes, while the design, economic viability and operating parameters of various options of MAR were warranted by experienced technology providers and enterprises.

This book summarises condensed years of research and practical experience of a broad multi-disciplinary international coalition. Despite the strong professionalism and sound scientific evidence, the broad implementation of water reuse practices, such as managed aquifer recharge, needs motivation, patience and perseverance. You can read, share and implement the provided recommendations, taking in view that each project is case specific and calls for prudent decisions regarding public health, environmental impacts, economic and financial concerns, as well as social and legal aspects.

Dr. Valentina Lazarova
Chair of the IWA Specialist Group on Water Reuse

Chapter 1

Introduction

Christian Kazner, Thomas Wintgens, Thomas Melin and Peter Dillon

1.1 THE IMPORTANCE OF MANAGED AQUIFER RECHARGE

Groundwater represents over 97% of all freshwaters available on earth and supplies water for about 75% of the EU citizens (European Communities, 2008). Furthermore groundwater is a main water source for agricultural irrigation and industry. Due to growing water consumption and unsustainable water management many countries and regions of the world are facing deterioration of groundwater quality and decreasing groundwater quantities. Integrated water cycle management and augmentation of available freshwater resources through alternative sources are one response to these challenges. Desalination of seawater and brackish water as well as water reclamation and reuse are technical water stress mitigation options applied up to now (Bixio *et al.* 2006; Fritzmann *et al.* 2007). This practice is particularly encouraged for many Mediterranean countries as outlined by the European Commission in their publications on the management of water scarcity and drought (European Commission, 2007). The already severe water stress conditions in the Mediterranean region are expected to aggravate as a probable consequence of climate change impacts (EEA, 2007).

In course of the implementation of the Water Framework Directive (2000/60/EC) the status of water resources has been analysed and evaluated with regard to achieving a "good ecological status" by 2015. European Union Member States had to assess the quality of the water bodies according to Article 5 of the Water Framework Directive (WFD). The investigation in the Member States has revealed that quite a number of water bodies in the European Union are at risk of failing to meet the quality objectives by 2015, as set in the WFD (Hochstrat *et al.* 2010). Between 30% and 75% of the groundwater bodies in UK, France, Spain, Malta and Cyprus are categorised "at risk", and an additional 15% to 50% were evaluated as being "probably at risk" (Figure 1.1).

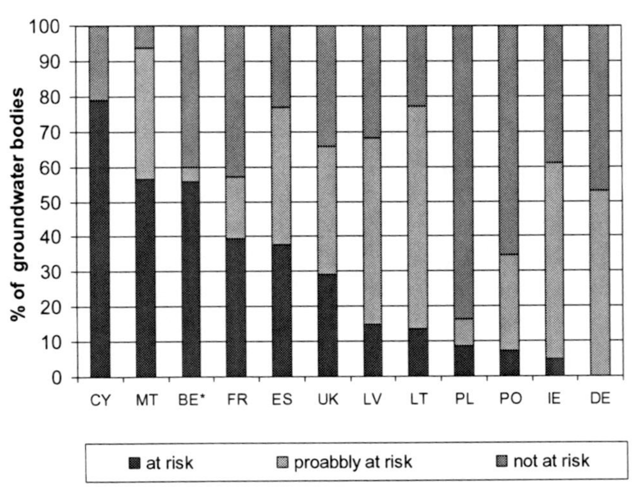

Figure 1.1 Status of groundwater in selected European countries (National reports on Art. 5 WFD accessed via CIRCA)

The types of pressures exerting a significant impact on water resources are either qualitative (e.g. diffuse or point pollution with nitrates and pesticides or seawater intrusion into coastal aquifers) or quantitative (e.g. over-abstraction of groundwater). In the scope of River Basin Management plans the water authorities will have to address the identified pressures and set up a programme of measures. The utilisation of alternative water resources or options such as aquifer recharge can be part of these programs. In the European policy water saving, e.g. supported and enforced by economical measures, has a priority. While water reuse is regarded as a possible water conservation measure, seawater desalination is regarded as a last resort option (European Commission, 2007).

Among the various beneficial uses of reclaimed wastewater, managed aquifer recharge (MAR) receives growing attention because it features advantages such as additional natural treatment, storage capacity to buffer seasonal variations of supply and demand as well as mixing of reclaimed water with natural water bodies which promotes the acceptance of further uses, particularly indirect potable use. Major concerns about the safety of this exploitation route of an alternative water source are connected to microbial and chemical contaminants occurring in wastewater, among which are emerging trace organics like endocrine disrupters and pharmaceuticals.

The concept of managed aquifer recharge offers potential for various subsequent uses like irrigation, industrial process water and augmentation of public water supplies. The latter, indirect potable use, is certainly one of the most challenging water reclamation and reuse applications with a high demand in terms of safety because of the potential use as drinking water and the general level of protection required for groundwater resources as laid down for example in the European directive on the protection of groundwater against pollution and deterioration (2006/118/EC). Even if drinking water supplementation is not explicitly foreseen in an aquifer recharge project, the provision of drinking water quality in the recovered product is a common bottom line in many applications, for example in Israel where the Dan Region Reclamation and Soil Aquifer Treatment scheme operated by the national water company Mekorot provides "accidental drinking water quality" on a large scale.

1.2 RESEARCH IN MANAGED AQUIFER RECHARGE

Managed aquifer recharge is not a new water management tool but it is one that is increasingly being applied in water recycling due to the storage, treatment and distribution functions it can fill as well as 'naturalising' water from a public perception perspective. It requires a sound understanding of the science of recharging, storing and recovering surface reclaimed waters in aquifers to make it safe for human health and the environment, and easier and more reliable to design, operate and regulate. That understanding has grown in recent years by an upsurge in the measurement capability of chemical, biological and physical processes occurring in water treatment trains, and especially including the aquifer. It has been assisted by a growing number of managed aquifer recharge projects that allow observations to be made under a very wide range of water types, treatment methods, aquifer types and end uses of the recovered water. Recycling via aquifers was identified as an important opportunity for the growth of water recycling in a preceding project, AQUAREC (Bixio and Wintgens, 2006) of the European Commission's 5th Framework Programme.

Important to the increasing knowledge is the ability to synthesise results across a variety of sites and the use of laboratory experiments to fill in gaps that are difficult to address under field conditions. Preceding multi-site studies managed aquifer recharge studies include Stuyfzand (1998), Dillon and Toze (2005), Clinton *et al.* (2007), Vanderzalm *et al.* (2009) and Sauter (2009). In addition there are a number of operational sites using reclaimed water that have been the focus of attention for numerous published research papers, generally involving a number of disciplines. These include groundwater replenishment via wells and basins at Orange County, California, soil aquifer treatment in Arizona, bank filtration and basins in Berlin, dune filtration at Amsterdam, and recycled stormwater and effluent recharge via wells in Adelaide.

The RECLAIM WATER project, which forms the main basis of this book, is different to the syntheses that have preceded it as its focus has been on reclaiming waters via aquifers by all methods (Figure 1.2). However the range of source waters and end uses has been very broad and it covers all aspects of projects from hydraulic design to water treatment process validation. Cases span from initially unintentional opportunistic reuse of water in Mexico, Italy and Spain for irrigation supplies, to intensively planned projects, where the aquifer is the recipient of highly treated water reclaimed from sewage effluent to produce drinking water supplies, for example in Belgium. RECLAIM WATER also includes a number of intermediate cases where there is planned partial dependence on the treatment processes that aquifers provide, including cases in Israel, South Africa and Australia. Significant work on the coupling of engineered and natural treatment processes has required studies of pathogens, trace organics and nutrient removals in engineered treatments, including activated carbon and membrane processes. Laboratory studies have enabled better interpretation of field measurements or substitution where these were not possible. Modelling of hydraulic, chemical and biological processes at the different sites have been undertaken using a variety of models to address specific local needs.

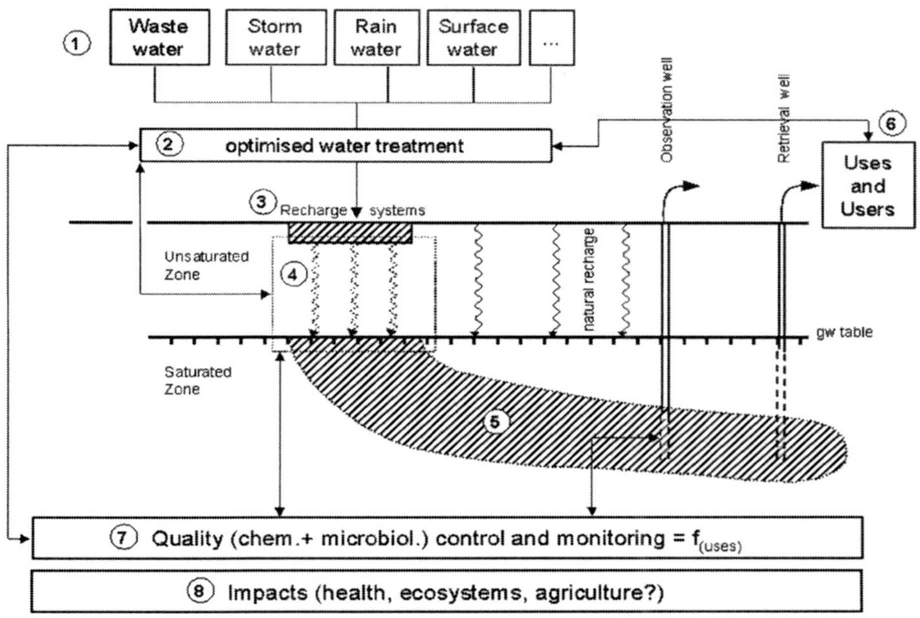

Figure 1.2 Concept of managed aquifer recharge and points of consideration in RECLAIM WATER

1.3 OBJECTIVES OF THIS BOOK

This book presents the results of European research activities in the field of managed aquifer recharge funded by the European Commission during the 6th Framework Programme for Research and Technological Development (FP6). It incorporates most of the project outcomes of the EU project RECLAIM WATER and selected highlights of the work from the EU project GABARDINE.

RECLAIM WATER was a Specific Targeted Research Project supported by the European Commission under the Thematic Priority 'Global Change and Ecosystems' of the 6th Framework Programme (Contract-No. 018309). The project started in October 2005 and ran till December 2008. The project consortium consisted of universities, research and technology institutes, as well as technology providers and small and medium size enterprises, with a total number of 20 partners from 16 countries coming from European Union Member States and Australia, China, Israel, Mexico, Singapore, South Africa, and Switzerland.

The strategic objective of RECLAIM WATER was to develop hazard mitigation technologies for water reclamation providing safe and cost effective routes for managed aquifer recharge. It assessed different treatment applications in terms of behaviour of key microbial and chemical contaminants such as organic micropollutans. The knowledge generated in the project and the technologies developed are also suited to the needs of developing countries, which have a growing need of supplementation of freshwater resources. The participation of partners from China, Australia, Singapore, South Africa and Mexico demonstrates the anticipation of the global dimension of the water reclamation and aquifer recharge issue.

The project embarked on these findings and focused its research objectives on the investigation of water reclamation technologies and their impact on key contaminants. The project aimed to provide new combinations of technologies to treat wastewater to appropriate water quality levels for sustainable groundwater recharge. RECLAIM WATER in its scope concentrates upon different issues indicated in Figure 1.2 following the water cycle in a water reclamation and aquifer recharge process:

(1) Improvement of municipal wastewater treatment processes as an important prerequisite for water reclamation;
(2) Optimised water reclamation technologies to produce feed water for aquifer recharge and other water reuse applications;
(3) Design and operation of actual recharge systems (e.g. infiltration-percolation systems, constructed wetlands, lagoons, well injection);
(4) The investigation of processes in the unsaturated soil zone and their dependence on the pre-treatment (steps 2 and 3);
(5) The investigation and modelling of the introduction of a reclaimed water phase in a natural aquifer using tracer compounds (like Boron);
(6) Anticipating the needs of various uses of recovered water (e.g. indirect potable use and non-potable uses) in terms of water quality requirements and definition of relevant target contaminants;

(7) Developing (if necessary) and applying the necessary analytical tools to monitor the relevant contaminants (microbial and chemical) including the investigation of contaminant fate in the processes;
(8) Collecting existing data on the behaviour of the target compounds and the impact of treatment options in case studies to provide a sound basis for human health and environmental risk assessment.

1.4 CHAPTER CONTENTS

The book is divided into five parts. Part A includes the description of the investigated eight international MAR case studies and the analytical results based on the selected standard parameters. Part B presents the analytical methods for the measurement of organic micropollutants, bulk organics and microbial contaminants as well as the results of the measurements. Part C looks into selected natural and engineered pre-treatment technologies for MAR. Part D contains key aspects for the design and management of MAR systems such as modelling tools, risk assessment and risk management as well as decision support for MAR planning. Part E provides the legal framework and a brief guidance for EU policies.

Part A: International MAR case studies

The project depicted water cycle schemes utilizing different alternative water sources and a range of pre-treatment options prior to infiltration or injection for aquifer recharge. Various recharge systems including direct infiltration through dug wells, infiltration basins and different geological settings (sandy to carstic aquifers) were investigated.

The majority of the data was generated in a set of globally distributed case studies. These also served as a basis to validate technical and managerial concepts, prove their practicability and effectiveness, take into account regional and use type variety as well as to assess a number of selected treatment processes. Table 1.1 gives an overview on the treatment technologies, recharge systems and end-uses relevant to the different case studies.

Table 1.1 Water reclamation schemes investigated in RECLAIM WATER.

	Site location	Scheme description
Europe	Sabadell (near Barcelona), Spain	Secondary treated wastewater effluent discharged into a river bed where it infiltrates and is recovered. The water is then disinfected (UV) and distributed for parks irrigation.
	Nardò, Salento Region, South of Bari, Italy	Secondary treated municipal effluent is transported to aquifer injection. Recharge acts as a salt intrusion barrier and resource is also used as drinking water source.
	Torreele (Wulpen), Belgium	Tertiary treated municipal effluent is upgraded by ultrafiltration and reverse osmosis, and then infiltrated via infiltration ponds to prevent salt intrusion and to recharge an aquifer used for drinking water production.
Asia	Shafdan, Tel Aviv Region, Israel	Secondary wastewater from the Tel-Aviv area is recharged to an aquifer via a soil aquifer treatment (SAT) system. Recovered water is primarily used for irrigation but has accidental drinking water quality.
	Gaobeidian, Beijing, China	Tertiary effluent is used for aquifer recharge. Treatment is provided by coagulation, filtration and ozonation (in test) prior to infiltration and recharge. Water is used for irrigation.
Australia	Salisbury, Adelaide Region, South Australia	Wetland treated urban stormwater injected into a brackish aquifer. Water recovered via separate recovery wells. Recovered water is intended for drinking supplies and until proven will be used for irrigation.
Africa	Atlantis (near Cape Town), South Africa	Urban stormwater run-off is collected via a series of detention basins, blended with secondary treated domestic wastewater and recharged up-gradient of a production well field for augmenting the water supply. The blend of natural groundwater and recharged water abstracted from the well field is used as potable water supply for the city of Atlantis.
America	Mezquital Valley, State of Mexico, Mexico	The raw wastewater from Mexico City mixed with stormwater and surface water is discharged to an irrigated area of more than 76,000 ha. About 40% of the irrigation water infiltrates into the aquifer. The water is recovered via separate wells and springs. 206 well systems, 31 springs, and 63 waterwheels are in operation. Recovered water is chlorinated and locally distributed for drinking water supply, industrial use, irrigation and other purposes (bathing, swimming, washing).

Part B: Water quality analysis in MAR – Methods and results

Microbial contaminants. Besides the standard indicator parameters for microbiological water quality such as Total bacteria counts, Total coliforms, *E. Coli*, Enteroccoci and Clostridium spores, an enhanced spectrum of parameters was considered including parasites, protozoea, pathogenic bacteria and viruses as well as antibiotic resistance genes. It could be confirmed that advanced treatment schemes, for example those utilising membrane filtration, could readily remove those compounds and that sub-surface processes can significantly attenuate those (Böckelmann *et al.* 2009; La Mantia *et al.* 2008).

Chemical contaminants. Particular attention has been paid to organic chemicals in the different stages of the investigated aquifer recharge schemes. Both bulk organic characterisation as well as analysis of trace organics has been carried out. The parameters included pharmaceutically active compounds and disinfection by-products as well as estrogenic activity. Analytical methods have been developed and adopted to the different matrices encountered in the sites (Fink and Ternes 2006; Krauss *et al.* 2008). Those methods have been used to elucidate the fate of the target compounds in different treatment processes (Schulz *et al.* 2008; Yu *et al.* 2008).

Part C: Water reclamation technologies in MAR

The effectiveness of both natural and intensive water reclamation technology options which can be applied prior to aquifer recharge options has been assessed. The investigations included constructed wetlands, slow sand filters, advanced primary treatment for a developing country context as well as advanced options such as membrane bioreactors (Abegglen *et al.* 2009), dense membrane processes (Joss *et al.* 2011), ozonation (Benner *et al.* 2008) and process combinations also for brine treatment, for example utilizing activated carbon adsorption plus membrane filtration as well as capacitive deionisation (Ng *et al.* 2008). A particular hybrid process investigated in the project included powdered activated carbon adsorption and nanofiltration (Kazner *et al.* 2007 and 2008). Joss *et al.* (2008) and Wintgens *et al.* (2008) have reviewed water reclamation technology options particularly with regard to micropollutants removal and suitability for indirect potable use.

Part D: Design and management of MAR systems

Sophisticated characterisation tools such as boron and lithium isotope monitoring has been utilised to characterise the hydrology and hydrogeochemistry of full scale aquifer recharge sites. Native and influenced groundwater could be clearly distinguished and input to flow modelling could be provided (Kloppmann *et al.* 2008; Kloppmann *et al.* 2009). Conceptual and numerical models have been set up to better describe and characterise hydrology and hydrogeochemistry in the investigated aquifer recharge sites.

State-of-the-art concepts for risk assessment such as Quantitative Microbial Risk Assessment (QMRA) and risk management (e.g. Hazard Analysis and Critical Control Points (HACCP) and Water Safety Plan) have been used to integrate the water quality and treatment efficiency data and to put the results into a perspective which is relevant for end-users and public perception. Comprehensive risk assessment approaches have been developed for selected full scale schemes involved in the project and the contribution of subsurface processes to risk control has been assessed (Dillon *et al.* 2008; Page *et al.* 2010a, 2010b; Ayuso-Gabella *et al.* 2011).

Part E: Promoting MAR systems for water recycling

The project outcomes have been summarised in a policy brief document directed to the European Commission and other decision makers in the water sector (Hochstrat *et al.* 2010). RECLAIM WATER has provided a comprehensive assessment of different water reclamation and managed aquifer recharge methods and technologies covering a broad range of water quality parameters (including pathogens, antibiotic resistance genes and emerging contaminants), different pre-treatment and infiltration options, different end-uses (potable and non-potable), raw water sources for replenishment (from untreated wastewater to highly treated tertiary effluent and purified stormwater). The investigations covered schemes on all continents and used a coordinated methodology for assessment. The project results underline that managed aquifer recharge can be a save and reliable climate change adaptation method. Technologies and methods can be tailored to the different socio-economic contexts. In developing countries' context extensive natural systems, in particular MAR, can also provide a decent water quality at very low cost.

REFERENCES

Abegglen C., Joss A., McArdell C. S., Fink G., Schlüsener M., Ternes T. A. and Siegrist H. (2009). The fate of selected micropollutants in a single-house MBR. *Water Research*, **43**(7), 2036–2046.

Ayuso-Gabella N., Page D., Masciopinto C., Aharoni A., Salgot M. and Wintgens T. (2011). Quantifying the effect of managed aquifer recharge on the microbiological human health risks of irrigating crops with recycled water. *Agricultural Water Management*, **99**(1), 93–102.

Benner J., Salhi E., Ternes T. A. and von Gunten U. (2008). Ozonation of reverse osmosis concentrate: kinetics and efficiency of beta blocker oxidation. *Water Research*, **42**, 3003–3012.

Bixio D., Thoeye C., De Koning J., Joksimovic D., Savic D., Wintgens T. and Melin T. (2006). Wastewater reuse in Europe. *Desalination*, **187**, 89–101.

Bixio D. and Wintgens T. (2006). Water reuse system management manual - AQUAREC. Office for Official Publications of the European Communities, Luxembourg, 2006.

Böckelmann U., Dörries H.-H., Ayuso-Gabella N., Salgot de Marçay M., Tandoi V., Levantesi C., Masciopinto C., Van Houtte E., Szewzyk U., Wintgens T. and Grohmann E. (2009). Quantitative real-time PCR monitoring of bacterial pathogens and antibiotic resistance genes in three european artificial groundwater recharge systems. *Applied and Environmental Microbiology*, **75**(1), 154–163.

CIRCA (2006). *CIRCA - Communication & Information Resource Centre Administrator*, http://forum.europa.eu.int/Public/irc/env/wfd/Library.

Clinton T. (ed.) (2007). Reclaimed Water Aquifer Storage and Recovery: Potential Changes in Water Quality. Report of Water Reuse Foundation Project WRF-03–009, http://www.watereuse.org/catalog/research-reports/natural-systems-groundwater-recharge-Storage.

Dillon P. and Toze S. (eds) (2005). Water Quality Improvements During Aquifer Storage and Recovery. American Water Works Assoc. Research Foundation Report 91056F, http://www.waterrf.org/Pages/WaterRFHome.aspx.

Dillon P., Page D., Vanderzalm J., Pavelic P., Toze S., Bekele E., Prommer H., Higginson S., Regel R., Rinck-Pfeiffer S., Purdie M., Pitman C. and Wintgens T. (2008). A critical evaluation of combined engineered and aquifer treatment systems in water recycling. *Water Science and Technology*, **57**(5), 753–762.

European Commission (2007). *Addressing the challenge of water scarcity and droughts in the European Union*. Communication from the Commission to the European Parliament and the Council. COM(2007) 414 Final.

European Environment Agency (2007). Climate Change and Water Adaptation Issues. Technical Report EEA 2/2007.

Fink G. and Ternes T. A. (2006). Simultaneous determination of psychoactive drugs and their metabolites in aqueous matrices by liquid chromatography mass spectrometry. *Environmental Science & Technology*, **40**(23), 7321–7328.

Fritzmann C., Löwenberg J., Wintgens T. and Melin T. (2007). State-of-the-Art of Reverse Osmosis desalination. *Desalination*, **216**, 1–76.

Joss A., Siegrist H. and Ternes T. A. (2008). Are we about to upgrade wastewater treatment for removing organic micropollutants? *Water Science and Technology*, **57**(2), 251–255.

Joss A., Baenninger C., Foa P., Koepke S., Krauss M., McArdell C. S., Rottermann K., Wei Y., Zapata A. and Siegrist H. (2011). Water reuse: >90% water yield in MBR/RO through concentrate recycling and CO_2 addition as scaling control. *Water Research*, **45**(18), 6141–6151.

Hochstrat R., Wintgens T., Kazner C., Jeffrey P., Jefferson B. and Melin T. (2010). Managed aquifer recharge with reclaimed water: approaches to a European guidance framework. *Water Science and Technology*, **62**(6), 1265–1273.

Kazner C., Fink G., Ternes T., Wintgens T. and Melin T. (2007). Removal of organic micropollutants by nanofiltration in combination with adsorption on powdered activated carbon for artificial groundwater recharge with reclaimed wastewater. *Proceedings of the 5th IWA Micropol & Ecohazard 2007 conference*, 17–20 June 2007, DECHEMA e.V., Frankfurt/Main, Germany, 259–265.

Kazner C., Lehnberg K., Kovalova L., Wintgens T., Melin T., Hollender J. and Dott W. (2008). Removal of endocrine disruptors and cytostatics from effluent by nanofiltration in combination with adsorption on powdered activated carbon. *Water Science and Technology*, **58**(8), 1699–1706.

Kloppmann W., Van Houtte E., Gaus I., Picot G., Vandenbohede A., Lebbe L., Guerrot C., Millot R. and Wintgens T. (2008). Monitoring reverse osmosis treated wastewater recharge into a coastal aquifer by environmental isotopes (B, Li, O, H). *Environmental Science & Technology*, **42**(23), 8759–8765.

Kloppmann W., Chikurel H., Gaus I., Picot G., Guttman Y., Aharoni A., Guerrot C., Millot R. and Wintgens T. (2009). B and Li isotopes as intrinsic tracers for injection tests in aquifer storage and recovery systems. *Applied Geochemistry*, **24**(7), 1214–1223.

Krauss M. and Hollender J. (2008). Analysis of nitrosamines in wastewater: exploring the trace level quantification capabilities of a hybrid linear ion trap/orbitrap mass spectrometer. *Analytical Chemistry*, **80**, 834–842.

La Mantia R., Masciopinto C., Levantesi C. and Tandoi V. (2008). Fate and transport of faecal contamination microbial indicators, pathogenic protozoa and Campylobacter in the artificially recharged fractured aquifer of Salento, Italy. *Water Science and Technology*, **57**(6), 849–856.

Ng H. Y., Lee L. Y., Ong S. L., Tao G., Viswanath B., Kekre K., Lay W. and Seah H. (2008). Treatment of RO brine-towards sustainable water reclamation practice. *Water Science and Technology*, **58**(4), 931–936.

Page D., Dillon P., Toze S., Bixio D., Genthe B., Jiménez Cisneros B. E. and Wintgens T. (2010a). Valuing the subsurface pathogen treatment barrier in water recycling via aquifers for drinking supplies. *Water Research*, **44**(6), 1841–1852.

Page D., Dillon P., Vanderzalm J., Toze S., Sidhu J., Barry K., Levett K., Kremer S. and Regel R. (2010b). Risk assessment of aquifer storage transfer and recovery with urban stormwater for producing water of a potable quality. *Journal of Environmental Quality*, **39**(6), 2029–39.

Sauter M. (2009). GARBADINE final activity report. http://www.gabardine-fp6.org/Publications.aspx.

Stuyfzand P. J. (1998). Simple models for reactive transport of pollutants and main constituents during artificial recharge and bank filtration. *Proceedings of the TISAR '98 Amsterdam*, Balkema, Rotterdam, 427–434.

Schulz M., Loffler D., Wagner M. and Ternes T. A. (2008). Transformation of the X-ray contrast medium lopromide in soil and biological wastewater treatment. *Environmental Science & Technology*, **19**, 7207–7217.

Vanderzalm J., Sidhu J., Bekele G.-G., Pavelic P., Toze S., Dillon P., Kookana R., Hanna J., Barry K., Yu X., Nicholson B., Morran J., Tanner S. and Short S. (2009). Water Quality Changes during Aquifer Storage and Recovery. Water Research Foundation, Denver, USA, http://www.waterrf.org/Pages/WaterRFHome.aspx.

Wintgens T., Salehi F., Hochstrat R. and Melin T. (2008). Emerging contaminants and treatment options in water recycling for indirect potable use. *Water Science and Technology*, **57**(1), 99–107.

Yu L., Fink G., Wintgens T., Melin T. and Ternes T. A. (2009). Sorption behavior of potential organic wastewater indicators with soils. *Water Research*, **43**(4), 951–960.

Part A

International MAR Case Studies

Chapter 2

Water reclamation for aquifer recharge at the eight case study sites: a cross case analysis

Kristell Le Corre, Avi Aharoni, Johan Cauwenberghs, Alma Chavez, Haim Cikurel, Maria Neus Ayuso Gabella, Bettina Genthe, Richard Gibson, Bruce Jefferson, Paul Jeffrey, Blanca Jimenez, Christian Kazner, Constantino Masciopinto, Declan Page, Rudi Regel, Stephanie Rinck-Pfeiffer, Miquel Salgot, Maronel Steyn, Emmanuel van Houtte, Gideon Tredoux, Thomas Wintgens, Cheng Xuzhou, Liang Yu and Xuan Zhao

2.1 INTRODUCTION

Water scarcity combined with the quality deterioration of freshwater due to the rapid augmentation of population and industrial development is a major concern in many parts of the world. For instance, more than 50% of European countries are being affected by water stress issues (Bixio *et al.* 2006) while under the current circumstances it is also predicted that by 2025, nearly 2.4 billion of the world population will be confronted to severe water stress situations with less than 1000 m^3/year of freshwater available per person (Zeman *et al.* 2006). In this context, the development of recycling options for the conservation or replenishment of groundwater levels is a must to face the 25–60% increase of the water demand that could be required in the next 100 years (Levine & Asano, 2004). The use of reclaimed municipal wastewater and excess stormwater runoff for aquifer recharge applications is particularly seen as a promising approach to improve the management of water resources as independently from climate changes, this technique allows the underground storage of surface and reclaimed water for future use (Asano & Cotruvo, 2004), but also provides an effective protection against salt water intrusion (Masciopinto *et al.* 2007), reduce the discharge of effluent to the sea (Wade Miller, 2006), can provide storage where surface storage is limited by land availability (Dillon *et al.* 2006) while improving the image of wastewater reuse projects as recycling takes place through a natural compartment (Bouwer, 1996).

Several examples of intentional groundwater recharge schemes with reclaimed wastewater or excess runoffs are available and applied for different water resources management purposes including groundwater level conservation; protection against salt water intrusion; management of surface and groundwater supplies for potable and non-potable uses such as agriculture or industry (Guendert, 2004; Masciopinto *et al.* 2007; Tredoux *et al.* 1999; Icekson-Tal *et al.* 2003). Although the number of wastewater and stormwater recycling schemes for aquifer recharge has developed significantly in the past few years, it is not common practise yet, especially in Europe. The reasons for this are the same reasons why water reuse in general is developing at a steady but slow pace which includes the lack of international regulations on reuse or the absence of global standards for recharge (Bixio *et al.* 2006).

Furthermore, concerns in water recycling through managed aquifer recharge are related to potential health risks arising from the eventual infiltration of pathogenic micro-organisms, minerals, heavy metals and organic substances into groundwater (Asano & Cotruvo, 2004), especially when recovery is meant for indirect potable reuse. Finally, recent studies have also highlighted the fact that the long term operation of managed aquifer recharge (MAR) sites could lead to reduction in treatment efficiency but also in quality of the abstracted water (Oren *et al.* 2007).

To date, a number of studies have looked at the performance of aquifer recharge sites regarding specific contaminants such as disinfection by products (Pavelic *et al.* 2006), natural organic matter (Vanderzalm *et al.* 2006) and microbial indictors of faecal contamination (La Mantia *et al.* 2008). However, the comparison of various aquifer recharge sites using reclaimed municipal wastewater and/or stormwater in terms of overall recharge performances, reclamation technologies and aquifer recharge practices is rarely available. The unique character of an aquifer recharge scheme depending on a combination of factors which include the performance, reliability and robustness of the treatment options applied prior recharge but also the recharge method used (i.e. indirect vs. direct injection), the hydrogeology of the aquifer effectively render the knowledge and experience gained at a specific site difficult to apply to another site (Dillon & Jimenez, 2008). If the knowledge gained from the operation of various sites might not be applicable to other sites a broad comparison of schemes could help to identify what quality of water would be acceptable for aquifer recharge.

To provide a better understanding of aquifer recharge, from 2005 to 2008, eight international pilot or full scale test sites (Figure 2.1) treating municipal wastewater effluent and/or stormwater for aquifer recharge took part in the European RECLAIM WATER project looking at water reclamation technologies for safe managed groundwater recharge. The main objective of this global study has been to assess the overall performance of these sites in recharging aquifers mainly for irrigation and potable water supply purposes by following contaminant fate throughout each scheme. The reclamation technologies used at the sites covered a range of pre-treatment technologies from conventional activated sludge to membrane technologies, and a suite of recharge methods including river bed filtration, dune filtration, sinkhole infiltration, infiltration ponds, soil aquifer treatment (SAT) and well injection. The characteristics of each site are summarised in Table 2.1. For a more detailed description of the schemes and performances obtained at each sites please refer to Chapter 3 to 10. The variety of the sites provided then an opportunity to compare different recharge methods under various geological and pre-treatment conditions, hence providing a broader knowledge of managed aquifer recharge (MAR) for sustainable groundwater recharge in both developing and developed countries.

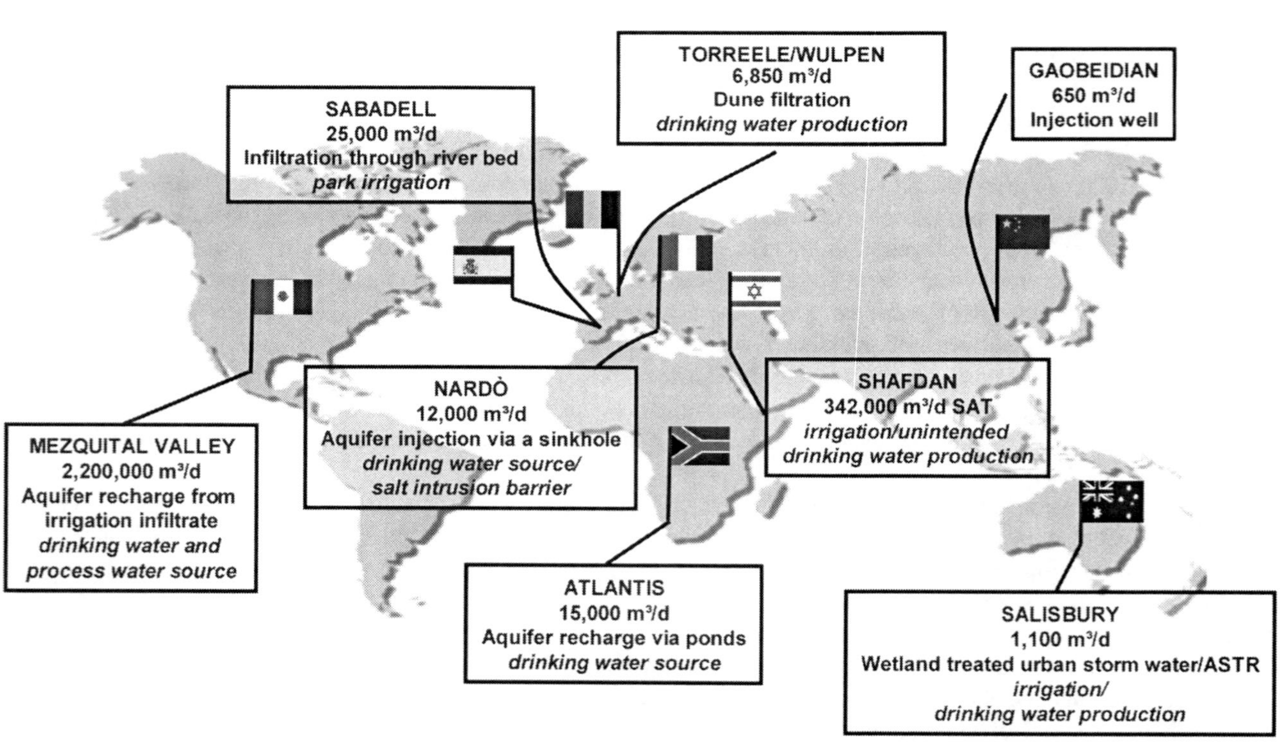

Figure 2.1 RECLAIM WATER case studies of MAR

Table 2.1 Characteristics of the eight RECLAIM WATER case study sites.

Test Site	Injected source	Recharge method	Recharge rate	Aquifer	Retention time (days)	Additional treatment	Reuse purpose
SPAIN Sabadell, Barcelona region	Secondary treated effluents (activated sludge + nutrient removal)	Infiltration through river bed	6.9 Mm^3/y (2006)	– Unconfined – Alluvial, mainly sand and gravel)	Not determined	UV and chlorination	Public park irrigation Street cleaning
ITALY Nardò, Salento region	Secondary treated effluents (conventional activated sludge + biological treatment plant + chlorination)	Injection via a sinkhole	4.4 Mm^3/y (average)	– Confined – Karstic and fractured aquifer (sandstone, limestone and loamy sand)	~60	None	Salt intrusion barrier + drinking water source
ISRAEL Shafdan, Tel Aviv region	Secondary treated effluents (conventional activated sludge incl. biological N and P removal) Tertiary treated effluent (Sec effl + UF-Pilot SAT site)	Soil Aquifer Treatment (SAT)	120 Mm^3/y (full site) ~120 m^3/d (pilot site)	– Unconfined – Mainly sand and sandstone	180–360 30–60	Intermediate chlorination	Irrigation (unintended drinking water quality)
CHINA Gaobeidian, Beijing	Tertiary treated effluents (secondary treatment + coagulation + sand filtration + ozonation + slow sand filter)	Well injection	350–500 m^3/d (pilot site)	– Mainly clay and sand	30	None	None*
BELGIUM Torreele/ Wulpen	Tertiary treated effluents (biological treatment (AS) + chemical P removal + UF + Chlorination + RO)	Dune filtration	2.5 Mm^3/y	– Unconfined – Mainly sand	~35	Chlorination, aeration, rapid sand filtration and UV prior to distribution	Sustainable groundwater management
AUSTRALIA Salisbury, Adelaide	Wetland treated urban stormwater (in-stream basins + holding storage basins + cleansing wetland)	Aquifer storage transfer and recovery (ASTR)	0.2 Mm^3/y (ASTR)	– Confined – limestones	~200	None	Intended irrigation, industrial use and drinking water production
SOUTH AFRICA Atlantis	Secondary treated effluent (incl. biological N removal) mixed with Urban stormwater runoffs	Infiltration ponds	2.7 Mm^3/y	– Unconfined – Mainly sand	150–210	Ion exchange + Chlorination	Potable water supply
MEXICO The Mezquital Valley (Unintentional aquifer recharge)	Untreated wastewater mixed with stormwater and natural surface water	Discharge to an irrigated area- (atypical SAT)	800 Mm^3/y	– Unconfined – Shallow alluvial deposits – Volcanic soil – Limestone soil	–	Chlorination	Industrial use, domestic use, Potable water production, Irrigation

2.2 METHODOLOGY

For this project, the different sites have been periodically monitored over two years except for the Mexican and South African sites, which have been monitored over one year. The monitoring covered basic wastewater parameters, microbiological parameters, trace elements and salinity related parameters (Table 2.2) that were analysed in secondary or tertiary treated effluent, prior to injection or soil aquifer treatment (SAT) application and in the abstracted water.

Table 2.2 Protocol for standard analysis of the eight RECLAIM WATER case study sites (Protocol 1).

Basic analysis	Microbiological analysis	Trace elements analysis	Salinity related analysis
Suspended solids, SS, BOD, COD, DOC, Ammonia, Nitrite, Nitrate, Total N, Phosphorus, Alkalinity, pH, Turbidity, Detergents, Mineral, oils, Phenols	Total bacteria count, Faecal coliforms, *E. coli*, Enterococci, *Clostridium* spores, Bacteriophages	Boron (inorganic tracer), Cadmium, Chromium, Cobalt, Copper, Fluoride, Iron, Lead, Manganese, Molybdenum, Nickel, Selenium, Barium, Cyanide, Zinc, Hexavalent chromium and sulfide	Chloride, Electrical conductivity, Sodium, Potassium, Calcium, Magnesium, Carbonate, Bicarbonate, Sulphate

Over the duration of the project, the data collected at each test site have been gathered in a common spreadsheet and analysed by determining, for each parameter data set, the average value, the minimum and maximum levels observed, and when applicable the median, the lower and upper quartile values.

The box and whisker diagrams plotted in the following sections serve as a support for the comparison of the sites performances. They display for each site the range of values measured per parameter in the "source" water (i.e. after secondary treatment; in the raw stormwater for Australia and South Africa, or the raw wastewater in Mexico), at the point of injection or infiltration (i.e. injectant) and in the water recovered after passage through soil (i.e. abstracted water). Table 2.3 gives a description of the type of source, recharged water and abstracted water at each site. On these diagrams, the line represents the limits of maximum and minimum values; the box the lower and upper quartile values.

Table 2.3 Sampling point description.

Location	Source	Recharge water	Abstracted water
ITALY	Secondary effluent	Secondary effluent after transportation via open channel	Water from well 500m away from injection point
SPAIN	Secondary effleunt	River water (mixed with secondary effluent)	Water recovered in a mine under the river bed
ISRAEL Conventional SAT site	Secondary effluent	Secondary effluent	Water abstracted via wells 300–1500 m from recharge basins
ISRAEL Pilot site (UF + rapid SAT)	Secondary effluent	UF effluent	Water abstracted in wells 7.5 and 17.5 m away from injection
BELGIUM	Secondary effluent	UF + RO treated effluent in infiltration pond	Water abstracted via wells after dune filtration
CHINA	Secondary effluent	Ozonated and sand filtrated effluent	Water monitoring via wells
AUSTRALIA	Stormwater	Wetland treated stormwater	Water recovered from set of 4 wells situated 50 m away from injection wells
SOUTH AFRICA	Secondary effleunt + stormwater	Secondary effluent + stormwater in infiltration basins	Water abstracted via wells
MEXICO*	Raw wastewater	Raw wastewater	Water recovered from wells, or dug wells, or forming springs

*The results presented are for the area where irrigation has been practised for 50 years

2.3 RESULTS

For clarity purposes, the results section and the comparative analysis of the different sites performance focuses on a selection of representative parameters for each category illustrated in Table 2.2, that is to say basic wastewater parameters, microbiological parameters, trace elements and salinity related parameters. All sites were able to reduce main basic parameters such as SS, COD, DOC. However some of the sites performed better than other despite differences of the reclaimed wastewater quality to be injected.

2.3.1 Basic wastewater parameters

Physical water parameters: suspended solids

Despite the diversity of the type of water used for recharge, levels in suspended solid (SS) were below 10 mg/L in all injectant waters except for Mexico where justifiably the raw wastewater concentrations in SS were high with values ranging from 266 to 383 mgSS/L (Figure 2.2). The lowest contents in suspended solid observed in injectant water throughout the sites were expectedly measured at sites practicing advanced pre-treatment with respectively below 5 mgSS/L after ozonation and sand filtration in China, below 4.0 mg/L in the infiltration water after dual membrane filtration at the Belgian site and 0.4 mgSS/L on average at the Israeli UF + SAT pilot site. However it is to notice that the blend of secondary treated effluent with stormwater used as injectant in South Africa also presented a low content in suspended solid (just above 5 mg/L on average) while at the Australian site the treatment of stormwater via wetland ensured a reduction in suspended solid from 11.4 down to 3.6 mg/L prior to well injection (ASTR). After passage through the aquifer, the average level of SS ranged below 5 mg/L in China down to 0.2 mg/L after conventional SAT treatment in Israel. An increase in SS in the abstracted water was only observed at the well located the furthest away from the injection point at the pilot site in Israel. As expected, in the case of site using advanced treatment, SS were mainly removed before injection, while in Italy the transportation of municipal effluent in an open channel reduced the concentration in SS by 72% on average in the injectant as a result of SS deposition at the bottom of the channel. Overall when looking specifically at the performance of the sites in removing suspended solid, the sites practising SAT (i.e. Israel and Mexico) achieved the highest SS removal as a result of filtration through the upper soil layer (Idelovitch *et al.* 2003).

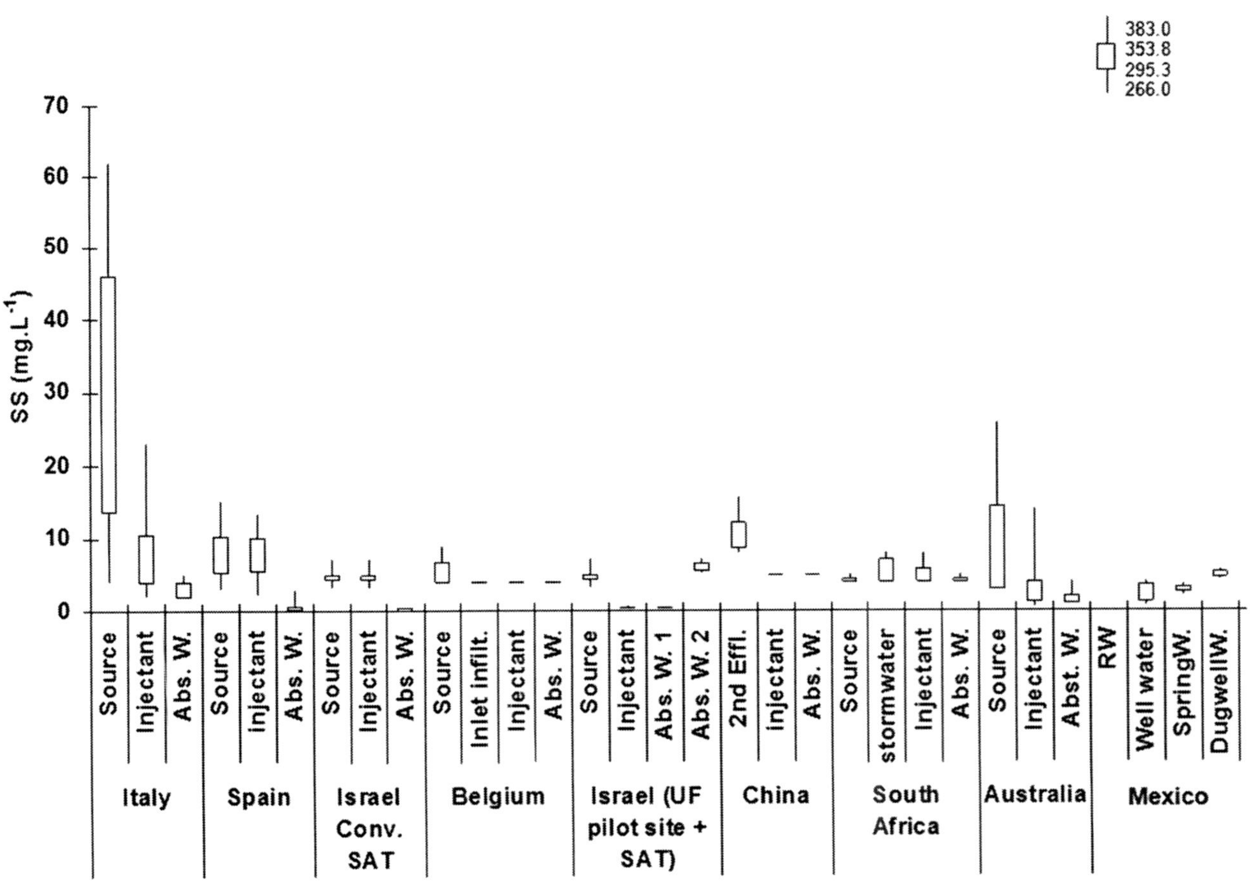

Figure 2.2 Comparison of SS levels in respectively the source, the injectant and the abstracted water the sites

Organics: COD, BOD and DOC

COD

When looking specifically at COD levels across the sites, levels for secondary effluent (i.e. source water) varied on average from 36 mg/L in Belgium to 61 mg/L in Italy, while the COD of the stormwater used as source water in Australia was 76 mg/L on average and the COD of the raw wastewater used for irrigation in Mexico varied from a minimum value of 536 mg/L to a maximum of 686 mg/L over the duration of the project (Figure 2.3).

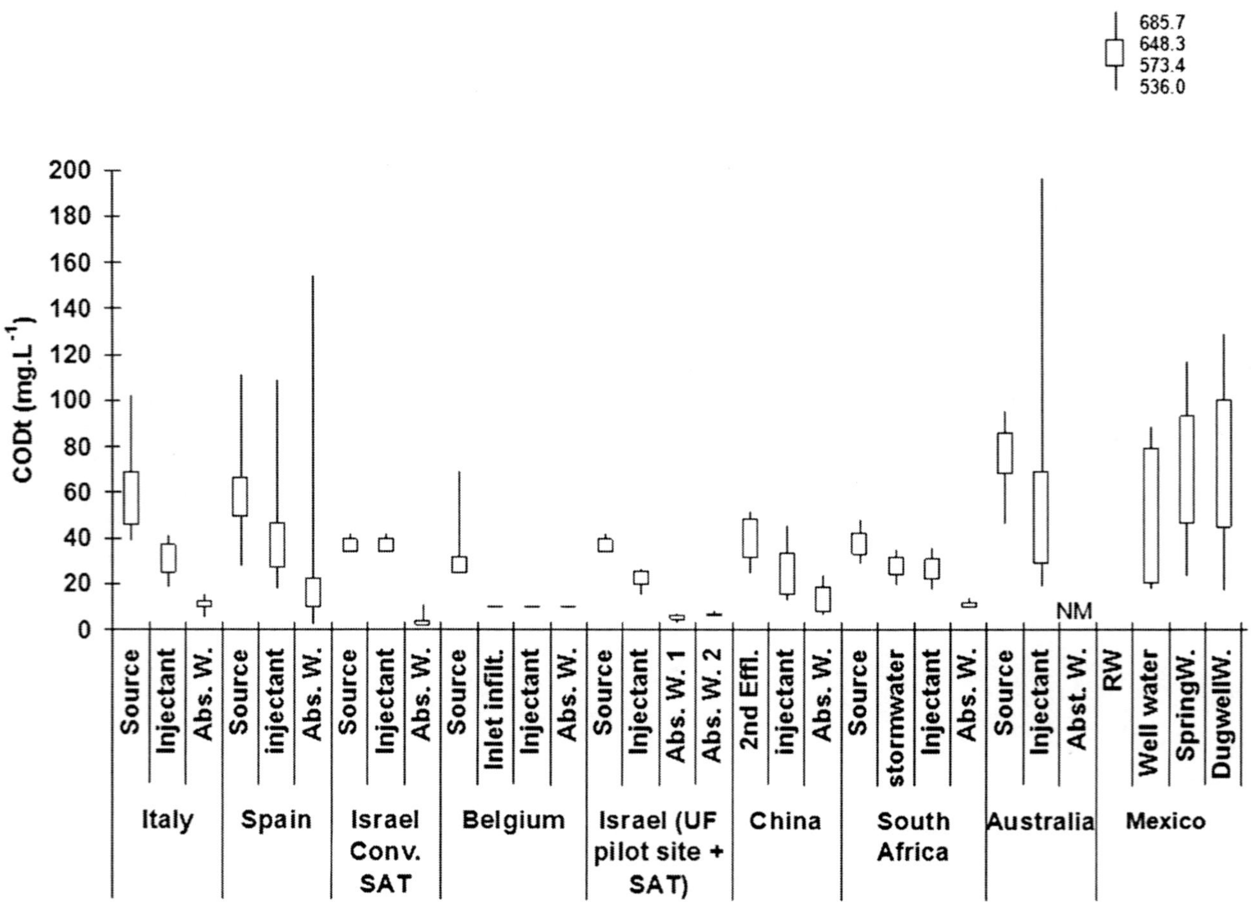

Figure 2.3 Comparison of COD levels in respectively the source, the injectant and the abstracted water the sites

The quality of the reclaimed water used for recharge (i.e. injectant) varied depending on the nature of the pre-treatment used prior recharge. Overall, the sites using advanced pre-treatment generally limited the injection of organic pollutant into aquifers with COD values below 10 mg/L in the infiltration pond water after dual membrane filtration in Belgium, 22 mg/L at the UF pilot site in Israel and 26 mg/L after ozonation + sand filtration in China, compared to values of 30, 37, 41, 55 and 611 mg/L on average in respectively Italy, Israel, Spain, Australia and Mexico (Figure 2.3). For sites practising indirect infiltration, the COD content in the sources used for recharge complied with the quality guidelines for indirect aquifer recharge suggested by Bixio and Wintgens (2006) for COD which is between 70 and 100 mg/L (Table 2.4), except in Mexico where the raw wastewater represents 6 times this recommended value.

After aquifer recharge and passage through soil, the sites using secondary effluent as injectant performed well, with COD levels comparable or even sometimes lower than for sites using advanced pre-treatment. To illustrate, the conventional SAT system in Israel was capable of reducing COD levels down to 4 mg/L which is comparable to standard values for drinking water (Table 2.4). The production of "unintended" drinking water at the Israeli site can be explained by the actual time the infiltrated water spends subsurface before being abstracted. Indeed, as mentioned by Idelovitch *et al.* (2003), the retention time of 6 to 12 month used at this site favours the degradation of persistent organic materials by a specific microbiological population. In Italy the passage of secondary effluent through the Nardò karstic aquifer also allowed a net decrease in organic compounds with an average reduction of 65%. Although the Mexican aquifer is able to reduce the high content in COD by more than 90%, final COD values at this site remained significantly higher than those observed at other sites with for example 75% of the values above 20.3 mg/L in the water abstracted from wells.

Table 2.4 Requirement for indirect groundwater recharge, proposed quality for direct groundwater recharge and drinking water quality requirements (Adapted from Bixio & Wintgens, 2006 and European Union, 1998).

Parameter	Unit	Polishing goal for indirect infiltration	Polishing goal for direct infiltration	Drinking water quality
pH		7–9	>6.5 and <9.5	>6.5 and <9.5
Conductivity	mS/cm	0.7	<0.7	<2.5
Alkalinity	$mgCaCO_3/L$			
TSS	mg/L			
COD	mg/L	70–100	<5	<5
TOC	mgC/L			
Total N	mgN/L	<25	<25	
NH_4^+	mgN/L	<0.2	<0.5	<0.5
NO_2^-	mgN/L		<0.5	<0.5
NO_3^-	mgN/L	<25	<25	<11.3
Total P	mgP/L			
PO_4^{--}	mgP/L			
SO_4^{--}	mgS/L	30	<30	<250
As	μg/L	5	<10	<10
B	mg/L	0.2	<1.0	<1.0
Cd	μg/L	3	<5.0	<5.0
Cl^-	mg/L	100	<250	<250
Cr	μg/L	25	<50	<50
Hg	μg/L	0.5	<1.0	<1.0
Pb	μg/L	5	<25 <10 in 2013	<25 <10 in 2013
Zn	μg/L			
Micropollutants	μg/L	<0.1	Compound specific	Compound specific
Faecal coliforms	CFU/100mL	$<10^4$	0	0
Faecal streptoc.	CFU/100mL		0	0
Helminth ova, Giardia	1/L	<1	0	0

BOD

Comparison of the data from the different injectant water used for recharge reveals the BOD to vary between 4.5 and 18.3 mg/L for sites using secondary effluent (i.e. Italy, Spain and Israel) and from below 1.0 mg/L to 4.6 mg/L for sites using advanced treatment (Israel, Belgium and China) (Figure 2.4). In Australia, the wetland treated stormwater used for injection had an average BOD content of 1.7 mg/L, while in South Africa an average BOD content below 10 mg/L was measured in the mix of secondary effluent. The raw wastewater used for irrigation purposes in Mexico presented BOD content varying between 227 and 530 mg/L (Figure 2.4) depending on the sampling time of the year with the lowest BOD values being observed during the wet season as a result of dilution effect. Indeed, the annual rainfall of 700mm in this area mainly occur during the wet season (NRC, 1995), June to September, when a single storm event can constitute 10 to 15% of the annual rainfall (Jimenez, 2008a).

A more detailed analysis of the nature of the organic components in the injectant sources used at the sites reveals as expected BOD:COD ratios ranging on average from 0 in Australia and in Israel (pilot site) to maximum 0.4 in Italy and Belgium while typical values for final effluent range from 0.1 to 0.3 (Metcalf & Eddy, 2003), hence suggesting that little biodegradation of organic material would occur in the aquifer. In Mexico the BOD:COD ratio of the untreated wastewater was 0.6 as reported for typical untreated municipal effluent (Metcalf & Eddy, 2003).

Comparison of the BOD data in the abstracted water across the sites reveals that after passage through the aquifer, a majority of the sites presented values≤5 mgBOD/L (i.e. Spain, Israel, China, Belgium) or ≤10 in South Africa hence meeting the most stringent standards for water reuse (USEPA, 2004). The only sites where BOD values were on average ≥10 mg/L were Italy and Mexico with respectively 16 mgBOD/L and 33 mgBOD/L in their abstracted water.

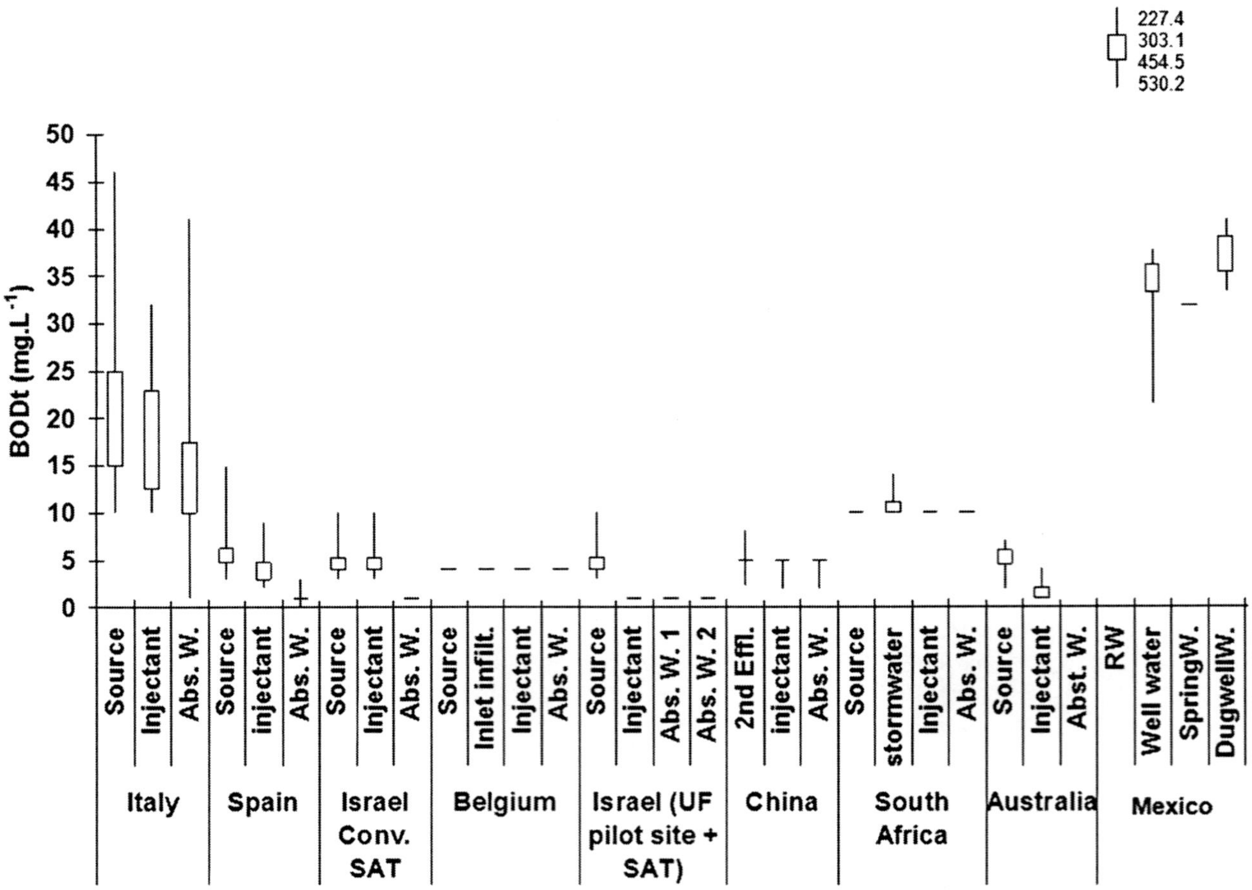

Figure 2.4 Comparison of BOD levels in respectively the source, the injectant and the abstracted water at all sites

The comparison of the capacity of the different sites to treat the biodegradable organic matter is delicate given the generally low BOD levels injected. However, these results illustrate the remarkable capacity of atypical SAT system in Mexico to remove biodegradable compounds with a 91% BOD decrease from an initial average value of 379 mg/L.

DOC

Comparing the sites' performances in terms of DOC is delicate as several factors such as pre-treatment options, recharge method, retention time, redox conditions, removal mechanisms (aerobic respiration, denitrification) and presence of oxidants (NO_3^-/O_2) can have an impact on the fate and removal of DOC within aquifers (Vanderzalm *et al.* 2006; Lin *et al.* 2008). Here the use of advanced treatment predictably offered a greater barrier to the injection of pollutants into aquifers as illustrated in Figure 2.5 which shows that the lowest organic contaminant levels (as DOC) in the injectant were found in Belgium with 2.1 mgDOC/L on average in the infiltration pond after RO + UF pre-treatment and in China with 3.0 mgDOC/L after ozonation (Figure 2.5). However, the aquifers subject to the highest contaminant loads, that is too say involving low levels of pre-treatment, proved capable of buffering high DOC concentrations and achieved an abstracted water quality comparable and sometimes higher to sites using high pre-treatment technologies. For example the lowest DOC content in abstracted water was measured in Israel where secondary treated effluent is used for recharge (Figure 2.5) with an average of 0.8 mgDOC/L in the abstracted water, while in Belgium, DOC levels were not affected and even slightly increased from 2.1 mgDOC/L to 2.9 mgDOC/L after pond infiltration into the dunes and mixing with native groundwater. When comparing the recharge methods used, the highest DOC removal were observed at sites using SAT (i.e. Israel). However, firsts results obtained in Australia after well injection of stormwater into a brackish aquifer showed significant reduction in DOC levels from 6.1 mgDOC/L in the wetland treated injectant to 1.6 mg/L in the water recovered from the ASTR system.

Macronutrients

Overall the recharge of reclaimed effluent at the different sites showed that macronutrient levels such as phosphorus and nitrogen compounds can be significantly reduced after passage through soil.

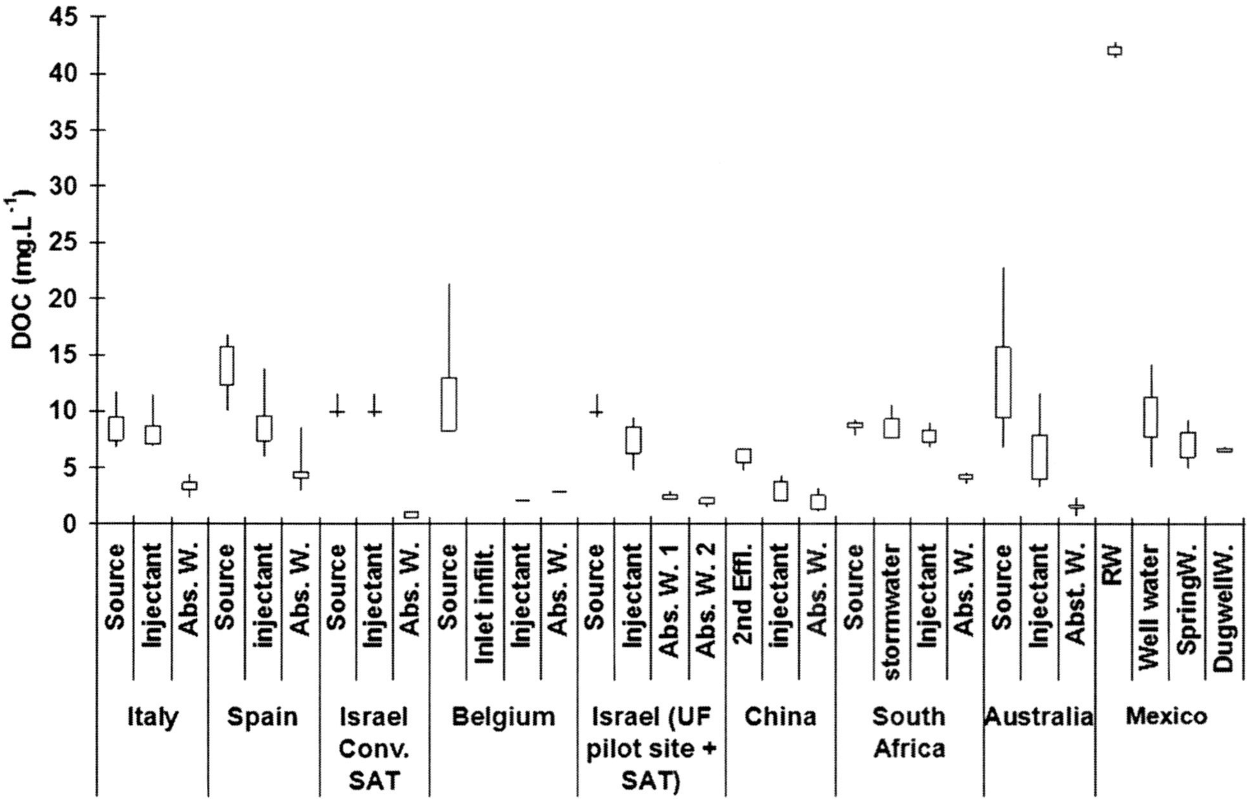

Figure 2.5 Comparison of DOC levels in respectively the source, the injectant and the abstracted water at all sites

Regarding phosphorus compounds, the analysis of the injectant (as total phosphorus (TP), Figure 2.6) showed values ranging between 0~0.1 mg/L in Belgium and Australia and 2.7 mg/L in China, except in Mexico where the raw wastewater used as injectant presented a TP value of 26.1 mg/L. After passage through the aquifer all values were very low with TP values all below 0.4 mg/L. If at the sites using advanced treatment the majority of the phosphorus removal occurred before aquifer recharge, the site using secondary treated effluent for recharge showed that independently from the recharge method and type of aquifers significant phosphorus removal, generally resulting from precipitation and adsorption mechanisms (Idelovitch *et al.* 2003), can be achieved. For example, TP content in secondary treated effluent in Italy was on average 1.8 mg/L to reach a TP content of 0.4 mg/L in abstracted water after passage through the karstic aquifer while in South Africa, infiltration of secondary effluent mixed with stormwater through the unconfined Atlantis aquifer mainly composed of sand led to a decrease from 2.6 down to 0.2 mg/L. It is to notice that at this latter site, the low content in phosphorus of the stormwater (i.e. ~1.0 mg/L) blended with the higher levels present in the secondary effluent (i.e. 8.6 mg/L) helped reducing the phosphorus load to be infiltrated in the aquifer (Figure 2.5). In Mexico the atypical soil aquifer treatment of raw wastewater caused a remarkable 98% reduction for phosphorus compounds with TP concentrations of 0.7 mg/L in the spring water and 0.4 mg/L in the water recovered from wells and dug wells, highlighting the obvious capacity of the volcanic and limestone soil to treat wastewater. However these results also suggest the accumulation of phosphorus in soil which at long terms could progressively lead to the saturation of phosphorus sorption sites in soils (Lin *et al.* 2006). However according to Lin and Banin (2006), who estimated P breakthrough at the Shafdan site, this phenomenon would be extremely long-term to occur, that is too say between many decades and centuries.

In 1991, the European Commission developed a directive for the protection of subsurface and groundwater against pollution caused by nitrates. This directive established that nitrates levels in groundwater should not exceed 50 mg/L, or 11.4 mg/L as NO_3-N (European Union, 1991). When looking at the nitrates levels (as NO_3-N) in abstracted water across the sites (Figure 2.7), all were significantly below that limit with values ranging from 0 in Australia to 5.8 mg/L in Israel. Overall the passage of the effluent used for recharge through the aquifer caused decreases in nitrates with for example nearly 38% removal at the Nardò site where the aquifer is karstified, to 85% nitrates removal in South-Africa, with an average final concentration of 0.6 mg NO_3-N/L in the abstracted water (Figure 2.7).

Overall, all sites' injectant sources, except the Chinese one, met the polishing step limit proposed by Bixio and Wintgens (2006) for aquifer recharge.

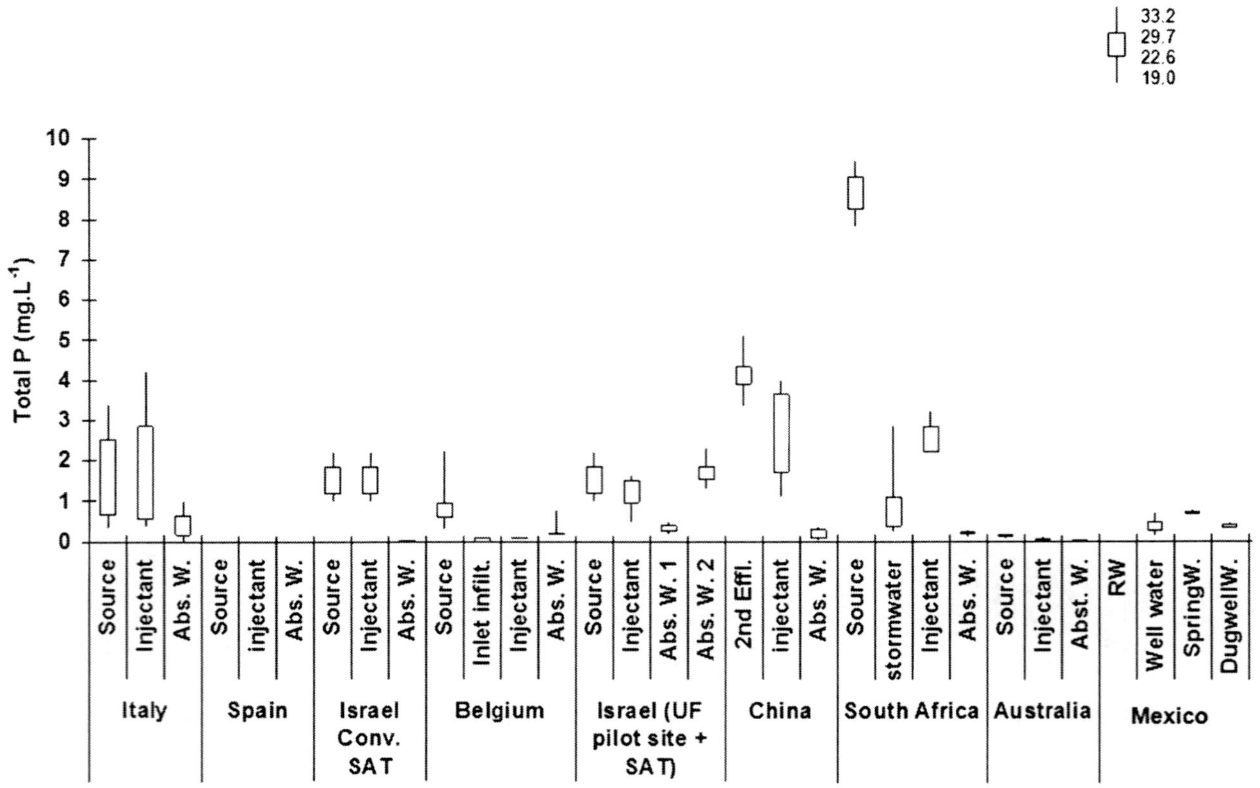

Figure 2.6 Total P levels in respectively the source, the injectant and the abstracted water at all sites

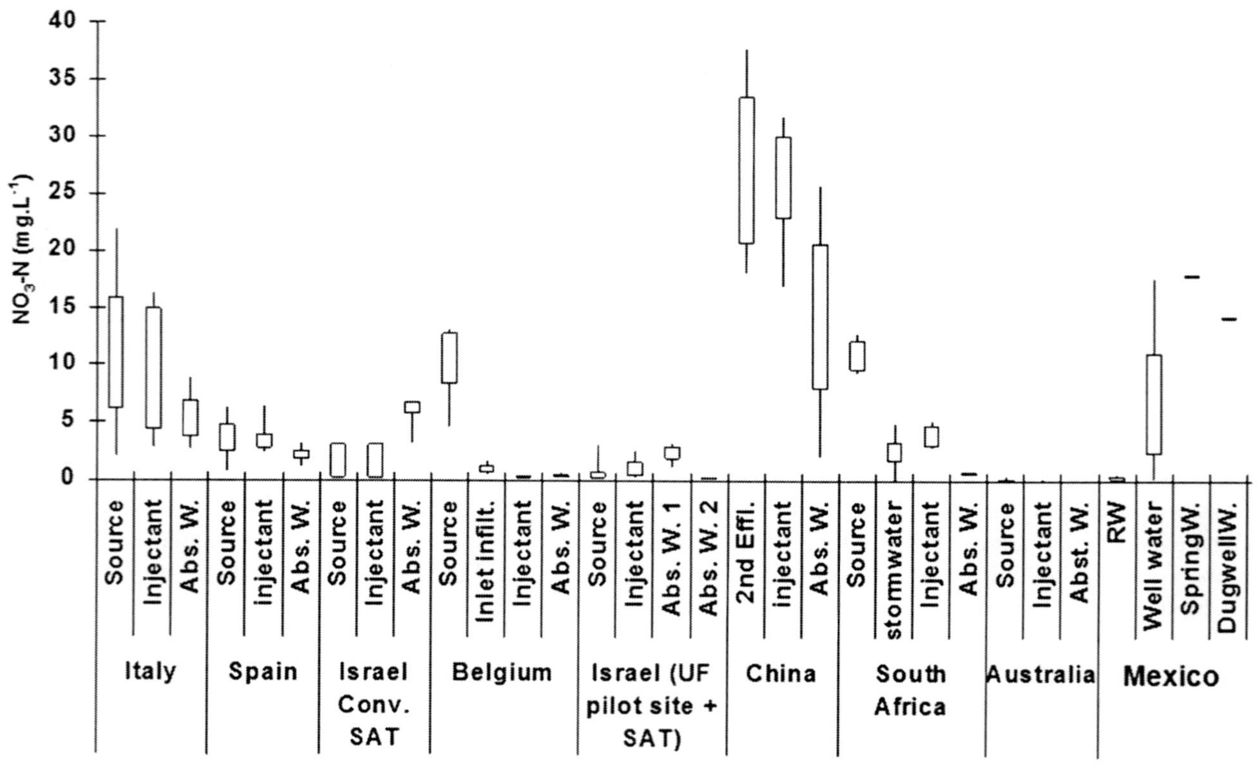

Figure 2.7 Nitrates levels in respectively the source, the injectant and the abstracted water at all sites

The only sites where NO_3-N concentrations increased after passage through the aquifer were the Mexican and the Israeli sites where recharge is achieved via managed soil aquifer treatment and unintended soil aquifer treatment respectively.

In Mexico the average nitrate levels in the well water were around 7.4 mg/L although they regularly exceeded the European limit as illustrated on Figure 2.7 with 25% of the concentration being above 11.1 mg/L. Furthermore, nitrates levels exceeded 14 and 17 mg/L on average in the water sampled from the dug well and spring water respectively. In fact, the atypical soil aquifer treatment of raw wastewater in Mexico resulted in significant leaching of NO_3-N as the injectant nitrate concentration (i.e. in the raw wastewater) in Mexico was only 0.2 mg/L. The long term irrigation with wastewater (50 years in the area considered here) in the Tula Valley is known to have contributed to an increasing amount of nitrogen not being used as nutrients by plants although not significantly accumulating in soils (Ramirez-Fuentes *et al.* 2002). This suggests reactions of nitrification are occurring subsurface, hence producing nitrates easily leached out as a consequence of regular irrigation and flooding events (Jimenez, 2008b; Ramirez-Fuentes *et al.* 2002).

In Israel, an average increase in NO_3-N from 0.8 to 5.7 mg/L was observed after passage of the secondary effluent through the aquifer at the long term SAT site. At the short-term SAT system, the UF treated effluent led to an increase from 1.1 to 2.3 mg/L in the well the closest to the injection point, while it decreased to 0.2 mg/L in the well the furthest away from the dug well. One of the main reasons to the net increase in nitrate levels at the conventional SAT in Shafdan is nitrification due to the prevailing aerobic conditions resulting from 1 day wetting/2–3 days drying infiltration regime (Idelovitch *et al.* 2003; Bixio & Wintgens, 2006). However, in Israel, changes of redox conditions thought to be due to anaerobic conditions that have developed in the saturated zone as a consequence of 30 years of SAT operation (despite the relaxation regime used), have resulted in manganese dissolution and clogging problems during irrigation by formation of Manganese oxides (MnO_2) (Oren *et al.* 2007). This manganese-reducing redox status of the Shafdan aquifer is also thought to contribute to the oxidation of ammonia to nitrates using labile manganese oxides as substrates (Petennati *et al.* 2008). At the pilot sites, the overall average increase in nitrates may have been also influenced by changes in redox conditions during the first sampling campaigns as the pilot site was originally designed to be operated in continuous flow, hence favouring anaerobic conditions.

Finally in China, high nitrates levels were measured across the scheme. After secondary treatment followed by ozonation and slow sand filtration, the injectant presented an effluent with a 26.1 mg/L concentration in NO_3-N, showing little impact of the pre-treatment on this parameter. The injection of this effluent into the Gaobeidian aquifer led to a 45% NO_3-N removal, although the concentration remained high in the abstracted water with 14.4mg NO_3-N/L.

2.3.2 Microbiological parameters

The removal of pathogens during aquifer recharge depends on a combination of factors including soil characteristics, residence time within the aquifer, hydrogeology, subsoil's microbial activity, water characteristics such as pH, oxygen levels (DO) but also the type of microbiological organisms (Asano & Levine, 1996; Jordan *et al.* 2004).

The analysis of the results obtained at the sites illustrated the good capacity of the different systems to eliminate microbiological parameters, with a wide range of micro-organisms being effectively removed after passage through soil independent of their actual concentration in the injectant source. For example, micro-organisms such as *E. coli* could be removed by 1 log units to 4 log units while Total Coliforms were reduced by 1 to 5 log units (Table 2.5 and 2.6).

Table 2.5 *E. coli* in respectively the source, the injectant and the abstracted water of the sites. [average (min-median-max); n = number of samples].

a) ***E. coli* (CFU/100 mL) Site ID**	**Source**	**Injectant**	**Abstracted water**
Italy	$1.3\ 10^3$ $(8.0 - 3.5\ 10^2 - 5.4\ 10^3)$ n = 7	$1.7\ 10^2$ $(4.0 - 2.6\ 10^1 - 9.2\ 10^2)$ n = 7	$8.1\ 10^1$ $(2.0 - 2.6\ 10^1 - 3.5\ 10^1)$ n = 7
Spain	$9.9\ 10^4$ $(2.2\ 10^4 - 5.9\ 10^4 - 3.3\ 10^5)$ n = 14	$6.4\ 10^4$ $(8.0\ 10^1 - 6.1\ 10^4 - 2.3\ 10^5)$ n = 14	$2.9\ 10^1$ $(2.2\ 10^{-1} - 3.7 - 2.1\ 10^2)$ n = 14
Israel Conv sat	$8.6\ 10^4$ $(4.0\ 10^2 - 3.7\ 10^4 - 2.7\ 10^5)$ n = 4	$8.6\ 10^4$ $(4.0\ 10^2 - 3.7\ 10^4 - 2.7\ 10^5)$ n = 4	0 (0 – 0 – 0) n = 6
Belgium	$3.3\ 10^4$ $(1.1\ 10^4 - 1.2\ 10^4 - 8.7\ 10^4)$ n = 6	$1.0\ 10^2$ $(0 - 4.8\ 10^1 - 3.6\ 10^2)$ n = 6	0 (0 – 0 – 0) n = 7

(Continued)

Table 2.5 *E. coli* in respectively the source, the injectant and the abstracted water of the sites. [average (min-median-max); n = number of samples] (*Continued*).

a) *E. coli* (CFU/100 mL) Site ID	Source	Injectant	Abstracted water
Israel UF Pilot Site	$8.6\ 10^4$ $(4.0\ 10^2 - 3.7\ 10^4 - 2.7\ 10^5)$ n = 4	$1.3\ 10^1$ $(1.0\ 10^1 - \times - 1.5\ 10^1)$ n = 2	*Close well:* 0 $(0 - \times - 0)$ n = 2 *Far well:* 0 (0 – 0 – 0) n = 3
China	$4.9\ 10^3$ $(4.5\ 10^1 - 2.2\ 10^2 - 2.4\ 10^4)$ n = 5	$1.3\ 10^1$ $(1.0 - 1.5\ 10^1 - 2.3\ 10^1)$ n = 6	1.0 $(1.0 - \times - 1.0)$ n = 2
Australia	$4.6\ 10^2$ $(0 - 4.6\ 10^1 - 2.8\ 10^3)$ n = 8	$3.4\ 10^1$ $(0 - 1.6\ 10^1 - 2.2\ 10^2)$ n = 34	0 (0 – 0 – 0) n = 14
South Africa	*Sec eff.* $1.2\ 10^4$ $(3.510^3 - 9.2\ 10^3 - 1.3\ 10^5)$ n = 4 – *Stormwater* $1.2\ 10^4$ $(1.4\ 10^3 - 2.5\ 10^3 - 4.1\ 10^4)$ n = 4	$1.7\ 10^3$ $(5.2\ 10^1 - 1.6\ 10^3 - 3.5\ 10^3)$ n = 4	0 (0 – 0 – 0) n = 4
Mexico*	$4.5\ 10^5$ $(3.5\ 10^5 - \times - 5.6\ 10^5)$ n = 2	$4.5\ 10^5$ $(3.5\ 10^5 - \times - 5.6\ 10^5)$ n = 2	*Well water* 1.4 $(3.3\ 10^{-1} - 1.0 - 3.0)$ n = 3 *Spring water* $1.9\ 10^1$ $(1.8\ 10^1 - \times - 2.1\ 10^1)$ n = 2 *Dug Well* $9.8\ 10^1$ $(5.6\ 10^1 - \times - 1.4\ 10^2)$ n = 2

* Values indicated are for the site irrigated for 50 years with raw wastewater; x: median not calculated for $n \leq 2$.

Table 2.6 Total Coliforms in respectively the source, the injectant and the abstracted water. [average (min-median-max); n = number of samples].

b) ***Total Coliforms* (CFU/100 mL) Site ID**	Source	Injectant	Abstracted water
Italy	$6.9\ 10^4$ $(1.7\ 10^4 - 2.8\ 10^4 - 1.6\ 10^5)$ n = 3	$1.2\ 10^4$ $(2.2\ 10^3 - 9.2\ 10^3 - 2.4\ 10^4)$ n = 3	$3.7\ 10^3$ $(9.2\ 10^2 - 9.2\ 10^2 - 9.2\ 10^3)$ n = 3
Spain	$1.3\ 10^6$ $(3.2\ 10^5 - 1.2\ 10^6 - 9.2\ 10^6)$ n = 14	$1.3\ 10^6$ $(5.3\ 10^4 - 8.7\ 10^5 - 4.9\ 10^6)$ n = 14	$1.1\ 10^3$ $(4.0 - 6.6\ 10^1 - 6.2\ 10^3)$ n = 14
Israel Conv Sat	$9.8\ 10^5$ $(8.0\ 10^5 - 9.0\ 10^5 - 1.3\ 10^6)$ n = 4	$9.8\ 10^5$ $(8.0\ 10^5 - 8.0\ 10^5 - 1.3\ 10^6)$ n = 4	4.0 $(0.0 - 6.6\ 10^1 - 5.0\ 10^{-1})$ n = 4

(*Continued*)

Table 2.6 Total Coliforms in respectively the source, the injectant and the abstracted water. [average (min-median-max); n = number of samples] (*Continued*).

b) ***Total Coliforms* (CFU/100 mL) Site ID**	**Source**	**Injectant**	**Abstracted water**
Belgium	$1.3\ 10^5$ ($3.5\ 10^4 - 6.4\ 10^4 - 3.3\ 10^5$) n = 6	$9.1\ 10^2$ ($2.4\ 10^1 - 3.9\ 10^2 - 2.6\ 10^3$) n = 6	0 (0 – 0 – 0) n = 7
Israel UF Pilot Site	$9.8\ 10^5$ ($8.0\ 10^5 - 9.0\ 10^5 - 1.3\ 10^6$) n = 4	$7.8\ 10^3$ ($7.0\ 10^1 - 1.2\ 10^3 - 2.2\ 10^4$) n = 3	*Close well:* $5.8\ 10^1$ ($0.0 - \times - 1.2\ 10^2$) n = 2 *Far well:* $5.5\ 10^1$ ($2.0 - 4.3\ 10^1 - 1.2\ 10^2$) n = 3
China *(as faecal coliforms)*	$6.0\ 10^2$ ($4.0 - 7.9\ 10^1 - 2.8\ 10^3$) n = 5	7.0 ($1.0 - 2.0 - 2.3\ 10^1$) n = 7	1.0 (1.0 – 1.0 – 1.0) n = 3
Australia *(as faecal coliforms)*	$4.6\ 10^2$ ($0 - 4.6\ 10^1 - 2.8\ 10^3$) n = 8	$3.6\ 10^1$ ($2.0 - 1.7\ 10^1 - 2.2\ 10^2$) n = 34	0 (0 – 0 – 0) n = 14
	Sec eff. $6.9\ 10^4$ ($5.0\ 10^3 - \times - 1.3\ 10^5$) n = 2	$4.3\ 10^4$	2.0
South Africa	– *Stormwater* $5.5\ 10^4$ ($5.0\ 10^3 - x - 1.0\ 10^5$) n = 2	($5.0\ 10^3 - \times - 8.2\ 10^4$) n = 2	($0 - \times - 4.0$) n = 2
Mexico* *(as faecal coliforms)*	$7.7\ 10^6$ ($1.3\ 10^6 - \times - 1.4\ 10^7$) n = 2	$7.7\ 10^6$ ($1.3\ 10^6 - \times - 1.4\ 10^7$) n = 2	*Well water* 8.0 ($3.3\ 10^{-1} - 3.9 - 2.3\ 10^1$) n = 4 *Spring water* $6.9\ 10^1$ ($2.0\ 10^1 - \times - 1.2\ 10^2$) n = 2 *Dug Well* $5.0\ 10^2$ ($2.1\ 10^2 - \times - 8.0\ 10^2$) n = 2

*Values indicated are for the site irrigated for 50 years with raw wastewater; x: median not calculated for $n \leq 2$.

As expected, advanced treatments preceding recharge offer a good and robust barrier to the injection of microbiological contaminants in aquifers. To illustrate in Belgium, *E. coli* and Total Coliforms have never been detected in the effluent after double membrane filtration (RO + UF), although a slight recontamination occurred prior to recharge in the infiltration pond. However, the dune filtration system is able to eliminate these residual pathogens as neither *E. coli* nor Total coliforms are detectable in the abstracted water after 35 days retention time in the sandy aquifer. Similarly the ozonation systems followed by slow sand filtration in China also ensure one of the highest pathogen removal as *E. coli* and Faecal coliforms prior recharge, with respective values <15 CFU/100 mL and <10 CFU/100 mL. This latest value is well below the Faecal coliform limit of 10^4 CFU/100 mL suggested by Bixio and Wintgens (2006) for aquifer recharge by indirect infiltration as a polishing step, and really close to the 0 CFU/100 mL limit for direct aquifer recharge as a polishing step.

In Israel, the UF filtration unit used to treat secondary effluent at the pilot site also reduced significantly contents in *E. coli* with up to 3 log units removal with an average concentration <15 CFU/100 mL in the injectant. The removal of Total coliforms seems on average lower, but this result was affected by heavy pollution of the feed and permeate tanks that deteriorated the microbiological quality during the summer season due to high temperatures (Aharoni & Cikurel, 2006; Le Corre *et al.* 2007). Although the UF system removed Total coliforms to a lesser extent, the passage through the aquifer after dug well filtration led to abstracted water containing 58 CFU/100 mL on average. When compared to the performance of the conventional SAT system using secondary treated effluents where no coliforms where found in abstracted water, the capacity of the aquifer to remove contaminants remaining in the injectant after UF is slightly lower. High removal of particles by ultrafiltration can be responsible for higher infiltration rates, hence lower removal of specific contaminants (Le Corre *et al.* 2007). Furthermore, the retention time of 10–12 months used at the conventional site, as opposed to 1–2 months at the pilot site, may also explain its higher capacity to remove contaminants.

Although advanced treatment occurred as a reliable way to limit risks of microbial contamination of aquifers, natural systems at the sites using secondary effluent as injectant can buffer variation in injectant quality and achieve an abstracted water quality similar to schemes involving advanced treatment processes. For example, *E. coli* levels measured in abstracted water were 0 CFU/100 mL after conventional soil aquifer treatment in Israel and after pond infiltration in South Africa. In Spain, despite little difference in *E. coli*, and Total Coliforms content between the secondary effluent and the river mixture, river bed filtration reduced by 3 to 4 log units microbiological contamination. In Australia, the wetland treatment of stormwater followed by ASTR performed really well as no *E. coli* nor faecal coliforms were detected. Finally, in Mexico, the study confirmed the capacity of the Tula valley atypical soil aquifer treatment to treat raw wastewater as 4–5 log units removal in *E. coli* and Total coliforms could be achieved with well, spring and dug well waters of quality suitable for irrigation purposes (See chapter 9).

Please note, that pathogens and resistance genes were further analysed in the frame of dedicated work package activities, and are reported in Chapter 11.

2.3.3 Trace elements

The elimination or attenuation of trace elements during the passage of effluent through soils relies on biological transformations, adsorption, precipitation and immobilisation mechanisms that are mainly influenced by the type of the trace element infiltrating the aquifer as well as subsoil biological and chemical reactions (Rivett *et al.* 2006; Lin *et al.* 2004; WHO, 2006; Lee *et al.* 2004), hence on the specific characteristics of a site used for recharge.

Trace element concentrations in sewage effluent are generally lower than limits for irrigation, unless mixed with drinking water or industrial effluents presenting naturally high concentrations of trace elements (Pescod, 1992). This was confirmed in the present study with trace element concentrations generally below regulatory limits for irrigation in the sources used for recharge. However some concerns arose at some of the sites for specific trace elements. For instance, the level of nickel and lead measured in the abstracted water in China were relatively high with respective average concentration of 27 ppb and 47 ppb, where at all other sites typical values were below 10 ppb.

One parameter that did cause problems at two of the sites is manganese. Indeed, the passage of secondary effluent through soil at the Israeli SAT and pilot sites and the river bed filtration scheme in Spain, caused manganese dissolution with for example an increase in magnesium concentration from 37.2 ppb in the injectant to 301.4 ppb in the abstracted water in Spain; or from 18 to 149 ppb at the Israeli pilot site while at the conventional SAT site almost no reduction in manganese was observed (Figure 2.8).

In Israel, an alternate infiltration regime including 1 day infiltration and 2–3 days relaxation is used to favour oxidation and maintain aerobic condition of the SAT system (Oren *et al.* 2007). However, after 30 years of irrigation, infiltration turned to an almost continuous mode in the unsaturated and saturated zones due to a "sink effect" meaning that the point were the hydraulic conductivity of the soil after 30 years of infiltration does not permit the infiltration of increasing amounts of effluents, the flow turns to a continuous flow and the conditions are anaerobic, that is to say no oxygen enters the saturated zone even at relaxation times since also the unsaturated zone is diminished (Cikurel & Aharoni, 2007). The continuous flow imposed at the pilot site in the first stages of experimentation was also responsible for changes in redox conditions that caused the reduction of manganese into soluble manganese oxides under suboxic conditions. (Le Corre *et al.* 2007).

In Spain, where aquifer recharge is practised via river bed filtration, the significant manganese dissolution could be linked to the occurrence of biochemical reactions in presence of DOC. Indeed, according to Petrunic *et al.* (2005), aquifers recharged via river bed filtration often undergoes geochemical changes due to the presence of DOC in the infiltrating water. The occurrence of manganese in abstracted water is thought to be the result of changes in redox conditions involving cation exchanges reactions but also reductive dissolution of manganese oxides influenced by the microbial population within the aquifer.

2.3.4 Salinity

This section focuses on the electrical conductivity (EC) data monitored at the sites as an indicator of the presence of salts in groundwater and the improvement of groundwater quality after recharged with reclaimed effluent. Indeed, salinity is an important parameter to control, especially when the reuse purpose is agricultural or urban irrigation. Indeed, excessive salinity levels can indirectly affect human health, while excess of salts can significantly reduce or stop plant growth and affect soils' hydraulic conductivity (Godfree & Godfrey, 2008; Toze, 2006). In the European Union, the Drinking Water Directive established an EC limit of 2,500 μS/cm while Bixio and Wintgens (2006) also proposed a conductivity requirement of 700 μS/cm for direct and indirect groundwater recharge as a polishing step. When applied for irrigation purposes, the Food and Agriculture Organisation guidelines suggest that EC levels below 700 μS/cm are not restrictive for reuse, EC levels between 700 μS/cm and 3000 μS/cm can cause slight to moderate issues, while the reuse of irrigation water containing EC levels above 3000 μS/cm should be severely restricted (Ayers & Westcot, 1985).

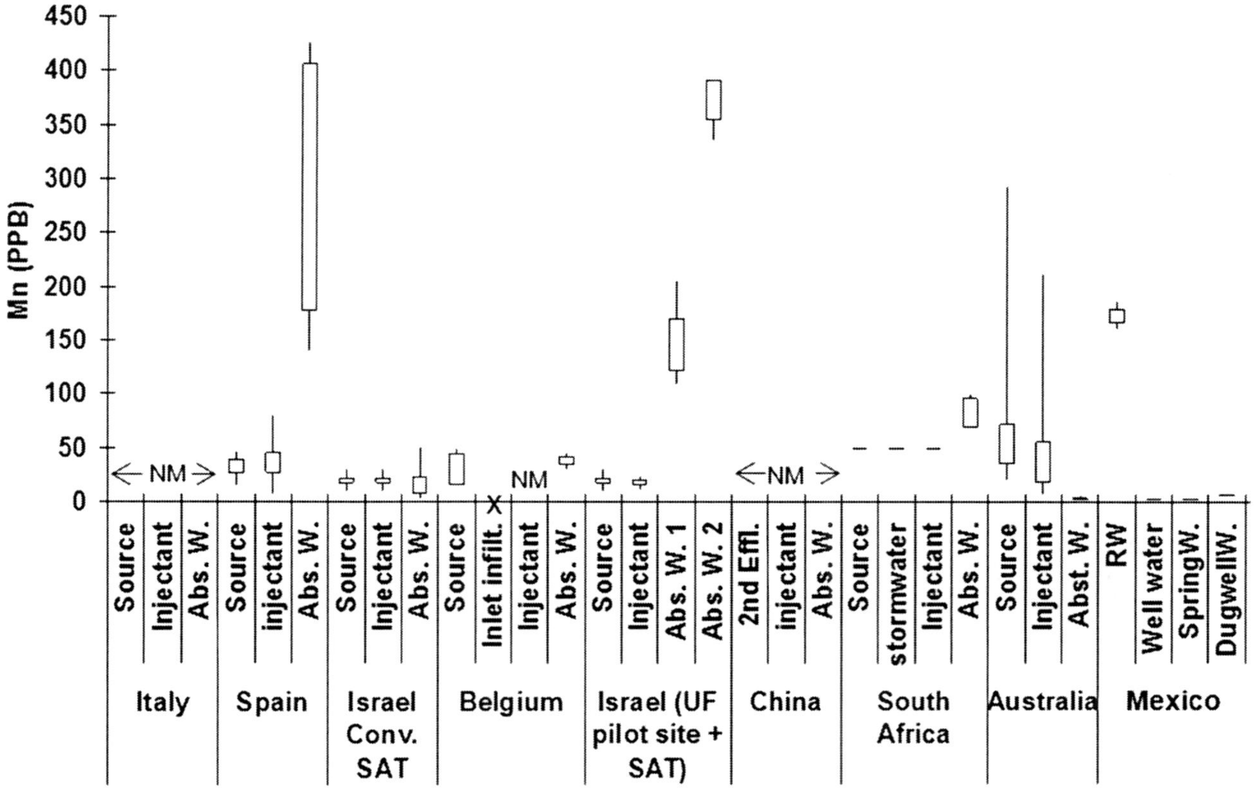

Figure 2.8 Manganese levels in respectively the source, the injectant and the abstracted water at all sites

Coastal aquifers are often subject to seawater intrusion in groundwater, the injection of reclaimed effluent into aquifers can then act as a hydraulic barrier against such intrusions. Based on the same principles, managed aquifer recharge can also help reducing salinity levels in brackish aquifer through mixing and dilution of water of higher quality with the native groundwater, hence providing further opportunities for recharge and reuse where aquifers are not exploitable (ex: Salisbury, Australia – for details on this site see Chapter 7).

In this project, the injection of reclaimed effluent has proved efficient as a barrier against salt water intrusion (i.e. Italy, South Africa) or for the remediation of brackish aquifers (i.e. Australia). Like many karstic aquifers located in costal areas (COST, 2005), the Nardò aquifer is affected by seawater intrusion. One of the principal motivations for recharge at this site was then the remediation of the salinised aquifer. In that context, following the injection of secondary treated effluent, a net reduction of the background groundwater salinity was observed with an average decrease in conductivity from 2092 μS/cm to 869 μS/cm. The quality of the brackish aquifer in Australia also improved significantly as a result of wetland treated stormwater injection. On average the treatment of stormwater via wetland produced an injectant source of low salinity, 242 μS/cm on average. As illustrated in Figure 2.9 the salinity of the water monitored at the recovery wells was 1645 μS/cm on average, representing a 54% decrease in salinity when considering that the native groundwater salinity was approximately 3600 μS/cm before the injection of 153 000 m^3 of wetland treated stormwater.

Overall, the injectant sources used at the sites presented a conductivity above the requirements of 700 μS/cm suggested by Bixio and Wintgens, 2006 and the European Union for the direct or indirect recharge of aquifer. The only exceptions to this were the South African site where the injectant source had on average a conductivity of 567 μS/cm and the Belgium site where the advanced pre-treatment option was able to remove nearly 98% of the salinity as EC prior infiltration through the Flanders dunes, hence resulting in a injectant source with an average EC of 54 μS/cm. In Spain the salinity levels remained relatively high and constant throughout the scheme with values ranging from 1947 μS/cm in the river mixture to 2010 μS/cm in the abstracted water (Figure 2.9). Although the reuse purpose at this site is urban irrigation (i.e. *municipal park*) and not agricultural irrigation, this is not a real issue at the moment, however irrigation with water containing this level of salinity could affect not only soil salinity but also the growth of plants in the park as well as their aesthetic aspect (Niu & Rodriguez, 2006). Finally, it is to notice that in Mexico the salinity levels measured over the duration of the project remain high with values ranging on average from 2607 μS/cm in the raw wastewater to 2445, 3411 and 3426 μS/cm in respectively the well water, spring water and dugwell water. According to the FAO guidelines suggested above (Ayers & Wescot, 1985), such values signify that when reused for irrigation purpose, such waters are synonymous of severe impact on crops.

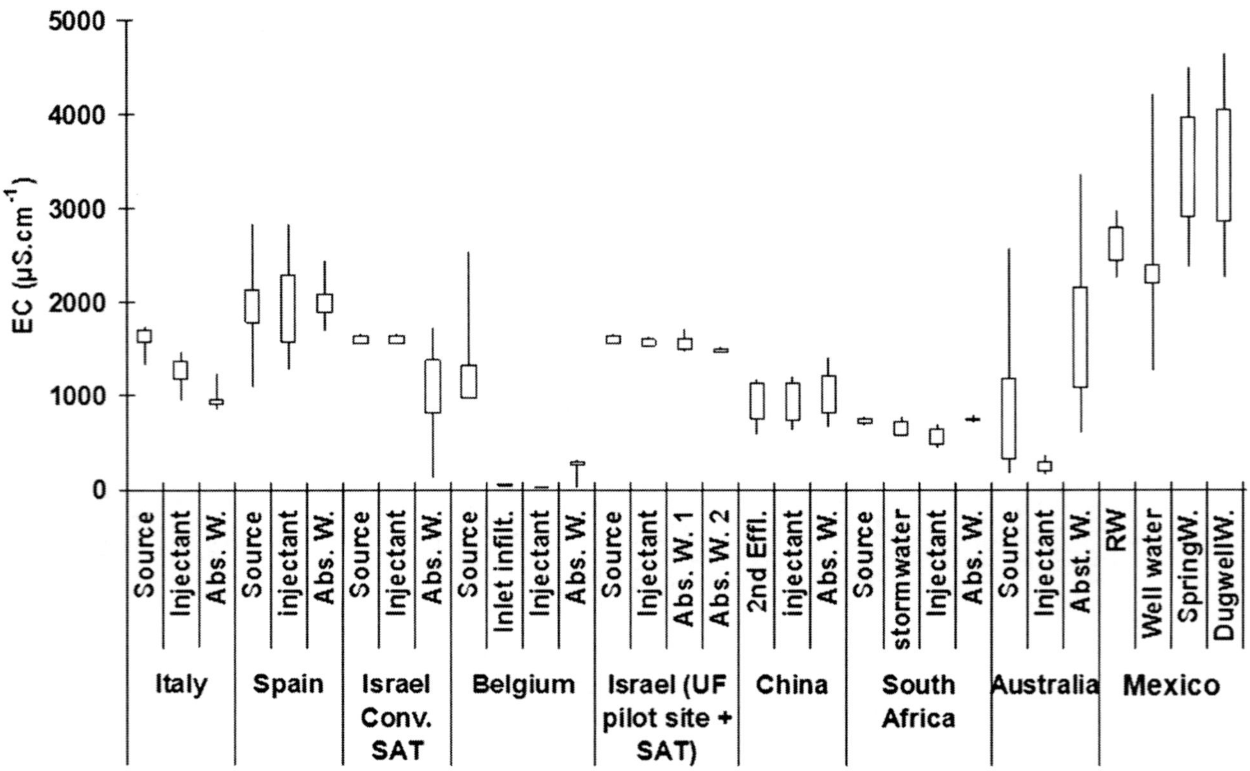

Figure 2.9 EC in respectively the source, the injectant and the abstracted water at all sites

2.4 CONCLUSIONS

The unique character of this study resides in the variety of the sites involved in term of sources used for recharge, pre-treatment options, aquifer recharge methods, aquifer types and reuse purposes. Comparing the sites in terms of their performances in recharging aquifers with reclaimed effluent for reuse is challenging as expectedly a multitude of factors are susceptible to influence their efficiencies.

The following factors affect the pre-treatment method choices for aquifer recharge of reclaimed effluent:

- Reuse purposed (quality and quantity requirements for uses such as agriculture, intrusion barrier etc.)
- Location (climate and land)
- Recharge method (surface spreading vs. direct injection)
- Reclaimed water source (quality, quantity, availability)
- Native groundwater quality
- Nature and type of aquifer to be recharged
- Economics and financing options

Indeed, as highlighted by Dillon *et al.* (2008) in their critical evaluation of combined engineered and aquifer treatment systems in water recycling, the source of reclaimed water to be used for recharge, hence its level of pre-treatment will depend on a combination of several factors which include the hydrogeology of the aquifer, confinement of the aquifer and the quality of the native groundwater. In his report on "*criteria for health related guidelines*" for aquifer recharge, Brissaud (2003) also suggested that the quality of the reclaimed water to be used for recharge should be adapted to the characteristics of the site (i.e. recharge method, hydraulic load, infiltration schedule) and the intended use of the abstracted water post recharge.

Despite an inevitable case by case approach when planning aquifer recharge as widely explained in the literature (Dillon & Jimenez, 2008; Bouwer, 1996, Franson 1988), the recharge and reuse practises studied here present technical aspects similar to a large number of schemes in the world. The performances of the sites tested and the difficulties encountered during the study should then provide valuable information for organisations planning or operating MAR sites.

Injectant sources used across the sites included raw wastewater at the unintentional soil aquifer treatment system in Mexico; secondary effluent in Italy, Spain, Israel (i.e. conventional site) and South Africa (blended with stormwater); advanced treated effluent in Belgium (RO + UF), Israel (UF), China (Coagulation-Ozonation-Sand Filtration) or wetland treated stormwater in Australia.

As illustrated above and in the subsequent chapters, the level and type of water reclamation technologies applied prior to recharge unsurprisingly determined the contaminant load needing polishing by the aquifer. However the aquifers preceded by lower levels of pre-treatment (i.e. secondary effluent, or raw wastewater) seem to be able to buffer high contaminant concentrations and achieve an abstracted water quality comparable to sites using high pre-treatment technologies. Though, when analysing the actual impact of pre-treatment levels on basic parameter removal rates after recharge (Figure 2.10), as expected removal rates after passage through soil typically decrease, when the level of pre-treatment increases. This corroborates the fact that independently from the recharge method and the type of aquifer used, the aquifer primarily acts as a treatment step when lower qualities of effluent are used, while it acts as a polishing and storage step when higher effluent quality is used.

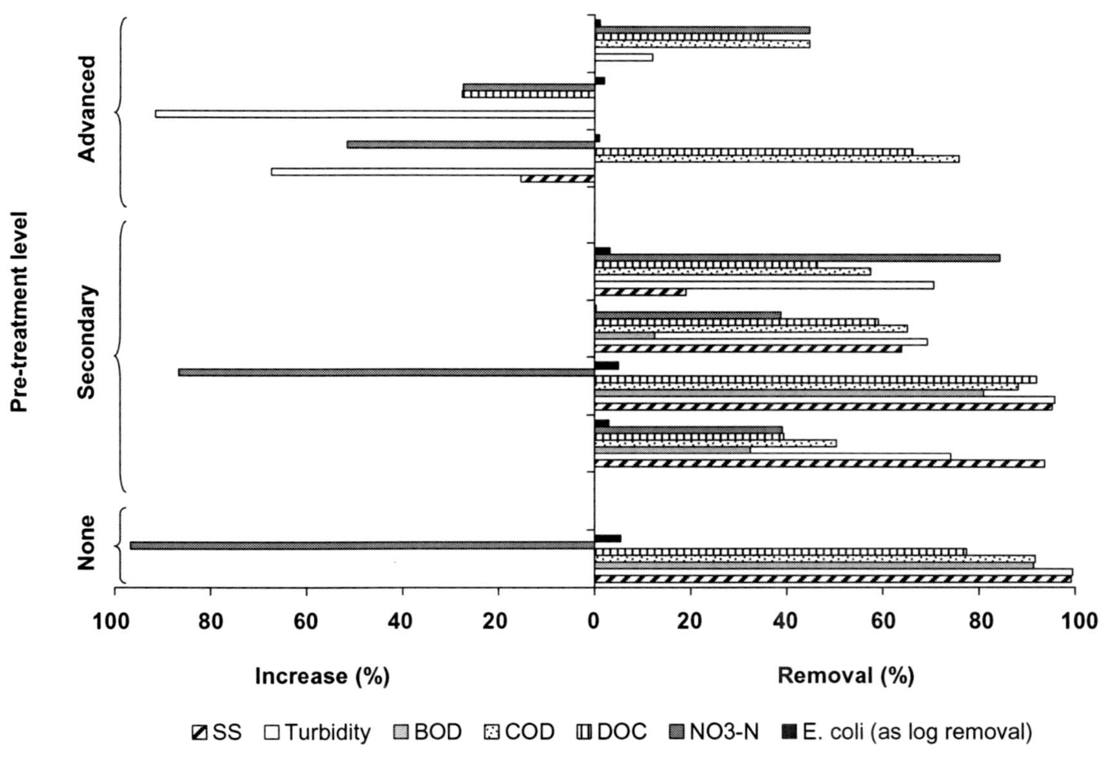

Figure 2.10 Influence of the injected/infiltrated water quality pre-treatment level on average percentage removals or increase after aquifer recharge at each site

The residence time that the injected/infiltrated water spends within the aquifer before recovery could also be one of the factors enhancing the capacity of the sites using secondary effluent to achieve removals comparable to sites using advanced technologies. For example, the Israeli conventional SAT site uses secondary effluent that travels within the aquifer for a minimum of 180 days to a maximum of 360 days, hence providing higher opportunities for removal than at the pilot Israeli site where highly treated effluent can lead to higher infiltration rates in the aquifer (Chilton & Alley, 2006), hence also reducing aquifer travel times and chances for quality improvements. Similarly, in Belgium the level of pre-treatment prior infiltration ensuring robust, reliable treatment of the effluent to very high quality standards prior recharge combined with the 35 days retention time leaves little opportunities for quality improvements during soil aquifer passage. Recharge of RO pre-treated water does not improve the water quality but can serve as additional barrier and storage. Overall, when comparing the performances of the site in terms of removal of basic wastewater parameters versus retention time in the aquifer independently of the recharge method used, no particular trend could be confirmed (Figure 2.11). However, as shown in Figure 2.11 for retention times below 60 days, removal performance seemed lower.

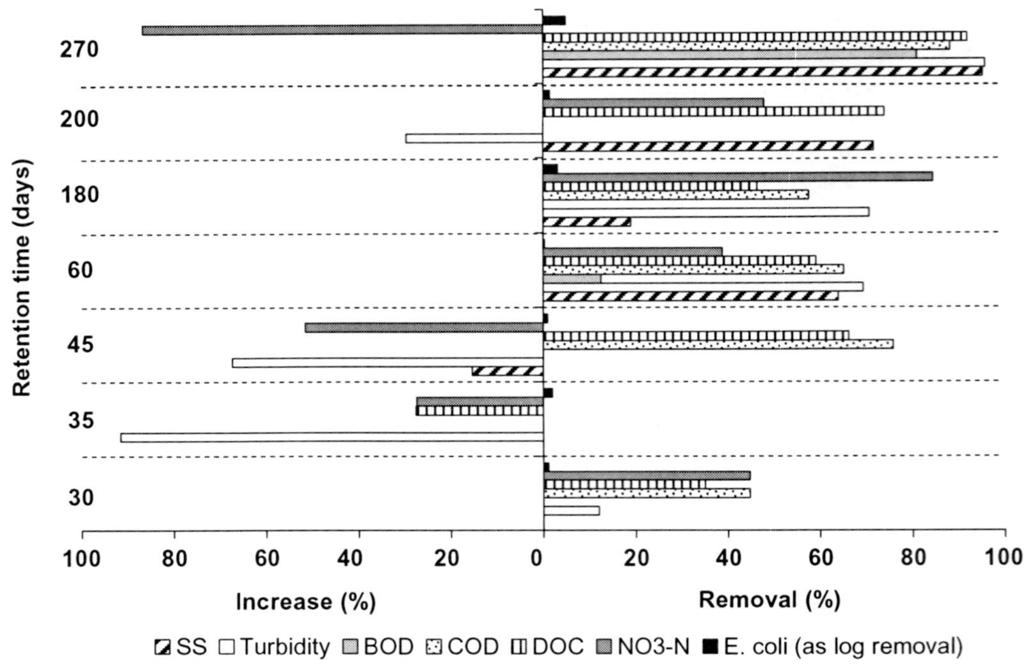

Figure 2.11 Average percentage removals or increase associated to the retention times of the sites studied

Case study	Source	Pretreatment level	Aquifer residence time	End use
Wulpen, B	sewage	3ry	35 d	Drinking
Atlantis, ZA	sewage + stormwater	2ry	> 150 d	Drinking
Salisbury, AUS	stormwater	wetland	200 d	Drinking
Mezquital Valley, MEX	sewage + surface water	dam	unknown	Drinking irrigation
Beijing, CN	sewage	3ry	30 d	Cleaning
Shafdan, ISR	sewage	2ry	180 – 360 d	Irrigation
Nardò, I	sewage	2ry	60 d	Irrigation
Sabadell, E	sewage	2ry	7 d	Irrigation cleaning

Figure 2.12 Treatment intensity and treatment trains for the respective end uses at each case study site

The particular case of the Australian site where wetland treated stormwater is used for recharge by means of ASR makes it difficult to compare to the other sites. Indeed, stormwater usually presents lower contents in dissolved solids, nutrients, oxygen demand and salinity and higher SS, heavy metals and bacteria level when compared to wastewaters (Barnett *et al.* 2000). However, this study showed the good capacity of wetland treatment to act as a reliable and robust polishing option for upgrading stormwater for groundwater recharge, hence helping to reduce the salinity of brackish groundwater and increasing options for reuse. Indeed, monthly injection rates of 16400 m^3 generally met Australian Drinking Water Guidelines (2004) except for turbidity, *E. coli*, Enterococci, faecal coliforms, and iron, while after conditioning of the aquifer the stored stormwater met drinking standards (exception for total iron).

Aquifer recharge and reuse schemes are highly complex undertakings involving multi-stage treatment processes utilizing a variety of physical, biological, and chemical interactions. Operational and climatic conditions strongly influence scheme performance and our suite of case study sites involved a variety of source waters and reuse applications. Figure 2.12 illustrates the relative importance of the pre-treatment and aquifer treatment in converting a specific source water to water fit for the intended use.

The detail of local climatic and operational circumstances coupled with uncertainties associated with aquifer dynamics and performance impose an iterative approach to the design and implementation of recharge/reuse schemes. The investigated case studies have all recognised this and have taken appropriately conservative approaches to scheme development. 'Appropriate' schemes might thereby be characterised as pursuing structured exploration of the recharge/ reuse system dynamics, focusing on issues of reliability and risk as elaborated in Chapter 19.

Over the duration of the project no major operational issues were observed although some problems remained regarding the control of redox conditions within the aquifer which could cause maintenance and clogging problems as illustrated in Israel and Spain. Overall, the project confirmed the multiple benefits of using reclaimed wastewater or stormwater for aquifer recharge with the ultimate objectives of combating water restrictions (Spain), preventing salinity and seawater intrusion (Italy, South Africa), and producing high water quality for irrigation (Israel), cleaning (China), and potable reuse purposes (Australia, Belgium, South Africa).

REFERENCES

Aharoni A. and Cikurel H. (2006). Mekorot's research activity in technological improvements for the production of unrestricted irrigation quality effluents. *Desalination*, **187**, 347–360.

Asano T. and Cotruvo J. A. (2004). Groundwater recharge with reclaimed municipal wastewater: health and regulatory considerations. *Water Research*, **38**(8), 1–14.

Asano T. and Levine A. D. (1996). Wastewater reclamation, recycling and reuse: past, present and future. *Water Science and Technology*, **30**(10-11), 1941–1951.

Ayers R. S. and Westcot D. W. (1985). Water quality for agriculture, FAO irrigation and drainage paper 29 rev. 1, Food and Agriculture Organization of the United Nations, Rome, Italy. Available at: http://www.fao.org/DOCREP/003/T0234E/T0234E00.htm (*accessed February 2009*).

Barnett S. R., Howles S. R., Martin R. R. and Gerges N. Z. (2000). Aquifer storage and recharge: Innovation in water resources management. *Australian Journal of Earth Sciences*, **47**(1), 13–19.

Bixio D. and Wintgens T. (eds) (2006). Water Reuse System Management Manual, European Comission Community Research. Aquarec Project Report, EKV1-CT-2002-00130, 676.

Bixio D., Thoeye C., De Koning J., Joksimovic D., Savic D., Wintgens T. and Melin T. (2006). Wastewater reuse in europe. *Desalination*, **187**(1-3), 89–101.

Bouwer H. (1996). Issues in artificial recharge. *Water Science and Technology*, **33**(10-11), 381–390.

Brissaud F. (2003). Groundwater recharge with recycled municipal wastewater: criteria for health related guidelines. In: Health Risks in Aquifer Recharge Using Reclaimed Water-State of the Art Report, R. Aertgeerts and A. Angelakis (eds). Water, Sanitation and Health, Protection of the Human Health, WHO Regional Office for Europe, Copenhagen, Denmark, 2003.

Canada health guidelines (2004). Guidelines for Canadian Drinking Water Quality, Supporting Documentation. *Enteric Viruses*, Section 6.2.1: Viral indicators in Untreated Surface Water or Groundwater. Prepared by the Committee on Drinking Water of the) Federal-Provincial-Territorial Committee on Health and the Environment, Health Canada Ottawa, Ontario, (www.hcsc.gc.ca/hecs-sesc/water/dwgsup.htm).

Chilton J. and Alley W. (2006). Hydrological management. In: Protecting Groundwater for Health: Managing the Quality of Drinking-Water Sources, O. Schmoll, G. Howard, J. Chilton and I. Chorus (eds). World Health Organization, 2006, pp. 678. (Available online at: http://www.who.int/water_sanitation_health/publications/protecting_groundwater/en/index.html).

Cikurel H. and Aharoni A. (2007). New Concepts Envisioned for SAT in Israel. In: *Proceedings of The Soil/Aquifer-Based Natural Systems for Drinking Water and Wastewater Treatment Workshop*, SWITCH Project, 28–29 November 2007, Tel Aviv, Israel.

COST Action 621 (2005). Groundwater management of coastal karstic aquifers. L. Tulipano (ed.), EU Publication office (OPOCE), 336p.

Dillon P., Page D., Vanderzalm J., Pavelic P., Toze S., Bekele E., Sidhu J., Prommer H., Higginson S., Regel R., Rinck-Pfeiffer S., Purdie M., Pitman C. and Wintgens T. (2008). A critical evaluation of combined engineered and aquifer treatment systems in water recycling. *Water Science and Technology*, **57**(5), 753–762.

Dillon P., Pavelic P., Toze S., Rinck-Pfeiffer S., Martin R., Knapton A. and Pidsley D. (2006). Role of aquifer storage in water reuse. *Desalination*, **188**(1-3), 123–134.

Dillon P. J. and Jimenez B. (2008). Water reuse via aquifer recharge: intentional and unintentional practises. In: Water Reuse, an International Survey of Practice, Issues and Needs, B. Jimenez and T. Asano (eds). Scientific and Technical Report No 20, 629, IWA publishing, London, UK.

European Union (1991). Council directive 91/676/ EEC of 12 December 1991 concerning the protection of waters against pollution caused by nitrates from agricultural sources. Official Journal of the European Communities.

European Union (1998). Council directive 98/83/EC of 3 November 1998 on the quality of water intended for human consumption. Official Journal Of The European Communities.

Franson J. W. (1988). Evaluating potential artificial recharge projects In: *Proceedings of The International Symposium on The Artificial Recharge of groundwater*, 23–27 August 1988, Anaheim, California, USA, ASCE Publishing.

Godfree A. and Godfrey S. (2008). Water Reuse criteria: environmental and health risk based standards and guidelines. In: Water Reuse, an International Survey of Practice, Issues and Needs, B. Jimenez and T. Asano (eds). Scientific and Technical Report No 20, 629, IWA publishing, London, UK.

Guendert D. (2004). Orange county's innovative water project to purify 70 Mgd in 2007. *Journal/American Water Works Association*, **96** (7), 58–63.

Icekson-Tal N., Avraham O., Sack J. and Cikurel H. (2003). Water reuse in israel–the dan region project: evaluation of water quality and reliability of plant's operation. *Water Science and Technology: Water Supply*, **3**(4), 231–237.

Idelovitch E., Icekson-Tal N., Avraham O. and Michail M. (2003). The long-term performance of soil aquifer treatment (SAT) for effluent reuse. *Water Science and Technology: Water Supply*, **3**(4), 239–246.

Jimenez B. (2008a). Incidental groundwater recharge of an aquifer used as a water supply in Mexico. *Water Reclamation and Aquifer Recharge - Experiences in the Americas*, 2nd Regional Reclaim Water Workshop, Mexico City, April 14Th - 16Th, 2008.

Jimenez B. (2008b). Unplanned reuse of wastewater for human consumption: the tula valley, Mexico. In: Water Reuse, an International Survey of Practice, Issues and Needs, B. Jimenez and T. Asano (eds). Scientific and Technical Report No 20, 629, IWA publishing, London, UK.

Jordan F. L., Sandrin S. K., Frye R. J., Brusseau M. L. and Maier R. M. (2004). The influence of system complexity on bacterial transport in saturated porous media. *Journal of Contaminant Hydrology*, **74**(1–4), 19–38.

La Mantia R., Masciopinto C., Levantesi C. and Tandoi V. (2008). Fate and transport of faecal contamination microbial indicators, pathogenic protozoa and campylobacter in the artificially recharged fractured aquifer of Salento, Italy. *Water Science and Technology*, **57**(6), 849–856.

Le Corre K., Aharoni A., Ayuso Gabella M. N., Cauwenberghs J., Chavez A., Cikurel H., De Heyder B., Genthe B., Jeffrey P., Jiménez B., Masciopinto C., Regel R., Rinck-Pfeiffer S., Salgot M., Steyn M., Tredoux G., Xuzhou C. and Zhao X. (2007). Milestone report M 4.2 integration of new treatment concepts and monitoring tools completed. WP4 Reclaim Water Project Report, 187p.

Lee S. Y., Lee J. U., Choi H. and Kim K. W. (2004). Sorption behaviors of heavy metals in SAT (soil aquifer treatment) system. *Water Science and Technology*, **50**(2), 263–268.

Levine A. D. and Asano T. (2004). Recovering sustainable water from wastewater. *Environmental Science and Technology*, **38**(11), 201A–208A.

Liebenberg and Stander (1976). Planning and Masterplan for the Stormwater Network, Outfall and Disposal System. Report GW/EduT/1961 PC. Cape Town, Liebenberg and Stander Consulting Engineers.

Lin C. and Banin A. (2006). Phosphorous retardation and breakthrough into well water in a soil-aquifer treatment (SAT) system used for large-scale wastewater reclamation. *Water Research*, **40**(8), 1507–1518.

Lin C., Eshel G., Negev I. and Banin A. (2008). Long-term accumulation and material balance of organic matter in the soil of an effluent infiltration basin. *Geoderma*, **148**, 35–42.

Lin C., Eshel G., Roehl K. E., Negev I., Greenwald D., Shachar Y. and Banin A. (2006). Studies of P accumulation in soil/sediment profiles used for large-scale wastewater reclamation. *Soil Use and Management*, **22**(2), 143–150.

Lin C., Shacahr Y. and Banin A. (2004). Heavy metal retention and partitioning in a large-scale soil-aquifer treatment (SAT) system used for wastewater reclamation. *Chemosphere*, **57**, 1047–1058.

Masciopinto C., La Mantia R., Carducci A., Casini B., Calvario A. and Jatta E. (2007). Unsafe tap water in households supplied from groundwater in the salento region of southern italy. *Journal of Water and Health*, **5**(1), 129–148.

Metcalf and Eddy, Inc. (2003). Wastewater Engineering – Treatment, Disposal and Reuse. In: G. Tchobanoglous, F. L. Burton, H. D. Stensel (eds). McGraw-Hill series in civil and environmental engineering, 4th edition. New York, USA.

Niu G. and Rodriguez D. S. (2006). Relative salt tolerance of selected herbaceous perennials and groundcovers. *Scientia Horticulturae*, **110**, 352–358.

NRC (National Research Council), AIC (Academia Nacional de la Investigación Científica, A.C.) and ANI (Academia Nacional de Ingeniería, A.C.) (1995). Mexico city's water supply: improving the outlook for sustainability. National Academy Press, Washington, D.C., USA.

Oren O., Gavrieli I., Burg A., Guttman J. and Lazar B. (2007). Manganese mobilization and enrichment during soil aquifer treatment (SAT) of effluents, the Dan Region Sewage Reclamation Project (Shafdan), Israel. *Environmental Science and Technology*, **41**(3), 766–772.

Pavelic P., Dillon P. J., Barry K. E. and Gerges N. Z. (2006). Hydraulic evaluation of aquifer storage and recovery (ASR) with urban stormwater in a brackish limestone aquifer. *Hydrogeology Journal*, **14**(8), 1544–1555.

Pescod M. B. (1992). Wastewater treatment and use in agriculture. *FAO Irrigation and Drainage Paper 47*, Food and agriculture organisation of the United Nations, Rome, Italy.

Petrunic B. M., Macquarrie K. T. B. and Al T. A. (2005). Reductive dissolution of Mn oxides in river-recharged aquifers: a laboratory column study. *Journal of Hydrology*, **301**, 163–181.

Pettenati M., Picot G., Kloppman W., Cikurel H., Aharoni A., Guttman Y. and Kraitzer T. (2008). ASTR modelling approaches: kinetic modelling of microbially-driven redox chemistry of SAT environments of Shafdan (Israel). *Water Reclamation and Aquifer Recharge Final Dissemination Workshop*, September 2008, Maribor, Slovenia.

Ramirez-Fuentes E., Lucho-Constantino C., Escamilla-Silva E. and Dendooven L. (2002). Characteristics, and carbon and nitrogen dynamics in soil irrigated with wastewater for different lengths of time. *Bioresource Technology*, **85**(2), 179–187.

Rivett M., Drewes J., Barrett M., Chilton J., Appleyard S., Dieter H. H., Wauchope D. and Fastner J. (2006). Chemicals: health relevance, transport and attenuation. In: Protecting Groundwater for Health: Managing The Quality of Drinking-Water Sources, O. Schmoll, G. Howard, J. Chilton and I. Chorus (eds). World Health Organization, 2006, 678 Pages, http://www.who.int/water_sanitation_health/publications/protecting_groundwater/en/index.Html.

Toze S. (2006). Reuse of effluent water – benefits and risks. *Agricultural Water Management*, **80**, 147–159.

Tredoux G., King P. and Cavé L. (1999). Managing urban wastewater for maximising water resource utilization. *Water Science and Technology*, **39**(10–11), 353–356.

USEPA (2004). Guidelines for Water Reuse. US Envitonmental Protection Agency, Washington, DC, Report EPA/625/R-04/108.

Vanderzalm J. L., Le Gal La Salle C. and Dillon P. J. (2006). Fate of organic matter during aquifer storage and recovery (ASR) of reclaimed water in a carbonate aquifer. *Applied Geochemistry*, **21**, 1204–1215.

Wade Miller G. (2006). Integrated concepts in water reuse: managing global water needs. *Desalination*, **187**(1-3), 65–75.

WHO (2006). Guidelines for the safe use of wastewater, excreta and greywater. *Volume 2: Wastewater Use in agriculture*, World Health Organization 2006.

Zeman C., Rich M. and Rose J. (2006). World water resources: trends, challenges, and solutions. *Re-Views in Environmental Science and Biotechnology*, **5**(4), 333–346.

Chapter 3

Indirect potable reuse via managed aquifer recharge in the Torreele/St-André project

Emmanuel van Houtte, Johan Cauwenberghs, Marjoleine Weemaes and Chris Thoeye

3.1 INTRODUCTION

3.1.1 Water management situation

The Intermunicipal Water Company of the Veurne Region (IWVA) is responsible for the distribution of drinking-water in the western part of the Flemish coastal plain. At the beginning of the 1990s the IWVA could no longer increase the groundwater extraction from its dune water catchments of St-André (123 ha) and the Westhoek (87 ha) to fulfil the increasing drinking-water demand (Figure 3.1). In the latter, salinity increased dramatically since the beginning of the 1980s, so a new approach for sustainable groundwater management was needed.

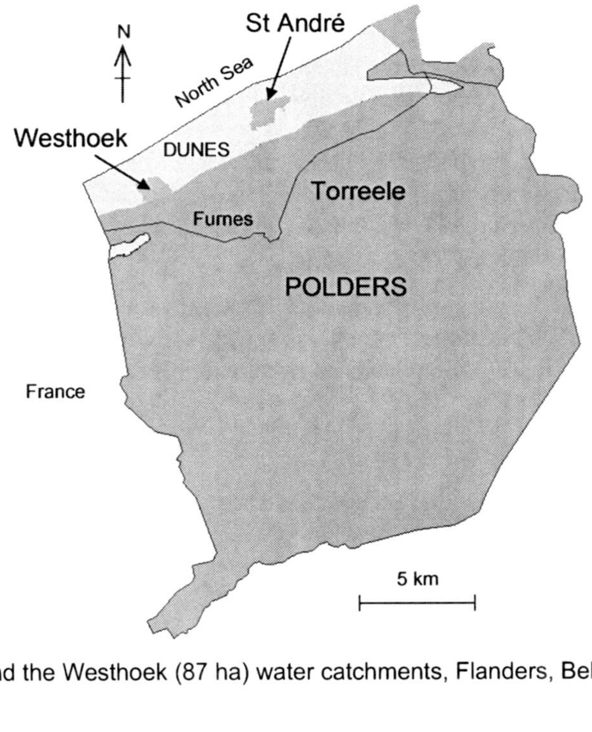

Figure 3.1 St-André (123 ha) and the Westhoek (87 ha) water catchments, Flanders, Belgium

Since wastewater treatment improved over the last decades, the effluent quality was of such a nature that it could be put to beneficial use, especially in water scarce regions. The IWVA region could be considered as such an area since freshwater is only available under the small dune ridge along the coast and with a limited capacity. Deeper aquifers are no alternative and surface water resources, e.g. drainage water from the polder area, are limited, often brackish, and only available during the winter period. However, due to tourism, most water is needed in the summer period. A project to integrate wastewater effluent into the existing drinking-water production was a logical choice for the IWVA. *Artificial recharge* of the unconfined aquifer of the dune water catchment St-André was the selected solution. It should restore the groundwater quality and enhance the ecological values of the dune areas as the extraction of natural groundwater reduced. Effluent from a nearby wastewater treatment plant was selected as the source for the production of infiltration water.

3.1.2 History of implementation

Ten years of research preceded the start of the project. The Flemish Institute of Nature Conservation performed an ecological study of the water catchment of St-André resulting in a delineation of the infiltration area, the indication of ecological conditions (Kuijken *et al.* 1993) and an ecological management plan for this area. Two infiltration tests, using groundwater from nearby sewage works, gave valuable information about the hydrogeological parameters of the dune aquifer in St-André (Lebbe, 1995). The impact of artificial recharge in St-André was simulated using MODFLOW (Van Houtte & Vanlerberghe, 1998). The model represented the unconfined aquifer in these dunes with the northern and southern borders of the model area, presenting the shore and the polder area respectively, introduced as constant head boundaries. The horizontal conductivity of the sands is 14 m/d with a mean natural infiltration rate of 280 mm/year. The mean annual hydraulic heads were calculated; seasonal effects were not taken into account. The calculation showed that the groundwater flow in St-André would not change very much. However, although the rise in hydraulic head would be small, the outflow of groundwater to the shore would increase by 10%.

From 1997 until 1999 pilot tests were performed on the effluent from the waste water treatment plant in Wulpen, which is operated by Aquafin, the Flemish wastewater treatment company. Different micro- and ultrafiltration systems were investigated for the pre-treatment (Van Houtte *et al.* 1998 and 2000) and two types of reverse osmosis membranes were tested for the desalination (Van Houtte & Vanlerberghe, 2001).

3.1.3 Treatment and implementation concept

The high ecological value of the dunes necessitated constant consideration especially when developing the infiltration project. A shallow depth, gentle slopes, bendy banks, an island in the middle and some wide bays increased the natural interest of the infiltration pond. The extraction wells were integrated within the existing infrastructure, minimizing the impact on the environment. The quality of infiltration water was subject to stringent standards (cf. Table 3). Both the salt and nutrient content had to be low and together with the hygienic safety of the water, this was the biggest challenge when integrating wastewater effluent into the drinking-water production.

The indirect potable reuse scheme via MAR was developed based on the multi-barrier principle and it is composed of an activated sludge plant, an advanced water reclamation facility, a groundwater infiltration unit, and a water treatment plant for the final treatment of the abstracted water (cf. Figure 3.2).

Based on the Orange County (USA) experience, where clarified secondary effluent was treated for groundwater injection as a barrier against saline water intrusion and where new pilot tests were performed (Leslie *et al.* 1996), a combination of *membrane filtration techniques* was chosen as advanced water treatement following the wastewater treatment to produce the infiltration water:

- Ultrafiltration (UF) as the first treatment step to remove suspended solids and bacteria from the effluent and
- Reverse osmosis (RO) as the final treatment, not only removing salts, nutrients and viruses, but also small organics (e.g. pesticides), endocrine disrupting compounds and pharmaceuticals.

3.1.4 Authorisation procedure

Prior to the development of the MAR scheme, the IWVA conducted pilot tests with infiltration in the dunes of St-André. These tests were executed in cooperation with the University of Ghent. The IWVA also had preliminary discussions with environmentalist specialists. As a result, at the start of the permitting procedure some ecological conditions were taken into account (Van Houtte & Vanlerberghe, 1998).

Authorisation for aquifer recharge or extraction is a regional, thus Flemish matter. An Environmental Impact Assessment was mandatory. It included a hydrogeological and ecological study of the area. In the final permit specific parameters had been set for infiltration water, especially regarding the nutrient content to avoid negative impact on the dunes. For many parameters the standards were even more stringent than for drinking-water (cf. Table 3.3). The most important ecological condition was that all recharged water should be re-extracted. The permit included a monitoring scheme for the quality of the infiltration water, the groundwater level and quality and the ecological value of the area. Since the start of the project, the IWVA has performed the monitoring dutifully and all requirements have been met. The information gained by monitoring is helpful and will lead to future adjustments and improvements.

3.2 DESCRIPTION OF THE TEST SITE

The Managed Aquifer Recharge (MAR) scheme of Torreele/St-André, located at the Flemish coast, has been developed to replenish the drinking water aquifer at St. André, *levelling out seasonal variations* of water availability and *preventing salt water intrusion*. Drinking water produced out of the dune aquifer is distributed to a resident population of 60,000 inhabitants. Because of tourism, daily consumption in summer can be as high as 2.5 times that of a normal day in winter. The scheme is operational since July 2002. The sandy unconfined dune aquifer is easy to exploit and the natural groundwater is of excellent quality. Capacity of the phreatic dune aquifer is limited due to the presence of salt water North (sea) and South (polder area). The schematic representation of the catchment is illustrated in Figure 3.2.

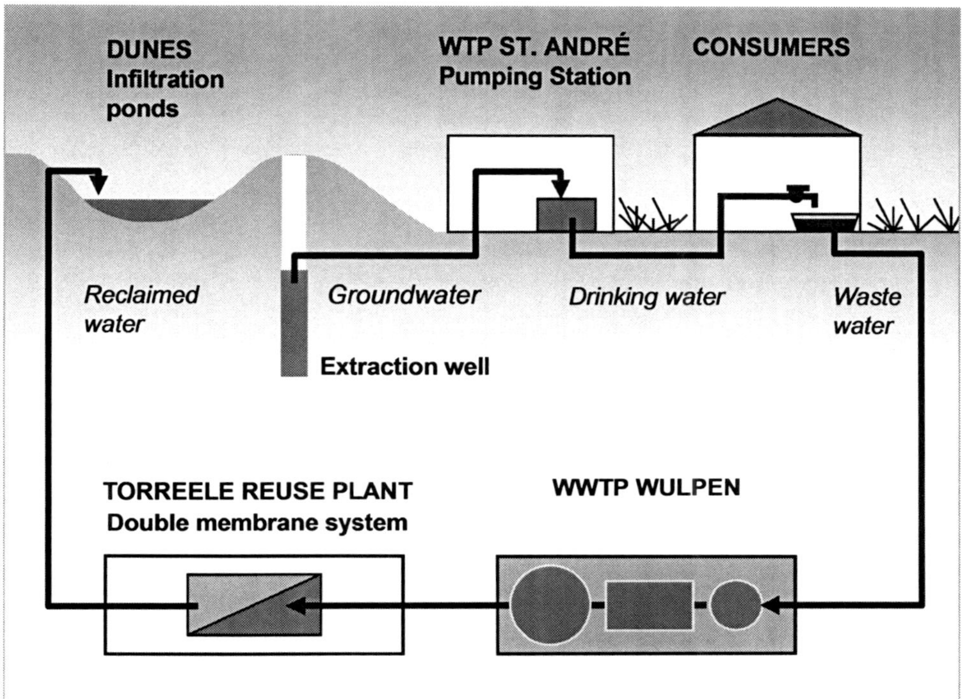

Figure 3.2 Concept of the closed loop system in Torreele/St. André

The infiltration area is part of the dunes protecting the inland from the sea and has a width varying between 2 and 2.5 km and a ground level from +6 to +35 m above sea level. The unconfined aquifer under the dune belt is formed by 25 to 35 m thick Quaternary sandy deposits, in which in some areas thin layers of fine grained sediments occur. South of the dunes, is the polder area consisting of two types of features being the more elevated sandy creek ridges (+3.5 to +5 m) and the lower marsh basins (+2.5 to +4 m). The creek ridges have sandy Quaternary sediments whereas under the marsh basins fine grained sediments (clay, silt and peat) occur. The polder area is drained by canals that discharge large amounts of surface water to the sea, especially during winter (Van Houtte and Verbauwhede, 2005). The Quaternary sediments lie upon Tertiary clay. This Ypresian clay, 110 m thick, separates the upper unconfined aquifer from the confined aquifer. This confined aquifer was formed by very fine silty green sands of Landenian age with a thickness of 20 m and lay on green clays.

3.2.1 Waste water treatment plant Wulpen

The waste water treatment plant (WWTP) Wulpen, built in 1987 for organic removal and operated by Aquafin, was extended in 1994 to comply with European standards for sensitive areas incorporating nutrient removal. The scheme has a nominal capacity of 74,700 population equivalent (at 60 gBOD/d). The sewerage network is of combined type, and the hydraulic capacity is of 800 L/s. Figure 3.3 shows an aerial picture of the WWTP.

Mechanical treatment consists of 2 step screens, a sand trap and rectangular primary clarifiers. Biological treatment is achieved by a conventional predenitrification activated sludge system. Phosphorus is removed by simultaneous chemical precipitation. Sludge-water separation is achieved by secondary settling tanks. WWTP Wulpen meets all the limits for sensitive areas of the EU Urban Waste Water Treatment Directive.

Figure 3.3 Wastewater treatment plant Wulpen (74,700 PE)

3.2.2 Advanced water treatment plant Torreele

The Torreele water reclamation plant, where IWVA produces the infiltration water, was built at the site of the Wulpen wastewater treatment plant. The influent is mainly of domestic nature and as large amounts of rainwater are drained to the sewers, the quality of the secondary effluent is variable. The process is shown in Figure 3.4 and summarized in Table 3.1.

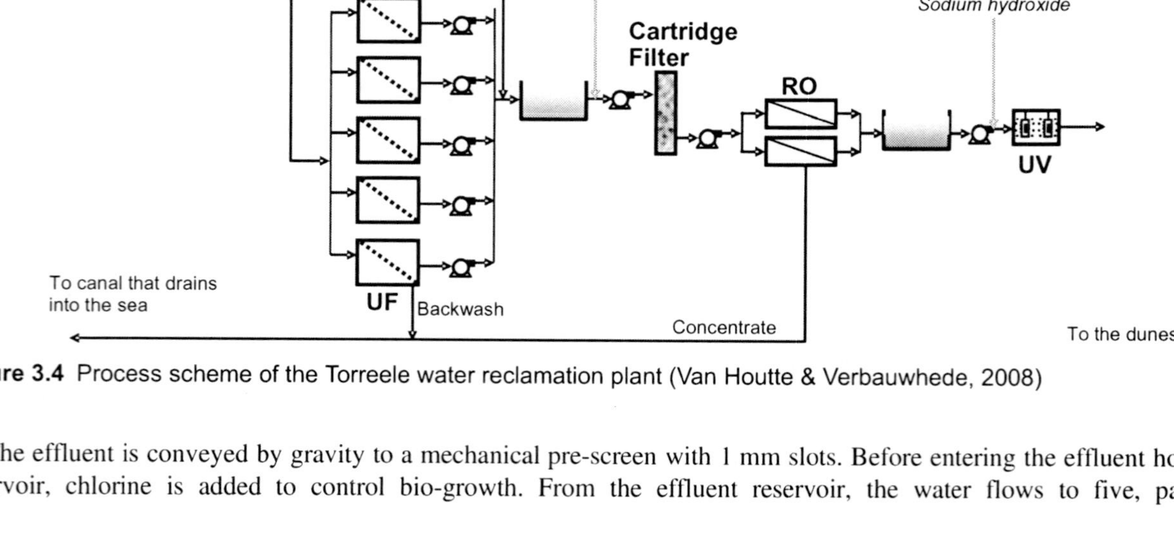

Figure 3.4 Process scheme of the Torreele water reclamation plant (Van Houtte & Verbauwhede, 2008)

The effluent is conveyed by gravity to a mechanical pre-screen with 1 mm slots. Before entering the effluent holding reservoir, chlorine is added to control bio-growth. From the effluent reservoir, the water flows to five, parallel ultrafiltration (UF) trains. UF filtrate enters a holding reservoir after being chloraminated. From there it is pumped to the reverse osmosis (RO) system. Both scale inhibitor and sulphuric acid are injected to control scaling. A third pump can dose $NaHSO_3$ to neutralize free chlorine. Before the RO system, the water first passes through a cartridge filter (pore size of 15 μm). The RO filtrate enters a reservoir from where it is pumped to the dunes, but first sodium hydroxide is dosed to adjust the pH to about 6.5. An extra treatment with UV disinfection system is possible as backup disinfection unit.

Table 3.1 Summary of the Torreele reclamation plant (Van Houtte & Verbauwhede, 2008).

History of project	1997–1999, pilot testing using different MF/UF and RO systems on effluent of Wulpen WWTP
Status of project	Operating since July 2002
Capacity	2,500,000 m^3/year
Source water	WWTP secondary effluent from Wulpen WWTP
Pre-treatment	Pre-screens with 1 mm slots
Membranes	UF: Zenon (ZW500C) RO: Dow (30LE-440)
Design flux	UF: max of 40 $L/h \cdot m^2$ recovery minimum 87% RO: max of 20 $L/h \cdot m^2$ recovery minimum 75%
Biofouling prevention	Chlorine to effluent, monochloramine to UF filtrate, sodiumbisulfite redox controlled to prevent damage of the RO membranes
Scaling prevention	pH adjustment using sulfuric acid and antiscalant to UF filtrate
Concentrate disposal	Drained to canal mixed with the part of the effluent that does not undergo advanced water treatment

The Torreele scheme has a hydraulic capacity of 80 L/s. The UF units are ZENON ZeeWeed® modules by (ZW500C), working at a maximum design flux of 40 $L/(m^2 \cdot h)$. The filtration time is 480 to 600 seconds, while backwash is 30 seconds. Every 30 to 35 backwashes chlorinated backwash is performed. The RO is composed by low energy membranes from DOW (30LE-440) in 2/1 configuration with an average recovery of 77%. The design flux is maximum 20 $L/(m^2 \cdot h)$. The RO permeate is then transported by a pipeline over a distance of about 2.5 km to the MAR site of St-André.

3.2.3 Groundwater infiltration at St. André

The infiltration occurs in an unconfined sandy aquifer. From the top to the depth of pumping, the sands are fine to medium in size containing shells. There are some small clayey and peaty layers present but their occurrence is irregular. The recharged water is recaptured using 112 pumping.

The maximum percentage of recycled water in the extracted well water is 70%. The horizontal separation between the infiltration pond and the abstraction wells varies between 33 m and 153 m, with an average of 59 m. All wells have screens between 8 and 12 meters depth. Figure 3.5 shows an aerial view of the groundwater infiltration site.

The extraction wells (Figure 3.6) were integrated within the existing infrastructure, minimizing the impact on the environment (Van Houtte & Verbauwhede, 2005).

The current groundwater flow regime of the wider area is described by Smith (2003). Under the current conditions of artificial recharge and recovery, due to the fact that extraction of natural groundwater has been decreased compared to the period before infiltration started, the outflow pattern towards the north and the south is reinforced. The infiltrated water forms a lens below the wider area of the infiltration ponds and increased flow occurs towards both the sea and the polder area. Evidence from recent isotopes analysis confirms the presence of this flow regime (Kloppmann *et al.* 2008).

After recharging the unconfined aquifer in St-André, the water is recaptured using 112 new wells with filter elements between 8 and 12 m depth. Due to varying distances between the wells and the pond (Table 3.2) residence times vary between 30 days and almost 5 year, whereby 50% of the recharged water reaches the extraction wells after 55 days (Vandebohede *et al.*, 2009). This underground passage ensured elimination of all pathogens.

Early 2004, after checking all distances, the infiltration pond was locally recalibrated to increase the distance to the nearest wells (Van Houtte & Verbauwhede, 2005).

The surface area of the infiltration pond is 18,200 m^2 and the infiltration capacity is 2,500,000 m^3/year. Annually 3,500,000 m^3 should be extracted, or 1.4 times the infiltration rate, to ensure that the infiltration water is totally recaptured. Additionally monitoring wells were installed to observe the quality of the groundwater body.

Figure 3.5 Groundwater infiltration site at St. André

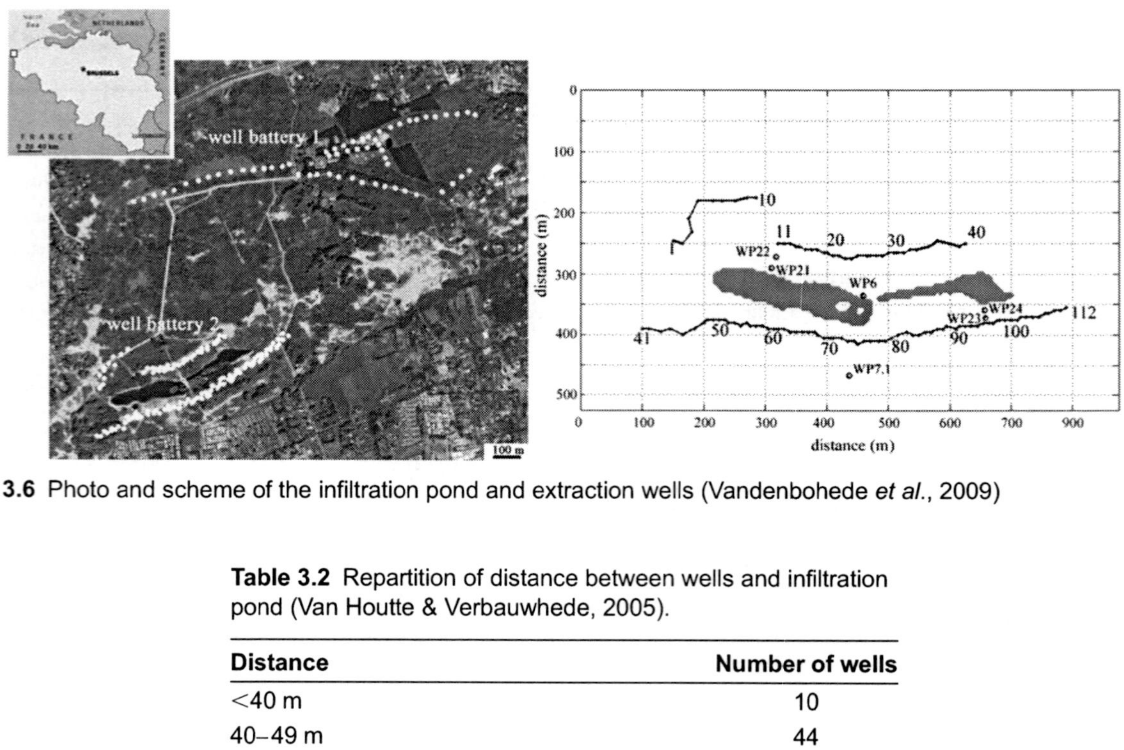

Figure 3.6 Photo and scheme of the infiltration pond and extraction wells (Vandenbohede *et al.*, 2009)

Table 3.2 Repartition of distance between wells and infiltration pond (Van Houtte & Verbauwhede, 2005).

Distance	Number of wells
<40 m	10
40–49 m	44
50–59 m	23
60–69 m	17
70–79 m	3
80–89 m	2
>90 m	13

3.2.4 Groundwater extraction and treatment facility at St. André

The groundwater is extracted by pumps and treated with the existing water treatment plant in St. André. The plant comprises aeration and rapid sand filtration. The final product is then supplied to the potable water distribution system of the cities of De Panne, Koksijde, Nieuwpoort, Veurne and Alveringem. Apart from the latter after its being used it is again collected and reaches as wastewater again the WWTP Wulpen.

3.2.5 Hydrogeology

Under the dunes a fresh water lens is formed by infiltrating rainwater. In natural conditions, groundwater flows from the dune area towards the polder area and the shore, where salt water infiltrates during the tides. This prevented salination of the fresh water lens under the dune belt. The groundwater extraction diminished the fresh water outflow from the dune area. Over-exploitation would lead to salt water encroachment by the inflow of salt water from the aquifer under the beach and polder areas. In the polder area, fresh water filled the upper and relict salt water the lower part of the aquifer. A small transition zone of brackish water lay in between. However under the creek ridges more fresh water occurs because of the higher infiltration rate of rainwater here compared with the marsh basins, where fine sediments reduce infiltration rates (Van Houtte & Vanlerberghe, 1998).

3.3 TREATMENT TARGETS AND REGULATORY FRAMEWORK

3.3.1 Regulated substances

The treatment goals set in the local permit for infiltration water quality are set out in Table 3.3. Based on the precautionary principle, infiltration water must already comply with the end product water quality standards (i.e. for drinking water). In other words, dilution and contaminant attenuation in both the unsaturated and saturated zones are considered as an additional safety factor. Table 3.3 also documents local standards for groundwater quality (Flemish Government, 2001), international benchmarks for drinking water standards as well as recently issued international benchmarks for indirect potable reuse. For the risk analysis, Aquafin adopted the most restrictive numerical targets, as the local regulation demands that "the water supplier has to carry out analyses on the chemicals and micro-organisms not covered by the limits but that can be present in the water and that can pose a risk to public health" (Flemish Government, 2002).

Table 3.3 Infiltration water permit at MAR Wulpen versus international benchmarks on selected parameters.

	Local permit	**VLAREM (2001)**		**DWD**	**WHO**	**USEPA**
	Unit (mg/L)	**G****	**MAX**	**(1998)**	**(1996)**	**(2007)**
Turbidity (NTU)	–			–	5	0.3–1*
Colour (Hazen units or mg/L Pt)	–	1	20	200	15	15
Conductivity (µS/cm)	1000	400			–	–
pH value	6.5–9.2	6.5–8.5			–	6.5–8.5
Total hardness (°F)	40			–		–
Total dissolved solids					1000	500
COD		2	5			
Total organic carbon					–	–
Min oils	0.005			–		
Calcium	–	100			–	–
Magnesium	50	30	50	–		
Sodium	150	20	150	200	200	–
Potassium	–	10	12			
Ammonium (NH_4^+)	1.5	0.05	0.5	0.5	1.5	–
Nitrate (NO_3)	15	25	50	50	–	44.3
Nitrite (NO_2)	0.1		0.1	0.5	0.2–3	3.3
Phosphorus (P_2O_5)	0.4	0.4	5			
Chloride (Cl^-)	250	25		250	250	250
Fluoride	1.5	1.5		1.5	1.5	4
Selenium	10		10	10		50
Silica (SiO_2)					–	–
Sulphate (SO_4)	250	25	250	250	250	250
Metals, metalloids and micro-pollutants (µg/L)						
Antimony	10		10	5		6
Arsenic	–		50	10***		10***
Aluminium	200	50	200	200	200	50–200

(Continued)

Table 3.3 Infiltration water permit at MAR Wulpen versus international benchmarks on selected parameters (*Continued*).

	Local permit	VLAREM (2001)		DWD	WHO	USEPA
	Unit (mg/L)	G**	MAX	(1998)	(1996)	(2007)
Barium	–	100			700	2000
Boron	–	1000		1000	900	–
Cadmium	–		5	5		5
Chromium (total)	50		50	50		100
Copper	100			2000	2000	1300
Iron	200	50	200		300	300
Lead	25	3000	50	10		15
Manganese	50	20	50	50	50	50
Mercury	1		1	1		2
Nickel	50		50	20		
Strontium					–	–
Zinc	200	5000		–	3000	5000
Individual pesticides	0.02			0.1		
Total pesticides				0.5		
Atrazine					2	3
Diuron						
Simazine					100	
PAC	0.02			0.1		
Surfactants (lauryl–sulphate)			200			
Phenols (C_6H_5OH)			500			
Residual Chlorine (Cl Total)		1			5	–
Bromate	–				25	10
Chlorite					200	1000
Total Trihalomethanes	200			–	–	80
Monochloramine					3	
Haloacetic acid						60
Bacteriolgical						
Hetero. plate count 22°C (/mm)	–	100				500
Hetero. plate count 37°C (/mm)	–	10				
T. Coliforms (Counts/100 mL)	–		0			5.0% +
F. Coliforms (Counts/100 mL)	–		0			0
E. Coli (Counts/100 mL)	–		–			0
Enterococci (Counts/100 mL)	–					–
Enteric viruses	–	–	–			4 logs
Giardia intestinalis	–	–	–	–		3 logs
Cryptosporidium parvum	–	–	–	–		2 logs

*filtration
**guideline
***refers to the LOQ. The excess lifetime skin cancer risk of 10^{-5} $p^{-1} \cdot y^{-1}$ is reached at 0.17 μg As/L.

3.3.2 Unregulated substances

At source there is possible microbiological and chemical contamination of a much larger number of compounds of anthropogenic origin. The European Water Framework Directive 2000/60/EEC counts 33 legally binding priority substances to be monitored in European basins ubiqituously present in wastewater (Commission of the European Communities, 2000). Their selection out of the hundreds or thousands that can be found took place based on their expected environmental concentrations, and their toxicity, persistence or bioaccumulation to the environment and to aquatic organisms rather than to humans. The link between public health protection and environmental exposure is not always well established.

3.4 WATER QUALITY MONITORING AND ASSESSMENT

3.4.1 Regular Monitoring Program

The list of parameters, the measuring points and frequency of measurements are set out in Table 3.4.

Table 3.4 Monitoring strategy per location as it is currently in place.

Location	On-line	Daily	Weekly	Monthly
WWTP effluent	Turbidity, pH, temperature	BOD, SS, NH_4N, NO_3N, PO_4P	BOD, COD, SS, NH_4N, KjN, NO_xN, TN, PO_4P, TP	Heavy metals
UF	Flow, conductivity, pH, temp., turbidity, pressures and energy consumption	SDI, temperature, pH, conductivity	plate counting, TC, *E. coli*, Enterococci, salts and nutrients	Compliance monitoring
RO	Temperature, pressure, conductivity, pH	Temperature, conductivity, pH	plate counting, TC, *E. coli*, Enterococci, salts and nutrients	Idem
Infiltration pond	Water level			plate counting, TC, *E. coli*, Enterococci, salts and nutrients, pH
Groundwater			Plate counting	
Aeration	n.a.	n.a.	n.a.	n.a.
Rapid sand filtration	Headlosses, turbidity	n.a.	n.a.	n.a.
UV	UV intensity	n.a.	n.a.	Compliance monitoring
Distribution	Turbidity, flow, pressure	Plate counting output at the tap	plate counting, TC, *E. coli*, Enterococci, salts and nutrients	Compliance monitoring

3.4.2 Measuring program of RECLAIM WATER

To gain additional insights and to produce data sets for additional risk and hydrological studies, the several sampling campaigns were conducted between February 2006 and November 2007. The analytical methods of the measuring campaigns are reported in Chapter 2. The sampling points along the treatment train are given in Figure 3.7.

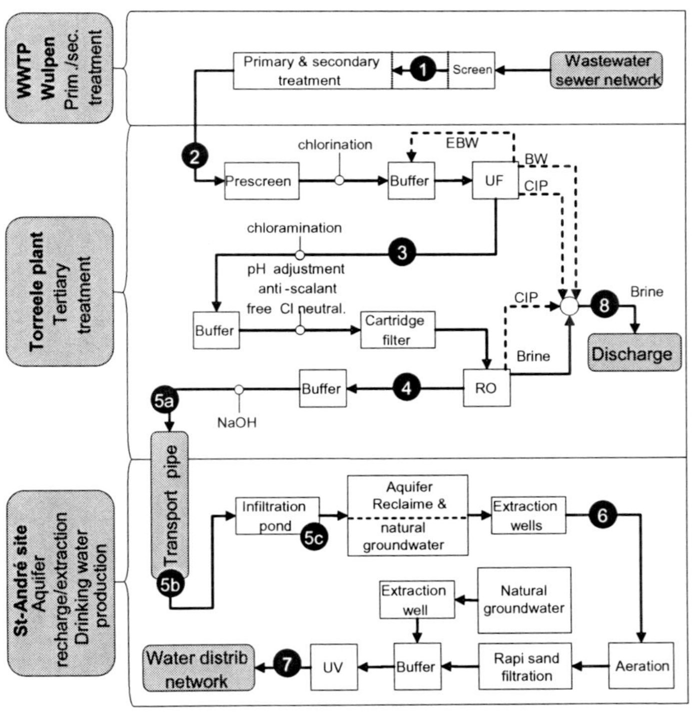

Figure 3.7 Schematic overview of the treatment concept at the Wulpen-Torreele MAR scheme with sampling points used for RECLAIM WATER

3.4.3 Results

Based on the data that were already available, Le Corre *et al.* (2007) concluded that advanced technologies like UF and RO improve significantly the quality of the reclaimed water used for recharge. As shown in Figure 3.9 UF succeeded in complete removal of bacteria while Böckelmann *et al.* (2009) found that UF succeeded in complete removal of the antibiotic resistance genes.

The quality of both UF and RO filtrate was as expected: Figures 3.8 and 3.9 demonstrate that nutrient and salt content is very low after RO and bacteria are absent. Consequently the infiltration water, composed of RO filtrate to which a small pH correction is done using sodium hydroxide, proved to be of very good quality. Most of the parameters that were controlled were below detection limit (Table 3.5). As the heterotrophic plate counts were constantly low, the UV disinfection that was installed prior to transport to the dunes, is no longer operational since September 2004 (Van Houtte & Verbauwhede, 2008).

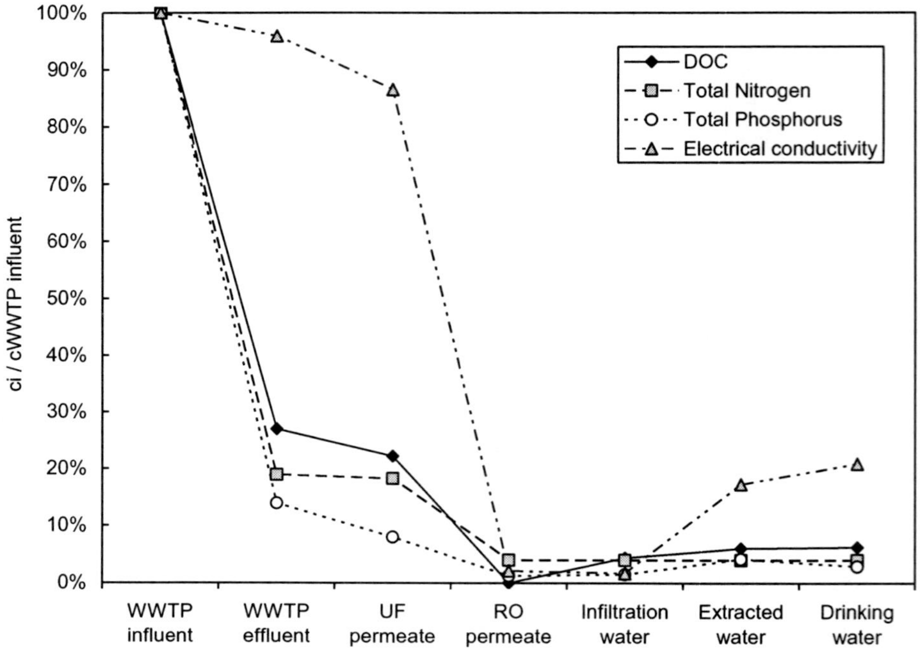

Figure 3.8 Fate of chemical constituents in the Wulpen-Torreele MAR scheme

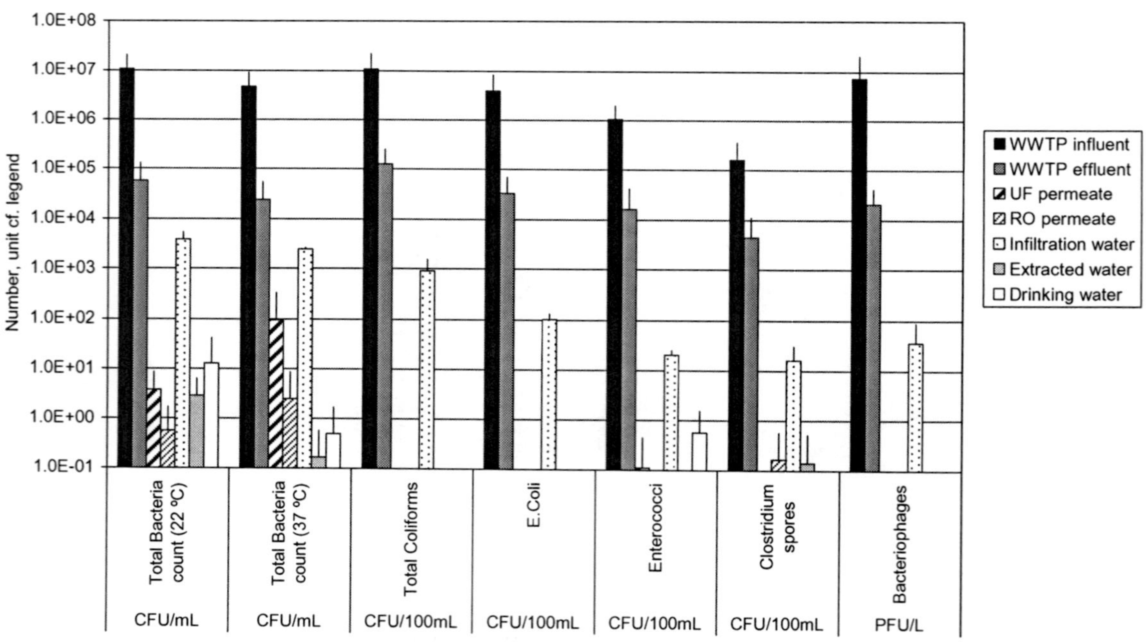

Figure 3.9 Fate of microbial contaminants in the Wulpen-Torreele MAR scheme

Table 3.5 Overview of quality in 2010 (Van Houtte & Verbauwhede, 2010).

Parameter	UF filtrate	RO filtrate	Infiltration water**
Conductivity (µS/cm)	1211 (364–1735)	25 (5–54)	45 (<10–89)
pH***	7.98 (6.94–8.33)	5.51 (4.96–6.0)	6.98 (6.28–8.02)
Total Organic Carbon (mg/l)	10.4 (6.8–12.8)	0.3 (0.1–1.1)	0.4 (0.1–1.1)
Total hardness (mg/l as $CaCO_3$)	28 (11–40)	<0.5	<0.5
Chlorides (mg /l)	220 (53–360)	5.1 (1.2–18)	3.2 (1.0–4.7)
Fluorides (mg/l)	0.17 (0.08–0.53)	<0.2	<0.2
Sulfates (mg/l)	75 (31–109)	14 (0.1–5)	<1
Nitrate (mg NO_3/l)	19 (6–32)	2.2 (0.1–5)	2.5 (<1–6.3)
Ammonia (mg NH_4/l)	1.9 (0.07–4.2)	<0.1	0.13 (0.03–0.38)
Phosphate (mg PO_4/l)	19.4 (6.9–26)	0.4 (0.07–1.1)	<0.1
Silicon (mg SiO_2/l)			0.3 (0.1–0.4)
Total trihalomethanes (µg/l)			3.8 (1.2–6.7)
Aluminum (µg/l)	65 (19–128)	2.6 (0.6–21)	12 (2–59)
Chromium µg/l)	1.8 (0.5–5)	<10	<2.5
Copper (µg/l)	32 (15–52)	<20	<5
Iron (µg/l)	44 (16–91)	<20	<75
Lead (µg/l)	2.6 (1–10)	<20	<5
Manganese (µg/l)	155 (35–236)	<5	<10
Mercury (µg/l)	16 (7–32)	<20	<0.2
Nickel (µg/l)		3.7 (1.4–6.4)	<3
Sodium (mg/l)		<20	10.5 (4.5–17.7)
Zinc (µg/l)			<20
Total coliform bacteria (counts/100 ml)	0–2	0	0
E. coli (counts/100 ml)	0	0	0
HPC 22°C (counts/ml)	22 (2–80)	2 (0–20)	<1 (0–10)

*Mean values are presented with minimum and maximum values provided in parentheses
**Infiltration water consisted of RO filtrate to which sodium hydroxide was dosed
***pH measured in the infiltration pond

Nitrosamines, often a concern with RO when chloramination is performed, tend to be very low in concentration. RO rejected about 50% of NDMA, while NMOR was rejected to a larger extent resulting in NMOR levels below DL (Krauss *et al.*, 2010). The results from different campaigns showed that the probability of significant human exposure to nitrosamines from the reclaimed water was low and that low levels of NDMA, when present in the RO filtrate, are removed in the aquifer (Krauss *et al.*, 2010).

3.4.4 Technology performance and contaminant monitoring

After two years of operation, it was decided to optimize the process in order to reduce operational costs. In the UF system the use of air has been reduced. Initially the aeration was 50% of the time, but it was reduced 30% resulting in a substantial reduction of energy consumption for a comparable amount of production.

In the RO system the recovery is no longer constant (initially 75%) but it is varied according to the conductivity of the UF filtrate. It was called 'recovery control' and means that the recovery is higher when the conductivity is lower. Lower conductivity means less risk for scaling and in this way the recovery respectively amounts around 77% on average without any additional dosing of chemicals (Van Houtte & Verbauwhede, 2008).

In the summer of 2005, in each stage of the RO system one PV was filled with new membranes, also resulting in lower energy consumption.

The chemical consumption was further reduced as chloramination was no longer continuous. The ratio between dosing and non-dosing varies according to the season, thus the temperature of the water. It resulted in a 20% reduction of sodium hypochlorite usage and 35% for ammonium chloride. No negative influence on the bio-fouling prevention was observed until now.

Thanks to artificial recharge, the extraction of natural groundwater could be reduced substantially, and this benefitted most from the dune water catchment in the Westhoek. In that area, since 10 years the IWVA gradually reduced the groundwater extraction. The salinity in the wells nearest to the polders declined from approximately 1.500 μS/cm in 1992 to 1.000 μS/cm in 2004.

In St-André the groundwater quality extracted outside the infiltration area did not change very much, but here problems with increasing salinity never occurred. There is a relative rise in groundwater levels resulting in an increased outflow of fresh water from the dunes. In the vicinity of the infiltration pond, where the water level is close to the surface, some valuable plants have been monitored in 2003 and 2004. In St-André there exists an ecological management of the dunes since 1994; around the infiltration pond this management was recently adjusted to enhance the positive development that was observed. Ecological management of the dunes of the Westhoek recently started.

3.4.5 Concentrate disposal

As UF and RO both produce concentrate, an important volume of water is discharged. 35% percent of the discharged volume is UF backwash water; 65% is RO concentrate. The discharge is into the adjacent canal with the part of effluent that has not been treated. The canal is brackish, so the salinity does not have a major negative effect on the quality.

The IWVA performed several tests to look for mitigating the effect of concentrate disposal (Van Houtte and Verbauwhede, 2010):

- Since October 2003 a subsurface flow reed bed (constructed wetland) was used; the nitrogen content reduced by 30%; organic content (TOC, COD) was removed only partially and phosphorous content did not change;
- Since April 2007 willows were used under the same conditions as reed and besides reducing nitrogen and organic load, the phosphorous content was lowered.

3.5 CONCLUSIONS

Reuse of wastewater via managed aquifer recharge resulted in a sustainable groundwater management of the Flemish dunes. The international state-of-the-art combination of UF and RO membranes guarantees an excellent quality of the infiltration water. Monitoring of parameters, both online, manually and analytical, safeguards a constant high performance as alarms can be generated when threshold values are exceeded and emerging problems can be quickly detected.

Water reuse opens opportunities in many regions. It could be a solution for inadequate water supply and for deteriorating water quality. The multiple barrier approach assures a safe solution when drinking water is at stake.

REFERENCES

Böckelmann U., Dörries H., Ayuso-Gabella M. N., Salgot de Marçay M., Tandoi V., Levantesi C., Masciopinto C., Van Houtte E., Szewzyk U., Wintgens T. and Grohmann E. (2009). Quantitative PCR monitoring of antibiotic resistance genes and bacterial pathogens in three european artificial groundwater recharge projects. *Applied and Environmental Microbiology*, **75**(1), 154–163.

Flemish Government. (2001). *Milieukwaliteitsnormen voor grondwater* (Environmental quality standards for groundwater), VLAREM II, version October 10, 2001.

Flemish Government. (2002). *Voorontwerp van decreet betreffende water bestemd voor menselijke aanwending* (Draft decree on water intended for human use).

Kloppmann W., Van Houtte E., Picot G., Vandenbohede A., Lebbe L., Guerrot C., Millot R., Gaus I. and Wintgens T. (2008). Monitoring reverse osmosis treated wastewater recharge into a coastal aquifer by environmental isotopes (B, Li, O, H). *Environmental Science and Technology*, **42**(23), 8759–8765.

Krauss M., Longree P., Van Houtte E., Cauwenberghs J. and Hollender J. (2010). Assessing the fate of nitrosamine precursors in wastewater treatment by physicochemical fractionation. *Environmental Science and Technology*, **44**(20), 7871–7.

Kuijken E., Provoost S. and en Leten M. (1993). *Oppervlakte-infiltratie in de Doornpanne, een verkennend onderzoek naar de ecologische implicaties*, (Surface infiltration in the nature reserve 'Doornpanne' – an exploratory study of the ecological implications). Instituut voor Natuurbehoud, A 93.69, 86.

Le Corre K., de Heyder B., Masciopinto C., Tandoi V., Aharoni A., Chikurel H., Zhao X., Rinck-Pfeiffer S., Regel R., Salgot M., Ayuso-Gabella M., Saperas N., Cartmell E., Jefferson B. and Jeffrey P. (2007). Preliminary results of managed aquifer recharge with reclaimed wastewater and stormwater in six operational case studies from around the world. *Proceedings of the 6th IWA Specialist Conference on Wastewater reclamation and Reuse for Sustainability*, 9–12 Octorber, 2007, Antwerp, Belgium.

Lebbe L., Tarhouni J., Van Houtte E. and De Breuck W. (1995). Results of an artificial recharge test and a double pumping test as preliminary studies for optimizing water supply in the western belgian coastal plain. *Hydrogeology Journal*, **3**(3), 53–63.

Leslie G. L., Dunivin W. R., Gabillet P., Conklin S. R., Mills W. R. and Sudak R. G. (1996). Pilot testing of microfiltration and ultrafiltration upstream of reverse osmosis during reclamation of municipal wastewater. In: *Proceedings of the American Desalting Associations Biennal Conference*, Monterey, California, August 1996, pp. 29–40.

Smith K. (2003). *Hydrogeologische studie te Koksijde: kunstmatige infiltratie als bijdrage tot duurzame waterwinning in kustgebieden*, (Hydrogeological study of the Koksijde: Artificial recharge as contribution to sustainable water abstraction in coastal areas). Faculty of Sciences, Department of Geology and Hydrogeology, University of Gent. Licentiate thesis.

Vandenbohede A., Van Houtte E. and Lebbe L. (2009). Water quality changes in the dunes of the western Belgian coastal plain due to artificial recharge of tertiary treated wastewater. *Applied Geochemistry*, **24**, 370–382.

Van Houtte E. and Vanlerberghe F. (1998). Sustainable groundwater Management by the integration of effluent and surface water to artificially recharge the phreatic aquifer in the dune belt of the western Flemish coastal plain. *IAH International Groundwater Conference, Groundwater: Sustainable Solutions*, Melbourne, Australia, pp. 93–99.

Van Houtte E., Verbauwhede J., Vanlerberghe F., Demunter S. and Cabooter J. (1998). Treating different types of raw water with micro- and ultrafiltration for further desalination using reverse osmosis. *Proceedings 'Membranes in Drinking and Industrial Water Production' Amsterdam, The Netherlands, Desalination*, **117**(1), 49–60.

Van Houtte E., Verbauwhede J., Vanlerberghe F. and Cabooter J. (2000). Comparison between different out-to-in filtration MF/UF membranes. *International Conference on 'Membrane Technology in Water and Wastewater Treatment'*, Lancaster 2000, 190–197.

Van Houtte E. and Vanlerberghe F. (2001). Preventing biofouling on RO membranes for water reuse - results of different tests. *AWWA Membrane Technology Conference*, San Antonio, 2001.

Van Houtte E. and Verbauwhede J. (2005). Artificial recharge of treated wastewater effluent enables sustainable groundwater management of a dune aquifer in Flanders, Belgium. *5th ISMAR International Symposium on Management of Aquifer Recharge Berlin*, Germany, 11–16 June 2005.

Van Houtte E. and Verbauwhede J. (2008). Operational experience with indirect potable reuse at the Flemish coast. *Desalination*, **218**, pp. 198–207.

Van Houtte E. and Verbauwhede J. (2010). Long time membrane experience at Torreele's water reuse facility in Belgium. *MIDW2010*, Trondheim Norway, June 2010.

Chapter 4

Managed aquifer recharge of a karstic aquifer in Nardó, Italy

Costantino Masciopinto, Rosanna La Mantia, Alfieri Pollice and Giuseppe Laera

4.1 INTRODUCTION

4.1.1 Historical background

Nardò is a town of about 30,000 inhabitants located in Apulia, Italy, 50 km south of Lecce. Apulia is one of the driest regions of Italy with an average annual rainfall of less than 600 mm (1950–2000). Its most southern part, known as Salento peninsula, forms the heel on the "boot" of Italy. Here the economy depends largely on farming, leading to a mainly agricultural land use with a large share of irrigated cropping due to the low precipitation. The total water consumption in Apulia is about 2400 Mm^3/year where of 58% (1400 Mm^3/year) are consumed by agriculture, 18% (430 Mm^3/year) by industry and 24% by urban users (580 Mm^3/year). As the Salento peninsula does not have any relevant surface water sources, groundwater has been traditionally the main water source. Due chronic water scarcity, the Apulian Aqueduct has been built already in 1906 (AQP, 2011) to provide additional water to the South from freshwater sources in the North of Apulia. Nowadays it provides about 440 Mm^3/year. The remaining annual water demand of 1960 Mm^3/yr is largely covered by groundwater. However, natural recharge does not refill the aquifers sufficiently, and overexploitation with consequent sea water intrusion into the water table is a severe problem at many locations. The number of the often private (and illegally drilled) wells amounts to around 140,000.

4.1.2 Motivations for recharge and use of abstracted water

Main problems in the Salento peninsula are related to the increase of water table depths and to the groundwater salinization due to over abstractions and sea water intrusion, as it is shown in Figure 4.2. At several places of the Salento coastal area the electrical conductivity of the groundwater already exceeds 15,000 μS/cm.

The groundwater flows under low pressure inside karstic fractures and it is affected by withdrawals for drinking and irrigation uses, as explained above. Artificial recharge has been considered in Nardò as a pilot study to counteract saltwater intrusion and overexploitation of the aquifer, and groundwater recharge operations have started in 1991. In Salento (Figure 4.1) natural recharge of coastal and fractured aquifer occurs via existing sinkholes replenished by small ditches. The natural sinkhole located near Nardò has thus been used for injection of reclaimed water via the Asso Channel. This channel drains local surface infiltration and runoff, the volume of which is approximately 0.6 Mm^3/year (Balice *et al.* 1989) and secondary effluents from wastewater treatment plants (WWTP) located nearby. The carbonate limestone of the Salento platform is very permeable because most of the joints and fractures caused by tectonic movements are karstified. The hydrological studies at the Nardò site show that the fractured aquifer, in which groundwater flows along preferential horizontal pathways, has a variable saturated thickness averaging 30 m. Here, for the last 20 years, treated secondary effluents from municipal treatment plants (towns of Aradeo, Galatone, Galatina, Sogliano and Maglie) have been injected into a natural sinkhole.

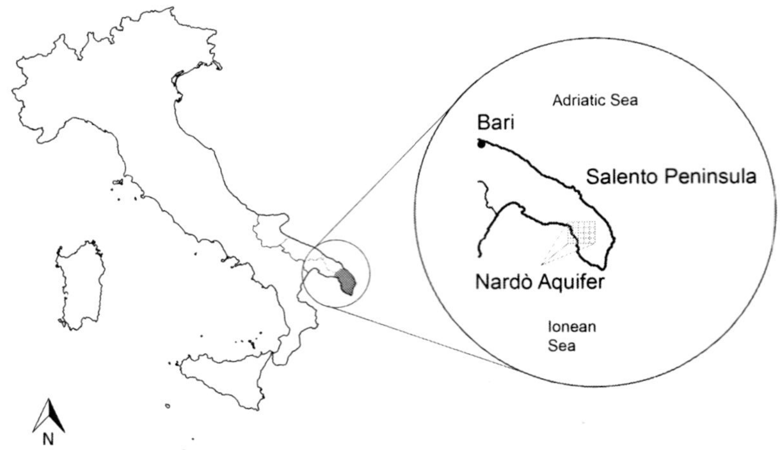

Figure 4.1 Location of the case study region

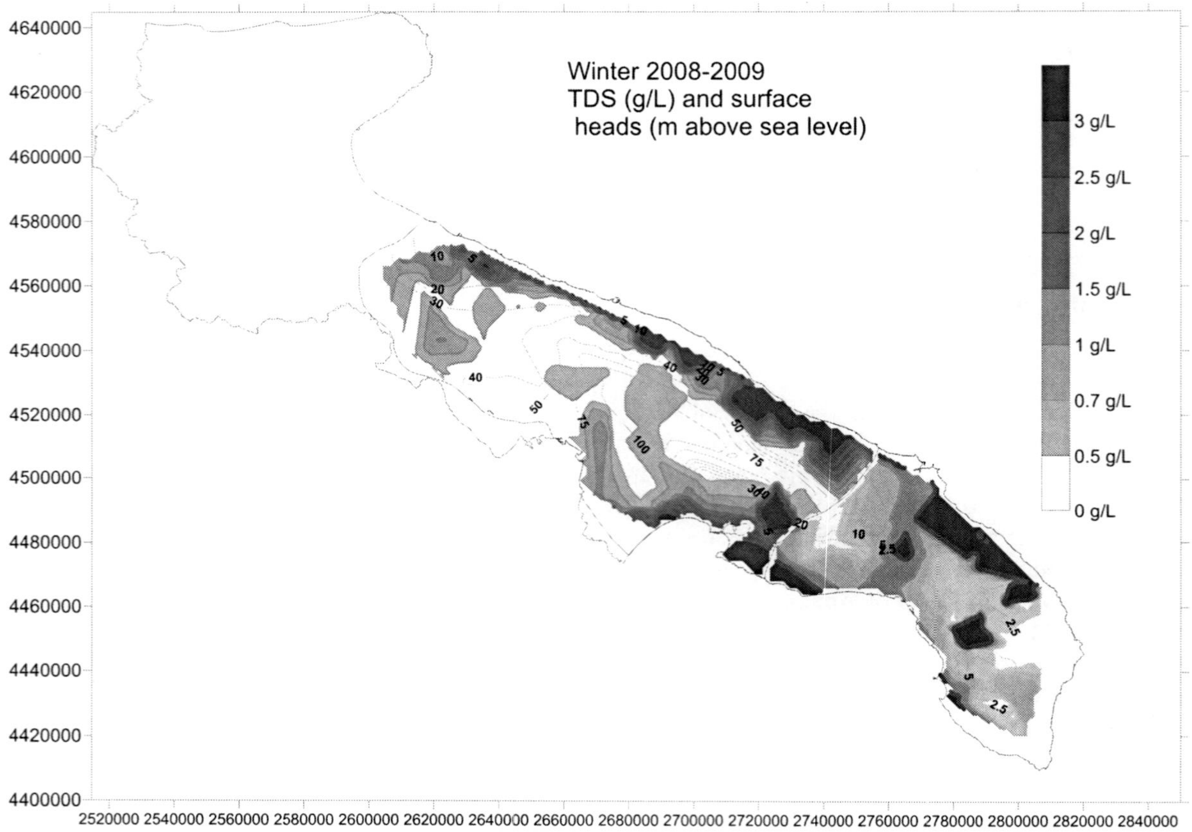

Figure 4.2 Salinity levels of the groundwater in the Salento peninsula given as TDS concentration in g/L (TIZIANO project, http://tiziano.regione.puglia.it/)

In 2002, by measuring the water velocity (1.4 m/s) at the outflow cross-section (0.8 × 25 m^2) of the Asso Channel (before the injection into the sinkhole), the annual average flowrate of artificial recharge was estimated as 140 L/s. More recent measurements (2006–2007) showed an increasing flowrate of up to 200 L/s (6.3 Mm^3/y), on average. This is due mainly to a second group of municipal wastewater treatment plants (towns of Cutrofiano, Muro Leccese, Scorrano and Neviano), which have also started to send their treated effluents into the Asso Channel. It should be noted that the treated wastewater discharged into the Asso Channel is authorised by the water management authority of the Apulia region. An increase in the electrical conductivity (1.24 mS/cm) of the water presently injected, compared to previous measurements (0.4 mS/cm, on average) (Masciopinto *et al.*, 2007), was also recorded. This change in water electrical conductivity is due to the difference in the salinity of the water supplied to the two groups of towns. The drinking water supplied to the first group is drawn from artificial lakes (Pertusillo and Sinni dams, southern Italy) and has a lower salinity compared with the water supplied to the second group of towns, which instead is derived from Salento groundwater.

The abstracted water is mainly used for irrigation (quality requirements are given in Table 4.1) but to some degree also for human purposes. The Salento karstic aquifer supplies 80% of the population (about 800,000 inhabitants) with 126 million m^3/yr of water for drinking purpose, and some concern has arisen regarding the groundwater quality. Public health concerns are particularly associated with ingestion of water contaminated with human or animal faeces, which can be a source of pathogenic bacteria, viruses, protozoa and helminths. The fate of these pathogens, associated with that of faecal contamination indicators has been investigated within Reclaim Water for establishing a minimum setback distance between effluent recharge and withdrawal that can occur in safe conditions. The results of this study, obtained by applying a new model for virus transport in fractures (Masciopinto *et al.* 2008), are presented in Chapter 11.

Table 4.1 Main quality parameters (D.Lgs n. 185/03) for agricultural reuse of municipal wastewater. Exceptions valid in Apulia are marked with asterix (*).

Parameter	Value
pH	6–9.5
Coarse solids (mg/L)	absent
TSS (mg/L)	10
COD (mg/L)	100–(50)*
BOD_5 (mg/L)	20–(10)*
Boron (mg/L)	1.0–(2.0)*
Chloride (mg/L)	250–(500)*
Sulphate (mg/L)	500
Electrical Conductivity (μS/cm)	3,000
Total Phosphorus (mg/L)	10
Total Nitrogen (mg/L)	35
Grease and Oil (mg/L)	10
Aldehydes (mg/L)	0.5
Surfactants	0.5
Chlorinated Pesticides	0.0001
Escherichia coli (CFU/100ml)	10
Salmonella (CFU/100ml)	absent
Sodium Adsorption Ratio	10

Notes: In addition to B also Al, As, Ba, Be,Cd, Cr, Cu, Fe, Hg, Mn, Ni, Pb, Se, Si, Th, V, Zn, THM, CN, SO_3, Benzene, Benzo(a)pyrene and other organics are considered.

4.1.3 Authorisation procedure

Seawater intrusion and overabstraction of aquifers are very relevant problems in Apulia as in many other Mediterranean coastal regions. Italian legislation for pollution prevention in natural water resources (D.Lgs. 152/1999) includes several measures for preserving the global groundwater balance by considering not only water demand and resource availability, but also potential sources for groundwater replenishment. The authorization procedures are regulated by D.Lgs. 152 (2006), which also establishes measures (art. 76–77), such as the creation of barriers, artificial recharge, etc., that regional authorities must apply to preserve and protect groundwater. Specifically, a subsequently national regulation (D.Lgs. n. 4/16–01, 2008) has required that all projects for the artificial recharge of phreatic aquifers with annual recharged volumes above 10 million cubic metres must be verified for their environmental impacts. Other proposed measures aimed at limiting seawater intrusion in groundwater consist of best management of groundwater volumes by reducing abstractions and by increasing seawater desalination plants. Several Italian Regional governments have planned groundwater remediation measures in Master Plans named "*Piani di Tutela delle Acque*" (PTA) (i.e. Water Regulation and Protection Plans). The PTA has already been published for the Apulia Region. In particular, the PTA according to Directive 2006/118/EU and the Italian legislation (D.Lgs. 30, 2009) establishes criteria for the assessment of both quantitative and qualitative (i.e. chemical) status of groundwater. These criteria must be applied from regional water management authorities to the entire region in order to ensure both good quality and quantity status of groundwater reservoirs before the end of 2015. Recommended threshold values to achieve the good standard of groundwater chemical quality are reported in Table 4.2:

Table 4.2 Threshold values to assess the good chemical status of Italian groundwaters (D.Lgs 30/2009).

	Threshold values (μg/L)	Threshold values (μg/L) for groundwater drained by surface water (SW)
Pollutants		
Nitrate	50 (mg/L)	
Total pesticide active compounds by including metabolites and degradation/reaction products	0.5	
Metals		
Antimony	5	
Arsenic	10	
Cadmium	5	0.08 (SW in class 1) 0.09 (SW in class 2) 0.15 (SW in class 3) 0.25 (SW in class 4)
Total Chromium	50	
Chromium VI	5	
Mercury	1	0.03
Nickel	20	
Lead	10	7.2
Selenium	10	
Vanadium	50	
Inorganic pollutants		
Boron	1000	
Cyanide	50	
Fluoride	1500	
Nitrite	**500**	
Chloride	**250 (mg/L)**	
Sulphate	**250 (mg/L)**	
Ammonia (ammonium ion)	500	
Organic aromatic pollutants		
Benzene	1	
Ethylbenzene	50	
Toluene	15	
Para-Xylene	10	
Polycyclic aromatic pollutants		
Benzo (a) pyrene	0.01	
Benzo (b) fluoranthene	0.1	0.03 (Summation of Benzo (b) and Benzo (k) fluoranthene
Benzo (k) fluoranthene	0.05	
Benzo (g, h, i) perylene	0.01	
Dibenzo (a, h) anthracene	0.01	
Indeno (1,2,3-c,d) pyrene	0.1	
Carcinogenic chlorinated pollutants		
Trichloromethane	0.15	
Vinyl chloride	0.5	
1,2 Dichloroethane	3	
Trichloroethylene	1.5	
Tetrachloroethylene	1.1	

(Continued)

Table 4.2 Threshold values to assess the good chemical status of Italian groundwaters (D.Lgs 30/2009) (*Continued*).

	Threshold values (μg/L)	Threshold values (μg/L) for groundwater drained by surface water (SW)
Esachlorobutadiene	0.15	0.05
Total halo-organic compounds	10	
Non-carcinogenic aliphatic chlorinated pollutants		
1,2 Dichloroethylene	60	
Aliphatic halogenated pollutants		
Dibromocromomethano	0.13	
Bromodichloromethano	0.17	
Chlorobenzene pollutants		
Monochlorobenzene	40	
1,4 Dichlorobenzene	0.5	
1,2,4 Trichlorobenzene	190	
Trichlorobenzenes (12002-48-1)		0.4
Pentachlorobenzene	5	0.007
Esachlorobenzene	0.01	0.005
Pesticides		
Aldrin	0.03	
Beta-esaclorocicloesano	0.1	0.2 (total)
DDT, DDD, DDE	0.1	DDT total: 0.025 p,p DDT: 0.01
Dieldrin	0.03	
Aldrin+dieldrin+entri+isodrin		0.01
Other pollutants		
PSDD + PCDF	4×10^{-6}	
PCB	0.01	
Total hydrocarbons (as n-hexane)	350	

Thus Italian law explicitly requires that groundwater cannot presents seawater intrusion problems and regional authorities must apply measures, such as groundwater replenishment, barriers, reduction of abstractions, desalination plants, and so on, in order to achieve the good qualitative and quantitative status of groundwater by the end of 2015.

4.2 DESCRIPTION OF THE TEST SITE

4.2.1 Study area

In the Nardò area, the aim of the experimental study carried out for the Reclaim Water project was to evaluate the efficiency of natural filtration through the fractured aquifers on the removal of toxic compounds and pathogens (biological treatment) occurring in the injected wastewater effluents. Indeed, the study allows an assessment of the groundwater quality when an aquifer is subjected to a continuous and prolonged artificial injection with water mainly derived from municipal wastewater treatment plants (activated sludge). As previously mentioned, the secondary treated effluents are transported via an open channel (the Asso Channel) to a sinkhole where the recharge flow rate under steady conditions and during drought periods is 140–200 L/s. The recharge flow rate increases to 450 L/s during winter due to surface runoff drained by the Asso Channel. Rainfalls usually occur consecutively for a few days (5–6) and 3–4 times per year during winter/spring seasons. However, the predominant steady conditions are during the drought periods (due to the arid climate) therefore a constant average recharge flow rate of 140–200 L/s into the sinkhole can be considered. The effluent infiltrates through a fractured and karstified aquifer. The water is then sampled through wells located 320 m and 500 m away from the sinkhole (Figure 4.3). The groundwater recharge acts as a salt intrusion barrier.

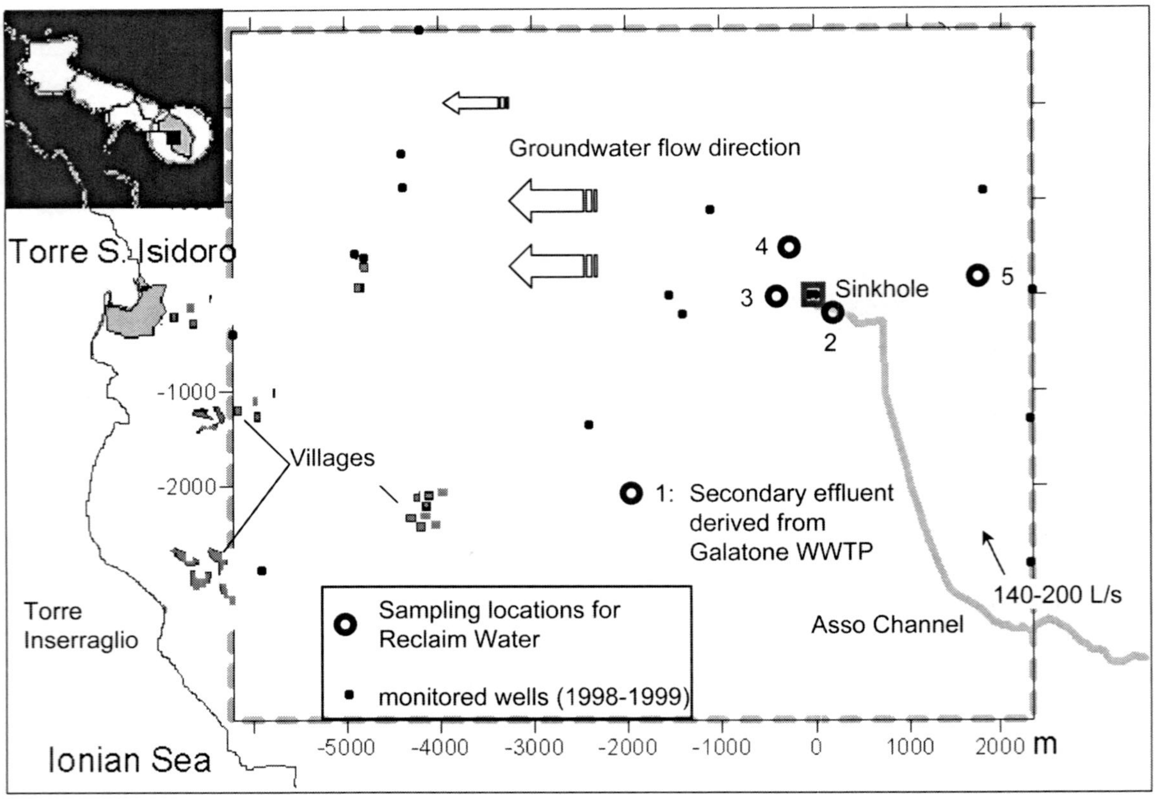

Figure 4.3 Study area and sampling locations: Sampling point 1 is the secondary effluent from the municipal treatment plant. Sampling point 2 is the aquifer injection point. Sampling points 3–4 are wells at 320 m and 500 m downgradient with respect to the groundwater flow. Sample point 5 is outside the recharge area and provides background values of groundwater quality

4.2.2 Hydrogeology

The Apulia carbonate (Mesozoic) stratum contains fresh water floating on sea water of continental intrusion. The formation of the Salento peninsula consists mainly of sandstone, limestone and dolomite deposits. In the study area (Figure 4.3–4.4) the groundwater flows under low pressure (<0.1 bar) inside the fractured Cretaceous (Senoniano) limestone which has a variable thickness between 20 and 40 m. Along the coastal area of the region, the Calcare di Altamura (limestone) is lapped by 5 to 10 m of Tertiary (Miocene) deposits known as Calcareniti di Gallipoli (sandstone) intercalated by lenses of terra rossa, calcspar and loamy sand. The underlying deposits are mainly limestone and dolomite.

The Nardò aquifer is very permeable (average hydraulic conductivity is 7.9×10^{-3} m^2/s), due to a large number of joints and fractures caused by tectonic movements, which are karstified. The oldest sediments in the Salento Peninsula belong to the limestone and dolomite formations. The average ground level is about 35 m above sea level, approximately 8 km from the Ionian Sea coast. The depth of the water table is about 32 m below ground with a piezometric head of 3 m above sea. The groundwater salinity increases with depth as freshwater floats over saline water originating from the Ionian Sea due to intrusion and it is confined within the fractured limestone. Fractures are interconnected and partly filled with calcspar or terra rossa. Groundwater flows along preferential pathways within the aquifer under a piezometric head of approximately 3 m above the mean sea level. The preferential pathways are mainly horizontal conduits. The Pleistocene sediments are also fractured but the interconnected fractures are orientated along the local bedding planes.

The Nardò aquifer supplies freshwater to numerous households along the Ionian Sea coastline. Consequently, the water quality of the Nardò aquifer should be maintained at acceptable levels. The presence of organic compounds (revealed by an oxidation level over 0.5 mg/L as O_2) together with an elevated concentration of chloride due to marine intrusion, makes the groundwater unfit for drinking purposes (Directive 2008/105/EU), without treatments.

4.2.3 Process design and operation

At the Nardò test site the managed aquifer recharge is based upon continuous direct injection in a sinkhole. Effluent from a municipal treatment plant (Galatone WWTP) and the surface draining water from its surrounding area are used for recharge. The water is collected and transported in an open channel (the Asso channel) to the sinkhole. The injection rate amounts to 12,000–17,000 m^3/d. This injection is stopped for about one-two months per year during May to September for maintenance purposes. During this period the water transported by the Asso Channel is diverted to another channel that flows towards other sinkholes or directly to the Ionian Sea (Figure 4.4). The reclaimed water is used for irrigation purposes and partly for domestic supply.

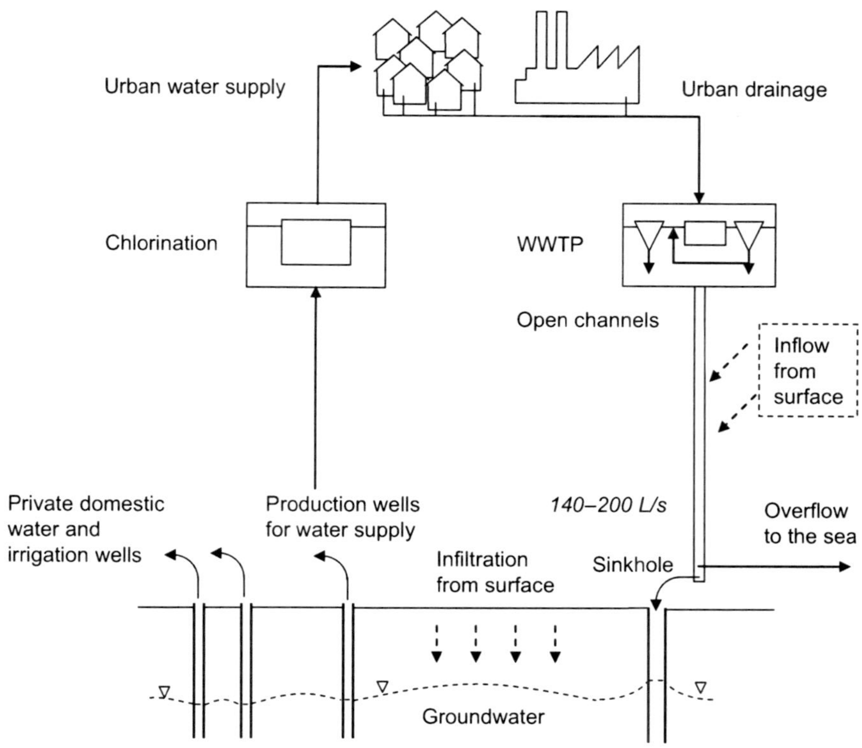

Figure 4.4 Schematic of the Nardò (South Italy) aquifer recharge site

4.2.4 Clogging

A full approach to the clogging theory has been developed by Perez-Paricio (2001), which includes operational guidelines to reduce clogging during artificial recharge. In particular, draying and cracking (mechanical treatment) in recharge basins or pumping in injection wells, are required respectively once or twice per year in basins and once or four times per month in wells, when there is a possibility for a slight (i.e. TSS < 10 mg/L; TOC < 10 mg/L) or moderate (i.e. TSS < 20 mg/L; TOC < 25 mg/L) clogging during artificial recharge. For possible strong clogging (i.e. TSS < 20 mg/L; TOC < 25 mg/L) water pre-filtration is required before recharge. The simplest theoretical approach presented by several authors (Taylor & Jaffe, 1990; Blazejewski & Murat-Blazejewska, 1997; Perez-Paricio, 2001; Skolasinska, 2006; Pavelic *et al.* 2007) for granular porous media, was here modified in order to estimate the period of time required to reach clogging in limestone fractured aquifers due to biological-physical colloidal material deposition into fractures. By considering the fractured media as compared by a 3D set of parallel fractures with variable apertures, the clogged mass (Chrysikopoulos *et al.* 2010)

$$\rho_s V_s = \left(\frac{M}{L^3} \times L^3\right) \quad (1)$$

is deposited at the constant water/rock surface and it can be considered as a function of changes in suspended solid concentration in filtered wastewater, i.e. $\rho_s\ V_s = V_w\ \Delta\ C_{ss}$. The latter mass can be derived from the conservation mass-balance, as follows:

$$d(\rho_s V_s) = dV_w \Delta C_{ss} = Q_w dt \times \Delta C_{ss} \quad (2)$$

where ΔC_{ss} ($M \cdot L^{-3}$) accounts for the reduction of concentration of suspended solids with density ρ_s into wastewater volume dV_w, which is filtered through limestone fractures at a constant flow rate Q_w (L^3/t). By assuming that the change on density of the attached solid ρ_s is negligible, Equation (2) can be integrated between instant $t = 0$, where deposited volume is $V_s = 0$ (and $C_{ss} = C_{ss0}$) and the generic instant t, to have:

$$\rho_s V_s = Q_w t[C_{ss0} - C_{ss}(t)] \quad (3)$$

or

$$t = \frac{\rho_s V_s}{Q_w[C_{ss0} - C_{ss}(t)]} \tag{4}$$

The volume of solids attached to the fracture walls will cause a reduction of the fracture apertures. By defining the clogged fraction of the saturated fractured media f_c as the n_c/n_e ratio, where n_e is the initial effective porosity and n_c is the clogged porosity, Equation (4) can be rewritten as:

$$t = \frac{f_c \rho_s}{Q_w[C_{ss0} - C_{ss}(t)]} \sum_{i=1}^{N_f} \overline{b_i} \times S_i \tag{5}$$

where N_f(-) is the number of the fractures of the parallel set, $\overline{b_i}$ is the average value of apertures on the fracture plane surface with area S_i (L^2). If all fractures of the set have the same surface area S_f and the same mean aperture $\overline{b}$, we can define the time for clogging as follows:

$$t = \frac{f_c \rho_s}{Q_w[C_{ss0} - C_{ss}(t)]} N_f \overline{b} \times S_f \tag{6}$$

It should be noted that Equation (6) implicitly accounts for both physical and biological clogging by means of the actual value of the suspended solid concentration, at assigned time t of filtration. Chemical clogging due to carbonate dissolution and precipitation was, instead, not included in this Equation.

The tested fractured set has then an hydraulic conductivity, as defined by,

$$K(C_{ss}) = (1 - f_c) n_e \frac{\overline{b}^2}{12} \frac{\gamma}{\mu} \tag{7}$$

where γ/μ can be set 10^7 ($m \cdot s^{-1}$) for wastewater at 18°C and by replacing f_c, given by Equation (6) into Equation (7), it produces:

$$K(C_{ss}) = \left[1 - Q_w \frac{[C_{ss0} - C_{ss}(t)]}{\rho_s N_f \overline{b} \times S_f} t\right] n_e \frac{\overline{b}^2 \gamma}{12 \mu} \tag{8}$$

It should be noted that to determine K from Equation (8) it is required $C_{ss}(t)$ at an assigned time of filtration. During lab tests, this value can be derived from measurements on water effluent at the end of filtration.

The application of the formula at the lab test carried out at IRSA laboratory for 5 years and to the Nardò wells 3 and 4 during Reclaim Water studies, has led to the values reported on Table 4.3. Thus results of the developed clogging theory suggest a complete clogging of the well 3 during 2008 and the clogging of well 4 during year 2016.

Table 4.3 Estimation of the clogging periods on the lab columns and Nardò site (well 3).

Filtration period (d)	ΔC_{ss} (mg/L of total SS)	f_c	K (cm/s) Estimated by Eq. (8)	K (cm/s) Lab/field estimations by using Darcy equation
		Filter #1 $\overline{b} = 1.08$ mm; $N_f = 16$		IRSA Laboratory
0*	0	0.00E+00	**4.85**	4.85±0.331 (2006)
230	14	7.04E-02	**4.51**	4.55±0.401 (2007)
3650 (10 y)	19.8	9.96E-01	**0.02**	–
		Filter #2 $\overline{b} = 1.23$ mm; $N_f = 16$		
0*	0	0.00E+00	**6.30**	6.31±0.595 (2006)
230	14	6.18E-02	**5.91**	6.02±0.780 (2007)
3650 (10 y)	19.8	8.74E-01	**0.80**	–

(Continued)

Table 4.3 Estimation of the clogging periods on the lab columns and Nardò site (well 3) (*Continued*).

Filtration period (d)	ΔC_{ss} (mg/L of total SS)	f_c	K (cm/s) Estimated by Eq. (8)	K (cm/s) Lab/field estimations by using Darcy equation
		Nardò site#1 (Well 3, Fioschi) $\bar{b} = 0.11$ mm; $N_f = 24$		Field estimations
0*	0.0	0	**3.43E-03**	–
3650 (10 y)	9.6	6.09E-01	**1.34E-03**	3.00E-03 (2001)
5840 (16 y)	9.6	9.75E-01	**8.60E-05**	8.46E-05 (2007)
5990 (16.4 y)	9.6	1.00E+00	**3.13E-08**	**Clogged**
		Nardò site#2 (Well 4, Fico) $\bar{b} = 0.11$ mm; $N_f = 24$		
0*	0	0.00E+00	**3.43E-03**	–
3650 (10 y)	9.9	4.00E-01	**2.06E-03**	3.0E-03 (2001)
5840 (16 y)	9.9	6.41E-01	**1.23E-03**	–
9113 (25 y)	9.9	1.00E+00	**8.86E-08**	**Clogged**

*0 is referred to the year 1991, when the artificial recharge operations started.

4.3 TECHNOLOGY PERFORMANCE AND CONTAMINANT MONITORING

During Reclaim Water, the case study site was investigated from January 2006 to September 2007. In this period, every three months a sampling campaign was conducted leading to a total of seven sampling campaigns. For each Reclaim Water sampling campaign, samples were taken in 3 monitoring wells (Figure 4.3) after 10 minutes of pumping to ensure the removal of stagnant water in pipelines. In wells 3 and 4 (Figure 4.3) water samples were collected by means of samplers located 0.5 to 1 m below the water table. The mean values of all samplings, according to the "Protocol 1" agreed in the project (standard chemical and microbiological parameters), are given in Table 4.4.

Besides the field activity carried out at the Nardò test site, bench scale experiments were also performed where the karstic aquifer was simulated by horizontal columns. These horizontal columns were composed of limestone slabs and reproduced wastewater velocities measured in the cracks of the Nardò subsoil (Carrieri and Masciopinto, 2000). To mimic groundwater recharge, the columns were fed with municipal wastewater effluents from a conventional activated sludge plant and a membrane bioreactor (MBR). Further details on the bench-scale experimental set-up are reported in Chapter 15. Experiments were aimed at comparing the two effluents having different quality features in terms of their effects on groundwater recharge both from the standpoint of microbiological safety and environmental pollution in general (La Mantia *et al.* 2008a). The main differences between the two tested effluents were mainly in the concentrations of suspended solids (absent in MBR), COD, nitrogen forms, and bacterial indicators. The membrane bioreactor provided effluents of higher quality with respect to CAS, with lower carbon concentration, completely nitrified, and with no bacterial contamination.

4.3.1 Wastewater chemical-physical parameters

Full scale results

Figure 4.5 reports the average values of the main parameters monitored at the test site at different sampling points. Relevant modifications across the different sampling points can be observed especially for SS and turbidity, while the nitrogen forms appear to be affected by nitrification and partial denitrification. As a matter of fact, nitrate concentrations in groundwater close the sinkhole were below the background values observed at well 5 (see Figure 4.5 and Table 4.4), and biodegradable organic carbon compounds (limiting factor) were almost absent in the injected wastewater.

Horizontal soil column experiments at the IRSA lab

The results of the bench scale activity showed that the horizontal columns could remove up to 85% of the total COD fed with the CAS effluent, while in case of the lower COD concentrations fed with the MBR effluent the removal observed was about 44%, due to the higher efficiency of the MBR in removing biodegradable substrates.

Table 4.4 Results of the Nardò monitoring programme during Reclaim Water (2006–2007, N = 7).

Sampling point		**Secondary effluent Sampling point 1 (SP1)**	**Injectant well SP2**	**Well at 320 m (well 3) SP3**	**Abstracted water well 4 SP4**	**Background groundwater, well 5 SP5**
		Average ± stdv	**Average ± stdv**	**Average ± stdv**	**Average ± stdv**	**Average ± stdv**
Basic parameters						
Suspended Solids	$mg \cdot L^{-1}$	29.9 ± 22.5	8.4 ± 7.7	9.9 ± 14.1	3.0 ± 1.4	2.9 ± 1.4
Turbidity	NTU	21.4 ± 10.7	6.0 ± 5.7	6.3 ± 5.8	1.8 ± 1.4	3.3 ± 3.1
COD Soluble	$mg \cdot L^{-1}$	42.7 ± 19.2	25.0 ± 6.7	13.4 ± 4.2	9.0 ± 2.2	6.4 ± 4.0
DOC	$mg \cdot L^{-1}$	8.5 ± 1.8	8.1 ± 1.7	3.9 ± 1.6	3.3 ± 0.7	1.3 ± 0.6
Ammonia NH_4-N	$mg \cdot L^{-1}$	1.1 ± 1.2	2.2 ± 2.0	0.39 ± 0.11	0.40 ± 0.13	0.39 ± 0.11
Total Kjeldahl Nitrogen	$mg \cdot L^{-1}$	3.6 ± 1.9	3.1 ± 1.9	0.99 ± 0.60	1.2 ± 0.83	0.68 ± 0.69
Nitrate NO_3-N	$mg \cdot L^{-1}$	10.9 ± 7.5	9.1 ± 6.0	2.7 ± 3.7	5.6 ± 2.3	33.3 ± 7.9
Nitrite NO_2-N	$mg \cdot L^{-1}$	0.28 ± 0.29	1.04 ± 1.17	0.07 ± 0.06	0.07 ± 0.06	0.05 ± 0.01
Total Phosphorus	$mg \cdot L^{-1}$	1.60 ± 1.28	1.8 ± 1.53	1.14 ± 1.12	0.44 ± 0.36	0.02 ± 0.01
PH	–	7.46 ± 0.21	7.85 ± 0.41	7.38 ± 0.32	7.58 ± 0.15	7.41 ± 0.20
Microbiological parameters						
Total bacteria count 22°C	CFU/mL	$4.3 \cdot 10^4 \pm 6.1 \cdot 10^4$	$7.3 \cdot 10^4 \pm 1.2 \cdot 10^4$	$2.9 \cdot 10^4 \pm 5.1 \cdot 10^4$	$1.3 \cdot 10^4 \pm 1.2 \cdot 10^4$	$6.4 \cdot 10^2 \pm 3.9 \cdot 10^2$
Total bacteria count 37°C	CFU/mL	$2.5 \cdot 10^4 \pm 4.1 \cdot 10^4$	$2.8 \cdot 10^4 \pm 3.6 \cdot 10^4$	$1.1 \cdot 10^4 \pm 1.1 \cdot 10^4$	$7.5 \cdot 10^3 \pm 7.7 \cdot 10^3$	$4.5 \cdot 10^2 \pm 3.8 \cdot 10^2$
Faecal coliforms	MPN/100mL	$3.0 \cdot 10^4 \pm 4.9 \cdot 10^4$	$4.5 \cdot 10^3 \pm 5.5 \cdot 10^3$	$5.7 \cdot 10^2 \pm 5.3 \cdot 10^2$	$3.1 \cdot 10^3 \pm 5.9 \cdot 10^3$	$3.3 \cdot 10^2 \pm 5.8 \cdot 10^2$
Total coliforms	MPN/100mL	$6.9 \cdot 10^4 \pm 8.0 \cdot 10^4$	$1.2 \cdot 10^4 \pm 1.1 \cdot 10^4$	$1.5 \cdot 10^4 \pm 3.4 \cdot 10^4$	$8.3 \cdot 10^3 \pm 6.3 \cdot 10^3$	$1.0 \cdot 10^3 \pm 1.2 \cdot 10^3$
E. coli	MPN/100mL	$1.3 \cdot 10^3 \pm 1.9 \cdot 10^3$	$1.7 \cdot 10^2 \pm 3.4 \cdot 10^2$	$1.4 \cdot 10^1 \pm 1.1 \cdot 10^1$	$8.1 \cdot 10^1 \pm 1.3 \cdot 10^2$	$8.0 \cdot 10^0 \pm 1.2 \cdot 10^1$
Faecal streptococci	MPN/100mL	$6.1 \cdot 10^3 \pm 8.7 \cdot 10^3$	$6.6 \cdot 10^2 \pm 8.3 \cdot 10^2$	$1.0 \cdot 10^2 \pm 1.0 \cdot 10^2$	$1.3 \cdot 10^2 \pm 1.9 \cdot 10^2$	$7.8 \cdot 10^1 \pm 1.3 \cdot 10^2$
Enterococci	CFU/100mL	$3.7 \cdot 10^3 \pm 3.6 \cdot 10^3$	$3.5 \cdot 10^3 \pm 7.0 \cdot 10^3$	$6.2 \cdot 10^2 \pm 1.2 \cdot 10^3$	$1.6 \cdot 10^3 \pm 8.3 \cdot 10^2$	$1.9 \cdot 10^2 \pm 2.3 \cdot 10^2$
Clostridium spores	CFU/100mL	$9.0 \cdot 10^3 \pm 8.7 \cdot 10^3$	$8.4 \cdot 10^2 \pm 1.1 \cdot 10^3$	$1.6 \cdot 10^3 \pm 3.0 \cdot 10^3$	$1.6 \cdot 10^2 \pm 2.1 \cdot 10^2$	$1.5 \cdot 10^2 \pm 1.1 \cdot 10^2$
Bacteriophages	PFU/100mL	$8.3 \cdot 10^3 \pm 6.9 \cdot 10^3$	$5.3 \cdot 10^3 \pm 5.9 \cdot 10^3$	$1.8 \cdot 10^3 \pm 2.2 \cdot 10^3$	$1.2 \cdot 10^2 \pm 2.5 \cdot 10^2$	$1.3 \cdot 10^2 \pm 1.8 \cdot 10^2$
Trace elements						
Zinc	$\mu g \cdot L^{-1}$	69.4 ± 38.8	47.3 ± 16.0	65.9 ± 23.9	29.3 ± 30.8	23.6 ± 25.9
Salinity						
Electrical conductivity	$\mu S \cdot cm^{-1}$	1616 ± 156	1256 ± 185	637 ± 384	966 ± 123	1147 ± 152

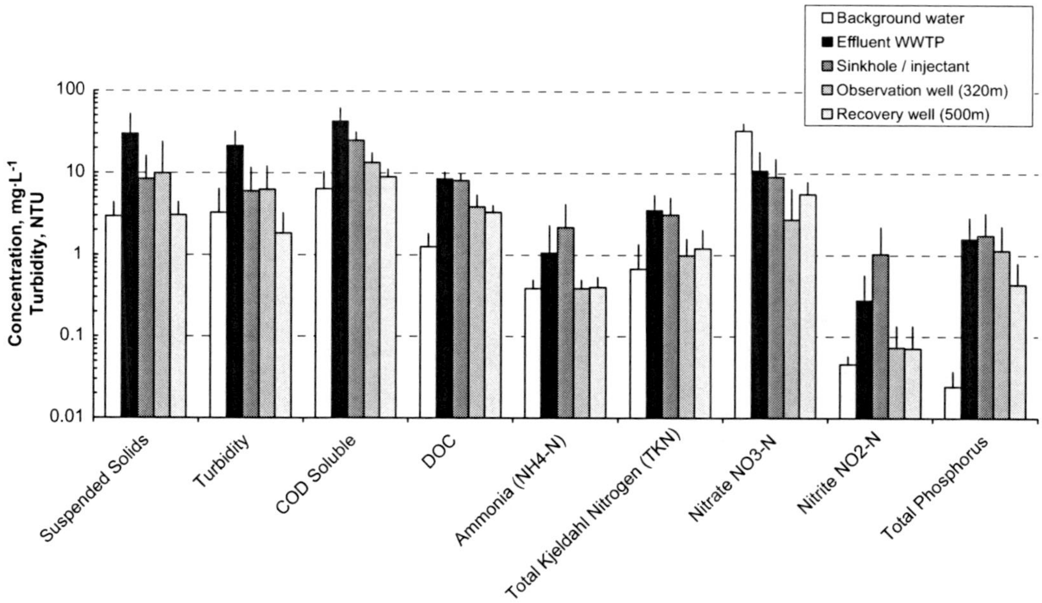

Figure 4.5 Development of chemical water quality during MAR in Nardò

The effects of passage through the two horizontal columns on the removal nitrogen species is reported in Figure 4.6. In the case of the CAS effluent (not nitrified, Figure 4.6A), it can be noted that the Filter RI completely removed the ammonium but no nitrate formation was observed. Therefore a nitrification/denitrification process was assumed and the low concentration of the nitrate observed during test indicates that the rate of denitrification was higher than that of nitrification. Although the COD removal process seemed to occur at a very high rate, due to both the filtration and the de-nitrification process, it should be noted that the mass balance was well correlated with the de-nitrification process (CODrem/Nrem = 4.2, in good agreement with the theoretical value). Differently, the batch reference for Filter RI showed complete nitrification and negligible de-nitrification, possibly due to the higher oxygen transfer. In the case of the MBR effluent (Figure 4.6B), nitrate removal only reached 35% possibly due to the lack of biodegradable COD. Furthermore, also in this case, the denitrification performance for the Filter RII was more effective than for the batch reference.

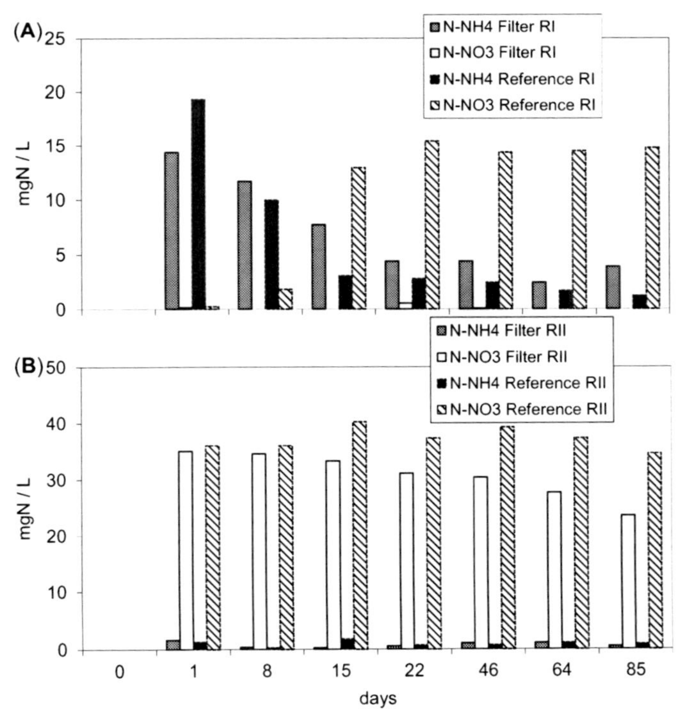

Figure 4.6 Nitrogen removal in Filter and Reference batch fed with CAS effluent (A) and MBR effluent (B)

The temporal trend of groundwater electrical conductivity (Figure 4.7), confirmed that during drought season artificial recharge causes a slight increases of the background salinity at the Nardò sinkhole, though correspondent estimated chloride remains about 50% below the threshold value (<250 mg/L). Similarly, the DOC trend (Figure 4.8) shows a slight increase of organic carbon compounds in groundwater.

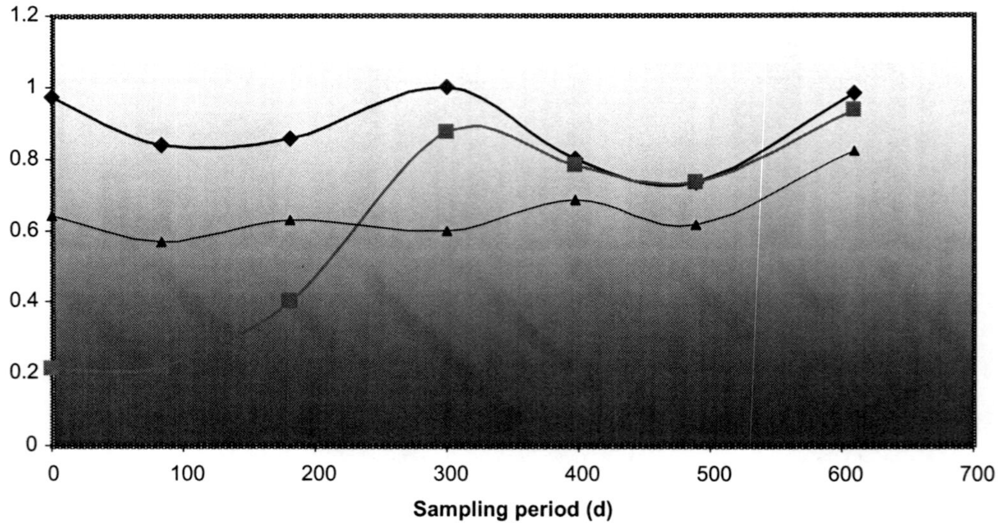

Figure 4.7 Trend of electrical conductivity (C/C0) in the injected water (diamonds) and groundwater (well 3, squares; well 4, triangles) during MAR in Nardò

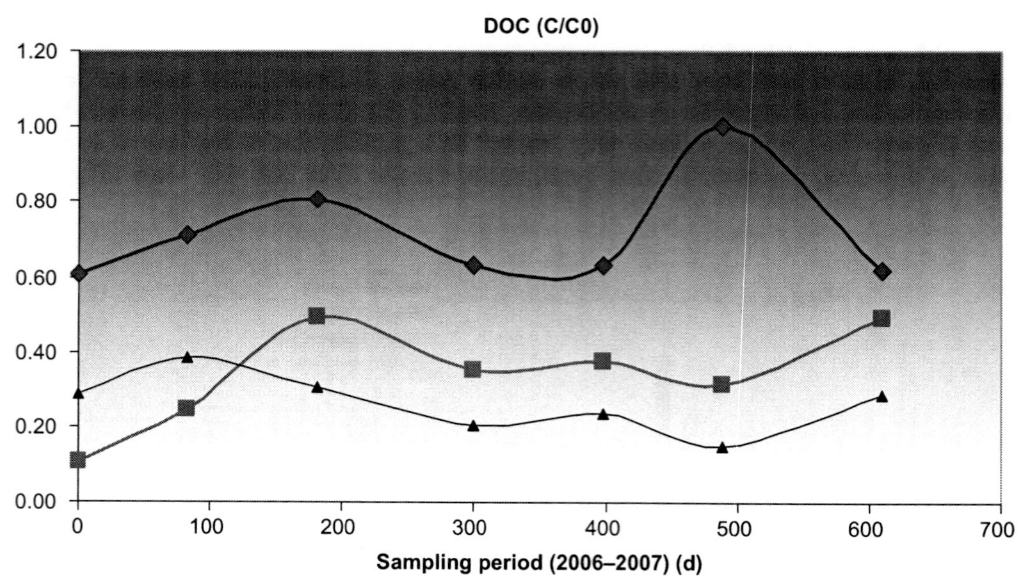

Figure 4.8 Temporal trend of DOC (C/C0) of injectant water (diamonds) and groundwater (well 3, squares; well 4, triangles) during MAR in Nardò

4.3.2 Microbiological parameters

Microbial indicators monitored at the Nardò site showed a low microbiological quality of the native groundwater. Indicator background concentrations in the groundwater ranged from $7.8 \cdot 10^1 \pm 1.3 \cdot 10^2$ MPN/100 mL of faecal streptococci to $4.5 \cdot 10^2 \pm 3.8 \cdot 10^2$ CFU/mL of TBC (37°C). In addition, the presence of specific microbial indicators of faecal origin, namely Enterococci ($1.9 \cdot 10^2 \pm 2.3 \cdot 10^2$ CFU/100 ml) and, at lower level, *E. coli* ($8.0 \cdot 10^0 \pm 1.2 \cdot 10^2$ MPN/100 mL) was an evidence of faecal contamination of the native groundwater (Figure 4.9).

Noteworthy in two sampling campaigns the *E. coli* level in well 4 was above the reported Italian threshold value for water reuse in agriculture (<100 CFU/100 mL, Italian law D.M. 185/2003). Nevertheless, a higher microbial contamination was found in the monitored groundwater wells affected by the recharged reclaimed water (well 3 and well 4) with respect to the background groundwater (well 5). In well 3 the indicator bacteria ranged from $1.4 \cdot 10^1 \pm 1.1 \cdot 10^1$ MPN/100 mL of *E. coli* to $1.1 \cdot 10^4 \pm 1.1 \cdot 10^4$ CFU/mL of TBC at 37°C. Similar values were obtained in well 4 where the indicator concentrations ranged from $1.3 \cdot 10^2 \pm 1.3 \cdot 10^2$ MPN/100 mL of faecal streptococci to $7.5 \cdot 10^4 \pm 7.7 \cdot 10^4$ CFU/mL of TBC at 37°C.

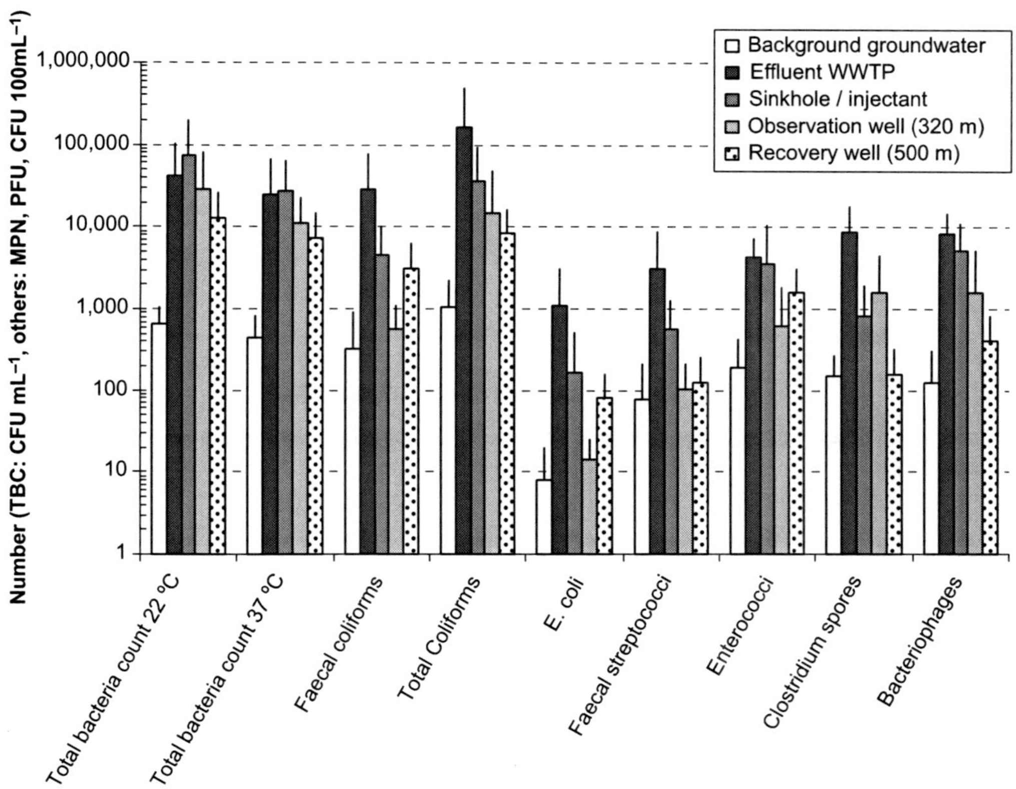

Figure 4.9 Development of microbiological water quality during MAR in Nardò

Overall each microbial indicator was more concentrated in well 3 and well 4 with respect to background well 5, with the exception of somatic coliphages, *Clostriudium* spores and faecal coliforms that were present at concentrations comparable to the background water at least in one of these wells. The level of *E. coli* in the groundwater in well 3 and well 4 was frequently (40%) above the limit (<10/100 mL) of Italian law (D.M. 185, 2003) for reuse of reclaimed water in agriculture (Figure 4.9).

During the Reclaim Water sampling campaigns only a slight reduction of microbial indicators from the injected (well 2) to the recovered water (well 4) was observed at this site (Table 4.4 and Figure 4.9). Higher decrease was observed between well 2 and well 4 with reference to somatic coliphages (Figure 4.10) (from $5.3 \cdot 10^3 \pm 5.9 \cdot 10^3$ to $1.2 \cdot 10^2 \pm 2.5 \cdot 10^2$ PFU/100ml), TBC 37°C (from $2.8 \cdot 10^4 \pm 3.6 \cdot 10^4$ to $7.5 \cdot 10^3 \pm 7.7 \cdot 10^3$ CFU/ml) and total coliforms (from $1.2 \cdot 10^4 \pm 1.1 \cdot 10^4$ to $8.3 \cdot 10^3 \pm 6.3 \cdot 10^3$ MPN/100 mL).

A 99% (or 2 Log disinfection credits) removal of *Giardia* cysts from the injection to well 4 groundwater stream was observed, on average, and all the groundwater samples tested negative for HAV. The concentration of *Cryptosporidium* oocysts decreased from injection to well 3 by 36% on average, although a high concentration (0.77 oocysts/L) was found at well 3 during sampling of May 2007. A removal higher than 94.8% of *Cryptosporidium* oocysts, on average, was also observed between the injection point to well 4. HAV has never been detected in the Salento groundwater, whereas Enteroviruses were tested positive in only four samples (out of the total 37 samples) by using PCR. Noroviruses were the most frequently retrieved viruses in the water samples analysed, even if they were only observed in the winter samples. This seasonal occurrence is confirmed by other data described in the literature. As a matter of fact, these viruses are one of the main causes of nonbacterial gastroenteritis (winter vomiting disease), which is common during cold seasons in developed countries. Of the 37 Nardò samples analysed, only eight (22%) tested positive for enteric viruses using PCR, and it was only possible to perform a quantitative estimation of genomic copies on 16 samples out of the total 148 determinations. In a previous study at the same Nardò site, only three samples (17%) tested positive for enteric viruses among the 18 water samples analysed using qRT-PCR (Masciopinto *et al.* 2007). Furthermore, in the present work it was also noted that when the sampled water tested positive for Enteroviruses, both the WWTP effluent and the samples from the injection point showed negative results. The maximum possible virus reduction in the Nardò groundwater was 2.8 Log disinfection credits for Enterovirus in well 3 and 3 Log credits for Norovirus in well 4.

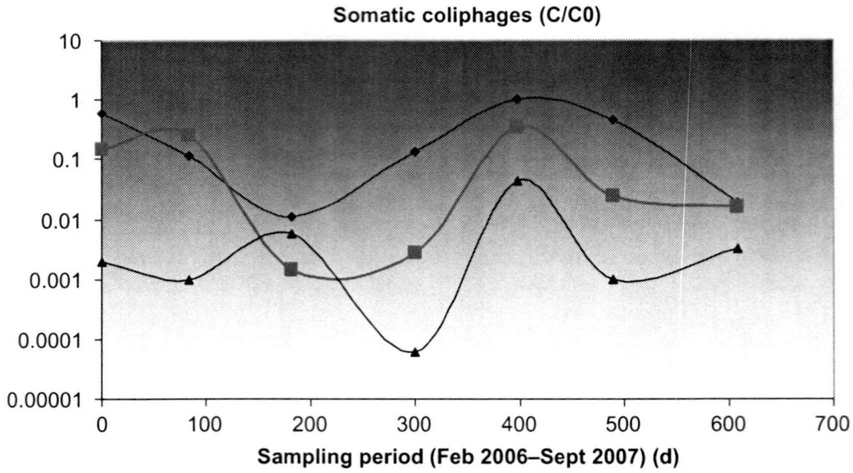

Figure 4.10 Development of somatic coliphages during MAR in Nardò: injectant water (diamonds) and groundwater (well 3, squares; well 4, triangles)

The apparent absence of Enteroviruses in samples of the injected water, despite contamination in wells downstream of it, could have been due to the presence of both dissolved organic compounds and salt ions. These could have interfered with the natural filtration process, by causing an increase of attachment of viruses to the soil and solid particles in the injected water, with subsequent particles sedimentation in the channel. Indeed, in the same injected water samples, in which enteric viruses were absent, both *Giardia* and *Cryptosporidium* were retrieved, together with high counts of all the faecal pathogen indicators. Furthermore, a second hypothesis should also be considered. According to Schmoll *et al.* (2006, p. 75), a decrease in the concentration of salts in the groundwater, such as would occur after a storm event, could have caused disaggregation-detachment of viruses and bacteria from fracture walls by increasing detachment rate coefficient. Redman et al. (1997) suggest that small changes in the concentrations of salts can dramatically affect the mobilisation of some of the organisms found in groundwater systems. The implication is that the concentration of salts in groundwater may be of great significance for pathogen transport, because this can affect attachment/detachment and deposition rate coefficients. Thus, the reduction in the concentration of salts due to stormwater runoff in February 2007 from 0.6 to 0.4 g/L in injected water samples and in the Nardò groundwater could explain why Enteroviruses were retrieved in the wells, but not in either the injected water or the WWTP effluent. In particular, the maximum Norovirus count (7600 gc/L) was found in well 3, 320 m from the injection point during sampling carried out only 2 days after an intensive (3-day) storm (11.4 mm) that occurred between 17 and 19 May 2007. During the same sampling campaign, the highest counts of *Giardia* and *Cryptosporidium* oocysts were also noted in the same well 3, whereas Norovirus was absent in WWTP effluent and injected water. This variability of virus loads suggests that a temporal variability of the initial conditions is required at the Nardò site in order to have realistic model simulations of the injected virus counts into the sinkhole, also by considering the distance of the injection site from the treatment plant (i.e. the survival time of viruses in surface water).

In conclusion, the virus reduction at a specific travel time (or distance) may mathematically be defined by the proposed Equation (see Chapter 7), which points to new findings on the theory of the time-dependent virus inactivation-deposition rate, because it accounts for a slowed down rate of virus reduction with time, due to the effects of detachment. The sensitivity analysis has shown that the attachment rate coefficient is the critical parameter for the model. Variations of 50% of this rate, lead to a change up to three times in the virus reduction rate. The spread of the time-dependent virus reduction rate can be reduced by decreasing groundwater temperature. The simulation results are in accordance with maximum Log disinfection credits (3 and 2.8) of the Norovirus and Enterovirus tested in the Nardò wells. For water pumped from a well 500 m from the injection point, a 99% removal of Giardia cysts and a higher than 94.8% removal of *Cryptosporidium* oocysts were experienced.

As for the bench scale experiments, the main microbial indicators measured are reported in Table 4.5. It should be noted that there was an initial contamination of the MBR effluent, due to pathogen indicators attached to the fracture walls into Filter RII. After few days of filtration, the pathogenic indicators decreased rapidly. The *E. coli* and Enterococci were zero after 18 days. After 18 days, the pathogenic indicators into the filtered CAS effluent (Filter RI) had higher counts with respect to the filtered MBR effluent, although the differences between somatic coliphage and Clostridium spore counts between the two filters appeared negligible.

Table 4.5 Microbiological indicators vs. time at the output section of the lab-filters during the 2nd filtration test in the laboratory.

Date	Days of filtration	Somatic coliphages PFU/mL	Total bacteria count at 22°C CFU/mL	Fecal coliforms MPN/100 mL	Enterococci CFU/100 mL	Clostridium spores CFU/100 mL
CAS effluent						
6-Mar	1	80	56000	27800	11000	3000
8-Mar	3	11	4900	542	500	350
12-Mar	7	8	2100	542	180	250
19-Mar	14	4.3	14900	34	19	220
26-Mar	21	2.6	680	33	5	90
2-Apr	28	1	390	5	4	190
27-Apr	53	0.00E+00	2660	0	0	65
4-Jun	91	0.00E+00				0
Static water CAS effluent						
5-Mar	0	1200	210000	160900	260000	45000
12-Mar	7	160	49000	54200	49000	8000
2-Apr	28	106	32000	348	30000	900
27-Apr	53	48	440	2	900	350
4-Jun	91				60	52
MBR effluent						
6-Mar	1	1	830	5	1	200
8-Mar	3	2	860	2	0	50
12-Mar	7	5	1560	2	4	150
19-Mar	14	3	1500	1609	0	90
26-Mar	21	5.00E-02	830	0	0	65
2-Apr	28	3.00E-02	1020	0	0	95
27-Apr	53	10.4				120
4-Jun	91	3.6				50
Static water derived from MBR effluent						
5-Mar	0	1.5	4100	4	0	20
27-Apr	53	1.25	2040	0	0	56
4-Jun	91	1.25		0	0	0

On the basis of these results, the adoption of MBR effluents instead of the existing CAS effluent for the artificial recharge could result in relevant improvements at the Nardò site. Membrane bioreactors produce effluents with pathogen indicators at counts lower than the background values usually monitored in the Nardò wells (La Mantia *et al.* 2008b). Also, based on the laboratory filtration tests, it was possible to estimate the inactivation constants, observing a first-order decay of the faecal pathogen indicators in filtered CAS effluent (La Mantia *et al.* 2008a). Moreover, in previous studies membrane filtration systems were reported to be an effective barrier against contamination by *Giardia* cysts and *Cryptosporidium* oocysts (Lonigro *et al.* 2006).

Finally, it should be considered that due to the low count of faecal pathogen indicators, MBR effluents require limited post-disinfection treatments, thus avoiding the potential formation of disinfection by-products in the injected water and then in the recharged aquifer. This contamination was shown to be relevant at the Nardò site, as demonstrated by monitoring the Adsorbable Organic Halogens (AOX, La Mantia *et al.* 2008a).

The above findings lead to the conclusion that the MBR technology should be strongly recommended when evaluating the aquifer recharge options, due to the effluent quality improvement in terms of chemical and microbial parameters and also in terms of clogging reduction (very low suspended solids content).

4.3.3 Salinity related parameters

Maps of the salinity contour line (Figure 4.11) show the improvement of groundwater quality subsequent to the injection into the sinkhole of 140 L/s, which was started in 1991. In 1999 the ground water salinity increased slightly in wells with the depth below the watertable. The salinity increased according to the range from 0.2 to 1.5 g/L, at 50 m below watertable. This information, together with the water temperature, pH and dissolved oxygen, was obtained in wells by means of the Ocean Service (Idronaut S.r.l.) probe, which allowed to draw water conductance, dissolved oxygen, and temperature profiles (Figure 4.12).

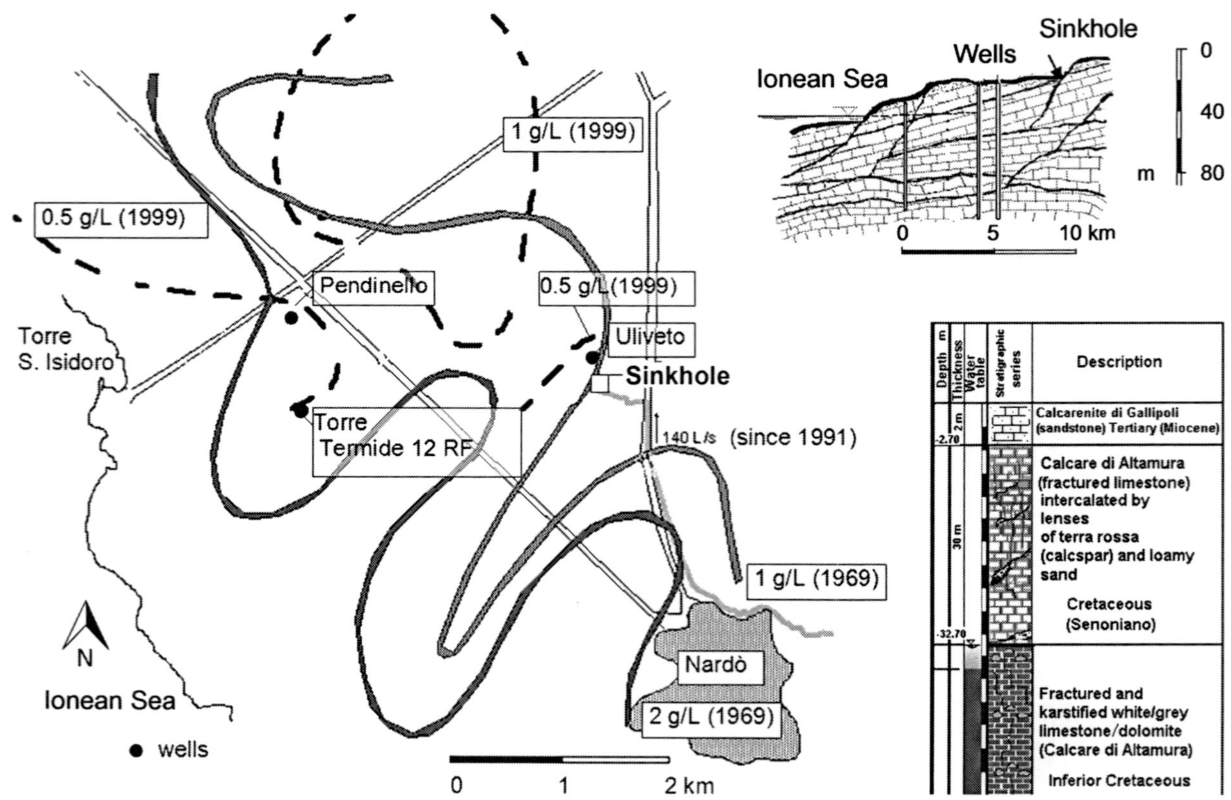

Figure 4.11 On the left: Salinity contour map (solid lines) in 1969 (i.e. before the injection into the sinkhole) reported by Cotecchia (1977) and contour salinities (broken lines) of water samples in wells in 1999, after 8 years of the injection of 140 L/s the into sinkhole. On the right: Geological cross-sections of the Nardò aquifer. Right below: Geological cross-sections of the Nardò aquifer

Groundwater was sampled in monitoring wells located in the area of Nardò, at different distances from the sinkhole, before and after 1991 when injection started (Table 4.6). The water samples were collected after ten minutes of pumping to ensure the removal of stagnant water in the pipelines. In wells without pumps, the water was collected by means of samplers located 0.5–1 m below the watertable. Ten wells were monitored from November 1953 to February

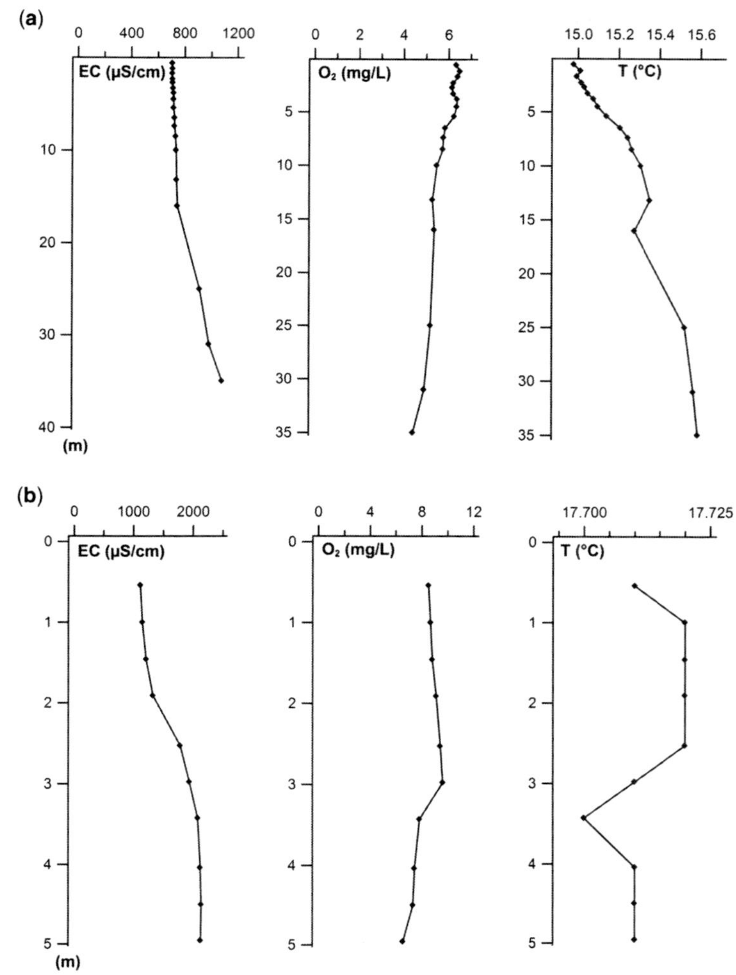

Figure 4.12 Vertical profiles (Ocean Service probe) into well Uliveto, 600 m from the sinkhole (a) (and 7600 form the coast) and well Torre Termide 12 RF (b) at 5800 m from the sinkhole (2600 m from the coast), during winter 1999

Table 4.6 Principal water quality constituents in monitoring wells with increasing distance from the injection, before and after artificial recharge started in 1991 (mean values).

Date (years)	Inject. time years	No. of wells	Specific conductance μS/cm	pH	COD mg/L	$N\text{-}NH_4^+$ mg/L	$N\text{-}NO_2^-$ mg/L	$N\text{-}NO_3^-$ mg/L	NaCl mg/L	TOC mg/L	TOX[a] μg Cl^- AOX	POX
Area 1 – distance from the injection = 3–4 km (7–8 km from the sea coast)												
2006/07	16	2	869	7.5	10.8	<0.3	0.05	4.5	243	3.3		
1998/99	10	7	1418	7.0	17.4	0	0.002	5.7	396	6.7	5.8	0.3
1979/80	–	6	2127	7.1	12.3	0.1	0.079	2.2	506	4.8		
1953/54	–	3	2093	7.5					513			
Area 2 – distance from the injection = 7–8 km (3–4 km from the sea coast)												
1998/99	10	7	1633	7.5	20.4	0	0	10.2	461	5.9	15.1	0.36
1979/80	–	2	2820	6.8	23.1	0.1	0.14	1.9	711	4.5		
1953/54	–	7	2352	7.4					584			

[a]TOX = sum of Purgeable (P) and Adsorbable (A) Organic (O) halogen compounds (X) (includes DBPs such as trihalomethanes)

1954, 8 wells during winter 1979/1980 and 14 wells from November 1998 to March 1999. Water sampled from these wells confirmed the reduction of groundwater salinity owing to the injection of the treated wastewater (see Table 4.6).

Furthermore, Table 4.6 shows a general improvement in groundwater quality due to the reduction of ammonia and nitrite (from 0.079 during 1979/80 to 0.05 mg/L during 2006/07), which was also coincident with the increased nitrates (Figure 4.13), chemical oxygen demand (COD) and Total dissolved Organic Carbon (TOC) in the wells near the

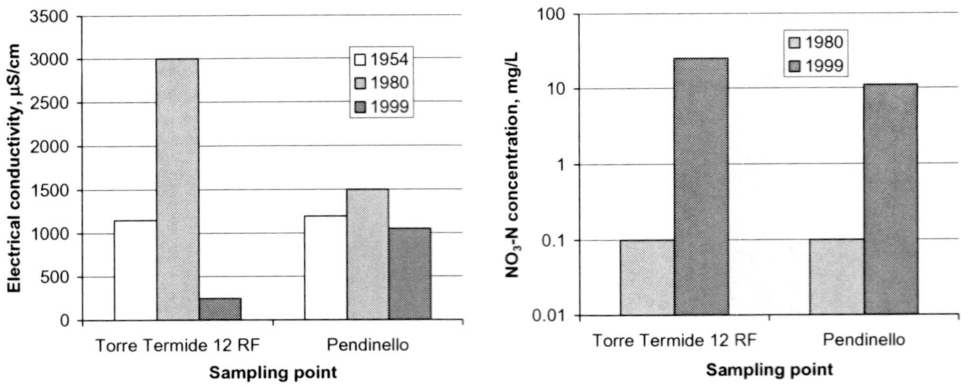

Figure 4.13 Electrical Conductivity (EC) (on the left) and N-NO3 concentration (on the right) vs. time in two wells (Torre Termide 12 RF and Pendinello) at respectively, 2600 m and 3200 m from the coast (i.e. 5400 and 5800 m from the sinkhole)

sinkhole (Table 4.6). The latter increase is due to the presence of weakly-biodegradable dissolved organic compounds. Moreover the lack of biodegradable organic carbon did not allow a complete denitrification of the groundwater.

Thus, the groundwater affected by municipal effluent recharge has shown salinity reduction of 58% (= 2092−869/2093) due to both the recession of seawater intrusion and the dilution of groundwater directly affected by injection.

The results of this study confirm also that water injection decreased the nitrite concentrations, that remained below the threshold value of good quality status (<0.05 mg/L). Also total nitrogen remained below the threshold limit (<50 mg/L), and similar behaviour was observed for the organic compounds (DOC) measured in groundwater with respect to the concentrations previously detected in wells close to the sinkhole during 1998/99.

4.3.4 Water quality changes during MAR

Quality of the injected water can be summarized as in the following: Flow rate: 140 L/s; Suspended solids 20–30 mg/L; COD: 30–40 mg/L; Water pH = 7–8; Electrical conductance: 1.1–1.5 mS/cm

Pollutant levels in the injected wastewater effluent are mainly due to the presence of pathogens, pharmaceuticals and disinfection by-products. For pharmaceuticals and micro-pollutants extensive studies carried out at the Nardò site have been described in Chapter 12. Experimental investigations have shown that groundwater is subject to faecal contamination and the presence of enteric viruses can be responsible of severe illnesses. Treatment technologies have to be combined with wellhead protection measures to reduce the risk of waterborne outbreaks. Canadian health guidelines (2004) and World Health Organization (WHO, 2006) established standard quality requirements and treatment technologies for drinking water that can ensure at least a 4-log (i.e. 99.99%) reduction credits and/or inactivation of pathogens. The experimental study carried out at Nardò established that 4-log reduction credits in a well located down gradient from injection can be achieved only after a filtration time over τ+85 d, where τ is (on average) the elapsed time required by wastewater to reach the recovery well positioned at a distance over 3000 m.

As a consequence, further water treatments before the injection in the Nardò site are recommended in order to reduce the risk of waterborne outbreaks. On the basis of the best water treatment technology proposed in chapter 15 and taking into account of the soil aquifer treatment into the fractured subsoil studied, appropriate additional treatments may be dense membrane filtration such as nanofiltration or reverse osmosis. These further treatments can, indirectly, also avoid fracture clogging which is expected to affect wells at a distance of 500 m from the injection by year 2016.

4.4 OPERATIONAL FEEDBACK

At the Nardò site, the improvement of injected water quality, i.e. the soil aquifer treatment (SAT), may be considered as a function of aquifer transmissivity and soil porosity and, obviously, the final quality of groundwater depend on the quality of the injected water. Hence, for one type of injected water, different types of subsoils may result in different abstracted water qualities. The hydraulic transmissivity of the Nardò aquifer was estimated to be elevated (860–8600 m^2/d).

At the Nardò coastal aquifer, in Southern Italy, the fractured limestone formation has been monitored to experience the change in groundwater quality after 17 years of artificial recharge (Masciopinto and Carrieri, 2002) with a constant rate of about 140 L/s (about 12,000 m^3/d).

As a main result of the groundwater water monitoring it can be stated that the natural inactivation of pathogens occurs within fractures of limestone aquifers. The model simulations and monitoring results show that the SAT could contribute to a reduction in treatment technology. The fitting of experimental data by using mathematical models suggests maximum Log disinfection credits ranging from 3 to of both the Norovirus and Enterovirus tested in the Nardò wells. For water pumped from a well 500 m from the injection point, a removal of 99% of *Giardia* cysts and higher than 94.8% of *Cryptosporidium* oocysts were experienced. All groundwater samples resulted negative to HAV and only 1 water sample, at the injection site, resulted positive to *Campylobacter*. Thus, even though the natural filtration in the fractures can be considered a "natural barrier" for some pathogens (i.e. *Campylobacter, Giardia* and *Cryptosporidium,* which were not detected or were only rarely present in the ground water samples at 500 m) a disinfection water treatment after pumping is still required to provide pathogen-free drinking water. Experimental investigations previously carried out at Nardò have shown that this groundwater is subjected to faecal contamination, which can lead to outbreaks of waterborne enteric viruses, responsible for severe illnesses. Therefore, treatment technologies have to be combined to wellhead protection measures to reduce the risk of such outbreaks. In fact, the monitoring results indicate that viruses are present in the studied water environments and the risk due to water contamination should be carefully evaluated, taking into consideration pathogenic indicators of water quality. Indeed, the quality of groundwater pumped at a distance of 500 m from the sinkhole does not meet the requirements for one of the two compulsory limits of the Italian rule (D.Lgs. 152/2006) of reclaimed water for irrigation: i.e. *E. coli* < 100 CFU/100 mL.

In the experimental investigations, it was determined that 22% of the groundwater samples were positive to enteric viruses. Virus recovery efficacy in the environmental water samples was affected by both salt ions (Ca^{2+}, Mg^{2+}, Fe^{2+} and Na^{+}) and dissolved organic compounds concentrations, which interfered with the water filtration process.

4.5 CONCLUSION

Attenuation of microbial contamination occurs within the fractured limestone aquifer. However the low level of microbiological water quality in well 4, obtained by indicators as well as by monitored pathogen (see Chapter 11), confirm that groundwater recovery must be done at a great distance from the injection source to protect human health. Natural inactivation of pathogens occurs within the fractured limestone aquifers and the quality of the water recovered in wells can be considered as a function of the aquifer transmissivity (and soil porosity) and, in particular, of the injected water quality. If the latter presents a quality inferior to that of the existent groundwater, the recycle of water may accumulate pollutants (such as nitrates, THMs, organic compounds weakly biodegradable) and salts that will produce unusable groundwater at recovery wells.

The experimental studies thus suggest the integration of the actual water treatments by including a new treatment plant before the injection into the sinkhole. The treatment can be selected by considering appropriate technologies and by taking into consideration the efficacy of the soil aquifer treatment of the Nardò subsoil, which allows log reduction credits in water disinfection and biodegradation, such as the water denitrification (when biodegradable organic carbon is available). The effectiveness of these processes was confirmed by the positive results of the bench scale simulations.

REFERENCES

AQP Acquedotto Pugliese (2011). http://www.aqp.eu, accessed May 2011.

Balice V., Vurro M., Di Fazio A. and Soria B. (1989). (In Italian) Qualità dell'acqua sotterranea in una zona della penisola salentina. *Ingegneria Ambientale*, **18**(2), 93–98.

Blazejewski R. and Murat-Blazejewska S. (1997). Soil clogging phenomena in constructed wetlands with subsurface flow. *Water Sci. Technol.*, **35**(5), 183–188.

Carrieri C. and Masciopinto C. (2000). Biological nitrogen removal during percolation of municipal treated wastewater in fractured subsoil. *Annali Di Chimica – Jour. of Analytical, Environmental and Cultural Heritage Chemistry, Roma*, **90**, 51–60.

Chrysikopoulos C. V., Masciopinto C., La Mantia R. and Manariotis I. D. (2010). Clogging of fractured formations by biocolloids suspended in reclaimed wastewater. *Environmental Science & Technology*, **44**, 971–977.

Cotecchia V. (1977). Studies and investigations on apulian groundwater and intruding seawaters (salento peninsula). *Quaderni IRSA*, **20**, Rome (I). 145–158.

Directive 2006/118/EU. Directive of the European parliament and of the council official journal of the European Union L 372/19 of 12 December 2006 on the protection of groundwater against pollution and deterioration.

Directive 2008/105/EC. Regulation of the European parliament and of the council of 16 December 2008 on standards of environmental quality.

D. Lgs. 152 (1999). (In Italian): Decreto legislativo 11 maggio 1999, recante disposizioni sulla tutela delle acque dall'inquinamento. Gazzetta Ufficiale n. 246 del 20 ottobre 2000 – Supplemento Ordinario n. 172. Rome (I).

D. Lgs. 152. (2006). (In Italian): Norme in materia ambientale. Supplemento ordinario alla "Gazzetta Ufficiale, n. 88 del 14 aprile 2006. Rome (I).

D. Lgs. 4/2008. (In Italian): Ulteriori disposizioni correttive ed integrative del D.Lgs. 3 aprile 2006, n. 152, recante norme in materia ambientale. Gazzetta Ufficiale, n. 24, S.O. del 29 gennaio 2008. Rome (I).

D. Lgs 30/(2009). (In Italian): Decreto marzo 2009, n. 30 Attuazione della direttiva 2006/118/CE, relativa alla protezione delle acque sotterranee dall'inquinamento e dal deterioramento. Gazzetta Ufficiale n. 79 del 4–4–2009. Rome (I).

D. M. 185 (2003). (In Italian): 12 giugno 2003, n. 185 regolamento recante norme tecniche per il riutilizzo delle acque reflue in attuazione dell'articolo 26, comma 2, del D.Lgs. 11 maggio 1999, n. 152. Gazzetta Ufficiale 23 luglio 2003, n. 169. Rome (I).

La Mantia R., Saturno D., Laera G., Masciopinto C. and Pollice A. (2008a). Suitability of MBR and CAS effluents for groundwater recharge in fractured aquifers: laboratory studies. SIDISA.08 – Intern. Symposium on Sanitary and Environmental Engineering, Firenze (Italy) 24–27 June 2008.

La Mantia R., Masciopinto C., Levantesi C. and Tandoi V. (2008b). Fate and transport of faecal contamination microbial indicators, pathogenic protozoa and campylobacter in the artificially recharged fractured aquifer of Salento, Italy. *Water Science & Technology*, **57**(6), 849–856.

Lonigro A., Pollice A., Spinelli R., Berrilli F., Di Cave D., D'Orazi C., Cavallo P. and Brandonisio O. (2006). *Giardia* cysts and *Cryptosporidium* oocysts in membrane-filtered municipal wastewater used for irrigation. *Applied and Environmental Microbiology*, Vol. 72, n. 12, December 2006, pp. 7916–7918.

Masciopinto C., La Mantia R. and Chrysikopoulos C. V. (2008). Fate and transport of pathogens in a fractured aquifer in the Salento area, Italy, Water. *Resour. Res.*, **44**, W01404, doi:10.1029/2006WR005643.

Masciopinto C., La Mantia R., Carducci A., Casini B., Calvario A. and Jatta E. (2007). Unsafe tap water in households supplied from groundwater in the Salento Region of Southern Italy. *Journal of Water and Health*, **5**(1), 129–148.

Masciopinto M. and Carrieri C. (2002). Assessment of water quality after 10 years of reclaimed water injection: the Nardò fractured aquifer (Southern Italy). *Ground Water Monitoring and Remediation, Winter*, **22**(1), 88–97.

Pavelic P., Dilon P. J., Barry K. E., Vanderzalm J. L., Correll R. L. and Rinck-Pfeiffer M. (2007). Water quality effects on clogging rates during reclaimed water ASR in a carbonate aquifer. *J. Hydrol.*, **334**, 1–16.

Perez-Paricio A. (2001). Integrated modelling of clogging processes in artificial groundwater recharge. Ph.D. Thesis. Department of Geotechnical Engineering and GeoSciences, Technical University of Catalonia, Barcelona, Spain.

Redman J. A., Grant S. B., Olson T. M., Hardy M. E. and Estes M. K. (1997). Filtration of recombinant Norwalk virus particles and bacteriophage MS2 in quartz sand: importance of electrostatic interactions. *Environ. Sci. Technol.*, **31**, 3378–3383.

Schmoll O., Howard G., Chilton J. and Chorus I. (2006). Protecting groundwater for health/managing the quality of drinking-water sources. IWA Publishing, Alliance House, 12 Caxton Street, London SW1H 0QS, UK.

Skolasinska K. (2006). Clogging microstructures in the vadose zone – laboratory and field studies. *Hydrogeology J.* 2006, **14**(6), 1005–1017.

Taylor S. W. and Jaffe P. R. (1990). Biofilm growth and the related changes in the physical properties of a porous medium 1. Experimental investigation. *Water Resour. Res.* 1990, **26**, 2153–2159.

WHO (2006). www.who.int/water_sanitation_health/wastewater/en/. Verified May 2011.

Chapter 5

Managed aquifer recharge via river bed in Sabadell, Spain

M. Neus Ayuso-Gabella and Miguel Salgot

5.1 INTRODUCTION

5.1.1 Historical background

Since the 18th century, the main activity of the city of Sabadell, which is located on the Ripoll River in the southern part of the *Vallès Occidental* district (Catalonia, Spain), is the textile industry. Factories have been exploiting the Quaternary and Miocene aquifers, as well as the Ripoll River, which crosses the city and which is an affluent to the Besós River (a river crossing Barcelona municipality). Over the years, the industrial activity led to a depletion of groundwater levels and also a high pollution of the river and the aquifers. Furthermore, river shores became an uncontrolled disposal site.

Parallel to the pollution of the area and the depletion of the aquifer levels, a flooding due to the inadequate river bed conditions in the 80s resulted in strong damages to many houses close to the river.

In view of this situation, the Environmental Department of Sabadell Municipality decided to restore the area and started a project to create a "Fluvial Park" in the shores of the river. The project is being developed jointly with CASSA, the water company exploiting the reuse system. Considering the situation of the area, the project main objectives were:

- The Ripoll River water improvement, as it was highly polluted due to the illegal disposals of the factories.
- Landscape ecological reclamation: the river shores were an illegal disposal site, which contributed to the deterioration of the area.
- Water savings: the aquifer recharge not only recovers the aquifer levels but also is an additional source of water which is used for parks irrigation and streets cleaning.
- Riverbank management (against floods): the bad conditions of the river bed enhanced the big flooding in the 80s.

Project planning

The project was divided in different units:

- Completion of the wastewater sewerage and control system: many companies were not connected to the general sewerage system, thus disposing their effluents directly to the river, and the Riu Sec WWTP had not capacity enough to treat the wastewaters from the new industrial area. Thus, it was necessary to construct a new WWTP, which is the Ripoll River WWTP. Nowadays, only few companies still dispose their effluents directly to the river, and are closely controlled by the local authorities. Besides, an online monitoring system is being implemented in the collectors close to the river, to monitor any disposal out of the quality limits set by the local authorities.
- Water reuse: the water recharged thanks to the river bed filtration is stored in the aquifer, the quality is improved thanks to the filtration and it can be utilised for non potable uses, as parks irrigation and streets cleaning. To fulfil this objective, emissaries were constructed to discharge the water into the river in 3 different points, as well as a pumping system to send the effluent upstream the river.
- Riverbank protection: the riverbank has been restored, carving a defined path for the river and reinforcing the walls with a metallic net.

- Wetland construction: wetlands in three different points of the river were planned to improve the quality of the river water, expecting a reduction in the organic and the microbial loads. These wetlands were not yet constructed at the time of the Reclaim Water project development, but one of them is in place nowadays.

The chronology of the project has been as follows:

- 1994: The project started.
- 2001: Ripoll River WWTP construction finished.
- 2002: Ripoll River WWTP came into operation.
- 2003: Pipes and transport of the reclaimed water from the WWTP to the river were constructed (1st stage)
- 2004: Water recovery from the mine. The old installations were updated, including the pumping system.
- 2005: Piping system came into operation, thus discharging the effluent of the Ripoll River WWTP to the Ripoll River.
- 2006: Pumping system to send WWTP effluent upstream the river came into operation.
- 2009: Wetland by the river construction. Wetland came into operation in 2010. They will treat- approximately 10% of the water discharged into the Ripoll River, enabling a nutrients reduction in the water that it is infiltrated through the riverbed to the alluvial aquifer.

Authorization procedure and regulations applicable to the system

MAR is considered as an environmental use of reclaimed water in the Spanish Water Reuse Royal Decree 1620/2007. Sabadell would be included in "5.1: Aquifer recharge by localized percolation through the soil". However, as the system is not interpreted as a "pure" artificial recharge scheme by the administration, no authorisation procedure for the recharge of the aquifer has been required to date. It is considered that the WWTP discharges the effluent in the Ripoll River, hence contributing to the preservation of its ecological flow, although it is inevitably recharging the alluvial aquifer at the same time.

Even though a permission is not required as a MAR system, regulations apply as a recycled water system. Thus, the same Spanish Water Reuse Royal Decree 1620/2007 has to be fulfilled in different ways in Sabadell. This Royal Decree defines the quality that the reclaimed water must have depending on its final reuse purpose. The water quality requirements of this law are summarized in Table 5.1. In Table 5.1, the rows highlighted in grey correspond to the reclaimed water uses found in Sabadell that would be abided by the RD 1620/2007.

Table 5.1 Royal Decree 1620/2007 requirements for water reuse.

Reclaimed water uses	Quality criteria (Maximum value permitted)				Other criteria
	Helminth eggs	E. coli (CFU/100mL)	SS (mg/L)	Turbidity (NTU)	
1. *Urban uses*					
1.1. Residential uses: Private gardens irrigation, toilet flushing.	1 egg/10 L	0	10	2	Legionella spp. (1) 100 CFU/L Other contaminants (2)
1.2. Urban services: Irrigation of open access landscape areas (parks, sport fields...), street cleaning, fire-fighting systems, cars washing (industrial)	1 egg/10 L	200	20	10	Legionella spp. (1) 100 CFU/L Other contaminants (2)
2. *Agricultural uses*					
2.1. Irrigation of raw consumed crops.	1 egg/10 L	100	20	10	Legionella spp. (1) 1000 CFU/L Other contaminants (2)
2.2. Irrigation of crops for canning industry and crops not raw consumed. Irrigation of pastures for milk or meat animals. Aquiculture.	1 egg/10 L	1,000	35	No limit established	Taenia saginata and T. solium (3) 1 egg/L Other contaminants (2)
2.3. Irrigation of industrial crops, nurseries, fodder, cereals and oleaginous seeds. Ornamental flowers.	1 egg/10 L	10,000	35	No limit established	Legionella spp. (1) 100 CFU/L Other contaminants (2)

(*Continued*)

Table 5.1 Royal Decree 1620/2007 requirements for water reuse (*Continued*).

Reclaimed water uses	Quality criteria (Maximum value permitted)				Other criteria
	Helminth eggs	**E. coli (CFU/100mL)**	**SS (mg/L)**	**Turbidity (NTU)**	
3. *Industrial uses*					
3.1.a. Process water and cleaning, except food industry. Other industrial uses.	No limit established	10,000	35	15	Legionella spp. 100 CFU/L Other contaminants (2)
3.1.b. Process water and cleaning in food industry.	1 egg/10 L	1,000	35	No limit established	Legionella spp. 100 CFU/L Other contaminants (2)
3.2. Cooling water.	1 egg/10L	0	5	1	Legionella spp. 0 CFU/L
4. *Recreational uses*					
4.1. Irrigation of golf courses.	1 egg/10 L	200	20	10	Legionella spp.(1) 100 CFU/L Other contaminants (2)
4.2. Impoundments, water bodies and streams for recreational uses in which public contact with the water is not permitted.	No limit established	10,000	35	No limit established	T. Phosphorus (stagnant water bodies): 2 mg/L Other contaminants (2)
5. *Environmental uses*					
5.1. Aquifer recharge by localized percolation through the soil.	No limit established	1,000	35	No limit established	TN: 10 mg/L NO^{3-}: 25 mg/L Other contaminants (4)
5.2. Aquifer recharge by direct injection.	1 egg/10L	0	10	2	TN: 10 mg/L NO^{3-}: 25 mg/L Other contaminants (4)
5.3. Irrigation of forested areas, landscape and green areas with restricted access	No limit established	No limit established	35	No limit established	Other contaminants (2)
5.4. Other environmental uses (habitat wetlands, enhancement of marsh and similar, ecological flow maintenance)	The quality required will be evaluated case by case.				

(1) If there is risk of aerosols creation.
(2) See RD 849/1986 Annex II (public disposals to the hydraulic system law), RD 907/2007 Annex IV (hydraulic planning law) and RD 606/2003 (environmental quality law, modification of RD 849/1986).
(3) Pastures for meat animals.
(4) See RD 849/1986 articles 257 to 259 (public disposals to the hydraulic system law).

As mentioned above, Sabadell case study is not considered as MAR site but specifically as an "ecological river flow maintenance" site (integrated in "5.4: Other environmental uses"). For this type of reuse application, the water quality or sampling routine required have not been specifically defined yet, the decree only indicates that "the minimum quality required will be studied case by case". In Sabadell, the Water Catalan administration gave the permit to discharge the WWTP effluent into the Ripoll River several years ago, just considering it as a release into the public hydraulic system. In this sense, the quality of Ripoll River WWTP effluent, which is discharged into the Ripoll River, is subjected to the requirements of the Royal Decree 509/1996 (Regulations for Urban Wastewater Treatment, see Table 5.2). The Ripoll River is declared as a sensitive area and the limits are stricter for the effluent quality.

Another use of the reclaimed water that can be found in Sabadell entails "1.2. Urban services: Irrigation of open access landscape areas (parks, sport fields,...), street cleaning,...". The recovered and disinfected water is used for park irrigation and street cleaning. Then, this reclaimed water should fulfil the requirements of the Royal Decree regarding this irrigation use, but as it is considered by the administration as a well exploitation, and the requirements are not set in any law, only permits are required. The Environmental Department of Sabadell Municipality was given a concession to exploit the well in 1994.

Figure 5.1 Geographical location of Sabadell (© from ICC, Catalan Institute of Cartography)

Table 5.2 Other laws applicable in Sabadell.

Other laws and requirements	RD 509/1996	Local administration
Scope of the law	WWTP effluent quality • BOD < 25 mg/L or red. > 70% • COD < 125 mg/L or red. > 75% • SS < 35 mg/L or a red. > 90%	Water used for parks irrigation and streets cleaning • Total bacteria 22°: 0 CFU/100 mL • Total bacteria 37°: 0 CFU/100 mL • Total Coliforms: 0 CFU/100 mL
Limits	Besides, at least one of the following limits must be assured: • TN < 10 mg/L or a red. > 70% • TP < 1 mg/L or a red. > 80%	• E. coli: 0 CFU/100 mL • Enterococci: 0 CFU/100 mL • Clostridium spores: 0 CFU/100 mL • Legionella < 50 CFU/L • Residual chlorine 0.5–2 ppm

At the end of 2009, a new law to regulate the use and protection of groundwater came into force, the Royal Decree 1514/2009. This Royal Decree is more focussed on groundwater used for potable uses. According to this law, MAR systems need to follow the RD 1620/2007 for Water Reuse.

Thus, considering the present situation, fulfilling the water reuse RD 1620/2007 limits would be enough, even though not required for some of the uses found in Sabadell, as it has been explained above.

Apart of the RD 1620/2007 and the RD 509/1996, CASSA (the water company exploiting the system) and Environmental Department of Sabadell Municipality (the local administration) can set different limits, more restrictive, according to the requirements of the system. The limits set by the local administration are presented in Table 5.2.

The limits imposed by the local administration concerning the microbiological quality of the water used for irrigation are very restrictive, asking for a quality as good or even better than for drinking water. If we compare the limits set by the local administration with the water reuse RD we can see that the RD is less restrictive; it permits up to 200 CFU/100 mL of *E. coli*. However, Sabadell is a special case-study, as the park being irrigated is close to a hospital, thus increasing the risk of illnesses, and the Environmental Department of Sabadell Municipality wanted to be cautious in this sense, being probably too conservative.

A further consideration on the indirect artificial recharge performed should be taken into account, but to date, the Spanish administration does not require a further permission neither does the Catalan administration.

5.1.2 Motivations for recharge and use of abstracted water

The primary motivation for the implementation of the aquifer recharge and water reuse scheme was water scarcity. Lately, the precipitation has drastically diminished in our study area – as it has in the rest of Catalonia and Spain – and solutions to mitigate the water stress are necessary. Specifically for Sabadell, the water stress drivers are:

- Very low precipitation in the recent years

- A high industrial water demand (even though the industrial volume has diminished in the recent years)
- The potable water in Sabadell comes from the Llobregat River which is overexploited as it supplies Barcelona as well as small municipalities.

Other motivations that must be taken into account for the implementation of the aquifer recharge and water reuse scheme are:

- The reduction of the utilisation of potable water in uses when it is not a necessity.
- To have an additional source of water available during the whole year, that can be utilised for non- potable uses as park irrigation or street cleaning.
- Maintenance of the ecologic flow of the river. The river flow is low due to the reduction of precipitations, and the river carries mainly effluents from different WWTPs. The ecologic flow of the river remains irregular despite the effluent discharges of the Ripoll River WWTP, but at least an additional flow gets to the river.
- To avoid restrictions in the irrigation of public green areas and in street cleaning during dry periods. These are the first recipients that experience water cuts during dry periods, as the "Catalan Drought Ordinance" asks for.
- Maintenance and use of the old installations already present in the area (the Ripoll River mine).

5.2 DESCRIPTION OF THE TEST SITE

5.2.1 Study area

Sabadell is a city located 20.6 km north from Barcelona (see Figure 5.1). Nowadays, Sabadell has slightly more than 200,000 inhabitants and occupies an area of 37.89 km^2. The average rainfall is of 600–700 mm per year, although in 2007, when the samplings for the Reclaim Water project took place, was less than 400 mm. Catalonia region has been experiencing a long lasting drought period until April 2008.

Sabadell is divided into two different areas concerning the collection of wastewater: half of the city sends its wastewaters to the Sec River WWTP and the other half to the Ripoll River WWTP. The current study focused on the Ripoll River reuse scheme.

5.2.2 Ripoll River WWTP

The Ripoll River WWTP has a treatment capacity of 200,000 population equivalent based on conventional activated sludge treatment including nutrient removal. The maximum hydraulic capacity is 30,000 m^3/d, while the average influent amounts were of 16,000 m^3/d. The wastewater collected in Sabadell is conveyed by gravity flow to a pumping station, where a pre-treatment is performed for solids removal: the coarse solids are separated and the small ones are sieved. After that, the water is pumped to the Ripoll River WWTP. After an optional additional pre-treatment, which is nowadays by-passed, the raw water may undergo an optional physico-chemical treatment if necessary. After this, the water enters the primary treatment with settling tanks and the secondary treatment with nutrient removal (N and P).

5.2.3 Ripoll River recharge and reuse scheme

The reuse system is based on the Ripoll River (Figure 5.2), which has a length of 40 km, seven of them within the limits of the municipality of Sabadell. The river has its source in 'Sant Llorenç del Munt i l'Obac' Natural Park, near Sant Llorenç Savall village.

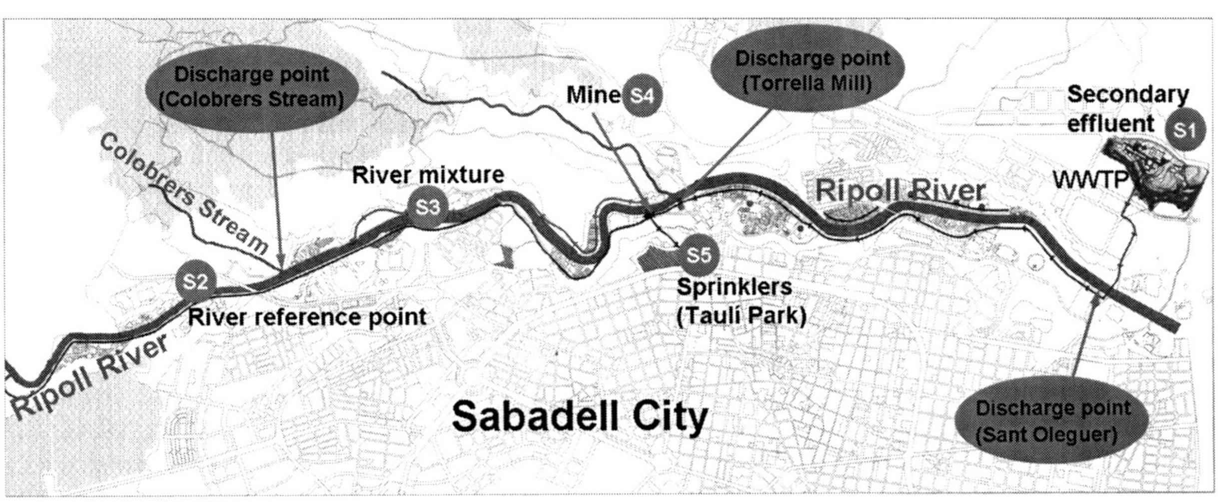

Figure 5.2 Scheme of the global reuse system in Sabadell and the sampling points

The secondary effluent of Sabadell WWTP is discharged into the Ripoll River in three different points (see section below on Process design and operation). Once in the river, part of the water infiltrates and reaches the aquifer, which is mainly formed by sand and gravel. The water is then recovered in a mine located in the Torrella Mill area, disinfected (UV + Chlorination) and used to date to irrigate the Taulí Park and trees along the river. In addition, the water is also used for streets cleaning.

There are two more WWTPs upstream of the study area: Castellar del Vallès and Sant Llorenç Savall. Castellar del Vallès has 20,000 inhabitants and Sant Llorenç Savall 2,200. Both plants discharge their effluents into the river, but the amount of treated water discharged into the river is smaller than the amount discharged by the Ripoll River WWTP. To allow a sound evaluation of the whole system, a sampling point upstream Colobrers Stream discharge area (the most distant discharge area in the Ripoll River, see Figure 5.2) was included to measure the water quality of the Ripoll River before arriving to the recharge area. Besides, we have to consider also other incomes into the Ripoll River: partially treated wastewater from a factory (this is a controlled disposal), rainfall, and stormwater and wastewater that can not be treated in the Ripoll River WWTP during heavy rain periods.

Considering the characteristics of the case study, the following sampling points were selected to monitor the water quality along the system (see Figure 5.2 and Figure 5.6):

- S1: Secondary effluent of the WWTP: the water discharged into the river. The sample is whether a grab sample or a composite sample. Composite samples are preferred regarding pollutants, but not for microorganisms quantification. Then, in all samplings a grab and a composite sample were taken. Microbiological parameters were measured with grab samples, and the other parameters with composite samples.
- S2: Ripoll River reference point (water from the Ripoll River upwards the Colobrers Stream discharge point): this sampling point has been selected because it is representative of the water that the river carries before arriving to the study area (mostly water discharged by the mentioned small WWTPs from other cities).
- S3: Ripoll River water after the Colobrers Stream discharge: this water is a mixture of the water that carries the Ripoll River itself and the secondary effluent of our WWTP. This mixture is the water that infiltrates into the aquifer.
- S4: Water recovered from the aquifer: water extracted in the mine. This water is representative of the recharged water through the riverbed plus the groundwater already present in the aquifer.
- S5: Water from the sprinklers of Taulí Park: this water is representative of the water used for irrigation purposes. The sprinklers water is recovered water from the aquifer (S4) that has been disinfected by UV and chlorination, and that has also been treated to reduce the amount of solids present.

5.2.4 Hydrogeology

In Sabadell the recharged aquifer is unconfined. Three different hydrogeological entities depending on the origin of the sediments in the area can be differentiated (cf. Figure 5.3):

- Alluvial aquifer (quaternary origin) corresponding to the old Ripoll River terraces: This aquifer is attached to an area mainly formed of sand and clay, with a little amount of gravels. It is a free aquifer. The recharge is mainly done by the Ripoll River, irrigation surpluses and losses from the distribution system, and rainfall to a minor extent (rains are scarce in the area).
- Alluvial aquifer (quaternary origin) corresponding to the new Ripoll River terraces: This aquifer is attached to an area mainly formed of sand and gravel, with little clay. It is a free aquifer too, with a maximum depth of 5 m, generally. The recharge is mainly done by the Ripoll River and irrigation surpluses, and rainfall to a minor extent (rains are scarce in the area).
- Miocene aquifer, under the alluvial aquifers: This aquifer is made of initial materials that were deposited in the Vallès-Penedès depression (Tortonian age). This aquifer is attached to an area mainly formed of conglomerate materials. It is a multi-layered aquifer, with different permeable levels. The recharge is done through the alluvial aquifers.

As the basin of the river is mainly formed of gravel and sand, with a much lower amount of clay, the river basin is highly permeable, and this highly influences the groundwater composition.

In the mine where the water is recovered for the Reclaim Water project (Mine: S4 - Figure 5.5), the infiltration of the water from the river to the alluvial aquifer is enhanced, because the groundwater level is 7 m below the river. In order to qualitatively understand the proportion of river water present in the groundwater recovered in the mine, samples representative of the Ripoll River water, the old alluvial aquifer and the miocene aquifer were taken. These samples were obtained from natural wells (fountains) and private wells (wells located in factories exploiting the aquifer), and formed part of a hydrochemical study (Franch, 2007). The well used to recover the water for the Reclaim Water project (Mine: S4) was not included in this hydrochemical study. To interpret the hydrochemical study, a Piper diagram was done, and is shown in Figure 5.4. The old alluvial aquifer samples taken for the hydrochemical study include: Font dels Capellans, Font de la Teula, Font de la Tosca, Font de Jonqueres, Font de Degotalls, Font de la Salut, Font del Quimet, Font de Llobateres, Font de Magí, Font de Petricó, Font de Pi i Maragall, Font Marià Feliu i Font Nicolau. The miocene aquifer samples taken for the hydrochemical study include: Pou Retintel, Pou Tintoretto, Pou Qualitex and Pou Mitjans. The Ripoll River samples taken for the hydrochemical study include: Ripoll N, Ripoll C and Ripoll S. Besides, in the Piper plot the samples from the Reclaim Water project have been included, namely the recovered water from the mine, S4, and the Ripoll River water, in this case, a mean of the samples taken at points S2 and S3.

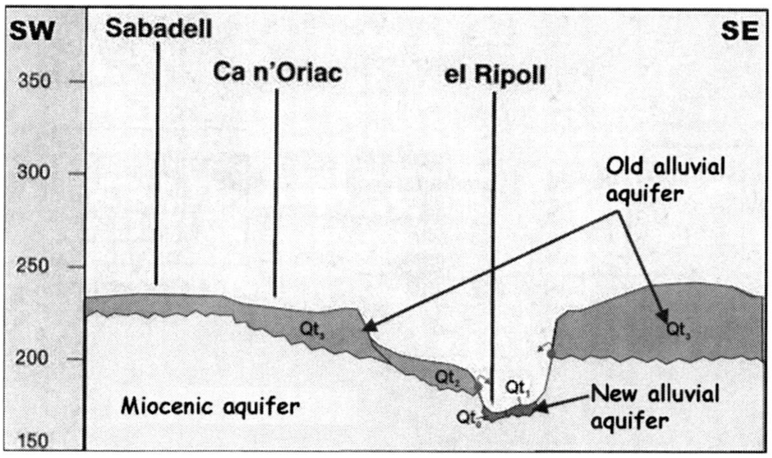

Figure 5.3 Cross section of the aquifer and river bed at the North area of the city, close to sampling point S2 (from Franch, 2007)

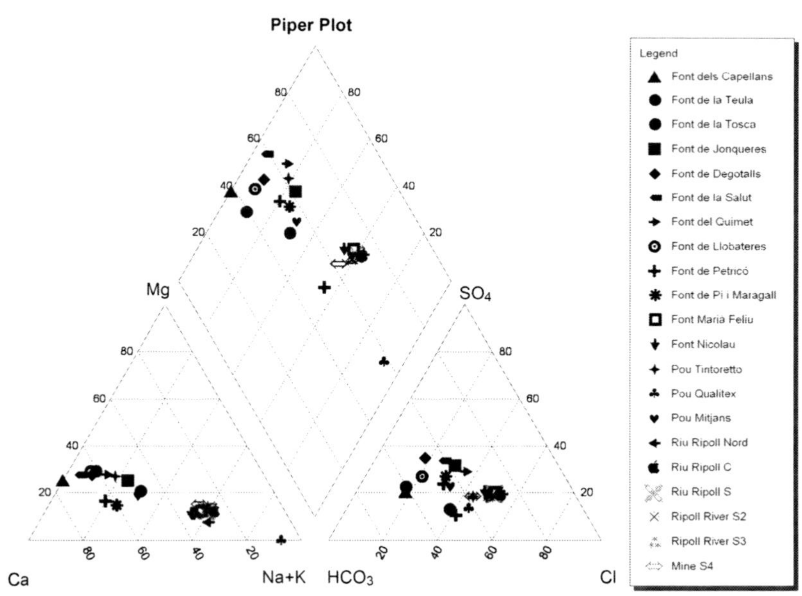

Figure 5.4 Piper diagram of hydrochemical study samples and Reclaim Water samples

This Piper plot (Figure 5.4) shows that the samples taken for the Reclaim Water project (Ripoll River: average between S2 and S3, and Mine: S4) are very close to each other, indicating that a very high amount of the water recovered in the mine comes from the Ripoll River water. The Reclaim Water project samples are also very close to the Ripoll River samples taken for the hydrochemical study (Ripoll N, Ripoll C and Ripoll S). By comparing the old alluvial aquifer samples with the samples of the Reclaim Water project it appears that the distance between the points in the diagram is rather significant. Besides, among the old alluvial aquifer samples there are clearly two groups, the samples taken in natural wells and the samples taken in factories, with different chemical composition, thus represented distantly in the plot. The old alluvial aquifer samples taken in factories are chemically closer to the river samples and the samples taken for Reclaim Water project than the old alluvial aquifer samples taken in natural wells.

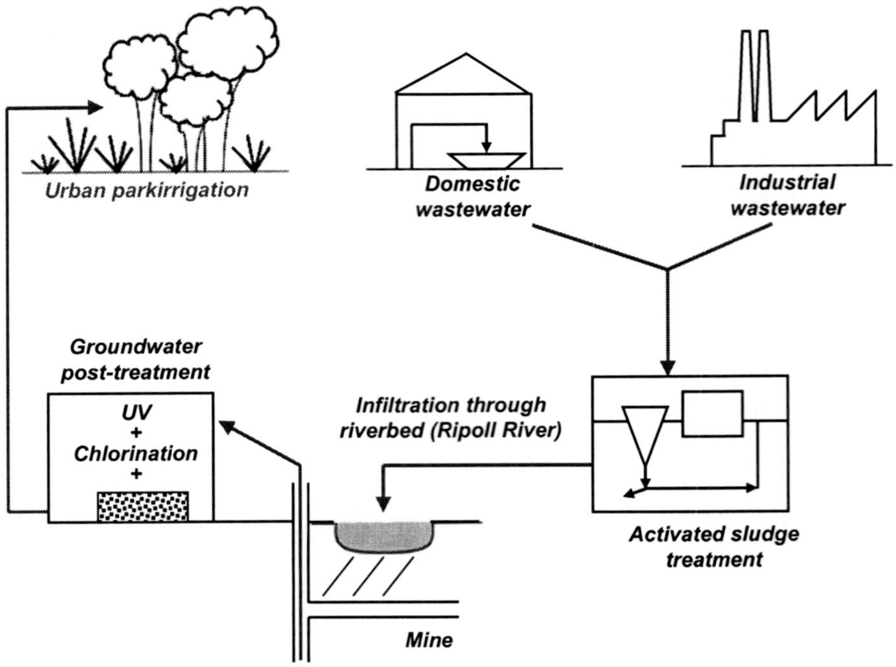

Figure 5.5 Schematic of the Sabadell water reuse system

When considering the miocene aquifer samples, a change of chemical characteristics can be noticed. The Qualitex sample is probably the most representative of the miocene aquifer, and Tintoretto's one is the closest to the old alluvial aquifer samples. Only one Miocene aquifer sample is rather close to the mine water sample of the Reclaim Water project, which is the sample from Retintel. This is in accordance to the high proximity of a factory to the site's well (the factory is located few meters away from the well). These results show that the alluvial aquifer and the miocene aquifer mix at some points.

Thanks to previous pumping studies, the hydraulic parameters relative to the site have been obtained. A recalculation of results obtained from past studies has been performed. The results show a good transmissivity of the alluvial aquifer (45–51 m^2/day), which enhances the recharge from the river basin, and a lower one for the Miocenic aquifer (5–7 m^2/day).

5.2.5 Process design and operation

The secondary effluent of the Ripoll River WWTP is disposed into the Ripoll River in three different points:

- Colobrers Stream area: up to 8000 m^3/d. This point is located upstream the Ripoll River WWTP, and the water is pumped from the Torrella Mill area. The pumping system started to work at the end of July of 2006.
- Torrella Mill area: up to 10,000 m^3/d. The mine where the water is extracted is located close to this area (250 m distance). This point is located upstream the Ripoll River WWTP and the water is pumped from the WWTP.
- Sant Oleguer Mill area: up to 12,000 m^3/d. This point is located close to the Ripoll River WWTP and the water gets there by gravity. This point is only used when there's a heavy raining period, and not all the water can be treated at the WWTP.

The WWTP was designed to treat 30,000 m^3/d, and during the Reclaim Water project sampling campaigns it was receiving between 15,000 to 16,000 m^3/d (the amount is reduced during summer). Circa 5,000,000 m^3 are sent to the river yearly. The treated wastewater is sent continuously to the river, and the recharge is continuous thanks to the riverbed filtration.

The water is recovered in a mine underneath the Ripoll River. In the treatment facility nearby the mine, the recovered water used to be disinfected by UV and then stored in a tank. In the storage tank, the water was chlorinated by means of a system that doses chlorine continuously depending on the levels detected by the probes. In addition, in the storage tank the remaining solids were removed by a small filter, also in a continuous closed circuit. Nowadays, the treatment after recovery has been improved, replacing the small sand filter by a big sand filter and a new UV treatment, that seemed to be much more effective in disinfection (data not shown).

After these treatment steps, the disinfected water is sent to another storage tank in the park, where it is used for irrigation. The irrigation is performed by a sprinkler system. The water is also used for irrigation of trees along the river. Another use of the disinfected water is street cleaning. Water trucks receive the flushing water from a facility nearby the park. About 70 % of the recovered water is used for irrigation and 30% for streets cleaning.

Figure 5.6 Sampling points selected for the Reclaim Water project monitoring

It must be considered that Sabadell Town Hall and CASSA (i.e. the water company exploiting the recharge system) are planning to use the recovered water for other uses, namely industrial processing water, recreational water and even drinking water (eventually as a solution for the water scarcity). These new uses require further treatment of the water, and the technologies to be used are being considered. The levels of the aquifer are monitored in 3 piezometers present in the area by the Catalan Water Agency and the trends show a remarkable increase during the last years. A longer monitoring is required to see the evolution of the aquifers of the area.

5.3 TECHNOLOGY PERFORMANCE AND CONTAMINANT MONITORING

During 2007 monthly sampling campaigns were performed to monitor the quality of the system. Table 5.3 summarizes all the results for Protocol 1 during 2007 sampling campaign. The sampling points selected have been described in the previous section. The analysis performed included basic wastewater parameters, nutrients, microbiological indicators, salinity related parameters and trace elements. Other more specific analyses for pathogens or pharmaceutical compounds were performed and appear in other chapters of the present book.

5.3.1 Basic wastewater parameters

Results of the wastewater parameters indicate that the quality of the water is mainly improved through aquifer recharge, and not through dilution effects of the WWTP effluent in the river, because the river quality before the discharges is very similar to the river quality after the discharges. Indeed, the river water is mainly a mixture of upstream WWTP effluents.

Suspended solids and turbidity are highly reduced after recharge, with reduction percentages ranging from 91% to 99%, and a final content below the detection limit. This is the parameter that experiences the highest reduction after aquifer treatment.

Table 5.3 Results from the monitoring within RECLAIM WATER – part 1.

	Unit	S1			S2			S3			S4			S5		
		Secondary effluent			River reference			River mixture			Mine			Sprinklers		
		min	av	max	min	av	max	min	av	max	min	av	max	min	av	max
Basic wastewater analysis																
Suspended Solids	$mg \cdot L^{-1}$	2.9	7.9	15	3.6	6.6	12	3.4	8.3	13	<0.10	0.28	0.97	<0.10	0.56	1.7
BOD Total	$mg \cdot L^{-1}$	3.0	6.2	15	1.0	4.9	10	2.0	4.5	9.0	0.00	1.2	3.0	0.00	1.4	4.0
COD Total	$mg \cdot L^{-1}$	27	54	76	8.2	25	42	16	34	55	2.3	14	24	0.00	9.2	18
COD Soluble	$mg \cdot L^{-1}$	21	42	55	4.7	21	33	13	25	40	1.2	11	20	0.00	8.1	18
Alkalinity	$mg \cdot L^{-1}$	194	270	323	318	365	410	235	327	386	334	357	388	289	355	395
pH	unit	6.7	7.2	7.6	7.5	8.0	8.6	7.1	7.9	8.6	6.8	7.2	7.7	7.3	7.7	8.2
Turbidity (NTU)	NTU	0.51	3.9	10	1.9	4.0	11	1.5	5.3	9.7	0.00	0.018	0.15	0.00	0.15	0.57
Detergents	$mg \cdot L^{-1}$	<0.20	0.41	0.79	<0.20	0.38	0.61	<0.20	0.41	0.77	<0.20	0.27	0.43	<0.20	0.27	0.41
Phenols	$mg \cdot L^{-1}$	0.63	1.1	1.4	0.43	0.72	1.1	0.44	0.78	1.1	0.24	0.46	0.65	0.27	0.42	0.57
Nutrients																
Ammonia (NH_4-N)	$mg \cdot L^{-1}$	0.59	3.6	12	0.29	1.9	5.0	0.60	2.2	4.0	<0.25	0.49	0.96	<0.25	0.45	1.1
DOC	$mg \cdot L^{-1}$	10	14	17	5.3	7.3	10	6.4	9.5	14	2.9	4.6	8.6	3.0	4.5	5.5
Nitrite (NO_2-N)	$mg \cdot L^{-1}$	<0.08	0.26	0.66	<0.08	0.30	1.2	<0.08	0.30	0.73	<0.08	0.08	0.09	<0.08	<0.08	<0.08
Nitrate (NO_3-N)	$mg \cdot L^{-1}$	0.71	3.6	6.2	2.30	3.7	6.6	2.7	3.8	6.4	1.2	2.2	3.2	1.2	2.1	3.1
Total Kjeldhal Nitrogen (TKN)	$mg \cdot L^{-1}$	2.3	6.3	15	0.87	3.3	11	1.5	4.7	12	0.58	1.6	3.9	<0.25	1.8	4.8
Phosphorus (PO_4-P)	$mg \cdot L^{-1}$	1.2	3.9	6.7	0.65	1.8	3.1	0.80	2.1	4.4	0.65	0.90	1.1	0.72	0.90	1.5
Microbiological analysis																
Total Bacteria 22°C	CFU/100 mL	9.5×10^{5}	3.2×10^{7}	$>3.0 \times 10^{8}$	4.7×10^{4}	3.0×10^{6}	$>3.0 \times 10^{7}$	7.0×10^{4}	2.9×10^{7}	$>3.0 \times 10^{8}$	4.1×10^{1}	5.6×10^{3}	3.3×10^{4}	3.3×10^{-1}	1.5×10^{2}	8.2×10^{2}
Total Bacteria 37°C	CFU/100 mL	1.6×10^{5}	5.7×10^{5}	1.6×10^{6}	1.2×10^{4}	7.1×10^{4}	1.8×10^{5}	3.0×10^{3}	4.9×10^{5}	2.6×10^{6}	1.4×10^{1}	1.3×10^{3}	8.1×10^{3}	$<5.0 \times 10^{-1}$	5.9×10^{2}	5.0×10^{3}
Total Coliforms	CFU/100 mL	6.7×10^{5}	1.3×10^{6}	2.6×10^{6}	5.7×10^{4}	2.6×10^{5}	1.1×10^{6}	5.3×10^{4}	1.1×10^{6}	3.7×10^{6}	1.5×10^{1}	9.6×10^{2}	6.2×10^{3}	2.0×10^{-1}	3.8×10^{1}	2.1×10^{2}
E.coli	CFU/100 mL	2.2×10^{4}	9.9×10^{4}	3.3×10^{5}	5.5×10^{2}	1.1×10^{4}	6.2×10^{4}	8.0×10^{1}	5.8×10^{4}	2.3×10^{5}	2.0×10^{-1}	1.7×10^{1}	8.0×10^{1}	$<2.0 \times 10^{-1}$	9.8×10^{-1}	7.0×10^{0}
Enterococci	CFU/100 mL	6.0×10^{3}	1.1×10^{4}	1.9×10^{4}	2.7×10^{2}	1.3×10^{3}	3.4×10^{3}	1.4×10^{2}	7.2×10^{3}	2.4×10^{4}	2.7×10^{-1}	5.6×10^{0}	2.1×10^{1}	2.5×10^{-1}	7.3×10^{0}	4.9×10^{1}
Clostridium spores	CFU/100 mL	1.8×10^{2}	1.1×10^{3}	2.8×10^{3}	7.0×10^{0}	6.9×10^{1}	4.1×10^{2}	4.0×10^{0}	2.7×10^{2}	8.8×10^{2}	$<1.0 \times 10^{-1}$	4.2×10^{-1}	1.9×10^{0}	$<1.0 \times 10^{-1}$	1.7×10^{-1}	1.0×10^{0}
Bacteriophages	PFU/100 mL	2.0×10^{2}	1.7×10^{4}	5.0×10^{4}	4.1×10^{2}	1.3×10^{4}	3.5×10^{4}	9.3×10^{2}	2.7×10^{4}	1.1×10^{5}	$<1.0 \times 10^{-1}$	1.7×10^{0}	9.6×10^{0}	$<1.0 \times 10^{-1}$	3.0×10^{-1}	1.2×10^{0}

The oxygen demand is also reduced by the system, as shown by the reduction in COD and BOD after recharge. However, an important part of the reduction in COD seems to be due to dilution effects in the river, although for most of the parameters tested the dilution effect does not exist.

Along the year, the pH is maintained in the neutral range, with a slight increase in the Ripoll River samples in summer. This is reflected analogously by changes in alkalinity, although alkalinity levels are much more variable than pH. Alkalinity is lower in the effluent of the Ripoll River WWTP than in the Ripoll River and aquifer samples.

Detergents are found in low levels in the system, with values always inferior to 1 ppm and at some cases under the quantification limit (0.2 ppm), and are slightly reduced by the recharge. Phenols are also reduced and the levels measured before recharge should be lower than the ones recorded. Indeed, the method used detects with a high specificity phenolic compounds that have a natural origin, e.g. coming from lignins. Results below the detection limits were obtained for all samples when using an alternative methodology (colorimetric methods).

5.3.2 Nutrients

Nitrogen, phosphorous and carbon concentrations in the different sampling points selected were monitored along the sampling campaigns. The presence of nutrients was tracked determining the presence of different species:

- Carbon: dissolved organic carbon (DOC)
- Nitrogen: ammonia, nitrite ($N\text{-}NO_2^-$), nitrate ($N\text{-}NO_3^-$) and total Kjeldahl nitrogen (which includes organic nitrogen plus ammonia)
- Phosphorus: total Phosphorus in phosphate form ($P\text{-}PO_4^{3-}$)

Nutrients are present in high concentrations in the untreated wastewater (data not shown), and are highly reduced by the treatments performed at the Ripoll River WWTP, especially in the secondary treatment by activated sludge. The Ripoll River WWTP has a secondary treatment for nutrients removal. In the case of nitrogen, the reduction thanks to the secondary treatment is very high, and the discharged water fulfils the requirements set by the administration, with removals always higher than 70%. However, the presence of phosphorus in the treated wastewater is still high sometimes, and contributes to the eutrophication in the Ripoll River. The performance of the secondary treatment for phosphorus removal was, in general, low and very variable, ranging from no removal to more than 80% removal, and not enough to reduce the phosphorus concentration to less than 1 mg/L, which is the amount set in the RD 509/1996.

Overall, as this law states to ensure the nitrogen and phosphorus requirements, the discharged water would fulfil this law. Regarding carbon, no requirements are set in the RD 509/1996.

Comparing the nutrients concentration in the Ripoll River WWTP effluent with the Ripoll River water nutrients concentration, a reduction can be appreciated. This is due to a little dilution effects in the river and the nutrients use by the algae present in the Ripoll River, which highly increase their presence during summer. Big macrophytes can be seen in the Ripoll River surface during that period. Thus, in the river water nutrients are reduced but for nitrate, that does not change its concentration.

On average, the nitrogen content is also reduced by the recharge, but oxidised compounds (nitrate and nitrite), which are found in similar levels have a tendency to be less reduced in the aquifer. During aquifer recharge, average reductions achieved are of 50% for carbon (DOC), 67% for ammonia, 49% for Total Kjeldahl Nitrogen, 44% for phosphorus, 37% for nitrate and 52% for nitrite. As nitrite is quickly oxidized to nitrate, it cannot be known exactly if the percentage reduction is due to aquifer treatment or to the oxidizing process of nitrite to nitrate.

5.3.3 Microbiological parameters

Microbiological quality of the recovered water was one of the first concerns regarding the case study. The recovered and disinfected water is used for urban parks irrigation, and the Taulí Park, located by the Taulí Hospital, is being irrigated.

The indicator microbiological results for 2007 are summarised in Table 5.3 and the calculated log reductions for the different treatment steps (dilution in the river, infiltration to the aquifer and the disinfection system) for every parameter are summarized in Table 5.4. Besides, Figure 5.7 represents the microbiological results, mean and average. When reporting the results, quite often there were no detections for a certain indicator and sample, especially regarding the disinfected water. Then, the limit of quantification was reported instead. This enabled the possibility to perform $\log_{10}$ reduction calculations.

For all sampling points a fluctuation in the microbiological concentrations of the different indicators during the sampling period was observed. This is due to the fact that the sampling period spans through a whole year, and the samplings were done monthly, thus covering the different seasons.

When discharging the secondary effluent (S1) to the Ripoll River (S3) the water does not suffer a marked dilution effect, and in some cases the microbiological load increases, especially in bacteriophages. Only for *Clostridium* spores there is always a dilution when the effluent is discharged into the Ripoll river. In Table 5.5, the calculated $\log_{10}$ reductions for each indicator and dilution in the river water are given.

Table 5.4 Results from the monitoring within RECLAIM WATER – part 2.

	Unit	S1			S2			S3			S4			S5		
		Secondary effluent			River reference			River mixture			Mine			Sprinklers		
		min	av	max	min	av	max	min	av	max	min	av	max	min	av	max
Trace elements analysis																
Boron	PPB	249	298	401	120	270	500	133	274	449	236	289	346	241	290	362
Cadnium	PPB	<0.10	0.17	0.75	<0.10	0.10	0.11	<0.10	<0.10	<0.10	<0.10	<0.10	<0.10	<0.10	<0.10	<0.10
Chromium	PPB	<10	32	202	<10	15	40	<10	11	18	<10	11	18	<10	95	1000
Cobalt	PPB	<5.0	<5.0	<5.0	<5.0	<5.0	<5.0	<5.0	<5.0	<5.0	<5.0	<5.0	<5.0	<5.0	<5.0	<5.0
Copper	PPB	4.8	8.8	13	3.4	6.7	8.7	4.7	6.9	9.6	4.2	5.5	6.8	3.7	5.7	6.8
Fluoride	$mg \cdot L^{-1}$	<0.50	<0.50	<0.50	<0.50	0.54	0.75	<0.50	0.56	1.1	<0.50	<0.50	<0.50	<0.50	<0.50	<0.50
Iron	PPB	56	115	171	20	43	71	29	58	123	<10	25	36	<10	33	152
Lead	PPB	0.42	0.90	3.1	0.25	1.2	7.4	0.32	0.70	3.2	0.24	0.40	0.70	0.13	0.33	1.1
Manganese	PPB	16	32	46	15	40	96	7.2	38	79	142	327	427	1.8	62	258
Molybdenum	PPB	1.1	20	214	1.2	3.5	18	1.4	3.6	17	2.1	2.6	5.5	2.0	2.7	6.6
Nickel	PPB	33	124	886	<10	17	49	<10	19	43	<10	13	30	<10	13	29
Selenium	PPB	<5.0	<5.0	<5.0	<5.0	<5.0	<5.0	<5.0	<5.0	<5.0	<5.0	<5.0	<5.0	<5.0	<5.0	<5.0
Barium	PPB	9.9	18	27	83	111	134	21	89	140	115	129	146	96	119	137
Cyanide	PPB	<10	<10	<10	<10	<10	<10	<10	<10	<10	<10	<10	<10	<10	<10	<10
Zinc	PPB	39	70	97	15	31	58	21	37	56	15	78	339	15	25	43
Salinity analysis																
Chloride	$mg \cdot L^{-1}$	161	316	419	166	365	579	156	325	491	169	295	373	256	316	362
Electrical conductivity	$\mu S/cm^{-1}$	1088	1883	2270	1224	2152	2850	1277	1953	2550	1770	2022	2440	1740	2035	2590
Sodium	$mg \cdot L^{-1}$	122	261	336	126	279	411	134	256	375	217	256	308	232	262	286
Potassium	$mg \cdot L^{-1}$	21	36	46	10	18	32	11	22	35	12	15	18	13	15	17
Calcium	$mg \cdot L^{-1}$	65	74	85	67	112	147	68	102	128	103	119	134	102	117	126
Magnesium	$mg \cdot L^{-1}$	15	19	25	21	34	45	18	31	40	31	37	41	30	36	39
Carbonate	$mg \cdot L^{-1}$	<2.5	<2.5	<2.5	<2.5	22	47	<2.5	20	93	<2.5	<2.5	<2.5	<2.5	6	25
Bicarbonate	$mg \cdot L^{-1}$	236	329	394	338	402	498	282	359	427	407	436	474	352	423	482
Sulphate	$mg \cdot L^{-1}$	100	164	223	94	184	286	97	172	269	119	170	207	131	175	211

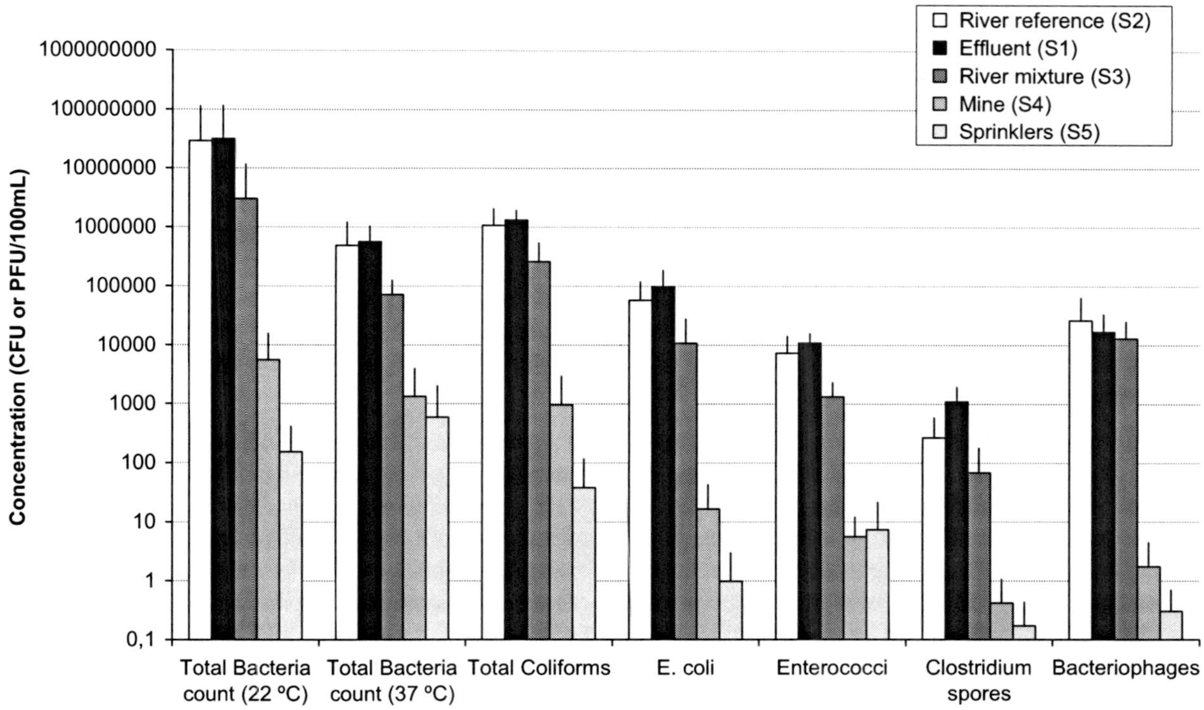

Figure 5.7 Results of the microbiological parameters from the 2007 monitoring campaign (n = 12; mean values and standard deviations)

Table 5.5 Microbiological log reduction by the SAT system and the disinfection.

Treatment Log10 reductions	**Dilution in the river**			**River bed filtration system**			**Disinfection**		
Minimum, average and maximum	**Log S1 – Log S3**			**Log S3 – Log S4**			**Log S4 – Log S5**		
	min	**av**	**max**	**min**	**av**	**max**	**min**	**av**	**max**
Total Bacteria count (22°C)	−0.64	0.36	1.5	2.6	3.7	4.9	−0.027	1.4	2.4
Total Bacteria count (37°C)	−0.33	0.48	1.9	0.74	2.9	4.5	−1.2	0.73	3.0
Total coliforms	−0.51	0.27	1.4	2.4	3.7	4.7	−0.61	1.7	4.1
E.coli	−0.61	0.54	2.8	2.6	3.6	4.5	0.00	1.1	2.4
Enterococci	−0.42	0.49	1.8	1.2	3.1	4.4	−1.7	0.20	1.8
Clostridium spores	0.24	0.88	2.2	1.6	2.7	3.5	0.00	0.23	1.2
Bacteriophages	−1.3	−0.15	1.2	2.0	4.1	5.1	−0.30	0.66	2.0

Opposite to what happens to the dilution in the river, all of the microbiological indicators tested were significantly reduced after aquifer recharge, with average reductions ranging from 2.7 to 4.1 log_{10} units. In this case, the indicators that suffer the highest reduction are bacteriophages and *Clostridium* spores the lowest. Removal values obtained for bacteriophages, ranging from 2.0 to 5.1 log_{10} units, are similar to values obtained by Sprenger *et al.* (2009), ranging from 3.3 to 4.8 log_{10} units removal in an Indian case study. For *Clostridium* spores, Weiss *et al.* (2005) obtained up to 4.5 log_{10} units in two river bank filtration sites in the US, and average 2.3 log_{10} units in another US site. The obtained values at Sabadell site are similar to the values obtained in both studies. A bit different are the values obtained for total coliforms, that seemed to be more highly reduced at one tested US site, with average reductions of 5.5 to 6.1 log_{10} units (Weiss *et al.* 2005).

Although the high removal rates obtained during the infiltration, bacteria are still present in the recovered water, so the aquifer itself is not enough to reduce the microbiological load of the recovered water to be reused. This is due to a short residence time in the aquifer and the bacterial load of the infiltration water. The disinfection step is necessary as part of the whole reuse system.

The disinfection treatment applied after recharge, UV + chlorination, is able to reduce between 0.2 and 1.7 log_{10} units the residual amount of bacteria present in the abstracted water on average. This reduction is relatively low probably due to a not very effective UV treatment. The highest disinfection reduction is observed in the Total coliforms and *E. coli*, but this reduction is smaller and nearly negligible for the other bacterial indicators. *Clostridium* spores, which are usually very resistant to UV and chlorination, are not reduced at all. It must be pointed out that the amount of *Clostridium* spores in the recovered water is very low or under the detection limit most of the times.

Additionally, CASSA analyses *Legionella* in an external laboratory, and this bacterium has never been found in the sprinklers water (<50 CFU/L). *Legionella* could pose a risk as it propagates by aerosols, which are inevitably formed due to the sprinklers system for irrigation. The Royal Decree for water reuse in parks irrigation (Table 5.1) states that *Legionella* must be analysed in case that aerosols are formed, and that the levels should be inferior to 100 CFU/L. So, the requirements are fulfilled in the case of Sabadell.

Overall, the quality of the final water is adequate for irrigation and street cleaning purposes, considering the requirements of the Royal Decree for water reuse in Spain (Table 5.1). Considering the local requirements set by the Environmental Department of Sabadell Town Hall, the recovered and disinfected water would not meet them. In this case, few considerations must be taken:

- As this was a research study, the volumes of water filtered and used to calculate the indicators concentration were much higher than the ones required. So, although it was required to have no presence of indicators in 100 mL, a higher volume was filtered when possible (see chapter 11 on microbiological analysis done for the Reclaim Water project). Then 500 mL or even 1 L was filtered in some cases, rendering positive results.
- Comparing the results obtained during the Reclaim Water sampling campaigns with the ones obtained by an external laboratory, this hypothesis is reinforced, as in most of the cases the results obtained by the external laboratory were no presence of indicators in 100 mL. Only during summer period there was growth of total bacteria at 22°C and total bacteria at 37°C, which is something reasonable considering that any kind of bacteria present in the soil particles or in the area and not coming from the irrigation water can grow with the rather unspecific culture media used for these indicators.
- In view of the occasional growth of indicators in summer, CASSA increased the chlorine dosage to disinfect the recovered water.

Then, although indicators could be detected in the disinfected water, it was considered that the quality was enough to irrigate the park.

In chapter 22 a risk analysis was performed, which underpins the suitability of the recovered and disinfected water for irrigation purposes.

5.3.4 Salinity related parameters

Salinity is high in the irrigation water, as it is also in the alluvial aquifer, the Ripoll River water and the effluent of the WWTP. A summary of the results for major ions and salinity measured in the five sampling points is given in Table 5.4. Major ions contributing to the high salinity and that are much higher than the recommended irrigation guidelines are chloride and sodium. The Ripoll River WWTP receives all the effluents from an industrial area, including textile factories. The use of NaCl in the textile industry is well known, thus releasing high amounts of sodium and chloride in their effluent. However, the analysis of the river water quality concerning these ions showed that the content of these ions is relatively similar before and after disposal of the effluent (Table 5.4). Overall the various anions and cations tested (chloride, carbonate, bicarbonate, sulphate, sodium, potassium, calcium and magnesium) are present in similar levels at all sampling points.

As chloride and sodium are responsible of foliar injury in crops, and also in other plants, it must be considered whether the grass and plants in the park present these symptoms after a period of time being irrigated with the water. Grass is especially sensitive to high amounts of these ions. Furthermore, the soil can suffer irreversible damage, making it not suitable for grass or plants anymore. However, as the water is not used for crop irrigation the risk posed by the high salinity of the irrigation water is low.

5.3.5 Trace elements analysis

Similarly to the salinity related parameters, trace elements are found at similar concentrations in all sampling points, but for manganese, molybdenum and nickel. A summary of the results for metals measured in the five sampling points is given in Table 5.4.

Abundance of toxic metals had been a concern in the area a long time ago, due to the illegal discharges from the companies present in the area. Nowadays, their presence in the Ripoll River and in the aquifers has considerably been reduced.

A peak of molybdenum and nickel was detected in secondary effluent, but lower levels were maintained in the recovered and disinfected water, thus not being a problem for irrigation. A point of concern is the high increase in manganese levels observed in the recovered water. This increase seems to be due to a change of redox conditions in the aquifer (see Section 5.3.6).

The other trace elements are under levels of detection or in levels safe enough not to cause problems to the aquifer neither to the reuse of the water for park irrigation or street cleaning.

5.3.6 Redox conditions

As far as aquifer recharge side effects, an increase in the manganese levels of the water recovered from the mine has been observed.

Releases of manganese or iron from the sediments are associated to the presence of organic matter in the water and its oxidation. During the organic matter degradation, the electron donor is the organic matter, and different electron acceptors can be used. The "natural" sequence, which is dominated by the energy required to perform the oxidation, is first to use the dissolved oxygen, then nitrates, after this manganese oxides and then iron oxides. These redox reactions are catalyzed by microorganisms, and the most efficient organisms dominate each stage, inhibiting the development of the ones below. Overall, this leads to a natural zonation. Thus, in the aquifer the organic matter is oxidised and the current electron acceptor is the manganese oxide.

The manganese concentration in the recovered water (S4) was already high in the first two sampling campaigns (143 ppb in January 2007 and 142 ppb in February 2007) comparing to the infiltration water close to the mine (S3; 27 ppb and 79 ppb, respectively) and the Ripoll River water before the discharges of the WWTP effluent (S2; 23 ppb and 61 ppb). From March 2007, there is an increase of manganese in the recovered water, to 333 ppb in March 2007, reaching a maximum of 427 ppb in August 2007. The concentration of the organic matter already present in the Ripoll River is high (S2; average DOC of 7.3 mg/L), and a bit higher (S3; average DOC of 9.5 mg/L) after the effluent (S1; average DOC 14 mg/L) discharges. Then, these high concentrations of DOC in the Ripoll River may have contributed to develop this shift in the redox conditions in the aquifer, increasing the manganese in the recovered water.

Part of the manganese present in the recovered water is oxidized by the chlorine during the disinfection, and it precipitates, creating black particles in the disinfected water. The manganese precipitation clogged the probes measuring the residual chlorine in the final water, and the sand filter installed previously to reduce particles in the final water needed to be cleaned more often.

5.4 OPERATIONAL FEEDBACK

Natural infiltration of the Ripoll River water through the riverbed occurs along the Ripoll River course through the municipalities of Sabadell, Barberà del Vallés and Ripollet. The recharge is enhanced close to the mine location by discharging the Ripoll River WWTP effluent upstream de mine area and close to the mine area, and the pumps sending the WWTP effluent upstream the Ripoll River came into operation in July 2006.

The groundwater abstraction has been performed since 2004, much before sending the WWTP effluent upstream the Ripoll River. This was due to strong needs to use this water for irrigation. The groundwater, in fact, has been used since a long time as process water for the factories located in the area. The operation of the system has experienced some problems regarding the disinfection of the water recovered and the pumping system implemented. The development of a biofilm which was observed initially in the storage tank that sends the water to the park was corrected by regular cleaning and disinfection. The pumping system experienced some start-up problems.

To date, the use of reclaimed wastewater for aquifer recharge using river bed filtration in Sabadell has proved to be valuable for its intended use, that is the irrigation of a local park. Indeed, in case of drought, the Catalan government typically requires the application of a "drought decree" which imposes primarily water restriction concerning the irrigation of parks like the Taulí Park in Sabadell. During the summers of 2006, 2007 and 2008 such decrees have been applied in Catalonia, however the utilisation of the aquifer recharge permitted not only to maintain the irrigation of the Taulí Park without problems but also to provide an additional source for street cleaning.

However, the system has experienced several failures/problems:

- March 2007: After a long extraction period, the redox processes in the mine seemed to have changed, as concentration in manganese increased. A black precipitate occurred in the recovered water and started to clog the chlorine probes and other small filters. The sand filter needed to be cleaned twice per week, instead of weekly.
- April 2007: Works were done in the pipes delivering the water to the sprinklers of the Taulí Park. During that period, the water from the sprinklers could not be sampled, so analyses were done directly after UV + chlorination disinfection.
- July 2007: The pipe sending the reclaimed water from the WWTP to the upper point of the river, the Torrent the Colobrers area, was broken. The breakage was close to the mine area, so water was arriving mainly to one discharge point and not to both, and the mine where the water is recovered was less recharged, as the water was delivered only to the point which is located a little downstream the mine (Molí Torrella).
- August 2007: Heavy rain fell between the 13th and the 19th of August, and a breakage in the collectors system was detected on 20th August. Part of the water of the collectors system (wastewater) reached the river, in an area after the

river mixture sampling point (S3). That week, the mine showed higher microbiological values (sampling not pertaining to the Reclaim Water project). However, due to the heavy rain, this wastewater was diluted, and the following week corresponding with a Reclaim Water sampling campaign (27th August) the results were similar as usual.

Another issue encountered over 2007 was a strong reduction in the volume of wastewater received by the WWTP. This means the recharge system starts with a lower volume of water to recycle and to recharge.

The Sabadell case study is an example of a system that can be easily operated and produce a final water appropriate for the intended use. However, the different aquifers and the area need to be closely monitored and a further hydrogeological study in more detail is required. It is also to consider that the WWTP secondary effluent does not worsen the quality of the Ripoll River water, as the quality of the river water itself is not the quality desired for a regular river. It is expected that the quality of the river water, thus the quality of the water that is being infiltrated, will improve considerably with the new wetland constructed close to the mine. Furthermore, it is expected that one of the WWTPs upstream the river will be upgraded in the near future, thus discharging an effluent with less nutrients.

If the organic matter and nutrients concentration is not reduced enough by the wetland pre-treatment, a post-treatment for manganese oxidation is recommended.

5.5 CONCLUSION

This study showed that, in Sabadell:

- The artificial recharge system is able to reduce substantially the suspended solids, COD and BOD while DOC, nitrogen and phosphorus are reduced to a lesser extent.
- The artificial recharge system is able to reduce nearly all the microbiological parameters by 3 to 4 $\log_{10}$ units.
- Salinity is similar at all sampling points, so the water recovered has a very high proportion of water infiltrated from the river basin (the aquifer that is being exploited is the alluvial aquifer).

The use of treated wastewater for aquifer recharge through river bed filtration prevents water restrictions and provides an additional source of water for street cleaning. However, manganese precipitation needs to be solved, and the organic matter in the infiltration water needs to be reduced. It is expected that the nowadays present wetland will reduce the nutrients and organic matter contents, thus improving the water quality for recharge. An improved quality of the groundwater offers the possibility of further reuse applications. However, the salinity is a question rather difficult to solve, as a specific treatment to reduce salt may need to be considered.

REFERENCES

Ayuso-Gabella M. N. (2012). Risk assessment in managed aquifer recharge systems. PhD thesis (in preparation).

Ayuso-Gabella N., Page D., Masciopinto C., Aharoni A., Salgot M. and Wintgens T. (2011). Quantifying the effect of managed aquifer recharge on the microbiological human health risks of irrigating crops with recycled water. *Agricultural Water Management*, **99**(1), 93–102.

Böckelmann U., Dörries H.-H., Ayuso-Gabella M. N., Salgot de Marçay M., Tandoi V., Levantesi C., Masciopinto C., Van Houtte E., Szewzyk U., Wintgens T. and Grohmann E. (2009). Quantitative PCR monitoring of antibiotic resistance genes and bacterial pathogens in three european artificial groundwater recharge systems. *Applied and Environmental Microbiology*, **75**(1), 154–163.

Franch D. (2007). Hydrogeological study on the alluvial and tertiary aquifers adjacent to the Ripoll River. Master thesis.

ICC (Catalan Institute of Cartography) (2011). Topographic map downloaded and available at www.icc.cat (accessed June 2, 2011).

Levantesi C., La Mantia R., Masciopinto C., Böckelmann U., Ayuso-Gabella M. N., Salgot M., Tandoi V., Van Houtte E., Wintgens T. and Grohmann E. (2010). Quantification of pathogenic microorganisms and microbial indicators in three wastewater reclamation and managed aquifer recharge facilities in Europe. *Science of the Total Environment*, **408**, 4923–4930.

Spanish Official Bulletin (BOE) (1996). Royal Decree 509/1996 of December 7, setting rules and regulations for urban wastewater treatment.

Spanish Official Bulletin (BOE) (2007). Royal Decree 1620/2007 of December 7, setting rules and regulations for treated wastewater reuse.

Spanish Official Bulletin (BOE) (2009). Royal Decree 1514/2009 of October 2, setting rules and regulations for protecting groundwater against pollution.

Sprenger C., Lorenzen G., Grunert A., Ronghang M., Mittal A., Selinka H.-C., Girones R. and Lopez-Pila J. M. (2009). The potential of river bank filtration for reducing chemical pollutants and pathogens from river water in mega-cities: The New Delhi experience. Proceedings of the 1st IWA Development Congress, Mexico, November 2009.

Vinyoles J., Quirante R. and Fos M. A. (2005). The benefits of water reuse inland Catalonia: Sabadell case study. Technical Workshop: The integration of reclaimed water in water resource management. Lloret de Mar, Spain.

Weiss W. J., Bouwer E. J., Aboytes R., LeChevallier M. W., O'Melia C. R., Le B. T. and Schwab K. J. (2005). Riverbank filtration for control of microorganisms: Results from field monitoring. *Water Research*, **39**, 1990–2001.

Chapter 6

Managed aquifer recharge for agricultural reuse in Shafdan, Israel

Haim Cikurel, Joseph Guttman and Avi Aharoni

6.1 INTRODUCTION

6.1.1 Water management situation

Israel is a semi-arid country with insufficient natural water resources. The current cumulative deficit in Israel's renewable water resources amounts to approximately 2000 Mm^3/yr, an amount equal to the annual consumption of the State. The natural water resources can be divided as: 58% ground water, 36% surface water and 6% runoff catchment (Foreign Ministry Report Israel, 2010). With the increase of sea water desalination (currently 15% of the total supplied water), the natural water resources are only 62% of all water resources while the rest besides the desalinated water comes from brackish water (5%) and high quality tertiary treated effluents (18%), (Israel Water Authority Report, 2010). The lack of natural water resources will not improve in the future as the population growth and the expanding modern economy will continue to increase the demand for water. Shortage of water in semi-arid countries like Israel might be accompanied by groundwater contamination, public health threats, aquifer depletion and sea and river pollution. Immediate and relatively cheap solutions like reducing the water consumption through economizing and increasing the availability of water through wastewater reuse could preserve available freshwater for potable uses.

Effluent reclamation for irrigation is important to water management in Israel and action was taken as early as the 1960s by planning the Dan Region Wastewater Treatment and Reclamation Project. The reuse of effluent for irrigation purposes is a reasonable solution that achieves two aims:

(1) Transfer of clean water that is allotted today for irrigation purposes to urban use, while supplying the irrigation needs with effluent, thus augmenting the water potential of the country
(2) Creation of an ecological acceptable disposal method to get rid of the treated effluents

The Dan Region Reclamation Project (known by its Hebrew acronym Shafdan) is the largest wastewater treatment and reclamation project in Israel and in Europe and the Middle East. Today it treats more than 140 million cubic meters per year Mm^3/yr from the Greater Tel-Aviv, that is, the Tel-Aviv Metropolitan area and several other neighbouring municipalities serving a population equivalent of over two million (Figure 6.1).

It also fulfils the aims mentioned above by further improving the effluent quality through the Soil Aquifer Treatment System to an almost drinking water quality and reclaiming and transporting the water for unrestricted irrigation to the Negev region (see Figure 6.8 for details of the "Third Line", the reclaimed water distribution system to the south of the country). This distribution system covers 125,000 acres of irrigated lands. Different crops that are irrigated unrestricted by this water are: Oranges, carrots, potatoes, lettuces, flowers, wheat.

6.1.2 Historical background

At the early days of the state of Israel, the wastewater (domestic and industrial) from all cities and settlements in Israel was disposed in septic tanks. Since septic systems need good maintenance in most cases the wastewater was percolated through the ground and reached into the aquifer used for drinking water production. In the 1950's most of the wastewater was discharged to the sea or the rivers. The authorities soon realized that this policy would lead to severe aquifer, sea and river pollution and that the only ecological way to dispose of wastewater was through adequate biological treatment.

Figure 6.1 Map of the Tel-Aviv Metropolitan area with the main sewage collection system and the Shafdan WWTP (Aharoni, 2007)

In 1955, an authority for the treatment of liquid and solid waste disposal was founded by 7 cities in the Dan Region: Tel-Aviv, Ramat Gan, Bat Yam, Holon, Petach Tikva, Givatayim and Bne Brak. The aim of the Dan Region Association of Towns was to prevent sea and river pollution by commonly treating the regional wastewater.

The national water company Mekorot Ltd. joined this project in order to reuse the quantities of effluent, resulting from the wastewater treatment, by reclaiming them for irrigation. The innovation in this project was conveying the water from the densely populated urban area in the center of the country to the dry South of Israel for further development of the agriculture in that region. Because of the large quantities of effluent and the long distance of conveyance, a decision was taken to bring the effluent to a quality close to potable water. This aim was achieved by a system developed by the project engineers. The system consists of Soil Aquifer Treatment that uses the special geological structure of the area surrounding the wastewater treatment plant for purifying the effluent by infiltration into the aquifer, recovering it, and conveying it by a pipeline to the South.

The Shafdan Wastewater Treatment Plant was planned to be built in the South of Tel-Aviv on the sand dunes of Soreq near the town of Rishon-Le Zion. The biological treatment mode was chosen to be oxidation ponds (Figure 6.2). In 1972 the first ponds were ready to receive 20 Mm^3/yr. wastewater (out of 80 Mm^3/yr produced that year). Since the beginning of operation, the wastewater has been treated biologically in re-circulated facultative oxidation ponds (Figure 6.2), followed by high pH lime-magnesium chemical treatment and final maturation in polishing ponds.

Figure 6.2 Shafdan WWTP with its first oxidation ponds (Aharoni, 2007)

In 1977, the recharge of effluents into the regional aquifer was started in order to comply with the above mentioned aims, that is, proper disposal of effluents and replacing potable water being used for agricultural irrigation with treated effluent, thereby freeing the potable water for urban use. The construction of the first infiltration site (Soreq) was delayed due to the stiff opposition from the Major of Rishon Le-Zion, the city where the project was constructed at the end. Figure 6.3(a, b) shows a typical infiltration field (Yavne 4) and Figure 6.6 shows the location and schema of the WWTP the Soreq site, and the later build Yavne sites. The advanced treatment plant meets all the EC Drinking Water Directive's limits.

Figure 6.3 (a) Recharge of the effluents in Yavne 4. (b) Dry Infiltration fields in Yavne 4 (Aharoni and Cikurel, 2011)

Since the oxidation ponds required a large surface area (200 hectares were required to treat the 20 Mm^3/yr of wastewater) planning for building a mechanical–biological activated sludge (MBAS) plant started mid 1970s. The WWTP was taken into operation in 1987 and extended in 1996. At the same time, in 1987 Yavne 1 and 1988 Yavne 2 (Figures 6.6 and 6.7a) infiltration fields were operational. Until 1999, 7% of the wastewater was still treated in the oxidation ponds. Starting at the end of 1999, the system of recirculated oxidation ponds were shut down and since then the total wastewater amount is treated in the mechanical–biological plant (Figure 6.6 and 6.7b). In 1996 the second stage of the MBAS plant was operated. At the same year the Yavne 3 fields were also operated to increase the infiltration capacity. With the operation of the second stage MBAS, in 1999, the operation of the oxidation and polishing ponds was stopped. In response to ever increasing available secondary effluents for infiltration and gradual clogging tendencies in the existing infiltration fields the last fields that were available were operated (Yavne 3 in 2003 and Soreq 2 in 2006, Figures 6.6 and 6.7a). Today about 140 Mm^3/yr are being treated by the SAT system.

Due to growing demand for "Third Line water" and the need to reclaim more available effluents new fields had to be opened in a heavily populated area. Today there are no more infiltration fields available due to growing population and housing construction around the available sites. The growing treated effluent quantities ready to be treated in the SAT system are infiltrated in a given area and this causes the "sink effect" with slower infiltration rates due also to gradual clogging. When the flow into the "sink" reaches the "passage capacity" overflow can occur. Due to clogging, the "passage capacity" decreases and the incoming flow has to be reduced. The same is valid for the SAT system:

(1) Gradual organic matter accumulation decreases the hydraulic conductivity (also due to hydrophobicity)
(2) Due to lack of new infiltration areas while the effluent application rate increases, the hydraulic conductivity limits the drainage capacity and the over all infiltered volume decreases.

In the actual SAT system (example Yavne 1) the area was prepared by taking sand from the original sand dunes and filling low areas (Figure 6.4) and also pressing to level the soil. This also caused local hydraulic conductivity problems. Due to these problems the infiltration rate is decreasing continuously. This is one of the main reasons to look for economically feasible alternative hybrid SAT technologies.

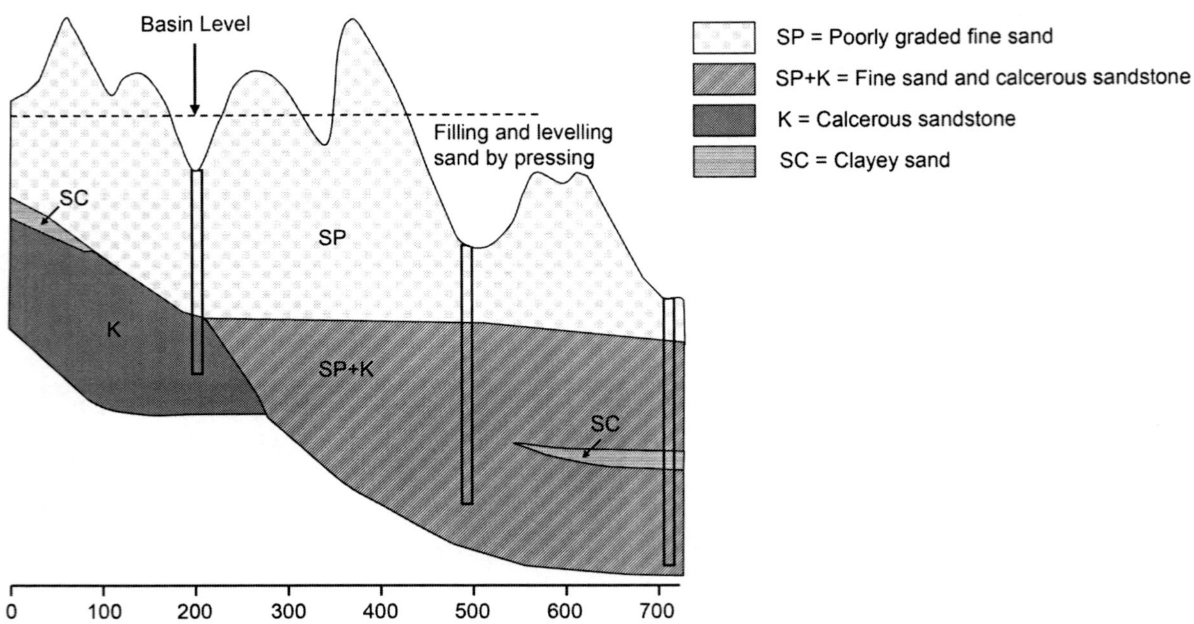

Figure 6.4 Yavne 1 infiltration field preparation (Cikurel & Aharoni, 2009)

6.1.3 Motivations for recharge and use of abstracted water

In the SAT method, dissolved organic matter is removed by combined biological, chemical, and physical processes mainly in the vadose zone:

- Suspended solids are filtered out.
- Biodegradable organic compounds are decomposed.
- Micro-organisms are adsorbed, strained out, or die because of competition with other soil micro-organisms.
- Nitrogen concentrations are reduced by denitrification.
- Synthetic organic compounds are adsorbed and/or biodegraded.
- Phosphate, fluoride, and heavy metals are adsorbed, precipitated, or otherwise immobilized.

Due to these effects, SAT or groundwater recharge can be an effective step in the treatment train for water reuse which offers many advantages: it is a relatively inexpensive technique suitable for countries with a warm climate; it consists of an open area that can be located far away from residential sites; it is a reliable process which is simple to operate; and it allows seasonal storage of reclaimed water during low water demand periods for future reuse during high water demand periods.

Motivation for improved SAT processes

During the last 50–60 years, Israel has been looking for technologies to save water, among them also recycling of treated sewage for irrigation of crops. Israel with 7 million inhabitants is consuming about 890 Mm^3 of water for domestic and industrial uses. The reuse of treated sewage for agriculture is 75–80%, one of the highest in the world. Today, the Soil Aquifer System depicted above, is treating and infiltering about 120–140 Mm^3/yr in area of about 110 hectare with a retention time of 6–12 months in the aquifer (long term SAT). The infiltration regime is mostly one day flooding and 2 days drying.

Substantial problems in the infiltration regime were encountered during the 30 years of infiltration (Aharoni and Cikurel, 2011):

- Deterioration of the infiltration rates (mostly in winter – rainy days and low temperature)
- Anoxic conditions in the aquifer that cause release of manganese and iron to the reclaimed water forming afterwards oxides and clogging the irrigation systems
- No more available lands for the increasing quantities of effluents to be infiltrated.
- Bio-fouling of effluent pipelines before and after SAT
- There are intentions to reduce the actual infiltration areas (to treat by UF-RO instead of SAT and use the areas close to the cities for housing). This will mean changing a natural process to an energy intensive process.

After 30 years of operation, the limits of the infiltration capacity using the existing infiltration basins are reached and in order to increase the recharged volume of water, a system of dug wells around the infiltrations basins closer to the extraction wells, leading to a shorter residence time, is considered.

6.1.4 Authorisation procedure

In Shafdan, the authorization procedure for infiltration of municipal effluent through soil aquifer treatment is well established and revised by the Infiltration Committee on a yearly basis depending on the performance of the aquifer recharge site. This infiltration committee consists of different water authorities and ministries including: the Israel Water Authority, the Health Ministry, Rishon-Le-Zion municipality (where the SAT system and Shafdan are located), the Ministry of Agriculture, the hydrological services, farmer representatives and Mekorot. This committee meets to decide on the quantities of treated effluents to be infiltrated, the quality required for infiltration and the quality of the reclaimed water, which are reported in a yearly document. In this document the quality of the water to be infiltered is set according to the requirements of the Ministry of Health. A monitoring plan is also mentioned in this document including the sampling points and the analyses to be performed (including chemical parameters and the bacteriology) on a yearly basis. The monitoring program includes also monitoring of drinking wells adjacent to the SAT area in order to see if there is any seepage of the infiltered – reclaimed water to the drinking water system. Every end of the year, Mekorot presents a book that includes all hydrological and analytical data that have to be discussed and approved. After approval, the new yearly program is issued and Mekorot has to work accordingly.

Preceding year's results are analyzed and ratified and a new yearly infiltration program and permit were issued for every infiltration zone (i.e. Yavne or Soreq area). This document includes: The name of the company in charge of the infiltration (i.e. Mekorot), the infiltration projects' name (i.e. Shafdan Project), the infiltration area (i.e. Yavne or Soreq), the year this infiltration will be performed (2009) and the maximum quantity that is allowed for infiltration (i.e. 25 Mm^3/yr for Yavne 1 area) and the quality.

The document includes also the technical description of the SAT: Infiltration of secondary effluents (after nitrification-denitrification) to infiltration basins in intermittent regime. The purpose of the SAT is to be "a slow sand filter to obtain water reusable in unrestricted agricultural irrigation and also a seasonal and multi-annual underground reservoir where water for agricultural irrigation is stored". After that, the technical process of infiltration is explained (it is allowed for example, to infiltrate 20,000–130,000 Mm^3 in 24 hectares area according to one day infiltration and 2 days drying). Also, the infiltration and sampling program is given (which points to be sampled, how many times a month or a year and which parameters). A minimum quality requirement for the infiltrated water is also mentioned. There are also some operational guidelines for the quality of the secondary effluent to be infiltrated in the SAT system: Turbidity <8 NTU and Total Suspended Solids <15 mg/L. The Dan Region secondary treatment produces effluent that complies with this water quality. The annual average turbidity of the effluents in 2007 was 2.7 NTU and the Total Suspended Solids is 6 mg/L.

The water obtained from the SAT treatment should comply with the extremely high quality requirements to be used for unrestricted agricultural irrigation and accidental drinking water quality. The quality of water after the SAT system is comparable to BOD <1 mg/L, TSS almost 0 mg/L, COD 10–20 mg/L, UVabs 20–50, $N\text{-}NH_4$ >0.2 mg/L, TN 5–10 mg/L, TP 0.01–0.03 mg/L, MBAS. 0.05–0.2 mg/L, Total coliforms 0 CFU/100 mL, Faecal Coliforms 0 CFU/100 mL, S.Fec. 0 CFU/100 mL, TB 30–2000 CFU/mL. These are only a few examples of the chemical and microbiological quality of the infiltered water. The effective monitoring program insures this quality.

6.2 DESCRIPTION OF THE TEST SITE

As explained above, the studies within RECLAIM WATER investigated the conventional SAT system and compared this to a novel short SAT system with a dug well infiltrating UF pretreated effluent.

The aim of Shafdan case study site was to look into alternative technologies to produce reclaimed water for unrestricted irrigation qualitatively comparable to the "accidental drinking water quality" of the "Third Line" water. The process proposed consists of an Ultra Filtration (UF) treatment of secondary effluents combined with a dug well injection at short term SAT (up to 2 month instead of nearly 12 month in the SAT system). This way more secondary effluents can be recharged at a given area. Figure 6.5 presents the conventional and the modified SAT systems to be compared.

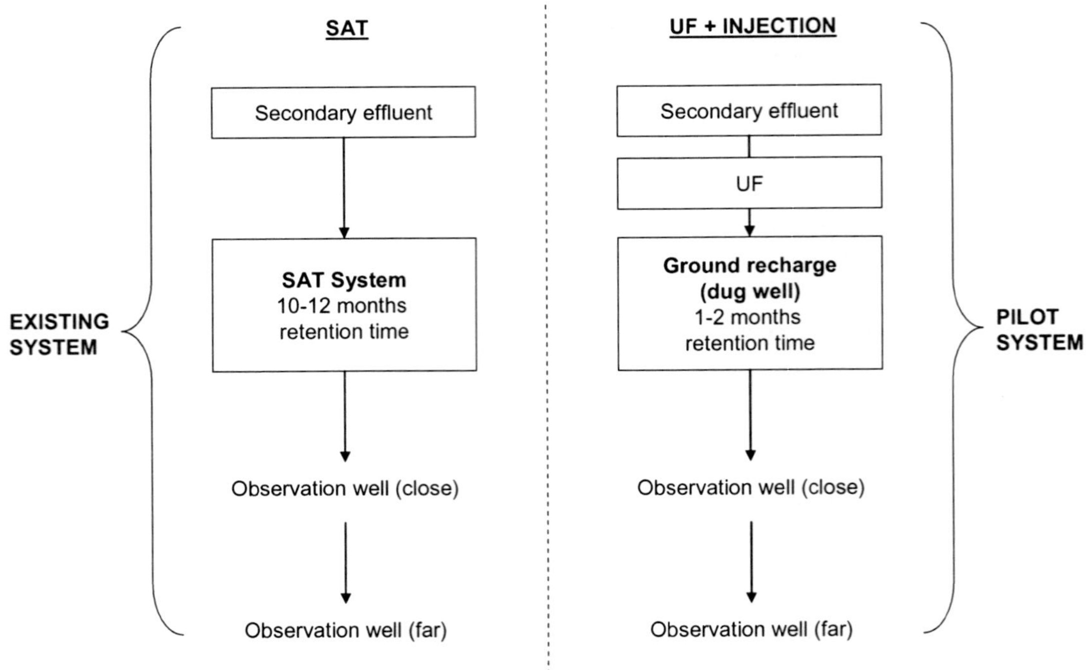

Figure 6.5 Recharge scheme in Shafdan: Conventional SAT versus pilot short SAT

6.2.1 Existing full-scale system

Currently the Dan Region Reclamation Project (Shafdan) reclaims approximately 135–140 Mm^3/yr of wastewater from the Tel Aviv Metropolitan area and several other neighbouring municipalities.

Shafdan wastewater treatment plant

A mechanical-biological activated sludge (MBAS) plant with nitrogen removal was put into operation in 1987 and was further expanded in 1996 (cf. Figure 6.7b). The wastewater treatment plant is an extended aeration plant consisting of pre-treatment (bar screens, grit removal and scum removal) and activated sludge tanks with aerobic and anoxic zones, so that simultaneous nitrification and denitrification occurs together with carbonaceous BOD removal, followed by secondary clarifiers. In 2010 the WWTP underwent another extension by primary clarifiers and a sludge treatment facility to improve the treatment quality and increase the energy efficiency of the plant.

Soil aquifer treatment scheme

After secondary treatment, the effluent is then further treated through recharge to the aquifer and subsequent recovery (from 80–100 m. depth) using a special Soil Aquifer Treatment (SAT) system. The very high quality (accidental drinking quality) water named also "Third Line water" is then pumped 100 km to the south of the country, where it is used to irrigate a variety of crops.

During the 30 years operation of the SAT system, due to organic matter accumulation in the upper layers the infiltration velocity has decreased down to 1 m/d (or hydraulic loading of 100–120 $m^3/m^2/yr$ taking into account the 1 day infiltration 2 days resting regime) in old fields causing more anaerobic conditions and, as a result, manganese dissolution has been detected in some wells. Figures 6.6 and 6.7a present a schematic and a map of the recharge facilities.

Agricultural reuse of the recovered water

The operation of the "Third Line" (Figure 6.8), the distribution system for the soil aquifer treated effluent from the Dan Region WWTP, started in November 1989. Since that time the reclaimed water of almost drinking water quality is used for unrestricted agricultural irrigation in the southern part of the country.

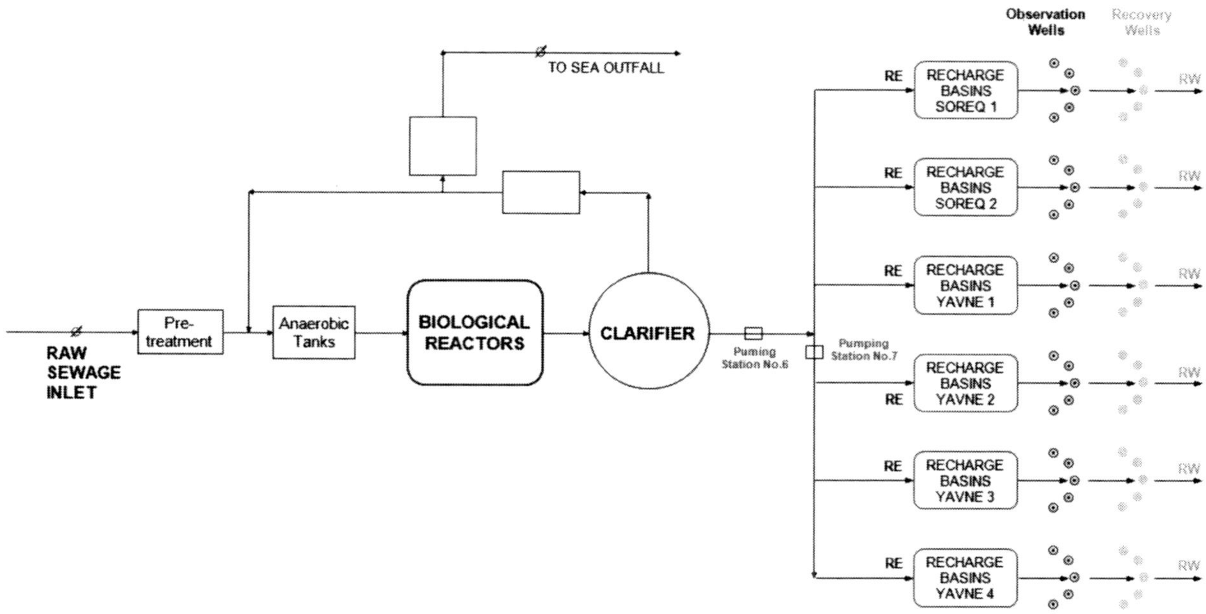

Figure 6.6 Scheme of the Dan Region WWTP, Soreq and Yavne infiltration fields (Cikurel & Aharoni, 2009)

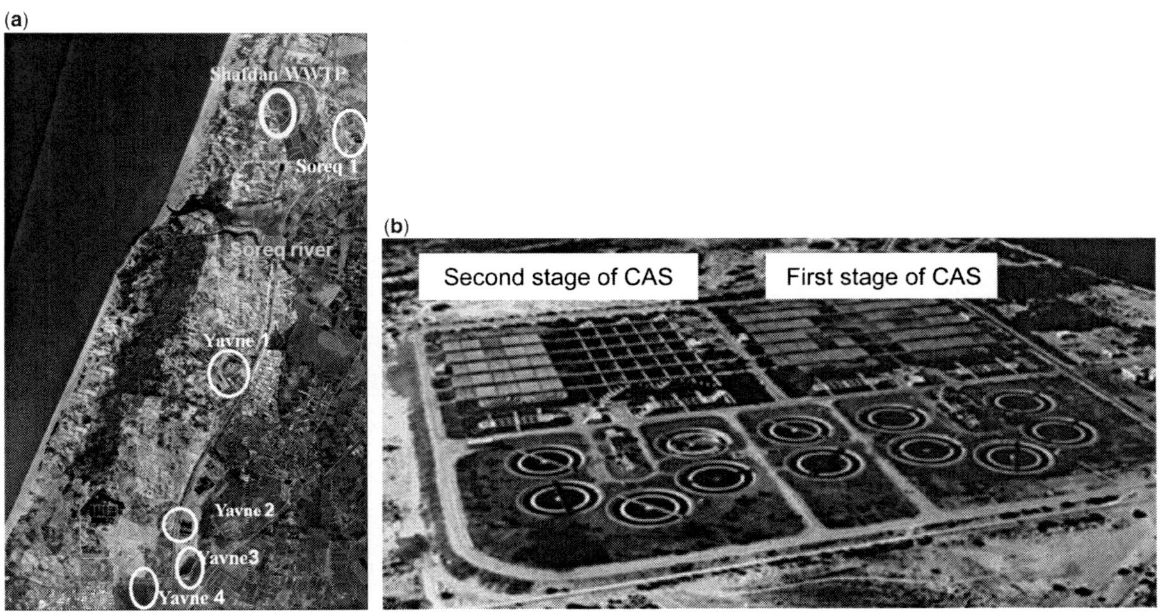

Figure 6.7 (a) Location of the Dan Region WWTP, Soreq and Yavne infiltration fields (Goren, 2005). (b) Aerial view of the first and second stage of the CAS WWTP (Aharoni, 2007)

6.2.2 Pilot-scale UF short SAT system

Study area

The Shafdan main facilities (including the UF pilot plant system) are located in one of the coastal aquifer depressions. At this depression there is a thick clay layer overlaid by a sandstone layer. The depression is bordered by two elongated sandstone ridges. This clay layer separates the regional aquifer (located below the clay layer) and a perch aquifer located above the clay layer, in the upper sandstone layers. From the geological point of view, the favorite site should be located outside of this depression, at a place where the clay layer is absent (zero thickness). At this point, there is a continuous sand sequence from the surface up to the water table.

For the UF-SAT pilot experiment, after preparing the structural and the isopach maps of the clay layer, four proposed places for the test site were chosen. At those places auger wells were drilled up to the regional water table in order to learn about the lithological sequence (Figure 6.9). The result of the drilling indicated, that in three places the clay layer is missing and there is a continuous sequence of sand from the surface up to the regional water table (Figure 6.9).

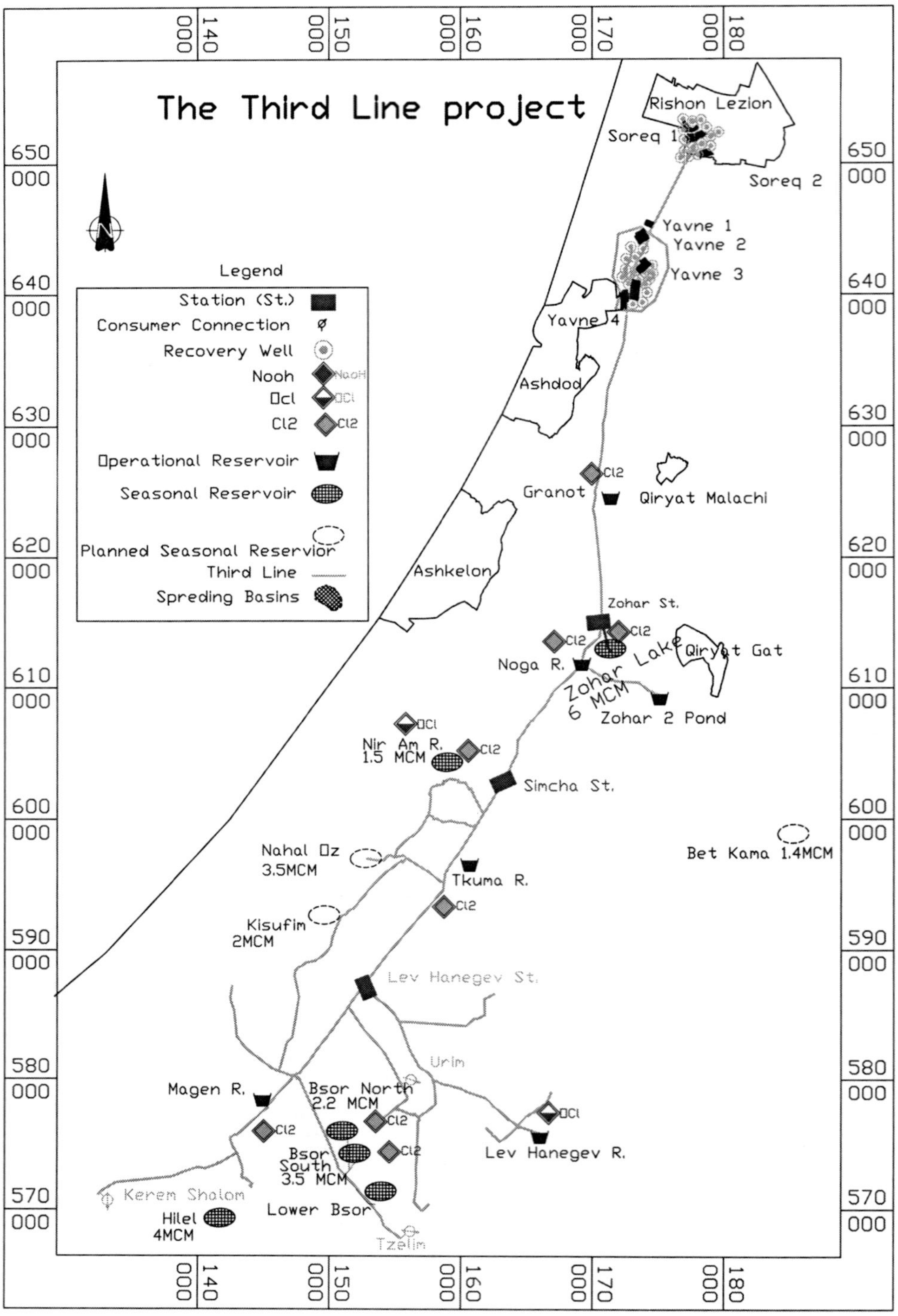

Figure 6.8 Map of the irrigation water pipeline "Third Line" to the South of Israel (1989) (Cikurel & Aharoni, 2005)

The water infiltrated through the dug well would undergo ultrafiltration as an extra treatment to guarantee the quality of recovered water and to compensate for the shorter residence time. As part of the work in RECLAIM WATER, a pilot dug well site including observation wells in the saturated zone and porous cups for water sampling in the unsaturated zone was set up in order to assess the performance and the resulting water quality in the system (Kloppmann *et al.* 2009).

In order to carry out the experiences a comprehensive system had been constructed at the site. It included:

- A recharge well called "Dug Well"
- Installation of three ceramic (porous) cups in the unsaturated zone.
- Drilling of three observation wells (in two phases).

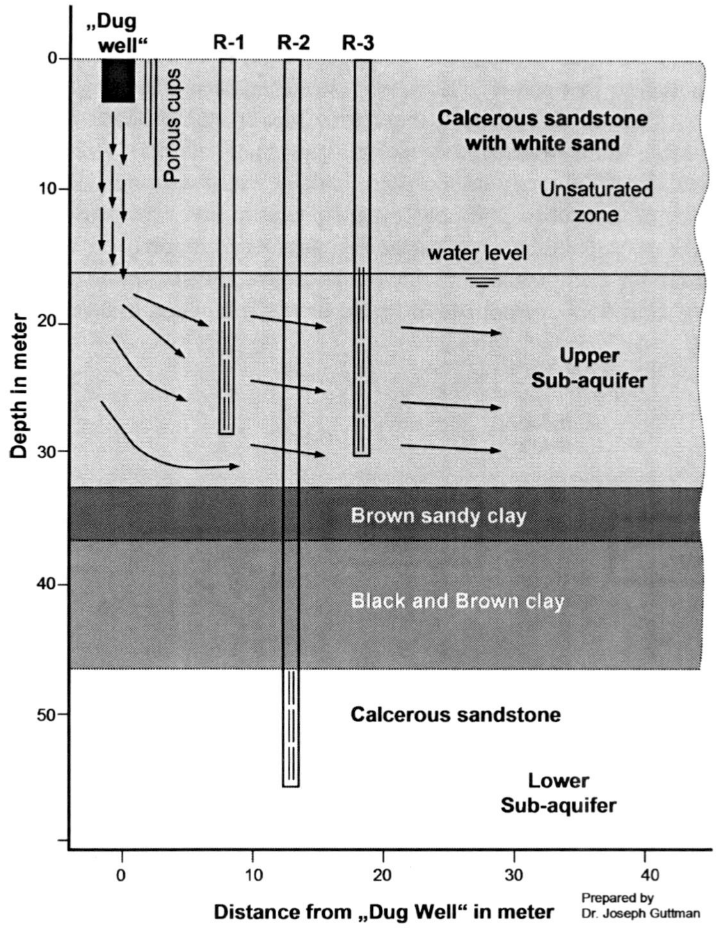

Figure 6.9 The results of the 4 bore holes. R-2 is the area where the Dug well was located (Guttman, 2007)

Hydrogeology

According to the drilling results, the most permeable section of the soil was in the uppermost part of the saturated zone at a depth between 14–21 m.

The "Dug well" (Figure 6.9) was a cylinder tube. Its size was calculated to enable the recharge of about 100–120 m^3/day of reclaimed water (after UF). The depth of the "Dug Well" was 2.5 m and an unsaturated zone of about 12–13 m was left. Close to the Dug well wall, three ceramic (porous) cups were installed in the unsaturated zone. The depths of the ceramic cups were: 5.5 m, 7.0 m and 8.5 m. These ceramic cups allowed sampling the infiltrated water in the unsaturated zone. About 7.3 m downstream of the "Dug Well" an observation well was drilled (well R-1) to a depth of 27.5 m. A deeper well (well R-2) was located 5.1 m downstream to well R-1 and was drilled to a depth of 54 m. Up to a depth of 32 m the lithological section was composed of calcareous sandstone with white sand similar to the section in well R-1. Below 32 m the section was composed of clay (sandy clay and black clay). This clayey layer ended at a depth of 46 m and separates two sand sequences. Below 46 m the section is composed of calcareous sandstone. This well was drilled to see if there is a clayey layer between depths of 27 and 46 m and the tracer tests that were conducted afterwards showed just that. The third well R3 which made up together with R1 the second experimental well (sampling was conducted in wells R1 and R3) was located 10.3 m downstream to well R-1 and 5.2 m downstream to well R-2. The well was drilled to a depth of 28.3 m, same layers as in well R-1. After completion of the observation wells, a tracer test was carried out. The concentration was measured in the ceramic cups and in the observation wells too. The results of the test showed a retention time of 9–10 days for the R1 well and 30–35 days for the R-3 well for the tracer injected in the dug well.

In conclusion, three wells were drilled at the site. The results showed that between 32 and 46 m there was a clay layer. This clay layer separated the sand sequence into two sub- aquifers. An upper sub-aquifer existed above the clay layer and a second one existed below it. The deepest well (well R-2) did not reach to the bottom of the lower sub-aquifer. There was a difference in salinity between the two sub-aquifers. The lower sub-aquifer showed lower salinity levels. The static water level in February 2007 (when measured) was at an elevation of about 1.9 m below sea level. The massive pumping in the Shafdan Reclamation plant was the reason why the water level was low (below sea level) although in that season the pumping was less then in August to September.

Process design and operation

As mentioned in the introduction part due to the growing demand for "Third Line water" and the need to reclaim more available effluents, new fields had to be opened in a heavily populated area. The aim of the Shafdan case study site in the RECLAIM project was to look at alternative technologies that would produce reclaimed water for unrestricted irrigation qualitatively comparable to the accidental drinking water quality of the "Third Line" water. In order to check the suitability of the reclaimed water the analyses included, besides classical chemical analyses (organic matter, N, P, salts, heavy metals, microbiology), also micropollutants (mainly neutral drugs and antibiotics, NDMA, estrogens, AOI and DOC characterization). The micropollutants results are not part of this report.

In this context, as described in the previous section, the pilot site was constructed at the Shafdan WWTP at the chosen location (Figure 6.10) where a short SAT system (up to 35 days) was compared to the conventional long retention time (around a year) SAT system.

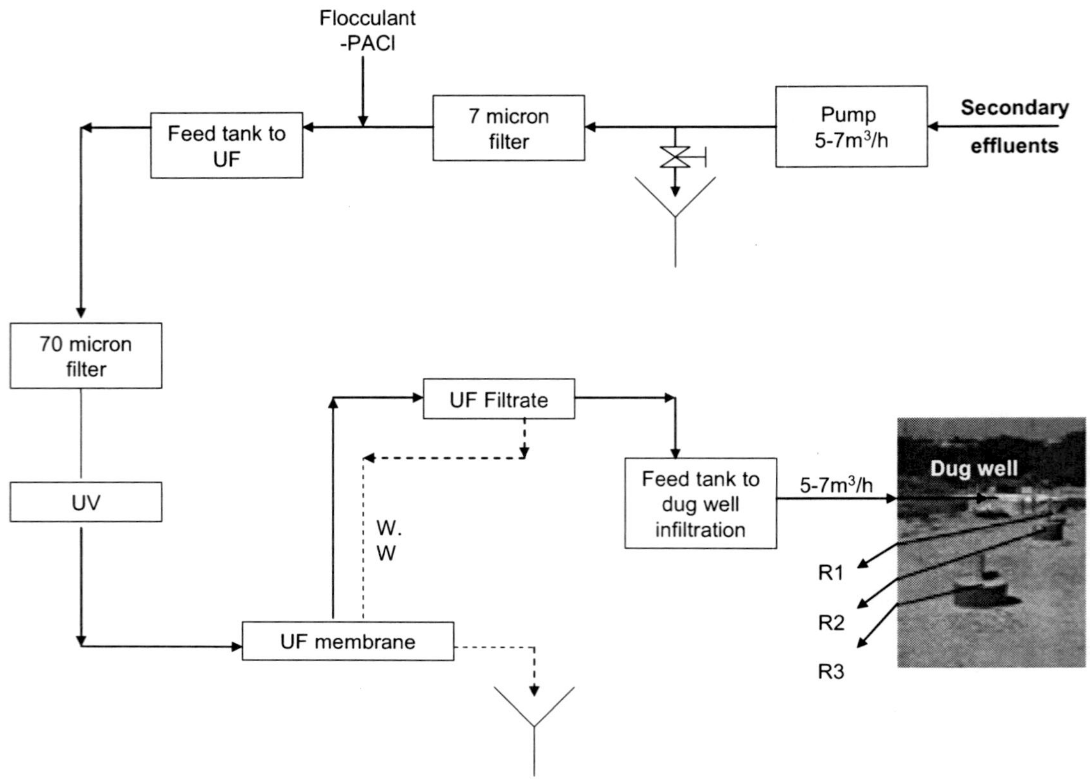

Figure 6.10 Shafdan Reclaim pilot plant final flow sheet (Cikurel & Aharoni, 2009)

The scheme was based on two stages: The system of dug-wells and observation wells are part of the second part of the system. At a first stage secondary effluents from Shafdan WWTP were UF filtered (Figure 6.10). The filtered effluents were then infiltrated at a high rate (10–12 m/d) by means of a dug-well at short term SAT (up to 35 days instead of nearly 12 month for the conventional SAT system). The infiltrated water was abstracted via the two observation wells at a depth of 20 m (R1: 9–10 days retention time and R3: 30–35 days retention time) described in the previous chapter. A deeper well (R2) was also drilled to investigate the effect of dilution by depth as also described in the previous chapter.

After the construction of the pilot plant and running of the tracer tests at the end of 2006, the whole pilot plant was operated with Third Line water (classical SAT water). The first sampling campaign to determine the micropollutants in the background water was performed in November and December 2006. The monitoring that included all classical analyses and micropollutants on the UF infiltrated water was carried out for the first time in May to June 2007 and another three campaigns were performed till May 2008. The different sampling points were: The secondary effluents, after UF filter, the observation wells R1 and R3 and a sampling point (recovery well Dan 9) which is part of the classical SAT (Figure 6.11).

The original set-up for the UF, which was operated in April 2007, included a pre-treatment based on a 400 micron pre-wire filter followed by a 2 micron fiber filter before a MP-UV system followed by 0.01 micron PS UF membrane operating at a dead-end mode at 75 LMH. This set-up resulted in easily clogged membranes and was gradually optimized. Due to the heavy organic matter containing secondary effluents (cf. Figure 6.12), the clogging of the fiber filter and the UF was high so that a change in the pre-filter and UF was performed.

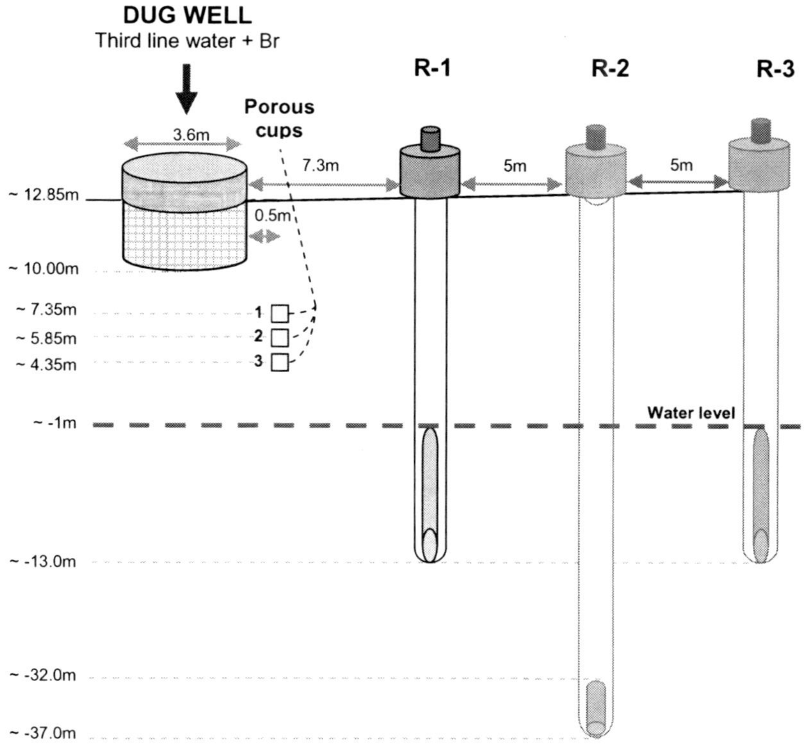

Figure 6.11 Sampling points at the Shafdan Reclaim pilot (Gaus *et al.* 2007)

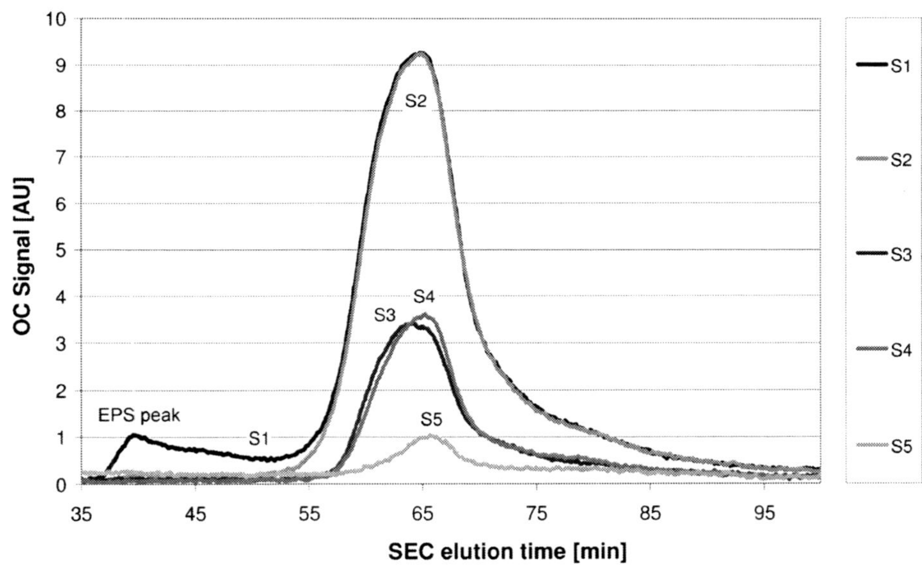

Figure 6.12 LC-OCD chromatograph of the secondary effluent (S1), UF filtrate (S2), water at observation well 1 and 2 of the pilot site (S3 and S4) and recovered water from full-scale SAT (S5)

From May to July 2007, 5 m^3/h effluents were pre-filtered at 7 micron and UF filtered at 55 LMH (3 membranes of 30 m^2 each) at 0.6–0.8 bar TMP. At this period, only 80% UF effluents (4 m^3/h) and 20% (1 m^3/h) conventional SAT water (Third line water) was mixed to obtain the desired 5 m^3/h flow.

In August 2007 another 70-micron filter was added to the feed tank before the UF, to prevent the biofilm growing in the tank (high temperatures, high sun radiation) prior UF. In August 2007 the mixing ratio of third line water/UF filtered water was 40% to 60%. In that period the UF was operated at 55 LMH and the Trans-membrane pressure (TMP) was around (0.6–0.7 bar).

At the beginning of September 2007 the TMP was 0.9–1.1 bar and a new working program based on 4 m^3/h of UF filtered secondary effluents was operated. Till the end of October the UF was operated at 0.6–0.7 bar at 45 LMH. The permeate turbidity was 0.04–0.05 NTU. The hydro-geological modeling was based on 5 m^3/d infiltration to the dug well so that an exact ratio of SAT water (2 m^3/hr) was mixed to the 3 m^3/d UF permeate actually produced to have the required amount of water for infiltration.

From September 2007 till end of February 2008 the membranes had to be operated at 3 m^3/h net flow and 2 m^3/h of Third Line water was mixed. From March 2008 till end of May 2008 a different UF filtration mode (cross-flow mode) of operation was used replacing the dead-end mode used up to that date. But at this stage it was decided that a new infiltration mode will be applied. Instead of the continuous mode infiltrating 10–12 m/d of 60–80% UF water/20–40% Third line water, only 16 hours infiltration was applied (at 10–12 $m \cdot h^{-1}$) and 8 hours of relaxation. On the average, 66–70 m^3/d of good quality (0.06-0.08 NTU) UF effluents were infiltered with no mixing of third line water. The infiltration rate was around 7 m/d (no water level in the dug-well). The results obtained showed a decrease of the manganese content that may be due to the more aerobic conditions in R1.

On 15 May 2008 the last sampling was performed. The last 5 months operation at intermittent mode enabled the introduction of more air into the SAT system, improving the oxic conditions at the closer well R1 and a decrease in soluble Manganese was obtained (Table 6.1).

Table 6.1 Manganese, iron and oxygen content in R1 well water at different periods (Cikurel & Aharoni, 2009).

Sampling date	15-08-06	12-09-06	29-11-06	15-05-07	15-05-07	04-12-07	15-05-08
Dilution	Background before infiltration	3rd line only	3rd line only	80% UF-20% 3rd line only	60% UF-40% 3rd line only	60% UF-40% 3rd line only	100% UF
DO ($mg \cdot L^{-1}$)			0.7	1.2	0.93	1.44	1.3
DOC_1 ($mg \cdot L^{-1}$)	2	1.4	1.1	3.2	2.5	2.1	2.9
Fe ($mg \cdot L^{-1}$)	1335	469	571	68	25	41	18
Fe_s ($mg \cdot L^{-1}$)				28	11	28	17
Mn ($mg \cdot L^{-1}$)	442	879	705	218	205	135	109
Mn_s ($mg \cdot L^{-1}$)				215	80	132	109

6.3 TECHNOLOGY PERFORMANCE AND CONTAMINANT MONITORING

6.3.1 Nutrients and bulk organics

The chemical parameter results in Figure 6.14 show that organic matter represented by COD and CODf was almost reduced in the Reclaim UF-short SAT pilot plant (at sampling well R3 which represents 30 days retention time in the aquifer) to the same level as in the classical SAT (12 months retention time in the aquifer), also the DOC was effectively reduced from a 10–12 mg/L level to 2 mg/L level in the R3 well (30–35 days retention time) although still less effective than the long term SAT (DOC 1–2 mg/L). Ammonia (Figure 6.14) and total N (Figure 6.13) were also removed effectively (around 0.5 mg/L of ammonia) but P was removed only partially (Figure 6.13).

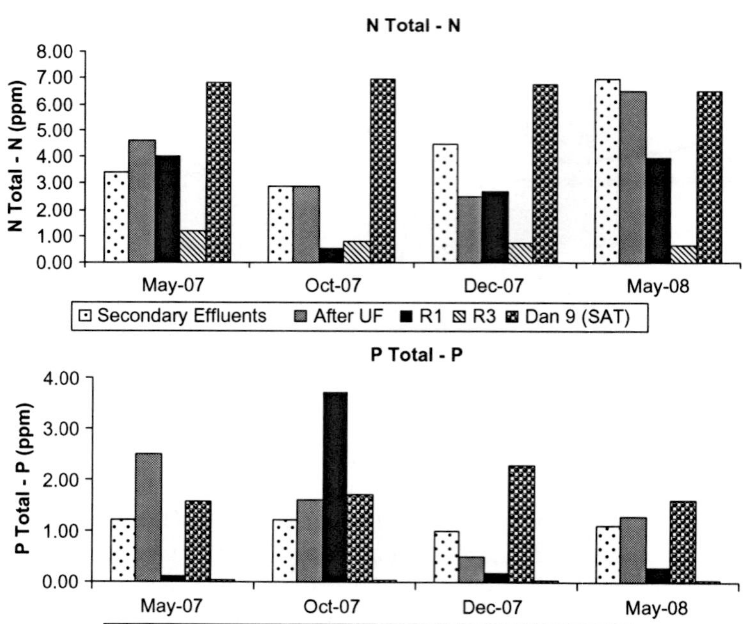

Figure 6.13 Chemical parameters Nt and Pt at different sampling points along the UF-short SAT system and a typical conventional SAT well (D9)

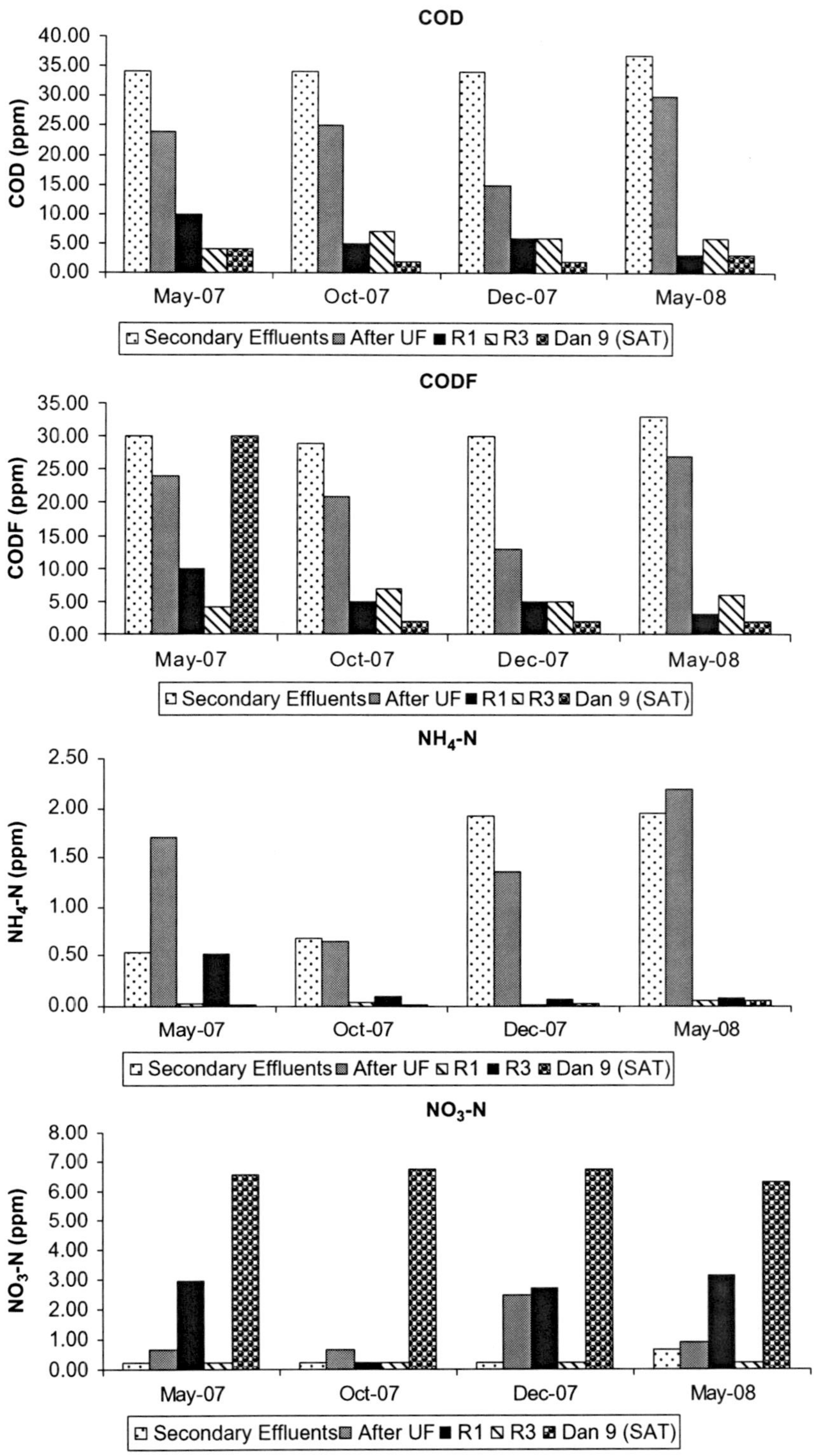

Figure 6.14 Chemical parameters COD, CODf, ammonia and nitrate results at different sampling points along the UF-short SAT system and a typical conventional SAT well (D9)

6.3.2 Dissolved oxygen, iron and manganese

Figure 6.15 shows the Dissolved Oxygen (DO) at different sampling points. The DO in the closer well R1 was around 1 mg/L while the DO in the furthest well R3 was less than 1 mg/L and this affected the redox conditions and the dissolution of manganese and iron (Table 6.1 and Figure 6.16).

The change from continuous infiltration to intermittent infiltration during the last sampling period (23/3–19/5/2008) affected the redox conditions. During this period only UF filtered water was infiltrated at intermittent rates of 16 hours and 8 hours rest.

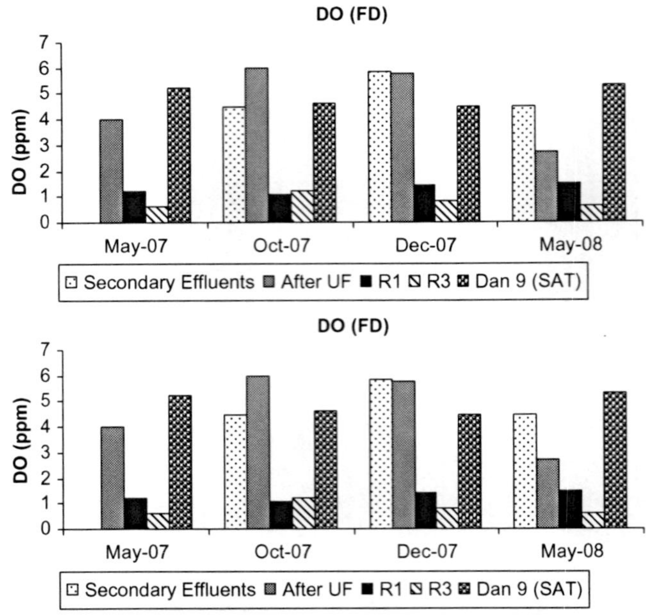

Figure 6.15 DO and EC at different sampling points along the UF-short SAT system and a typical conventional SAT well (D9)

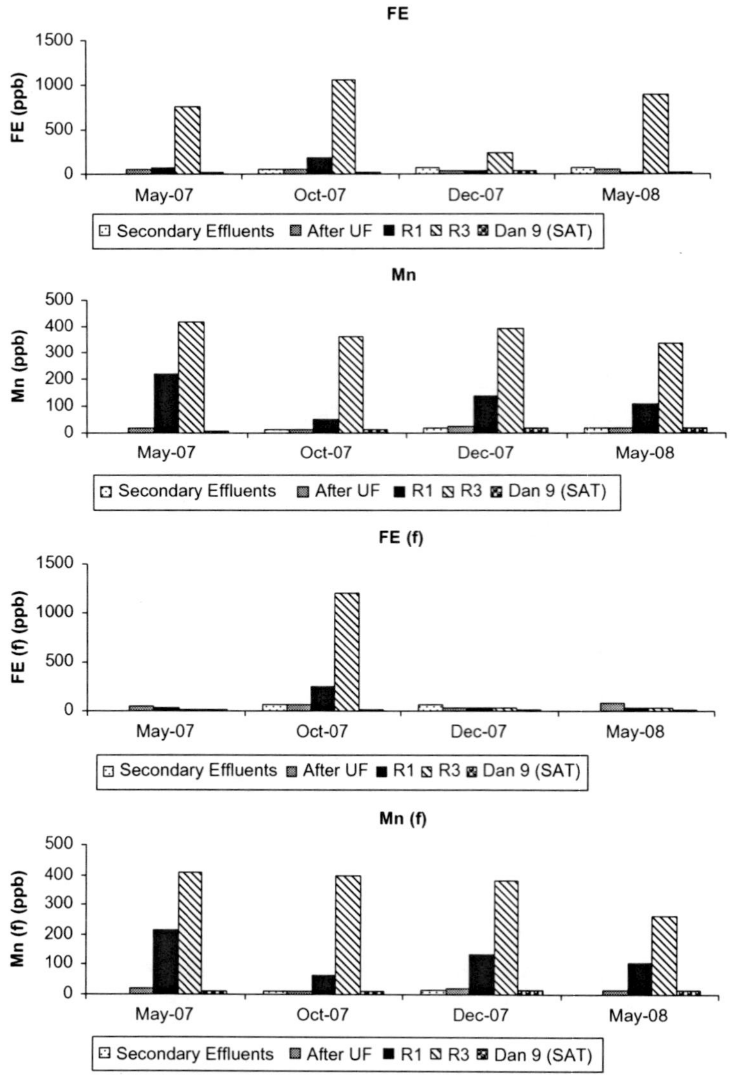

Figure 6.16 Non-soluble and soluble manganese and iron concentrations in different sample points along the UF-short SAT system and a typical conventional SAT well (D9)

One of the positive results of the pilot site was that in R1, probably due to the positive redox conditions, the soluble manganese rate was the lowest observed up to now in one and a half years of infiltration. The lowest dissolved manganese results in well R1 had been obtained with these experiments which allowed a better introduction of oxygen to the unsaturated zone and also the saturated zone (only in R1 whereas in R3 there was no difference in soluble oxygen). As a result of more oxygen-rich conditions, manganese precipitation increased, hence leading to less soluble manganese ending up in well R1. R3 did not show a serious decrease in soluble manganese. In this well, where the conditions are still anoxic, the manganese dissolution rate did not change significantly.

These results also can strengthen the postulate that was mentioned before, that by introducing organic free and oxygen rich water the redox conditions of the ground can be improved (in that case even the R3 could be oxygen rich since organic matter will not deplete the oxygen).

6.3.3 Microbiological contaminants

The bacteriological quality of the water at the furthest well R3 (30–35 days retention time) shows that (Figures 6.17 and 6.18) there were no or very low numbers (less than 5) of Faecal Coliforms, *E. Coli*, Enterococci, Enteroviruses, Bacteriophages and Clostridium spores.

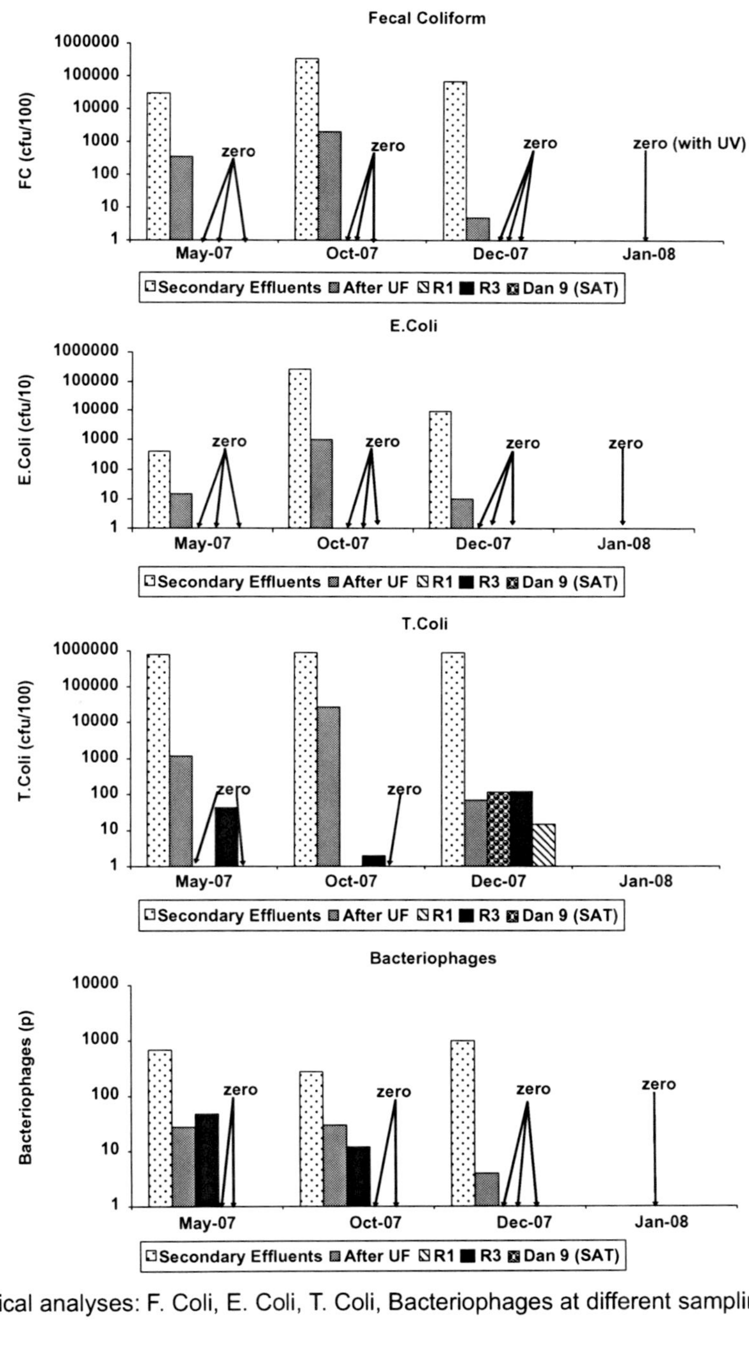

Figure 6.17 Bacteriological analyses: F. Coli, E. Coli, T. Coli, Bacteriophages at different sampling points

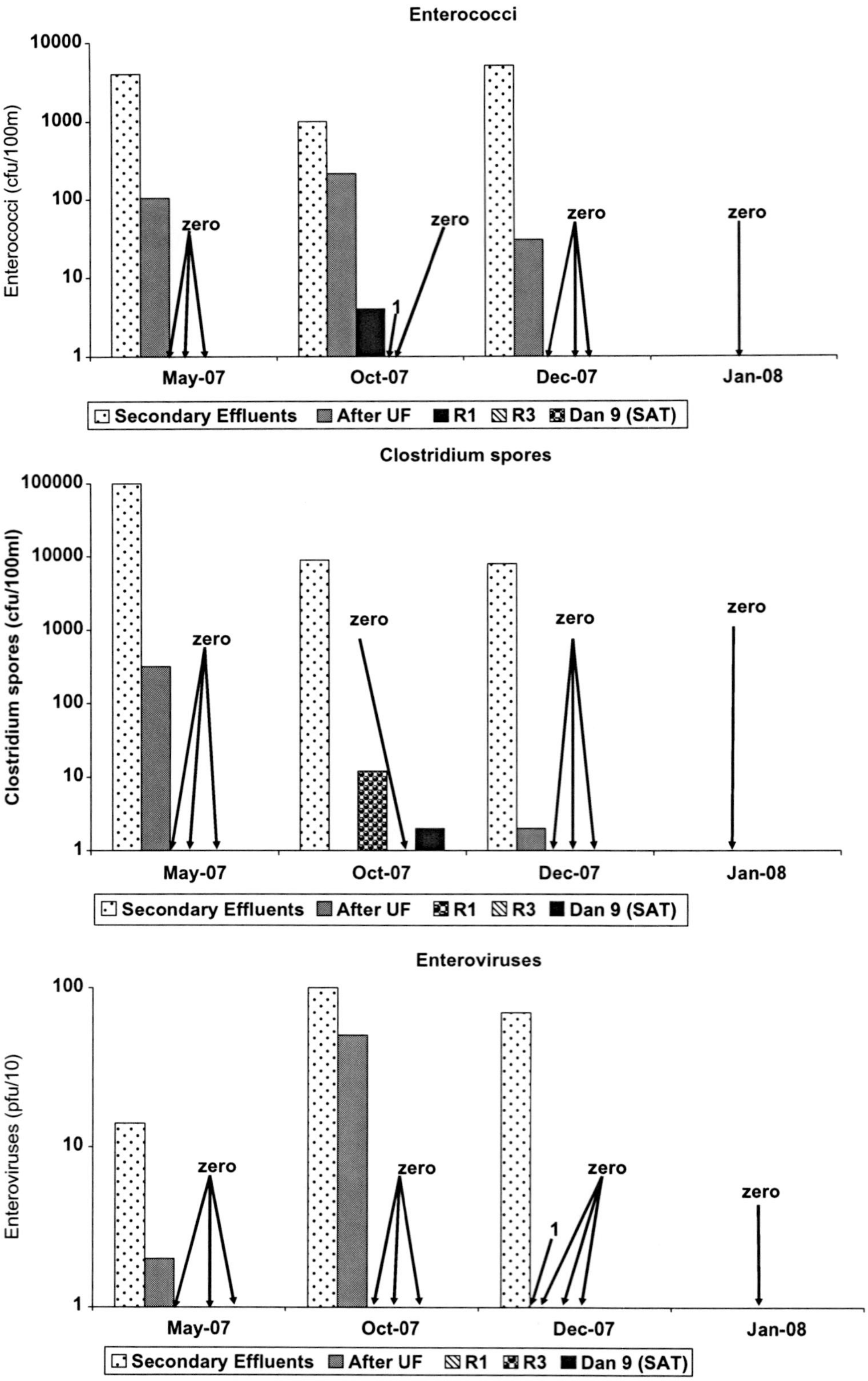

Figure 6.18 Bacteriological analyses: Enterococci, Clostridium spores, Enteroviruses at different sampling points

This shows that water clean from micro-organisms can be obtained by the slow passage of UF filtered secondary effluents at much shorter soil retention times then the conventional SAT.

6.3.4 Summary removal capacity UF-short SAT

Table 6.2 and Figure 6.19 show the different removal steps of various contaminants for the UF-short SAT process. The contaminants are mainly removed by the activated sludge process but the second polishing biotreatment process occurs during the passage of the effluents through the upper layers of the soil in the Soil Aquifer treatment. Organic matter, nitrogen, phosphorous and pathogenic bacteria are removed in the first 60–100 cm thick sand layers, which act as biofilters (Sharma *et al.* 2007). So that there is no significant additional DOC removal between 10 and 30 days retention time (Table 6.2).

Table 6.2 Removal steps of different contaminants for the UF-short SAT pilot (Absolute values and removal efficiencies).

		Absolute values				
		1 Capture zone	2 sec. effl.	3 sec effl after UF	10 days r.t.	30 days r.t
Sampling point description		**WWTP influent (2009)**	–	**Infiltration water**	–	**Extracted water**
Basic wastewater analysis		**AV**	**AV**	**AV**	**AV**	**AV**
Suspended solids (105)	$mg \cdot L^{-1}$	424.00	4.76	0.37	0.43	6.08
BOD total	$mg \cdot L^{-1}$	407.00	5.25	1.00	1.00	1.00
BOD soluble	$mg \cdot L^{-1}$	170.00	1.50	1.00	1.00	1.00
COD total	$mg \cdot L^{-1}$	896.00	37.38	22.00	5.33	6.75
COD soluble	$mg \cdot L^{-1}$	274.00	31.88	20.00	5.00	6.50
DOC	$mg \cdot L^{-1}$	86.30	10.22	7.40	2.50	2.10
Ammonia	$mg \cdot L^{-1}$	44.11	1.61	0.82	0.03	0.14
NKjT-N	$mg \cdot L^{-1}$	66.40	3.78	1.99	0.53	0.60
Nitrite NO_2	$mg \cdot L^{-1}$	0.00	1.00	0.38	0.20	0.00
Nitrate NO_3	$mg \cdot L^{-1}$	0.00	0.79	1.13	2.33	0.23
Total nitrogen - N	$mg \cdot L^{-1}$	20.00	4.90	3.50	3.07	0.83
Total phosphorus - P	$mg \cdot L^{-1}$	11.40	1.46	1.17	0.33	1.73
Electrical conductivity (LB)	$\mu S/cm^{-1}$	1,874.00	1,602.80	1,571.75	1,564.00	1,488.00

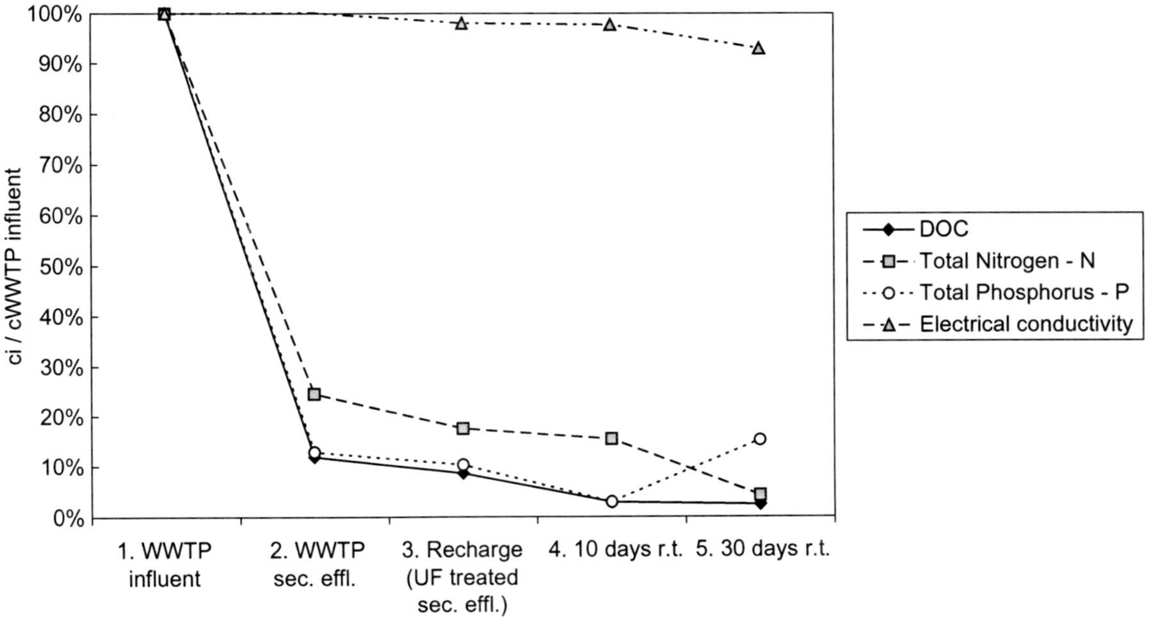

Figure 6.19 MAR scheme Shafdan: Average reduction in pilot UF-short SAT

6.3.5 Summary removal capacity conventional SAT

Figure 6.20 and Table 6.3 show the data for the conventional 12 months retention time SAT. The pattern for organic matter removal is the same but there is some secondary removal of hardly biodegradable matter due to the longer retention time. There was an increase in nitrates and not complete denitrification.

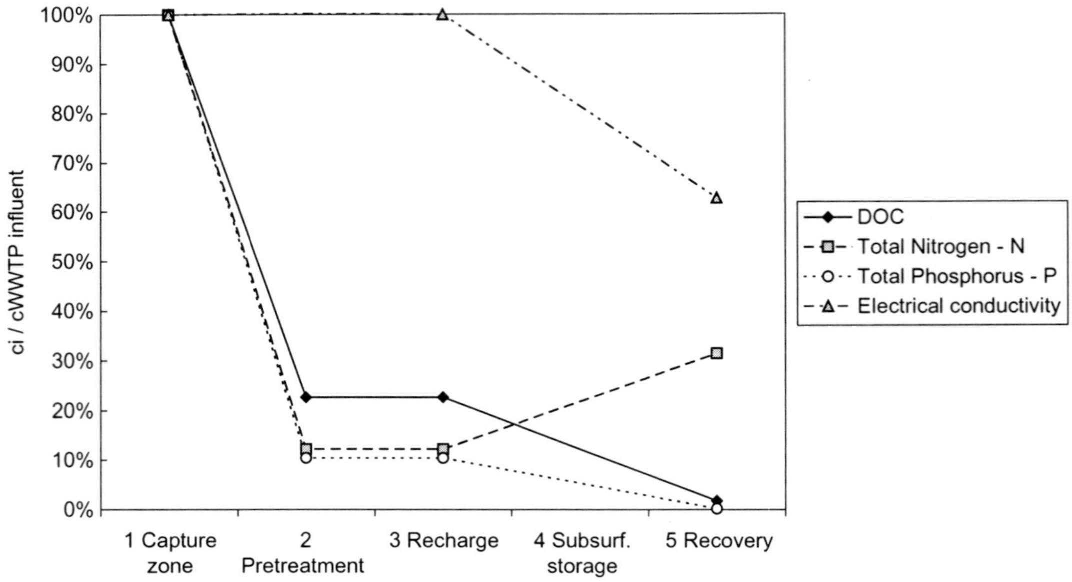

Figure 6.20 MAR scheme Shafdan: Average reduction in existing full scale SAT

Table 6.3 Removal steps of different contaminants for the long term SAT (Absolute values and removal efficiencies).

		Absolute values			
		1 Capture zone	2 sec. effl.	Infiltered effluents	Extraction
Sampling point description		**WWTP influent**			**Long SAT**
Basic wastewater analysis		**AV**	**AV**	**AV**	**AV**
Suspended solids (105)	$mg \cdot L^{-1}$	424.00	4.76	4.76	0.23
BOD total	$mg \cdot L^{-1}$	407.00	5.25	5.25	1.00
BOD soluble	$mg \cdot L^{-1}$	170.00	1.50	1.50	1.00
COD total	$mg \cdot L^{-1}$	896.00	37.38	37.38	4.00
COD soluble	$mg \cdot L^{-1}$	274.00	31.88	31.88	4.00
DOC	$mg \cdot L^{-1}$	86.30	10.22	10.22	0.81
Ammonia	$mg \cdot L^{-1}$	44.11	1.61	1.61	0.50
NKjT-N	$mg \cdot L^{-1}$	66.40	3.78	3.78	0.30
Nitrite NO_2	$mg \cdot L^{-1}$	0.00	1.00	1.00	0.03
Nitrate NO_3	$mg \cdot L^{-1}$	0.00	0.79	0.79	5.84
Total nitrogen - N	$mg \cdot L^{-1}$	20.00	4.90	4.90	12.57
Total phosphorus - P	$mg \cdot L^{-1}$	11.40	1.46	1.46	0.03
Electrical conductivity (LB)	$\mu S/cm^{-1}$	1,874.00	1,602.80	1,602.80	1,005.25

The conclusions from the analysis of the results are:

(1) The long-term DOC was <1 mg/l in all cases on the long SAT showing a very good removal of organic matter.
(2) Ammonia was very low in the long SAT.
(3) There was no good denitrification in the long SAT (high nitrate content) although there was denitrification in the short term SAT.
(4) The phosphate content in the long SAT was <1 mg/L (0.02–0.04 mg/L) which is not the case for the short SAT (around 1.5–2.5 mg/L).

(5) The EC and chloride content of the long term SAT was lower than the short term SAT.
(6) Redox conditions in the short and long term SAT (D.O., Mn and Fe). The DO was always high in the long SAT sample since due to the sampling conditions there was always an entrance of external oxygen before the sampling. In the pilot plant this could be done much more carefully showing the real DO in the different samples.

It was noticed that in the first well R1 (10 days retention time) there was 1.1–1.5 mg/L DO while in R3 it was always <1 (0.6–0.8 mg/L). This affected the Mn dissolution rate (Figure 6.16). It was observed that Mn in R1 decreased from 218 mg/L to 109 mg/L (50% decrease) while in R3 due to anaerobic conditions in one year (May 07–May 08) this decreased from 419 mg/L to 336 mg/L (around 20%). Taking into account that the distances were not very high (10 days compared to 35 days) this could be due to the more aerobic conditions prevailing in R1.

The soluble iron content was (except in the October 07 sample in R3) relatively low, showing that there are not yet adequate anaerobic conditions for iron dissolution. After analysis of different bacteriological parameters like *T. coli*, Fecal Coli and *E. coli*, Enterococci, Clostridium spores, bacteriophages and Enteroviruses, it was observed that in the long term SAT (Dan 9) there are no Faecal coliforms and *E. Coli*, no Enterococci, Clostridium spores, bacteriophages and Enteroviruses. This is even true for the 35 days retention time short SAT (R3). After UF and after UV, no Faecal, *E. coli*, Enterococci, Clostridium spores, bacteriophages and Enteroviruses were found either.

In two out of three analyses there were no total coli in the long SAT and in one case it was 15 CFU/100 ml. The total bacteria varied from 340 to 2400 CFU/100 ml in the long retention time SAT.

6.4 CONCLUSION

6.4.1 Operational results from short SAT

The Reclaim pilot was designed especially to save SAT infiltration area. This was obtained by increasing the infiltration velocity from 2 m/d (obtained by conventional SAT experiments conducted at the same site with secondary effluents without pre-filtration) to 12 m/d by pre-treating the secondary effluents by UF and infiltrating them in a dug-well.

The UF filtration does not affect the soluble organic matter so DOC, N and P were not reduced after filtration. Only suspended solids were removed at a 0.01 micron level by the UF and this helped increasing the infiltration rate.

Also high infiltration rates helped introducing high oxygen concentrations affecting the redox condition. On the other hand, since the retention time in the unsaturated zone (14–16 m.) of the UF-filtered secondary effluents was shortened, the aerobic reactions less effectively removed certain chemicals. Organic matter was mostly removed in the first 2 m of soil so that a relatively good organic matter removal could be obtained in a short SAT. This was also true for ammonia removal while P is less effectively removed and in a future application there is a need for pre-treatment for P removal.

Since the main reason for soil clogging is linked to the presence of polysaccharides that are removed by the UF (Figure 6.12) the UF is clogged and needs a very effective washing. In the pilot plant the chosen membrane and the operation mode (hollow fibre, dead-end mode) did not enable an easy operation at high fluxes. In case that such a system will be operated a suitable membrane and operation mode is required.

The experiment (as mentioned above) was performed in pristine sand area and accordingly the infiltration rates were high.

It has to be seen when such a system is built in a mature infiltration field if the infiltration velocity will be as high as it was obtained in the Reclaim pilot experiment.

6.4.2 Removal capacity of short SAT

Analysis of the final results with secondary effluent treatment by UF and short SAT (35 days) compared to conventional SAT results can be summarised as follows:

- Efficient removal of microorganisms can be obtained by the short SAT similar to conventional SAT.
- Efficient removal of total DOC (from 10 mg/L to 2–3 mg/L) was obtained in the short SAT, but slightly less effective then in the conventional SAT system (reaching 1–2 mg/L DOC).
- Efficient removal of nitrogen (from around 6–7 mg/L total N to <0.5 mg/L) almost similar to conventional SAT was obtained in the short SAT.
- Relatively low phosphorous removal (from 2–4 mg/L Total P to 1.3–1.5 mg/L).

6.4.3 Outlook

Based on the last results for the conditions to prevent Manganese dissolution, in the context of a new research, it is planned to enrich the UF filtered secondary effluent with oxygen. The idea is to produce conditions for minimum bacterial activity and maximum oxygen to improve the Redox conditions. A new research project that investigates the efficiency of the dug-well infiltration of UF filtered effluents as compared to sand filtered and non-filtered effluents at almost field conditions (columns filled with undisturbed soil from the infiltration fields) started lately. The decision for the project was made based on the RECLAIM WATER research results.

REFERENCES

Aharoni A. (2007). "SAT as practiced historically in israel: principles, operation and problems" communications EU SWITCH FP6 Project (No. 18530) SAT process workshop, Tel-Aviv, Israel (Nov. 2007).

Aharoni A. and Cikurel H. (2011). "Experience with SAT systems in Israel and Future Plans", presentation in : "The Future of Urban Water: Solutions for Liveable and Resilient Cities", EU SWITCH FP6 project (No. 18530), January 24–26, 2011 UNESCO Paris, France.

Cikurel H. and Aharoni A. (2009). "New Concepts Envisioned for SAT in Israel", presentation, EU SWITCH FP6 Project (No. 18530), SAT technology workshop, Accra, Ghana, 26 February 2009.

Cikurel H. and Aharoni A. (2005). EU FP5 AQUAREC Project (EVK1-CT-2002-00130) Communications.

Gaus I., Cikurel H., Picot G., Aharoni A., Guttman Y., Azaroual M. and Kloppmann W. (2007b). Soil aquifer treatment (SAT) of secondary effluents using a combination of UF and short term SAT at the Shafdan site (Israel): first results on water quality changes. In: IWA 6th Conf. Wastewater Reclamation and Reuse for Sustainibility WRRS, Antwerpen, Belgium, October 9–12, 2007.

Guttman J. (2007). "The hydrogeology at the reclaim site at the Shafdan area-Israel – description of the hydrogeology works", Mekorot, communications.

Kloppmann W., Cikurel H., Picot G., Guttman J., Pettenati M., Aharoni A., Guerrot C., Millot R., Gaus I. and Wintgens T. (2009). "B and Li isotopes as intrinsic tracers for injection tests in aquifer storage and recovery systems", Applied Geochemistry (2009), doi: 10.1016/j.apgeochem.2009.03.006.

Israel Foreign Ministry report (2010). "Israel's chronic water problem". www.mfa.gov.il.

Israel Water Authority report (2010). "Water situation in Israel". www.water.gov.il.

Oren O., Gavrieli I., Burg A., Guttman J. and Lazar B. (2005). Manganese mobilizaton and enrichment during soil aquifer treatment (SAT) of effluents, the Dan Region Sewage Reclamation Project (Shafdan). Presentation in Hebrew for Mekorot.

Oren O., Gavrieli I., Burg A., Guttman Y. and Lazar B. (2007). Manganese mobilization and enrichment during soil aquifer treatment (SAT) of effluents, the Dan Region Sewage Reclamation Project (Shafdan), Israel. *Env. Sci. and Tech.*, **41**, 766–772.

Sharma S. K., Katukiza A. and Amy G. (2007). Effect of water quality and process parameters on the removal of effluent organic matter (EfOM) during soil aquifer treatment. *Proceedings of the ISMAR6 Conference (28 October – 2 November 2007)*, Phoenix, USA, 272–284.

Chapter 7

The aquifer storage, transfer and recovery project in Salisbury, South Australia

Declan Page, Karen Barry, Rudi Regel, Sarah Kremer, Paul Pavelic, Joanne Vanderzalm, Peter Dillon, Stephanie Rinck-Pfeiffer and Colin Pitman

7.1 INTRODUCTION

7.1.1 Historical background

The South Australian capital, Adelaide is the driest capital city in Australia with an average rainfall of 585 mm and a range of 257–882 mm per year. Monthly rainfall greater than 50 mm occurs during May to September (ABS, 2008). Traditionally, Adelaide and regional towns have relied on the River Murray and the Mount Lofty Ranges watershed for potable and non-potable water uses, with an Adelaide consumption of $\sim$180 million m^3/year. Summer evaporation rates are also high. Reduced rainfall in South Australia and in the catchment of the River Murray has led the community to place a greater value on water, increased efficiencies in water use, domestic and agricultural water restrictions, desalination and an increase in sewage and stormwater recycling.

The major sewage treatment plants in metropolitan Adelaide incorporate significant reuse schemes for municipal, agricultural and viticulture irrigation. Stormwater capture and reuse has also been widely practised in northern Adelaide incorporating water sensitive urban design principles and MAR schemes. End uses of stormwater have included municipal irrigation, ornamental lake replenishment and the supply of water for dual reticulation. For example, the Mawson Lakes dual reticulation housing scheme comprising of 11,000 residents receives up to 500,000 m^3 of stormwater and 500,000 m^3 of reclaimed water per year (Carr & Kolhagen, 2005) and is reliant on the Parafield Stormwater Harvesting Facility to freshen the reclaimed sewage prior to irrigation (Marks *et al.* 2005).

7.1.2 Motivations for recharge and use of abstracted water

The role of Managed Aquifer Recharge (MAR), and particularly Aquifer Storage and Recovery (ASR), has been growing in importance in sewage and stormwater reuse schemes in South Australia. The hydrogeology of aquifers below Adelaide is suitable for MAR and offers several environmental, technical, community perspective and financial benefits in the context of sewage and stormwater reuse including:

- large storage capacity in Tertiary 2 aquifer, not currently used due to brackish quality;
- storage of sewage during winter production for later use in summer;
- augmentation of summer sewage production circumventing the need for plant capacity upgrades;
- flexibility in terms of small surface footprints and site locations;
- circumventing the need to build above ground storages;
- reducing losses through evaporation;
- ability to remove microbiological pathogens and selected chemical compounds;
- ability to store short and intense precipitation events all year round;

- integration into existing schemes;
- public perception and provision of an additional storage or treatment step.

The electrical conductivity of the native groundwater in the Tertiary 2 aquifer ranges between 3,390 and 3,770 μS/cm (an average salinity of 2,100 mg/L TDS). The brackish groundwater quality makes the aquifer unsuitable for beneficial uses such as drinking water or irrigation without further treatment. The acceptability of ASR with recycled water increases from a regulatory perspective as the potential detrimental affects of storing recycled water are lower when the groundwater quality is poorer. By comparison the salinity of the T1 aquifer is 900 mg/L TDS and consequently ASR schemes require more regulation and oversight.

The Aquifer Storage Transfer and Recovery (ASTR) concept of turning stormwater into drinking water was borne from the knowledge and experiences gained from the Bolivar ASR reclaimed water trials (DWLBC, 2005; Dillon *et al.* 2003; Dillon & Toze, 2005) and the Parafield ASR scheme involving stormwater ASR (Marks *et al.* 2005). The ASTR concept differs from ASR in that it involves the use of separate production bores for injection and abstraction and the target aquifer interval is narrower with greater homogeneity (Figure 7.1). The separate bores introduce a horizontal component which should provide a longer and more uniform residence time and travel distance to ensure more reliable inactivation of pathogens and biodegradation of trace organics (Pavelic *et al.* 2004).

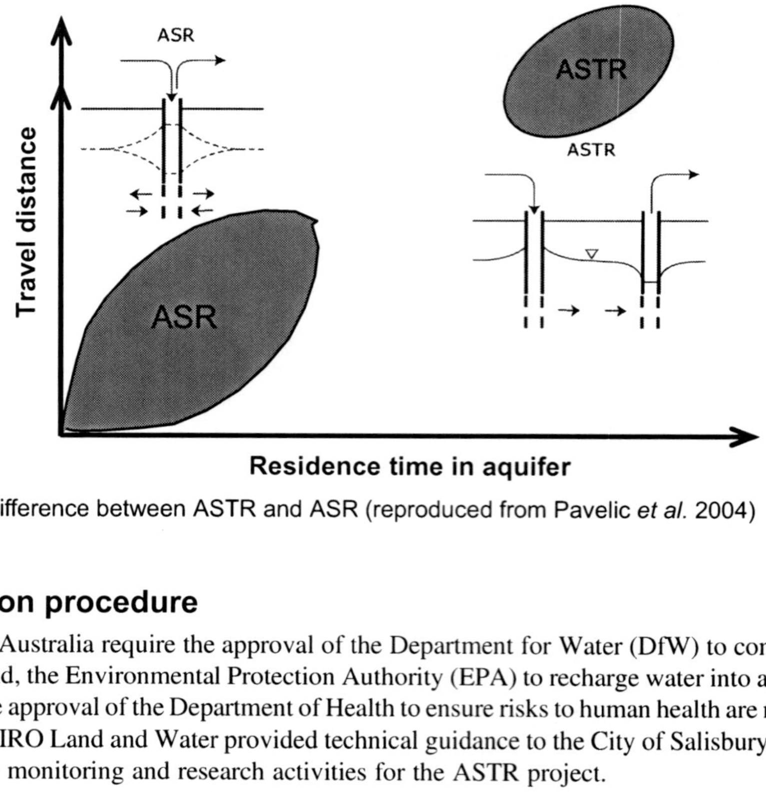

Figure 7.1 Schematic difference between ASTR and ASR (reproduced from Pavelic *et al.* 2004)

7.1.3 Authorisation procedure

MAR projects in South Australia require the approval of the Department for Water (DfW) to construct a bore and extract water from an aquifer and, the Environmental Protection Authority (EPA) to recharge water into an aquifer. The extraction of water also requires the approval of the Department of Health to ensure risks to human health are minimised. United Water International and the CSIRO Land and Water provided technical guidance to the City of Salisbury for their operations, and performed water quality monitoring and research activities for the ASTR project.

7.2 DESCRIPTION OF THE TEST SITE

7.2.1 Study area

The ASTR project trial encompasses the Parafield Stormwater Harvesting Facility located at the Parafield Airport and the ASTR bore-field located at the Parafield Gardens Reserve, within the City of Salisbury, South Australia (Figure 7.2). The stormwater harvesting facility was commissioned in 2003 and is operated by the City of Salisbury (Marks *et al.* 2005). It collects stormwater from a 1,600 ha catchment area consisting of residential and industrial uses. The harvesting facility covers an area of 11.2 hectares and has a design yield of 1.1 Mm^3/year. The scheme also incorporates two existing and operational ASR bores, known as the Parafield ASR bores. These bores are drilled to a depth of 160–180 m and have a combined injection rate of 35 L/s and are used for non-potable purposes.

The ASTR well field consists of 6 production wells spaced at 50 m, arranged in a quadrilateral configuration with the outer four wells used for injection and the inner two wells used for abstraction (Figure 7.2). The well field was designed using FEFLOW software (refer to Pavelic *et al.* 2004). Three piezometers are located between IW1 and RW1 and are used for water quality monitoring purposes. The production bores and piezometers target the Tertiary 2 aquifer at approximately

160–180 m (Table 7.1) for reasons given in Pavelic *et al.* (2004) including provision of adequate travel time from injection to recovery wells and retaining low salinity in recovered water.

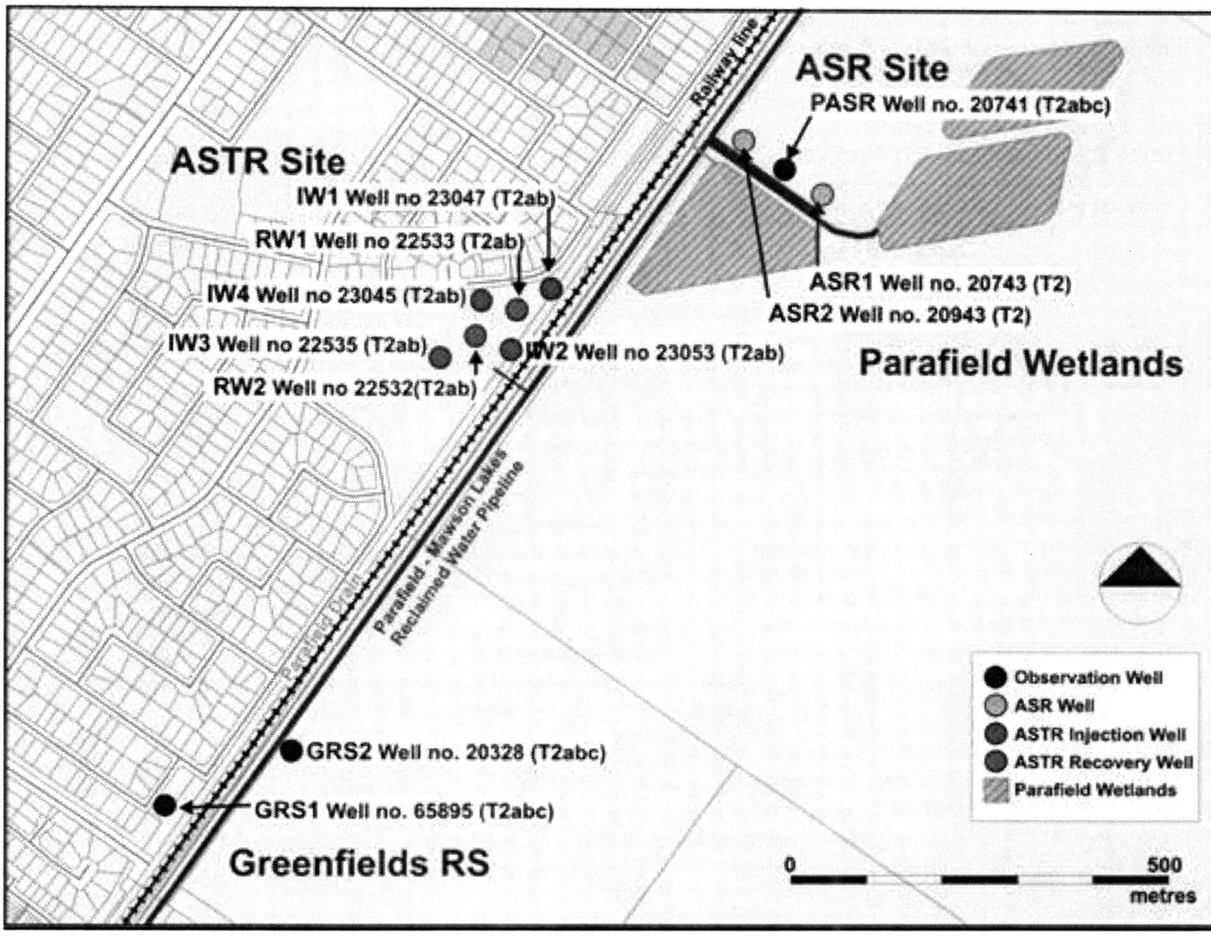

Figure 7.2 City of Salisbury water harvesting facilities, reproduced from Kremer *et al.* (2010)

Table 7.1 Summary of ASTR bores.

Unit number* 6628-	Permit number*	Construction	Max. depth drilled (m)	Open hole interval (m)	Conductivity (μS/cm)	Yield (L/s)	ASTR/Reclaim Water nomenclature	Purpose
22532	117712	12/05/2006	180	164–180	3,730	12.5	RW1	Abstraction
22533	119379	29/05/2006	183	168–183	3,770	12.5	IW3	Injection
22535	117711	22/05/2006	182	165–182	3,770	12.5	RW2	Abstraction
23045	126355	05/02/2007	184	165–184	3,450	18.0	IW4	Injection
23047	126353	15/01/2007	184	165–184	3,390	18.0	IWI	Injection
23053	126354	22/01/2007	184	165–184	3,450	18.0	IW2	Injection
24537	149448						PP1	Observation
24538	149450						PP2	Observation
24539	149449						PP3	Observation

*The South Australian Government – Department for Water, is responsible for bore construction permits, and granting MAR approvals.

7.2.2 Hydrogeology

The local T2 aquifer was characterized using geophysical and lithological logs and combined pump tests and electromagnetic (EM) flowmeter profiling. The hydrogeological units are summarised in Table 7.2. The T2 aquifer is

60 m thick, from 160 to 220 m in depth and is divided into three sub-aquifers T2 a, b and c with thicknesses of 12, 15 and 33 m, respectively (AGT, 2007). Depth-averaged measures of transmissivities (T) and storage coefficient (S) were determined from pump tests as reported in Gerges (2005) and AGT (2007) and are summarised in Table 7.3. Transmissivity ranged from 188 to 203 m^2/d in fully penetrating wells (T2 a,b,c) to an average of 46 m^2/d in the ASTR production wells (T2a,b). The small differences in transmissivity between IW1-IW4, RW1, RW2 of 38–53 m^2/d indicate low lateral heterogeneity (Kremer *et al.* 2010). Average storage coefficient approximated 2.6×10^{-4} across all 6 production bores.

Table 7.2 Lithology of ASTR aquifer.

Interval (m)	Lithology	Aquifer	Stratigraphic name
0–3	Top soil & clay	Confining bed	Recent to quaternary
3–6	Clay & calcrete		
6–27	Clay – contains Q2 aquifer	Confining bed	
27–33	Clay & sand	Q3 aquifer	
33–42	Clay (olive-grey)	Confining bed	
42–54	Clay & sand	Confining bed	
54–66	Clay	Confining bed	
66–90	Fine sand	Q4 aquifer	Carisbrook Sand
90–114	Fine sand & grey silt	T1a sub-aquifer	Hallet Cove Sandstone &
114–148	Limestone	T1b sub-aquifer	Dry Creek Sand, Upper Port Willunga Formation
148–156	Clay	Confining bed	Munno Para Clay
156–168*	Limestone well cemented interbedded with sand/silt	T2a sub aquifer – low permeability	Lower Port Willunga Formation
168–187*	Limestone moderately cemented interbedded with sand/silt	T2b sub-aquifer – moderate permeability	
187~220	Sand highly fossilferous inter-bedded with thin limestone layers	T2c sub-aquifer – high permeability	
~220–222	Clay-marl hard drilling	Confining bed	Ruawarung Member

*target aquifer

Table 7.3 Summary of transmissivity (T) and storage coefficients (S) determined from pump tests.

Bore	Aquifer	T (m^2/day)	S (-)
RW1	T2a,b	53	
RW2		47	1.9 to 2.7×10^{-4}
IW1		45	
IW2		38	
IW3		53	1.8 to 2.8×10^{-4}
IW4		41	
Average		46	2.6×10^{-4}
GRS1	T2a,b,c	188	
GRS1		203	
Average		195	

EM flow metering was performed at four production wells (IW1-IW2, RW1-RW2) between 2006 and 2007 with intervals of T2 a,b (Figure 7.3). A heterogeneous distribution was determined with two orders of variation in hydraulic conductivity K with values from less than 0.1 to greater than 3 (Kremer *et al.* 2010) which is consistent with previous findings of the stratified T2 aquifer (Pavelic *et al.* 2004). Collectively, the findings indicate that the aquifer is a 2 layered structure consisting of a relatively high layer at 164-173 m and a low transmissivity layer at 173-182 m (Kremer *et al.* 2010).

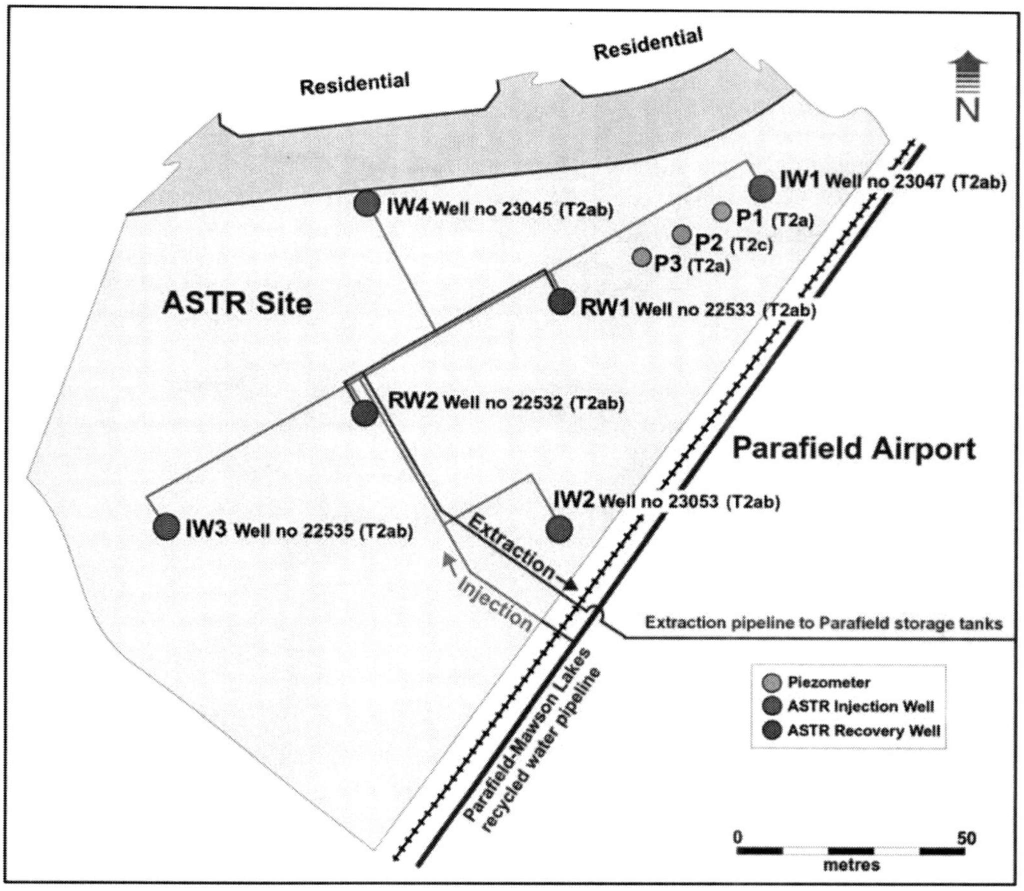

Figure 7.3 Location of injection and recovery wells and piezometers at ASTR site in the City of Salisbury, reproduced from Kremer *et al.* (2010)

7.2.3 Process design and operation

The Parafield Stormwater Harvesting Facility receives rainfall generated stormwater from the Parafield and Ayfield catchments (Figure 7.4). The stormwater is collected in the Parafield drain, diverted via a weir into the in-stream basin, from where it is pumped to a holding storage. The detention time varies and is dependent upon storage availability and rainfall intensity. From the holding storage, the stormwater gravitates into the reedbed which has a hydraulic retention time of ~10 days (Page *et al.* 2010a). Once stormwater has been treated by the reedbed, it can either be pumped (Figure 7.5)

- into the ASTR well-field;
- into the Parafield ASR bores or
- distributed to users (either for municipal irrigation or blended with reclaimed sewage at the third pipe system at Mawson Lakes).

The ASTR project is considered in terms of three phases of operation: conditioning, injection and abstraction (Figure 7.6). The conditioning phase involves pumping stormwater into the T2 aquifer to displace native brackish groundwater and create a freshwater lens below the well-field. The inner two production bores, RW1 and RW2 were initially recharged to create a freshwater bubble. Following the injection phase commenced by pumping stormwater into the outer four production bores, IW1-IW4. The abstraction phase involved pumping stored stormwater out of the inner two production bores, RW1 and RW2.

The main pump station located at the Parafield Stormwater Harvesting Facility is used to pump stormwater into the T2 aquifer via ASTR production bores at a pressure of ~450 KPa. The RW1 and RW2 production bores contain pumps deployed at ~100 m depth to abstract stored storm water during the abstraction phase.

To ensure the conditioning phase of the ASTR scheme was not delayed when there was little or no rainfall, the ASTR well-field received stormwater which had been stored during previous years in the Parafield ASR bores. This stormwater had undergone the same process of reedbed treatment, followed by injection/abstraction via Parafield ASR bores and a second process of reedbed treatment before injection into the ASTR bore-field.

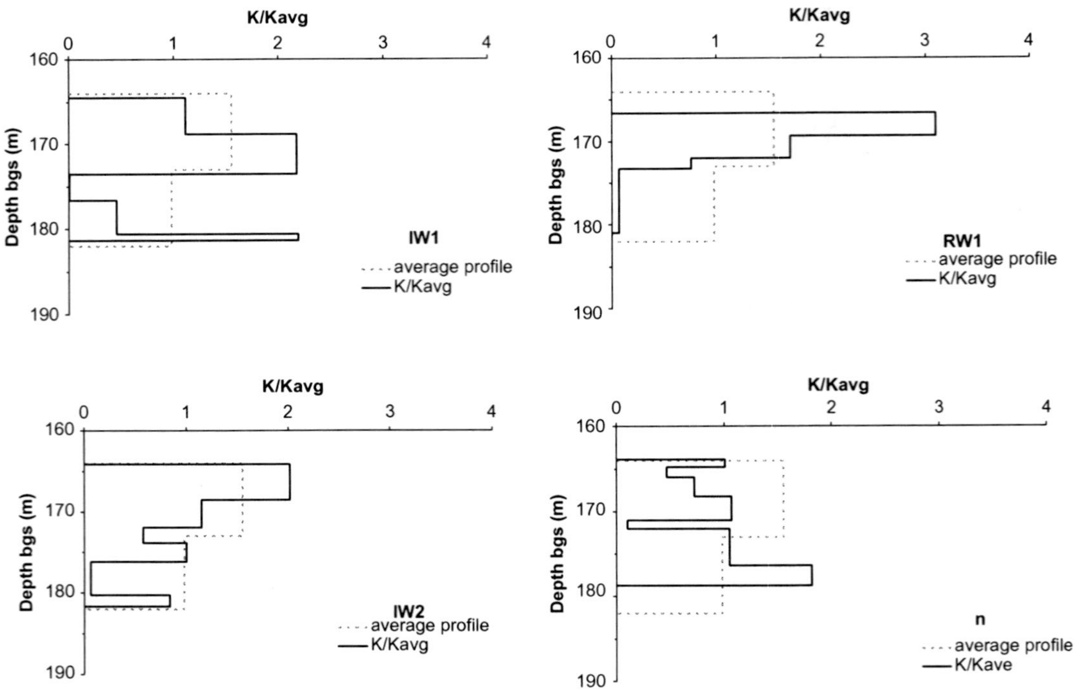

Figure 7.4 Hydraulic conductivity, K/Kavg interpreted from EM flowmeter analysis versus depth at RW1, RW2, IW1 and IW2 wells, showing comparison average 2-layered K/Kavg profile (dashed line). Reproduced with permission from Kremer *et al.* (2010)

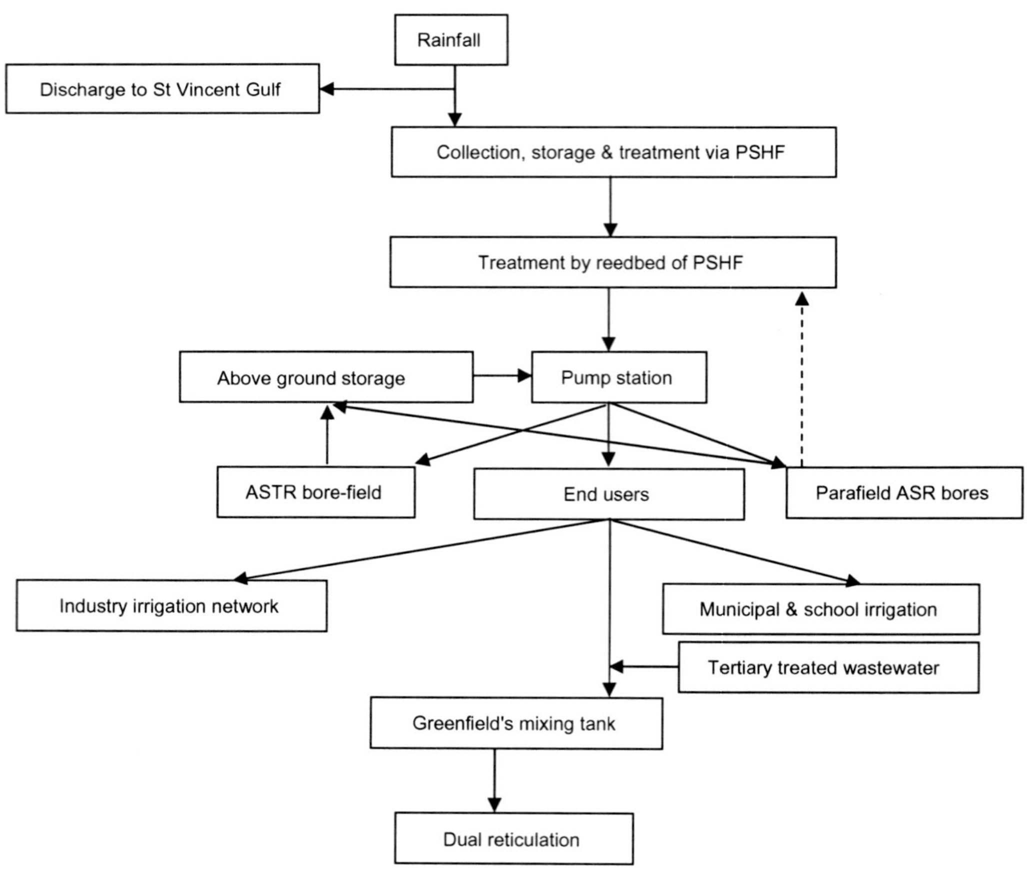

Figure 7.5 Process schematic for ASTR

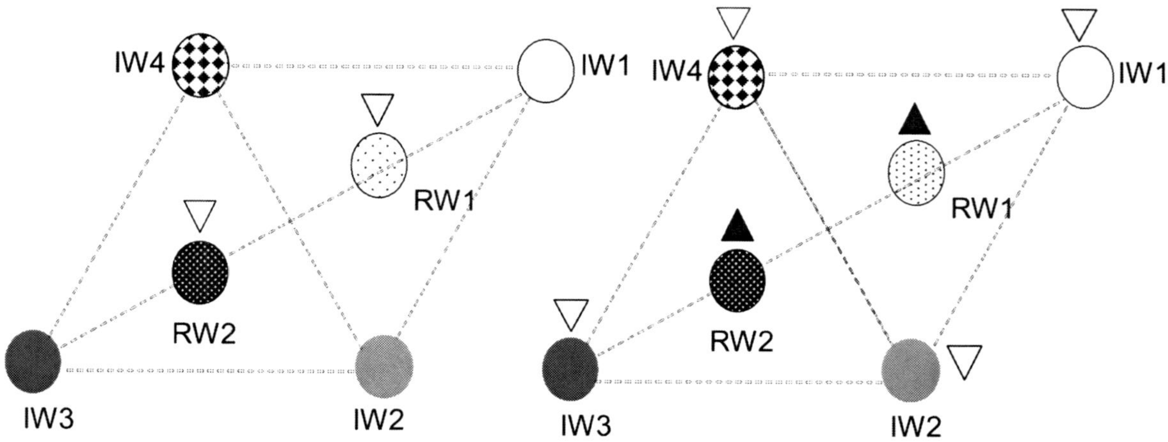

Figure 7.6 Difference in production bore injection and recovery bores for the aquifer conditioning and injection/recovery phases of the scheme

Abstraction from the Parafield ASR bores previously (2005–2008) involved pumping to the reedbed before distribution. Modifications now enable abstracted water to be pumped to concrete header tanks adjacent to the main pump station. Abstracted water from the ASTR trial will be pumped to the header tanks before distribution with the main pump station.

The main factors which have influenced the schedule of injection and abstraction at the ASTR well-field include:

- rainfall and stormwater runoff quantity;
- quantity of stormwater stored within Parafield ASR bores;
- quality (salinity) of stormwater stored within Parafield ASR bores;
- customer demand at Mawson's Lake dual-reticulation housing scheme;
- operation of Bolivar Wastewater Tertiary DAFF Reuse plant;
- quantity of stormwater stored within ASTR bores and
- quality (salinity) of stormwater stored within ASTR bores.

7.3 TECHNOLOGY PERFORMANCE AND CONTAMINANT MONITORING

The technology performance of the ASTR trial is broadly assessed according to:

(1) Hydrogeological characterisation parameters (transmissivity, modelling);
(2) Volume of stormwater captured and pumped into the T2 aquifer;
(3) Aquifer conditioning (i.e. mixing fraction);
(4) Salinity of target aquifer;
(5) Water quality of stormwater pumped from reedbed or header tanks (*Protocol 1* parameters);
(6) Water quality of abstracted water from ASTR production bores (*Protocol 1* parameters);
(7) Degradation of microbiological pathogens in reedbed and aquifer;
(8) Degradation of chemicals hazards (pesticides, herbicides) and
(9) Ecotoxicological experiments.

Performance indicator #1 is addressed in further detail in Pavelic *et al.* (2004) and Kremer *et al.* (2010). Indicators #7 and #8 are addressed in Work Packages 2 and 6, respectively and reported in Sidhu *et al.* (2010) and Page *et al.* (2010b, 2010c, 2010d), while #9 will be addressed in a follow up CSIRO technical report (Kumar *et al.* 2011).

7.3.1 Rainfall, stormwater capture and reuse

Monthly rainfall ranged from 0 for the months of October 2006, February 2007, and March 2008 to 83 mm for April 2007 and approximately 60 mm in May and August 2008 (Figure 7.7).

The stormwater flow into the PSHF was dependent upon rainfall, resulting in an average and median daily flow of 1.94 and 0.03 × 1000 m^3/day and a maximum of 36.16 × 1000 m^3/day (Figure 7.8). The daily flow exiting the reedbed was relatively constant with an average, median and maximum of 2.23, 2.03 and 6.59 × 1000 m^3/day, respectively. The reedbed volume was maintained at a constant level with an average and median daily volume of 21.25 and $21.36 \times 1000\ m^3$ (Figure 7.8). The salinity of the stormwater leaving the reedbed ranged between 133 to 372 μS/cm (Figure 7.9).

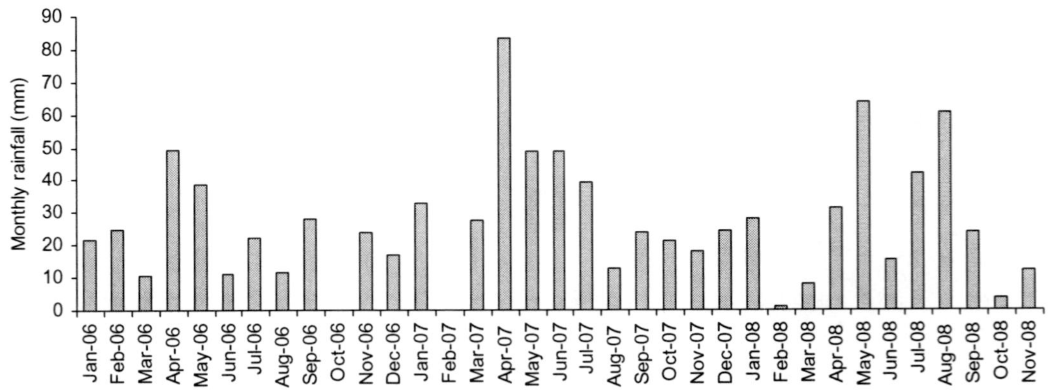

Figure 7.7 Monthly rainfall data from the Parafield Airport adjacent to PSHF. *Source*: Bureau of Meteorology

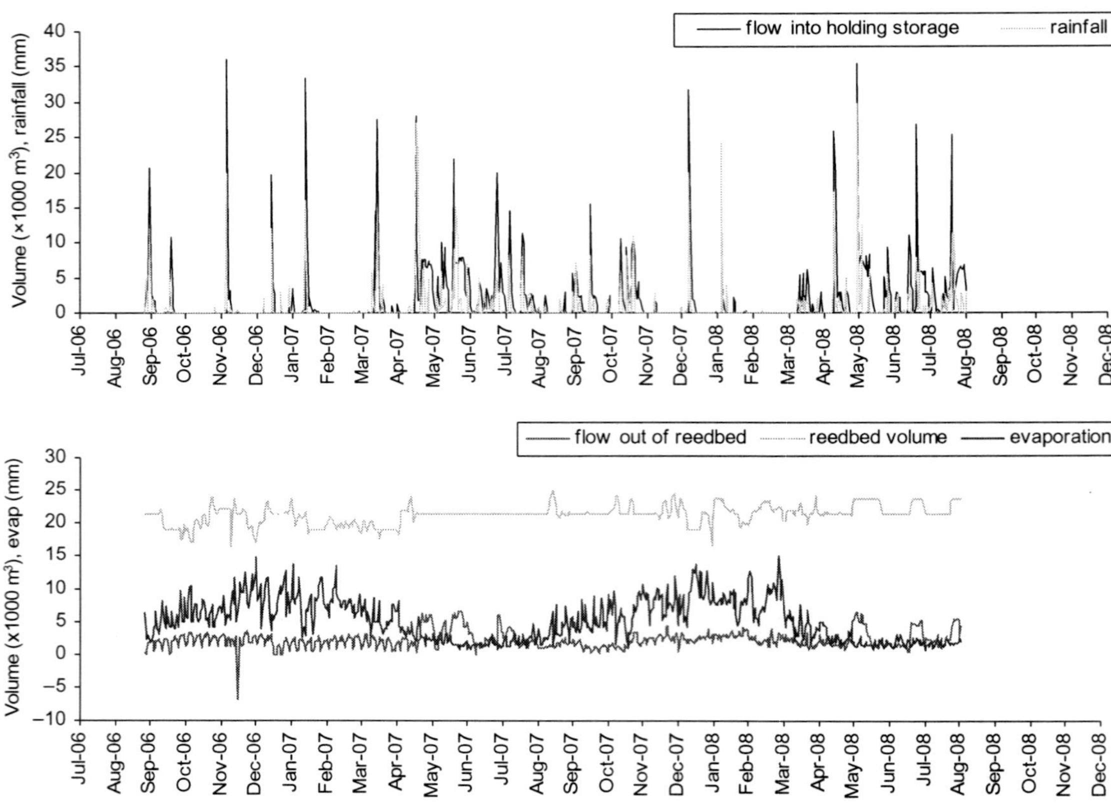

Figure 7.8 Rainfall, evaporation and flow into the PSHF holding storage and out of reedbed during the conditioning phase

7.3.2 ASTR well-field aquifer conditioning

The ASTR well field aquifer conditioning phase involved pumping stormwater into the RW1 and RW2 bores. Pumping started on 7 September 2006 and ended on the 30 of June 2008 (Figure 7.10). The average, median and maximum flow rates for RW1 were 4.8, 5.2 and 9.2 L/s, respectively and for RW2, 4.3, 4.5 and 8.3 L/s, respectively. The volume of stormwater pumped into RW1 and RW2 was 197,200 and 179,700 m^3 culminating in a total of 376,900 m^3 with a monthly injection rate of 16,400 m^3. No well development was performed during conditioning. The total volume of stormwater injected for conditioning was 377,000 m^3. Hydrogeological modelling using flow and solute transport observation data revealed that an effective volume of stormwater had been injected to flush and condition the

T2 aquifer, to ensure appropriate recovery efficiency after the first phase of injection and abstraction (Kremer *et al.* 2010). Injection ceased while the three piezometers were constructed between July and August 2008. On the 9th of September injection recommenced as part of the 1st phase of injection.

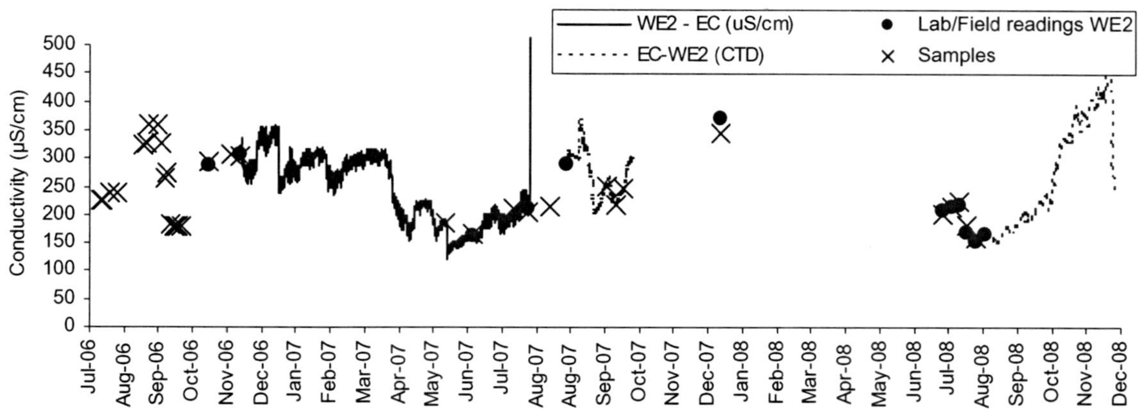

Figure 7.9 Salinity (electrical conductivity µS/cm) of stormwater exiting the PSHF reedbed during the conditioning phase. Conductivity was measured with two different instruments and in the field and via grab samples

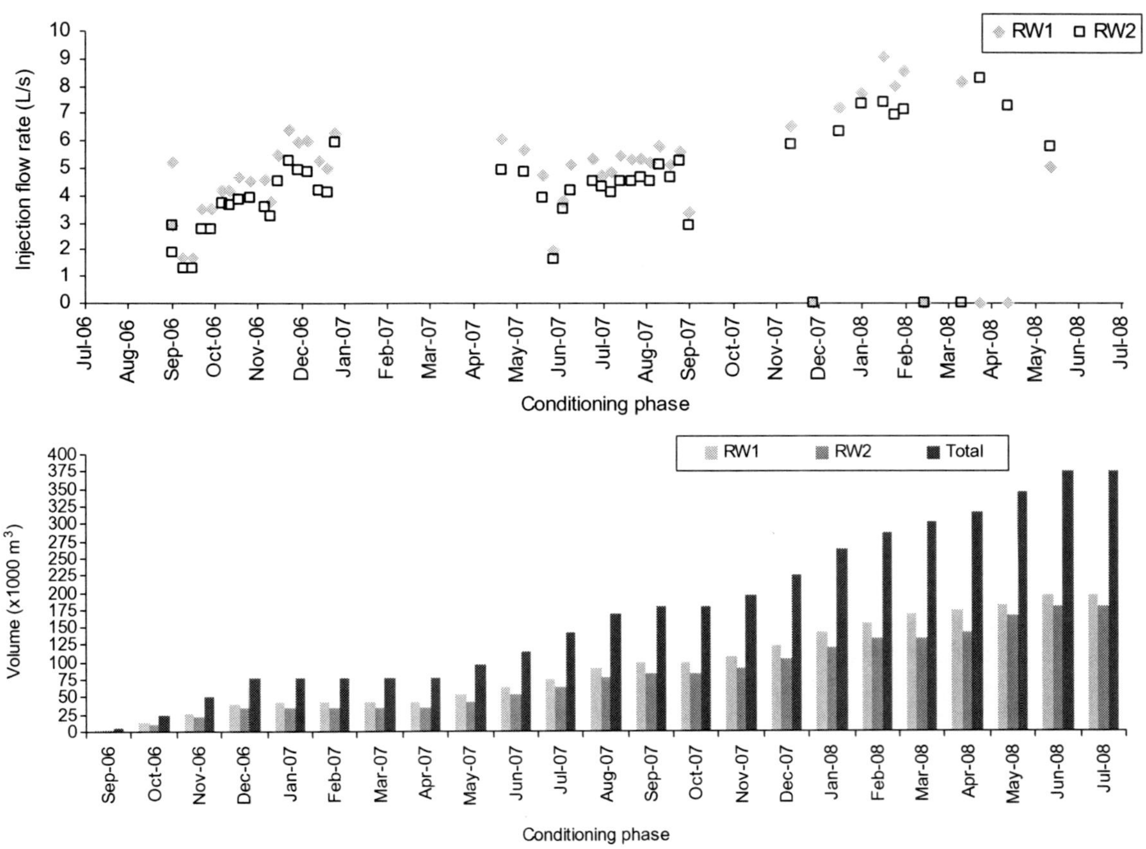

Figure 7.10 Flow rate and volume of stormwater injected into RW1 and RW2 bores during conditioning phase

Aquifer conditioning was assessed via routine down-hole profiling, *in situ* CTD sondes and water quality sampling. Down-hole profiling was undertaken between 2006 and 2008 when IW1-IW4 bores were used as observation monitoring bores. The most comprehensive dataset is for IW3 bore and monitoring demonstrated that as the volume of stormwater increased, the salinity progressively decreased, falling to 1,100 µS/cm in April 2008 (Figure 7.10). The CTD sondes deployed in IW1 and IW3 bores also demonstrated the substantial decrease in salinity and gradual decrease in aquifer temperature over time. The calculated mixing fraction through a measure of the proportion of stormwater to native groundwater was found to increase at least to 0.8 (Figure 7.11). Water quality sampling results for conductivity of the stormwater, production bores and piezometers are outlined in Protocol 1 results below (Figure 7.12).

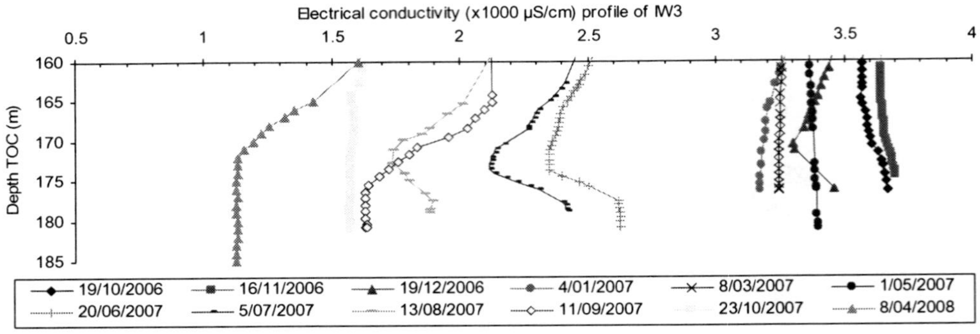

Figure 7.11 Salinity profile of IW3 bore during conditioning phase

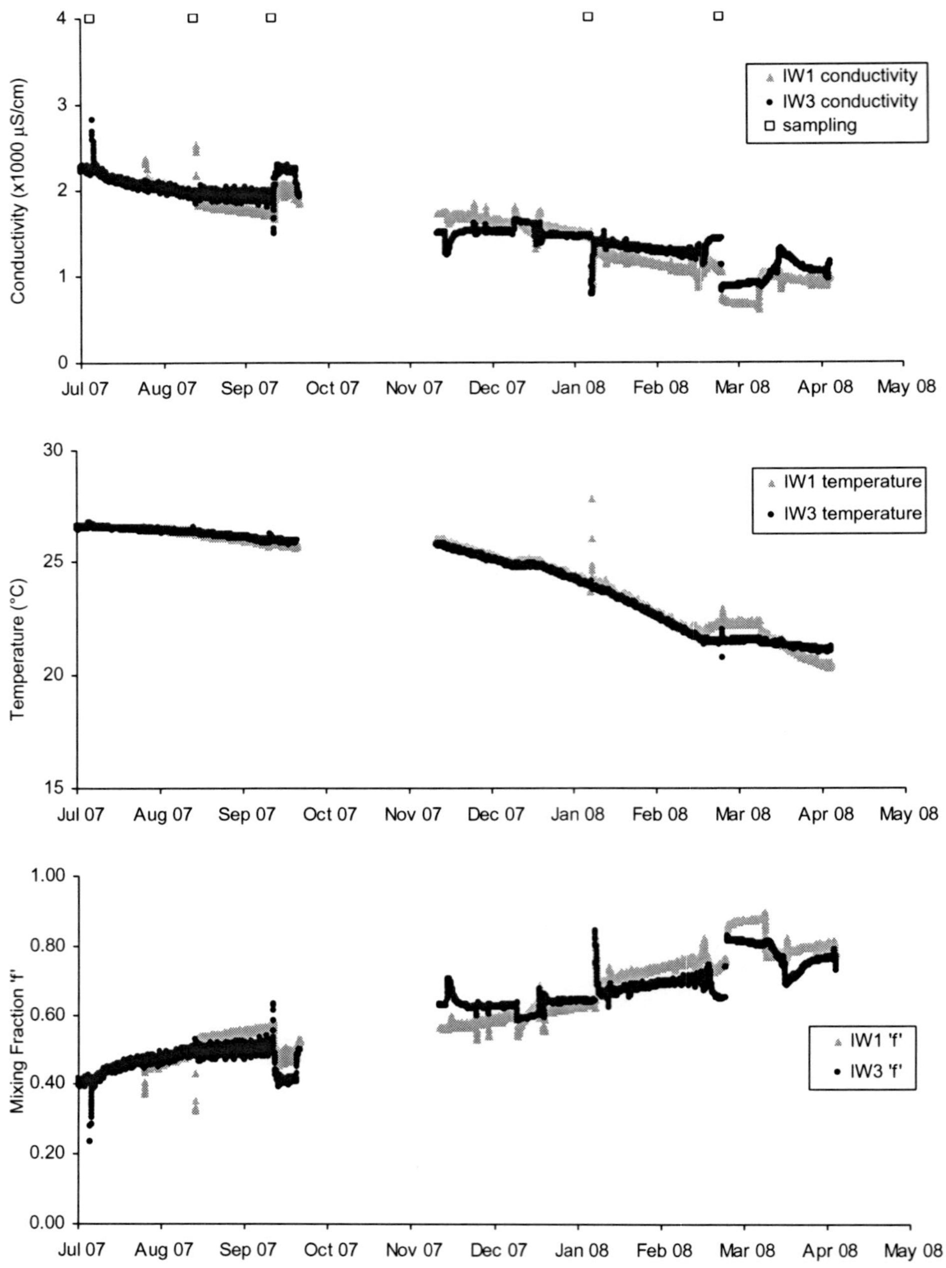

Figure 7.12 Conductivity, temperature and calculated mixing fraction during conditioning phase for bores IW1 and IW3. Conductivity and temperature data were measured hourly from CT sondes deployed at 175 m in each bore

7.3.3 ASTR well field, first injection phase

Stormwater injection commenced into IW1, IW2, IW3 and IW4 bores on 9 of September 2008 and the total volume of stormwater pumped into IW bores equated to 15,600 m^3 at the end of October 2008 (Figure 7.13). A small volume of stormwater was abstracted from RW1 and RW2 in September equating to ~1,000 m^3. During this period stormwater pumped tended to be higher in electrical conductivity ranging between 180–460 μS/cm which indicates that it had been abstracted from the Parafield ASR bores and pumped to the concrete header tanks (Figure 7.9). Previously, the electrical conductivity of any abstracted Parafield ASR water would have been diluted by passage through the reedbed before being pumped to the ASTR well-field. The header tank is more efficient hydraulically but circumvents any potential dilution through the reedbed.

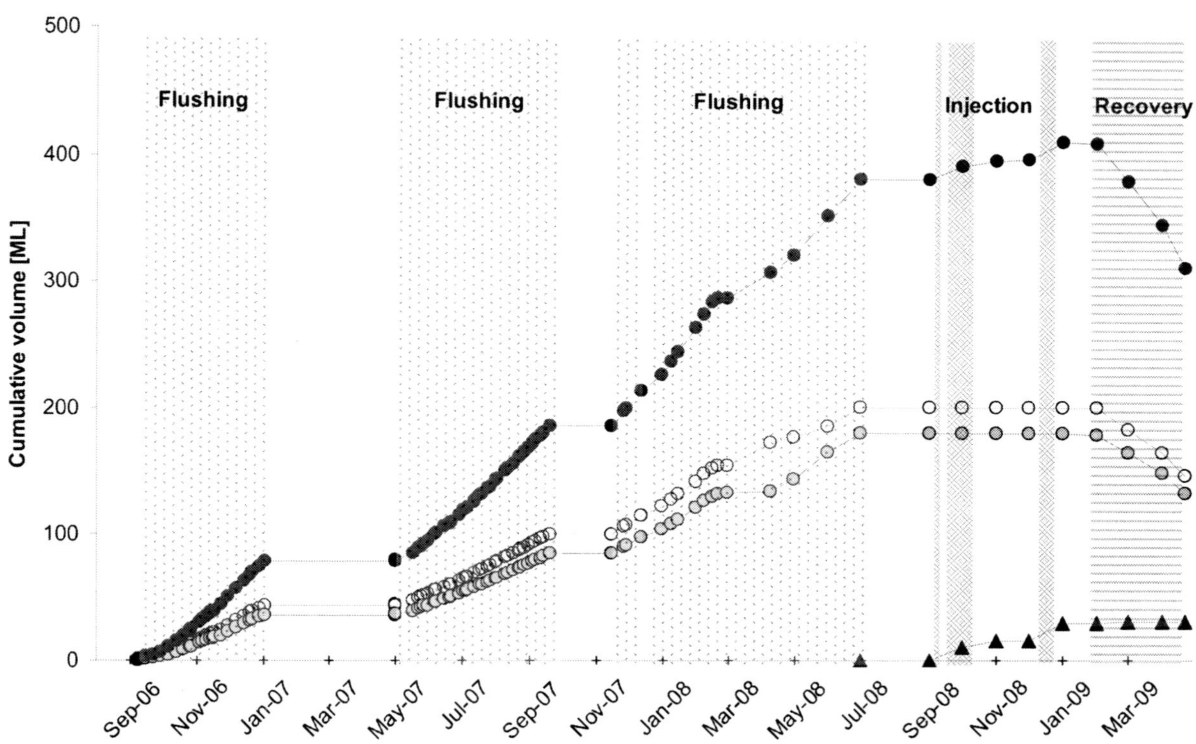

Figure 7.13 Volume of stormwater injected into IW1-IW4 bores and abstracted from RW1 and RW2 bores during the injection phase

7.3.4 Water quality assessment based on protocol 1

The water quality monitoring program is linked to the HACCP risk assessment and management plan (Swierc *et al.* 2005; Page *et al.* 2008b; 2009; 2010c). The broad objectives of the monitoring have been to identify hazards present in the stormwater and assess the treatment performance of the surface and groundwater components. Water quality monitoring commenced in June 2006 and lasted until April 2009. The 2006 and 2007 sampling campaigns focussed on the quality of stormwater and surface waters including the performance of the Parafield Stormwater Harvesting Facility (Page *et al.* 2008b). In 2008, groundwater sampling intensified to gain an understanding of the change in water quality and whether or not abstracted water will meet drinking water standards (Page *et al.* 2010c).

Sampling locations and frequency

Water samples were collected throughout the Parafield Stormwater Harvesting Facility and ASTR scheme according to critical control points and quality control points identified in the preliminary risk assessment (Swierc *et al.* 2005) including: the in-stream basin, holding storage, reedbed and injection and recovery bores (Table 7.1). WE2 represents the exit of the reedbed and relates to the quality of stormwater to be injected into the aquifer. Samples have also been collected from the production bores.

The frequency of sampling was related to targeted investigations and/or was event-driven. Events included for example: native groundwater before stormwater injection, rainfall and stormwater runoff, targeted experiments such as pathogen decay diffusion chamber studies and prior to injection or abstraction. Table 7.4 shows a summary of the sampling dates for different locations within the ASTR scheme.

Table 7.4 Sampling dates for Protocol 1 water quality monitoring program during conditioning and injection phases.

	In stream		Reedbed		Production at well head		Production bores from aquifer						Piezometers inlet		
	Inlet IS1 A1	Exit IS2	Inlet WE1 A5	Exit WE2 A7b	RW1	RW2	IS1 A1	IS2	WE1 A5	WE2 A7b	RW1	IW4 A9	IS1 A1	IS2	WE1 A5
Jun-06							27	22			26				
Jul-06	31		17	17, 18, 25, 31											
Aug-06	22, 25, 28			22, 25, 28											
Sep-06	4, 7, 13, 15		11, 18, 20, 22, 25	4, 7, 11, 13, 15, 18, 20, 22, 25											
Oct-06				19		19									
Nov-06				7, 16		30									
Dec-06						19, 28									
Jan-07						2									
Feb-07									12	8		10			
Apr-07		30				30					20				
May-07				14											
Jun-07			7	7											
Jul-07			18	12, 25		5			5		5				
Aug-07				13					13		13				
Sep-07			25	25	11				11	11	11	11			
Oct-07			2, 9, 16	2, 9, 16											
Jan-08				8					8		8				
Feb-08									27		27				
Jul-08			21, 28	21, 28											
Aug-08			4, 11, 18	4, 11, 18			11	11		11	11	11			
Sep-08							2		1, 23				23	23	23
Oct-08													16	16	16
Nov-08															11
Feb 09*							2	2							

*Represents the first sample during abstraction phase.

Water quality of stormwater exiting reedbed

The stormwater exiting the reedbed originates from either recent stormwater runoff or stored stormwater abstracted from the Parafield ASR bores and pumped back through the reedbed. The outflow from the reedbed forms the injectant for the ASTR scheme. The quality of the stormwater and the performance of the Parafield Stormwater Harvesting Facility during 2006 were investigated by Page *et al.* (2008b; 2009; 2010a). The 2007–2009 data is summarised in Table 7.5 and generally the reedbed outflow meets Australian Drinking Water Guidelines (2004). The parameters which do not meet drinking water guidelines include: turbidity, *E. coli*, Enterococci, thermotolerant coliforms, iron (total) and colour.

Salinity of production bores and piezometers during aquifer conditioning and injection

Table 7.6 summarises the change in salinity of the production bores and piezometers during the conditioning and injection phases. The samples were collected by pumping out 3 bore volumes before taking a sample. Ambient salinity decreased from 3,600 μS/cm to ≤500 μS/cm in the central RW1 and RW2 bores and was reduced to 242 μS/cm in IW2 bore on 23/09/08 at the end of aquifer conditioning. The P3 piezometer, closest to IW1, was also reduced to ~350 μS/cm at the beginning of the first phase of injection.

Table 7.5 Selected water quality monitoring results for WE2 (reedbed outlet).

	DWG	Mean < guideline	2006				2007				2008			
			Min	Max	Mean	SD	Min	Max	Mean	SD	Min	Max	Mean	SD
Conductivity (μS/cm)	–	√	179	361	262	64	167	251	214	27	161	345	222	65
pH (pH units)	6.5–8.5	√	6.9	7.7	7.3	0.2	6.8	7.5	7.0	0.2	6.6	6.9	6.7	0.1
SS (mg/L)	–	√	<1	6	2.7	1.3	<1	14	5.8	4.5	2	8	4.8	2.3
Turbidity (NTU)	5	×	1.3	6.2	3.4	1.5	1.4	13	4.3	3.8	2	10	5.4	3.0
True Colour (HU)	15	×	12	86	47	26								
Alkalinity as $CaCO_3$	–	√	25	118	76	31	43	64	54	7.2	44	109	68	29
Bicarbonate	–	√	30	144	93	38	53	78	66	8.7	54	132	83	35
Sulphate	250	√	7.8	20.4	11.0	3.0	7.8	10.8	8.8	1.1	9	15.3	11.1	2.44
Chloride	250	√	12	46	26	9.0	18	38	29	5.6	17	41	27	8.2
Fluoride	1.5	√	0.13	0.29	0.19	0.06	0.1	0.14	0.12	0.01	0.12	0.22	0.15	0.04
Calcium	–	√	15.2	36.1	24.2	8.7	14.7	23.6	18.6	2.9	15.1	32.9	20.5	6.4
Magnesium	–	√	3.2	10.6	5.6	2.1	2.9	4.5	3.9	0.51	2.6	6.7	3.7	1.5
Potassium	–	√	2.4	7.8	4.4	1.5	2.5	4	3.0	0.48	2.6	3.5	3.2	0.3
Sodium	180	√	12.2	27.9	18.5	4.5	13.9	22	17.9	2.7	12.8	29.5	18.3	5.9
Thermotol. coliforms (cfu/100 mL)	0	×	4	140	37	39	2	77	24	24	4	220	49	84
Total Nitrogen	–	√	0.26	0.81	0.46	0.17	0.22	0.56	0.35	0.11	0.28	0.86	0.38	0.23
Total Phosphorus	–	√	0.021	0.095	0.055	0.024	0.024	0.097	0.050	0.026	0.023	0.042	0.030	0.007
DOC	–	√	3.8	11.7	6.9	2.5	3.4	8.2	5.7	2.0	3.7	4.7	4.2	0.4
Arsenic (t)	0.007	√	<0.001	0.006	0.001	0.002	<0.001	0.001	<0.001	<0.001	0.0005	0.0020	0.0009	0.0006
Copper (t)	1	√	<0.030	<0.030			<0.001	0.002	0.002	0.001	<0.001	0.003	0.002	0.001
Iron (t)	0.3	×	0.247	1.200	0.633	0.310	0.258	0.743	0.476	0.165	0.203	0.484	0.366	0.112
Lead (t)	0.01	√	<0.0005	0.0036	0.0007	0.0008	<0.0005	0.0012	0.0005	<0.0005	<0.0005	0.0012	0.0006	0.0004
Manganese (t)	0.1	√	0.003	0.213	0.056	0.053	0.008	0.056	0.032	0.016	0.0131	0.0823	0.0281	0.0269
Zinc (t)	3	√	<0.01	0.069	0.028	0.017	0.004	0.020	0.012	0.006	0.007	0.031	0.016	0.009

DWG = drinking water guideline, taken from Australian Drinking Water Guidelines

Table 7.6 Electrical conductivity (µS/cm) of reedbed and ASTR production bores during conditioning and injection phases.

Date	Location								
	RW1	**RW2**	**IW1**	**IW2**	**IW3**	**IW4**	**P1**	**P2***	**P3**
22/06/06		3,630							
26/06/06					3,620				
27/06/06	3,650								
12/02/07			3,420	3,480		3,280			
20/04/07					3,340				
05/07/07			2,630		2,690				
13/08/07			2,240		1,900				
11/09/07			1,790	1,810	1,500	1,190			
09/01/08			1,100		1,080				
27/02/08			874		961				
12/08/08	462	470		815	709	598		1,790	
01/09/08			845						
02/09/08		415							
23/09/08				242			439	2,610	333
16/10/08							417	1,930	373
11/11/08									369

*P2 conductivity data is high due to problems with well construction and data is not considered representative of the aquifer at this location.

Water quality of production bores during aquifer conditioning

Table 7.7 summarises the change in water quality of production bores during conditioning. Ambient groundwater prior to injection is represented by bores IW1, IW2 and IW4 in early February 2007, whereas the second and third data sets represent progressive volumes of stormwater being injected. After bore construction in February 2007, water quality of the native groundwater did not meet Australian Drinking Water Guidelines for turbidity, sulphate, chloride, sodium, iron or total dissolved solids. However, the stored stormwater met drinking standards in January 2008 and August 2008, with the only exception being total iron.

Water quality of abstracted water

Water was abstracted from the T2 aquifer on the 2nd of February 2009 which represented the first sample of stormwater injection, storage and abstraction for the ASTR project (Table 7.8). The samples were taken from the inner production bores RW1 and RW2. Water was pumped out for approximately 45 minutes before samples were taken. The abstracted water meets the Australian Drinking Water Guidelines (2004) with the exception of colour, turbidity, iron and manganese.

The abstracted water is currently being used for municipal irrigation in the City of Salisbury and for blending with reclaimed water of the Bolivar wastewater treatment plant to achieve a TDS of <900 mg/L for use in the Mawson Lakes dual-reticulation housing scheme.

Post-treatment options will be considered in the future and may encompass aeration and rapid sand filtration, followed by UV disinfection. UV disinfection has also been proposed by Page *et al.* (2009; 2010b) as a means of providing an additional barrier to remove viruses, which may persist for up to 180 days (Sidhu *et al.* 2010; Page *et al.* 2010; NRMMC, EPHC and NHMRC (2009)).

7.4 OPERATIONAL FEEDBACK

The ASTR project has proven valuable for the local project partners as it has enabled all partners to increase their hydrogeological understanding of such systems and to consider the major risks with stormwater harvesting. Turning stormwater into drinking water is possible and further advances will depend upon the cost of post-treatment, the severity and frequency of drought and the cost of alternative water supplies. The ASTR project has contributed to the momentum of reuse activities with reclaimed water and stormwater in South Australia. Currently a sequiel project is underway; the Managed Aquifer Recharge and Stormwater Use Options project.

Table 7.7 Selected water quality monitoring results for ASTR production bores during different stages of aquifer conditioning: IW well flushing phase 2007, IW well flushing phase 2008 and RW injection phase 2008–2009.

	2007 – Flushing phase – IW wells						2008 – Flushing phase – IW wells						2008–2009 – Injection phase – RW wells					
	n	Min	Max	Median	Mean	SD	n	Min	Max	Median	Mean	SD	n	Min	Max	Median	Mean	SD
Physical characteristics																		
EC (μS/cm)	12	1156	3270	2426	2326	729	8	591	1121	891	890	181	3	413	484	464	454	37
Temperature (°C)	12	19.8	28.7	22.2	23.7	3.2	8	18.0	23.5	21.5	21.1	2.1	3	13.1	13.6	13.3	13.3	0.3
pH (−)	12	6.93	7.85	7.46	7.39	0.32	8	7.43	8.22	7.86	7.81	0.29	3	7	7.98	7.84	7.61	0.53
DO (mg/L)	11	<0.01	0.77	0.01	0.17	0.29	8	<0.01	0.53	0.02	0.08	0.24	3	0.07	2.86	0.09	1.01	1.61
Eh (mV SHE)	11	−56	96	−32	−7.4	50	8	−236	6	−5.5	−33	82	3	−55	29	−15	−14	42
Suspended Solids (mg/L)	10	1	5	2	3	2	4	1	2	1	1.3	0.5	3	9	48	25	27	20
Turbidity (NTU)	12	3.9	22	10	12	6.7	8	1.5	3.1	2.1	2.2	0.6	3	5.2	32	15	17	14
Microbiological (cfu/100 mL)																		
E. coli	11	0	0	0	0	0	8	0	0	0	0		3	0	1	0	0	1
Thermotolerant coliforms	11	0	0	0	0	0	8	0	0	0	0		3	0	1	0	0	1
Nutrients (mg/L unless otherwise indicated)																		
Nitrate+Nitrite as N	12	<0.005	0.006	0.003	0.003	0.001	8		<0.005				3	<0.0005	0.21	0.0003	0.07	0.1
Ammonia as N	12	0.024	0.048	0.03	0.032	0.007	8	0.018	0.029	0.023	0.023	0.004	3	3.4	4.6	4.4	4.1	0.6
Kjeldahl Nitrogen (t)	10	0.05	0.14	0.10	0.09	0.03	8	0.09	0.33	0.14	0.16	0.09	3	3.5	5.6	5.3	4.8	1.1
Nitrogen (t)	11	0.06	0.14	0.10	0.09	0.03	8	0.10	0.34	0.14	0.17	0.09	3	3.7	5.6	5.3	4.8	1.0
Phosphorus (t)	11	0.006	0.019	0.013	0.012	0.004	8	0.008	0.014	0.0105	0.011	0.002	3	0.19	0.32	0.30	0.27	0.07
Organic Carbon (d)	12	0.6	1.8	1.5	1.3	0.4	8	1.6	3.1	2.3	2.2	0.5	3	7.3	12.5	9.5	9.8	2.6
Metals and metalloids (mg/L)																		
Arsenic (t)	12	0.004	0.008	0.006	0.0059	0.00	8	0.004	0.006	0.005	0.005	0.001	3	<0.001	0.008	0.006	0.005	0.004
Iron (t)	12	0.45	1.13	0.72	0.77	0.25	8	0.33	0.39	0.35	0.35	0.02	3	5.4	11.0	8.7	8.4	2.8
Manganese (t)	12	0.002	0.007	0.004	0.004	0.00	8	0.002	0.003	0.002	0.002	0.001	3	0.39	0.73	0.44	0.52	0.19
Zinc (t)	7	0.003	0.035	0.011	0.015	0.01	7	0.007	0.023	0.012	0.014	0.01	3	0.004	0.032	0.006	0.014	0.02

Table 7.8 Recovered water quality (RW wells) in 2009.

	DWG	Mean < guideline	Average RW 2 Feb 09	RW1 5 Mar 09
Physical characteristics				
Conductivity (μS/cm)	–		470	
pH (pH units)	6.5–8.5	√	7.3	7.6
Suspended Solids (mg/L)	–		9	
Turbidity (NTU)	5	×	6.7*	1.0
True Colour (HU)	15	×	28	26
Major Ions (mg/L)				
Alkalinity as $CaCO_3$	–		216	137
Bicarbonate	–		264	167
Bromide	–		0.11	0.12
Sulphate	250	√	<1.5	11.4
Chloride	250	√	19.5	32
Fluoride	1.5	√	0.20	0.21
Calcium	–		69.7	46.9
Magnesium	–		6.6	6.3
Potassium	–		5.2	4.6
Sodium	180	√	13.8	22.9
Microbiological				
E.coli (cfu/100 mL)	0	√	0	0
Thermotolerant coliforms (cfu/100 mL)	0	√	0	0
Nutrients (mg/L)				
Total Nitrogen	–		5.705*	0.3
Total Phosphorus	–		0.2325*	0.035
Dissolved Organic Carbon	–		6.5*	4.2
Metals and metalloids (mg/L)				
Antimony (t)	0.003	√	0.0013	
Arsenic (t)	0.007	√	0.0015	0.002
Copper (t)	1	√	<0.001	
Iron (t)	0.3	×	5.67‡ *	0.379
Lead (t)	0.01	√	<0.0005	
Manganese (t)	0.1	√	0.331*	0.0565
Zinc (t)	3	√	0.01	
Radiological (mBq/L)				
Gross alpha activity	500			
Gross beta activity (K-40 corrected)	500	√	<10	

*initial purging of recovery well – not considered representative of recovered water.
‡adjusted for analytical error (reported soluble > total).

7.5 CONCLUSION

Between 2006 and 2009, stormwater was captured from an urban catchment, treated in two detention basins and reedbed, injected into and abstracted from the Tertiary 2 aquifer in Salisbury, South Australia. Some of the key learning's of the project include:

- Transmissivity can be variable within the T2 aquifer – depending on drilling depth (46 m^2/day);
- Stormwater injection rates in the T2 a,b sub-aquifers approximate 5–6 L/s;

- Conductivity and temperature are useful tracers in assessing aquifer conditioning and the status of stormwater injection;
- Hydrodynamic modelling using FEFLOW can be used to design ASTR well field schemes;
- Hydrodynamic modelling using FEFLOW can reliably simulate observed changes in conductivity and be used to develop operational injection and recovery strategies with end-use constraints (i.e. recovered water <300 mg/L TDS);
- The water exiting the Parafield Stormwater Harvesting Facility can be of a high quality and with the exception of iron, colour and selected microbiological parameters, meet Australian Drinking Water Guidelines (2004) (Page *et al.* 2009, 2010c);
- Trace organics (mainly simazine and atrazine) were sporadically found using conventional water quality techniques in the Parafield Stormwater Harvesting Facility (Page *et al.* 2009);
- Advanced sampling techniques which are still in a development phase, detected very small concentrations (nanogram levels) of trace organics in the Parafield Stormwater Harvesting Facility (Page *et al.* 2010a);
- HACCP risk assessment plans can be used to structure water quality monitoring plans and be used proactively to assess performance and risk (Page *et al.* 2010b).
- Quantitative risk assessments (QRA) advance HACCP plans in that a final measure of risk can be determined for both chemical and microbiological hazards (Page *et al.* 2009, 2010c).

Based on 2006 water quality data, the water exiting the Parafield Stormwater Harvesting Facility reedbed was found to meet drinking water standards except for high concentrations of iron, colour and microbiological indicators. Water sampled from a Parafield Stormwater Harvesting Facility ASR bore in 2007 and subjected to advanced treatment (granular activated carbon and microfiltration) was found to meet Australian Drinking Water Guidelines (2004). The water was bottled and labelled *'Recharge Water'*. This activity indicated that ASTR and ASR have the potential use within the treatment train for producing high quality water. In 2009, stored stormwater was abstracted from the ASTR production bores, RW1 and RW2 for the first time and preliminary results indicate that with the exception of colour, turbidity, iron and manganese, Australian Drinking Water Guidelines (2004) can be met.

The role of ASTR in producing constant reliable supplies of drinking water will require further assessment, such as ongoing water quality monitoring and risk management including quantitative risk assessments and several cycles of injection and recovery. However, this raises the question that if stormwater can be reliably and safely turned into drinking water with minimal risk, how could the water be supplied for drinking or alternatively, how could it be supplied to industries which need high quality water (e.g. food industry)?

To consider the above, a number of questions are raised that will be addressed in future research,

(1) If recovered bore water requires additional advanced treatment to meet various use options, what are the appropriate technologies and their cost and performance (e.g. granular activated carbon, microfiltration, disinfection/chlorination, sand filtration)?
(2) What are the impacts of reticulating recovered bore water +/− additional advanced treatment in potable water supply mains?
(3) How technically, financially and environmentally feasible are various urban stormwater use options that fully account for the safety of those uses?
(4) What is the level of public acceptance of various options for use of stormwater via aquifers?

REFERENCES

ABS (2008). Australian Bureau of Statistics. www.abs.gov.au.

AGT (2007). Parafield ASTR Wells Completion Report and Aquifer Test Results. Australian Groundwater Technologies Report # 2077/22. Technical Report.

Carr J. and Kohlhagen L. (2005). Recycled water in an urban environment – managing the risks. American Water Works Association Annual Conference, 11-15 June 2006, San Antonio.

Dillon P. and Toze S. (2005). Water Quality Improvements During Aquifer Storage and Recovery, Vol. 2. Compilation of information from ten sites. AWWARF Project 2618, Final Report, pp. 203.

Dillon P., Martin R., Rinck-Pfeiffer S., Paveic P., Barry K., Vanderzalm J., Toze S., Hanna J., Skjemstad J., Nicholson B. and Le Gal La Salle C. (2003). Aquifer storage and recovery with reclaimed water at Bolivar, South Australia. Proc. AWA SA Regional Conference, Glenelg, 6 Aug 2003, and Proc. 2nd Australian Water Recycling Symposium, Brisbane 1-2 Sept 2003.

DWLBC (2004). Aquifer Storage Capacities of the Adelaide region. Todd Hodgkin. Report DWLBC 2004/47 Parts 1-3. Department for Water Land and Biodiversity Conservation.

DWLBC (2005). Bolivar Reclaimed Water Aquifer Storage and Recovery Research Project – Final Report. R. Martin and P. Dillon (eds), Report DWLBC 2005/45. Department for Water Land and Biodiversity Conservation.

Gerges N.Z. (2005). Greenfield Railway Station ASTR Project Briefing on Aquifer Tests. Technical report.

Kremer S., Miotlinski K., Barry K., Dillon P. and Levett K. (2010). Revised Flow and Solute Transport Modelling for ASTR Operations, South Australia, CSIRO: Water for a Healthy Country National Research Flagship, http://www.clw.csiro.au/publications/waterforahealthycountry/2010/wfhc-ASTR-flow-solute-modelling.pdf retrieved 29 Apr 2011.

Kumar A., Doan H., Woods M., Gonzago D., Barry K., Page D. and Dillon P. (2011). Preliminary Ecotoxicological Assessment of Water Before and After Treatment in a Reed-bed and Aquifer at Salisbury Stormwater ASTR project, CSIRO Water for a Healthy Country Report.

Marks R., Chapman F., Lane S. and Purdie M. (2005). Parafield urban stormwater harvesting facility. *Water, Journal of the Australian Water Association*. **32**(5), 42–45.

NRMMC, EPHC and NHMRC (2009). Australian Guidelines for Water Recycling, Managing Health and Environmental Risks, Volume 2C–Managed Aquifer Recharge. Natural Resource Management Ministerial Council, Environment Protection and Heritage Council National Health and Medical Research Council, Jul 2009, 237p. http://www.ephc.gov.au/taxonomy/term/39

Page D., Barry K., Pavelic P., Dillon P. and Chassagne A. (2008a). Preliminary Quantitative Risk Assessment for the Salisbury Stormwater ASTR Project. Technical Report. CSIRO Land and Water, Adelaide, Australia.

Page D., Vanderzalm J., Barry K., Levett K. and Dillon P. (2008b). Revised Quantitative Risk Assessment for the Salisbury Stormwater ASTR project.

Page D., Vanderzalm J., Barry K., Levett K., Kremer S., Ayuso-Gabella M. N., Dillon P., Toze S., Sidhu J., Shackleton M., Purdie M. and Regel R. (2009). Operational Residual Risk Assessment for the Salisbury Stormwater ASTR project, CSIRO Water for a Healthy Country Report.

Page D., Dillon P., Mueller J. and Bartkow M. (2010a). Quantification of herbicide removals in a constructed wetland using composite water quality monitoring and passive samplers, *Chemosphere*, **81**, 394–399.

Page D. W., Dillon P., Toze S. and Sidhu J.P.S. (2010b). Characterising aquifer treatment for pathogens in managed aquifer recharge, *Water Science and Technology*, **62**(9), 2009–2015.

Page D., Vanderzalm J., Barry K., Levett K., Sidhu J., Toze S., Kremer S. and Dillon P. (2010c) Managed Aquifer Recharge (MAR) staged risk assessment framework applied to the Parafield Aquifer Storage Transfer Recovery system to assess human and environmental risks, *Journal of Environmental Quality*, **39**(6), 2029–2039.

Page D., Dillon P., Toze S., Bixio D., Genthe B., Jiménez Cisneros B. E. and Wintgens T. (2010d). Valuing the subsurface pathogen treatment barrier in water recycling via aquifers for drinking supplies, *Water Research*, **44**, 1841–1852.

Pavelic P., Dillon P. and Robinson N. (2004). Groundwater Modelling to Assist Well-field Design and Operation for the ASTR Trial at Salisbury, South Australia. CSIRO Land and Water Technical Report 27/04.

Sidhu J. P., Toze S., Hodgers L., Shackelton M., Barry K., Page D. and Dillon P. (2010) Pathogen inactivation during passage of stormwater through a constructed reedbed and aquifer transfer, storage and recovery, *Water Science and Technology*, **62**(5), 1190–1197.

Swierc J., Page D., van Leeuwen J. and Dillon P. (2005). Preliminary Hazard Analysis and Critical Control Points Plan (HACCP) – Salisbury Stormwater to Drinking Water Aquifer Storage Transfer and Recovery (ASTR) Project. CSIRO Land and Water Technical Report 20/05.

Vanderzalm J. L., Page D. W., Barry K. E. and Dillon P. J. (2010) A comparison of the geochemical response to different managed aquifer recharge operations for injection of urban storm water in a carbonate aquifer, *Applied Geochemistry*, **25**, 1350–1360.

Chapter 8

Managed aquifer recharge for potable reuse in Atlantis, South Africa

Gideon Tredoux, Bettina Genthe, Maronel Steyn and Jason Germanis

8.1 INTRODUCTION

8.1.1 Historical background

The town of Atlantis with a population exceeding 60 000 is located 50 km north of the centre of the City of Cape Town and presently forms part of the metropolitan area. Previously, this area resorted under the authority of the Cape Divisional Council and in 1976, the new Atlantis development was declared a National Growth Point, under the government's "decentralisation initiative". As it was planned to become an industrial growth point various incentives were introduced for attracting industries to the area. As the west coast is semi-arid, the importation of surface water from the Berg River system, located some 70 km distance away, was planned as the permanent water supply source. However, in 1972 the Department of Water Affairs and Forestry started developing local groundwater as an interim measure allowing for the postponement of the pipeline construction. In the early phases of development all water supply was provided by the perennial springs at Silwerstroom. These produced an average flow of about 1×10^6 m^3 per year. A weir was constructed to capture the springs' flow in a small dam, from which it was pumped to the growing town. Subsequently two wellfields were developed and over the next 25 years the local groundwater proved itself as a sustainable water resource providing the bulk of the water demand. Only in 2000 limited augmentation of the water supply by surface water was introduced via an alternative route.

More than three decades ago, at the time the Atlantis Water Resource Management Scheme was developed the Water Act No 54 of 1956 was in force in South Africa. That act did not have any sections dealing with the authorisation and regulation of artificial groundwater recharge nor with the recycling of water. In 1979 the CSIR concluded pilot scale artificial recharge studies using treated wastewater in the Cape Flats (Tredoux *et al.*, 1980). Hence the Regional Office of the Department of Water Affairs and Forestry requested the local authority to set up a monitoring programme for the Atlantis water supply scheme in collaboration with the CSIR.

The National Water Act (Act No 36 of 1998) replaced the Water Act of 1956 and this act specifically controlled artificial groundwater recharge through a permitting system. However, it is a gradual process, and whereas new systems will only be permitted once pilot studies are completed and strict requirements are met, the permitting of existing operations has not been completed. Nevertheless, in 2002 an audit was recommended covering all legal compliances, licences, and authorisations required to legalise the Atlantis water system (Tredoux and Cavé, 2002).

Artificial recharge has played a key role in the augmentation of the groundwater supplies at Atlantis. Initially, indirect recycling of wastewater and urban stormwater was considered an economic means of wastewater disposal as it eliminated the need for a marine outfall for such discharges. Soon water conservation became a key feature of the scheme and various combinations of urban stormwater and treated wastewater from sources in the town have been infiltrated into the aquifer over the years to maximise the amount of available groundwater. At the same time, several refinements have been made to the artificial recharge system to ensure that any potential deterioration of water quality does not jeopardise the scheme. In this way the Atlantis Water Resource Management Scheme has pioneered the application of artificial groundwater recharge as a water management tool for bulk water supply in southern Africa.

8.1.2 Motivations for recharge and use of abstracted water

In the South African coastal areas marine discharge of wastewater used to be the preferred disposal option. Strict marine water quality criteria promulgated by the Department of Water Affairs and Forestry in the 1980s and enforced by means of a permitting process required detailed monitoring programmes for marine water quality and filter feeders near such marine outfalls. The associated costs gradually became prohibitive and outfall designs had to be refined to ensure thorough blending and dilution with seawater to achieve the microbiological criteria within a set distance from the outfall. Therefore, the initial motivation for managed aquifer recharge at Atlantis was to obviate a marine outfall for wastewater disposal. Recycling of the water was, however, a secondary objective. Similarly the case was made to add the urban stormwater runoff to the wastewater recharge system instead of disposal into the ocean.

Once the aquifer was studied more intensively, it emerged that the natural yield of the aquifer is too little to sustain the water supply to Atlantis. This need for water recycling introduced a new perspective to the water management system at Atlantis and the management of water quality throughout the system became a key issue from that time. Domestic and industrial wastewater is treated separately and only the domestic wastewater is recycled. Similarly the peak flow and base flow in the stormwater system are channelled to different recharge basins to maintain good quality water in selected areas of the aquifer. Summarizing the Managed Aquifer Recharge (MAR) scheme in Atlantis serves several purposes:

- Augment local groundwater supplies by recycling (about 25–30% of current water demand)
- Prevent seawater intrusion into the aquifer
- Sensible stormwater and treated wastewater disposal, obviating need for costly marine discharge

8.1.3 Authorisation procedure

More than three decades ago, at the time the Atlantis Water Resource Management Scheme was developed, the Water Act No 54 of 1956 was still in force in South Africa. The act did not have any sections dealing with the authorisation and regulation of artificial groundwater recharge or the recycling of water, and all the Regional Offices of the Department of Water Affairs required was a general monitoring programme for the water supply scheme. The standard requirements for a potable water supply (SANS-241:2006) were applied to the water distributed in the town of Atlantis.

In South Africa, the National Water Act (Act No 36 of 1998) replaced the Water Act of 1956 and this act specifically controls artificial groundwater recharge through a permitting system. However, it is a gradual process, and whereas new systems will only be permitted once pilot studies are completed and strict requirements are met, the permitting of existing operations has not been completed.

8.2 DESCRIPTION OF THE TEST SITE

8.2.1 Study area

Atlantis is located along the semi-arid to arid west coast of South Africa (Figure 8.1) and has a Mediterranean climate, with most of the 450 mm mean annual rainfall received from April to September. As a result of the sandy surface over most of the area, recharge percentages of 15 to 30% of the annual rainfall are generally experienced, the higher recharge occurring in the unvegetated dune area.

The aquifer covers an area of about 130 km^2, stretching inland from the Atlantic Ocean to the town itself in the east. It pinches out against the Malmesbury Group shales and Cape granite outcrops to the north and east. The thin aquifer slopes steeply in a southwesterly direction from a maximum elevation of about 160 m in the north down to sea level in the west. A small part of the aquifer extends below sea level in the Witzand and Silwerstroom areas. The granite outcrops of Dassenberg, Kanonkop and Mamre-Darling near Mamre form the highest points in the area, at heights of 210 to 410 m above sea level (van der Merwe, 1983).

8.2.2 Hydrogeology

The primary coastal aquifer system in the Atlantis area is formed of unconsolidated Cenozoic sediments of tertiary to recent age, overlying Malmesbury Group bedrock consisting of greywacke and phyllitic shale (Van der Merwe, 1983). Granite plutons have intruded the bedrock. The Cenozoic successions in the area consist of quartz sands belonging to the Sandveld Group. These comprise a lower unit of shallow marine origin, the Varswater Formation, and an upper, primarily aeolian unit, the Bredasdorp Formation (Rogers, 1980). The Bredasdorp Formation is subdivided into the Springfontein, Mamre and Witzand Members. The base of the Springfontein Member is a peaty sand bed, while the rest of the member consists of relatively well-sorted quartz sand, free of shelly material, ranging from fine- to coarse-grained. The Witzand Member overlies the central part of the area and consists of calcareous quartz sand, shell fragments and discontinuous calcrete layers (Figure 8.2 and 8.3). The total sand cover reaches a thickness of 60 m in the central area, with an average thickness of 25 m (Van der Merwe, 1983).

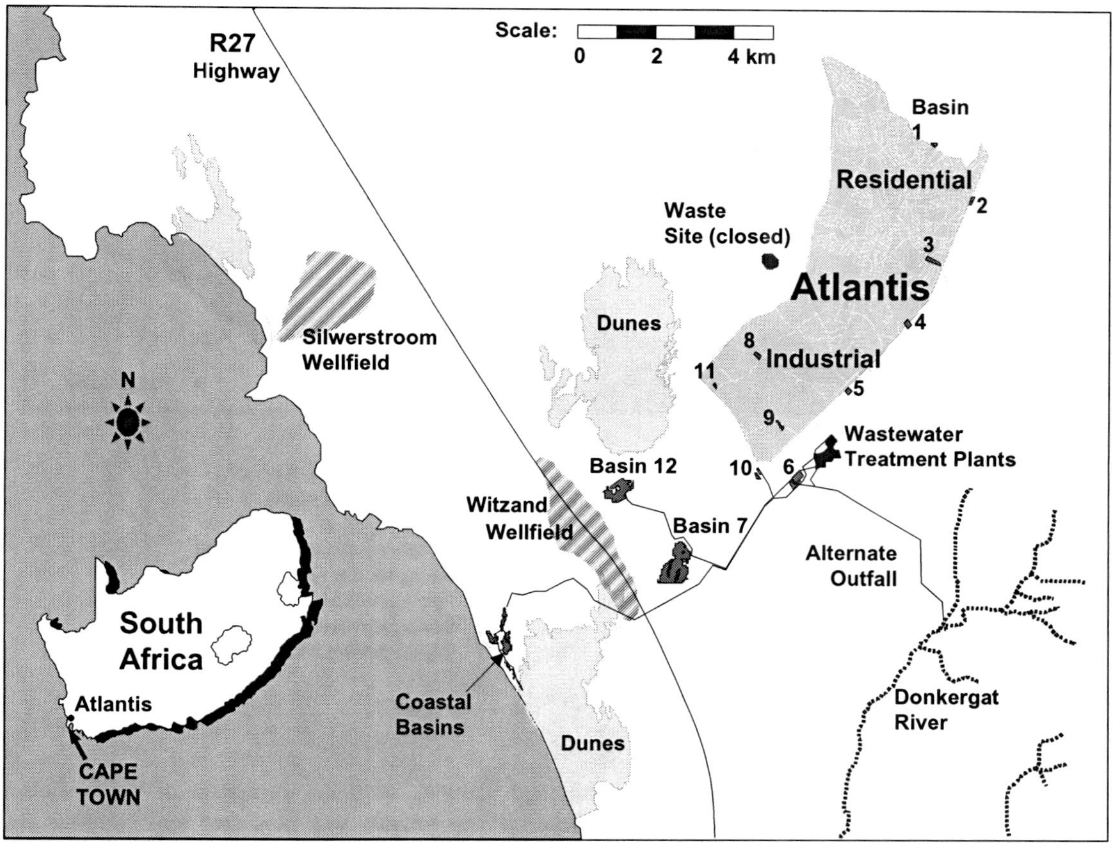

Figure 8.1 Location and layout of Atlantis water supply system

The bedrock also contains groundwater, but the weathered upper zone of the shale forms an impervious clay layer preventing any exchange of groundwater. Sandy-peat layers of varying thickness and calcrete lenses are often interbedded in the aquifer or interfinger laterally with clean sands, producing local heterogeneities in aquifer properties.

Groundwater flows westwards to southwestwards and discharges along the coast in areas where the aquifer dips below sea level. The groundwater table has a relatively steep gradient (approximately 1:58) towards the coast. The saturated thickness varies considerably, but seldom exceeds 35 m. Groundwater is abstracted in two wellfields, Silwerstroom in the north and Witzand in the south (Figure 8.1). Due to topographic constraints, artificial recharge is only practised near the Witzand wellfield in the south. The overlying Witzand Member is a calcareous quartz sand succession containing shell fragments and cemented calcrete horizons.

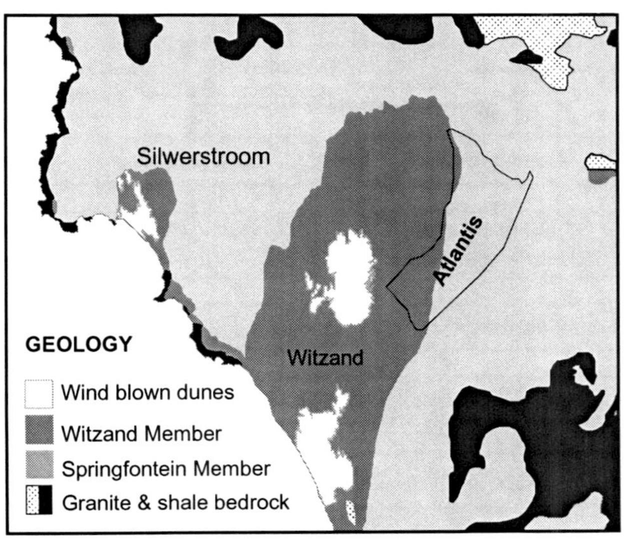

Figure 8.2 Geology of the Atlantis area

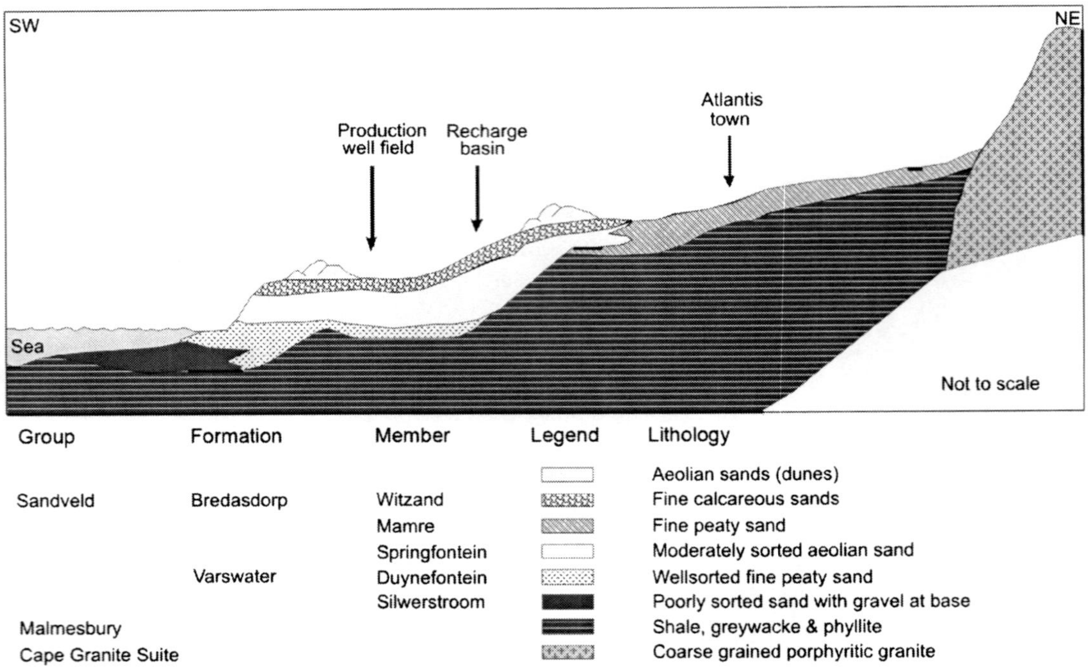

Figure 8.3 Geological cross section through the Atlantis aquifer

The Atlantis area is practically devoid of surface drainage features, with the exception of the Buffels River at Silwerstroom. The Donkergat and Sout rivers flow to the south of the Atlantis area in winter, while surface drainage to the north and east of Atlantis contributes to the catchment areas of the Modder and Diep rivers respectively. All these rivers are non-perennial, drying up in summer.

Perennial springs feeding the Buffels River near the coast at Silwerstroom have been used for water supply since 1976. There is also a spring at Mamre and a minor spring at Groot Springfontein.

8.2.3 Process design and operation

The various components of the Atlantis Water Resources Management Scheme are shown schematically in Figure 8.4. The natural characteristics of the aquifer material affect the groundwater quality, for example, imparting high hardness to the water, significant dissolved organic carbon and, in certain parts, measurable dissolved iron, or high salinity to the water. The hardness is reduced using ion exchange softening part of the flow. In the process iron is also partly removed but the organic carbon remains in solution.

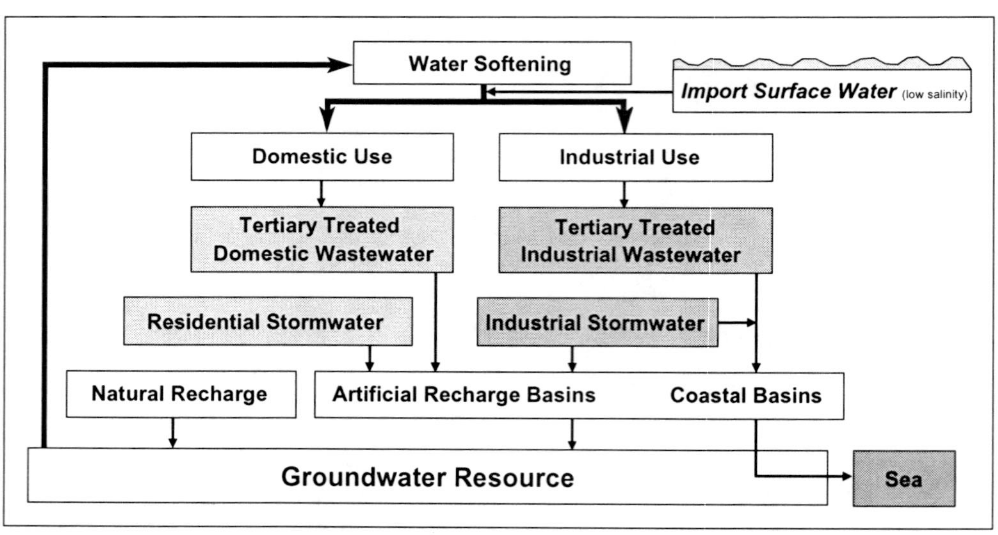

Figure 8.4 Components of the Atlantis Water Resource Management Scheme

The large volumes of stormwater runoff that would be generated after urbanisation and the associated hardening of surfaces was seen as a valuable water source for augmenting water supplies and it prompted the construction of an appropriate stormwater collection system. This consists of twelve detention & retention basins and the necessary interconnecting pipelines with peak flow reduction features (Liebenberg & Stander, 1976). The stormwater system at Atlantis was redesigned to allow flexibility for controlling water flows of differing salinity and to collect the best quality water for infiltration into the aquifer. Low salinity flows are channelled into two large spreading basins, basins #7 and #12, for artificial recharge up-gradient of the Witzand wellfield. Higher salinity baseflow is diverted to the coastal basins or to the Donkergat River in the south (Figure 8.1). Discharges during storm events can reach up to 72 000 m^3/d at Atlantis, while summer baseflow, averages 2 160 m^3/d (Wright, 1994). The baseflow is mostly groundwater entering the stormwater pipelines in areas where these are below the water table.

Initially all wastewater was treated in a single wastewater treatment plant and all the treated effluent used for artificial recharge. In 1986 this practice was discontinued due to water quality considerations and a separate treatment plant was constructed for domestic wastewater treatment. This came on line in the mid 1990s. The domestic wastewater undergoes full secondary treatment with nitrification-denitrification steps (anaerobic-anoxic-aerobic). The effluent from the secondary settling tanks is polished in a series of maturation ponds (Figure 8.5). The maturation pond effluent is blended with the urban stormwater runoff before discharge into the main recharge basins #7 and #12. The more saline treated industrial wastewater is discharged into the coastal recharge basins and seeps into the ocean through the subsurface.

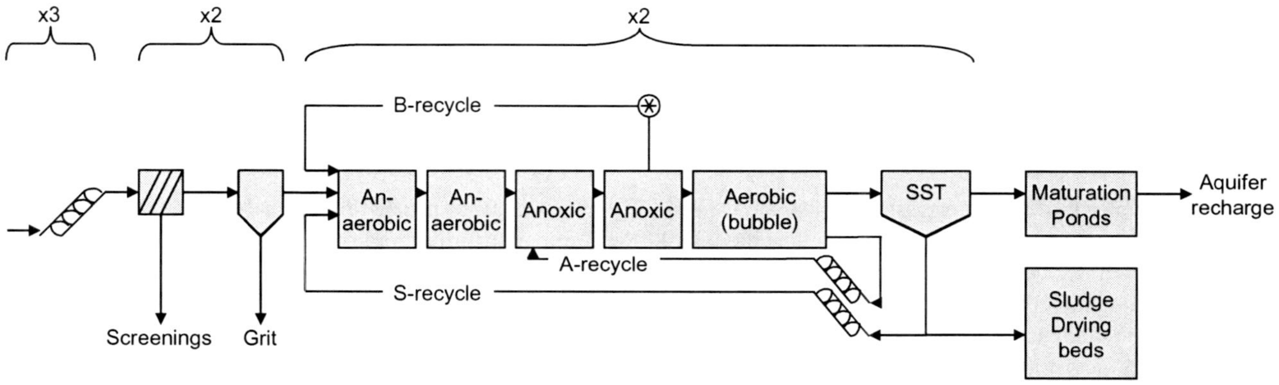

Figure 8.5 Schematic diagram of domestic wastewater treatment plant at Atlantis

The characteristics of the recharge facilities consisting of various large basins are shown in Table 8.1. Basin #7 is the largest and is mostly wet due to the larger volume of water reaching this basin, the shallow groundwater table, and the low infiltration rate. Basin #12 is mostly dry as it receives predominantly peak flow stormwater, has a deeper water table, and a higher infiltration rate. The three coastal basins operate in series with the primary basin always full and some overflow in one secondary basin. It is estimated that on average approximately 7 500 m^3/d of stormwater and wastewater is recharged up-gradient of the well field (Basins #7 and #12, Figure 8.6) augmenting the water supply by more than 2.7 Mm^3/a. Some 4 000 m^3/d higher salinity industrial wastewater is treated and discharged into the coastal basins down gradient of the well field close to the ocean.

Table 8.1 Recharge facility characteristics (average values).

Facility	Area (ha)*	Unsaturated zone (m)	Infiltration rate (m/day)**
Basin 7	28.3	1.5	0.01
Basin 12	16.8	4.5	0.16
Coastal Basins	12.5	10.5	0.11

* Total basin area when full; Basin 12 mostly dry.
** Rate based on total basin area.

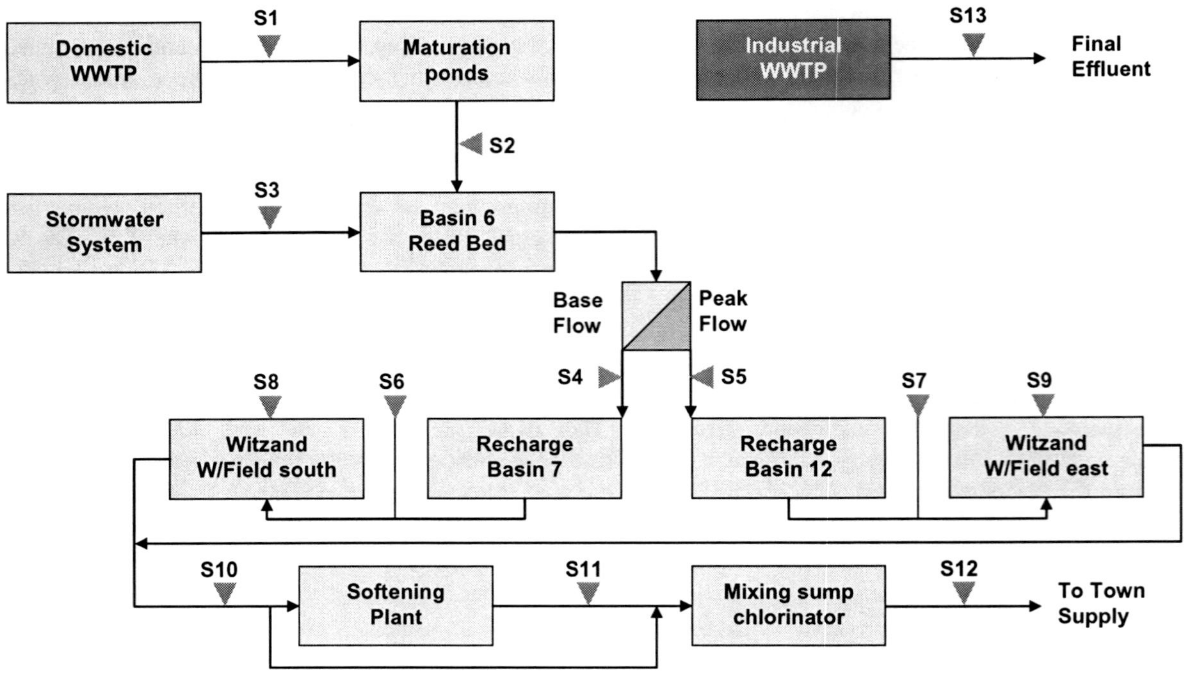

Figure 8.6 Schematic layout of the Atlantis system showing the monitoring points

8.3 TECHNOLOGY PERFORMANCE AND CONTAMINANT MONITORING

The Atlantis water resource management scheme has been fully operational for nearly three decades and during that time several upgrades and modifications were implemented. In 2006 a number of further improvements and modifications were identified. However, due to the participation of the CSIR in the Reclaim Water Project and the selection of the scheme as a study site for the project it was decided not to implement any modifications for the duration of the project and only to carry out the routine maintenance and monitoring.

The CSIR joined the Reclaim Water Project in January 2007 (i.e. month 16 of the Reclaim Water project which started in October 2005) which is during summer in the southern hemisphere. For the case study at Atlantis the key aspect was the evaluation of the performance of the system from a water quality perspective, that is, the removal of nutrients, heavy metals, endocrine disruptors and other organic micro-components such as pharmaceuticals, but also bacteria, viruses and other microorganisms. This data was needed for determining the efficiency of removal of the various pollutants and the health risks related to the remaining contaminants. Hazard mitigation is envisaged for ensuring the longer term viability of the system. A mass balance-based sustainability analysis for the Atlantis artificial groundwater recharge system was attempted.

For determining the efficiency of removal of the contaminants listed above the following groups of samples were taken at the indicated points, that is, the input, in the aquifer, and product at the indicated sampling points shown schematically in Figure 8.6.

(i) Sampling points 1 to 5: The feed water sources are: secondary treated domestic wastewater, before and after the maturation ponds, the final stormwater, as well as the blend after passing through the reed bed that is used for recharging the aquifer, both at peak flow and base flow.

(ii) Sampling points 6 to 10: The groundwater blend was sampled down gradient of each of the recharge Basins #7 and #12, approximately halfway to the production well field. Groundwater was sampled at two production boreholes influenced by the respective recharge operations at the two basins. Finally, the groundwater blend from all production boreholes was sampled immediately before the softening process.

(iii) Sampling points 11 to 12: The product water after softening was sampled in order to determine the effect of softening on the general water quality. The last sample represented the final chlorinated water before pumping to the town for determining the effect of the chlorination on the overall water quality.

(iv) Sampling point 13: The treated industrial wastewater which has a higher salinity and is discharged into the coastal recharge basins, it therefore does not form part of the recycling system. Water quality analyses from industrial wastewater samples provides an indication of the possibility to make better use of this source instead of being discharged into basins in the coastal dunes and seeping into the ocean.

Sampling at Atlantis followed the Reclaim Water standardised Protocol 1 and Protocol 2 (cf. chapter 2). Sampling according to Protocol 1 was completed four times while Protocol 2 sampling was completed twice (Table 8.2). The testing programme also included characterisation of natural organic matter and effluent organic matter. This was done together with the partners at TUB and IHE UNESCO due to a lack of suitable local facilities and expertise. Sampling for pharmaceuticals and contrast media took place in June 2008 in collaboration with the project partners of BfG. A second sampling using the SPE method was completed in October 2008 to confirm the results of the first survey.

Table 8.2 Sampling programme completed at Atlantis.

Sampling	Season/Basin	Dates
Protocol 1	Autumn	May 2007
	Winter	August 2007
	Spring	October 2007
	Summer	February 2008
Protocol 2	Winter	August 2007
	Summer	March 2008
Pharmaceuticals and contrast media	Winter Spring	August 2008 October 2008
Soil sampling	Basin #12	July 2008
	Basin #7	September 2008

The soil and unsaturated zone sampling in the two recharge basins, #7 and #12 took place in July and September 2008 as shown in the site schedule (Table 8.2). This was not ideal due to the fact that it was in the midst of the rainy season but it had to be organised when the auger drill was available for the unsaturated zone sampling (Figure 8.7). For this purpose all recharge water was at first diverted to Basin #7. Afterwards, all the water was diverted to Basin #12 for sampling in Basin #7. In the case of Basin #7 only soil sampling was possible due to the shallow groundwater table. The main purpose of the soil sampling was to attempt a mass balance for the various constituents (contaminants) in the recharged water. The soil sample of the first 50 cm seen in the split spoon sampling unit in Figure 8.7 (right), clearly shows the filtering effect of the infiltration surface in the basin.

Figure 8.7 Auger drilling for unsaturated zone sampling in Basin #12; split spoon core (right)

The four sets of results for Protocol 1 (standard parameters) and two sets for Protocol 2 (micro-contaminants) presented in Tables 8.4 and 8.5 allow a qualitative evaluation of the overall performance of the water recycling scheme with regard to chemistry and microbiology. The Protocol 1 sampling results are discussed below. From a water supply point of view the general inorganic chemistry of the water supply at Atlantis is of key importance due to the salinity and hardness of the groundwater. The average values of the macro chemical results at the different sampling points are illustrated by means of Stiff diagrams in Figure 8.8. In each diagram the cations are shown on the left and the anions on the right. For direct comparison the concentrations of the parameters are presented in milliequivalents per litre (meq/L).

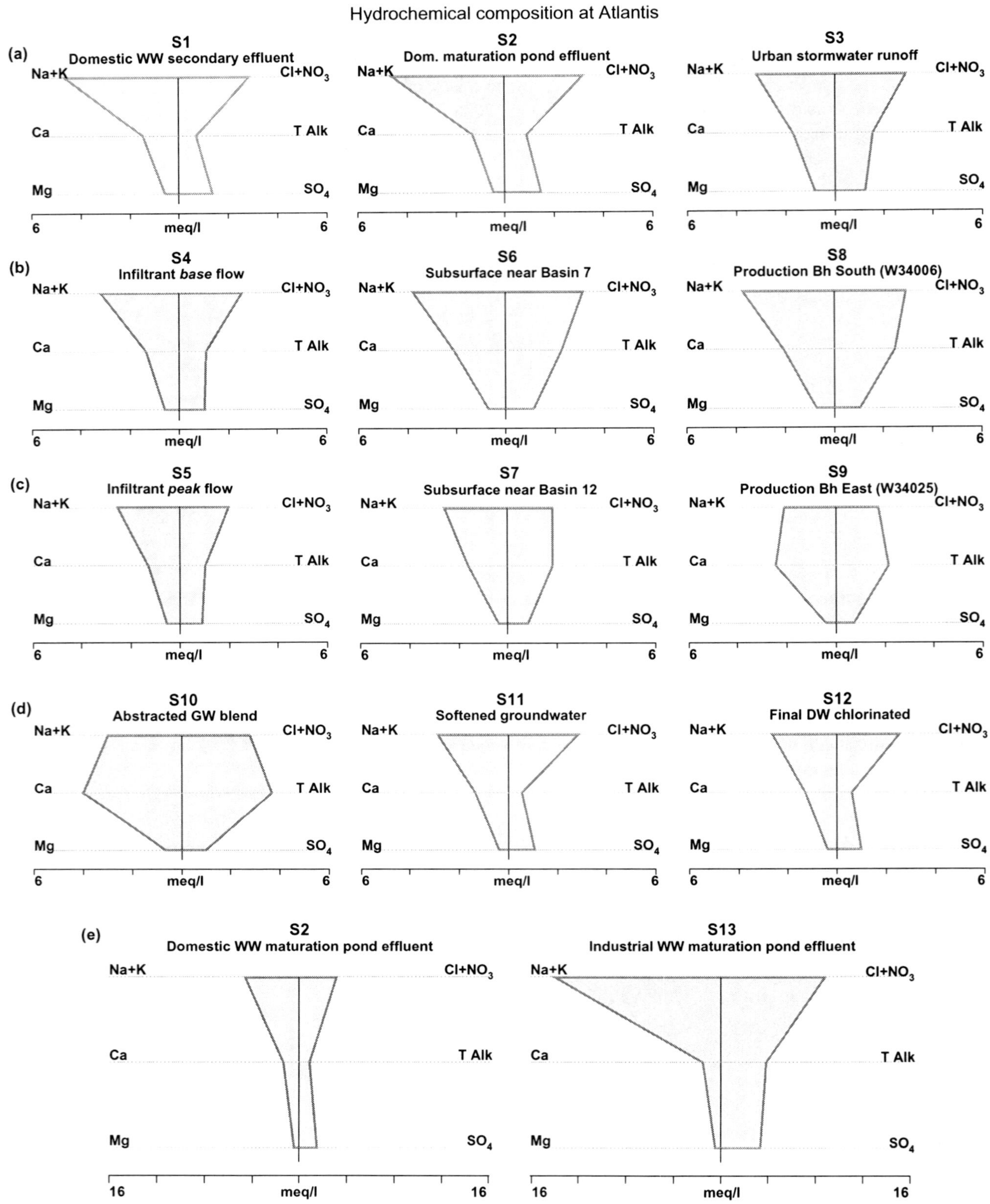

Figure 8.8 Hydrochemistry and salinity in the Atlantis recycling system (concentration scale is different in Series (e))

In the case of the first four sets of diagrams a) to d) the full scales for the cations and anions are both 6 meq/L while for the last set of two diagrams e) the full scale is 16 meq/L. On the top axis of each diagram, sodium and potassium are shown opposite chloride and nitrate, in the middle calcium opposite total alkalinity, and on the bottom axis magnesium opposite sulphate.

The series a) to d) in Figure 8.8 shows the sampling points S1 to S12 which represent the various stages in the recycling scheme. Series a) represents the source water and it is evident that the maturation pond effluent (S2) is a sodium chloride water which is chemically very similar to the secondary effluent (S1). Inspection of the data shows that only the nitrate decreased slightly during the residence in the maturation ponds. The urban stormwater (S3) has a slightly different composition with a little less sodium and chloride but slightly higher calcium, magnesium, and bicarbonate (total alkalinity).

In series b) the blend of maturation pond effluent and stormwater used for recharge in Basin #7 is shown at S4. As the water progresses through the subsurface past sampling point S6 to the nearest production borehole (W34006, sampling point S8) both calcium and bicarbonate increase to some extent. These changes together with the slight increase in sodium and chloride are ascribed to the dissolution of calcium carbonate from the aquifer and blending with slightly more saline groundwater. Series c) represents the lower salinity parallel flow path from Basin #12 past observation point S7 to the closest production borehole (W34025, sampling point S9) in the wellfield. In this case calcium and bicarbonate also increase but both sodium and chloride decrease due to blending with low salinity natural groundwater. The impact of the lower salinity groundwater is clearly visible by comparison of the two production boreholes at S8 and S9.

Series d) shows the groundwater blend from the whole wellfield (S10), the softened water (S11) and the final chlorinated water (S12). It is evident that sodium and chloride levels in the blend (S10) are similar to those at the production borehole S8 but higher than at production borehole S9. The blend also has a significantly higher calcium and bicarbonate content and this represents the impact of the natural groundwater in the aquifer which is unaffected by artificial groundwater recharge and is relatively hard. After softening, calcium, magnesium, and particularly bicarbonate are significantly lower (S11) and the composition remains the same after chlorination (S12). During use in the town the sodium, chloride and sulphate concentrations increase notably when considering the treated domestic effluent (S1) and comparing it with the final chlorinated water (S12).

In series e) the domestic and industrial wastewater are compared after the maturation ponds (S2 and S13). The diagrams in series e) were redrawn on a different scale to accommodate the higher salinity of the industrial wastewater hence the diagram for S2 in series e) looks different to that in series a). The large increase in most parameters (compare S13 with S12) during use in the industries is noteworthy and underlines the necessity from a salinity viewpoint to divert the industrial wastewater from the recycling system.

8.3.1 Dissolved organic carbon

Dissolved organic carbon (DOC) levels are relatively high both in the treated wastewater and in the stormwater (Figure 8.9). The stormwater shows the largest variation in DOC levels at S3, the outlet from the stormwater system and before the reed

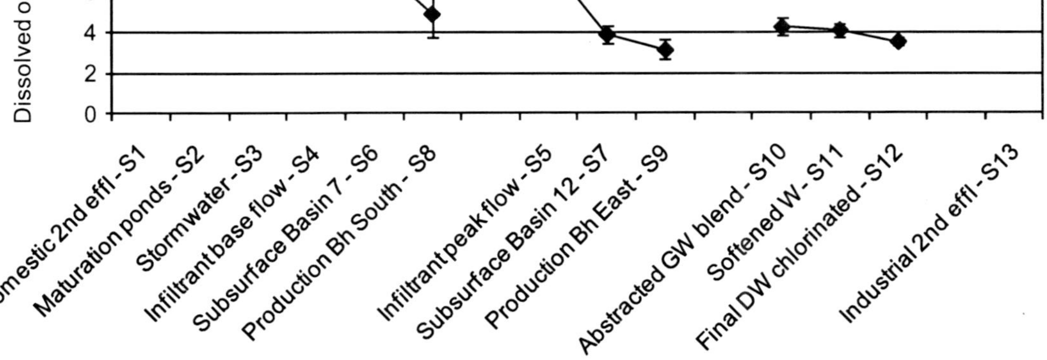

Figure 8.9 Dissolved organic carbon levels at various points in the system at Atlantis

bed, possibly due at times to intensive wash-off from roads and other surfaces in town.

Via both recharge basins (S6 and S8 representing the Basin #7 route, and S7 and S9 representing the Basin #12 route) there is a significant reduction in DOC but the level in the abstracted groundwater is still approximately 4 mg/L. In the case of Basin #12 receiving mainly peak flow stormwater and a thicker unsaturated zone the resultant groundwater has a lower

DOC when comparing production borehole S9 with S8. The presence of natural organic matter in the groundwater elevates the DOC level in the final water. The DOC is of concern due to the potential mobilization of iron from the geological material which causes clogging of the production boreholes. Furthermore, the final product water is chlorinated leading to the formation of trihalomethane compounds. The much higher DOC level in the industrial wastewater (S13) is noteworthy and supports the decision to exclude the industrial wastewater from recycling.

The Technical University in Berlin, Germany, analysed the dissolved organic carbon by means of liquid chromatography and the results are shown in Table 8.3. These sampling runs took place after the data series shown in Figure 8.9. In broad terms the results in Table 8.3 support the earlier conclusions but also add perspective with regard to the reduction in dissolved organic carbon during domestic wastewater treatment. The DOC results for the production boreholes in the Silwerstroom wellfield shows the background level of natural organic matter in groundwater which is unaffected by artificial recharge of treated wastewater and stormwater runoff.

Table 8.3 Dissolved organic carbon results by passive LC-OCD (DOC in mg/L).

Sampling point	June 2008 (Winter)	October 2008 (Spring)
S0 Raw domestic wastewater	–	17.8
S1 Domestic wastewater secondary effluent	7.7	7.7
S2 Maturation pond effluent	7.3	8.1
S3 Urban stormwater runoff	6.5	–
S4 Infiltrant blend	6.4	6.0
S8 Production borehole near Basin 7	4.7	5.0
S9 Production borehole near Basin 12	2.9	3.0
S11 Softened groundwater blend	–	4.0
S12 Chlorinated drinking water from reservoir	–	2.1
Silwerstroom production bh W34018 (no recycled water)	1.9	–
Silwerstroom production bh G29757 (no recycled water)	2.6	–

8.3.2 Electrical conductivity

The salinity of the groundwater is of importance for the Atlantis Water Resource Management Scheme as the aim is to keep the electrical conductivity (EC) of the town water supply below 700 μS/cm. This is generally achieved (Figure 8.10).

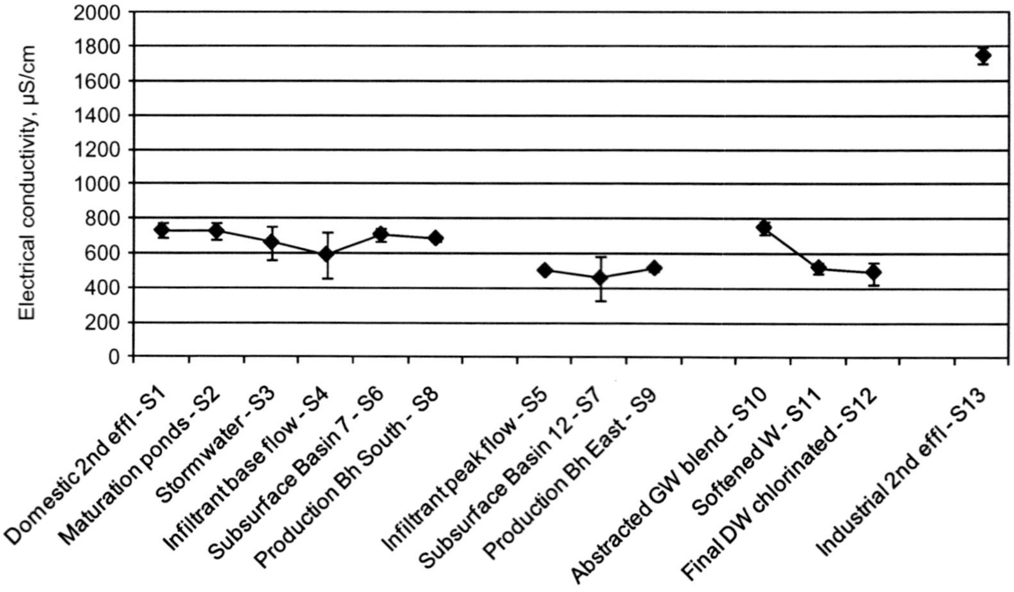

Figure 8.10 Electrical conductivity at various points in the Atlantis water recycling system

The salinity of the domestic wastewater is generally slightly above 700 μS/cm compared to the average of the water supply which is just below 500 μS/cm. On average the salinity of the stormwater runoff is lower than that of the wastewater and this helps to offset the increase in salinity in the subsurface. The salinity difference between Basin #7 (sampling points S6 and S8) and the low salinity Basin #12 (sampling points S7 and S9) is evident from the graph (Figure 8.10). The effect of the low salinity peak flow water recharged in Basin #12 during winter is clearly visible at the monitoring point (S7) down gradient of the basin. The wellfield include boreholes outside the influence of the recharge basins and hence the overall EC of the abstracted groundwater blend (S10) is higher than the two production boreholes (S8 and S9) shown in the graph. This is mainly due to the increase in calcium (bicarbonate) (Figure 8.13) but by ion exchange softening both calcium and bicarbonate is removed. Sampling point S13 shows that the treated industrial wastewater has EC levels of the order of 1700 to 1800 μS/cm which would make treatment for recycling for potable purposes very costly.

8.3.3 Sulphate

The sulphate concentration in the water recycling system at Atlantis is shown in Figure 8.11. The domestic wastewater has a sulphate concentration of between 70 and 80 mg/L while that of the stormwater is sometimes significantly lower. Sulphate reduction in the deeper layers of the aquifer is a natural phenomenon and this together with dilution with natural groundwater lower in sulphate helps to reduce the sulphate concentration at the production boreholes. The effect of the input of low salinity stormwater into Basin #12 is clearly evident at sampling point S7. The treated industrial wastewater has significantly higher sulphate concentrations (S13).

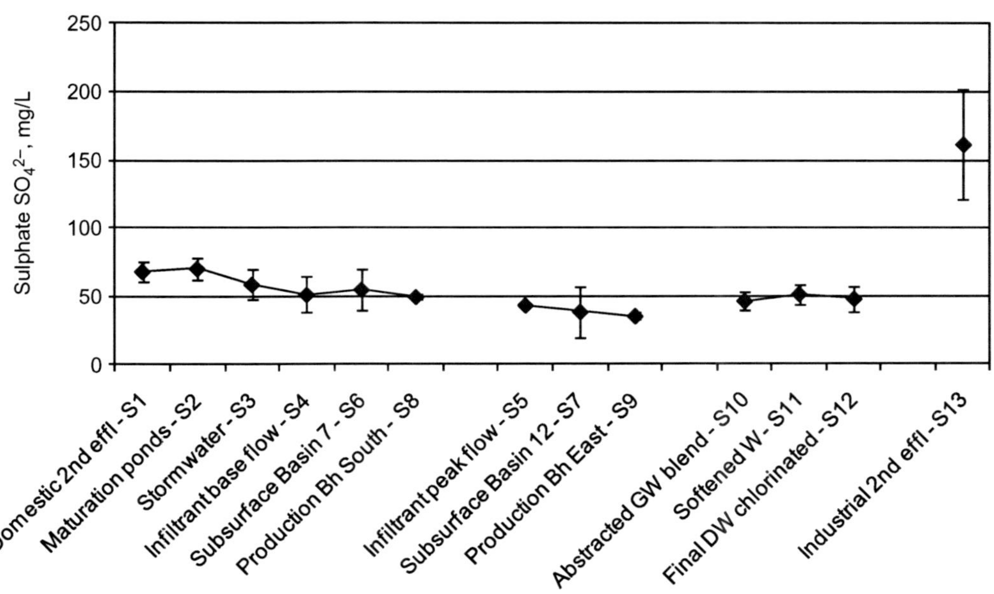

Figure 8.11 Sulphate concentrations at various points in the Atlantis water recycling system

8.3.4 Potassium and calcium

The usefulness of potassium as a tracer for wastewater is shown by the high levels for this parameter in the treated domestic and industrial wastewater (Figure 8.12). Stormwater also contains potassium but only slightly above background levels. In the subsurface potassium is removed by ion exchange on clay particles and the abstracted water has low potassium levels approaching the natural background in the aquifer. As the clay content of the aquifer material is very low, it is expected that potassium will break through eventually and reach higher levels in the boreholes affected by the recycling.

The calcium concentration is shown in Figure 8.13. In contrast to salinity, the water supply and the domestic wastewater have very similar calcium concentrations. Also, the industrial wastewater does not show any significant increase in calcium during use in the various industries. However, even the stormwater has a higher calcium content, perhaps caused by leaching of the cement pipes conducting the stormwater. The calcium increases significantly in the subsurface from the input values (S4) to the production boreholes (S8 and S9) due to leaching of calcium carbonate from the aquifer. At this stage leaching down-gradient of Basin 12 is more pronounced than at Basin 7 which has been in operation for much longer. The levels at these production boreholes are still lower than the mean value for the wellfield as the abstracted groundwater blend (S10) is significantly higher confirming the need for softening.

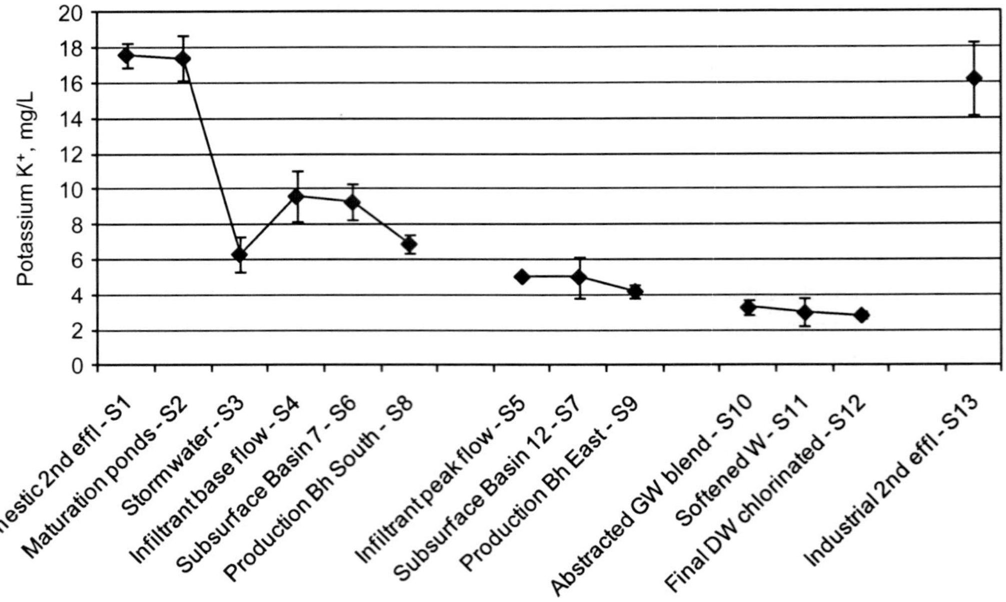

Figure 8.12 Potassium concentrations at various points in the Atlantis water recycling system

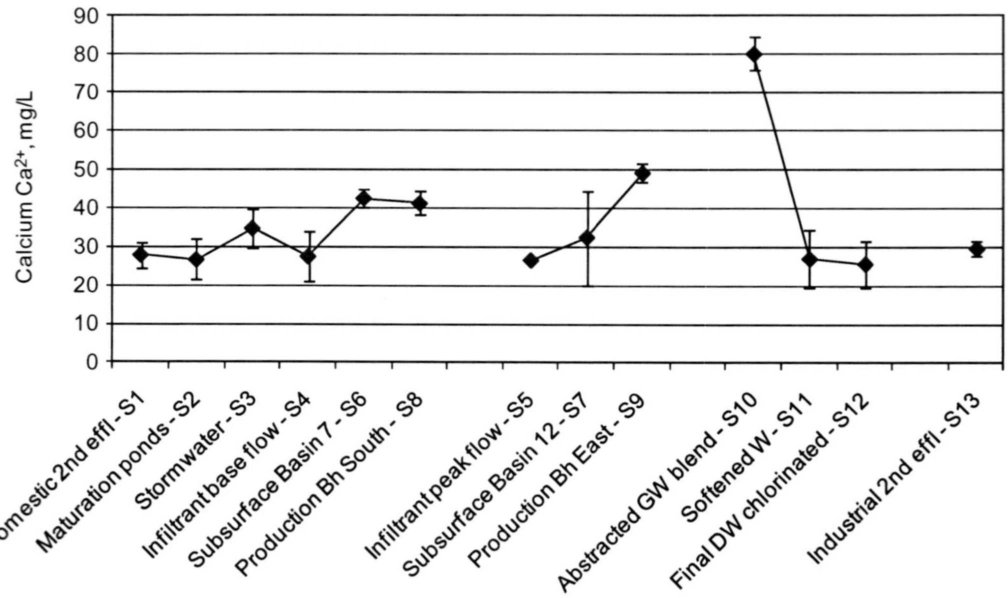

Figure 8.13 Calcium concentrations at various points in the Atlantis water recycling system

8.3.5 Boron

The boron values measured at Atlantis were generally close to the detection limit of 25 μg/L reported by the laboratory that carried out the analyses. Nevertheless, it was evident that the wastewater contained traces of boron originating from detergents. The industrial wastewater contained higher levels of boron (Figure 8.14). It is noteworthy that the boron was not removed in the subsurface but only diluted by blending with the natural groundwater.

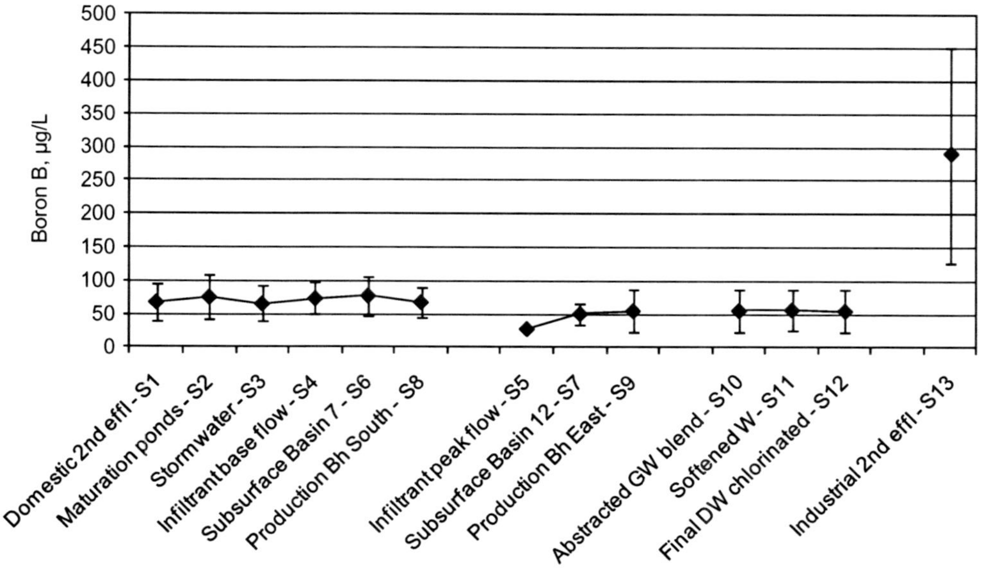

Figure 8.14 Boron concentrations at various points in the Atlantis water recycling system

8.3.6 Redox conditions

The concentrations of certain chemical species affected by oxidation-reduction conditions in the aquifer are shown in Figure 8.15. Both the domestic and industrial effluent contain nitrate and after circulation through the maturation ponds (S2 and S13) the mean nitrate-N concentrations in the two effluents are very similar at about 8 mg/L. The nitrate concentration in the stormwater (S3) is much lower and the blend with treated domestic wastewater after the reed bed at Basin 6 (S4 and S5) is even lower. This means that the input into the recharge basins was only in the order of 3.5 to 4.5 mg/L as NO_3-N. In the subsurface (S6) between Basin 7 and the wellfield nitrate decreased significantly and by the time groundwater reached the wellfield nitrate had virtually disappeared (S8). Although the exact redox potentials were not determined, it is evident that reducing conditions prevailed in the subsurface and that denitrification took place. Considering the situation at Basin 12 it is evident that nitrate reduction was also taking place in that part of the aquifer (S5 to S7) but to a lesser extent as even at the production borehole (S9) nitrate was still present. This would imply that compared to Basin 7 the redox potential was not as low in the area down-gradient of Basin 12. This also relates to DOC levels that are higher in the area down-gradient of Basin 7 (S6).

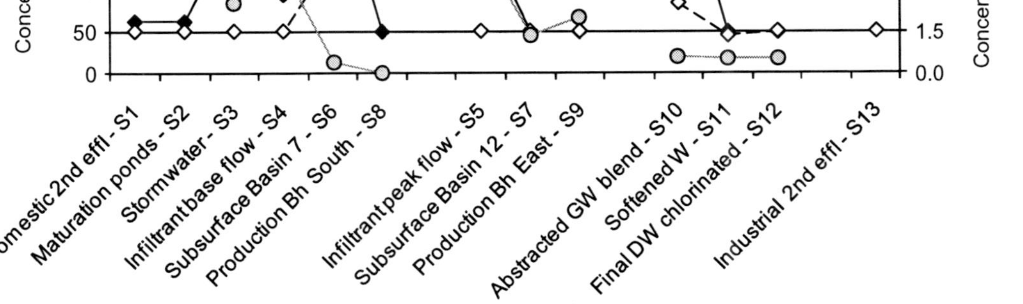

Figure 8.15 Iron, manganese and nitrate as indicator of Redox conditions along the recycling system at Atlantis

Ammonium-N concentrations were generally at or just above detection levels except in the stormwater system (S3) where the mean concentration was 2.2 mg/L (Table 8.4). Together with other forms of nitrogen as measured by Total Kjeldahl Nitrogen (TKN) significant oxidation to nitrate and removal took place during the passage through the reedbed (between S3 to S4 and to S5). In the subsurface near Basin 7 (S6) there was a slight increase in ammonium but by the time the water reached the wellfield (S8) the ammonium had disappeared. This may relate to the presence of higher levels of oxygen and variable redox conditions near the wellfield due to the ingress of oxygenated water from shallow depth.

The presence of iron in the groundwater at Atlantis causes significant production borehole clogging problems in all areas, not only where artificial groundwater recharge is practised. During the initial stage of aquifer development at Atlantis the presence of hydrogen sulphide indicated that sulphate reduction occurred in the deeper layers of the aquifer, generally at a depth greater than 20 m. At the low redox potential in this part of the aquifer Fe^{2+} concentrations would also seem to be stable. From Figure 8.15 it is evident that stormwater (S3) is the main source contributing iron to the feedwater. In the subsurface (S6) down-gradient of Basin #7 the iron concentration was at its highest while manganese was also being mobilised. For manganese this process continued up to the production borehole (S8) while the iron concentration decreased at the production borehole. This may imply that the soluble Fe^{2+} was converted to insoluble Fe^{+3} at the production borehole (S8) while manganese remained in solution in the reduced form despite the likely increase in the redox potential. Down-gradient of the peak flow recharge basin (S7) denitrification takes place but the redox potential is not low enough to keep iron and manganese in solution. The lower concentration of organic compounds in the form of DOC in this part of the aquifer also favours less reducing conditions. It was evident that the blend of abstracted groundwater originating from a large number of production boreholes including several others not covered by the graph in Figure 8.15 with higher iron concentrations indicated stronger reducing conditions at production boreholes in certain parts of the aquifer. Both iron and manganese are removed in the ion exchange softening process (S11).

8.3.7 Microbiological parameters

The bacterial counts at the various points in the system illustrate the importance of the subsurface passage as a safety barrier in the system (Figure 8.16). It was found that not only the indicator organisms follow this pattern but also the pathogens, including viruses follow a similar pattern of log-reductions to provide the necessary safety margins for the recycling system.

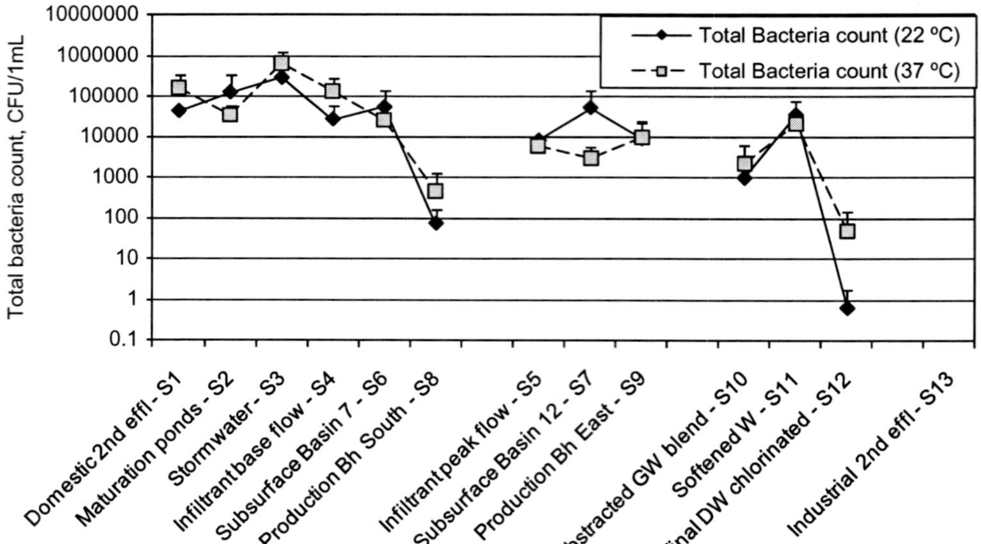

Figure 8.16 Bacteria removal in the recycling system at Atlantis

As expected, following the total general bacterial numbers, total coliforms were found in the highest numbers (except for enterococci once during winter), followed by *E. coli*, *Clostridium*, enterococci and lastly coliphage (Figures 8.16 and 8.17). The most notable factor observed is the low level of bacteria, if any, and surrogate pathogens (*Clostridium* and bacteriophages) in the groundwater and the final water before distribution to the town, that is, sampling points S6-S12. The highest counts in the infiltrant (S4 and S5) were observed in the winter which is the rainy season in Atlantis. Maximum total coliforms were in the order of $10^5/100$ mL. The absence of microbes or low microbial numbers found in the groundwater may be a result of filtration of the poorer quality recharge water. It may also be a result of natural die-off of the microorganisms.

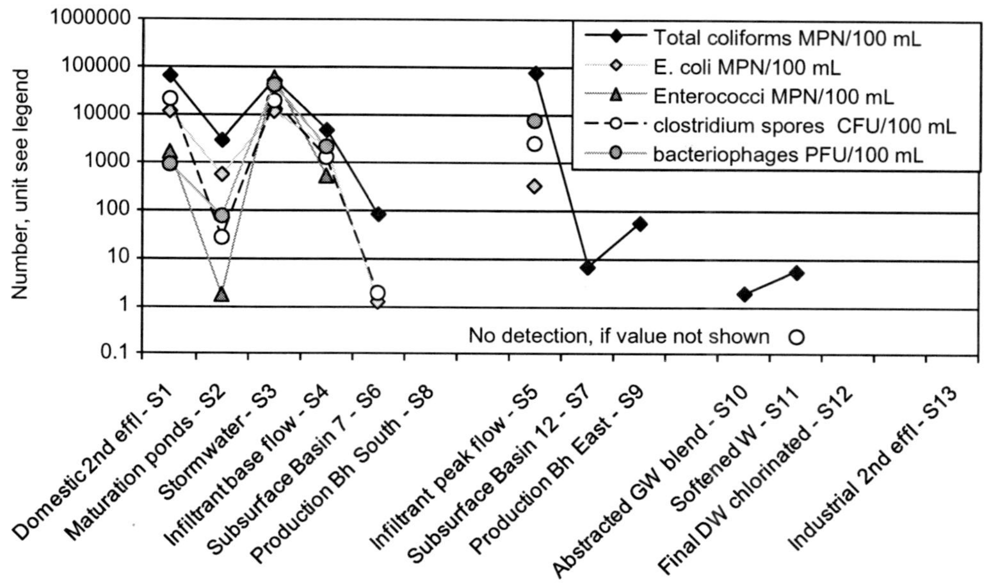

Figure 8.17 Bacterial counts for all sampling protocols

Another important factor is that the stormwater (S3) tended to have higher microbial counts than the treated wastewater effluent. This has implications for the management of the water scheme.

These results were used in a human health risk assessment and management programme within the Reclaim Water project to further understand how to best manage the recharge scheme to protect human health. Results are given in Chapter 19. Surrogate pathogen counts, namely E. coli representing bacterial pathogens, Clostridium representing parasite pathogens and coliphages representing viral pathogens, were used in addition to the protocol 2 parameters for the risk assessment.

8.3.8 Organic micropollutants

With the assistance of the Bundesanstalt für Gewässerkunde tests were carried out for a wide range of about forty trace organic compounds known to occur in domestic wastewater. Two sampling campaigns were organised in winter (June 2008) and spring (October 2008). At first the wastewater and stormwater inputs to the system were sampled as well as the production boreholes nearest to the recharge basins. A total of twelve of these compounds were detected in the domestic wastewater secondary effluent while nine were detected in at least one of the production boreholes at nanogram per litre levels. In the wastewater some compounds reached the microgram per litre level (Figure 8.18a and b). The compounds included antibiotics, antiepileptic and psychoactive drugs, compounds used as contrast media, and anti-inflammatory medication in the group of "acidic compounds". As these compounds occur at very low levels triplicate samples were taken but this limited the number of sampling points as four SPE extractions are required on each sample plus blanks and spiked samples to increase confidence limits. For the second round it was considered important to extend the range of samples by including the raw domestic sewage and the final drinking water after softening and chlorination. This was achieved by taking only duplicate samples. By including the raw wastewater the number of compounds detected in the second round increased to fifteen. This time a total of eight compounds were detected in at least one of the production boreholes (Figure 8.18a and b). Traces of only two compounds were detected at nanogram level in the final chlorinated drinking water. These were the antiepileptic medication carbamazepine and its metabolite dihydrodihydroxycarbamazepine with concentrations between 10×10^{-9} and 20×10^{-9} g/L. Such very low concentrations are well within the Australian National Guidelines for Water Recycling (2006) and can be considered as being safe for human consumption.

The highest concentrations of the trace organic compounds occurred in the wastewater and sometimes the treated effluent. In four cases the wastewater treatment effected a reduction in the levels as shown by sulphamethoxazole, diclofenac, ibuprofen and iopromide (Figure 8.18). In the case of some of the other compounds it would seem that the concentrations detected in the raw wastewater was lower than in the treated water. This may be correct but it is also possible that the grab samples were taken at a time when the concentrations in the wastewater were actually lower. During the wastewater treatment process the water is mixed well, also due to the various recycling loops. Comparing the maturation pond effluent (S2) to the domestic wastewater secondary effluent (S1) shows that the concentrations generally remain largely unaltered during the fourteen day residence time in the maturation ponds although carbamazepine, dihydrodihydroxycarbamazepine, and ibuprofen showed a decrease. The concentrations of the trace organic compounds were generally low in the urban stormwater runoff (S3) with the exception of carbamazepine, dihydrodihydroxycarbamazepine, and ibuprofen. The ibuprofen concentration in the stormwater is even higher than in the secondary effluent as it would seem that the wastewater treatment significantly reduces the ibuprofen concentration. Blending the secondary effluent with the stormwater causes a reduction in the concentration of the trace organic compounds in the feed water (S4) to the recharge basins. In the subsurface further decomposition and dilution with natural groundwater take place and very low concentrations of most of the eight compounds shown were found in the production boreholes.

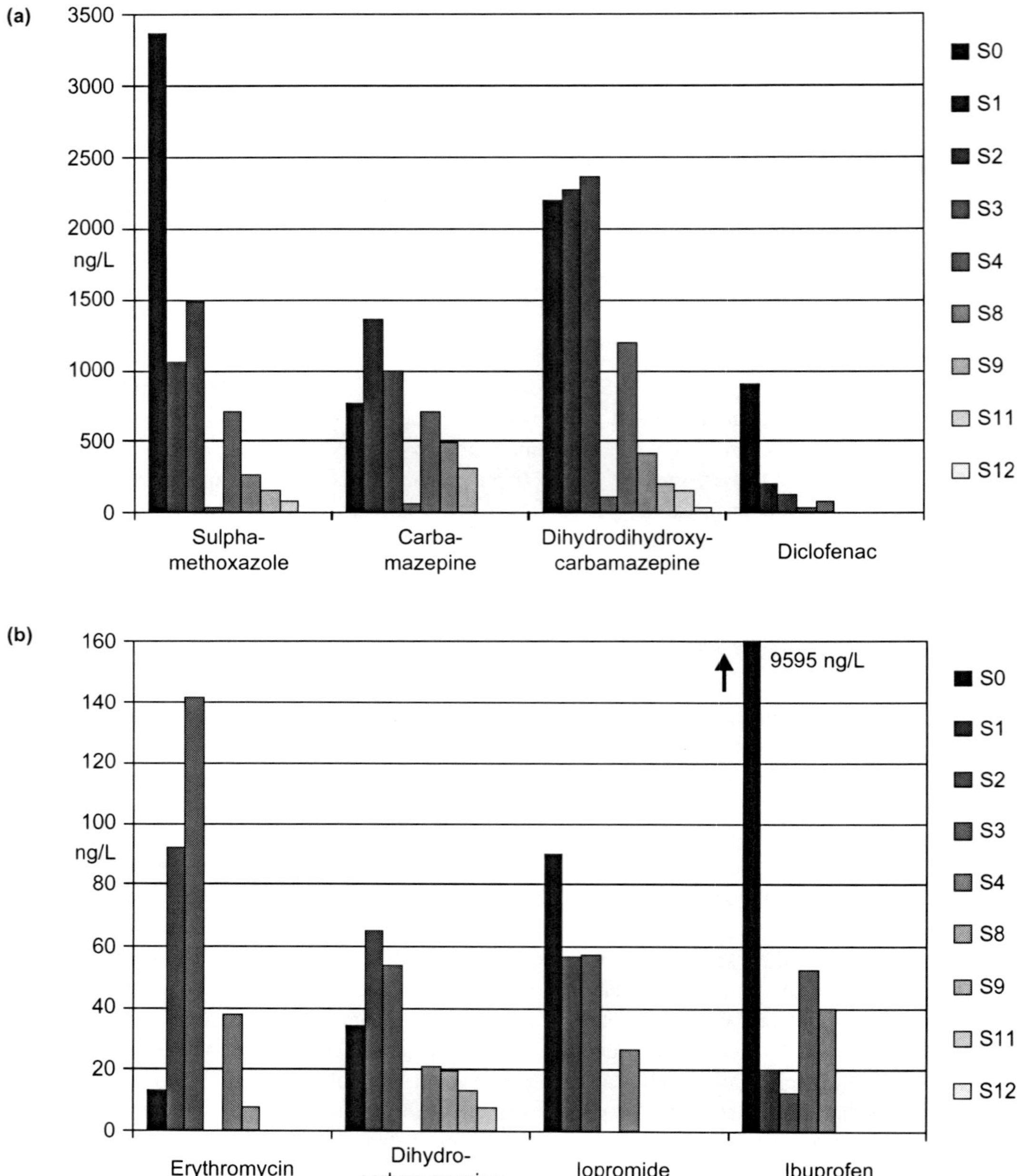

Figure 8.18 Occurrence of trace organic compounds in various parts of the water recycling system (S0 represents raw domestic wastewater, other sampling points as before)

8.3.9 Summary of water quality monitoring

The following Tables 8.4 and 8.5 present an overview of the monitoring results at Atlantis during the years 2007 and 2008 with the mean values of the main quality parameters. Barium, cadmium and lead were consistently below the respective detection limits and were not included in the table.

Table 8.4 Overview of the water quality monitoring results with regard to wastewater analyses and microbioloby at Atlantis (mean values).

Sampling point description		Domestic 2nd effl	Maturation ponds	Stromwater	Infiltrant base flow	Subsurface Basin 7	Production Bh South	Infiltrant peak flow	Subsurface Basin 12	Production Bh East	Abstracted GW	Softened W	Final DW chlorinated	DWG SANS 241 2006	Industrial WW - S13
Number		**SA1**	**SA2**	**SA3**	**SA4**	**SA6**	**SA8**	**SA5**	**SA7**	**SA9**	**SA10**	**SA11**	**SA12**		**SA13**
Basic WW analysis															
Suspended solids	$mg \cdot L^{-1}$	<4.0	5.1	5.7	5.7	<4.0	4.3	<4.0	<4.0	<4.0	<4.0	<4.0	<4.0	–	4
COD total	$mg \cdot L^{-1}$	38	31	28	27	19	18	26	15	12	11	14	11	–	53
COD soluble	$mg \cdot L^{-1}$	53	26	24	26	39	12	23	12	9	9	11	11	–	40
DOC	$mg \cdot L^{-1}$	8.8	8.6	8.8	8.1	7.4	4.9	7.4	3.9	3.1	4.2	4.1	3.5	<10	14
Ammonium (NH_4–N)	$mg \cdot L^{-1}$	0.2	0.2	2.2	0.1	0.4	0.1	0.1	0.1	<0.1	<0.1	<0.1	<0.1	<1.0	0.5
Nitrate NO_3–N	$mg \cdot L^{-1}$	11	7.6	2.5	3.7	0.4	<0.1	4.5	1.4	2.0	0.6	0.6	0.6	<10	8
Total nitrogen (TKN)	$mg \cdot L^{-1}$	2.3	4.4	4.7	1.1	1.0	0.8	1.9	0.8	0.7	0.8	0.8	0.8	–	3
Total phosphorus	$mg \cdot L^{-1}$	8.6	5.9	1.0	2.5	1.3	0.2	2.7	<0.2	0.4	0.2	<0.2	<0.2	–	2
Alkalinity	$mg \cdot L^{-1}$	40	45	77	57	113	123	53	91	107	184	28	30	–	190
pH	unit	7.0	7.6	7.4	7.6	7.9	8.0	7.2	7.9	7.8	7.8	7.6	7.3	5.0–9.5	7.8
Turbidity	NTU	8.4	1.5	7.4	2.5	1.0	0.2	12.8	1.6	0.3	1.5	0.2	0.2	<1	5
Detergents	PPB	293	180	840	170	135	175	400	180	133	128	150	113	–	167
Mineral oils	PPB	<200	<200	<200	<200	<200	<200	–	<200	<200	<200	<200	<200	–	265
Phenols	PPB	<4.0	<4.0	<4.0	<4.0	<4.0	<4.0	<4.0	<4.0	<4.0	<4.0	<4.0	<4.0	<10	<4
Microbiological															
Total bacteria count (22°C)	CFU/1 mL	4.5×10^4	1.3×10^5	3.0×10^5	2.8×10^4	5.6×10^4	76	8300	5.5×10^4	9400	1000	3.7×10^4	0.67		–
Total bacteria count (37°C)	CFU/1 mL	1.7×10^5	3.6×10^4	6.8×10^5	1.4×10^5	2.5×10^4	500	6200	3100	1.0×10^4	2400	2.2×10^4	50	<100	–
Total coliforms	MPN/100 mL	6.9×10^4	3.0×10^3	5.5×10^4	5000	87	0	8.2×10^4	7.0	58	2	5.7	0	<5 95% of samples	–
E. coli	MPN/100 mL	1.2×10^4	550	1.2×10^4	2200	1.3	0	327	0	0	0	0	0	<1	–
Enterococci	MPN/100 mL	1700	1.8	6.3×10^4	538	0	0	0	0	0	0	0	0	<1	–
Clostridium spores	CFU/100 mL	2.3×10^4	28.5	2.1×10^4	1250	2	0	2600	0	0	0	0.25	0	<1	–
Bacteriophages	PFU/100 mL	910	75	4.5×10^4	2140	0	0	7900	0	0	0	0	0	<1	–

Table 8.5 Overview of the water quality monitoring results with regard to trace elements and salinity at Atlantis (mean values).

Sampling point description		Domestic 2nd effl	Maturation ponds	Stromwater	Infiltrant base flow	Subsurface Basin 7	Production Bh South	Infiltrant peak flow	Subsurface Basin 12	Production Bh East	Abstracted GW Blend	Softened W	Final DW chlorinated	DWG SANS 241 2006	Industrial WW - S13
Number		**SA1**	**SA2**	**SA3**	**SA4**	**SA6**	**SA8**	**SA5**	**SA7**	**SA9**	**SA10**	**SA11**	**SA12**		**SA13**
Trace elements															
Boron	$\mu g \cdot L^{-1}$	68	76	66	74	78	68	29	51	56	56	57	56	–	291
Fluoride	$\mu g \cdot L^{-1}$	154	130	102	119	198	299	<100	206	146	202	206	179	<1000	392
Iron	$\mu g \cdot L^{-1}$	63	63	177	95	253	<50	174	<50	<50	264	<50	<50	<200	235
Manganese	$\mu g \cdot L^{-1}$	<50	<50	<50	<50	141	352	<50	<50	<50	83	<50	<50	<100	<50
Zinc	$\mu g \cdot L^{-1}$	55	<50	<50	<50	<50	<50	<50	<50	<50	<50	<50	<50	<5000	111
Salinity															
Chloride	$mg \cdot L^{-1}$	104	112	102	97	112	104	72	64	62	98	100	90	<200	313
Electrical conductivity	$\mu S/cm^{-1}$	730	725	660	590	708	685	500	458	515	750	520	494	<1500	1750
Sodium	$mg \cdot L^{-1}$	95	96	70	72	82	82	57	56	46	67	64	59	<200	313
Potassium	$mg \cdot L^{-1}$	18	17	6.3	9.6	9.3	6.9	5.1	5.0	4.2	3.3	3.0	2.9	<50	16
Calcium	$mg \cdot L^{-1}$	28	27	35	28	43	41	27	32	49	80	27	26	<150	30
Magnesium	$mg \cdot L^{-1}$	5.9	5.7	10.0	7.0	8.6	8.8	6.8	4.2	5.1	7.7	4.8	4.9	<70	6
Bicarbonate	$mg \cdot L^{-1}$	49	54	94	70	138	150	64	111	131	224	35	37	–	232
Sulphate	$mg \cdot L^{-1}$	69	71	59	51	55	50	44	39	36	47	52	48	<400	162

8.3.10 Operational feedback

The water reclamation scheme at Atlantis is operated by the City of Cape Town and the CSIR is involved in an advisory capacity. An important feature incorporated into the Atlantis recharge system is the separation of the runoff and wastewater into components with different qualities, mainly with respect to salinity (Tredoux and Cavé, 2002). This allows the recharge of lower salinity water in parts of the aquifer where the natural groundwater salinity is lower. Hence two recharge basins were constructed with Basin #7 intended for higher salinity water and Basin #12 for lower salinity water. The more saline industrial wastewater and high salinity stormwater is diverted to the coastal recharge basins and is not used for recycling. During the water quality monitoring and inspection visits the operational performance of the various units were observed and several matters noted for improvement. The main difficulty experienced in the operation is the borehole clogging ascribed to biofouling and the natural occurrence of iron in the groundwater. This seriously affects the well field production capacity and borehole rehabilitation is being planned. However, no modifications to the system were implemented during the Reclaim Water Project period in order not to complicate the interpretation of the results. One result is that larger volumes of surface water from other water resources of the City of Cape Town are being imported. Hence the groundwater (and recycled water) content of the final water supply may be as low as 50% at this stage. It means that the aquifer is not optimally utilised but an important spin off is the decrease in overall salinity of the system. This is brought about by the imported surface water of low salinity which is recharged into the aquifer after use and which blends with or even displaces more saline groundwater.

The Atlantis scheme faces several water quality management challenges, ranging from the control of saline water encroachment to industrial pollution threats and biofouling of production boreholes. The relatively thin unsaturated zone limits the attenuation capacity of the soil-aquifer system during basin recharge.

8.4 CONCLUSIONS

Managed aquifer recharge ensured the sustainability of the Atlantis groundwater supply over more than two decades and will continue to play a key role.

- Indirect recycling of stormwater and treated domestic wastewater via the aquifer as a means to augment supplies found public acceptance.
- Water quality management is the dominant issue regarding water supply at Atlantis. Separation of source water into different quality fractions allowed recharge of the highest quality water in the areas of importance for the production wellfield.
- Diverting the industrial wastewater from the recycling system has been shown to be the correct approach considering its high salinity and organic carbon content.
- When managed carefully the wastewater and stormwater recycling system is robust with regard to consistent water quality.
- While the importation of low salinity surface water lowered the overall salinity in the system calcium carbonate is gradually leached from the aquifer matrix.
- The presence of iron in the groundwater causes clogging of the production boreholes and, although the deeper groundwater is naturally reducing, organic carbon added through recycling may be contributing to the intensity of such conditions.
- The heavy metal concentrations in the system are low and in the abstracted water the heavy metal content, except for iron, is negligible.
- Microbiologically the subsurface passage plays a decisive role as a safety barrier. The presence of organic carbon needs to be controlled as this may affect the bacteria removal in the subsurface.
- Trace organic carbon compounds (e.g. pharmaceutical compounds) are effectively removed and diluted to such an extent that the nanogram levels remaining are well within internationally accepted norms.

The Atlantis groundwater scheme provides a cost-effective water supply option when coupled with strict management of the resource. The importation of limited quantities of low salinity surface water has enhanced the viability of the recharge scheme and also allows the utilisation of slightly more saline groundwater. The future introduction of membrane processes for water softening and partial desalination will further enhance the scheme, even allowing the possible export of potable water to other residential areas.

ACKNOWLEDGEMENTS

The authors thank the City of Cape Town for access to the site and support of the study as well as our colleagues at CSIR, Pannie Engelbrecht, Jac Wilsenach, and the CSIR chemical laboratories for analytical support, the Bundesanstalt für Gewässerkunde for the trace organic analysis, and Guido Fink for the preparation of samples.

REFERENCES

Australian Guidelines. (2006). Australian National Guidelines for Water Recycling: Managing Health and Environmental Risks.

Department of Water Affairs. (2007). South Africa's Artificial Recharge Strategy. Accessed August 2011. http://www.artificialrecharge.co.za/03strategy.html

Department of Water Affairs. (2010a). The Atlantis Water Resource Management Scheme: 30 years of artificial ground water recharge.

Department of Water Affairs. (2010b). Strategy and Guideline Development for National Groundwater Planning Requirements. Water Banking: A practical guide to using Artificial Groundwater Recharge, dated November 2010.

Liebenberg and Stander (1976). Planning and Masterplan for the Stormwater Network, Outfall and Disposal System. Report GW/EduT/1961 PC. Cape Town: Liebenberg and Stander Consulting Engineers.

Rogers J. (1980). First report on the cenozoic sediments between Cape Town and Elandsbay. Open File Report 1980 – 0249. Pretoria: Geological Survey of South Africa.

SANS 241:2006 (South African National Standard. Drinking Water). (2006). Part 1: Microbiological, physical, aesthetic and chemical determinants. Edition 6.1. SABS Standards Division, Pretoria.

Tredoux G., Ross W. R. and Gerber A. (1980). The potential of the Cape Flats aquifer for the storage and abstraction of reclaimed effluent (South Africa). *Z. dt. geol. Ges.*, **131**, 23–43.

Tredoux G. and Cavé L. C. (2002). Atlantis aquifer: a status report on 20 years of groundwater management at Atlantis. *Report Submitted to the City of Cape Town*. CSIR Report No ENV-S-C 2002-069.

Van der Merwe A. J. (1983). Exploration, development and evaluation of groundwater in the sand deposits in the Atlantis area for water supply to the Atlantis growth point. M.Sc thesis. Bloemfontein: University of the Orange Free State.

Wright A. (1994). Artificial recharge of urban wastewater, the key component in the development of an industrial town on the arid west coast of South Africa. In: *Water Down Under '94. Proceedings of the IAH Congress: Vol.2, Part A*, 39–41. Adelaide: International Association of Hydrogeologists.

Chapter 9

Unplanned aquifer recharge in El Mezquital/Tula Valley, Mexico

Blanca Jiménez, Alma Chávez, Richard Gibson and Catalina Maya

9.1 INTRODUCTION

9.1.1 Water management situation

Reuse of the large amounts of wastewater generated by modern societies has advantages such as reducing the pressure on other water sources and helping to avoid direct discharge into and contamination of natural water bodies (Toze *et al.*, 2009). Irrigation of farmland with wastewater has a long history (Tzanakakis *et al.*, 2007) and many cities around the world without adequate infrastructure to treat wastewater use it as a relatively inexpensive management option. However infiltration can occur and water is often extracted from aquifers without awareness of its origin.

One prominent example of this type of water management that has resulted in unplanned aquifer recharge occurs around Mexico City. Wastewater is sent from the city to the Tula Valley located 100 km to the north in the southern part of the State of Hidalgo (Figure 9.1). The wastewater, presently at a mean annual rate of approximately 52 m^3/s (Jiménez and Chávez, 2004), is transported to irrigation districts in the valley using a complex system of six wastewater and three freshwater storage dams, over 677 km of main and lateral distribution canals, and 565 km of agricultural drains; some of these dams and main distribution canals are shown in Figure 9.1. It has been estimated that half of the water is infiltrated, a value that is more than 13 times the natural rate (BGS, 1998). This infiltration has resulted in a large increase in the flow of the Tula River, which is partially fed from the aquifer (only from 1.6 m^3/s in 1945 to more than 12.7 m^3/s by 1995). New artesian wells and springs with flows from 40 up to 600 L/s started to appear from 1964, a result of the water table rising from a level of 50 m below the surface (BGS, 1998). Currently, 6 m^3/s of water extracted from the subsoil is used in the valley, 17% for human consumption, 38% for agriculture, 33% for industry, and 12% for other uses (Jiménez and Chávez, 2005), while 21 m^3/s still flows to the neighbouring basin for reuse in agriculture.

The Tula Valley varies in altitude from 1700 m in the northern part to 2 100 m in the southern part of the valley and its climate is classified as BS1kw(i')gw" (Garcia, 1988), that is, the most humid of the arid zones. It is a mild climate with a mean annual temperature of 16.7°C and annual rainfall from 435 to 618 mm, which almost exclusively occurs between May and October. The valley is surrounded by hills and mountains, largely composed of volcanic rocks of varying age and composition, and calcareous sediments (BGS, 1998). According to the PAHO classification (PAHO, 1988), soils in the valley are: (a) rendzic and melanic Leptosols, (b) calcic and haplic Phaeozems; and (c) eutric Vertisols (Siebe, 1994). Subsoil in the valley is composed of 89% shallow to slightly sandy soils, 10% clay soils, and 1% sand (Jiménez and Chávez, 2005).

The geohydrology of the Tula Valley is mainly influenced by the structure of its rocks, geological faults, and the permeability and porosity of its various deposits. These deposits form an interlinked sequence of aquicludes and aquifers in several areas. Aquifers in the area are formed by three formations; the lava flows and coarse sediments in the Tarango formation, the tertiary lava flows, and the Lower Cretaceous limestones.

Quaternary lava flows (in the Tarango formation) and tertiary lava flows form the most important aquifer in the area in which most groundwater accumulates. A multilayered aquifer is formed due to the low permeability of the rocks associated with the fracturing and faulting of rocks. This gives rise to formations of variable thickness interlayered with formations of less permeable deposits. Most of the aquifers are semi-confined with the exception of some unconfined aquifers close to the surface (BGS, 1998).

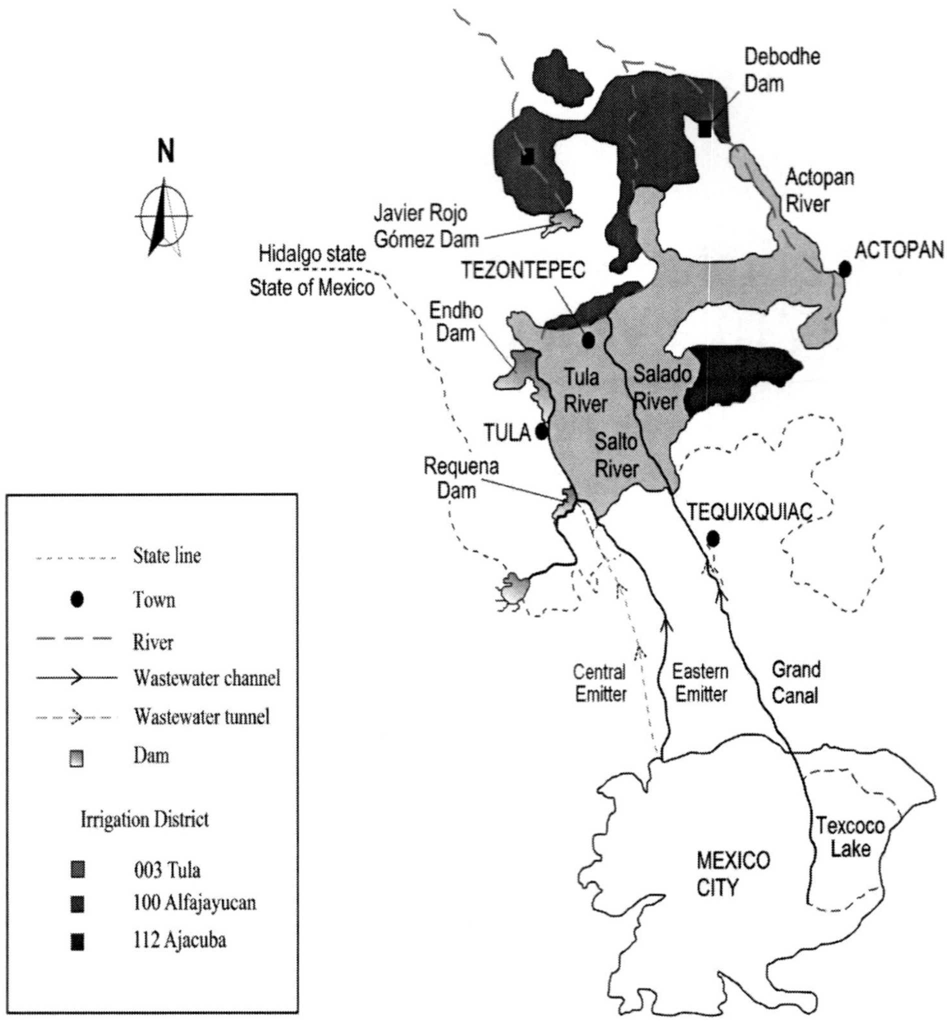

Figure 9.1 Map of Mexico City and the irrigation districts in Hidalgo State

The coarse sediments within the Tarango formation also house an important groundwater resource. They exhibit considerable thickness over the whole valley with moderate to high intergranular permeability of the coarser sediments interbedded with low permeability sediments and lavas. This formation has low transmissivity because of the large thickness of clay and clayey deposits.

A groundwater resource also exists in the lower Cretaceous limestones. These are present at a considerable depth beneath the valley floor. Knowledge of their hydraulic characteristics is limited due to the small number of boreholes that have been sunk into these layers. Outcrops of the El Doctor limestone typically have high permeability, probably resulting from the development of solution features. Where clay sequences are present, they give rise to zones of low permeability. In many areas of the valley there are differences in hydraulic head with depth, indicating a multilayered aquifer sequence containing interbedded aquicludes. Adjacent boreholes of different depths may have static levels that differ by as much as 30 m (BGS, 1998).

Although the local geological formations of the Mezquital area are complex with a complicated distribution of aquifers, it is possible to define the general groundwater flow regime using groundwater monitoring data with reasonable confidence. Two semi-independent groundwater flow units exist. One of these is situated in the western part of the Salado River with a flow direction from south to north. The other flows first eastwards and then to the north into the Actopan Valley.

Groundwater recharge is possible in the mountains to the south of the Tula Valley via infiltration of storm runoff at the valley margins and subsurface inflow. During the summer limited direct infiltration occurs, due to occasional heavy storms on the valley floor.

The relative depths of the aquifers vary greatly and the infiltration of irrigation water is understood only in general terms. The predominant flow through the upper aquifer is lateral, while direct downward infiltration to the second aquifer is very slow except when the water encounters an extrusion or fracture, allowing relatively rapid infiltration. Although recharge of the aquifers still occurs naturally through infiltration on the floor and at the margins of the valley and through subsurface inflow from the valley flanks, this natural recharge has been overtaken by artificial recharge from wastewater sent from Mexico City. Artificial recharge occurs as a result of the infiltration of excess irrigation water, as well as from leakage from both the distribution channels and drainage systems.

In the Tula Valley there are 283 water sources including wells, springs, and dugwells supplying approximately 500 000 inhabitants (Jiménez *et al.*, 2005). Most wells have been bored into the second and third aquifers. In the towns, water is derived from deep boreholes and chlorinated before delivery to the mains distribution system, while in rural areas shallower boreholes with less sanitary protection are often used. Although the quality of the extracted water complies with the majority of the established regulations for drinking water, there is a worry about the presence of some contaminants such as some microorganisms and organic micropollutants that could affect the health of the population.

The objective of this study was to assess the impact of the use of wastewater for irrigation on the quality of the sources of water supply in the Tula Valley, specifically measuring the removal of pathogens and micropollutants. The role of the soils in protecting the aquifers from contamination was also examined by direct measurement of micropollutant concentrations and through a column study of the fate of the micropollutants in different soil profiles.

9.1.2 History of the development of the site

Artificial removal of wastewater and excess rainwater from Mexico City to the Tula Valley began at the end of the nineteenth century. For this, three conducts have been constructed as part of the General Drainage System. The first was the Eastern Interceptor (in 1889), then the Gran Canal (in 1898), and more recently the Central Sewer (in 1975). The Gran Canal Sewer is 47 km long and is the main part of the drainage system that drains water from the lowest parts of the city with the help of twelve pumping stations along its length. The Central Sewer operates from between 30 and 220 meters below ground level, is not affected by the terrain above, and operates efficiently purely by gravity flow without any use of pumps.

The area under irrigation in the Tula Valley has increased in tandem with the increase in amounts of wastewater generated by the growth of Mexico City. In 1926 the irrigated area was 14 000 ha. By 1950 it had increased to 28 000 ha, to 42 460 ha by 1965, and reached a maximum area of 84 500 ha in 1992. At the present time, due mainly to farmer migration to the United States, this area has reduced to 76 119 ha. Currently the area is 80 880 ha (2009). It is distributed among six irrigation districts: 003-Tula Valley, 100-Alfajayucan, 025-Ixmiquilpan, 112-Ajacuba, 088-Chiconautla and 073-La Concepción the last two in the State of Mexico (Jiménez, 2005). The most important district in terms of surface area irrigated is 003-Tula Valley, which occupies an area of 45 000 ha (Figure 9.1).

9.1.3 Technical set-up and operational experiences

Three different regions of the Tula Valley were identified where the irrigation was predominantly either with raw wastewater soon after its arrival in the valley (zone 1), raw wastewater diluted with rainwater or previously stored in dams (zone 2), and raw wastewater mixed with spring or well water (zone 3). From each area, samples of water were taken from wells (water extracted by a pump), dugwells (water extracted by hand), and springs. In total, infiltrated water was sampled from 12 wells, 4 springs, and 2 dugwells. Wastewaters were sampled from the Central Sewer (the principal canal that delivers wastewater from the drainage system of Mexico City to the Tula Valley) and distribution canals from the three different zones. A map of the Tula Valley and the sampling points are presented in Figure 9.2. Samples (all in triplicate) were collected in two campaigns lasting approximately six weeks each, one in the wet season (summer, 2007) and one in the dry season (spring, 2008). The water was sampled directly into clean glass bottles, transferred to the laboratory and stored at 4°C overnight before processing the next day. Each sampling point was assigned a tag that identified the study zone, the type of water, and also an individual number. Z corresponds to the zone (1–3), W to well (1–4), S to spring (1–2), D to dugwell, UWW to untreated wastewater, and WWM for wastewater mixed with other waters (Table 9.1).

Microbiological content was carried out in triplicate with a total of 10 microorganisms determined for each sample. The work included total bacterial counts (at 5°C, 22°C, 37°C, and 55°C), faecal coliforms, *Escherichia coli*, *Streptococcus*, Clostridium spores, somatic bacteriophages, *Giardia spp.* and helminth eggs. Six organic micropollutants, ibuprofen, naproxen, carbamazepine, 4-nonylphenols, triclosan, and bisphenol-A, were analyzed. In addition, measurements of basic water and salinity parameters and metals were made.

Field soil sampling in each of the three zones was carried out seven days after the previous irrigation by the farmers. A composite sample (1 kg) was made from the A horizon (approximately the top 20–30 cm) from eight points chosen randomly in each zone. The soils were partially air-dried at room temperature under atmospheric conditions then sieved to 2 mm. The physicochemical characteristics of the soils sampled are detailed in Table 9.2. Soil columns were packed with 10 cm of soil from A, transition, and B horizons, from profiles of Phaeozem and Leptosol soils which had been irrigated for different lengths of time. They were irrigated with wastewater (250 mL; equivalent to 13 cm of water which is the typical amount of water used in each irrigation event) fortified with the micropollutants (5000 ng of each) five times at four weeks intervals (a programme that equates approximately to one crop cycle). The leachates were collected and processed after each irrigation. The columns were left to stand for a further four weeks after the final irrigation before being cut into three equal sections. Samples were air-dried sufficiently to allow them to be sieved to 2 mm and then stored at −15°C before analysis. Soils were labelled SL for soils and Z for the different zone from which soil and water was collected. Z1SL1, Z1SL3, Z2SL1, and Z3SL1 soils were used for the column study.

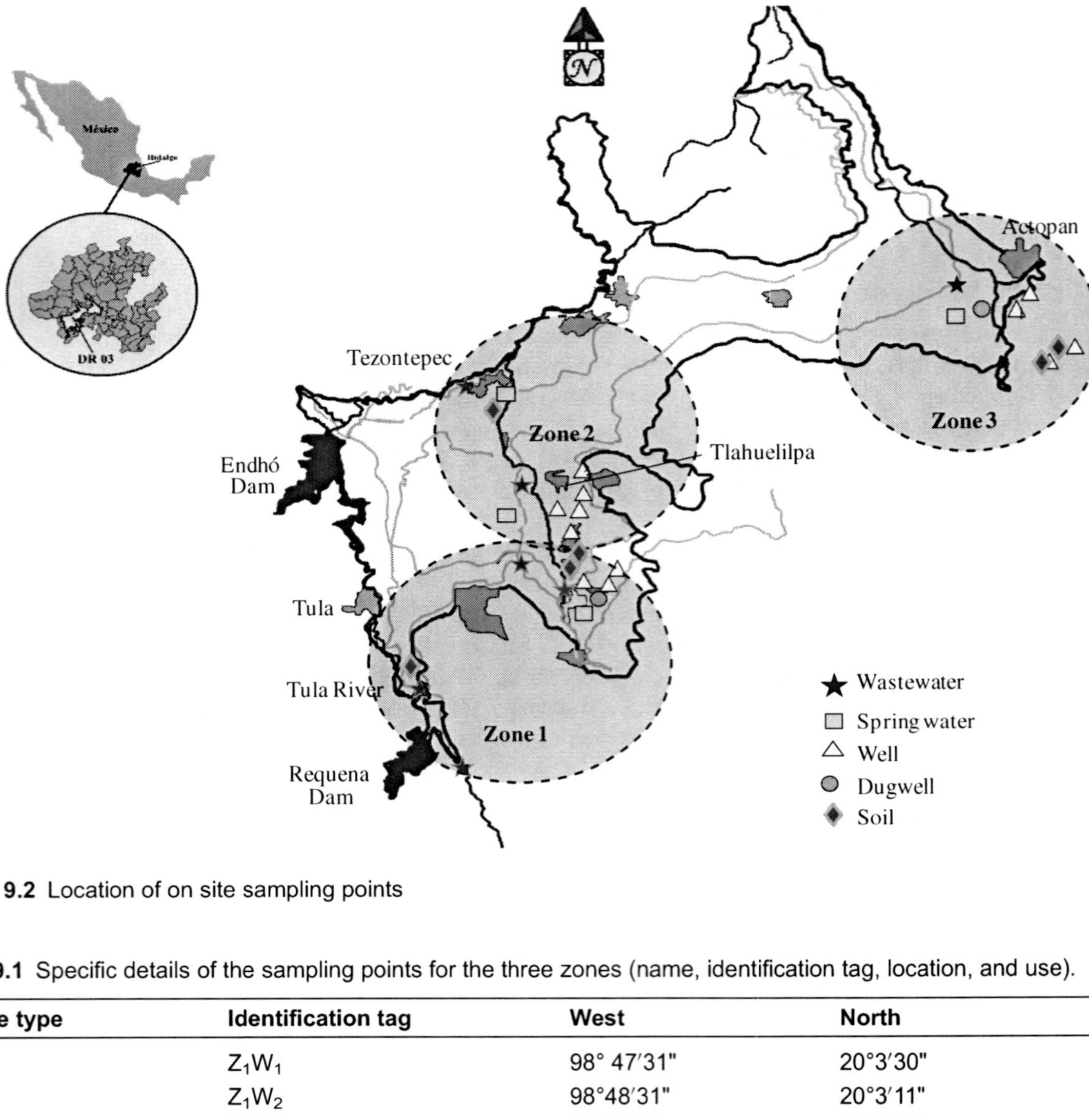

Figure 9.2 Location of on site sampling points

Table 9.1 Specific details of the sampling points for the three zones (name, identification tag, location, and use).

Sample type	Identification tag	West	North	Uses
Well 1	Z_1W_1	98° 47′31"	20°3′30"	WS
Well 2	Z_1W_2	98°48′31"	20°3′11"	WS
Well 3	Z_1W_3	98°47′11"	20°2′53"	WS
Well 4	Z_1W_4	98°47′7"	20°7′47"	WS
Springwater 1	Z_1S_1	98°46′40"	20°2′22"	RC
Dugwell 1	Z_1D_1	98°47′32"	20°3′29"	WS
E. Central	Z_1UWW_1	99°17′00"	19°5′80"	–
WW canal 1	Z_1UWW_2	99°12′45"	20°7′36"	–
WW canal 2	Z_1UWW_3	99° 12′45"	20°4′24"	–
WW canal 3	Z_1UWW_4	99° 15′ 11"	20° 4′ 2"	–
Soil 1	Z_1SL_1	98°42′49"	20°10′57"	–
Soil 2	Z_1SL_2	–	–	–

(*Continued*)

Table 9.1 Specific details of the sampling points for the three zones (name, identification tag, location, and use) (*Continued*).

Sample type	Identification tag	West	North	Uses
Soil 3	Z_1SL_3	20°07′37″	20°07′36′	–
Well 1	Z_2W_1	98°44′60"	20°6′40"	WS
Well 2	Z_2W_2	98°45′30"	20°8′14"	WS/Ind
Well 3	Z_2W_3	99°13′21"	20°8′21"	WS
Well 4	Z_2W_4	99°13′16"	20°7′42"	WS
Springwater 1	Z_2S_1	98°44′16"	20°7′13"	WS/Rc
Springwater 2	Z_2S_1	98°43′15"	20°11′28"	WS/Rc
WW canal 1	Z_2WWM_1	99°14′6"	20°6′37"	–
WW canal 2	Z_2WWM_2	99°15′57"	20°8′44"	–
Soil 1	Z_2SL_1	98°44′60"	20°6′40"	–
Soil 2	Z_2SL_2	98°45′30"	20°8′14"	–
Soil 3	Z_2SL_3	98°44′49"	20°04′26"	–
Well 1	Z_3W_1	99°3′34"	20°13′53"	WS
Well 2	Z_3W_2	99°2′43"	20°16′25"	WS
Well 3	Z_3W_3	99°3′33"	20°12′21"	IR
Well 4	Z_3W_4	99°3′46"	20°13′7"	IR
Springwater 1	Z_3S_1	99°1′44"	20°16′25"	WS
Dugwell 1	Z_3D_1	98°57′30"	20°15′54"	WS
WW canal 1	Z_3WWM_1	98°58′10"	20°15′13"	–
Soil 1	Z_3SL_1	99°03′49"	20°12′31"	–
Soil 2	Z_3SL_2	99°04′37"	20°13′37"	–

Note: WS = water supply, Rc = Recreation, IR = Irrigation, Ind = Industrial

Table 9.2 Soil classification, pH, CEC, and organic matter.

Soil i.d.	Classification WRB, 2006	Wastewater type (years under irrigation)	pH	CEC (cmol/kg)	Organic matter (%)	Texture			Textural classification
						Sand	Silt	Clay	
Z_1SL_1	Leptosol	UWW (10)	6.75	42	6.3	48	25	26	Sandy clay loam
Z_1SL_2	Cambisol	UWW (40–60)	7.31	46	4.2	41	33	26	Loam
Z_1SL_3	Phaeozem	UWW (100)	6.60	38	4.5	13	38	48	Silty clay loam
Z_2SL_1	Phaeozem	WWM (40–60)	6.63	47	7.0	39	35	26	Loam
Z_2SL_2	Vertisol	WWM (40–60)	6.79	50	6.0	16	39	45	Clay
Z_2SL_3	Leptosol	WWM (90)	6.87	43	5.5	24	31	45	Clay
Z_3SL_1	Leptosol	WWM (30–40)	8.09	47	1.8	54	35	11	Sandy loam
Z_3SL_2	Fluvisol	WWM (10)	7.96	39	1.5	59	27	14	Sandy loam

Note: CEC = cation exchange capacity

9.1.4 Analytical methodology

For all biological methods (see Table 9.3 for details) the limit of detection was zero organisms or not detected except for helminths eggs which was defined as 0.2 egg/L with a recovery efficiency of 82% (Maya *et al.*, 2006).

For the analysis of organic micropollutants, all water samples were stored at 4°C overnight and extracted within 24 hours. Wastewaters were passed through 1.2 μm filters and acidified before extraction while supply waters were simply acidified. The methodology for the analysis has been described elsewhere (Gibson *et al.*, 2007). Briefly, the micropollutants were isolated from the water samples on Oasis HLB cartridges then the acidic pharmaceuticals and carbamazepine were eluted with a mixture of acetone and 0.1 M sodium hydrogen carbonate (adjusted to pH 10) followed by elution with acetone. After evaporation and derivatization, the samples were separated and analyzed using an Agilent 6890 gas chromatograph fitted with a 30 m HP5-MS fused silica capillary column (30 m × 0.25 mm, 0.25 μm film thickness), connected to an Agilent 5973 mass selective detector. The soils were analyzed according to a procedure described in Durán-Álvarez *et al.*, 2009. Briefly, the sample (10 g) was mixed with diatomaceous earth (2 g) then extracted with acetone:hexane:acetic acid (49:49:2) using a DIONEX® 200 accelerated solvent extraction system under the following conditions: pre-heat (0 min), heating time (5 min), static time (5 min), 50% flush, 2 cycles. After extraction, the samples were evaporated to a volume of approximately 3–4 mL. Water (20 mL) was added before being passed through an Oasis HLB cartridge (200 mg). The acidic compounds and carbamazepine were eluted as described for the water samples while the phenols were eluted with acetone:dichloromethane (1:1, 6 mL). Evaporation, derivatization, and final analysis were similar to the procedure described above for water samples. Limits of detection are recorded in Table 9.4. Basic parameters including total suspended and dissolved solids, dissolved organic carbon (DOC), chemical oxygen demand (COD), nitrates, some trace elements, and parameters related to salinity (chlorides, electrical conductivity, carbonates, hydrogen carbonates, and sulphates) were measured using methods described by the American Public Health Association (APHA *et al.*, 1998).

Table 9.3 Analytical Methods used for the Tula Valley Study.

Group	Parameter	Methods	Preservative	Time between sample and analysis
Microorganisms				
	Total bacterial counts faecal coliforms *Escherichia coli* *Streptococcus* *Clostridium* spores	Membrane filtration 0.45 µm diameter by APHA *et al.*, 1998, Membrane method 9222C Membrane method 9222D Membrane method 9222G Membrane method 9230C Membrane method 9222B	Ice	24 h
	Somatic bacteriophages	Concentration by centrifugation Analysis : Host Culture: *E. coli* Detection of phage by dual layer methods according Adrams, 1959	Ice	24–48 h
	Giardia spp.	Method 9711B APHA *et al.*, 1998 (concentration & staining with Iodine)	Ice	24–48 h
	Helminth eggs	Dibasic phase, Methods USEPA, 1992	Formaldehyde	3–5 d
Micropollutants				
	Ibuprofen, naproxen, carbamazepine, 4-nonylphenols, triclosan, and bisphenol-A.	Water-extraction by solid phase, final analysis by GC-MS Soils – accelerated solvent extraction, solid phase clean-up and final analysis by GC-MS		24 h
Basic wastewater analysis				
	pH	Electrometric method 4500 H B. APHA *et al.*, 1998	Analyzed "*in situ*"	Immediately
	Temperature	Thermometer, 2550 B. APHA *et al.*, 1998	Analyzed "*in situ*"	Immediately
	Redox potential	Electrometric, 2580 B. APHA *et al.*, 1998	Analyzed "*in situ*"	Immediately
	Turbidity	Nephelometric method 2130 B. APHA *et al.*, 1998	Analyzed "*in situ*"	Immediately
	Hardness	Method 2340, APHA *et al.*, 1995	Ice	24 h
	Biological Oxygen Demand (BOD, total and soluble)	Nephelometric method 2130 B. APHA *et al.*, 1998	Ice	24 h
	Chemical Oxygen Demand (total and soluble, COD)	High temperature combustion Method 5310 B APHA *et al.*, 1998	pH less 2 with sulfuric acid	48 h
	Dissolved Oxygen Carbon (DOC)	Method 10129 Hach. U.S. Patent 6,368,870	Ice	24 h

(*Continued*)

Table 9.3 Analytical Methods used for the Tula Valley Study (*Continued*).

Group	Parameter	Methods	Preservative	Time between sample and analysis
	Nitrate-N	Cadmium reduction method. Method 8117 Hach. Procedure is equivalent to Standard Method 4500-NO_3^- E., APHA *et al.*, 1998		24 h
	Nitrite-N	Method No. 8507 Hach. Federal Register, 44(85), 25505 (May 1, 1979)	Ice	24 h
	Ammonia	Nessler methods No 8038 Hach. Adapted APHA *et al.*, 4500-NH_3 ByC	pH less 2 with sulfuric acid	48 h
	Total Kjeldahl Nitrogen (TKN)	Nessler method, 8075 adapted from Hach *et al.*, 1987 and Standard Method 4500-N org A., APHA *et al.*, 1998	pH less 2 with sulfuric acid	48 h
	Phosphorus	Molybdovanadate method. Method 8114 Hach. Adapted from Standard Method 4500-P C, APHA *et al.*, 1998	Ice	48 h
	Total Suspended Solids (TSS)	Method 2540-D. APHA *et al.*, 1998	Ice	24 h
	Total Dissolved Solids (TDS)	Method 2540-C. APHA *et al.*, 1998	Ice	24 h
	Sulphate	Gravimetric method with ignition of residue 4500-SO_4^{2-} C, APHA *et al.*, 1998	Ice	48 h
Salinity related analysis				
	Chloride	Method 4500-B, APHA *et al.*, 1998	Ice	24 h
	Electrical conductivity (EC)	Method 2520 B, APHA *et al.*, 1998	Analyzed "*in situ*"	Immediately
	Sodium, Potassium	EPA 6010b-1996		
	Carbonate/ bicarbonate	Titration method 2320 B. APHA *et al.*, 1998	Ice	24 h
	Sulphate	Gravimetric method with ignition of residue 4500-SO_4^{2-} C, APHA *et al.*, 1998	Ice	48 h
Trace elements analysis				
	Aluminum, cobalt, copper, iron, manganese, molybdenum, nickel, and zinc.	EPA 6010b-1996	pH less than 2 with sulfuric acid	
	Barium, cadmium, chromium, lead, selenium, boron and fluoride. Cyanides	EPA 2010B-1996		

Table 9.4 Limits of detection for the organic micropollutants in wastewater, supply water, and soils.

Micropollutant	LOD (ng/L)		LOD (ng/g) soil
	Wastewater	Supply water	
Ibuprofen	50	0.20	0.1
Naproxen	50	0.50	0.2
Carbamazepine	n/a	0.05	0.5
4-nonylphenols	100	50	25
Triclosan	10	1.00	1.0
Bisphenol-A	20	2.50	2.0

LOD: Limit of detection

9.2 WATER QUALITY ASSESSMENT

9.2.1 Irrigation water in the Tula Valley

Microorganisms

The wastewater sent via the drainage system that serves Mexico City arrives in the Tula Valley having received no treatment and therefore it was found to contain a high concentration of faecal coliforms (between 10^3 and 10^8 CFU/100 mL; generally greater than 10^6), Clostridium spores (between 10^3 and 10^5 CFU/100 mL), Streptococcus (between 10^2 and 10^6 CFU/100 mL), and somatic bacteriophages (greater than 10^3 PFU/100 mL); data shown in Figure 9.3 for each zone and season. Although it was expected that in general concentrations would be lower in the wet season due to rainfall dilution, this occurs intermittently as the wet season is characterized by irregular intense storms. An additional complication is that the wastewater can be sent directly to the fields, stored in dams, or mixed with stored rainwater, all depending on water flows and the demands of the farmers which in turn depend on the crops they are cultivating. Therefore it was not unexpected to see only sporadic differences in wastewater concentrations between seasons and little clear trend in concentrations as the wastewater moved through the valley (from zone 1 to zone 3).

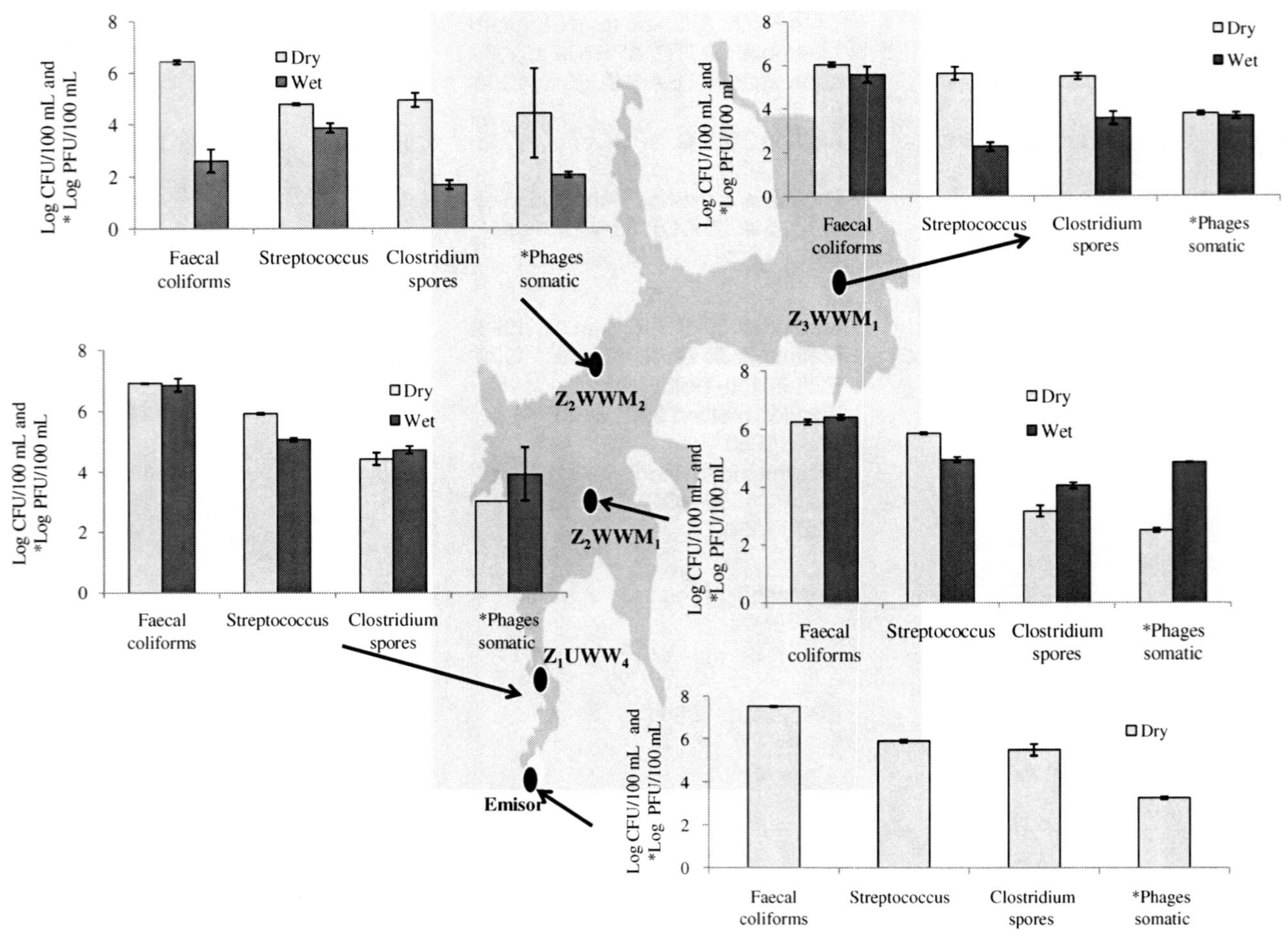

Figure 9.3 Microorganisms in the irrigation water (Error bars show standard deviation)

The two main indicator organisms considered in this study were *Giardia spp.* and helminths eggs (Figure 9.4). The concentration of *Giardia spp.* varied from 70 to 5567 cysts/L in zone 1, from 0 to 2000 cysts/L in zone 2, and from 277 to 3233 cysts/L in zone 3. The decrease between zone 1 and 2 could be explained by dilution of the wastewater with other types of water. For the increased number of cysts in zone 3, one explanation is the direct contribution of discharges from the population to irrigation channels, which are in a larger number than in zones 1 or 2. In the dry season, there was a large reduction in the *Giardia spp.* content of the wastewater in all zones. The helminth eggs content ranged from 8 to 14 eggs/L in zone 1, from 8 to 12 eggs/L in zone 2, and from 16 to 18 eggs/L in zone 3. There was no consistent difference between the wet and dry seasons but again values were greater for zone 3, suggesting a direct contribution from wastewater discharges from the local population. These concentrations of microorganisms are similar to those reported previously for untreated wastewater in Mexico (Jiménez *et al.*, 1999; Jiménez and Chávez, 1997; Jiménez *et al.*, 2008) and are similar to untreated wastewaters in other countries (e.g. Blumenthal *et al.*, 1996; Modella *et al.*, 2008) but much greater than concentrations seen in effluents of wastewater treatment plants.

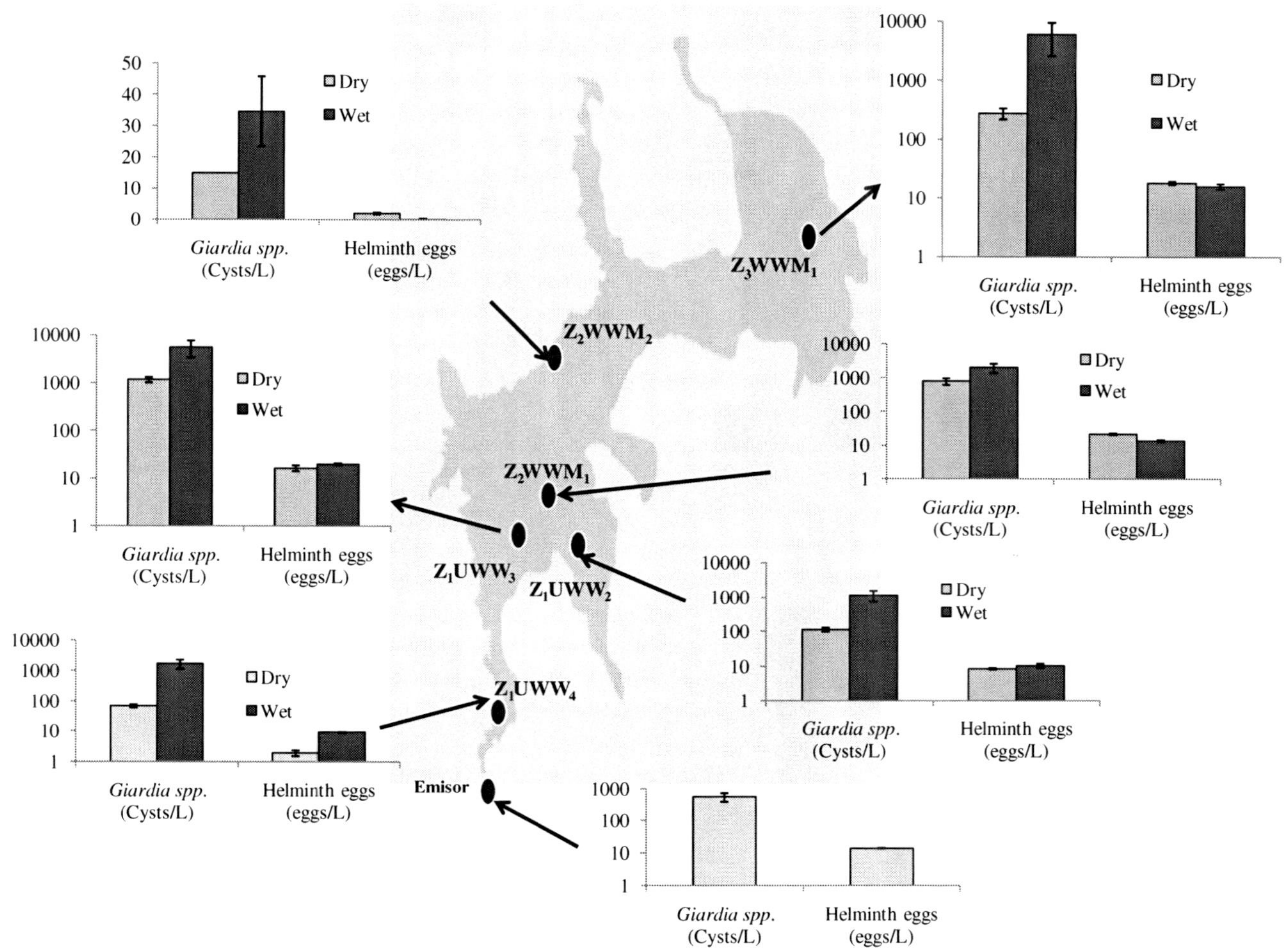

Figure 9.4 *Giardia spp.* and Helminth eggs in the irrigation water (Error bars show standard deviation)

Organic micropollutants

The concentrations of the organic micropollutants were measured in the Central Sewer and in canals in each of the three zones, in both the wet and dry season (Figure 9.5). Concentrations varied from 327 to 4 700 ng/L for ibuprofen, 3 114 to 16 336 ng/L for naproxen, 5 499 to 33 798 ng/L for 4-nonylphenols, 341 to 1 821 ng/L for triclosan, and from 432 to 3 177 ng/L for bisphenol-A. Consistent measurement of carbamazepine in the wastewater was not possible by GC-MS but concentrations of 200 ± 35 (Z1UWW1) and 203 ± 32 ng/L (Z1UWW3) were measured in two different wastewater canals in zone 1 by LC-MS-MS. Overall, values were within typical reported ranges for untreated wastewaters with notably high concentrations of naproxen and 4-nonylphenol. Concentrations of naproxen, 4-nonylphenols, and triclosan were generally lower in the wet season, probably due to dilution effects. Ibuprofen and bisphenol-A were significantly greater in the Central Sewer in the wet season; this may relate to use patterns or less degradation in the wet season due to a shorter travel time from the city to the sampling site. There was a general decrease in concentrations from zone 1 to 3 for most of the compounds in both the wet and dry season, probably due to degradation and sedimentation processes during transport.

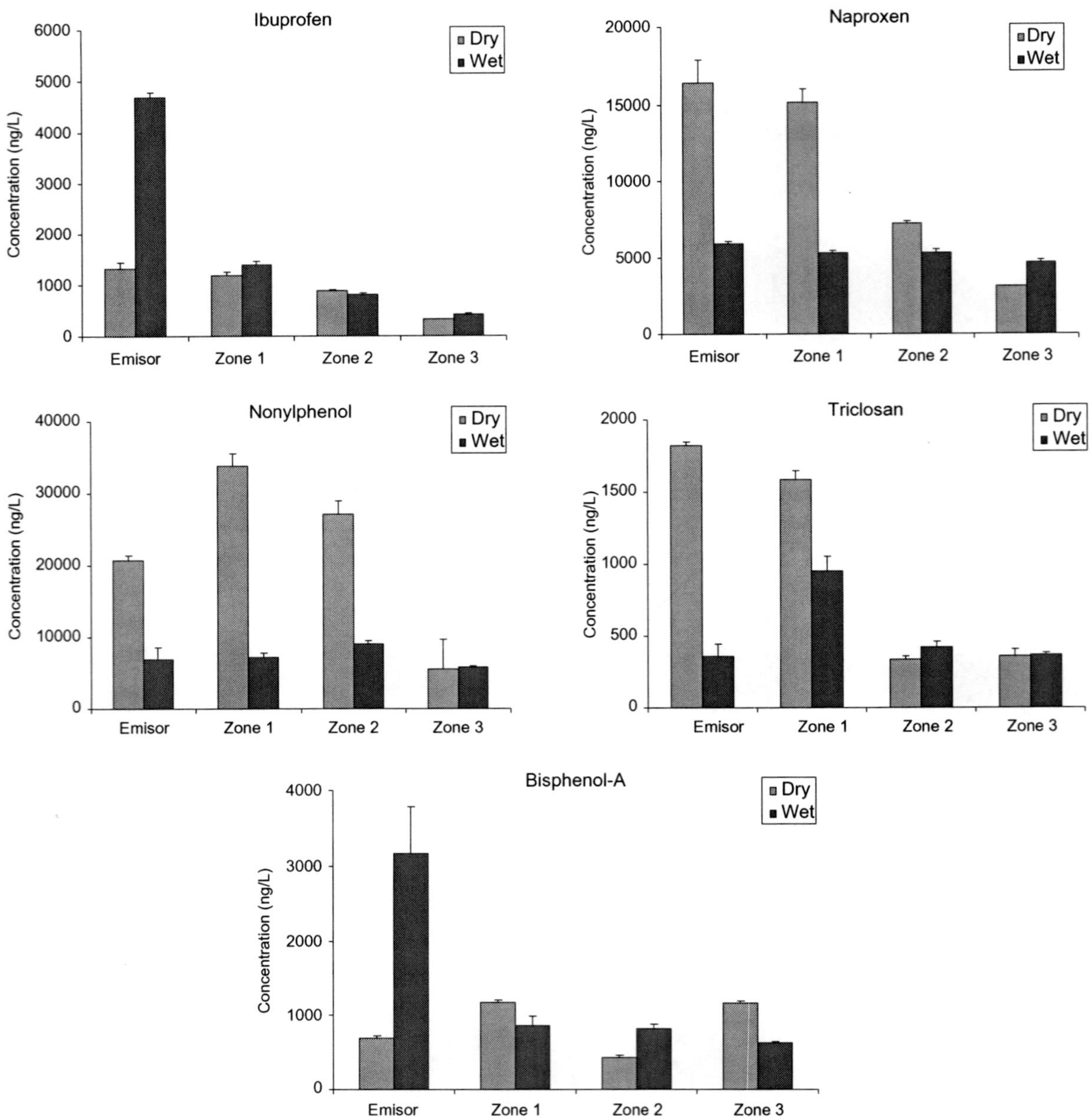

Figure 9.5 Micropollutant concentrations (ng/L) in the Emisor Central and wastewater canals in each zone in the dry and wet season (Error bars show standard deviation)

9.2.2 Supply water

Microorganisms

There was a large reduction in the microorganisms content of all sources of supply water when compared to the irrigation water. No obvious difference was noted between the wet and dry seasons (before disinfection). Bacteriophages were also present in some of the sources, with more frequency in springs and dugwells than wells. There was a general trend for more microorganisms in the dugwells than in the springs and wells (Figures 9.6 a to f). This is probably a result of the level of protection afforded to the source rather than the quality of the water emerging from the aquifer, as well as to the discontinuities in the strata of the aquifers (BGS *et al.*, 1998). The greater concentrations in the wet season could be caused by solids being carried in the water rather than being sedimented, and the poor protection of these water sources. The risk of microorganism contamination of aquifers from wastewater is well recognized (Fent *et al.,* 2006; Toze *et al.*, 2009), yet there are relatively few reports of actual microorganism contamination of groundwater on infiltration. Where described there has often been an unusually rapid infiltration of surface water that has some wastewater content, for example, after extreme rainfall events (Katz and Griffin, 2008; Fong *et al.*, 2007) as a result of the movement of microorganisms from the non-saturated zone towards the aquifer (Jean, 2000; Miossec *et al.,* 2000; Cronin *et al.,* 2003; Taylor *et al.,* 2004).

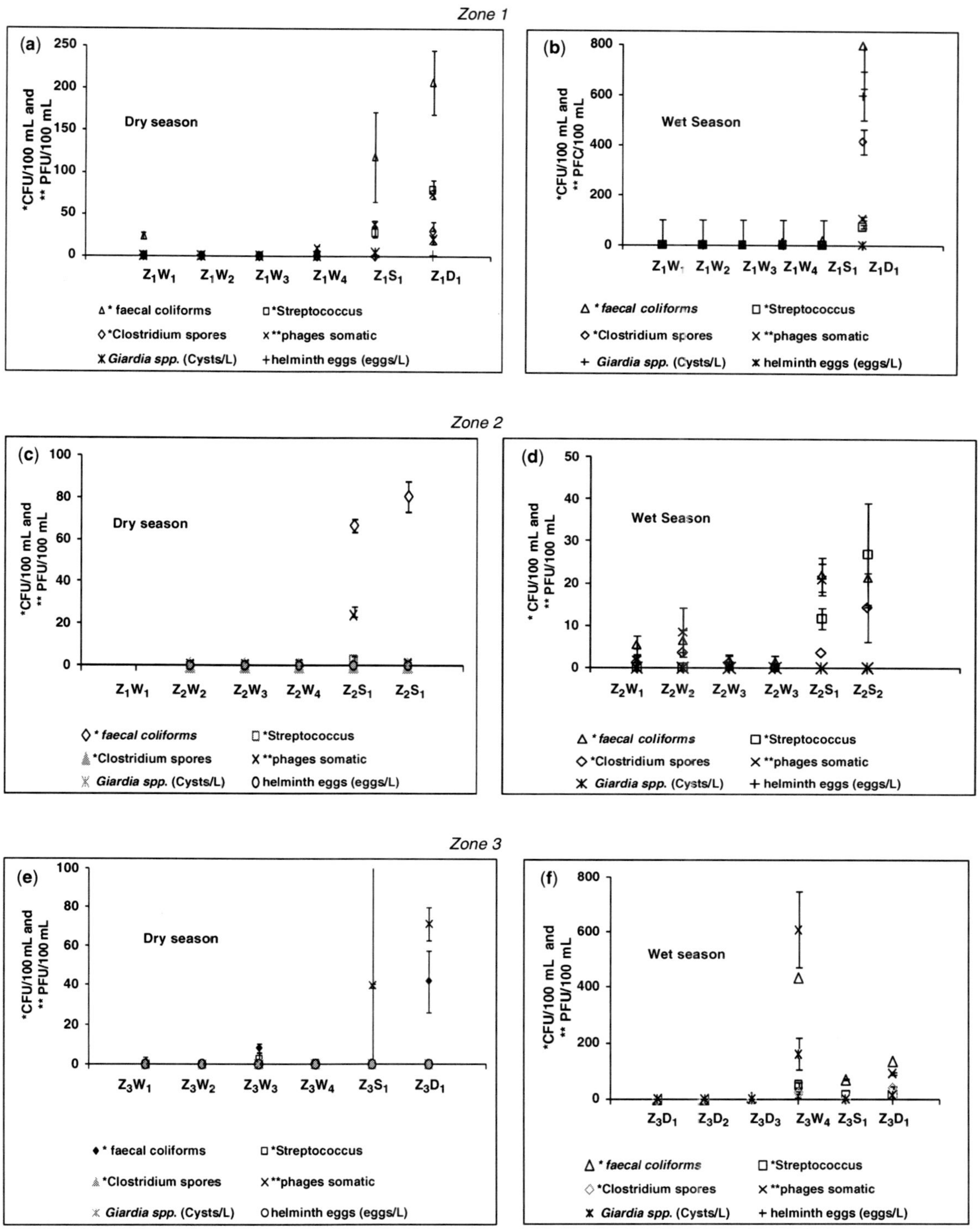

Figure 9.6 (a–f) Microorganism content in the Water Supply, Tula Valley 2007–2008

Generally, the water sources in zone 1 were less contaminated than those of zone 2, which in turn were less contaminated than zone 3. This may be because of the quality of the wastewater or possibly the fact that in zone 3 the water does not infiltrate to any great depth but flows through an upper stratum. The majority of the wells contained faecal coliforms and Clostridium spores but because of the variability of the data it was not possible to draw overall conclusions for a particular zone from the individual wells.

It appears that the soil currently removes microorganisms efficiently by the processes of filtration, adsorption, and natural death (Gerba and Bitton, 1984; Zelikson, 1994). The majority of microorganisms are retained easily in the soil, presumably in the first 10 cm as has been reported by various authors (Gerba *et al.*, 1975; Yavuz-Corapcioglu and Haridas, 1985; Craun, 1985; Yates, 1988; Modella *et al.*, 2008). In general terms, the Tula Valley could be considered as an unplanned SAT (soil-aquifer treatment) system, which removes between 4.0 and 5.8 (mean 4.8) log units of faecal coliforms, 1.7 to 3.4 (mean 2.7) log units of somatic bacteriophages, between 93.8 and 99.8% (mean 97.8%) of *Giardia spp.* and between 98.1 and 100% (mean 99.3%) of helminths eggs.

Under certain hydrogeological conditions, particularly where there are superficial aquifers or aquifers with fractures, viruses and bacteria could possibly penetrate significantly into the aquifer. However, in general the attenuation in the vadose zone (from 2 to 5 metres depth) is very effective at elimination of pathogens, simulating a tertiary treatment (Foster and Chilton, 2004).

Organic micropollutants

Despite the continual application of untreated wastewater containing high concentrations of the organic micropollutants using a flood irrigation system, the concentration of 4-nonylphenols, bisphenol-A, ibuprofen, and naproxen in the source water was nearly always below or very close to detection limits.

Triclosan was present in the spring water in zone 1 (Z1S1) in the dry season at 22.6 ng/L, but was close to or below the limit of detection at all other sites. This indicates that, in general, adsorption and degradation processes in the soil and subsoil protect the aquifer from contamination with these micropollutants. The exception was carbamazepine where concentrations were above the detection limit for the majority of the sites in the dry season and occasionally in the wet season (Figure 9.7). Greatest concentrations were observed at two sites in zone 1 164.4 ng/L (Z1W3) and 193.4 ng/L (Z1S1). Comparing these concentrations with those measured in the wastewater suggests that in these locations little removal of carbamazepine is occurring. There were large differences in carbamazepine concentrations in different source waters within zones, indicating a more direct influence from the wastewater at some sites compared to others.

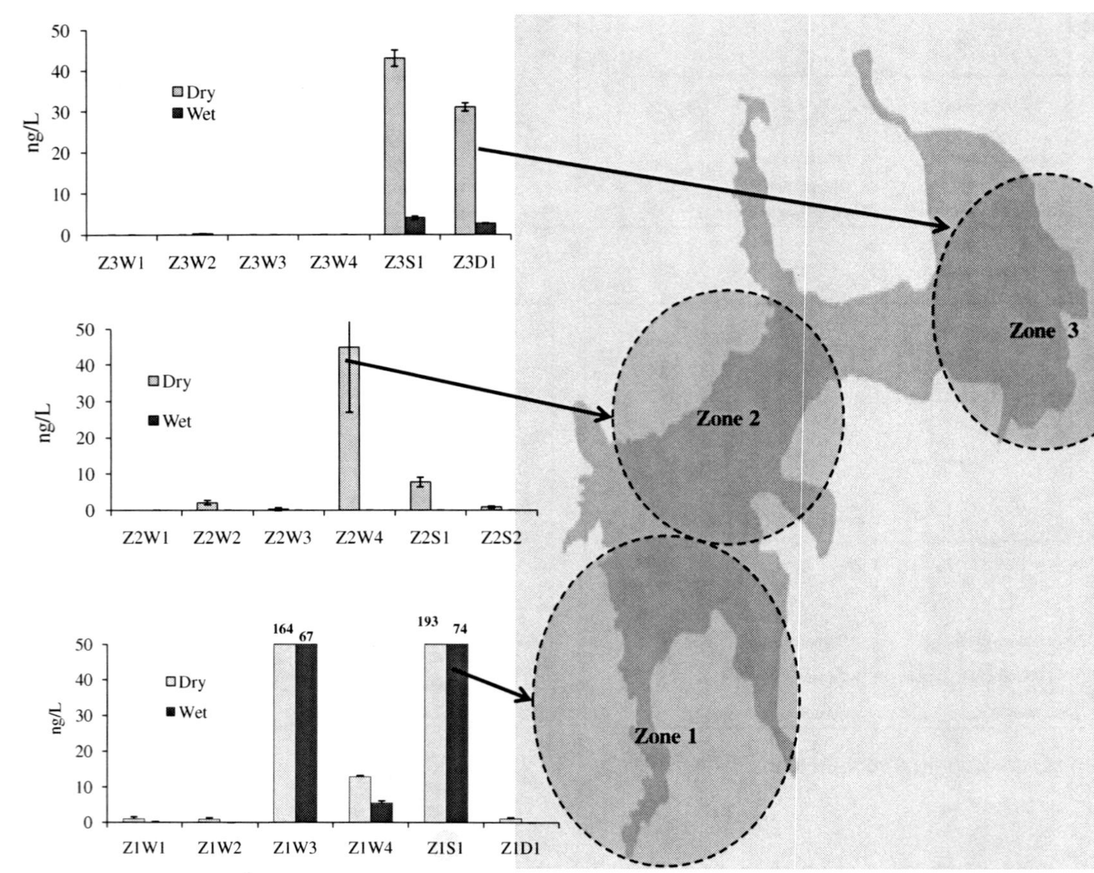

Figure 9.7 Carbamazepine concentrations in wells, springs and dugwells in each zone in the dry and wet season (Error bars show standard deviation)

Basic parameters

Basic parameters for the wastewaters are recorded in Table 9.5 and generally reflect the fact that the water is untreated. Total dissolved solids (TDS) were high as were parameters related to salinity, such as chloride, sulphate, and electrical conductivity. Trace elements in the wastewater were below regulatory limits, except for cyanide which was above the limit in a single canal during the dry season (data not shown). Data for the well water is also recorded in Table 9.5.

Table 9.5 Basic parameters, trace elements, and salinity related measurements for the wells.

Parameter		Untreated and canal wastewater (mean)		Zone 1		Zone 2		Zone 3		Limits	
		Wet	Dry	Wet (min-max)	Dry (min-max)	Wet (min-max)	Dry (min-max)	Wet (min-max)	Dry (min-max)	Mexico	(European Union)
Basic parameters (mg/L)	Total DS	792	673	865–1 372	858–1 455	941–1 199	1 147–1 291	1 098–1 586	893–1 487	1 000	–
	COD	29	31	5–14	7–12	1–5	5–7	2–3	3-7	–	2
	Nitrate (N)	0.24	0.03	18–18	5–18	10–17	10–13	NR	17–32	10	11.3
Trace elements (μg/L)	Aluminium	3 654	9 264	56–262	14–19	ND	ND	ND	18–29	200	200
	Fluorine	715	307	197–405	98–1809	502	587–1 309	119–128	163–392	1 500	1 500
	Boron	891	570	587–1 042	650–1 262	2 056	665–2 233	60–219	66–495	–	1 000
Salinity related parameters (mg/L)	Chloride	146	185	245–674	84–279	185–270	182–254	250–469	188–334	250	250
	EC[1]	2 502	1 496	1 262–2 195	2 387–4 187	1 643–1 985	3 130–3 560	1 554–2 265	2 773–4 150	–	2 500
	Sodium	102	142	115–179	127–296	355	183–361	109–119	121–296	–	–
	Hardness[2]	141	138	299–649	306–372	194–353	645–1 006	188–579	192–352	500	–
	CO_3^{2-}	87	94	32–262	24–96	60–96	52–84	102–120	44–84	–	–
	HCO_3^-	394	364	296–590	334–546	494–652	432–610	254–594	212–514	–	–
	Sulphates	480	747	300–579	245–348	276–671	1 221–3 025	613–1 028	190–319	400	250

Notes: [1]μmhos/cm, [2]mgCaCO_3/L, [3]NOM–127–SSA1–1994, [4]EU (European Union) (1998)
DS = dissolved solids, COD = chemical oxygen demand
ND = not detected; NR = not recorded

The unplanned soil aquifer treatment efficiently removes some pollutants, for example, for wells between 28 and 100% of COD (although the European Union defined limits were still regularly exceeded), most trace elements, between 60 and 100% of BOD, and between 98 and 100% of phosphorus (data not shown). However, concentrations of total dissolved solids were not dissimilar to those recorded for the irrigation water, and most exceeded the limit stipulated in the Mexican drinking water regulation (1 000 mg/L). Nitrates regularly exceeded stipulated limits (both Mexican and the European Union) while fluorine and aluminium were close to national permissible limits in some locations. Data related to salinity indicated high salt content, with little change from values recorded for the irrigation water, and it frequently exceeded national and international limits. The farmers use high irrigation rates deliberately to wash salts through the soil therefore it was not unexpected that salt concentrations in the supply waters were similar to or even greater than in the irrigation water.

9.2.3 Soil and soil column analysis

Soils in the Tula Valley

Eight soils taken to a depth of approximately 30 cm were analyzed in total from the three zones, covering a range of soil types and years under irrigation. The pH, texture, and notably the organic matter content are known to be important in influencing the retention of organic compounds. The pH of the soils in the Tula Valley are generally neutral therefore the acidic pharmaceuticals are likely to be in an ionized form and capable of movement whereas carbamazepine, 4-nonylphenols, triclosan, and bisphenol-A will tend to associate with organic material of the soil and be less likely to migrate downwards (Trolls, 2001; Topp, 2008; Williams *et al.*, 2006; Kinney *et al.*, 2006). Also, the fine texture of clays and loams favour the retention of micropollutants in micropores that can be an effectively irreversible process (Krahe *et al.*, 2006). The concentration of micropollutants in the soils (Table 9.6) varied from <LOD to 0.34 ng/g (ibuprofen), <LOD to 2.9 ng/g (naproxen), <LOD to 21.0 ng/g (carbamazepine), 33 to 316 ng/g (4-nonylphenols), <LOD to 30 ng/g (triclosan), and 2.0 to 71 ng/g (bisphenol-A). Based on concentrations in the wastewater it is possible to estimate the amount of the compounds added to the soil in a typical year, knowing that typical irrigation rates are approximately 2 m/yr (Table 9.7). This data suggests that there is no build up of ibuprofen, naproxen or bisphenol-A in the soil as concentrations were in general well below the estimated amounts that would be added over a year of irrigations. Concentrations of triclosan and 4-nonylphenols were also relatively low indicating degradation processes at work; however concentrations were generally greater than that which would be added over a year of irrigation, suggesting slower degradation. The concentration of carbamazepine in soils from zone 1 and 2 was up to 10 times the estimated annual addition, suggesting persistence. There did not appear to be any relation between length of time under irrigation and concentration of compounds, but concentrations were generally lower in zone 3 where wastewater mixed with well water was used for irrigation therefore lower amounts of the contaminants were added. Also, soil organic matter was much lower, suggesting less possibility for retention in the soil.

Table 9.6 Organic micropollutants in the soil (ng/g).

Soil i.d.	Ibuprofen		Naproxen		Carbamazepine		4-NPs		Triclosan		Bisphenol-A	
	Wet	Dry	Wet	Dry	Wet	Dry	Wet	Dry	Wet	Dry	Wet	Dry
Z_1SL_1	0.20 (±0.3)	0.10 (±0.01)	2.9 (±0.74)	0.6 (±0.1)	9.2 (±0.6)	6.6 (±0.02)	203 (±6)	77 (±10)	30 (±0.9)	8 (±1)	5.7 (±0.8)	2.8 (±0.4)
Z_1SL_2	0.16 (±0.04)	0.17 (±0.09)	1.2 (±0.02)	2.0 (±0.03)	3.0 (±0.1)	6.5 (±0.3)	172 (±17)	316 (±22)	20 (±2)	30 (±3)	71 (±8)	8.6 (±1.2)
Z_1SL_3	0.25 (±0.04)	0.13 (±0.02)	0.6 (±0.01)	1.2 (±0.09)	6.5 (±0.6)	13.0 (±0.7)	203 (±21)	63 (±6)	8.8 (±2.3)	4.6 (±0.1)	5.2 (±1.5)	2.0 (±1.3)
Z_2SL_1	0.15 (±0.02)	0.10 (±0.05)	0.9 (±0.1)	0.23 (±.09)	17.0 (±4.5)	4.6 (±0.5)	160 (±22)	64 (±21)	4.8 (±1.5)	3.4 (±0.2)	2.6 (±0.3)	2.6 (±0.3)
Z_2SL_2	0.34 (±0.15)	0.24 (±0.02)	1.5 (±0.2)	1.5 (±0.04)	21.0 (±3.5)	6.5 (±0.9)	275 (±77)	300 (±20)	21 (±8)	26 (±1.7)	8.6 (±2.4)	5.9 (±0.9)
Z_2SL_3	<LOD (–)	0.10 (±0.01)	0.48 (±0.11)	0.73 (±0.2)	1.7 (±0.1)	5.1 (±0.5)	120 (±13)	145 (±9)	3.4 (±0.4)	18.8 (±1.1)	6.7 (±1.2)	15.7 (±3.2)
Z_3SL_1	<LOD (–)	<LOD (–)	0.2 (±0.09)	<LOD (–)	<LOD (–)	<LOD (–)	77 (±3)	93 (±11)	1.9 (±0.4)	<LOD (–)	7.0 (±0.9)	5.6 (±0.5)
Z_3SL_2	<LOD (–)	<LOD (–)	<LOD (–)	<LOD (–)	<LOD (–)	<LOD (–)	33 (±5)	84 (±7)	<LOD (–)	<LOD (–)	3.2 (±0.5)	11.4 (±1.7)

Note: LOD = Limit of detection

Table 9.7 Analytes in the soils expressed as a percentage of estimated annual addition by irrigation with wastewater in the Tula Valley (Source: Duran *et al.*, 2009).

Analyte	Estimated annual addition (μg/kg)	Percentage of annual addition			
		Pha10	Pha90	Lep10	Lep90
4-Nonylphenols	52	166	139	108	210
Naproxen	41	0.7	1.1	1.3	1.5
Bisphenol-A	6.2	95	45	67	157
Ibuprofen	3.9	0.0	0.0	0.0	0.0
Triclosan	2.0	519	760	539	858
Carbamazepine	0.8	603	766	640	942

Note: Pha = phaeozem, Lep = Leptosol, 10 and 90 represent the lengths of time [years] that the soil has been irrigated with wastewater

Column study of the fate of organic micropollutants

The results for the soil columns were considered in terms of recovery of the compounds applied during irrigation, which consisted of the amounts present in the wastewater, 5000 ng from fortification, and the amounts present in the soil (which were generally low by comparison). The total amounts recovered from the leachates and soils of control columns irrigated with distilled water varied between <LOD and 0.05% of the amounts applied to the columns irrigated with fortified wastewater. Mean recoveries for the five leachates and three soil horizons for each of the four profiles irrigated with wastewater are shown in Figure 9.8a–b. Recoveries in the leachates varied from 5.9 to 37.2% for ibuprofen, 3.8 to 31.2% for naproxen, and from 0.9 to 16.1% for carbamazepine (Figure 9.8a). For all three, amounts in leachates were greatest in profile P4 and lowest in profiles P1 and P2, suggesting that retention in the soil during this type of periodic irrigation is directly related to the organic matter content, that is to say the more organic material the less leaching of compounds. Leaching of 4-nonylphenols (3.1 to 18.6%), triclosan (0.2 to 2.2%), and bisphenol-A (0.4 to 7.9%) was lower than for the acidic pharmaceuticals but again profile P4 showed the greatest vulnerability (Figure 9.8b).

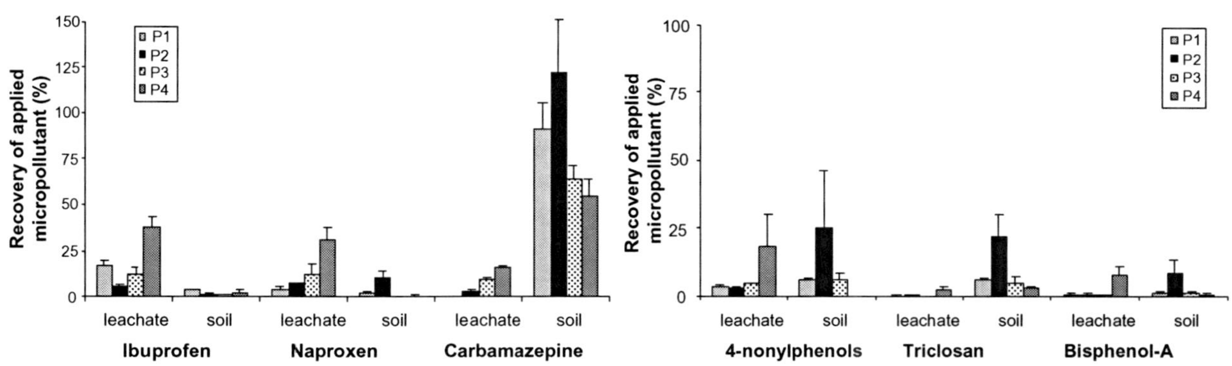

Figure 9.8 a–b Total recovery (%) of micropollutants applied to the columns in the leachates and soils (Error bars show standard deviation)

Recoveries of ibuprofen (0.5 to 3.5%) and naproxen (0.4 to 10.2%) in the soil suggested degradation was relatively rapid and prevented any build-up of the acids over time. The same was also broadly true for 4-nonylphenols (<LOD to 24.9%), triclosan (2.9 to 22.1%), and bisphenol-A (0.6 to 8.5%) although recoveries were generally greater for these phenols. In all cases, residual amounts of these compounds were greatest in profile P2, the profile with high organic matter content and lowest in profiles P3 and P4 where organic matter content was low. It is possible that adsorption to organic matter may result in the analytes being less available for degradation. Recovery of carbamazepine (54.9 to 123.1%) reflected the resistance to degradation and its probable build-up in soils over time, as well as the slow but continual release of this relatively polar and persistent compound into groundwater.

This corroborates the data from soils and supply waters generated in the sampling campaigns in Tula Valley. There was some difference between the profiles, with recoveries again being greater from soils with greater organic matter content. Table 9.7 shows the amounts of the analytes in the soils expressed as a percentage of the estimated annual amount added in the Tula Valley. The soil columns suggested rapid degradation of the acidic pharmaceuticals, and this was borne out in the field samples where there was no sign of accumulation, concentrations being less than 2% of annual addition. The evidence for build up of carbamazepine in the soil columns was replicated in the field samples where carbamazepine concentrations were between 640 and 942% of the annual addition, indicating some persistence in the soil. Concentrations of triclosan and carbamazepine were also greater in the soils that had been irrigated for 90 years. These soils have received many more irrigations, and thus greater additions of triclosan and carbamazepine. This has resulted in greater residual concentrations in the soils (Duran *et al.* 2009).

9.3 CONCLUSIONS

The case study of the Tula Valley is different from the other case studies being considered in Reclaim Water, because the aquifer recharge is an incidental result of the use of untreated wastewaters for irrigation of farmland, and operates on a far larger scale. The wastewater entering the Tula Valley from Mexico City is almost entirely without treatment, and therefore it contains a high concentration of a variety of contaminants including microorganisms, organic micropollutants, salts, and various metals. The atypical natural soil aquifer treatment receiving highly contaminated wastewater removes the vast majority of most contaminants. The soils are important in the removal of microorganisms, metals, and organic micropollutants. The column study and field soil sample analysis demonstrated that most organic micropollutants are removed and degraded, with the exception of carbamazepine which was persistent and polar enough to move through the soil and contaminate the aquifer. The infiltrated water complies with most potable water criteria. However there are concerns over the quality of the water, for example:

- Most of the microorganisms evaluated were detected with a relatively high frequency in wells, springs, and dugwells.
- Most of the parameters related to salinity, such as sodium, potassium, sulphates, chlorides, and electrical conductivity, indicate that the supply water is at least as saline as the wastewater and in some cases more so.
- Carbamazepine, a persistent and relatively polar micropollutant, was shown to have accumulated in soils and was occasionally present in wells and springs. This may indicate a vulnerability of specific sites to contamination with other micropollutants with similar properties.
- Nitrates and total dissolved solids exceeded regulatory limits in some sources of water; the same was true of fluorine and aluminium.
- Metal accumulation in the soils is occurring although regulatory limits have not yet been exceeded.

A simple measure that could be applied quickly would be better protection of the sources of supply (particularly dugwells, and springs) to avoid incidental contamination with microorganisms. Improvement in the water to reach the standards necessary for potable supply could be achieved using an advanced treatment such as nanofiltration which would remove most of the contaminants and would result in a high degree of confidence in the quality of the water produced. Studies at laboratory and pilot level would be necessary to compare different membrane processes, investigate the efficiency of the membranes and the costs of the process, and most importantly to ensure that the resulting water was of a standard fit for human consumption. Finally, it is important to evaluate the population health risks associated with the consumption of the "naturally treated" water, and water quality indicators should be proposed that serve to verify the functioning of the mechanisms of potablization of the water in the region, both in a form of monitoring and also periodic evaluation.

ACKNOWLEDGEMENTS

We would like to acknowledge the Secretaría de Medio Ambiente y Recursos Naturales and the CONACyT of the Mexican government (No. SEMARNAT-2002-01-0519) and the European Commission (RECLAIM WATER; project number: 018309) for supporting this research.

Many thanks to the staff of the treatment and reuse group, Juan Carlos Duran, Claudia Hernández, Lucina Mendoza, Soledad Lucario, Erika Becerril, Karina León and Alfonso Soriano, for their support in the laboratory during the conduct of this research.

REFERENCES

Adams M. H. (1959). Coliphages. Intercience Publisher, Inc., New York.

American Public Health Association (APHA), American Water Works Association and Water (AWWA), and Water Environment Federation (WEF). (1998). Standard methods for the examination of water and wastewater, 20th edition. American Public Health Association, Washington D.C.

Blumenthal U., Mara D., Ayres R., Cifuentes E., Peasey A., Scott R., Lee D. and Ruiz-Palacios G. (1996). Evaluation of the WHO nematode egg guidelines for restricted and unrestricted irrigation. *Water Science and Technology*, **33**(10–11), 277–283.

BGS, LSHTM and UB. (1998). Effects of wastewater reuse on groundwater in the Mezquital Valley, Hidalgo State, Mexico. Final Report- November 1998. BGS Technical Report WC/98/42.

Craun G. F. (1985). A summary of waterborne illness transmitted through contaminated groundwater. *J Environ Health*, **48**, 122–7.

Cronin A. A., Taylor R. G., Powell K. L., Barrett M. H., Trowsdale S. A. and Lerner D. N. (2003). Temporal variations in the depth-specific hydrochemistry and sewage related microbiology of an urban sandstone aquifer, nottingham, United Kingdom. *Hydrogeology Journal*, **11**, 205–216.

DOF, Diario Oficial de la Federación. (2000). Modificación a la norma oficial mexicana NOM-127-SSA1-1994, salud ambiental. Agua para uso y consumo humano. Límites permisibles de calidad y tratamientos a que debe someterse el agua para su potabilización. Secretaría de Salud, Publicada el 20 de junio de 2000. *In Spanish.*

Durán-Álvarez J. C., Becerril-Bravo E., Silva Castro V., Jiménez B. and Gibson R. (2009). The analysis of a group of acidic pharmaceuticals, carbamazepine, and potential endocrine disrupting compounds in wastewater irrigated soils by gas chromatography-mass spectrometry. *Talanta*, **78**, 1159–1166.

EU. (1998). The Quality of Water Intended for Human Consumption. Council Directive 98/83/EC. 3 November 1998.

Fent K., Weston A. A. and Caminada D. (2006). Review: ecotoxicology of human pharmaceuticals. *Aquatic Toxicology*, **76**(2), 122–159.

Fong T.-T., Mansfield L. S., Wilson D. L., Schwab D. J., Molloy S. L. and Rose J. B. (2007). Massive microbiological groundwater contamination associated with a waterborne outbreak in Lake Erie, South Bass Island, Ohio. *Environmental Health Perspectives*, **115**(6), 856–864.

Foster S. and y Chilton P. (2004). Downstream of downtown: urban wastewater as groundwater recharge. *Hydrogeology Journal*, **12**, 115 –120.

García E. (1988). Modification to the Köppen climate classification to adapt it to Mexico, Offset Larios, S.A. 4th edition. México, D.F. 217. *In Spanish.*

Gerba C. and Bitton G. (1984). Microbial pollutants: their survival and transport pattern to groundwater. In: Groundwater Pollution Microbiology, G. Bitton and C. y Gerba (eds), Wiley & Sons, NY, pp. 65–88.

Gerba C. P., Wallis C. and Melnick J. L. (1975). Fate of wastewater bacteria and viruses in soil. *Journal of the Irrigation and Drainage Division*, **101**, 157–174.

Gibson R., Becerril-Bravo E., Silva-Castro V. and Jiménez B. (2007). Determination of acidic pharmaceuticals and potential endocrine disrupting compounds in wastewaters and spring waters by selective elution and analysis by gas chromatography – mass spectrometry. *Journal of Chromatography A*, **1169**(1–2), 31–39.

Jean J. S. (2000). Reply to comments by P. Gale and others on "Outbreak of enteroviruses and groundwater contamination in Taiwan: concept of biomedical hydrogeology (Jean 1999)". *Hydrogeology Journal*, **8**, 350–353.

Jiménez B. (2005). Treatment technology and standards for agricultural wastewater reuse: a case study in Mexico. *Irrigation and Drainage*. **54**(1), 23–35.

Jiménez B. and Chávez A. (2004). "Quality assessment of an aquifer recharged with wastewater for its potential use as drinking source: El Mezquital Valley case". *Water Science and Technology*, **50**(2), 269–273.

Jiménez B. and Chávez A. (2005). Water quality in an aquifer recharged with wastewater and its possible use for drinking purposes in Mexico. *Latin American Journal of Hydrology (in Spanish)*, **5**, 111–116.

Jiménez B., Chávez A. and Hernández C. (1999). Alternative wastewater treatment intended for agricultural use. *Water Science and Technology*, **40**(4–5), 355–362.

Jiménez B. and Chávez A. (1997). Treatment of Mexico city wastewater for irrigation purposes. *Environmental Technology*, **18**(7), 721–736.

Jiménez B., Chávez A., Gibson R. and Maya C. (2008). Water reclamation technologies for safe artificial ground water recharge reclaim water. CASE STUDY Tula Valley. *Final_Deliverable*, 35 pages, December 5, 2008. European Commission (Research Directorate - General) Reclaim Water. Contract No. GOCE-018309. 24.

Katz B. and Griffin D. W. (2008). Using chemical and microbiological indicators to track impacts from the land application of treated municipal wastewater and other sources of groundwater quality in a karstic springs basin. *Environmental Geology*, **55**(4), 801–821.

Kinney C. A., Furlong E. T., Werner S. L. and Cahill J. D. (2006). Presence and distribution of wastewater-derived pharmaceuticals in soil irrigated with reclaimed water. *Environ. Toxicol. Chem.*, **25**, 317–326.

Krahe S., Düring R. A., Huisman J. A., Horn A. L and Gäth S. (2006). Statistical modeling for the partitioning of nonylphenol in soil. *Water, Air and Soil Pollution*, **172**, 221–237.

Maya C., Jimenez B. and Schwartzbrod J. (2006). Comparison of techniques for the detection of helminth ova in drinking water and wastewater. *Water Environment Research*, **78**(2), 118–124.

Miossec L., Le Guyader F., Haugarreau L. and Pommepuy M. (2000). Magnitude of rainfall on viral contamination of the marine environment during gastroenteritis epidemics in human coastal population. *Journal of Epidemiology and Public Health (in French)*, **48**(2), 2562–2571.

Modella P., Blanco I., Ansola G. and de Luis G. (2008). Removal of wastewater pathogen indicators in a constructed wetland in leon, spain. *Ecological Engineering*, **33**(3–4), 252–257.

PAHO. (1988). Soil Map of the World, Revised Legend. Reprinted with Corrections. World Soil Resources Report 60, FAO, Rome.

Taylor R., Cronin A., Pedley S., Barker J. and Atkinson T. (2004). The implications of groundwater velocity variations on microbial transport and wellhead protection-review of field evidence. *FEMS Microbiol Ecology*, **49**(1), 17–26.

Toop E., Hendel J. G., Lapen D. R. and Chapma P. (2008). Fate of the nonsteroidal anti-inflammatory drug naproxen in agricultural soil receiving liquid municipal biosolid. *Environmental Toxicology and Chemistry*, **27**(10), 2005–2011.

Toze S., Bekele E., Page D., Sidhu J. and Shackleton M. (2009). Use of static quantitative microbial risk assessment to determine pathogen risks in an unconfined carbonate aquifer used for managed aquifer recharge. *Water Research*, **44**(4), 1038–1049.

Trolls J. (2001). Sorption of veterinary pharmaceuticals in soils: a review. *Environmental Science and Technoloy*, **35**(17), 3397–3406.

Tzanakakis V. E., Paranychianaki N. V. and Angelakis A. N. (2007). Soil as a wastewater treatment system: historical development. *Wat. Sci. and Tech.: Water Supply*, **7**(1), 67–75.

United States Environmental Protection Agency. (1992). Control of Pathogens and Vector Attraction in Sewage Sludge. Report EPA/625/R-92/013. Environmental Regulations and Technology, Washington, DC.

Williams C. F., Williams C. F. and Adamsen F. J. (2006). Sorption-desorption of carbamazepine from irrigated soils. *Journal of Environmental Quality*, **35**, 1568–1574.

Yates M. and Yates S. (1988). Modeling microbial fate in the subsurface environment. *Crit. Rev. Env. Control.*, **17**, 307–44.

Yavuz-Corapcioglu M. and y Haridas A. (1985). Microbial transport in soils and groundwater: a numerical model. *Advances in Water Resources*, **8**, 188–200.

Zelikson R. (1994). Microorganisms and viruses in groundwater. In: Groundwater Contamination and Control, U. Zoller (ed.), Marcel Dekker Inc., New York, pp. 425–436.

Chapter 10

Managed aquifer recharge by enhanced direct injection-well recharge in Gaobeidian/Beijing, China

Cheng Xu Zhou and Xuan Zhao

10.1 INTRODUCTION

In China, the total annual quantity of water resources is $2.88 \cdot 10^{12}$ m^3 or a per-capita average of 2220 m^3yr^{-1}, about 1/4 of the world average value. The water resource in China is unevenly distributed both temporally and spatially. The water quantity is comparatively small in the Yellow River, the Huaihe River, the Haihe River and the Liaohe River, and comparatively great in the Yangtze River, the Pearl River and the Songhua River. It is exiguous in the North West China and abundant in the South East areas of China. The distribution of the water resources is not in consistence with the land resource availability and the social economic development levels.

The total water withdrawal in 2005 were estimated at 563.3 cubic kilometres, with the demand coming from agriculture (69%), industry (23%) and urban (8%) sectors, accounting for around 70% of the total available water resource in China. According to the national long-term development plan, in 2030 the social, economical and ecological water demand will reach 700 cubic kilometres or a per-capita average of 1700 m^3. In addition to the stressing water supply situation, water pollution adds to water shortage as all the main rivers, 90% of the urban surface water bodies and 97% of the urban groundwater bodies are to some extent polluted. The enhancement of integrated management, the improvement of utilization efficiency, and the building of scientific supporting systems for water resources are urgently needed.

As the cities of north China suffer from groundwater overexploitation with dropping groundwater levels, land subsidence, and salt water intrusion especially in the coastal areas, groundwater recharge is gaining more and more attention to relieve these problems.

10.1.1 Historical background

The Gaobeidian Wastewater Treatment Plant is the largest of 15 operational or planned wastewater treatment plants of the Beijing Municipality Master Plan (Figure 10.1). Gaobeidian has a capacity of 1 million m^3/d equivalent to about 40% of the total sewage discharge of Beijing. The plant provides wastewater treatment for a catchment area of 9661 hectares reaching from the city centre to the eastern suburbs. Its serviced population is more than 2.4 million. The plant itself covers an area of 68 hectares.

Considering the inadequacies of financing and sewage network in the upper reaches of the catchment, Gaobeidian WWTP was constructed in two phases. Phase I, with a capacity of 500,000 m^3/day, was started in 1990 and completed in 1993, as an investment of RMB 524 million Yuan (~US$ 63 million). During maximum load operation effluent quality meets the discharge standards. Phase II, with a further capacity of 500,000 m^3/day was started in 1995 and completed in September 1999, as an investment of RMB 1.12 billion Yuan (~US$ 134 million).

The treatment process of Gaobeidian WWTP is a conventional activated sludge secondary treatment (Figure 10.2). The primary treatment is composed of a bar screen, pumping station, aerated grit chamber and rectangular horizontal flow primary sedimentation tank. The secondary treatment is a conventional activated sludge process with biological nutrient removal (anaerobic-anoxic-aerobic). Secondary mesophilic anaerobic digestion is employed in sludge treatment. The sludge cake from dewatering plants are trucked out and used as fertilizer in farmland and grassland. About 20% of the electricity used in the plant is generated by biogas produced from anaerobic digestion.

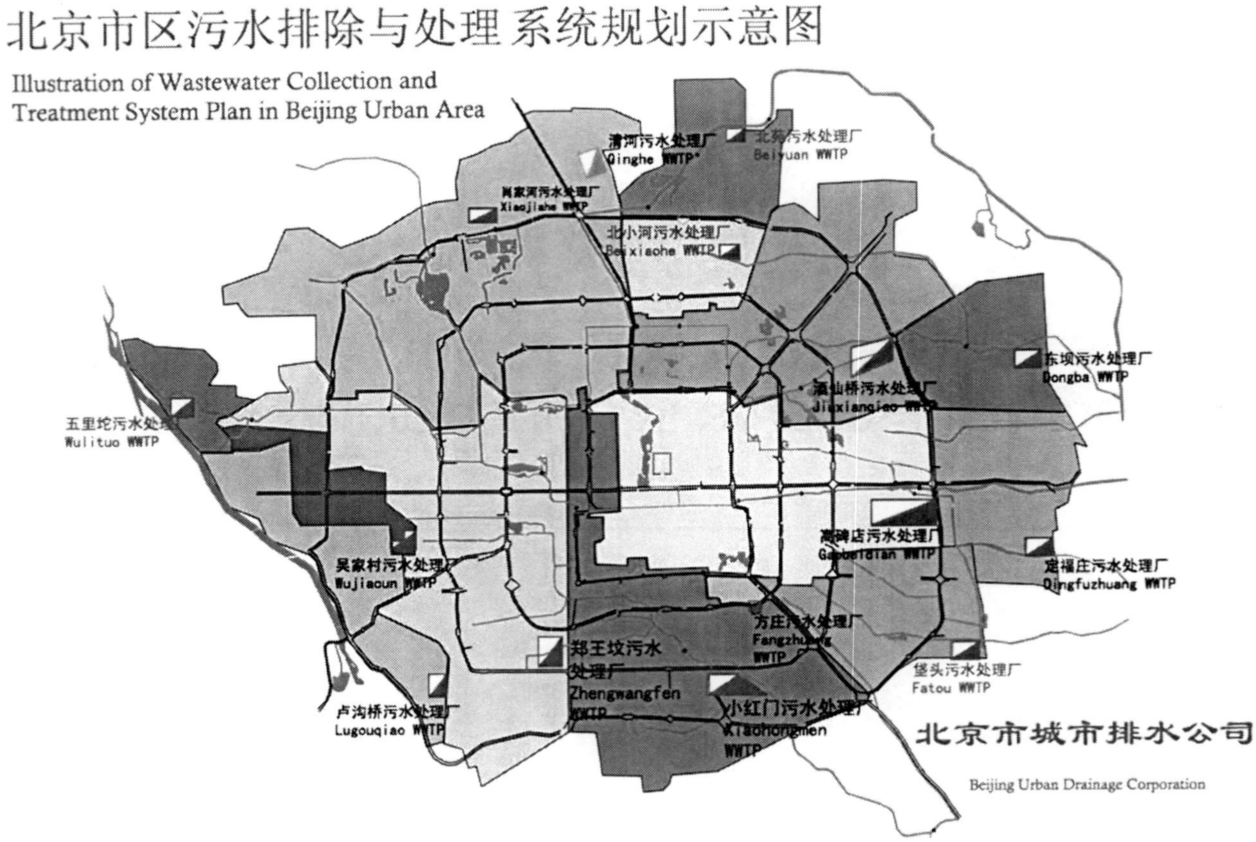

Figure 10.1 Wastewater collection and treatment system master plan for the Beijing urban area

Figure 10.2 Aerial view of the Gaobeidian WWTP

In addition, the plant avails of advanced treatment facilities consisting of flocculation with polyaluminium chloride (PACl), sedimentation and rapid sand filtration with a capacity of 10,000 m^3/d. The effluent from the advanced treatment process is used for washing the belts of the dewatering process, and watering the grassland in the plant. The plant complies with ISO9002, ISO14001 and OHSAS 18001 standards.

The construction and stable operation of the plant has significantly contributed to the water pollution control in the centre and eastern suburbs of Beijing, and has an effective function in improving water quality of the Tonghui River. It also played a very important role for the improvement of the water environment of the Haihe River.

10.1.2 Motivations for recharge and use of abstracted water

Beijing is a city of very severe water shortage. The groundwater is heavily over-exploited. Secondary effluent may serve as an alternative water source with reuse via managed aquifer recharge. However, as groundwater recharge was not very well understood and recognized, it was decided to build a demonstration plant to gain experience for the implementation of full scale managed aquifer recharge projects.

10.1.3 Legal framework and authorisation procedure

Concerning municipal wastewater recycling and reuse a series of national standards has been issued since 2003, including reuse of urban recycling water standards for urban miscellaneous water consumption, scenic environmental use, industrial and agricultural uses, and for groundwater recharge. The national guideline "The reuse of urban recycling water – water quality standard for artificial groundwater recharge" (GB/T 17992-2005) was issued in 2005 by the Ministry of Construction.

Water quality requirements

The recharge water quality standard GB/T 17992-2005 was the first standard of such kind in China. The standard describes basic water quality parameters, and selected parameters and their corresponding limits for surface spreading and well injection. Municipal wastewater should be treated to meet the limits of the basic parameters and site specific parameters before being recharged into the aquifer. This standard also describes the monitoring points, sampling frequency and analysis methods of all the water quality parameters. The standard also gives provision on the aquifer characteristics and retention time of recharged water in the aquifer before recovery.

Approval practice and responsible administration

Concerning artificial recharge projects, the Ministry of Water Resources, the Ministry of Construction, the Water Bureau, the National Development and Reform Commission, and the Ministry of Environmental Protection and others share the function of administration. As mentioned above, the Ministry of Construction had issued the water quality standard of reclaimed municipal wastewater for artificial recharge, at the same time the Ministry of Water Resources is now initiating artificial groundwater recharge and conservation projects with storm water or surface water.

The permission of the construction and operation of the Gaobeidian pilot reclamation and recharge scheme was granted by the Beijing Drainage Group Ltd in 2002 when the first artificial recharge pilot plant was built at the Gaobeidian Wastewater Treatment Plant.

10.2 DESCRIPTION OF THE TEST SITE

10.2.1 Study area

The managed aquifer recharge demonstration plant is located along the Western side of Gaobeidian Wastewater Treatment Plant in Beijing. The recharge area is about 200 meters long and 30 meters wide and secondary effluent from the Gaobeidian WWTP is used to recharge the aquifer after pre-treatment.

Before construction of the pilot plant, a field investigation was carried out including a survey of historic data, geological drilling, analysis of physical, chemical and hydraulic properties of soil samples at different depths, and analysis of groundwater samples. Table 10.1 summarises the results from the field drilling investigation.

Table 10.1 Hydrogeological characteristics of the test site.

Depth (m)	Lithology description	Aquifers
0–9.6	Surface layer is filling soil, down layer is powdery clay, viscid powdery soil with a thin layer powdery sand	
9.6–14.0	Fine sand, fine-mid sand	Aquifer I
14.0–17.5	Round gravel	
17.5–24.8	Powdery clay, sandy powdery soil	
24.8–28.0	Silver sand	Aquifer II
28.0–33.2	Medium, coarse sand in upper layer and cobble stone in lower layer	
33.2–35.0	Clay	
35.0–39.4	Silver sand, medium silver sand	Aquifer IIa

(Continued)

Table 10.1 Hydrogeological characteristics of the test site (*Continued*).

Depth (m)	Lithology description	Aquifers
39.4–45.0	Powdery clay	
45.0–50.5	Mainly round gravel with silver sand in top and bottom	Aquifer III
50.5–56.0	Clay and powdery soil alternant	
56.0–62.0	Mainly silver sand with a cobble stone layer in the lower part	Aquifer IV
62.0–77.8	Clay with a thin layer of powdery sand	
77.8–80.0	Silver sand (not drilled to the bottom)	

10.2.2 Process design and operation

Due to very high land prices in Beijing and low permeability of the upper soil layer, managed aquifer recharge could not be achieved by soil aquifer treatment and had to be performed via well injection. However, expensive pre-treatment such as double membrane filtration, often required for well injection, is not acceptable from a financial point of view. In search of an alternative solution, a special recharge concept called "enhanced direct injection-well recharge (EnDir)" was developed which first treats the reclaimed water in a spreading basin with 1 m deep sand layer simulating the processes in the vadose zone and then recharges the aquifer via an injection well.

Tertiary treated effluent was used at this site for aquifer recharge. The secondary effluent (conventional activated sludge treatment with biological nutrient removal) from the Gaobeidian WWTP was firstly dosed with 30 mg PACl/L, and then went through sedimentation and rapid sand filtration for SS removal. The effluent was then ozonated with a dosage of 10–15 mg/L before being diverted to the slow sand filtration basin and conveyed into wells for recharge of an alluvial aquifer mainly composed of alternating sand and clay layers. The information of the Gaobeidian pilot reclamation plant is summarized in Table 10.2.

Table 10.2 Summary of the Gaobeidian pilot reclamation plant.

Status of project	Operating since 2006
Capacity	Tertiary treatment 10,000 m^3/d Ozonation + EnDir* MAR 350–500 m^3/d
Source water	WWTP secondary effluent from Gaobeidian WWTP
Tertiary treatment	Coagulation (30 mg/L PACl) + sedimentation + rapid sand filtration
Advanced treatment	Ozonation (10–15 mg/L) + slow sand filtration ($A = 400\ m^2$, $h_{basin} = 1$ m) with percolation velocity of 0.05 m/h, intermittent flooding (2–3 days flooding, 1 day drying)
MAR	ASTR: direct injection in aquifer I (−9.6 to −17.5 m) minimum travel time: 20 days

*EnDir = Enhanced direct injection-well recharge.

The scheme and the sampling points are shown in Figure 10.3.

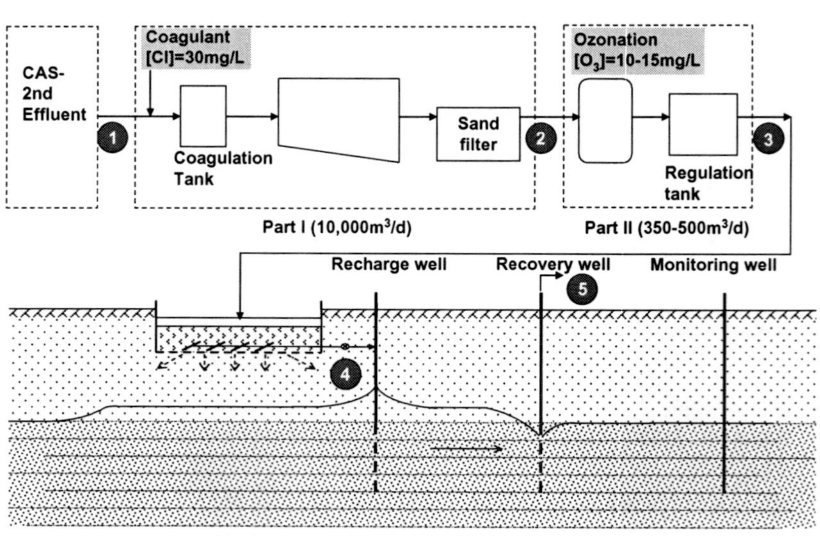

Figure 10.3 Scheme of the Gaobeidian case study site

The ozonation plant (cf. Figure 10.4) is of 6.5 kW with an ozone generating rate of 200g/h. The pre-treatment plant and EnDIR MAR system commenced operation in 2006 (Figures 10.5, 10.6 and 10.7).

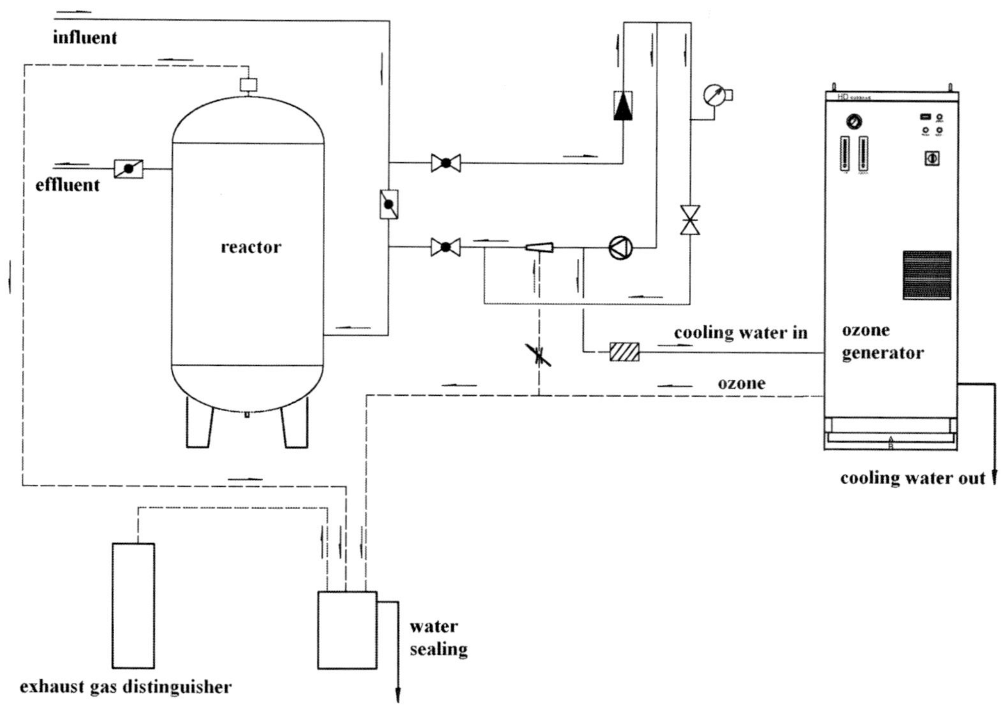

Figure 10.4 Ozonation facilities

Original spreading basins were transformed into a sand infiltration basin for further water quality improvement after pre-treatment. The sand infiltration basin has a 1 meter deep sand layer and water distribution – collection pipes system. The water enters the recharge site and passes a sand layer. Then drainage pipes divert the water to the infiltration well with a continuously screened section between 9.6 and 17 m. The drainage pipes (Figure 10.5) are covered with a felt layer to prevent clogging of pipes by sand particles. The basin is filled with treated water up to around 20 cm above the sand surface. Then the water level decreases until it reaches the height of the effluent pipe. Figure 10.5 and 10.6 show the slow sand filtration basin being built and in operation.

Figure 10.5 Slow sand filtration basin during construction and after completion

The scheme was in stable operation since June 2006 except for the frozen period in winter. The slow sand infiltration basin was filled with 1 m deep gravel and quartz sand providing a 1 day retention time. Regular cleaning of the sand basin was carried out during the operation period. The injection rate through the recharge well was 20 m^3/h. No clogging was found in the recharge well during the 3 years operation.

Figure 10.6 Sand filtration basin in operation in wet and dry weather

Figure 10.7 The recharge well and sampling wells

10.3 TECHNOLOGY PERFORMANCE AND CONTAMINANT MONITORING

Over the duration of the Reclaim Water project, 11 sampling campaigns relative to basic wastewater parameters and 3 micropollutant sampling campaigns have been carried out in Gaobeidian. Some additional analyses were carried out including GC/MS, EEM, AOC, BDOC and molecular weight distribution analysis in parallel with sampling and analysis done for the Reclaim Water project.

10.3.1 Water quality

Table 10.3 to 10.4 give an overview of the average results obtained during the 2nd year of the project (2007) when the system had reached stable operation. The treated effluent before recharge has a DOC of about 5.8-5.2 mg/L. ozonation removes then around 15% of the DOC (0.8 mg/L) As the final polishing step before the recharge, the slow sand infiltration contributed to additional 2 mg/L DOC removal through biodegradation. As shown in Table 10.3 the recovered water had a DOC of 1.3 mg/L in average. The total DOC removal from the secondary effluent to the recovered water was 78%.

Table 10.3 Basic water analysis (average concentrations during 2007, n = 5).

Sampling point		Secondary effluent 1	After flocculation + filtration 2	After ozonation 3	After slow sand (recharged water) 4	Recovered water 5
Suspended Solids	mg/L	11.3 ± 3.2	<5	<5	<5	<5
BOD Soluble	mg/L	<5	<5	<5	<5	<5
COD Soluble	mg/L	41.0 ± 8.6	40.4 ± 9.0	33.5 ± 6.4	29.8 ± 13.4	17.2 ± 5.6
DOC	mg/L	5.8 ± 0.8	5.2 ± 0.6	4.4 ± 0.9	2.4 ± 0.5	1.3 ± 0.2
Ammonia (NH_4-N)	mg/L	0.76 ± 1.44	0.71 ± 1.32	0.11 ± 0.09	0.03 ± 0.01	0.03 ± 0.01

(Continued)

Table 10.3 Basic water analysis (average concentrations during 2007, n = 5) (*Continued*).

Sampling point		Secondary effluent 1	After flocculation + filtration 2	After ozonation 3	After slow sand (recharged water) 4	Recovered water 5
Nitrate NO_3-N	mg/L	31.3±6.3	28.8±3.4	28.3±1.7	28.7±3.1	13.8±11.3
Total Nitrogen	mg/L	34.9±3.0	33.4±2.2	32.5±2.0	31.9±2.2	17.1±11.8
Total Phosphorus	mg/L	4.3±0.5	2.1±1.6	2.6±2.1	2.7±1.4	0.2±0.1
Alkalinity	mg/L	151.9±110.0	126.0±75.7	122.2±66.2	136.9±55.0	140.7±90.7
pH	unit	7.6±0.2	7.6±0.2	7.5±0.3	7.9±0.2	7.9±0.0

During the pre-treatment process the ammonia was almost completely removed. The nitrate nitrogen as well as the total nitrogen showed little variance during the pre-treatment steps, but about half the nitrate was generally denitrified or diluted during travel and storage in the aquifer. The total P was also removed mainly during the aquifer passage presumably due to adsorption to the aquifer materials. As given in Table 10.4, quite variable pathogen concentrations were observed. The aquifer passage proved to be a robust barrier since no faecal coliforms or *E. coli* were found in the recovered water. The total bacteria were also significantly reduced.

Table 10.4 Microbiological parameters (average concentrations during 2007, n = 5).

Sampling point		Secondary effluent 1	After flocculation + filtration 2	After ozonation 3	After slow sand (recharged water) 4	Recovered water 5
Total Bacteria count (37°C)	CFU/100mL	122400±155500	74800±42600	38800±9900	49600±43800	22000±8700
Faecal coliforms	CFU/100mL	600±1230	430±830	54±78	9±10	ND
E. coli	CFU/100mL	4930±10700	4530±9770	92±91	15±10	ND

Table 10.5 presents the values of salinity related parameters and trace elements. The groundwater has a relatively high concentration in chloride (i.e. between 84 mg/L for the secondary effluent and 79 mg/L in the abstracted water) and other inorganic compounds. The concentrations of nickel and lead in the abstracted water, 42 μg/L and 25 μg/L, respectively, did not comply with drinking water standards. Without further treatment, the recovered water could thus only be used for non-potable purposes.

Table 10.5 Analysis of salinity related parameters and trace elements (average concentrations during 2007, n = 5).

Sampling point		Secondary effluent 1	After flocculation + filtration 2	After ozonation 3	After slow sand (recharged water) 4	Recovered water 5
Salinity related analysis						
Chloride	mg/L	84.3±4.3	82.8±4.2	83.6±2.5	79.8±1.1	78.8±1.2
Electrical conductivity	μS/cm	880±279	810±228	803±221	836±230	989±335
Sodium	mg/L	114.4±37.4	91.0±28.6	96.4±33.1	98.4±32.2	109.8±18.2
Potassium	mg/L	10.9±1.4	11.1±1.9	11.1±2.2	9.8±1.4	6.4±5.7
Calcium	mg/L	74.2±16.8	71.0±23.0	71.3±22.1	81.4±20.9	85.3±25.1
Magnesium	mg/L	35.9±5.3	33.7±5.4	33.6±6.0	34.8±10.1	48.8±17.3
Carbonate	mg/L	0	0	0	0	0
Bicarbonate	mg/L	136.4±93.7	124.6±79.1	126.2±62.4	134.0±56.0	136.1±76.4
Trace elements						
Cadmium	μg/L	1.75±1.27	1.37±1.19	1.43±0.97	1.33±0.80	1.78±1.23

(*Continued*)

Table 10.5 Analysis of salinity related parameters and trace elements (average concentrations during 2007, n = 5) (*Continued*).

Sampling point		**Secondary effluent** 1	**After flocculation + filtration** 2	**After ozonation** 3	**After slow sand (recharged water)** 4	**Recovered water** 5
Chromium	μg/L	2.00±1.67	2.41±2.33	0.99±0.42	1.30±0.87	1.53±1.02
Cobalt	μg/L	6.5±6.8	7.3±7.5	6.8±7.1	5.9±6.2	6.0±6.0
Copper	μg/L	4.9±5.3	15.3±19.5	11.5±14.2	6.1±6.7	2.6±1.7
Fluoride	μg/L	115.0±15.4	116.2±17.2	108.4±13.8	190.2±19.5	231.9±16.6
Iron	μg/L	39.5±4.9	13.1±9.0	14.0±4.9	8.4±4.9	11.2±9.7
Lead	μg/L	34.8±19.7	20.3±11.9	21.8±11.5	20.2±10.4	24.9±17.3
Nickel	μg/L	39.5±28.7	39.6±29.5	32.9±20.6	28.6±16.4	41.8±33.6
Cyanide	μg/L	<2.0	<2.0	<2.0	<2.0	<2.0
Zinc	μg/L	34.0±32.2	34.4±32.2	33.3±31.3	38.9±35.9	33.6±32.3

Figure 10.8 shows the shift of molecular weight distribution via ozonation and slow sand infiltration. The DOC levels for the 3 water samples were 6.68 mg/L, 5.06 mg/L and 2.17 mg/L, respectively. The parts of DOC of secondary effluent in the range of 5~10 KDa and 500 Da~1 KDa increased during ozonation and were relatively more efficiently removed during slow sand infiltration step. Ozone showed preferential removal of smaller MW fraction of <500 Da to the fraction of >10 KDa which were less removed during sand infiltration. The ozonation and sand infiltration showed mutual beneficial DOC removal properties.

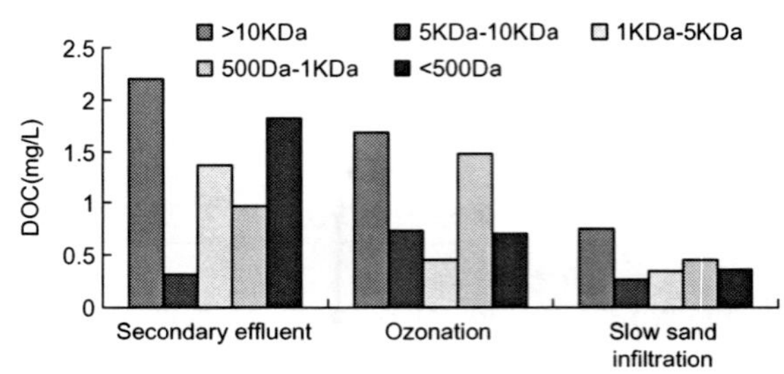

Figure 10.8 Molecular Weight distribution analyses

The Fluorescence EEM spectroscopy results give the differentiation between the humic like and protein like substances in the secondary effluent and reclaimed wastewater after the slow sand infiltration step (Figure 10.9).

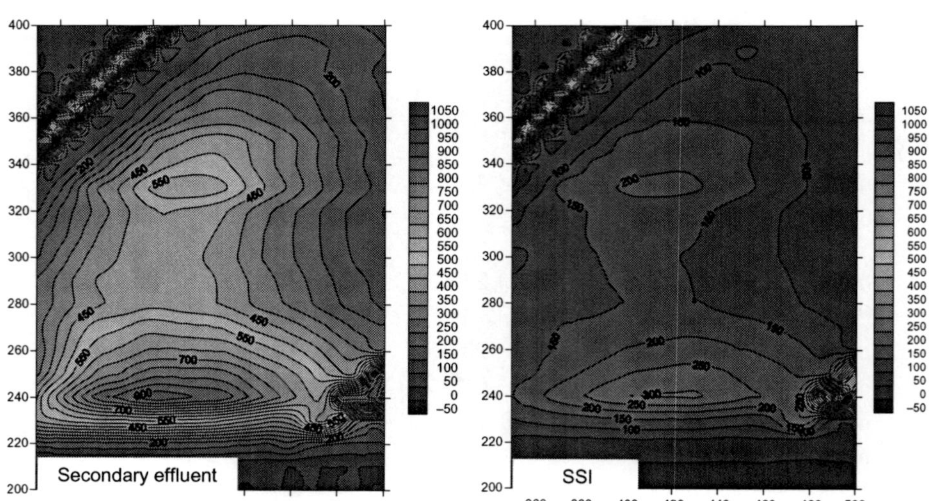

Figure 10.9 EEM spectra showing removal efficiency of the combination of ozonation and slow sand infiltration (SSI) at Gaobeidian site

As we could observe, the humic-like substances and the fulvic acid-like substances showed nearly the same ratio in DOC between secondary effluent and SSI effluent. This impression was also acquired from a rough volume calculation of corresponding areas of the EEM spectra. The SMP-like and aromatic protein-like substances were removed preferentially to humic-like and fulvic acid-like substances. This meant that in this case the added organics in the secondary effluent have a better degradability than the original natural organic matter.

BDOC tests were also carried out to characterise slow sand infiltration (SSI). The ratio BDOC/DOC dropped from 35.8% to 9.24% through SSI which again verified the effectiveness of slow sand infiltration as an artificial vadose zone.

The removal of DOC by ozonation was not significant considering the high dosage but it contributed to biodegradability improvement and control of microbiological parameters. The analysis on ozonation such as MW distribution and EEM analysis further supported this concept.

10.3.2 Operational feedback

During the three years of operation from 2006 to 2009, the surface sand infiltration basin was regularly cleaned to remove the upper sediment layer and grasses and no clogging was encountered. Figure 10.10 shows the basin before and after cleaning of the surface.

Figure 10.10 The slow sand infiltration basin before and after cleaning

As we can see from the above results, the natural system combined with gravity well recharge, as illustrated in Figure 10.11, demonstrated the feasibility to implement economical artificial recharge projects with reclaimed wastewater especially in areas of low surface soil permeability.

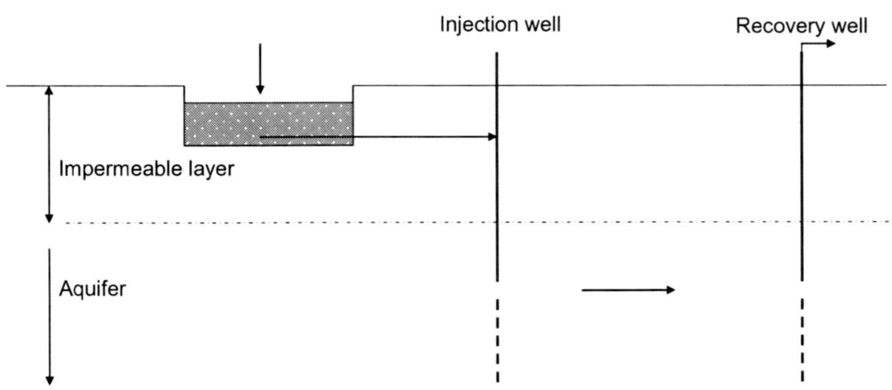

Figure 10.11 Concept of slow sand infiltration in combination with gravity well recharge

10.4 CONCLUSION

The employed MAR method of enhanced direct injection-well recharge (EnDir) proved to be a good efficient solution for the boundary conditions in Beijing with limited land availability. The ozonation contributed to biodegradability improvement while the slow sand infiltration further removed DOC. Overall the product water had very low DOC values below 1.5 mg/L. The pre-treatment system comprising PACl coagulation, ozonation, and slow sand infiltration ensured stable operation of the Gaobeidian site.

The Gaobeidian recharge scheme demands less space and less pre-treatment than soil aquifer treatment, but at the same time provides comparable treatment. The combination of slow sand infiltration and well recharge showed that clogging was managed by cleaning the infiltration basin only and this combination would be especially applicable on areas with low surface infiltration rates for implementing managed aquifer recharge projects.

Insufficient removal was observed for several heavy metals which would prevent potable reuse of the water. Parallel to the investigated pre-treatment with ozone, tests with GAC filtration (Xuan *et al.* 2009) and with nanofiltration (Wu *et al.* 2009, 2010) have been conducted which have been previously published.

REFERENCES

Ministry of Construction (2005). The Reuse of Urban Recycling Water – Water Quality Standard for Artificial Groundwater Recharge (GB/T 17992-2005).

Wu L. L., Zhao X., Zhang M. and Xu Zhou C. (2009). Desalination of reclaimed water by nanofiltration in an artificial groundwater recharge system. *Journal of Water Supply: Research and Technology—AQUA*, **58.7**, 463–469.

Wu L. L., Zhao X. and Zhang M. (2010). Studies on the integration of nanofiltration and soil treatment for municipal effluent reclamation as a groundwater supplement. *Water Environ. Res.*, **82**, 20–26.

Wu L. L., Zhao X. and Zhang M. (2011). Removal of dissolved organic matter in municipal effluent with ozonation, slow sand filtration and nanofiltration as high quality pre-treatment option for artificial groundwater recharge. *Chemosphere*, **83.5**, 693–699.

Xue S., Zhao Q. L., Wei L. L. and Ren N. Q. (2009). Behavior and characteristics of dissolved organic matter during column studies of soil aquifer treatment. *Water Res.*, **43**, 499–507.

Zhang M. (2008). Study on the Enhanced Direct-well Recharge. Master Thesis Tsinghua University, Beijing, pp. 36–40.

Zhao X., Zhang M. and Xuzhou C. (2009). Bulk organic matters and nitrogen removal from reclaimed water during groundwater recharge by enhanced direct injection-well. *Water Environ. Res.*, **81**, 69–76.

Zhao X., Xuzhou C. and Zhang M. (2009). A case study: bulk organic matters and nitrogen removal from reclaimed water by enhanced direct injection-well groundwater recharge system. *Water Science & Technology*, **60.3**, 745–749.

Part B

Water Quality Analysis in MAR – Methods and Results

Chapter 11

Water quality analysis – microbiological hazards

Valter Tandoi, Caterina Levantesi, Simon Toze, Uta Böckelmann, Maurizio Divizia, Neus Ayuso-Gabella, Miquel Salgot, Rosanna La Mantia and Elisabeth Grohmann

11.1 INTRODUCTION

The risks associated with pathogens in wastewater are important to consider in water reuse to protect human health (Asano and Cotruvo, 2004). The assessment of pathogen fate in water reuse schemes can provide relevant additional information to that provided by the traditional microbial parameters (e.g. faecal indicators) as survival capability of pathogens often differs from that of faecal indicators and the presence of "indicator" organisms is often not correlated with that of pathogens. Pathogen monitoring can:

(1) Provide data, on the presence and abundance of waterborne pathogens in the country of interest. The results are useful inputs to quantitative microbial risk assessment (QMRA) studies.
(2) Improve the knowledge on the efficiency of water and wastewater treatment processes to eliminate pathogens characterised by different resistances and survival capacities.
(3) Allow, in comparison with microbial indicator data, evaluation of the capacity of each microbial marker to show the diffusion and removal of each specific pathogenic microorganism.

In this context the development and optimization of advanced analytical methods for fast and specific detection of pathogens is of major importance. Recently, molecular methods, that allow the identification and quantification of specific microorganisms in complex environmental matrices, have been developed (Gauthier *et al.* 2010; Rothrock *et al.* 2009; Yáñez *et al.* 2009; Linke *et al.* 2010). These methods are usually simple and highly specific and are very promising for the rapid monitoring of pathogens in different samples (Cook & Britt, 2007; Jyoti *et al.* 2010; Park *et al.* 2011).

Besides the presence of pathogens the emergence and spread of antibiotic resistance genes in bacteria is a major public health issue (Auerbach *et al.* 2007; Levy, 2002) that should also be evaluated in water reclamation processes. The input of antibiotic resistant bacteria into the environment seems to be an important cause of these resistances. However, the impact of antibiotic resistant bacteria on the spreading of antibiotic resistance in the environment is not yet well understood (Kümmerer, 2004). In particular, research is warranted to determine the possible role of water reclamation processes in spreading antibiotic resistance genes.

Within the Reclaim Project, the occurrence and fate of 15 waterborne pathogens and six antibiotic resistance genes were monitored during 1 year in three reclamation sites: Belgium (Wulpen/Torreele), Spain (Sabadell), and Italy (Nardò). Molecular methods were developed, improved and adapted for the detection of microorganisms and antibiotic resistance genes in environmental water samples. These methods were then applied to detect pathogens and antibiotic resistance genes in water samples from the selected reclamation sites.

In managed aquifer recharge (MAR) improvement of microbial water quality is achieved during soil filtration, retention in the aquifer, decay in the aquifer, dilution in the aquifer, etc. The knowledge on the survival capacity of human pathogens in the aquifer is therefore required to predict their fate in MAR processes. To obtain data on the specific *in situ* decay rates of pathogenic microorganisms, diffusion cell chamber technology was used (Page *et al.* 2010a). Using this method the reduction of pathogenic microorganisms can be followed over time to determine an *in situ* decay rate for inclusion into Quantitative Microbial Risk Assessments (QMRA).

11.1.1 Overview on selected water quality parameters, relevance

Pathogens in water reclamation and reuse

A wide variety of pathogenic microorganisms, classified into four broad groups (viruses, bacteria, protozoa and helminths) can be transmitted through water contact and ingestion to humans (Jiménez, 2003; WHO, 2006; Asano *et al.* 2007). The type and abundance of waterborne pathogens varies highly among countries, especially between developed and developing, and it is therefore fundamental to identify the microorganisms that can be of health concern in water reuse cases in the specific region of interest (Jiménez, 2003). Waterborne pathogens are very diverse in their sizes, infection doses, and capacity for survival in the environment and thus are removed at different rates by water treatment processes. In the RECLAIM WATER project, representatives of the four major classes of waterborne pathogens, were selected: protozoa (*Giardia* and *Cryptosporidium*), bacteria (*Salmonella* spp., *Campylobacter, Yersinia enterocolitica, Mycobacterium avium* subsp. *paratuberculosis, Helicobacter pylori*), viruses (enterovirus, Norovirus group I and II, HAV) and helminths (*Ascaris lumbricoides, Trichuris trichiura, Ancylostoma duodenale*). In Table 11.1 a short overview of the recent available data on abundance and frequency of the target pathogens in environmental water is reported.

Table 11.1 Pathogens frequency and abundance in environmental waters.

Pathogen	Wastewater	Secondary/tertiary effluent	Surface water
Giardia cysts/L	10–10^2 Rose *et al.* (1996) 1,007–1,814 Jiménez *et al.* (2001) 9,300-65,000 Bonadonna *et al.* (2002) 10^3–10^4 Cacciò *et al.* (2003) 10–10^3 Briancesco[C] (2005) 10^2–10^3 Harwood *et al.* (2005) 300-3,900 Ottoson *et al.* (2006)	0–10 (83%) Rose *et al.* (1996) 66–1100 Bonadonna *et al.* (2002) 10^2–10^3 Cacciò *et al.* (2003) 30–90 Briancesco[C] (2005) 10–100[A] (80%) Harwood *et al.* (2005) 0–2.11[A] (43%) Ryu *et al.* (2005) 2–10 Ottoson *et al.* (2006) 79–2469[A] Castro-Hermida *et al.* (2008) 10–10^3 Li *et al.* (2009)	0–100 (71%) Briancesco[C] (2005) 0–>250 (52%) Carmena *et al.* (2007) 0–10^3 (80%) Di Benedetto *et al.* (2005) 0–3.6 (25%) Wilkes *et al.* (2009)
Cryptosporidium oocysts/L[F]	0–10^2 (67%) Rose *et al.* (1996) 1–35 Bonadonna *et al.* (2002) 0 Cacciò *et al.* (2003) 3–21 (74%) Harwood *et al.* (2005) 3–220 Ottoson *et al.* (2006)	0–10 (42%) Rose *et al.* (1996) 0.2–1.8 Bonadonna *et al.* (2002) 0.13–3.45[A] (83%) Gennaccaro *et al.* (2003) 10–50 Briancesco[C] (2005) 0–133 (87%) Di Benedetto *et al.* (2005) <50/100 L[A] (70%) Harwood (2005) 2–10 Ottoson *et al.* (2006) 0–15.9 (16%) Ryu *et al.* (2005) 0.79[AB] (94%) Costán-Longares *et al.* (2008) 2–390[A] Castro-Hermida *et al.* (2008)	0–5 (48%) Briancesco[C] (2005) 0–10 (80%) Di Benedetto *et al.* (2005) 0–17 (63%) Carmena *et al.* (2007) 0–8 (40%) Wilkes *et al.* (2009)
Salmonella MPN-CFU/100 mL[F]	10^5–10^6/100 mL Jiménez *et al.* (2001) 600 gc/100 mL Lee *et al.* (2006) 10^4 gc/100 mL (100%) Wéry *et al.* (2008)	0 gc/100 mL Lee *et al.* (2006) 1000 gc/100 mL (16%) Wéry *et al.* (2008)	(6.2%) Johnson *et al.* (2003)
Campylobacter MPN-CFU/100 mL[F]	26.2–7900 Jones (2001) 0 gc/10 mL Alexandrino *et al.* (2004) 0–10^3 gc/mL (83%) Wéry *et al.* (2008)	0 gc/60 mL Alexandrino *et al.* (2004) 0/100 mL Wéry *et al.* (2008)	>1/400 mL (30%) Pianetti *et al.* (1998) 0–11 Savill *et al.* (2001) >1 Abulreesh *et al.* (2005)

(Continued)

Table 11.1 Pathogens frequency and abundance in environmental waters (*Continued*).

Pathogen	Wastewater	Secondary/tertiary effluent	Surface water
Enterovirus PFU/L[F]	180–1500 Bonadonna *et al.* (2002) 99–1141 Lodder[E] (2005)	0–100 Bonadonna *et al.* (2002) 3–49 Lodder[E] (2005)	0.1–2 Lodder[E] (2005) 0–3.2 × 10^3 gc/L[D] Noble *et al.* (2006)
Norovirus GI and GII (PDU/L)	444–80,00000 Lodder[E] (2005)	57–71,000 Lodder[E] (2005)	0.2–46,000 Lodder[E] (2005)

Note: Range or average value of pathogen concentrations, in brackets occurrence frequencies; MPN: Most probable number; PFU Plaque forming units; PDU: PCR Detectable Units; gc: gene copies. [A]tertiary effluent; [B]viable *Cryptosporidium*; [C]Briancesco and Bonadonna (2005); [D]urban watershed; [E]Lodder and de Ronda Husman (2005); [F]if not differently stated.

The protozoa *Giardia* and *Cryptosporidium* are indicated by the WHO (2008) as waterborne pathogens of primary concern and are reported as one of the most common causes of waterborne gastroenteritis outbreak (Asano *et al.* 2007, Craun *et al.* 2006). Their infective/resting stage, cysts (*Giardia*) and oocysts (*Cryptosporidium*), are approximately 8–14 μm and 4–6 μm in diameter respectively and are secreted in the faeces of infected mammals in very high number (10^{10}/day in patients). *Cryptosporidium parvum* and *Giardia lamblia* are species infectious for humans. A large range of animals are reservoirs of *C. parvum*, but humans and livestock, particularly young animals, are the most significant source of human infectious organisms (WHO, 2008). Animals, such as beavers, have also been reported as reservoir and source of *G. lamblia* in the environment (Hibler and Hancock, 1990). However, to date no human infection with *G. lamblia* has been directly linked to an animal host and it is still controversial whether animal produced *G. lamblia* strains are infectious for humans (Gerba 2009).

Characteristic of *Giardia* and *Cryptosporidium* resting stages are their elevated survival capacity (John & Rose, 2005) and resistance to disinfection treatments (Asano *et al.* 2007). Cysts of *G. lamblia* have been documented to survive for up to 77 days in distilled water at 8°C (Bingam *et al.* 1979) and *C. parvum* has been reported to survive for weeks in surface waters (Johnson *et al.* 1997). Furthermore very high Ct values for chlorine inactivation (99%) of *G. lamblia* cysts (Ct 119–192 at 5°C, pH 8) and *C. parvum* oocysts (>7200 at 25°C, pH 7) in distilled water, were reported (Gerba, 2009). The *Giardia* and *Cryptosporidium* infective dose is low, being 1–10 oocysts and 10–25 cysts respectively; thus a low number may pose a risk of infection in water recycling via aquifers. Oocysts are commonly found in WWTP effluent as well as surface waters but very variable numbers of their frequency and abundance are reported (Table 11.1). Low numbers of these pathogens have also been found in groundwaters, effluent of water treatment processes and treated drinking water (Di Benedetto *et al.* 2005; Carmena *et al.* 2007).

Viruses are obligate, host specific, intracellular parasites ranging from 0.01–0.3 μm in size. Enteric viruses, that infect the gastrointestinal tract of humans, are mostly obligate human pathogens and are potentially transmitted by water contaminated with human faeces (Asano *et al.* 2007). Enterovirus, Norovirus group I and II and HAV were monitored within the RECLAIM WATER project. Although Enterovirus are readily found in faecally contaminated recreational and drinking water, waterborne transmission of these viruses has been only rarely confirmed (Gerba 2009; Asano *et al.* 2007). On the contrary *Norovirus* was identified as the etiological agent of numerous waterborne outbreaks associated with both drinking and recreational waters in the USA (Asano *et al.* 2007). Similarly, strong evidences support the waterborne transmission of HAV (Asano *et al.* 2007). The presence of viruses in reclaimed water is of high concern due to their relatively low infectious dose (1–10 median infectious dose N_{50}) and their elevated survival in the environment (Asano *et al.* 2007; WHO, 2003). An average life of 3 months in the specific environment such as water, wastewater and soil has been estimated for infective Enterovirus and Hepatitis A viruses (Feachem *et al.* 1983). According to the Guidelines for drinking water quality (WHO, 2008) the effectiveness of treatment processes used to remove viruses will require validation. Nevertheless it is believed that enteric viruses are more resistant to chlorine than most bacteria. Reported Ct values for chlorine inactivation (99%) of Enterovirus in distilled water ranged from 0.56 to 2.16 (5°C, pH 7.8) (Gerba, 2009). Enteric viruses are usually present in large numbers in wastewater: 10–1,000/L in untreated sewage and 10–100/L in secondary treated effluent (WHO, 2003). However their concentration is highly variable (Table 11.1) and is linked to the season and the age distribution of the population (Jiménez, 2003).

Bacteria are unicellular prokaryotic organisms of 0.2–5.0 μm in length. Most bacterial pathogens potentially transmitted by water infect the gastrointestinal tract and are excreted in the faeces of infected humans and other animals (WHO, 2008). Pathogenic bacteria are usually characterised by an elevated infectious dose (10^2–10^9 cells) and are less resistant than protozoa and viruses to traditional disinfection treatments. Five pathogenic bacteria known as waterborne or emerging waterborne pathogens were selected for the analysis in RECLAIM WATER: *Salmonella* spp., *Campylobacter* spp. *Yersinia enterocolitica*, *Helicobacter pylori* and *Mycobacterium avium* subsp. *paratuberculosis:* Long-term survival in the environment is well documented for all the selected bacterial pathogens (Ashbolt, 2004; Bellack *et al.* 2006; Lehtola *et al.* 2007; Moreno *et al.* 2007; Queralt and Araujo, 2007; Torvinen *et al.* 2007; Whittington *et al.* 2005; Winfield & Groisman, 2003; Hanninen *et al.* 1998).

More than 200 of the *Salmonella* serovars are pathogenic to humans being the etiologic agents of typhoid fever and gastroenteritis. *Salmonella* is one of the most common pathogenic bacteria detected in wastewater (Jiménez *et al.* 2001; Jiménez, 2003) and was reported as etiological agent in 4% of waterborne outbreaks in the USA (Gerba *et al.* 1994). This microorganism can enter the aquatic environment directly with faeces of infected humans or animals or indirectly, e.g., via sewage discharge or agricultural land runoff. Overall *Salmonella* spp. and subspecies can be found in a large number of vertebrates comprising wild and domestic animals (Wray & Linklater 2000; Wray & Davies 2000). Plants, fresh vegetable and fruits, insects and algae were also found capable of harbouring *Salmonella* spp. and might be implicated in the transmission of this enteric pathogen (Guo *et al.* 2002; Byappanhalli *et al.* 2009). In addition to its widespread occurrence, elevated survival capacity in non-host environment, was reported for *Salmonella* (Winfield & Groismann 2003). However, standard disinfection procedures used in drinking water treatment processes are active against salmonellae (Cicmanec *et al.* 2004).

Campylobacter spp. are important pathogens causing gastroenteritis and mild to severe diarrhoea, and several thermotolerant species such as *C. jejuni* are occasionally invasive. *C. jejuni*, *C. coli*, and *C. lari* account for more than 99% of the human isolates (*C. jejuni* 90%). *Campylobacter* infections are usually reported through foods but they also have a role in waterborne infections (Hanninen *et al.* 2003). Large waterborne *Campylobacter* outbreaks were reported in Norway, Denmark and Sweden (Sharma *et al.* 2003) while in the USA *Campylobacter*, together with *Salmonella* and pathogenic *E. coli*, has been estimated to cause three million waterborne illnesses per year (Asano *et al.* 2007). Wild and domestic animals, especially poultry, wild birds and cattle, are important reservoirs and possible sources of *Campylobacter* contamination in aquatic environments (WHO, 2008). The low infectious dose of this microorganism, in respect to other enteric bacteria, increases the possibility of infection by drinking water containing only a few hundred viable cells (Nachamkin, 1997). *Campylobacter* spp. can survive for several weeks in aquatic environments at temperatures around 4°C (Hazeleger *et al.* 1998).

Helicobacter pylori, *Mycobacterium avium* subsp. *paratuberculosis*, and *Yersinia enterocolitica* are reported as emerging waterborne pathogens (Sharma *et al.* 2003) and little data are available on their role in water reuse systems.

Yersinia enterocolitica is reported as the cause of various clinical manifestations ranging from self-limited gastroenteritis to more invasive syndromes like terminal ileitis and mesenteric lymphadenitis. Domestic and wild animals are the principal reservoir for *Yersinia* spp. and pigs are the major reservoir of human pathogenic *Y. enterocolitica* (Waage *et al.* 1999). Thus the most likely source of pathogenic *Yersinia* spp. in water is human or animal waste. *Y. enterocolitica* waterborne outbreaks have occurred in the United States but such documented outbreaks are very rare (Craun, 1979; Gerba, 2009). *Y. enterocolitica* can persist for long periods in a water environment and it was shown to survive for several weeks in natural river water at both 6 and 16°C (Karapinar and Gonul, 1991; Tashiro *et al.* 1991; Terzieva and McFeters, 1991). These organisms are sensitive to usual disinfection processes (WHO, 2008).

The main relevant species of the genus *Mycobacterium* that can occur as opportunistic pathogens are *M. avium*, *M. intracellulare*, *M. kansasii*, *M. marinum*, *M. simiae*. Paratuberculosis, a chronic enteric disease in ruminants also known as Johne's disease, is caused by *Mycobacterium avium* subsp. *paratuberculosis* (MAP) (Chiodini *et al.* 1984). Epidemiological studies have associated water sources with infection with mycobacteria but the precise mode of transmission of MAP remains unknown (Burns *et al.* 1991). These bacteria can multiply in oligotrophic water environment and are more resistant that than other enteric bacteria to disinfection by chlorination (Collins *et al.* 1984, Falcao *et al.* 1993, Gerba *et al.* 2003). Gerba *et al.* (2003) reported high Ct values (ranging from 106 to 204 at 23°C, pH 7) for 99% chlorine inactivation of different *M. avium* strains in distilled water. The risk that MAP may infect humans makes its control particularly important in water recycling applications.

Helicobacter pylori, a spiral, Gram-negative microaerophilic bacterium, colonizes at least half of the world's human population and is recognized as a major cause of chronic gastritis, peptic ulcer, and an important risk factor for gastric cancer (Parsonnet *et al.* 1991; The Eurogast Study Group, 1993, Uemura *et al.* 2001). On the basis of various epidemiological studies, *H. pylori* has been classified as a class I carcinogen in humans by a Working Group of the World Health Organization International Agency for Research on Cancer (International Agency for Research on Cancer Working Group, 1994). Humans appear to be the primary host of *H. pylori*. Person to person spread is probably the dominant source of transmission of this pathogen, but infections may also be spread by ingesting contaminated food or water (Gerba, 2009). *H. pylori* has been found to survive for 3 weeks in biofilms and up to 20–30 days in surface waters (WHO, 2008) and a greater risk of *H. pylori* infections was associated to the use of untreated drinking water (Klein *et al.* 1991).

The ability of *Y. enterocolitica* (Smith *et al.* 1994), *H. pylori* (Adams *et al.* 2003; Moreno *et al.* 2007; Rowe *et al.* 2006) *M. avium* (Archuleta *et al.* 2005; Queralt *et al.* 2007; Whittington *et al.* 2004), *Salmonella* (Roszak *et al.* 1984; Smith *et al.* 1994) and *Campylobacter* to persist for a long time in a dormancy or viable but non culturable (VBNC) state in aquatic environments has been reported. Bacteria in a VBNC state are not recovered by culture dependent methods, therefore culture-independent, rapid and specific tests are required for the detection of these pathogens.

Worldwide, enteric worms are one of the principal causative agents of human disease. It is estimated that 133 million people suffer from high intensity intestinal helminths infections, which often lead to severe consequences such as cognitive impairment, massive dysentery, or anaemia. These infections cause around 9,400 deaths every year (WHO, 2008). The burden of helminthiasis is much larger in developing countries, up to 25–33% of the population is infected, (Bratton & Nesse, 1993; Wani & Chrungoo, 1992) while in developed countries helminths are present in no more than 1.5% of people (WHO, 1997). Helminths include the phyla Nematoda (roundworms), Trematoda (flatworms) and Cestoda (segmented worms). The phylum Nematoda includes some well-known members associated with human infections: the large roundworm *Ascaris lumbricoides*, the whipworm *Trichuris trichiura*, the hookworms *Necator americanus* and *Ancylostoma duodenale*, and the threadworm *Strongyloides stercoralis*. Ascariasis is considered to be the most prevalent parasitic infection worldwide (Tchobanoglous *et al.* 2003). Helminth eggs, which range in size from ~10 μm to >100 μm, can be removed by many wastewater-treatment processes such as sedimentation, coagulation and flocculation, filtration, and stabilization ponds (Jiménez *et al.* 2000; Jiménez, 2003).

However, some Helminth eggs are extremely resistant to environmental stresses and may survive conventional wastewater treatment and disinfection procedures, becoming a potential risk to human health; that is, *Ascaris* eggs were reported to remain infective for several years in the environment (Smith, 1998). Viruses and bacteria are less persistent than nematodes, and immunity generally develops readily in early life (Campos, 2006). Thus, monitoring the presence of helminth eggs in water and wastewater is very important especially in developing countries where these microorganisms are more common, and raw wastewater is often used untreated for irrigation.

Antibiotic resistance genes (ARG)

Antibiotics are released daily into the natural environment in treated wastewater effluent and through use in animal husbandry (Auerbach *et al.* 2007). Many of these antibiotics can now be detected in water resources (Lindsey *et al.* 2001; Kolpin *et al.* 2002; Yang & Carlson, 2003) leading to increasing concerns with regard to their contribution to the abundance and persistence of resistance in populations of pathogenic, commensal and non-pathogenic microorganisms (Auerbach *et al.* 2007). These concerns are partly derived from a lack of critical information with respect to the transfer of resistance genes within and between bacterial populations in the environment (Isaacson & Torrence, 2002).

The input of antibiotic resistant bacteria into the environment seems to be an important source for these resistances in the broader population pool but the extent of the impact of antibiotic resistant bacteria on the environment is not yet known. To understand the ecology of antibiotic resistance, characterisation of the occurrence, fate and transport of antibiotic-resistant microorganisms is required. A first step in this process is the identification of major sources of antibiotics and antibiotic resistance in the environment. Wastewater treatment plant (WWTP) effluent is an important source of antibiotics to surface waters (Auerbach *et al.* 2007; Yang & Carlson, 2003). Hence, for the RECLAIM WATER project six different antibiotic resistance genes with different mechanism of action, were selected due to their abundance and reported increase of resistant microorganisms in the environment (Auerbach *et al.* 2007; Chee-Sanford *et al.* 2001; Esiobu *et al.* 2002; Ferreira da Silva *et al.* 2006; Kümmerer, 2004; Sengelov *et al.* 2003; Smith *et al.* 2004).

The following resistance genes were chosen:

(1) *ampC* (ampicillin resistance): confers resistance to third generation or extended spectrum cephalosporins, e.g. ceftazidime and cefotaxime; additionally it confers resistance to penicillins.
(2) *mecA* (methicillin resistance): *mecA* mediates resistance to methicillin and oxacillin. MRSA (All methicillin-resistant *Staphylococcus aureus*) strains encode the *mecA* gene.
(3) *blaSHV-5* (extended β-lactam resistance): encodes an extended spectrum β-lactamase, conferring resistance to expanded spectrum cephalosporins and aminoglycosides. It is often found in multi-resistant *Klebsiella pneumoniae* strains.
(4) *ermB* (erythromycin resistance): resistance gene conferring macrolide resistance, it is often present in streptococci, e.g. in *S. pneumoniae*.
(5) *tet*(O) (tetracycline resistance): conferring resistance to tetracyclines, commonly found, including in environmental samples; tetracycline resistance genes are present in both Gram-positive and Gram-negative bacteria.
(6) *vanA* (vancomycin resistance): conferring resistance to vancomycin, the *vanA* gene is responsible for vancomycin resistance in VRSA (vancomycin resistant *S. aureus*) strains.

All of the resistance genes are highly transmissible.

11.1.2 Detection and quantification methods

Pathogens and ARG detection methods

Traditional and standardised methods as well as advanced molecular techniques were applied for pathogens detection, these include:

(1) Available Polymerase Chain Reaction (PCR) and Real Time PCR (also known as quantitative PCR, qPCR) methods were adapted to environmental water samples and applied for the analysis of pathogenic bacteria

(*Salmonella*, *Campylobacter* spp. *Yersinia enterocolitica*, *Helicobacter pylori* and *Mycobacterium avium* subsp. *paratuberculosis*.) and viruses (Enterovirus, HAV, Norovirus GGI and GGII).

(2) A Peptidic Nucleic Acid Fluorescence In Situ Hybridization (PNA FISH) protocol was adapted for the determination of the human pathogenic species of *Campylobacter* (*C. coli, C. lari* and *C. jejuni*) in environmental waters samples.

(3) Traditional approaches were utilised for the detection of protozoa (method 1623; EPA 2001), helminth eggs (concentration and morphological identification according to Thienpont *et al.* 1986), Enterovirus (cultivation on BGM cells) and *Campylobacter* spp. (ISO 17995 method; ISO, 2005).

Within the RECLAIM WATER Project qPCR was also applied for the monitoring of six antibiotic resistance genes in the studied water reuse systems. To this scope new Taq-man qPCR assays were developed for quantitative detection of four antibiotic resistance genes (Böckelmann *et al.* 2009).

Traditional methods for the detection of waterborne pathogenic bacteria are based on the selective cultivation of the target microorganism and the identification of obtained isolates by different tests such as biochemical characterization, immunofluorescence and staining. Usually enrichment under non selective conditions is introduced before the selective cultivation for the retrieval of pathogenic cells with low activity. An advantage of these methods is that only the viable bacteria, i.e. those that might represent a health risk for the population, are enumerated. However, the cultivation-identification approach is complex and time consuming: 48–72 h are usually required for a presumptive identification with even longer times necessary for species specific identification. Furthermore some enteric pathogens are not cultivable or difficult to grow, requiring complex growth condition (e.g. microaerophilic growth of *Campylobacter*) and the cultivation approach may also fail to detect bacteria in the viable but nonculturable state (VBNC).

Molecular methods are based on the detection, by hybridization with fluorescent probes (Fluorescent *In Situ* Hybridization: FISH) or amplification (PCR and qPCR), of DNA/RNA sequences specific to the targeted microorganism. Using these methods, the problem of no or limited pathogen cultivability is overcome hence rapid and specific responses on the presence or absence of pathogens in the environment can be easily obtained. A major drawback in the use of these techniques is that they cannot clearly differentiate active infective pathogens from non viable ones. In particular PCR and qPCR can also amplify DNA from dead cells and/or extracellular remains of DNA that is persistent in the environment. Further research is necessary to fully understand the correlation between positive results obtained by molecular methods and potential human health risks. However, as suggested by Viau & Peccia (2009), in spite of the specific difference between cultivation and qPCR quantitative pathogens detection data, the higher concentration values obtained by molecular approach indicate that qPCR results can be used as a conservative parameter for pathogen monitoring. Furthermore in contrast to cultivation methods, PCR and qPCR can also detect VBNCs and extracellular virulence genes that when eventually transformed into viable organisms can be of human health concern.

Two biomolecular methods were chosen for the microbial contaminant detection in RECLAIM WATER, qPCR and FISH.

PCR targets nucleic acid sequences present in a complex sample and can selectively amplify DNA by the use of specific primers. Quantitative PCR (qPCR), has the added ability to monitor the progress of the PCR as it occurs (i.e. in real time). Data are therefore collected throughout the PCR process, rather than at the end of the PCR and quantitative data are obtained. In real-time PCR, reactions are characterized by the time point during cycling when amplification of a target is first detected rather than the amount of target accumulated after a fixed number of cycles. The higher the starting copy number of the nucleic acid target, the sooner a significant increase in fluorescence is observed.

Quantitative detection of specific nucleic acid sequences is possible using the fluorogenic 5′ nuclease assay. To this purpose, a fluorogenic probe complementary to the target sequence is added to the PCR reaction mixture. The probe consists of an oligonucleotide with a reporter and quencher dye attached (=TaqMan probe). During PCR, if the target of interest is present, the probe anneals specifically between the forward and reverse primer sites. The nucleolytic activity of the polymerase cleaves the probe, which results in an increase in the fluorescence intensity of the reporter dye. This process occurs in every cycle and does not interfere with the accumulation of the PCR product. To induce fluorescence during PCR, laser light is distributed to the samples via a multiplexed array of optical fibres. The resulting fluorescent emission returns via the fibres and is directed to a spectrograph with a charge-coupled device camera. To ensure that the PCR works and to verify that a negative amplification result is not due to inhibiting components, an internal activity control (IAC) is used. This IAC is composed of template DNA, two specific primers and a specific TaqMan probe and should give a positive result when added to the environmental sample. Quantitation data for the *blaSHV-5* standard are shown in Figure 11.1 as an example.

In the RECLAIM WATER project TaqMan real-time assays for six different antibiotic resistance genes were applied. Of the six assays, four of them were newly developed or modified as part of the RECLAIM WATER project, namely the assays for detection of *ermB*, *mecA*, *blaSHV-5*, and *tetO*.

For quantitative detection of the pathogens, *Salmonella*, *Yersinia enterocolitica*, *Helicobacter pylori* and *Mycobacterium avium* subsp. *paratuberculosis* previously published assays were applied as described in Böckelmann *et al.* (2009) and Levantesi *et al.* (2010). With the exception of the *vanA* and *H. pylori* all the bacterial pathogens and ARG could be detected in environmental samples. In particular the presence of *Y. enterocolitica*, *H. pylori*, and

M. avium subsp. *pseudotuberculosis*, not detected at quantifiable level in RECLAIM WATER samples from Sabadell (Spain), Nardò (Italy) and Wulpen/Torreele (Belgium), was also investigated in samples from Mézquital Valley (Mexico). Influent samples from the Mexican site showed the presence of *M. avium* subsp. *pseudotuberculosis* at concentrations between 1.6×10^2 and 2.6×10^3 gene copies/100 mL (U. Böckelmann and S. Hergett, unpublished data). *Y. enterocolitica* was also present in 50% of the influent samples from this site. However, the concentration was below the limit for accurate quantification (10 gene copies/reaction). Corroborating previous observation in the European reclamation sites *H. pylori* was not detected in any of the Mexican samples.

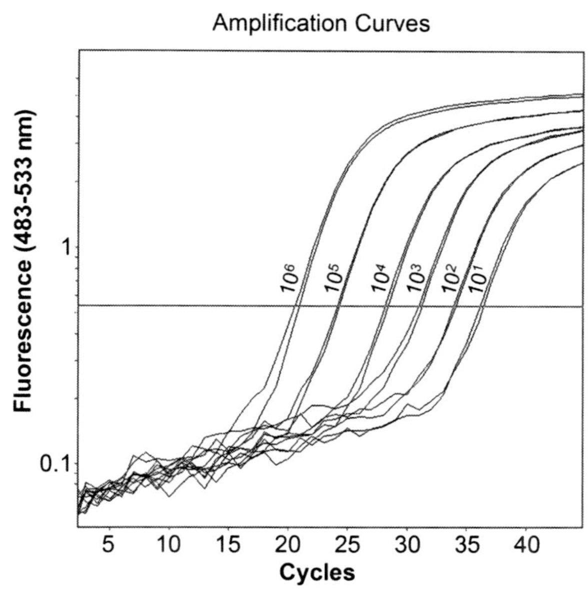

Figure 11.1 TaqMan technology principle, quantitation data for the *blaSHV-5* standard (qPCR amplification curves of the standard for the *blaSHV-5* gene quantification assay)

FISH is an identification approach based on the use of fluorescently labelled probes that specifically bind to the cells of a targeted taxon (Figure 11.2). This culture-independent technique is a reliable and validated method for the enumeration of whole bacterial cells in mixed populations. The procedure allows the observation and quantification of identified cells in their natural habitat (Figure 11.2).

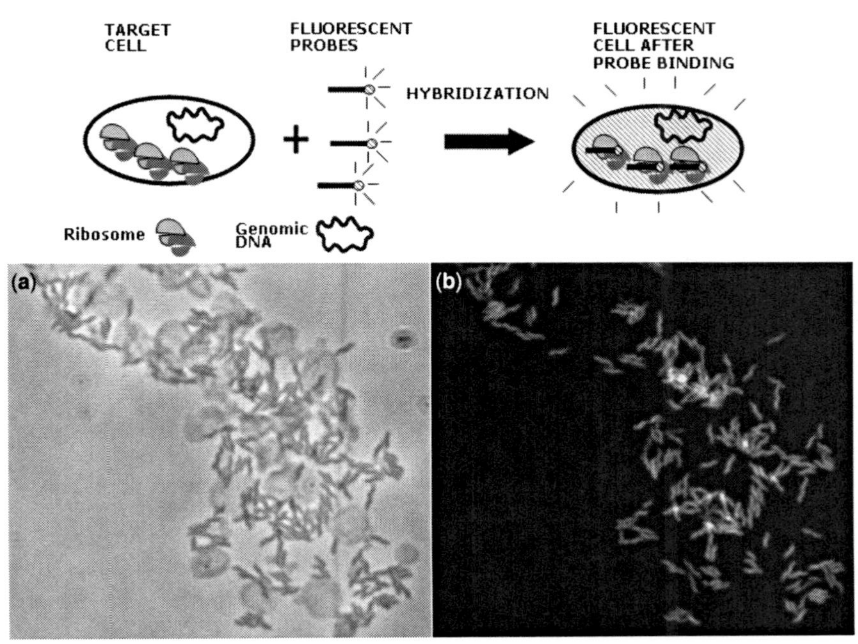

Figure 11.2 The principles of FISH, specific detection of *Campylobacter* cells (a) phase contrast 1000X, (b) FISH fluorescent *Campylobacter* cells upon probe hybridization in the same microscopic field of an enrichment culture

Briefly the FISH probes are short oligonucleotides complementary to the rRNA sequences of the microorganism of interest; upon hybridization the cells of the targeted taxon will appear fluorescent due to binding of the labelled probe to their ribosomal RNA. The high copy numbers of bacterial RNA (ribosomes 10^3–10^5 per bacterium) provide a naturally amplified target and afford greater assay sensitivity. An alternative to the classical FISH, the PNA-FISH utilizes probes of peptidic nucleic acids (PNA) instead of oligonucleotide probes. PNA is a molecule that mimics the DNA structure and can hybridize to complementary DNA/RNA sequences. The negatively charged sugar phosphate backbones of nucleic acid are replaced in PNA by an achiral neutral polyamide (usually formed by N-(2-aminoethyl)-glycine) backbone. Because of their neutral backbone PNA probes are characterized by a higher binding affinity and therefore increased specificity in binding complementary DNA; a PNA/DNA mismatch is less stable than DNA/DNA ones. Other advantages of PNA in respect to DNA probes are their higher resistance to degradation and the possibility to perform hybridization at low salt concentrations, conditions that reduce the formation of rRNA secondary structures and therefore increase the accessibility of rRNA binding sites.

Within the RECLAIM WATER project a PNA-FISH protocol developed by Lehtola *et al.* (2005) was adapted for the determination of the human pathogenic species of *Campylobacter* (*C. coli, C. lari* and *C. jejuni*) in environmental water samples as described in La Mantia *et al.* (2008a). The PNA-FISH protocol is simple, can be performed by non-specialised operators and *Campylobacter* cells can be detected in the enrichment broth (24 h) within three hours. At present the major drawback of this approach are the elevated cost of PNA probes (10 times higher than oligonucleotides probes) and the fluorescent background of blood proteins, present in the enrichment broth, that require a more careful and time consuming observation of FISH slides.

Pathogen decay studies

For QMRA studies to assess human health risks in water recycling via MAR, understanding of the decay inactivation of these pathogens in the subsurface is needed. Risk assessment and management of aquifer recharge systems can be improved by knowledge on the rate of decay of pathogens in natural systems such as aquifers and wetlands. Decay studies are best undertaken *in situ* using processes such as diffusion chambers (Toze *et al.* 2010). The *in situ* studies provide information on the rate of pathogen decay, usually as a decay rate ($\log_{10}$ per day) or as a time for 1.0 $\log_{10}$ removal (90%) of the target microorganism(s) (otherwise known as T90).

The *in situ* pathogens decay rates and their abundance numbers allow an estimation of the water treatment potential for natural systems such as aquifers or wetlands. Data on the rate of decay of pathogens in the environment can be important for the design of natural treatment systems and the subsequent operational management protocols or to determine if additional treatment steps are needed to bring any residual risk from microbial pathogens to within acceptable guideline limits (e.g., see Toze *et al.* 2010).

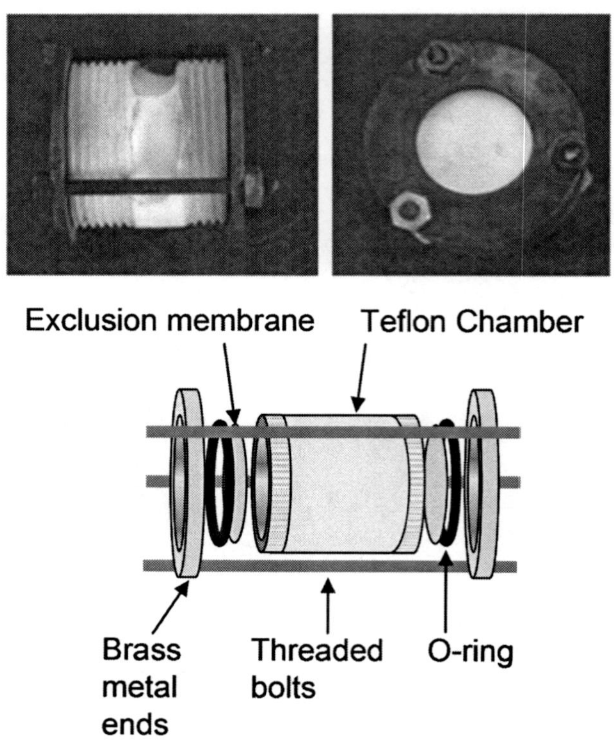

Figure 11.3 Design of diffusion chambers used to study pathogen decay in groundwater at the ASTR site

Diffusion chambers approach

The diffusion chambers were first developed by Pavelic *et al.* (1998) and have since been modified and improved to suit changing experimental requirements. The diffusion cells permit target microbial pathogens to be contained within the chamber while allowing passage of water across the membranes located at either end of the chambers (Toze *et al.* 2010). An experiment usually involves the construction of series of chambers filled with water or groundwater collected from the study site and seeded with equal numbers of the pathogen(s) of interest. A schematic diagram of the most common version of the diffusion chamber design is given in Figure 11.3.

The diffusion chambers were sealed at both ends with a confining membrane that has a pore size that is sufficiently small to prevent the loss of the target pathogens from the chamber but allows passage of fresh groundwater across the membrane. Commonly when working with enteric viruses a pore size of 0.025 μm (e.g. Toze *et al.* 2010) is used but membranes with larger pore sizes can be used for larger microorganisms. Groundwater collected from the study site is seeded with microbial pathogens. This seeded groundwater is then added to a number of chambers, sufficient to collect three replicate chambers on each pre-determined sampling event. The completed chambers are then strung together on a steel line and suspended in a monitoring well at the study location. To enable a determination of the loss of pathogens over time, individual chambers are collected on different sampling occasions and an assessment made on the number of viable or detectable pathogens present within the chamber at that time. The number of pathogens present within the chambers at each sampling time is then log transformed and plotted with respect to time. The resulting slope is then used to determine the decay rate.

11.2 CASE STUDIES

The removal efficiency of microorganisms was investigated over a year at three of the RECLAIM WATER case study sites:

(1) Wulpen/Torreele (Belgium): This case study exemplifies the most advanced treatment train including ultrafiltration and reverse osmosis prior to infiltration ponds with infiltration rate of 6,850 m^3/d: water is withdrawn and distributed after further treatment for drinking water supplies (cf. chapter 3).
(2) Sabadell (Spain): At this site treated domestic wastewater is discharged into the Ripoll River and infiltrates into groundwater by bank filtration with a flow rate of 16,000 m^3/d. The product water is used for irrigation of public parks after UV and Cl_2 treatment. Due to the vicinity of the park to a hospital and the applied spray irrigation, pathogen contamination of the irrigation water is a potential concern (cf. chapter 5).
(3) Nardò (Italy), where treated domestic wastewater is transported in an open channel and introduced in a sinkhole into a karstic aquifer with an average flow rate of 12,000 m^3/d: water is withdrawn at a distance of 3 km from the sinkhole for irrigation. This case study implies the need to assess pathogen transport and decay in a karstic aquifer (cf. chapter 4).

The microbial water quality in the injectant and in the abstracted water was investigated in the three MAR sites. In Wulpen/Torreele the performances of traditional and advanced wastewater treatments were also evaluated. In this chapter the results of pathogens and antibiotic resistance genes monitoring is summarized and compared to the data of microbial indicator surveys described in Chapters 3, 4 and 5. The results of pathogens and antibiotic resistance genes surveys in Wulpen/Torreele, Nardò and Sabadell sites, were described in detail by La Mantia *et al.* (2008a), Böckelmann *et al.* (2009), Levantesi *et al.* (2010), and Masciopinto *et al.* (2011). The role of the subsurface as a pathogen treatment barrier was also discussed for four sites, Wulpen, Atlantis, Salisbury and Tula Valley (the sites with intended potable use of recovered water) by Page *et al.* (2010c).

11.2.1 Belgium (Wulpen/Torreele)

At the Wulpen/Torreele site, pathogens and antibiotic resistance genes were monitored during three sampling campaigns (January, July and October 2007). Viruses were only analysed in July and October 2007. To evaluate removal of hazards by tertiary treatments and SAT, samples were collected at five sampling points: the effluent from the WWTP (SP2), water after ultrafiltration (SP3); the infiltration pond (SP5), groundwater after infiltration (SP6) and drinking water (SP7). Helminth eggs (*Ascaris lumbricoides, Trichuris trichiura, Ancylostoma duodenale*), protozoa (*Giardia* and *Cryptosporidium*) and viruses (Enterovirus, Norovirus group I and II, HAV) were also analysed in the wastewater influent (SP1) to assess their removal by conventional wastewater treatments. Enteric bacteria (*Salmonella* spp., *Campylobacter, Yersinia enterocolitica, Mycobacterium avium* subsp. *paratuberculosis, Helicobacter pylori*) and antibiotic resistance genes (*ampC, ermB, mecA, blaSHV-5, vanA* and *tetO*) were instead monitored in the infiltration water before transport (SP4) to asses reverse osmosis removal. A diagram of the Wulpen/Torreele site is reported in Figure 3.5 of chapter 3.

Occurrence and removal of pathogenic microorganisms

Wulpen/Torreele site was characterised by the lowest pathogen numbers and diversity of the three case study sites. Bacteria and viruses were never found in Wulpen/Torreele samples during the whole monitoring period, and, overall, pathogens were never detected in groundwater (SP6) or in the final effluent (SP7). *Giardia* cysts and *Cryptosporidium* oocysts were frequently found in dirty waters, wastewater and secondary effluents in this site as shown in Table 11.2.

Table 11.2 *Giardia* cysts and *Cryptosporidium* oocysts concentrations in the Wulpen/Torreele site (after Levantesi *et al.* 2010).

Sampling points	*Giardia* (cysts/L)			*Cryptosporidium* (oocysts/L)		
	February 2007	July 2007	October 2007	February 2007	July 2007	October 2007
SP 1 wastewater	302	713.6	289.6	–	1.6	–
SP 2 secondary effluent	0.26	27.64	2.43	0.17	0.17	–
SP 3 ultrafiltration permeate	–	–	–	–	–	–
SP 5 infiltration pond	0.06	–	–	–	–	–
SP 6 groundwater after infiltration	–	–	–	–	–	–
SP 7 UV treated groundwater	–	–	–	–	–	–

–: below detection limit. Sample volumes analysed varied depending upon the treatment stage and the amount of water that could be filtered. Approximately 1 L of wastewater, 10–30 L of secondary effluent, 30–60 L of surface water, and 50–115 L of ultra-filtration (UF) filtrate, groundwater and water for final use were filtered.

Giardia was the most common pathogen (39% of Wulpen/Torreele samples); it was detected in the WWTP influent and effluent in all the sampling campaigns at 289.6–713.6 and 0.26–27.64 cysts/L respectively. Although not detected in the ultrafiltration effluent, *Giardia* cysts were also observed once (January 2007) in the infiltration pond (0.06 cysts/L). *Cryptosporidium* oocysts were rarer than *Giardia* cysts (17% of Wulpen/Torreele samples) and were only detected in WWTP influent and effluent (Table 11.2). *Giardia* cysts and *Cryptosporidium* oocysts concentrations in wastewater and secondary treated effluent at Wulpen/Torreele were in the low range reported in literature (See Table 11.1).

In January-February 2007 oocysts were detected in the secondary treated effluent sample but not in the untreated wastewater sample. The increase of oocysts from wastewater to secondary effluent, observed in the Torreele site, was also described by other authors (Gennaccaro *et al.* 2003; Ryu *et al.* 2005) that addressed process factors, such as recovery and sensitivity of the method in different types of water to explain this result. In our case the inherent variability in water quality of the samples as well as the lower detection limit in secondary effluent (0.1–0.03 instead of 1 oocyst $\cdot L^{-1}$ in wastewater) should also be considered.

Even with the limited number of sampling events (n = 3), ultrafiltration seemed to be a useful barrier to remove protozoa as *Giardia* cysts and *Cryptosporidium* oocysts, still present in the secondary effluent, were never detected after membrane filtration.

Helminth, namely *Ascaris lumbricoides*, eggs were present in 17% of the samples. They were found twice in wastewater samples (July and October 2007) and once in the infiltration pond (October 2007), always at low level (<2 eggs/20 L). Consistently with the evidence showing an increased concentration of microbial indicators from the infiltration water before transport (SP4) to the infiltration pond (SP5) (Chapter 3), helminth eggs and *Giardia* cysts presence in SP5 suggested that this pond can be a source of recontamination probably from wildlife. However, helminth eggs and *Giardia* cysts detected in this sampling point were not detected further down the water treatment train.

Antibiotic resistance genes (ARG)

In Wulpen/Torreele the resistance genes *ampC*, *bla*, *SHV-5*, *mecA* and *vanA*, were absent from all the analysed samples. Only two antibiotic resistance genes *tetO* and *ermB* were detected in this site respectively in 22% and 5% of the samples (Table 11.3).

A net reduction of *tetO* and *ermB* resistance genes concentration was observed from the secondary effluent to the ultrafiltration permeates in the Wulpen/Torreele plant indicating that this barrier can effectively reduce the spreading of ARG in the environment. Nevertheless in October 2007 *tetO* was also found in low concentration in the infiltration water before transport and in the groundwater after infiltration.

11.2.2 Italy (Nardò)

The water filtration which takes place in the fissured aquifer of Nardò was monitored by sampling water at assigned five locations: SP1-WWTP effluent, SP2-injection point (the sinkhole collecting Galatone WWTP secondary effluent and surface water flows), SP3-well (320 m from the sinkhole), SP4-well (500 m from the sinkhole) and SP5-well reference groundwater (located 1,000 m upstream by considering natural groundwater flow direction and 1,000 m away from the sinkhole). At Nardò the wells at 320 and 500 m distances from the infiltration sinkhole were selected for monitoring to allow an estimation of the removal of pathogens in the subsurface. Protozoa, helminths and bacteria were monitored during four sampling campaigns (November 2006, February 2007, May 2007, September 2007) while ARG and viruses were analysed during the last three campaigns. A diagram of Nardò site is reported in Figure 4.3 chapter 4.

Table 11.3 Antibiotic resistance genes detected at Wulpen/Torreele (copy number/100 mL water) (after Böckelmann *et al.* 2009).

ARG		***tetO***		***ermB***
Sampling points		**January 2007**	**October 2007**	**January 2007**
SP 2 secondary effluent	m	1.05×10^7	4.35×10^6	1.92×10^5
	sd	3.54×10^6	5.59×10^5	1.06×10^4
SP 3 ultrafiltration permeate	m	0	0	0
	sd	0	0	0
SP 4 the infiltration water before transport	m	0	5.92×10^3	0
	sd	0	1.39×10^3	0
SP 5 infiltration pond	m	0	0	0
	sd	0	0	0
SP 6 groundwater after infiltration	m	0	3.13×10^3	0
	sd	0	1.52×10^3	0
SP 7 UV treated groundwater	m	0	0	0
	sd	0	0	0

Note: m: median; sd: standard deviation of replicate, of the same sample; only the sampling campaigns with positive samples are reported.

Occurrence and removal of pathogenic microorganisms

Nardò was characterised by the highest pathogen numbers and diversity of these three case study sites. All the analysed pathogens were detected in at least one sample, with the exception of *Helicobacter pylori*, *Mycobacterium avium* subsp. *pseudotuberculosis* and HAV. Protozoa, helminth eggs, *Salmonella* and viruses (Norovirus and Enterovirus) were also found in the groundwater samples.

Three different enteric bacteria were detected in the Nardò system. *Campylobacter* was only detected in 5% of the samples, it was present in the sinkhole water (SP2) of February 2007 sampling campaign. *Yersinia enterocolitica* was only detected in one sample (secondary effluent February 2007) at a level below the quantification limit (10 gene copies per reaction). *Salmonella* gene copies were detected in 10% of the samples, this pathogen was present in May 2007 and November 2006 sampling campaigns respectively in the WWTP effluent (SP1 at 9.1×10^3 gene copies/100 mL) and in the groundwater (SP4 at 1.1×10^3 gene copies/100 mL). *Salmonella* numbers in the secondary treated effluent in Nardò were very similar to those previously reported, and obtained by the same method (Wéry *et al.* 2008). Wéry *et al.* (2008) detected this bacterium in WWTP effluent samples at concentrations of 3.3×10^3 gene copies/100 mL. The presence of *Salmonella* spp. in groundwater is of particular relevance at the Nardò site because in Italy <1/100 mL *Salmonella* (Italian law D.M. 185/2003) are allowed for water used in agriculture. However, qPCR cannot distinguish between viable and non viable pathogenic microorganisms and therefore detection of *Salmonella* gene copies is not a direct evidence of the presence of viable *Salmonella*. In any case at the Nardò site, the molecular detection of *Salmonella* genes together with the presence of fecal indicators, described in Chapter 4, indicate a poor level of microbial quality of groundwater at 500 m from the sinkhole. This confirms that water recovery must be done at greater distance from the infiltration source to protect human health as also shown by modelling of pathogen decay and safe recovery distance (Masciopinto *et al.* 2011).

Nardò was the only European site in the project where pathogenic viruses were detected (Table 11.4).

Enterovirus and/or Norovirus Group I and II, were detected in 42% of samples and mainly in groundwater (50% of groundwater samples). Table 11.4 shows the presence of viruses in February 2007; Norovirus Group I and II were detected in all the sampling points while Enterovirus were only detected in the groundwater samples. Enterovirus were never found by isolation on cell culture but only by RT-PCR suggesting that the detected viral particles were inactive.

Norovirus were the most commonly detected viruses, they were present in 42% of the analysed samples and were mainly detected in the winter period (February 2007). Consistently Norovirus, like many other respiratory and gastrointestinal viruses, are known to exhibit wintertime seasonality in temperate climates (Mounts *et al.* 2000).

All samples were negative for HAV. The absence of HAV positive samples is not surprising considering the start of the vaccination program (1998–2006) where the incidence of infection dropped from 22.8 cases/100,000 during the epidemic period in 1998 to 0.7 cases/100,000 in 2006, with a reduction of the environmental circulation of the virus (Lopalco *et al.* 2008).

Table 11.4 Presence of Enterovirus, HAV and Norovirus in Nardò site (after Masciopinto *et al.* 2011).

Sampling campaigns	Sampling points	Enterovirus		HAV	Norovirus GGI	Norovirus GGII
		Viable test	RT-PCR	RT-PCR	RT-PCR	RT-PCR and qRT-PCR
February 2007	SP 1 secondary effluent	Negative	Negative	Negative	Positive	2×10^3/L
	SP 2 sinkhole	Negative	Negative	Negative	Positive	1×10^3/L
	SP 3 groundwater, well 300 m	Negative	Positive	Negative	Positive	2×10^1/L
	SP 4 groundwater, well 500 m	Negative	Positive	Negative	Positive	Positive
May 2007	SP 1 secondary effluent	Negative	Negative	Negative	Negative	Negative
	SP 2 sinkhole	Negative	Negative	Negative	Negative	Negative
	SP 3 groundwater, well 300 m	Negative	Negative	Negative	Positive	7.6×10^3/L
	SP 4 groundwater, well 500 m	Negative	Negative	Negative	Negative	Negative
September 2007	SP 1 secondary effluent	nd	Negative	Negative	Negative	Negative
	SP 2 sinkhole	nd	Negative	Negative	Negative	Negative
	SP 3 groundwater, well 300 m	nd	Negative	Negative	Negative	Negative
	SP 4 groundwater, well 500 m	Negative	Negative	Negative	Negative	Negative

Viruses were analysed in 1L of secondary effluent, 4 L of sinkhole water and 20 L of groundwater. Enterovirus were detected by RT-PCR and isolation on cell cultures, while only RT-PCR was applied for the remaining viruses. Quantitative analysis was also performed by qRT- PCR on Norovirus genomic groups II. nd: not determined.

Helminth eggs (*Trichuris* sp.) were found in 15% of the Nardò samples, namely two secondary effluent samples (November 2006 and February 2007) and one groundwater sample (SP3 February 2007). Similarly to Wulpen/Torreele the numbers were always low (1 egg/20 L), however, by contrast to the Belgian site they were present in 50% of the secondary treated effluent samples.

Protozoa were detected in 76.4% of Nardò samples (Table 11.5).

Table 11.5 *Giardia* cysts and *Cryptosporidium* oocysts numbers at the Nardò site (after Levantesi *et al.* 2010).

Sampling points	*Giardia* (cysts/L)				*Cryptosporidium* (oocysts/L)			
	November 2006	February 2007	May 2007	September 2007	November 2006	February 2007	May 2007	September 2007
SP 1 secondary effluent	4.8	5.2	27.3	75.4	0.33	0.1	–	–
SP 2 sinkhole	0.4	0.8	6.14	0.24	0.43	0.075	0.4	–
SP 3 groundwater, well 320 m	0.26	0.7	1.8	0.24	0.77	–	–	–
SP 4 groundwater, well 500 m	0.01	–	–	–	–	–	–	–
SP 4a groundwater background well	nd	nd	nd	–	nd	nd	nd	–

Sampling campaigns –: below detection limit. Sample volumes analysed varied depending upon the treatment stage and the amount of water that could be filtered. Approximately 1 L of wastewater, 10–30 L of secondary effluent, 30–60 L of surface water, and 50–115 L of ultra-filtration (UF) filtrate, groundwater and water for final use were filtered. nd: not determined

Giardia cysts were commonly detected (76.4% of total samples), and they were present in 55.5% of groundwater samples ranging from 0.01–1.8 cysts/L. Cysts were also present in all secondary treated effluent and sinkhole samples with an average of 28.2 and 2.4 cysts/L respectively. *Cryptosporidium* oocysts were detected more rarely than *Giardia* cysts; they were present in 35.3% of samples and numbers were lower (from 0.01 to 0.43 oocysts/L). Oocysts were found only in one occasion in the groundwater (well at 320 m in May 2007). The numbers of *Giardia* and *Cryptosporidium* (oo)cysts in the Nardò secondary treated effluent were the highest of the three sites, however, they were lower in comparison to published data (Table 11.1). A decrease of cysts numbers was observed from the sinkhole (2.4×10^{-1}–6.14 cysts/L) to the well at 500 m (0–10^{-2} cysts/L) suggesting removal with water passage through the aquifer.

Antibiotic resistance genes

Antibiotic resistance genes were monitored at Nardò site during three sampling campaigns (February, May and September 2007) and only two of the five targeted ARG, *tetO* and *ermB*, were detected, similar to the observation in Wulpen/Torreele. The results are reported in Table 11.6.

Table 11.6 Antibiotic resistance genes (copy number / 100 mL water) detected at Nardò (after Böckelmann *et al.* 2009)

Sampling points		*tetO*			*ermB*
		February 2007	May 2007	September 2007	February 2007
SP 1 secondary effluent	m	8.65×10^4	2.67×10^5	3.76×10^6	1.35×10^5
	sd	9.19×10^3	1.68×10^5	0	2.69×10^4
SP 2 sinkhole	m	3.20×10^4	1.01×10^4	1.22×10^4	4.00×10^3
	sd	1.81×10^4	1.50×10^3	8.38×10^3	3.82×10^2
SP 3 groundwater, well 300 m	m	1.39×10^5	2.19×10^4	1.07×10^4	4.54×10^4
	sd	1.41×10^4	1.33×10^4	7.06×10^3	1.17×10^4
SP 4 groundwater, well 500 m	m	1.27×10^5	8.74×10^3	0	3.87×10^4
	sd	3.25×10^4	1.92×10^3	0	1.91×10^3
SP 4a groundwater background well	m	9.99×10^2	1.21×10^3	1.93×10^4	0
	sd	4.69×10^2	4.04×10^2	3.54×10^2	0

Note: m: median; sd: standard deviation of replicates of the same sample; only the sampling campaigns with positive samples are reported.

tetO genes were detected in all the sampling campaigns at the Nardò site. In February and May 2007 it was present in the groundwater of SP4 at 1.27×10^5 and at 8.74×10^3 gene copies/100 mL respectively. However, this ARG was also present, at a similar level, in the groundwater not affected by the reclaimed water recharge (background well SP5) ranging from 9.99×10^2 to 1.93×10^4 gene copies/100 mL. *ermB* was detected only in the February sampling campaign from SP1 to SP4 but not in the groundwater of the background well (SP5). A net reduction of ARGs during the flow in the aquifer (from SP2 to SP4) was only observed for one gene in one sampling campaign (*tetO* in September 2007).

As shown in details in Böckelmann *et al.* (2009) the abundance of Enterococci in SP4 showed some correlation with the concentration of ARG: a two fold increase of enterococcal numbers was observed in February and May 2007 sampling campaigns in comparison with that of September 2007 when ARG were not detected. This is a noteworthy result, as it has been reported that Enterococci can exchange a variety of ARG (Grohmann *et al.* 2003; Ferreira da Silva *et al.* 2006).

11.2.3 Spain (Sabadell)

At Sabadell, pathogens and ARG were monitored at five different sampling points, (SP1) secondary effluent from the Ripoll River WWTP, (SP2) Ripoll River reference point (sample of the river water before WWTP effluent discharge), (SP3) infiltration water (sample of the river water after WWTP effluent discharge), (SP4) recovered groundwater and (SP5) disinfected groundwater ready to use (park irrigation and street cleaning). Microbial attenuation by bank filtration (SP3-SP4) and disinfection (SP4-SP5) were mainly evaluated. Antibiotic resistance genes were analysed during four sampling campaigns (January, March, June and October 2007). Protozoa and bacteria were analysed during three sampling campaigns, from March to October 2007 and viruses only in June and October 2007. A diagram of the Sabadell site is presented in Figure 5.2 of Chapter 5.

Occurrence and removal of pathogenic microorganisms

At Sabadell a notable improvement of water quality was achieved during aquifer recharge but pathogens were still detected in the final effluent (*Salmonella* gene copies) and in groundwater (cysts and oocysts).

Pathogenic bacteria were more commonly observed at Sabadell during June 2007: *Campylobacter* was detected in the river water, not influenced by wastewater discharge, while *Salmonella* gene copies were found at all the analysed sampling points. *Salmonella* concentration ranged from 1.2×10^2 to 2.1×10^3 gene copies/100 mL with the highest level associated with river water, both before (3.9×10^2 gc/100mL) and after effluent discharge (2.1×10^3 gene copies/100 mL), and groundwater (4.4×10^2 gene copies/100mL). *Salmonella* was also found in the secondary treated effluent in March 2007 but the concentration was below quantification limit. As described above qPCR can detect also the DNA of non viable *Salmonella* cells and remains of extracellular *Salmonella* DNA; therefore the amplification of *Salmonella* genes in sprinkler water, even if at high level, does not imply a direct hazard for the public.

At Sabadell, MAP, *Yersinia enterocolitica, Helicobacter pylori*, viruses and helminth eggs were not detected, while *Giardia* cysts and *Cryptosporidium* oocysts were commonly found (80% and 66% of the total samples respectively). In Table 11.7 numbers of *Giardia* cysts and *Cryptosporidium* oocysts in Sabadell water are reported.

Table 11.7 *Giardia* cysts and *Cryptosporidium* oocysts numbers at the Sabadell site (after Levantesi *et al.* 2010)

Sampling points	*Giardia* (cysts/L)			*Cryptosporidium* (oocysts/L)		
	March 2007	June 2007	October 2007	March 2007	June 2007	October 2007
SP 1 secondary effluent	5.5	0.54	3.3	0.12	0.09	6.5
SP 2a river background	1.22	3.9	1.9	0.01	0.1	0.09
SP 2 river after wastewater discharge	6.2	0.4	0.69	0.11	0.1	–
SP 3 groundwater after SAT	0.07	0.01	–	0.01	–	–
SP 4 disinfected groundwater	–	–	–	–	–	–

–: below detection limit, sample volumes analyzed varied depending upon the treatment stage and the amount of water that could be filtered. Approximately 1 L of wastewater, 10–30 L of secondary effluent, 30–60 L of surface water, and 50–115 L of ultra-filtration (UF) filtrate, groundwater and water for final use were filtered.

(Oo)cysts were detected in 66% of groundwater samples at Sabadell: in March and June 2007 cysts were present at respectively 0.07 and 0.01 cyst/L, *Cryptosporidium* oocysts were only found in March 2007 at 0.01 oocysts/L. Furthermore (oo)cyst were never detected in the sprinkler water.

In addition to groundwater, cysts and oocysts were both detected in all secondary effluent samples and all but one sample of river water after wastewater discharge, where only *Giardia* was found. Cysts and oocysts were also detected in the river water not influenced by effluent discharge (SP2a), at numbers comparable to or greater than that of the river after wastewater discharge (Table 11.7). Similarly to the Belgian and Italian sites (oo)cysts numbers in Sabadell secondary treated effluent were in the low range reported in literature (Table 11.1). Similarly in the Sabadell surface water, cysts and oocysts concentrations were low with values in the lower range reported in literature (Carmena *et al.* 2007; Di Benedetto *et al.* 2007).

Pathogens analysis showed a very low river water quality at the river reference point (SP2a), in particular a prevalence of pathogens was observed in June 2007. The presence of *Salmonella*, *Campylobacter* and parasites in the river water (SP2a) is likely due to the discharge of two WWTPs upstream of this sampling point. Only a minor effect on reduction of pathogens was observed at Sabadell by mixing and dilution of the secondary treated effluent with the river water (SP1-SP2). Cysts and oocyst reduction from SP1 to SP2 was observed only during two sampling campaigns and usually at low level ($<1\ \log_{10}$ removal). Natural attenuation by bank filtration appeared to have a role in parasite reduction at Sabadell: from SP2 to SP3 a several orders of magnitude decrease in (oo)cysts numbers was observed in all the sampling campaigns.

Antibiotic resistance genes

In Sabadell the highest level and diversity of antibiotic resistance genes of the three sites were found.

tetO was the most common ARG detected at Sabadell, it was present in 70% of total samples, it was found at all sampling points of three out of four sampling campaigns (Table 11.8a and b). *ermB* was found in 50% of the samples in this site, it was present in January (in all the samples except the disinfected water), in March (in all the samples) and in June (only in SP1) sampling campaigns. *mecA* was only detected in January 2007 (25% of total samples). A decrease of *tetO*, *ermB* and *mecA* concentrations was observed along the treatment train at Sabadell (SP1-SP4), however a net reduction of ARG during bank filtration (SP3-SP4) was only observed for *tetO* and *ermB* in respectively two (March and October 2007) and one (January 2007) sampling campaigns. *bla, SHV-5* was found in 30% of the samples; it was detected in all the samples except the disinfected water in March, with a similar number of copies, and in January it was only detected in SP4. *ampC* was detected only in 10% of the samples; it was found once in SP2 (January 2007) and once in SP1 (June 2007).

By comparing the results from the different sampling campaigns across the year with consumption of antibiotics it appears that there is a correlation with the occurrence of ARG. Higher consumption rates of antibiotics in winter are consistent with the higher concentrations of ARGs in the water during that period (January and March). Within the RECLAIM WATER project antibiotics were monitored in the Sabadell case study, but only erythromycin is linked to the ARGs monitored. The highest concentrations were detected in Sabadell WWTP effluent, with a maximum concentration of 463 ng/L (McArdell *et al.* 2008). However, this is far below concentrations increasing gene transfer frequencies (Böckelmann *et al.* 2009). Ohlsen *et al.* (2003) measured erythromycin concentrations up to 0.027 mg/L and ciprofloxacin concentrations up to 0.051 mg/L in sewage. These low-level concentrations of antibiotics did not increase plasmid transfer frequencies of resistance plasmids of staphylococci (Ohlsen *et al.* 2003).

Table 11.8a Antibiotic resistance genes *tetO* and *ampC* (copy number/100 mL water) at Sabadell (after Böckelmann *et al.* 2009).

ARG		*tetO*			*ampC*	
Sample		**March 2007**	**June 2007**	**October 2007**	**January 2007**	**June 2007**
SP 1 secondary effluent	m	7.23×10^5	2.77×10^6	1.59×10^6	0	6.88×10^3
	sd	3.18×10^5	1.07×10^6	3.25×10^5	0	5.69×10^3
SP 2 river background	m	9.18×10^6	9.16×10^4	1.67×10^6	0	0
	sd	7.85×10^5	9.12×10^3	4.03×10^5	0	0
SP 3 river after wastewater discharge	m	2.62×10^6	2.83×10^4	3.22×10^6	3.85×10^3	0
	sd	8.49×10^4	1.05×10^4	3.32×10^5	1.34×10^3	0
SP 4 groundwater after bank filtration	m	6.67×10^5	4.27×10^3	2.11×10^4	0	0
	sd	8.49×10^4	7.02×10^2	4.38×10^3	0	0
SP 5 disinfected groundwater	m	6.23×10^3	0	2.39×10^4	0	0
	sd	1.03×10^3	0	1.41×10^2	0	0

Only the sampling campaigns with positive samples are reported, m: median; sd: standard deviation.

Table 11.8b Antibiotic resistance genes *ermB*, *mecA*, and *blaSHV-5* (copy number/100 mL water) at Sabadell (after Böckelmann *et al.* 2009).

ARG		*ermB*			*mecA*	*blaSHV-5*		
Sample		**January 2007**	**March 2007**	**June 2007**	**January 2007**	**January 2007**	**March 2007**	**June 2007**
SP 1 secondary effluent	m	1.35×10^5	9.51×10^4	2.00×10^4	5.65×10^4	0	4.72×10^3	1.14×10^4
	sd	2.69×10^4	1.69×10^4	8.49×10^2	1.56×10^3	0	2.48×10^3	0
SP 2 river background	m	1.87×10^5	2.64×10^6	0	1.35×10^5	0	2.19×10^3	0
	sd	3.04×10^4	4.38×10^5	0	1.06×10^4	0	3.75×10^2	0
SP 3 river after wastewater discharge	m	2.28×10^5	2.03×10^5	0	4.40×10^4	0	1.07×10^3	0
	sd	4.95×10^3	1.48×10^4	0	3.61×10^3	0	3.75×10^2	0
SP 4 groundwater after bank filtration	m	1.78×10^4	1.10×10^5	0	1.19×10^5	1.60×10^2	1.45×10^3	0
	sd	1.41×10^3	8.49×10^3	0	9.90×10^3	2.83×10^1	3.22×10^2	0
SP 5 disinfected groundwater	m	0	9.50×10^2	0	3.04×10^3	0	0	0
	sd	0	7.07×10^1	0	2.26×10^2	0	0	0

Only the sampling campaigns with positive samples are reported; m: median; sd: standard deviation.

11.2.4 Coherence of pathogen and indicator presence

Results of microbial fecal indicator monitoring at the three sites, described in chapters 2, 3 and 4, were compared to pathogen data to evaluate the efficiency of the analysed microbial indicators as *index* of the pathogens presence. According to drinking water quality guidelines an *index* organism is a microorganism that can be used as reliable index of the presence/absence of pathogens or of fecal contamination in water (WHO, 2008). The percentage of false positives (pathogen absent and indicator present) and false negatives (pathogen present and indicator absent) given by each of the analysed microbial indicators for the detection of *Giardia* cysts, *Cryptosporidium* oocysts, helminth eggs and *Salmonella* spp. in the three MAR sites were determined (Table 11.9).

Table 11.9 Coherence of detection of pathogens and indicators in %, in the three MAR sites.

	Giardia cysts			*Cryptosporidium* oocysts			Helminth eggs			*Salmonella* spp.		
	Total	FN	FP	Total	FN	FP	Total	FN	FP	Total	FN	FP
Clostridium spores	49	2 (3.2)	10	48	0	35	48	2 (16)	69	53	3.7 (28.6)	62
E. coli	49	2 (3.2)	16	48	0	42	48	0	70	53	1.8 (14)	64
Enterococci	49	4 (6.4)	16	48	0	39	48	2 (16)	70	53	1.8 (14)	62
Somatic coliphages	44	2.2 (3.7)	32	43	0	58	43	0	83	48	1.8 (14)	68

Total: total samples number; FN: False negative samples that are negative for indicators (<1 CFU-PFU $\cdot$ 100 ml^{-1}) and positive for pathogens expressed as % of total samples; FP: false positive samples that are positive for indicators and negative for pathogens expressed as % of total samples. In brackets: false negative results expressed as % of pathogen positive samples.

False positive results (presence of indicator in the absence of pathogen) were very common for all the analysed indicators/pathogen couples. The fraction of false positive samples ranged from 10-32% for *Giardia* cysts to 69-83% for helminth eggs. False negative results (presence of pathogen in the absence of indicators) were also obtained for almost all the indicator-pathogenic microorganism combinations. However, these results were quite rare being <4% of the total samples. The lowest level of false negative results was obtained with the indicator *E. coli* independently of the pathogen analysed, confirming the suitability of this microorganism as reliable index of pathogen presence in the analysed sites.

The data on *Clostridium* spores showed the best agreement with the presence of *Giardia* cysts showing the lowest number of false positive (2% of the samples) and false negative results (10% of the pathogen positive samples). False negative results were not obtained for any *Cryptosporidium* oocysts-indicator combination nor for the combinations of helminth eggs with *E. coli* and somatic coliphages.

11.2.5 Results of pathogen decay studies

An example of the use of *in situ* pathogen diffusion chambers studies to assess the potential for pathogen inactivation in the aquifer was undertaken at an Aquifer Storage, Transfer and Recovery (ASTR) scheme in Adelaide, Australia (Page *et al.* 2010a). Two diffusion chamber experiments were undertaken to determine the decay rates of enteric pathogens and microbial indicators both in captured urban stormwater within a constructed reedbed, used as a pre-treatment step prior to the ASTR scheme, and in the aquifer, where the captured stormwater was injected and stored. As described in Sidhu *et al.* (2010) the microorganisms studied in both the reed bed and in the aquifer experiments were the fecal indicators *E. coli*, *Enterococcus faecalis*, and the enteric pathogens *Salmonella enterica*, adenovirus and *Cryptosporidium parvum* oocysts. In the aquifer experiment *Campylobacter jejuni*, coxsackievirus and rotavirus were also analyzed.

An example of *in situ* pathogens decrease over time, as measured in the diffusion chamber studies in the reedbed and the aquifer at the Salisbury ASTR site, Australia is shown in Figure 11.4. The decay rates determined for microbial indicators and pathogens in the wetland and aquifer experiments are reported in Table 11.10a and b.

In order to evaluate the influence of autochthonous groundwater microorganisms on the decay of some of the analysed pathogens, diffusion chamber experiments in the aquifer were performed also with filtered groundwater depleted of autochthonous microorganisms (achieved by passing the groundwater sample twice through sterile 0.2 μm nitrocellulose filters). The decay rate of *E. coli*, adenovirus, coxsackievirus, rotavirus and *Cryptosporidium* was determined; results are reported in Table 11.11 in comparison with the data of microbial decay in the non-filtered groundwater.

The results of the wetland study indicate that pathogens do decay in stormwater within the Parafield Stormwater Harvesting Facility, but the actual rate of decay is specific to the type of microorganism (Sidhu *et al.* 2010). The results show that the bacteria all had T90 times of <6 days while the T90 times for Adenovirus and *Cryptosporidium* were >30 days and >80 days respectively. As the hydraulic residence time of the captured stormwater in the reedbed has been calculated at ~10 days (Page *et al.* 2010b), this suggests that the reedbed should be capable of removing ~2 $\log_{10}$ of the common bacterial contaminants such as *E. coli* and enterococci. The reedbed is much less capable, however, of removing viruses and protozoa with ~0.3 $\log_{10}$ and 0.1 $\log_{10}$ removal, respectively. The low removal rates for adenovirus and *Cryptosporidium* are likely to be attributable to the low water temperatures in the reedbed during the winter season (Sidhu *et al.* 2010). The results obtained by Sidhu *et al.* (2010) have shown that the reedbed cannot be relied upon as a sole treatment barrier for viruses and protozoa. This places greater emphasis on the aquifer component of the ASTR scheme or post treatment of recovered water for the removal of these pathogens.

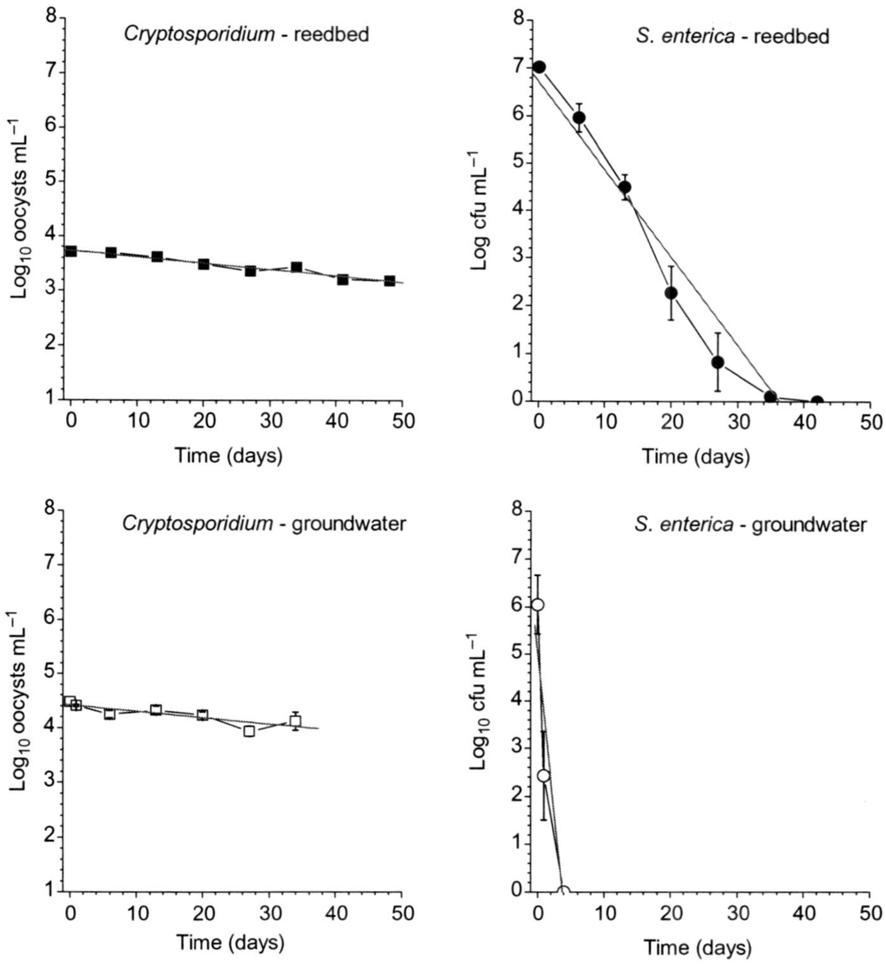

Figure 11.4 Decay of *S. enterica* cells and *C. parvum* oocysts in constructed reedbeds and in the aquifer

Table 11.10a Decay rate and 1 $\log_{10}$ reduction times (T90) for tested microorganisms in the reedbed (after Sidhu *et al.* 2010).

Microorganism	Decay rate ($\log \cdot day^{-1}$)	T90 (days)
E. coli	−0.235	4
E. faecalis	−0.161	6
S. enterica	−0.212	5
Adenovirus	−0.0307	33
Cryptosporidium oocysts	−0.0115	87

Table 11.10b Decay rate and 1 $\log_{10}$ reduction times (T90) for tested microorganism in the aquifer (after Sidhu *et al.* 2010).

Microorganism	Decay rate ($\log \cdot day^{-1}$)	T90 (days)
E. coli	−10	0.1
E. faecalis	−0.4	2.5
S. enterica	−1.42	0.7
C. jejuni	−5	0.2
Coxsackievirus	−0.009	111
Adenovirus	−0.017	59
Rotavirus	−0.0054	185
Cryptosporidium oocysts	−0.012	83

Table 11.11 Influence of groundwater autochthonous microbial population on enteric pathogens and indicators as *in situ* decay rates in the aquifer.

Microorganism	Decay rate ($log \cdot day^{-1}$) in non-filtered groundwater	Decay rate ($log \cdot day^{-1}$) in filtered groundwater
E. coli	−10	−5
Coxsackievirus	−0.009	−0.001
Adenovirus	−0.017	−0.0085
Rotavirus	−0.0054	−0.0036
Cryptosporidium oocysts	−0.012	−0.002

Similarly, Sidhu *et al.* (2010) showed that the bacteria had T90 reduction times in the aquifer of <2 days while the viruses and *Cryptosporidium* oocysts had T90 times of 58 days or longer. Of particular note is the relatively slow decay of rotavirus and coxsackievirus (Table 11.10b). These results indicate that, with a calculated hydraulic residence time of ~200 days, the aquifer can not be relied upon for the complete removal of these viruses. It is likely that a post-recovery treatment step, for example UV, may be needed to ensure the water meets Australian guideline values for acceptable risk from viral pathogens (Page *et al.* 2010c).

The slower decay for the viruses and *Cryptosporidium* in the diffusion chambers where the autochthonous groundwater microorganisms had been removed by filtration (Table 11.11) demonstrated that, despite the low temperatures within the ASTR scheme, the autochthonous groundwater microorganisms appear to have a major influence on the decay of the tested enteric pathogens in the groundwater as has been previously observed (Gordon and Toze, 2003; Toze and Hanna, 2004). Additionally, the reduction in virus and protozoan numbers are due to the actual decay of these microorganisms and not due to factors such as adsorption of the microorganism to the walls of the chamber or to the surface of the exclusion membranes filters as evidenced by the lack of decrease in numbers in the chambers containing the filtered groundwater.

11.3 CONCLUSIONS

11.3.1 Pathogen contamination numbers

Within RECLAIM WATER, useful data on the occurrence of pathogenic viruses, protozoa, bacteria and helminths in three European countries (Italy, Spain and Belgium) and in different types of environmental waters, from wastewater to groundwater were obtained. Low pathogen numbers and a limited diversity of pathogenic microorganisms were detected with the highest numbers observed in Italy and the lowest in Belgium. The presence of pathogens in the groundwater (Nardò) and in the final effluent (Sabadell) confirmed, however, the importance of a careful examination of the microbial quality of water reuse systems and the need for risk management.

Giardia and *Cryptosporiudium* are among the most common causes of waterborne outbreaks in developed countries (Craun *et al.* 2006; Dale *et al.* 2007; Schuster *et al.* 2005). Accordingly, protozoa were the most common pathogens in the three sites and were present in 62% of the samples. In particular, *Giardia* cysts were frequently found in wastewater secondary effluents and recharge waters but they were also detected in groundwater. Similar to previous research (Bonadonna *et al.* 2002; Cacciò *et al.* 2003; Cantusio *et al.* 2006; Carmena *et al.* 2007), *Cryptosporidium* oocysts were detected less frequently and in smaller numbers than *Giardia* cysts in the analysed samples. In spite of their frequent occurrence, (oo)cysts were usually present at low numbers with their concentrations close to the lower level of the previously reported ranges (Table 11.1). Moreover a reduction of protozoa was observed in the case study sites and these pathogens were only detected once, at low level (0.01 cyst/L in Nardò SP4), in the final reclaimed water.

In accordance to previous data from developed countries (EPA 1992; Schwartzbrod *et al.* 1989) RECLAIM WATER results showed that helminth eggs are not very common in wastewaters and secondary effluents of reuse schemes in Europe (Levantesi *et al.* 2010). Helminth eggs were detected in 12% of the samples always at low level, however they were also found in groundwater samples.

Overall enteric pathogenic bacteria were found in 19% of RECLAM WATER samples. *Campylobacter* spp. were only found in 6% of samples and were not detected in wastewaters or secondary treated effluent in this study nor by Alexandrino *et al.* (2004) and Wéry *et al.* (2008). Results obtained for the other bacterial pathogens are also consistent with previously reported data. Alexandrino *et al.* (2004) did not detect *Y. enterocolitica* in wastewater samples and recovered water from two different constructed wetlands in Germany by multiplex PCR and Shannon *et al.* (2007) did not detect *H. pylori* in any stage of municipal wastewater treatment in Canada using real-time qPCR. However *H. pylori* was detected in wastewater samples in Mexico (Mazari-Hiriart *et al.* 2001). A nested PCR study in water systems for human use in Mexico City resulted in the absence of *H. pylori* in dam water used as water sources but 17% of the treated wastewater samples contained the pathogen (Mazari-Hiriart *et al.* 2001). For *M. avium* comparable molecular studies on WWTPs are not available.

Consistently with previous observation (e.g. Hu and Gibbs, 1995; Jiménez *et al.* 2001) *Salmonella* was the most common enteric bacterium detected in the analysed waters. Most of the literature data on the distribution and amount of *Salmonella* in water were obtained by traditional cultivation methods and cannot be directly compared to the data reported in this Chapter; only recently qPCR was applied for the monitoring of *Salmonella* in the aquatic environment (Lee *et al.* 2006; Shannon *et al.* 2007; Ahmed *et al.* 2008; Wéry *et al.* 2008). In contrast with these results, Shannon *et al.* (2007) could not detect this bacterium in wastewater and secondary effluents. However, both Lee *et al.* (2006) and Wéry *et al.* (2008) found high numbers (respectively 4.6×10^2 and 8.5×10^4 gene copies/mL) of the latter in wastewater. Wéry and colleagues (2008) detected this bacterium also in one of the WWTP effluent samples (3.3×10^3 copies/100 mL) in a concentration very similar to the concentrations reported in this chapter.

Salmonella was the only pathogenic bacterium detected in the recovered water at the reclamation sites. *Salmonella* genes were present, at levels comparable to those observed in secondary treated effluent, in the final Sabadell water (SP4) and in the recovered groundwater (SP4) in Nardò. Comparable qPCR studies on *Salmonella* presence in groundwater are not available. qPCR cannot differentiate between viable and non viable cells; at present the only available study (Wéry *et al.* 2008) on the difference between viable *Salmonella* counts and qPCR data, in environmental waters, showed a difference of qPCR values 4 to 5 orders of magnitude higher than viable counts. Similar studies targeting different pathogens and indicators in environmental waters and sludge also showed concentrations obtained by qPCR typically higher than those obtained by traditional cultivation methods (He and Jiang 2005; Yaradou *et al.* 2007). Further research is necessary to fully understand the correlation between positive results obtained by molecular methods and the human health risks associated with water reuse.

Viruses were rarely detected in the case study sites, they were not detected at Wulpen/Torreele and Sabadell. At Nardò they were most frequently found in groundwater samples. Norovirus were detected in secondary effluent, recharge water as well as groundwater samples while Enterovirus were only found in groundwater. By contrast human enteric viruses such as Enterovirus have been detected in wastewaters and in WWTP effluents; however their reported occurrence is extremely variable (Jiménez, 2003) . The presence of these pathogens in the Nardò groundwater could be due to phenomena of microbial attachment and detachment from the karstic subsoil (La Mantia *et al.* 2008b; Masciopinto *et al.* 2007; Masciopinto *et al.* 2011) in association to the virus elevated survival capacity in groundwater, also shown by the diffusion chamber studies at the ASTR site in Australia (Sidhu *et al.* 2010).

The data indicate that the detected Enterovirus were not active as they did not grow on cell lines; furthermore no data are available on Norovirus viability and infectivity. However considering the low decay rates (Table 11.11), their presence in reclaimed water is an indication of possible risk to human health. At Nardò the data of protozoa and viruses presence, together with microbial indicators monitoring results, allowed the determination of a minimum distance of 3000 m from the sinkhole necessary to predict pathogen removal to an acceptable risk level (Masciopinto *et al.* 2011).

11.3.2 Treatment performances of MAR processes

In agreement with microbial indicators monitoring, pathogen analysis showed that, in the investigated sites, MAR was potentially a good although not necessarily a complete barrier for pathogen removal. A reduction of several orders of magnitude of *Giardia* cysts and *Cryptosporidium* oocysts level was observed from the injectant (river water) and the recovered (mine water) at the Spanish site, and at the Italian site a decrease of cysts concentration occurred from the sinkhole to the well at 500 m in spite of the fractured nature of the karstic aquifer (Levantesi *et al.* 2010). In the Wulpen/Torreele sites pathogens were detected in the infiltration water (SP4) but not in the recovered water after SAT (SP6) (Levantesi *et al.* 2010). Many factors such as soil properties, residence time, dilution factor, hydraulic conditions, organic matter content, dissolved oxygen concentrations, indigenous groundwater microorganisms and the type and number of microbes can influence microorganisms removal in MAR (Gordon & Toze, 2003; John & Rose, 2005). However the role of these individual factors in amelioration of water quality on a site specific basis remains difficult to evaluate. The results indicate that at sites that can be characterised by having low pathogen numbers in the recharged water, MAR might provide a valuable active contribution to microbial pathogen inactivation complimentary with engineered treatments (Levantesi *et al.* 2010; Page *et al.* 2010a, b; Toze *et al.* 2010). The importance of the multiple barrier approach for the inactivation of pathogens and indicators with different resistance to water treatment was also confirmed (Levantesi *et al.* 2010; Page *et al.* 2010c). Different processes (e.g. conventional activated sludge treatment, ultrafiltration, disinfection, MAR) contributed to the removal of microorganisms in the studied MAR systems. Furthermore, overall indicators (Chapters 2, 3, 4) and pathogens monitoring results (this chapter) suggested that treated water can be re-contaminated, (i.e. in the open infiltration pond, Belgian site), and that if a specific stage of the process (i.e. disinfection of groundwater in Sabadell, Chapter 2) fails the resulting removal performances will be lower than expected.

11.3.3 Pathogens and indicators relationships

The results suggest that microbial indicators of fecal contamination, and in particular *E. coli*, are reliable indexes of the presence of a variety of pathogens in the water reuse schemes. The frequency of samples containing a pathogen in the absence of microbial indicators was low ($\leq$4% of total samples). Similar values were obtained by Wilkes *et al.* (2009) that reported a low level of false negative results with microbial and viral markers in a survey of *Giardia* cysts and *Cryptosporidium* oocysts in surface waters. Other authors (Harwood *et al.* 2005; Costán-Longares *et al.* 2008) reported a much higher frequency of samples containing protozoan pathogens in the absence of detectable microbial fecal indicators in water reclamation processes. In agreement with previous observations (Harwood *et al.* 2005; Costán-Longares *et al.* 2008; Wilkes *et al.* 2009), it was found that a large number of pathogen free samples containing microbial indicators were detected. As described by Berg (1978) a microorganism can be considered a valuable index of pathogen presence even if it is detected when the pathogen is absent (false positive results). However these results might cause an increase in costs related to excessive treatment of water that represents no increased risk to human health. Fast specific molecular methods for the detection of pathogens, in spite of their limits to assess the infectivity of the detected microorganism have an advantage in this context. They can supply important information on the presence of pathogens and thereby avoid excessive treatment of the reclaimed water. The pathogen decay studies do provide a warning that in aquifer pathogen indicators decay rates are higher than those for pathogens and the incidence of false negatives is projected to increase in affected groundwater.

11.3.4 ARG in reclaim water sites

Only two resistance genes, *tetO* and *ermB* were detected at all of the three case study sites. Chen *et al.* (2007) and Cauwerts *et al.* (2007) found that the abundances of the *erm* ARG often appeared to be positively correlated with those of the *tet* ARG. Martins da Costa *et al.* (2006) investigated ARG in *Enterococcus* spp. isolated from municipal WWTPs in Portugal and found that 34.6% of the isolates were resistant against tetracycline and 24.8% against erythromycin. Wastewater treatment in the Portuguese plants showed a decrease of enterococci up to 4.0 $\log_{10,}$ nevertheless, more than 4.4×10^5 CFU/100 mL were present in the outflow of the plants (Martins da Costa *et al.* 2006). Lower concentrations of enterocci were obtained in the final water in all the investigated RECLAIM WATER sites resulting in a maximum of 1.27 log CFU/100 mL for Sabadell and a maximum of 3.42 log CFU/100 mL for Nardò.

WWTP effluent can be a source for ARG in the environment (Ferreira da Silva *et al.* 2006, Koike *et al.* 2007). The best investigated ARG are the tetracycline resistance genes including *tetO*. Ferreira da Silva *et al.* (2006) showed that tetracycline-resistant enterococci were not eliminated by secondary wastewater treatment. These data are consistent with the results from Sabadell and Nardò. At both sites secondary WWTP effluent is polished by dilution with surface waters and soil passage. Both sites resulted in a release of enterococci and in a temporary release of *tetO* and *ermB* genes in the environment through the irrigation water. Volkmann *et al.* (2004) performed real-time qPCR analysis on the presence of *ampC*, *vanA* and *mecA* in municipal wastewater in Germany. *vanA* was detected in 21% of the wastewater samples, *ampC* in 78%. *mecA* was not found in municipal wastewater but in two hospital wastewater samples. The detection of *ampC* in wastewater is in agreement with data from Sabadell where *ampC* was found in two of the four sampling campaigns. However, *vanA* was never detected. The presence of *mecA* in all samples from the Sabadell January 2007 sampling campaign could be explained by Sabadell WWTP receiving wastewater from the local hospital. For the resistance gene *blaSHV-5* no comparable analyses have been performed.

No clear trend for the extent of ARG detections in the course of the year was observed for the three sites. Consistent for all sites is that the lowest ARG contamination was found in the summer sampling campaign: no ARG were detected in the July samples of Wulpen/Torreele, in Sabadell in the irrigation water in June, and in Nardò in the groundwater (well at 500 m) in September 2007.

The release of *ermB* and *tetO* from the Nardò and Sabadell MAR systems and of *mecA* from the Sabadell site may contribute to the ARG pool in the environment.

The three systems demonstrated different removal capacity for ARG. Ultrafiltration and reverse osmosis in Wulpen/Torreele proved to be efficient barriers for the elimination of ARG whereas aquifer passage followed by UV treatment and chlorination at Sabadell and the fractured and permeable aquifer at Nardò posed only partial barriers for ARG. It is concluded that *tetO*, *ermB* and *mecA* can occur in groundwater derived from MAR.

11.3.5 *In situ* pathogen decay rate

Knowledge of the rate of pathogen attenuation in aquifers and other natural treatment systems has advantages for the design and operation of water reclamation systems such as MAR and to determine if any additional inactivation credits may be required. Pathogens have been shown to have decay rates that are dependent on factors such as pathogen type, aquifer conditions and the characteristics of the recharged water (Toze *et al.* 2010). The decay studies to-date have been designed to assess the decay of WHO index pathogens that pose the greatest risk to human health. Pathogen decay studies in conjunction with hydrogeological evaluation testing and a QMRA assessment have been used to demonstrate the design of a MAR scheme for water recycling (Toze *et al.* 2010).

REFERENCES

Abulreesh H. H., Paget T. A. and Goulder R. (2005). Recovery of thermophilic Campylobacters from pond water and sediment and the problem of interference by background bacteria in enrichment culture. *Wat. Res.*, **39**(13), 2877–2882.

Adams B. L., Bates T. C. and Oliver J. D. (2003). Survival of *Helicobacter pylori* in a natural freshwater environment. *Appl. Environ. Microbiol.*, **69**(12), 7462–7466.

Ahmed W., Huygens F., Goonetilleke A. and Gardner T. (2008). Real-time PCR detection of pathogenic microorganisms in roof-harvested rainwater in southeast Queensland, Australia. *Appl. Environ. Microbiol.*, **74**(17), 5490–5496.

Alexandrino M., Grohmann E. and Szewzyk U. (2004). Optimization of PCR-based methods for rapid detection of *Campylobacter jejuni, Campylobacter coli* and *Yersinia enterocolitica* serovar 0:3 in wastewater samples. *Wat. Res.*, **38**, 1340–1346.

Archuleta R. J., Hoppes P. Y. and Primm T. P. (2005). *Mycobacterium avium* enters a state of metabolic dormancy in response to starvation. *Tuberculosis* (Edbn), **85**, 147–158.

Asano T. and Cotruvo J. A. (2004). Groundwater recharge with reclaimed municipal wastewater: health and regulatory considerations. *Water Res.*, **38**, 1941–1951.

Asano T., Burton F. L., Leverenz H. L., Tsuchihashi R. and Tchobanoglous G. (2007). Water Reuse – Issues, Technologies and Applications, 1st edn, McGraw-Hill, New York.

Ashbolt N. J. (2004). Microbial contamination of drinking water and disease outcomes in developing regions. *Toxicol.*, **198**, 229–238.

Auerbach E. A., Seyfried E. E. and McMahon K. D. (2007). Tetracycline resistance genes in activated sludge WWTPs. *Wat. Res.*, **41**, 1143–1151.

Berg G. (1978). The indicator system. In: Indicators of Viruses in Water and food, G. Berg (ed.), Ann Arbor Science Publisher, Ann Arbor, pp. 1–13.

Bellack N. R., Koehoorn M. W., MacNab Y. C. and Morshed M. G. (2006). A conceptual model of water's role as a reservoir in *Helicobacter pilori* transmission: a review of the evidence. *Epidemiol. Infect.*, **134**, 439–449.

Bingham A. K., Jarroll E. and Meyer E. (1979). *Giardia* spp.: physical factors of excystation *in vitro*, and excystation vs eosin exclusion as determinants of viability. *Exp. Parasitol.*, **47**, 284–291.

Böckelmann U., Dörries H. H., Ayuso-Gabella M. N., Salgot de Marçay M., Tandoi V., Levantesi C., Masciopinto C., Van Houtte E., Szewzyk U., Wintgens T. and Grohmann E. (2009). Quantitative real-time PCR monitoring of bacterial pathogens and antibiotic resistance genes in three european artificial groundwater recharge systems. *Appl. Envir. Microbiol.*, **75**, 154–163.

Bonadonna L., Briancesco R., Cataldo C., Divizia M., Donia D. and Pana A. (2002). Fate of bacterial indicators, viruses and protozoan parasites in a wastewater multi-component treatment system. *New Microbiol.*, **25**, 413–420.

Briancesco R. and Bonadonna L. (2005). An italian study on *Cryptosporidium* and *Giardia* in wastewater, fresh water and treated water. *Environ. Monitor and Assess.*, **104**, 445–457.

Bratton R. and Nesse R. (1993). Ascariasis: An infection to watch for in immigrants. *Postgraduate Medicine*, **93**, 171–178.

Byappanahalli M. N., Sawdey R., Ishii S., Shively D. A., Ferguson J. A., Whitman R. L. and Sadowsky M. J. (2009). Seasonal stability of *Cladophora*-associated *Salmonella* in lake michigan watersheds. *Water Research*, **43**(3), 806–814.

Burns D., Wallace R. J. Jr., Schultz M. E., Zhang Y., Zubairi S. Q., Pang Y., Gibert C. L., Brown B. A., Noel E. S. and Gordin F. M. (1991). Nosocomial outbreaks of respiratory tract colonization with *Mycobacterium fortuitum*: demostration of the usefulness of pulse field electrophoresis in an epidemiological investigation. *Am. Rev. Respir. Dis.*, **144**, 1153–1159.

Cacciò S. M., De Giacomo M., Aulicino F. A. and Pozio E. (2003). *Giardia* cysts in wastewater treatment plants in italy. *Appl. Environ. Microbiol.*, **69**, 3393–3398.

Campos C. (2006). New perspectives on microbiological water control for wastewater reuse. Proceedings from the AQUAREC final meeting. Barcelona.

Cantusio Neto R., Santos J. U. and Franco R. M. B. (2006). Evaluation of activated sludge treatment and the efficiency of the disinfection of *Giardia* species cysts and *Cryptosporidium* oocysts by UV at a sludge treatment plant in campinas, south-east brazil. *Water Sci. Technol.*, **54**(3), 89–94.

Carmena D., Aguinagalde X., Zigorraga C., Fernández-Crespo J. C. and Ocio J. A. (2007). Presence of *Giardia* cysts and *Cryptosporidium* oocysts in drinking water supplies in northern spain. *J. Appl. Microbiol.*, **102**(3), 619–629.

Castro-Hermida J. A., García-Presedo I., Almeida A., González-Warleta M., Correia Da Costa J. M. and Mezo M. (2008). Contribution of treated wastewater to the contamination of recreational river areas with *Cryptosporidium* spp. and *Giardia duodenalis*. *Wat. Res.*, **42**(13), 3528–3538.

Cauwerts K., Decostere A., De Graef E. M., Haesebrouck F. and Pasmans F. (2007). High prevalence of tetracycline resistance in Enterococcus isolates from broilers carrying the *erm(B)* gene. *Avian. Pathol.*, **36**, 395–399.

Chee-Sanford J. C., Aminov R. I., Krapac L. J., Garrigues-Jeanjean N. and Mackie R. I. (2001). Occurrence and diversity of tetracycline resistance genes in lagoons and groundwater underlying two swine production facilities. *Appl. Environ. Microbiol.*, **67**, 1494–1502.

Chen J., Yu Z., Michel F. C. Jr., Wittum T. and Morrison M. (2007). Development and application of real-time PCR assays for quantification oferm genes conferring resistance to macrolides-lincosamides-streptogramin bin livestock manure and manure management systems. *Appl. Environ. Microbiol.*, **73**, 4407–4416.

Chiodini R. J., van Kruiningen H. J. and Merkal R. S. (1984). Ruminant paratuberculosis (Johne's disease): the current status and future prospects. *Cornell Vet.*, **74**, 218–262.

Cicmanec J. L., Smith J. E. Jr, and Carr R. (2004). Control of zoonotic diseases in drinking-water in waterborne zoonoses identification, causes, and control. In: World Health Organization, J. A. Cotruvo, A. Dufour, G. Rees, J. Bartram, R. Carr, D. O. Cliver, G. F. Craun, R. Fayer and V. P.J. Gannon (eds), IWA Publishing, Alliance House, 12 Caxton Street, London SW1H 0QS, UK.

Collins C. H., Grange J. M. and Yates M. D. (1984). Mycobacteria in water. *J. Bacteriol.*, **57**, 193–211.

Cook K. L. and Britt J. S. (2007). Optimization of methods for detecting *Mycobacterium avium* subsp. *paratuberculosis* in environmental samples using quantitative, real-time PCR. *J. Microbiol. Methods.*, **69**(1), 154–60.

Costán-Longares A., Montemayor M., Payán A., Méndez J., Jofre J., Mujeriego R. and Lucena F. (2008). Microbial indicators and pathogens: removal, relationships and predictive capabilities in water reclamation facilities. *Wat. Res.*, **42**(17), 4439–4448.

Craun G. F. (1979). Waterborne disease- a status report emphasizing outbreaks in ground-water systems. *Ground Water*, **17**, 183–193.

Craun G. F., Calderon R. L. and Craun M. F. (2006). Waterborne disease outbreaks: their causes, problems and challenges to treatment barriers. Waterborne Pathogens, 2nd edn, American Water Works Association, Denver, CO.

Dale K., Kirk M., Sinclair M., Hall R. and Leder K. (2007). Reported outbreaks of gastrointestinal disease in australia are predominantly associated with recreational exposure. *Australian and New Zeland Journal of Public Health*, **10**(34), 527–530.

Di Benedetto M. A., Di Piazza F., Maida C. M., Firenze A. and Oliveri R. (2005). Occurrence of *Giardia* and *Cryptosporidium* in wastewater, surface water and groundwater samples in palermo (sicily). *Ann. Ig.*, **17**(5), 367–75.

EPA. (2001). *EPA Method 1623: Cryptosporidium and Giardia in Water by Filtration/IMS/FA United States*. Office of Water Environmental Protection 4603 April 2001, United States Environmental Protection Agency (EPA-821-R-01-025), Washington DC, USA.

EPA. (1992). *Manual of Guidelines for Water Reuse*. Washington DC, United States Environmental Protection Agency (EPA/625/R-92/004), Washington DC, USA.

Esiobu N., Armenta L. and Ike J. (2002). Antibiotic resistance in soil and water environments. *Intern. J. Environ. Health Res.*, **12**, 133–144.

Falcao D. P., Valentini S. R. and Leite C. Q. F. (1993). Pathogenic or potentially pathogenic bacteria as contaminants of fresh water from different sources in araraquara, Brazil. *Water Research*, **27**, 1737–1741.

Feachem R. G., Brandley D. J. and Garelick H. (1983). Sanitation and disease: health aspects of excreta and waste management, Wiley, New York.

Ferreira da Silva M., Tiago I., Verissimo A., Boaventura R. A. R., Nunes O. C. and Manaia C. M. (2006). Antibiotic resistance of enterococci and related bacteria in an urban WWTP. *FEMS Microbiol. Ecol.*, **55**, 322–329.

Gauthier D. T., Reece K. S., Xiao J., Rhodes M. W., Kator H. I., Latour R. J., Bonzek C. F., Hoenig J. M. and Vogelbein W. K. (2010). Quantitative PCR assay for *Mycobacterium pseudoshottsii* and *Mycobacterium shottsii* and application to environmental samples and fishes from the chesapeake bay. *Appl. Environ. Microbiol.*, **76**(18), 6171–6179.

Gennaccaro A. L., McLaughlin M. R., Quintero-Betancourt W., Huffman D. E. and Rose J. B. (2003). Infectious Cryptosporidium parvum oocysts in final reclaimed effluent. *Appl. Environ. Microbiol.*, **69**, 4983–4984.

Gerba C. P., Rose J. B. and Haas C. N. (1994). "Waterborne disease who is at risk?" Proocedings of the american water works association's water quality technology conference, S. Francisco, American Water Works Association, Denver, CO.

Gerba C., Nwachuku N. and Riley K. (2003). Disinfection resistance of waterborne pathogens on the united states environmental protection agency's contaminant candidate list (CCL). *Journal of Water Supply Research and Technology AQUA*, **52**(2), 81–94. IWA Publishing.

Gerba C. (2009). Environmentally transmitted pathogens. In: Environmental Microbiology, R. M. Maier, I. L. Pepper and C. Gerba (eds), Accademic Press, Elsevier.

Gordon C. and Toze S. (2003). Influence of groundwater characteristics on the survival of enteric viruses. *J. Appl. Microbiol.*, **95**(3), 536–544.

Grohmann E., Muth G. and Espinosa M. (2003). Conjugative plasmid transfer in gram-positive bacteria. *Microbiol. and Mol. Biol. Rev.*, **67**(2), 277–301.

Guo X., van Iersel M. W., Chen J., Brackett R. E. and Beuchat L. R. (2002). Evidence of association of salmonellae with tomato plants grown hydroponically in inoculated nutrient solution. *Appl. Environ. Microbiol.*, **68**, 3639–3643.

Hazeleger W. C., Wouters J. A., Rombouts F. M. and Abee T. (1998). Physiological activity of *Campylobacter jejuni* far below theminimal growth temperature. *Appl. Environ. Microbiol.*, **64**, 3917–3922.

Hanninen M. L., Haajanen H., Pummi T., Wermundsen K., Katila M. L., Sarkkinen H., Miettinen I. and Rautelin H. (2003). Detection and typing of *Campylobacter jejuni* and *Campylobacter coli* and analysis of indicator organisms in three waterborne outbreaks in finland. *Appl. Environ. Microbiol.*, **69**(3), 1391–1396.

Hanninen M., Niskanen M. and Korhenen L. (1998). Water as a reservoir for *Campylobacter jejuni* infection in cows studied by pulsed-field gel electrophoresis. *J. Veterinary Medicine*, **45**, 37–42.

Harwood V. J., Levine A. D., Scott T. M., Chivukula V., Lukasik J., Farrah S. R. and Rose J. B. (2005). Validity of the indicator organism paradigm for pathogen reduction in reclaimed water and public health protection. *Appl. Environ. Microbiol.*, **71**, 3163–3170.

He J. W. and Jiang S. (2005). Quantification of enterococci and human adenoviruses in environmental samples by real-time PCR. *Appl. Environ. Microbiol.*, **71**, 2250–2255.

Hibler C. and Hancock C. (1990). Waterborne giardiasis. In: Drinking water microbiology, G. McFeters (ed.), Springer-Verlag, New York, pp. 271–293.

Hu C. J. and Gibbs R. A. (1995). A comparison of culture methods for the detection of *Salmonella* in wastewater sludge. *Water Sci. Technol.*, **31**, 303–306.

ISO. (2005). ISO 17995: 2005-06-15-Water quality – detection and enumeration of thermotolerant *Campylobacter* species. Geneva, Switzerland: International Organization for Standardization.

International Agency for Research on Cancer Working Group. (1994). Schistosomes, liver flukes and *helicobacter pylori*. IARC working group on the evaluation of carcinogenic risks to humans. *IARC Monogr. Eval. Carcinog. Risks Hum.*, **61**, 1–124.

Isaacson R. E. and Torrence M. E. (2002). The role of antibiotics in agriculture. American Academy of Microbiology, Washington, DC, USA.

Italian law. (2003). D. M. 185/2003 Regolamento recante norme tecniche per il riutilizzo delle acque reflue in attuazione dell'articolo 26, comma 2, del D.Lgs. 11 maggio 1999, n. 152. (G.U. 23 luglio 2003, n. 169).

Jiménez B. (2003). Health risk in aquifer recharge with recycled waters. In: State of the art report on health risk in aquifer recharge using reclaimed waters, R. Aertgeerts and A. Angelakis (ed.), WHO, Geneva, pp.16–122.

Jiménez B., Maya-Rendon C. and Salgado-Velazquez G. (2001). The elimination of helminth ova, faecal coliforms, *Salmonella* and protozoan cysts by various physicochemical processes in wastewater and sludge. *Water Sci. Technol.*, **43**(12), 179–182.

Jiménez B., Chavez A., Leyva A. and Tchobanoglous G. (2000). Sand and synthetic medium filtration of advanced primary treatment effluent from mexico city. *Water Research*, **34**(2), 473–480.

John D. E. and Rose J. B. (2005). Review of factors affecting microbial survival in groundwater. *Environ. Sci. Technol.*, **39**, 7345–7356.

Jones K. (2001). Campylobacter in water sewage and environment. *Journal of Applied Microbiol.*, **90**, 68S–69S.

Johnson D. C., Enriquez C. E., Pepper I. L., Davies T. L., Gerba C. P. and Rose J. B. (1997). Survival of *Giardia, Cryptosporidium*, polioviruses and *Salmonella* in marine waters. *Water Science and Technology*, **35**, 261–268.

Johnson J. Y. M., Thomas J. E., Graham T. A., Townshend I., Byrne J., Selinger L. B. and Gannon V. P. J. (2003). Prevalence of *Escherichia coli* O157:H7 and *Salmonella* spp. in surface waters of southern alberta and its relation to manure sources. *Can. J. Microbiol.*, **49**(5), 326–335.

Jyoti A., Ram S., Vajpayee P., Singh G., Dwivedi P. D., Jain S. K. and Shanker R. (2010). Contamination of surface and potable water in south Asia by Salmonellae: culture-independent quantification with molecular beacon real-time PCR. *Sci. Total Environ.*, **408**(6), 1256–1263.

Karapinar M. and Gonul S. A. (1991). Survival of *Yersinia enterocolitica* and *Escherichia coli* in spring water. *International Journal of Food Microbiology*, **13**, 315–319.

Klein P. D., Graham D. Y., Gaillour A., Opekun A. R. and Smith E. O. (1991). Water source as risk factor for *Helicobacter pylori* infection in peruvian children. gastrointestinal physiology working group. *Lancet*, **337**(8756), 1503–5306.

Koike S., Krapac I. G., Oliver H. D., Yannarell A. C., Chee-Sanford J. C., Aminov R. I. and Mackie R. I. (2007). Monitoring and source tracking of tetracycline resistance genes in lagoons and groundwater adjacent to swine production facilities over a 3-year period. *Appl. Environ. Microbiol.*, **73**, 4813–4823.

Kolpin D. W., Furlong E. T., Meyer M. T., Thurman E. M., Zaugg S. D., Barber L. B. and Buxton H. T. (2002). Pharmaceuticals, hormones, and other organic wastewater contaminants in US streams, 1999–2000: a national reconnaissance. *Environ. Sci. Technol.*, **36**, 1202–1211.

Kümmerer K. (2004). Resistance in the environment. *J. Antimicrob. Chemother.*, **54**, 311–320.

La Mantia R., Masciopinto C., Levantesi C. and Tandoi V. (2008a). Fate and transport of faecal contamination microbial indicators, pathogenic protozoa and *Campylobacter* in the artificially recharged fractured aquifer of salento, Italy. *Water Sci. Technol.*, **57**(6), 849–856.

La Mantia R., Saturno D., Laera G., Masciopinto C. and Pollice A. (2008b). Suitability of MBR and CAS effluents for groundwater recharge in fractured aquifers: laboratory studies. In Proc. SIDISA Conference, 24–27 June 2008, Florence, Italy.

Lee D. Y., Shannon K. and Beaudette L. A. (2006). Detection of bacterial pathogens in municipal wastewater using an oligonucleotide microarray and real-time quantitative PCR. *Journal of Microbiol. Methods*, **65**, 453–467.

Lehtola M. J., Torvinen E., Kusnetsov J., Pitkänen T., Maunula L., von Bonsdorff C. H., Martikainen P. J., Wilks S. A., Keevil C. W. and Mietttinen I. T. (2007). Survival of *Mycobacterium avium, Legionella pneumophila, Escherichia coli*, and caliciviruses in drinking water-associated biofilms grown under high-shear turbulent flow. *Appl. Environ. Microbiol.*, **73**, 2854–2859.

Lehtola M. J., Christopher T., Loades J. and Keevil W. C. (2005). Advantages of peptide nucleic acid oligonucleotides for sensitive site directed 16S rRNA fluorescence in situ hybridization (FISH) detection of *Campylobacter jejuni*, *Campylobacter coli* and *Campylobacter lari*. *Journal Microbiol. Methods*, **62**, 211–219.

Levantesi C., La Mantia R., Masciopinto C., Böckelmann U., Ayuso-Gabella M. N., Salgot M., Tandoi V., Van Houtte E., Wintgens T. and Grohmann E. (2010). Quantification of pathogenic microorganisms and microbial indicators in three wastewater reclamation and managed aquifer recharge facilities in europe. *Sci. Total Environ.*, **408**, 4923–4930.

Levy S. B. (2002). The Antibiotic Paradox: How the Misuse of Antibiotics Destroys their Curative Powers, Perseus Publishing, Cambridge, MA.

Li D., Craik S. A., Smith D. W. and Belosevic M. (2009). Infectivity of *Giardia lamblia* cysts obtained from wastewater treated with ultraviolet light. *Wat. Res.*, **43**(12), 3037–3046.

Lindsey M. E., Meyer T. M. and Thurman E. M. (2001). Analysis of trace levels of sulphonamide and tetracycline antimicrobials in groundwater and surface water using solid-phase extraction and liquid chromatography/mass spectrometry. *Anal. Chem.*, **73**, 4640–4646.

Linke S., Lenz J., Gemein S., Exner M. and Gebel J. (2010). Detection of *Helicobacter pylori* in biofilms by real-time PCR. *Int. J. Hyg. Environ. Health.*, **213**(3), 176–182.

Lodder W. J. and de Ronda Husman A. M. (2005). Presence of noroviruses and other enteric viruses in sewage and surface waters in The Netherland. *Appl. Environ. Microbiol.*, **71**, 1453–1461.

Lopalco P. L., Prato R., Chironna M., Germinario C. and Quarto M. (2008). Control of hepatitis a vaccination of adolescents, Puglia, Italy. *Emerg. Infect Diseases*, **14**(3), 526–528.

Masciopinto C., La Mantia R., Carducci A., Casini B., Calvario A. and Jatta E. (2007). Unsafe tap water in households supplied from groundwater in the salento region of southern italy. *J. Wat. And Health*, **5**(1), 129–148.

Masciopinto C., La Mantia R., Levantesi C., Tandoi V., Divizia M., Donia D., Gabrieli R. and Petrinca A. R. (2011). An analytical solution for the modelling of the natural time-dependent reduction of waterborne viruses injected into fractured aquifers. *Environ. Sci Technol.*, **45**(2), 636–642.

Martins da Costa P., Vaz-Pires P. and Bernardo F. (2006). Antimicrobial resistance in *Enterococcus* spp. isolated in inflow, effluent and sludge from municipal sewage water treatment plants. *Wat. Res.*, **40**, 1735–1740.

Mazari-Hiriart M., López-Vidal Y. and Calva J. J. (2001). *Helicobacter pilori* in water systems for human use in mexico city. *Water Sci. Technol.*, **43**, 93–98.

McArdell C. S., Asmin J., Ayuso-Gabella M. N., Balest L., Fink G., Hollender J., Mascolo G., Salgot M. and Ternes T. A. (2008). Fate of micropollutants in two artificial recharge sites with reclaimed wastewater. Paper presented at the IWA World Water Congress, Vienna, 7–12 September 2008.

Moreno Y., Piqueres P., Alonso J. L., Jimenez A., Gonzalez A. and Ferrus M. A. (2007). Survival and viability of *Helicobacter pylori* after inoculation into chlorinated drinking water. *Wat. Res.*, **41**, 3490–3496.

Mounts A. W., Ando T., Koopmans M., Bresee J. S., Noel J. and Glass R. I. (2000). Cold weather seasonality of gastroenteritis associated with norwalk-like viruses. *J Infect. Dis.*, **181**, Suppl 2, S284–S287.

Nachamkin I. (1997). *Campylobacter jejuni*. In: Food Microbiology: Fundamentals and Frontiers. M. P. Doyle, L. R. Beuchat and T. J. Montville (ed.), American Society for Microbiology, Washington, D.C, pp. 159–170.

Noble R. T., Griffith J. F., Blackwood A. D., Fuhrman J. A., Gregory J. B., Hernandez X., Liang X., Bera A. A. and Schiff K. (2006). Multitiered approach using quantitative PCR to track sources of fecal pollution affecting santa monica bay california. *Appl. Environ. Microbiol.*, **72**, 1604–1612.

Ohlsen K., Ternes T., Werner G., Wallner U., Löffler D., Ziebuhr W., Witte W. and Hacker J. (2003). Impact of antibiotics on conjugational resistance gene transfer in *Staphylococcus aureus* in sewage. *Environ. Microbiol.*, **5**(8), 711–716.

Ottoson J., Hansen A., Bjorlenius B., Norder H. and Stenstrom T. A. (2006). Removal of viruses, parasitic protozoa and microbial indicators in conventional and membrane processes in a wastewater pilot plant. *Wat. Res.*, **40**, 1449–1457.

Page D. W., Dillon P., Toze S. and Sidhu J. P. S. (2010a). Characterising aquifer treatment for pathogens in managed aquifer recharge. *Water Sci. Technol.*, **62**(9), 2009–2015.

Page D., Dillon P., Mueller J. and Bartkow M. (2010b). Quantification of herbicide removals in a constructed wetland using composite water quality monitoring and passive samplers. *Chemosphere*, **81**, 394–399.

Page D., Dillon P., Toze S., Bixio D., Genthe B., Jiménez Cisneros B. E. and Wintgens T. (2010c). Valuing the subsurface pathogen treatment barrier in water recycling via aquifers for drinking supplies. *Wat. Res.*, **44**, 1841–1852.

Park S. H., Hanning I., Jarquin R., Moore P., Donoghue D. J., Donoghue A. M. and Ricke S. C. (2011). Multiplex PCR assay for the detection and quantification of *Campylobacter* spp., *Escherichia coli* O157:H7, and *Salmonella* serotypes in water samples. *FEMS Microbiol. Lett.*, **316**(1), 7–15.

Parsonnet J., Friedmann G. P., Vandersteen D. P., Chang Y., Vogelman J. H., Orentriech N. and Sibley R. K. (1991). *Helicobacter pylori* infection and the risk of gastric carcinoma. *N. Engl. J. Med.*, **325**, 1127–1131.

Pavelic P., Ragusa S. R., Flower R. L., Rink-Pfeiffer S. M. and Dillon P. J. (1998). Diffusion chamber method for *in situ* measurement of pathogen inactivation in groundwater. *Wat. Res.*, **32**, 1144–1150.

Pianetti A., Baffone W., Bruscolini F., Barbieri E., Biffi M. R., Selvaggio L. and Albano A. (1998). Presence of several pathogenic bacteria in the metauro and foglia rivers (pesaro-urbino italy). *Wat. Res.*, **32**(5), 1515–1521.

Queralt N. and Araujo R. (2007). Analysis of the survival of *H. pylori* within a laboratory-based aquatic model system using molecular and classical techniques. *Microb. Ecol.*, **54**, 771–777.

Rose J., Dickson L., Farrah S. and Carnahan R. (1996). Removal of pathogenic and indicator microorganisms by a full-scale water reclamation facility. *Wat. Res.*, **30**(11), 2785–2797.

Roszak D. B., Grimes D. J. and Colwell R. R. (1984). Viable but nonrecoverable stage of *Salmonella enteritidis* in aquatic systems. *Can. J. Microbiol.*, **30**, 334–338.

Rothrock M. J. Jr, Cook K. L. and Bolster C. H. (2009). Comparative quantification of *Campylobacter jejuni* from environmental samples using traditional and molecular biological techniques. *Can. J. Microbiol.*, **55**(6), 633–641.

Rowe M. T. and Grant I. R. (2006). *Mycobacterium avium* ssp. *paratuberculosis* and its potential survival tactics. *Lett. Appl. Microbiol.*, **42**, 305–311.

Ryu H., Alum A. and Abbaszadegan M. (2005). Microbial characterization and population changes in nonpotable reclaimed water distribution systems. *Environ. Sci. Technol.*, **39**, 8600–8605.

Savill M. G., Hudson J. A., Ball A., Klena J. D., Scholes P., Whyte R. J., McCormick R. E. and Jankovic D. (2001). Enumeration of Campylobacter in New Zealand recreational and drinking waters. *J. Appl. Microbiol.*, **91**, 38–46.

Schuster C. J., Ellis A. G., Robertson W. J., Charron D. F., Aramini J. J., Marshall B. J. and Medeiros D. T. (2005). Infectious disease outbreaks related to drinking water in Canada, 1974–2001. *Canadian Journal of Public Health*, **96**(4), 254–258.

Schwartzbrod J., Stein J. L., Bouhoum K. and Balenx B. (1989). Impact of wastewater treatment on helminth eggs. *Water Sci. Technol.*, **21**, 295–297.

Sengelov G. Y., Agerso Y., Halling-Sorensen B., Baloda S. B., Andersen J. S. and Jensen L. B. (2003). Bacterial antibiotic resistance levels in danish farmland as a result of treatment with pig manure slurry. *Environ. Internat.*, **28**, 587–595.

Shannon K. E., Lee D. Y., Trevors J. T. and Beaudette L. A. (2007). Application of real-time quantitative PCR for the detection of selected bacterial pathogens during municipal wastewater treatment. *Sci. Total Environ.*, **382**(1), 121–129.

Sharma S., Sachdeva P. and Virdi J. S. (2003). Emerging water-borne pathogens. *Appl. Microbiol. Biotechnol.*, **61**, 424–428.

Sidhu J. P., Toze S., Hodgers L., Shackelton M., Barry K., Page D. and Dillon P. (2010). Pathogen inactivation during passage of stormwater through a constructed reedbed and aquifer transfer, storage and recovery. *Water Sci. Technol.*, **62**(5), 1190–1197.

Smith J. J., Howington J. P. and McFeters G. A. (1994). Survival, physiological response, and recovery of enteric bacteria exposed to a polar marine environment. *Appl. Environ. Microbiol.*, **60**, 2977–2984.

Smith H. V. (1998). Detection of parasites in the environment. *Parasitology*, **1998**(117), S113–S141.

Smith M. S., Yang R. K., Knapp C. W., Niu Y., Peak N., Hanfelt M. M., Galland J. C. and Graham D. W. (2004). Quantification of tetracycline resistance genes in feedlot lagoons by real-time PCR. *Appl. Environ. Microbiol.*, **70**, 7372–7377.

Tashiro K., Kubokura Y., Kato Y., Kaneko K. and Ogawa M. (1991). Survival of *Yersinia enterocolitica* in soil and water. *Journal of Veterinary Medical Science*, **53**, 23–27.

Tchobanoglous G., Burton F. L. and Stensel H. D. (2003). Wastewater Engineering: Treatment and Reuse, 4th edn. (revised). Metcalf & Eddy Inc., McGraw-Hill, NewYork.

Terzieva S. I. and McFeters G. A. (1991). Survival and injury of *Escherichia coli, Campylobacter jejuni*, and *Yersinia enterocolitica* in stream water. *Canadian Journal of Microbiology*, **37**, 785–790.

The Eurogast Study Group. (1993). An association between *Helicobacter pylori* infection and gastric cancer. *Lancet*, **341**, 1359–1362.

Thienpont D., Rochette F. and Vanparijs O. F. J. (1986). Diagnosis helminthiasis through coprological examination. Beerse: Janssen Research Foundation.

Torvinen E., Lehtola M. J., Martikainen P. J. and Miettinen I. T. (2007). Survival of Mycobacterium avium in drinking water biofilms as affected by water flow velocity, availability of phosphorus, and temperature. *Appl. Environ. Microbiol.*, **73**, 6201–6207.

Toze S. and Hanna J. (2002). The survival potential of enteric microbial pathogens in a treated effluent ASR project. In: Management of Aquifer Recharge for Sustainability, P. Dillon (ed.), Balkema Publishers, Australia, pp. 139–142.

Toze S., Hanna J., Smith T., Edmonds L. and McCrow A. (2004). Determination of water quality improvements due to the artificial recharge of treated effluent. *Wastewater Reuse and Groundwater Quality*, IAHS Publication, **285**, 53–60.

Toze S. and Bekele E. (2009). Determining the requirements for managed aquifer recharge in western australia. Final Report Water Foundation Project No. 022/04G CSIRO. http://www.clw.csiro.au/forms/publications/details.aspx?ID=1234.

Toze S., Bekele E., Page D, Sidhu J. and Shackelton M. (2010). Use of static quantitative microbial risk assessment to determine pathogen risks in an unconfined carbonate aquifer used for managed aquifer recharge. *Water Research*, **44**, 1038–1049.

Uemura N., Okamoto S., Yamamoto S., Matsumura N., Yamaguchi S., Yamakido M., Taniyama K., Sasaki N. and Schlemper R. J. (2001). *Helicobacter pylori* infection and the development of gastric cancer. *N. Engl. J. Med*, **345**, 784–789.

Viau E. and Peccia J. (2009). Survey of wastewater indicators and human pathogen genomes in biosolids produced by class A and class B stabilization treatments. *Appl. Environ. Microbiol.*, **75**, 164–174.

Volkmann H., Schwartz T., Bischoff P., Kirchen S. and Obst U. (2004). Detection of clinically relevant antibiotic-resistance genes in municipal wastewater using real-time PCR (TaqMan). *J. Microbiol. Methods*, **56**, 277–286.

Waage A. S., Vardund T., Lund V. and Kapperud G. (1999). Detection of low numbers of pathogenic *Yersinia enterocolitica* in environmental water and sewage samples by nested polymerase chain reaction. *J Appl Microbiol.*, **87**(6), 814–821.

Wani N. A. and Chrungoo R. K. (1992). Biliary ascarisis: surgical aspects. *World Journal of Surgery*, **16**, 976–979.

Wéry N., Lhoutellier C., Ducray F., Delgenès J. P. and Godon J. J. (2008). Behaviour of pathogenic and indicator bacteria during urban wastewater treatment and sludge composting, as revealed by quantitative PCR. *Wat. Res.*, **42**(1–2), 53–62.

Whittington R. J., Marsh I. B. and Reddacliff L. A. (2005). Survival of *Mycobacterium avium* subsp. *paratuberculosis* in dam water and sediment. *Appl. Environ. Microbiol.*, **71**, 5304–5308.

Whittington R. J., Marshall D. J., Nicholls P. J., Marsh I. B. and Reddacliff L. A. (2004). Survival and dormancy of Mycobacterium avium subsp. paratuberculosis in the environment. *Appl. Environ. Microbiol.*, **70**, 2989–3004.

Wilkes G., Edge T., Gannon V., Jokinen C., Lyautey E., Medeiros D., Neumann N., Ruecker N., Topp E. and Lapen D. R. (2009). Seasonal relationships among indicator bacteria, pathogenic bacteria, *Cryptosporidium* oocysts, *Giardia* cysts and hydrological indices for surface waters within an agricultural landscape. *Wat. Res.*, **43**(8), 2209–2223.

Winfield M. D. and Groisman E. A. (2003). Role of nonhost environments in the lifestyles of *Salmonella* and *Escherichia coli. Appl. Environ. Microbiol.*, **69**, 3687–3694.

WHO. (1997). Amoebiasis – an expert consultation. world health organization. Weekly Epidemiological Record, No. 14. Geneva, April.

WHO (World Health Organization). (2003). Assessing microbial safety of drinking water: improving approaches and methods. A. Dufour, M. Snozzi, W. Koster, J. Bartram, E. Ronchi, L. Fewtrell (ed.), Published on behalf of the World Health Organization and the Organisation for Economic Co-operation and Development by IWA Publishing Alliance London UK.

WHO (World Health Organisation). (2006). Guidelines for the safe use of wastewater, excreta and greywater. Volume 2, Wastewater use in agriculture. Geneva, 2006.

WHO. (2008). Guidelines for drinking-water quality third edition incorporating the first and second addenda. Volume 1 Recommendations World Health Organisation, Geneva.

Wray C. and Linklater K. A. (2000). *Salmonella* infections in sheep. In: Salmonella in Domestic Animals, C. Wray and A. Wray (ed.), chapter 12, CABI Publishing, Oxon.

Wray C. and Davies R. H. (2000). *Salmonella* infections in cattle. In: Salmonella in Domestic Animals C. Wray and A. Wray (ed.), Chapter 10, CABI Publishing, Oxon.

Yang S. and Carlson K. (2003). Evolution of antibiotic occurrence in a river through pristine, urban and agricultural landscapes. *Wat. Res.*, **37**, 4645–4656.

Yáñez M. A., Barberá V. M., Soria E. and Catalán V. (2009). Quantitative detection of *Helicobacter pylori* in water samples by real-time PCR amplification of the cag pathogenicity island gene, *cagE*. *J. Appl. Microbiol.*, **107**(2), 416–424.

Yaradou D. F., Hallier-Soulier S., Moreau S., Poty F., Hillion Y., Reyrolle M., André J., Festoc G., Delabre K., Vandenesch F., Etienne J. and Jarraud S. (2007). Integrated real-time PCR for detection and monitoring of *Legionella pneumophila* in water systems. *Appl. Environ. Microbiol.*, **73**, 1452–1456.

Chapter 12

Water quality analysis: Detection, fate, and behaviour, of selected trace organic pollutants at managed aquifer recharge sites

Mathias Ernst, Arne Hein, Josef Asmin, Martin Krauss, Guido Fink, Juliane Hollender, Thomas Ternes, Claus Jørgensen, Martin Jekel and Christa S. McArdell

12.1 INTRODUCTION

In treated municipal wastewater, residual organic compounds are of high relevance especially if water recycling and potable water reuse is envisaged. After biological treatment, such as the activated sludge process, some organic compounds remain that are either non-biodegradable, or are minimally biodegradable. If these chemicals are polar, they are commonly poorly absorbable, and are therefore identified as persistent polar organic compounds (also persistent polar pollutants, PPPs).

In the last decade, there have been important analytical improvements in detecting trace levels of pollutants, and within the water reuse community, new "hazardous" compounds are frequently discussed. This includes consideration of which organic residuals are really of health concern, which transformation products can be generated, and what is their human and environmental impact? Within the present chapter relevant PPPs and their fate during (advanced) wastewater treatment and managed aquifer recharge are identified and discussed as results of measuring campaigns at technologically different demonstration sites within the European research project RECLAIM WATER. Such PPPs mainly belong in the group of pharmaceuticals but also industrial chemicals.

Here antibiotics such as the macrolides and sulfonamides are of particular concern, because of the eco-toxicological potential of these parent micropollutants, and the potential threat posed by the build-up of antibiotic resistance genes. In addition to known multi-resistant bacteria such as *Staphylococci*, multi-resistant genes have recently been identified in the intestinal bacteria *Citrobacter*, *Enterobacteriaceae* and *Escherichia coli* (Patoli *et al.* 2010; Tao *et al.* 2010). Other organic substances of interest are the very persistent tracer compounds, such as the antiepileptic carbamazepine and the iodinated contrast media (ICM) diatrizoate (WateReuse report, 2008; Rauch-Williams *et al.* 2010). Such compounds can act as markers for anthropogenic activity, and be used to calibrate models of ground water flows; as reduction of concentration in the aquifer can only be explained by dilution (Gaser *et al.* 2011; Massmann *et al.* 2008). Another relevant group of substances are 1H-benzo-1,2,3-triazole (Benzotriazole, BTr) and its methylated analogues (tolyltriazole, TTri), these compounds are used as corrosion inhibitors in many industrial applications, in dishwashing agents, and in de-icing fluids for aircraft. After activated sludge treatment of sewage, the concentrations range from 7–18 µg/L BTri and 1–5 µg/L TTri (Giger, 2006; Reemtsma, 2010).

Within the "organic contaminants" work package of the RECLAIM WATER project (www.reclaim-water.org), a list of PPPs has been selected, analysed and assessed at managed aquifer recharge demonstration sites. There were two main goals for the "organic contaminants" study: (1) measure and identify relevant organic compounds, and (2) apply the new analytical protocol (protocol II) at 5 demonstration sites to study the fate and behaviour along the different treatment pathways. In conjunction with work package 1 (assessment and development of water reclamation technology), results from protocol II should identify the intensity of treatment needed to reduce levels of micropollutants. The four analytical partners BfG (German federal institute of hydrology), Eawag (Swiss Federal Institute of Aquatic Science and Technology, UNESCO-IHE (Delft institute for water education) and Technische Universität Berlin (TUB), developed the methods of protocol II and applied these to the 5 managed aquifer recharge sites (MAR). UNESCO-IHE however measured no specific micropollutant, but provided the methodology for characterisation of bulk organic matter (see Chapter 13).

Selected water quality parameters

The analytical program focused on more than 60 different organic compounds, most of which are known to be present in wastewater effluents at concentrations ranging from a few ng/L to several μg/L. These organic compounds commonly show limited biodegradability in the environment, together with a weak sorption tendency onto soil, thus they can often penetrate into unconfined aquifers.

The selected compounds can be divided into seven groups: (i) antibiotics, (ii) antiepileptics (neutral drugs), (iii) iodinated contrast media, (iv) antipholgistics, analgesics, and lipid regulators (acidic drugs), (v) estrogens, (vi) nitrosamines, and (vii) other micropollutants. A list of the determined compounds within each of the seven groups, and the abbreviations used in the text, is given in Table 12.1.

Table 12.1 Groups of compounds measured in the RECLAIM WATER project.

Group	Measured compounds
i. Antibiotics	Clarithromycin (CLA), erythromycin (ERY), anhydro-erythromycin (ERY-H_2O), roxithromycin (ROX), azithromycin, sulfadiazine, sulfapyridine, sulfamethazine (SMZ), sulfadimethoxine, sulfamethoxazole (SMX), N-acetyl-sulfamethoxazole (N-Ac-SMX), trimethoprim (TMP)
ii. Antiepileptics (neutral drugs)	Carbamazepine, primidone
iii. Iodinated contrast media (ICM)	Iopamidol, iomeprol, iopromide, iohexol, diatrizoate, ioxitalamic acid, adsorbable organic iodine (AOI as bulk parameter)
iv. Antipholgistics, analgesics, lipid regulators (acidic drugs)	Ibuprofen, diclofenac, clofibric acid, naproxen, bezafibrate, fenoprofen, mephenamic acid, paracetamol
v. Estrogens	Estrone (E1), estradiol (E2), ethinylestradiol (EE2)
vi. Nitrosamines	N-nitrosodimethylamine (NDMA), -methylethylamine (NMEA), -diethylamine (NDEA), -di-*n*-propylamine (NDPA), -di-*n*-butylamine (NDBA), -diphenylamine (NDPhA), -morpholine (NMOR), -piperidine (NPIP), -pyrrolidine (NPYR)
vii. Other micropollutants	Benzotriazole (BT), 4-tolyltriazole (4TT), 5-tolyltriazole (5TT), bisphenol-A

Within the antibiotics, the sulfonamide sulfamethoxazole (SMX) is a persistent compound. In Germany, SMX is used in both human medicine (53.6 tons / year, MUNLV-NRW 2007), and in veterinary medicine. Together with SMX, trimethoprim (TMP) is often applied in a ratio of 5:1, resulting in an *in vitro* synergistic antibacterial effect (cotrimoxazol). SMX is excreted in high amounts by patients, and its metabolite N-acetylsulfamethoxazole (N-Ac-SMX) can be converted back to SMX during wastewater treatment (Göbel *et al.* 2005). Therefore, both SMX and N-Ac-SMX need to be analyzed for correct mass balance calculations. The environmental fate of antibiotics and other pharmaceuticals during bank filtration have recently been discussed by Maeng *et al.* (2011) and the Water Research Foundation (2010).

Among the antiepileptics, carbamazepine is a refractory compound, often detected in the ng/L range in groundwater, drinking waters respectively (Sacher *et al.* 2001; Massmann *et al.* 2006).

The iodinated contrast media (ICM) in general is the group with the highest single compound concentration in treated municipal wastewater, even if μg/L concentration can be accessed for ethylene diamino tetraacetate (EDTA), benzotriazoles, diclofenac and sometimes carbamazepine as well (Reemtsma *et al.* 2006). The ICM are applied in high dosages for medical diagnostics, and are released from the patient by urine, nearly unchanged, within several hours of administration. The single ICM compounds often can be found in wastewater treatment effluent in concentrations ranging from several hundred ng/L to several μg/L. In particular, iopamidol and diatrizoate are quite resistant to natural removal mechanisms. As ICM contain iodine atoms, the sum content of iodine can be measured by the bulk organic parameter of absorbable organic iodine (AOI). AOI is determined to be 50% of the content of the ICM in source material. If this share decreases, the AOI allows distinction of whether the target compound was truly mineralized, or just transformed into new iodine containing organic molecules (Putschew, 2006).

In the group of estrogens, endocrine disruption compounds (EDC) are frequently discussed within the scientific community and have received considerable public attention, due to news reports on feminisation of male fish. The present study focussed on the natural and synthetic hormones estrone (E1), estradiol (E2), and ethinylestradiol (EE2). However bisphenol-A (see group of other compounds) can show endocrine disruption effects too. Bisphenol A is often contained in plastics as a conditioner.

The acid drug group, includes the well-known pain relieving drugs ibuprofen and diclofenac, which are often found in municipal effluents. Information on their behaviour during soil passage is given by Heberer *et al.* (2002) and Scheytt *et al.* (2007).

In the group of disinfection by-products, the nitrosamines and related compounds are of major concern. N-nitrosodimethylamine (NDMA) is carcinogenic at very low concentrations, and is difficult to analyse. A new, reliable and accurate, method to detect nine nitrosamines was developed within the RECLAIM WATER project (Krauss & Hollender, 2008).

The corrosion inhibitors benzotriazole, 4-tolyltriazole and 5-tolyltriazole, were measured; these compounds are found in high amounts in surface waters and also in groundwater (Giger *et al.* 2006; Reemtsma *et al.* 2006).

12.2 METHODS

12.2.1 Sampling, storage and processing at the demonstration sites

Prior to laboratory and demonstration site sampling, the analytical partners (BfG, Eawag and TUB) established and optimised five analytical methods for organic compound analysis (Table 12.2). At each demonstration site, five sampling locations were selected, and three to four sampling programs were conducted over at least 12 months were to show seasonal fluctuation.

Table 12.2 Analytical methods established for organic compound analysis in the RECLAIM WATER project.

Method	Application
1	Antibiotics, neutral drugs and other micro-pollutants: Clarithromycin (CLA), anhydro-erythromycin (ERY-H_2O), roxithromycin (ROX), azithromycin, sulfadiazine, sulfapyridine, sulfamethazine (SMZ), sulfadimethoxine, sulfamethoxazole (SMX), N-acetyl-sulfamethoxazole (N-Ac-SMX), trimethoprim (TMP), carbamazepine (CBZ), primidon (PMD), bisphenol-A, benzotriazole (BTri), 4-tolyltriazole (4-TT), 5-tolyltriazole (5-TT)
2	Acidic drugs and ICM: ibuprofen, diclofenac (DCF), clofibric acid, naproxen (NPX), bezafibrate, mephenamic acid, paracetamol, iopamidol, iomeprol, iopromide, iohexol, ioxitalamic acid, diatrizoate
3	Estrogens: estrone (E1), estradiol (E2), ethinylestradiol (EE2)
4	Nitrosamines: N-nitrosodimethylamine (NDMA), -methylethylamine (NMEA), -diethylamine (NDEA), -di-n-propylamine (NDPA), -di-n-butylamine (NDBA), -diphenylamine (NDPhA), -morpholine (NMOR), -piperidine (NPIP), -pyrrolidine (NPYR)
5	Adsorbable organic iodine (AOI)

Sample volume for ground water and surface water was 1.0 L, and for effluents of sewage treatment plants (STP) and waters with high organic content, 100 mL. Samples of treated wastewater effluent were usually taken as 24h-composite samples. Further processed samples prior to infiltration, and samples from infiltrated waters (groundwater, saturated or unsaturated zone), were taken as grab samples. The samples were usually stored at −20°C; or in exceptional cases (when sample processing was done within the following two days) at 4°C.

Solid phase extraction (SPE) was performed at each site by local laboratory staff. Before the SPE enrichment, the samples were filtered with glass fibre filters of Schleicher and Schuell (GF6) or with glass fibre filters of Whatman (GF/F). At least one blank sample (bottled drinking water, non-carbonated) was included as an additional sample for each sample batch. The recoveries were determined in the various matrices by spiking samples with 100 ng/l to 2000 ng/l of analytes prior to the SPE. After the SPE enrichment, the cartridges were sent from the sampling sites in a cooling box (blue ice, 3–5°C) to the analytical partners (BFG, Eawag, TUB). The enriched cartridges were sent according to these conditions: (i) SPE–cartridges were completely dried by N_2 before storage, (ii) cartridges were sealed with aluminium foil on both sites, (iii) during transport, the sample cartridges were kept as cool as possible (temperature should never exceed 35°C).

For analysis of nitrosamines and AOI, liquid samples were sent to the analytical partners Eawag and TUB.

Samples were processed according to five different methods, listed in Table 12.2, and described in detail in the following methods section.

Detailed information on the demonstration sites can be found in the result section, and in the book Chapter 2. The sites were sampled according to the following schedule:

- **Nardo**, Italy: (i) 13 November 2006, (ii) 19 February 2007, (iii) 22 May 2007, (iv) 18 September 2007, the secondary effluent sample was also a grab sample
- **Sabadell**, Spain: (i) 19 March 2007, (ii) 16 July 2007, (iii) 19 November 2007
- **Shafdan**, Israel: (i) 17 June 2007, (ii) 14 October 2007, (iii) 12 December 2007
- **Gaobeidian**, China: (i) 6 December 2006, (ii) 1 June 2007, (iii) 23 July 2007, (iv) 8 October 2007
- **Torrele/Wulpen**, Belgium: (i) 29 January 2007, (ii) 23 July 2007, (iii) 15 October 2007

All pharmaceuticals and internal standards (see Table 12.4) were analytical grade, and purchased from one of the following companies: Abbott Laboratories, Cambridge Isotope Labs (CID), Dr. Ehrenstorfer GmbH, Fluka, Riedel-de Haën, Sigma-Aldrich, Toronto Research Chemicals (TRC). Bayer Schering Pharma AG is acknowledged for the donation of DMI. Methanol, acetonitrile, and water for chromatography, were all of HPLC- grade (Acros Organics).

Mixtures of internal standards were prepared for each method separately, and added to the samples after filtration (surro-Mix A: 200 ng each for method 1, surro-Mix B1: 800 ng for contrast media in method 2, surro-Mix B2: 450 ng for acidic drugs in method 2, surro-Mix C: 25 ng each for method 3).

12.2.2 Method 1: antibiotics, neutral drugs, and other micropollutants

An analytical method was established which enables the simultaneous determination of sulfonamide antibiotics, macrolide antibiotics, the neutral drugs carbamazepine and primidone, benzotriazol and bisphenol-A in wastewater. The method comprises the addition of surrogate standards (usually isotopic labelled standards) to the liquid samples, a solid phase extraction with OASIS HLB and detection via LC electrospray tandem MS. A summary of the procedure is shown in the following scheme:

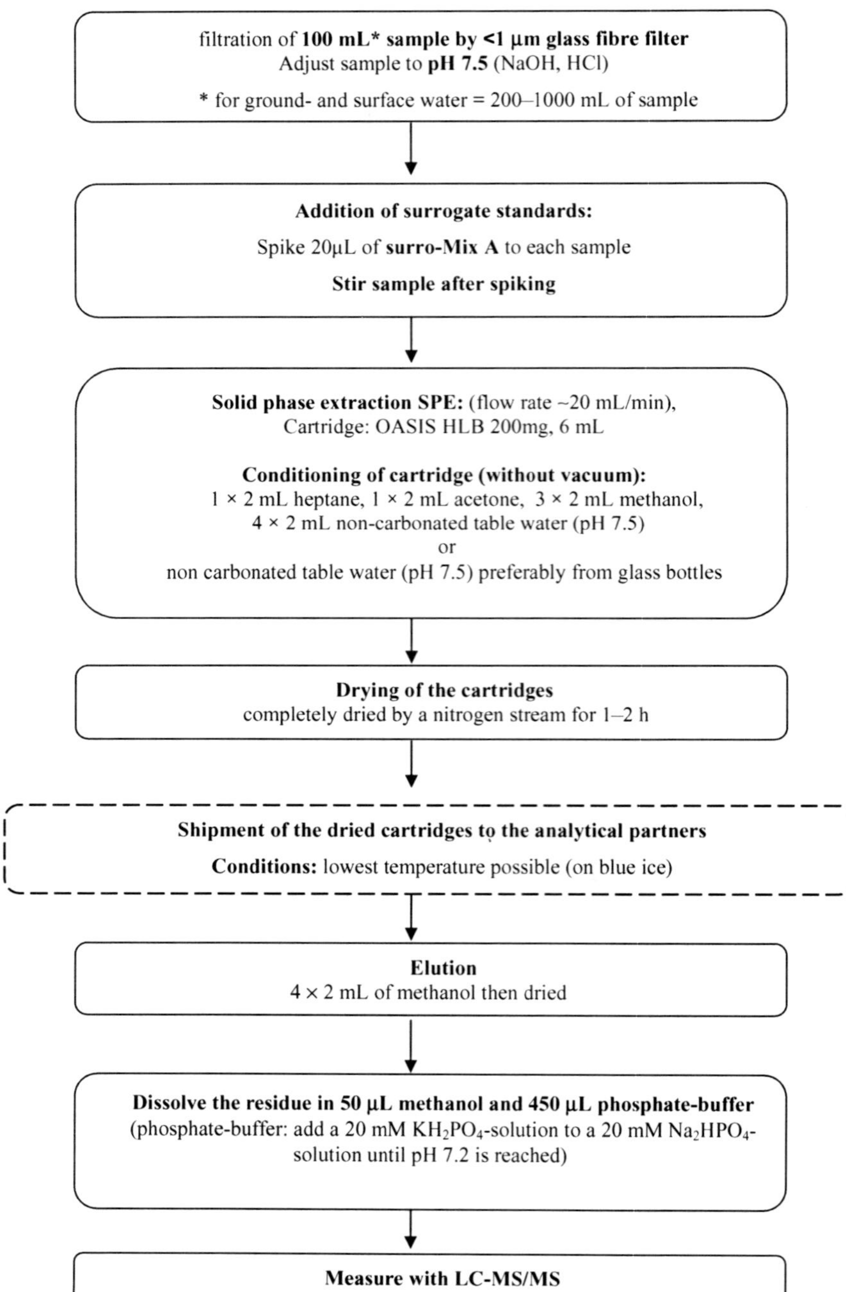

12.2.3 Method 2: acidic drugs and ICM

A second analytical method was developed which enables the simultaneous enrichment of iodinated contrast media (ICM) as well as acidic drugs in wastewater. The method includes the addition of surrogate standards, a combined solid phase extraction with OASIS MCX and Isolute ENV+, separation of the coupled extraction cartridges, elution of the individual cartridges for contrast media and acidic drugs, respectively, and detection with LC/MS/MS:

Filtration of **100 mL* sample** with **<1 µm glass fibre filter**
Adjust sample to **pH 2.8** with 3.5 mM H_2SO_4
* for ground- and surface water = 200–1000 mL of sample

↓

Addition of Surrogate standards:
Contrast media: First spike 40µL of **Surro-Mix B1** to each sample
Acidic drugs: Then spike 30µL of **Surro-Mix B2** to each sample
Stir sample after spiking

↓

Solid phase extraction: (flow rate 10 mL/min, ~600 mbar)
In any case 10 mL/min should NOT be exceeded!!!!
Cartridge 1: **Waters Oasis MCX 3cc 60 mg, 3 mL for acidic drugs**
Cartridge 2: **Isolute ENV+ 200 mg, 3 mL for contrast media**
Conditioning of cartridge (without vacuum):
1 × 2 mL heptane, 1 × 2 mL acetone, 3 × 2 mL methanol, 4 × 2 mL non carbonated table water **(pH 2.8)** preferably in glass bottles
Connect cartridge 1 and 2

↓

Drying of the cartridges: completely by a nitrogen stream for 2 h

↓

Shipment of the dried cartridges to the partners
Conditions: lowest temperature possible

Acidic drugs branch:

Elution acidic drugs. (cartridge 1)
4 × 1 mL of acetone

↓

Evaporation acidics
to 100 µL (NOT to dryness)
by a nitrogen stream

↓

Addition of 300 µL methanol and
evaporation to 200 µL by a nitrogen stream

Fill up with 300 µL 0,01 M formic acid

↓

Contrast media branch:

Elution contrast media (cartridge 2)
4 × 1 mL of methanol

↓

Evaporation contrast media:
to dryness by a nitrogen stream

↓

Dissolve the residue in 50 µL methanol and
450 µL phosphate-buffer

↓

Measure with LC-MS/MS

12.2.4 Method 3: estrogens

A third analytical method was established which enables the detection of estrogens in wastewater.

Acidification of sample directly after sampling to **pH 2–3** with 3.5 mM sulfuric acid

↓

Filtration of **1L sample** with **<1 μm glass fibre filter**

↓

Addition of Surrogate standards:
Spike 50μL of **Surro-Mix C** to each sample
Stir sample after spiking

↓

Solid phase extraction: (flow rate ~ 20 mL/min, ~ 200 mbar)
Cartridge: **Baker C18 500mg, 3 mL**

Conditioning of cartridge (without vacuum):
1 × 2 mL heptane, 1 × 2 mL acetone, 3 × 2 mL methanol, 4 × 2 mL groundwater
or
non carbonated table water **(pH 3)** preferably in glass bottles

↓

Drying of the cartridges: Completely by a nitrogen stream for 1 h

↓

Shipment of the dried cartridges to the partners
Conditions: lowest temperature possible

↓

Elution: 4 × 1 mL of acetone

↓

Evaporation to 200 μl

↓

Clean up
Silicagel 60: dry at 150°C over night, deactivate with 1.5 % water
Eluent: 65 ml n-hexane + 35 ml acetone
a) condition 1 g Silicagel with eluent and put into in cartridge
b) add the 200 μL acetone extract to the cartridge, rinse cartridge with 4 mL of the eluent

↓

Evaporation: evaporate to dryness by a gentle nitrogen stream

↓

Dissolve the residue in 50 μL methanol and add 450 μl of Milli Q water

↓

Measure with LC-MS/MS

12.2.5 Method 4: nitrosamines

Nitrosamine analysis followed the method of Krauss & Hollender (2008). 500 mL of sewage samples were spiked with internal standards (50 ng of deuterium labelled nitrosamines), filtered through a glass fiber filter (GF/F; 0.7 mm, Whatman, Brentford, UK), and solid-phase extracted using a combination of Oasis HLB (200 mg; Waters, Milford, MA) and Bakerbond Carbon (1000 mg; Mallinckrodt-Baker, Philippsburg, NJ). Nitrosamines were eluted using 15 mL of dichloromethane and this solvent was exchanged by a water: methanol mixture (95:5, v/v), which was transferred to an autosampler vial. Nitrosamines were separated using reversed-phase HPLC. A Waters X-Bridge C18 column

(100 × 2.1mm, 3 mm particle size) and a gradient elution with water and methanol, each containing 0.4% (v/v) acetic acid was used. After electrospray ionization in positive mode nitrosamines were quantified using an LTQ Orbitrap hybrid mass spectrometer in selected-ion-monitoring mode (molecular ions) or selected reaction monitoring mode (MS/MS product ions).

12.2.6 Method 5: AOI

The analytical method to detect AOI was according to Oleksy-Frenzel *et al.* 2000.

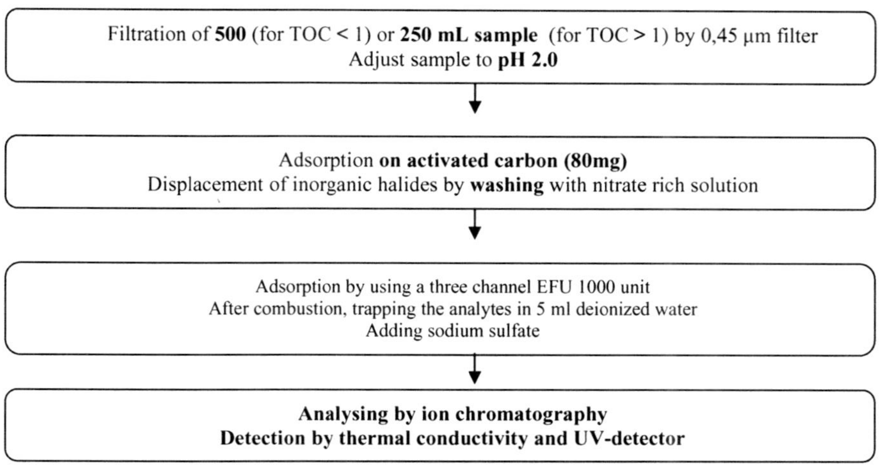

12.2.7 Quality assurance

Recovery studies were conducted by spiking at least three samples of each sample matrix. For most of the analytes, isotope labelled internal standard were available which were added prior to solid phase enrichment (SPE) to account for possible losses during the analytical procedure. For each site a mean relative recovery over the entire procedure was calculated as an average for all analysed samples (Table 12.3). Recoveries were generally in the range of 80–120%, and out of this range usually only for analytes without labelled internal standard. Results were corrected with the corresponding recovery rates obtained in the same matrix and sample batch if no labelled internal standard was available. Sample-based limit of quantification LOQ were defined as concentrations in a sample matrix resulting in peaks with signal-to-noise ratios (S/N) of 10. When samples contained analytes, the concentration corresponding to the defined S/N was determined by downscaling, using the measured concentration and the assigned S/N of the peaks assuming a linear correlation through zero. Usually they were in the range of 10–20 ng/L. Details of the MS/MS analysis and settings are given in Table 12.4. For each substance, two transitions of the precursor ion were monitored (product ion I and II). Together with the retention times, they were used to ensure correct peak assignment and to evaluate peak purity.

In general better recoveries for diluted samples could be observed. Effluents and groundwater samples were enriched without dilution to avoid a decrease of the LOQ but influent samples (only taken at site Wulpen) were not quantifiable without previous dilution due to strong matrix effects.

12.3 RESULTS AND DISCUSSION

The results of the measurements at the different sites are now presented according to the technologies applied, beginning with sites with simple infiltration of treated wastewater to groundwater, and continuing to sites with sophisticated technical treatments. The sites are presented in the following order: Nardo (Italy), Sabadell (Spain), Shafdan (Israel), Gaobeidian (China) and Wulpen/Torrele (Belgium). The location is briefly described, boundary conditions presented, and the fate of micropollutants are given. Finally a cross-demonstration site discussion points out resulting similarities and outcomes.

Raw water quality was a decisive factor for resulting water qualities, and varied greatly between the demonstration sites. Infiltrated water, aquifer soil material, aquifer depths, redox conditions, hydraulic retention times and local background water quality differ among sites and govern the resulting water quality.

Table 12.3 Mean recovery and limit of quantification (LOQ) for all analytes at each site. N: amount of samples spiked for calculation of mean recovery. '–' denotes that analyte was not analysed at this site.

	Wulpen (BE)			Nardo (IT)			Sabadell (ES)			Shafdan (IL)			Gaobeidan (CN)		
	n	Mean recovery [%]	LOQ [ng/L]	n	Mean recovery [%]	LOQ [ng/L]	n	Mean recovery [%]	LOQ [ng/L]	n	Mean recovery [%]	LOQ [ng/L]	n	Mean recovery [%]	LOQ [ng/L]
Method 1															
Clarithromycin	23	117 ± 53	2–10	25	96 ± 18	10	9	103 ± 36	10	6	78 ± 15	50	6	41 ± 2	50
Erythromycin	23	59 ± 28	2–10	25	93 ± 20	10	9	107 ± 31	10–20	6	14 ± 7	50	6	8 ± 2	50
Roxithromycin	23	113 ± 52	2–10	25	88 ± 19	10	10	86 ± 31	10	6	107 ± 18	50	6	57 ± 24	50
Sulfamethoxazole	23	110 ± 23	2–10	24	80 ± 29	10	12	76 ± 37	10	6	54 ± 12	25	6	100 ± 66	25
Sulfadiazine	–	–	–	3	39 ± 8	10	–	–	–	–	–	–	–	–	–
Sulfamethazine	–	–	–	25	66 ± 24	10	15	89 ± 28	10–20	6	81 ± 33	25	6	118 ± 84	25
Sulfadimethoxine	–	–	–	25	112 ± 22	10	10	110 ± 24	10–20	–	–	–	–	–	–
Sulfapyridine	–	–	–	5	65 ± 20	10	–	–	–	–	–	–	–	–	–
Trimethoprim	–	–	–	5	59 ± 26	10–30	5	38 ± 39	10	6	40 ± 17	25	6	67 ± 22	25
Azithromycin	–	–	–	5	103 ± 7	10–50	–	–	–	–	–	–	–	–	–
Primidone	23	93 ± 33	2–10	19	96 ± 27	10	8	53 ± 20	10–20	–	–	–	–	–	–
N-Ac-SMX	20	183 ± 64	10–25	25	97 ± 10	10–30	15	82 ± 35	10–20	5	93 ± 16	50	7	128 ± 28	50
Carbamazepine	22	108 ± 14	2–10	25	94 ± 34	10	9	60 ± 38	10	–	–	–	–	–	–
Benzotriazole	15	141 ± 69	10–50	5	112 ± 61	0.1	5	87 ± 7	0.1	15	103 ± 20	0.1	15	109 ± 22	0.1
Bisphenol-A	19	104 ± 34	5–25	25	73 ± 26	10–120	10	75 ± 20	10–40	–	–	–	–	–	–
Method 2															
Ibuprofen	11	106 ± 8	2–20	20	84 ± 15	20–30	4	45 ± 9	40	3	109 ± 6	20	6	97 ± 6	20
Diclofenac	11	107 ± 27	5–50	19	70 ± 14	10–20	9	99 ± 59	10–20	3	94 ± 9	20	6	104 ± 9	20
Clofibric acid	11	114 ± 27	5–50	20	110 ± 43	10–30	4	273 ± 116	10	3	102 ± 6	20	6	105 ± 10	20
Naproxen	11	90 ± 30	2–20	19	97 ± 35	10	9	67 ± 29	20–40	3	104 ± 19	20	6	116 ± 22	20
Bezafibrate	11	97 ± 41	10–100	19	81 ± 27	10	9	90 ± 32	10	3	110 ± 20	20	6	148 ± 20	20
Mephenamic acid	–	–	–	–	–	–	5	113 ± 36	40	–	–	–	–	–	–

(*Continued*)

Table 12.3 Mean recovery and limit of quantification (LOQ) for all analytes at each site. N: amount of samples spiked for calculation of mean recovery. '–' denotes that analyte was not analysed at this site (*Continued*).

	Wulpen (BE)			Nardo (IT)			Sabadell (ES)			Shafdan (IL)			Gaobeidan (CN)		
	n	Mean recovery [%]	LOQ [ng/L]	n	Mean recovery [%]	LOQ [ng/L]	n	Mean recovery [%]	LOQ [ng/L]	n	Mean recovery [%]	LOQ [ng/L]	n	Mean recovery [%]	LOQ [ng/L]
Iopamidol	12	64 ± 20	2–20	18	79 ± 28	10–20	15	101 ± 30	20–80	4	130 ± 5	50	6	127 ± 35	50
Iomeprol	12	82 ± 27	2–20	24	77 ± 19	10–30	15	122 ± 22	40–120	4	99 ± 4	50	6	103 ± 13	50
Iopromide	8	109 ± 31	2–20	25	85 ± 25	10	11	102 ± 24	20–60	4	85 ± 6	50	6	90 ± 12	50
Iohexol	13	66 ± 20	10–20	10	56 ± 17	10–50	14	122 ± 23	30–110	4	104 ± 10	50	6	116 ± 20	50
Ioxitalamic acid	–	–	–	4	58 ± 18	10	5	78 ± 29	–	–	–	–	–	–	–
Paracetamol	–	–	–	18	62 ± 28	10	12	92 ± 32	–	–	–	–	–	–	–
Diatrizoate	9	133 ± 57	10–20	22	95 ± 42	10–20	11	117 ± 49	10–40	4	142 ± 9	50	6	121 ± 27	50
Method 3															
Estrone E1	15	91 ± 12	1–10	10	106 ± 9	1.0	7	95 ± 18	1.0	6	81 ± 10	1.0	–	–	–
Estradiol E2	7	80 ± 12	1–10	–	–	1.0	–	–	1.0	6	73 ± 15	2.0	–	–	–
Ethinylestradiol EE2	7	70 ± 21	1–10	–	–	1.0	–	–	1.0	6	78 ± 16	2.0	–	–	–
Method 4															
NDMA	5	67 ± 34	0.7	3	109 ± 17	0.7	2	108 ± 0	0.7	4	82 ± 40	0.7	–	–	–
NDEA	5	76 ± 19	0.3	3	106 ± 9	0.3	2	75 ± 6	0.3	4	74 ± 15	0.3	–	–	–
NDPA	5	101 ± 3	1.1	3	106 ± 4	1.1	2	94 ± 2	1.1	4	105 ± 7	1.1	–	–	–
NDBA	5	99 ± 29	3.0	3	75 ± 2	3.0	2	91 ± 25	3.0	4	86 ± 22	3.0	–	–	–
NPYR	5	72 ± 32	3.0	3	96 ± 2	3.0	2	98 ± 19	3.0	4	87 ± 17	3.0	–	–	–
NPIP	5	94 ± 7	3.2	3	87 ± 7	3.2	2	108 ± 3	3.2	4	84 ± 21	3.2	–	–	–
NMOR	5	103 ± 20	1.3	3	84 ± 22	1.3	2	107 ± 0	1.3	4	86 ± 31	1.3	–	–	–

Table 12.4 Substance specific MS/MS settings for analytes and internal standards from the analytical lab at Eawag and BfG and TU Berlin (DP: declustering potential, CE: collision energy, CXP: cell exit potential). If no labelled internal standard was available, a surrogate was used (*italic*).

Analyte							Internal Standard						
	Pre-cursor m/z	Product ion I/II m/z	Quantifier m/z	DP (V)	CE (V)	CXP (V)		Pre-cursor m/z	Product ion I/II m/z	Quantifier m/z	DP (V)	CE (V)	CXP (V)
Method 1													
N-Ac-Sulfamethoxazole	296.0	134.1/198.1	134.1	56	30	12	d5-N-Acetyl-Sulfamethoxazole	301.1	139.2/203.1 (111.4;157.0)	139.2	40	32	12
Sulfamethoxazole	254.0	108.0/155.9	155.9	48	26	18	d4-Sulfamethoxazole	258.4	111.9/160.2	160.2	41	25	10
Sulfadiazine	250.6	108.1/156.1	156.1	44	24	18	d4-Sulfadiazine	255.1	112.0/160.2	160.2	48	26	10
Sulfamethazine	278.7	124.0/186.1	186.1	58	27	12	$^{13}C_6$-Sulfamethazine	285.2	123.9/186.1	186.1	56	27	12
Sulfadimethoxine	311.2	108.1/156.1	108.1	61	31	10	d4-Sulfadimethoxine	315.0	156.2/160.1	156.2	45	32	15
Sulfapyridine	250.1	184.2/156.1	184.2	50.7	23.8	10.0	*$^{13}C_6$-Sulfamethazine*						
Trimethoprim	290.7	230.2/261.1	230.2	65	34.3	20.7	Trimethoprim-d9	300.9	235.1/122.9	235.1	46	35	20
Azithromycin	749.5	591.2/573.5	591.2	45/59	39/47	16/16	Azithromycin-d3	753.6	595.4/577.2	595.4	85/86	41/47	16/16
Clarithromycin	748.6	158.2/590.4	158.2	60	31	15	Clarithromycin-N-methyl-d3	752.5	594.3/161.2	594.3	70/66	27/39	16/9
Erythromycin	734.6	576.5/158.1	576.5	54/55	27/40	16/10	$^{13}C_2$Erythromycin	737.6	579.3/160.2	579.3	60/60	27/39	16/16
Erythromycin-H_2O	716.8	540.5/558.2	558.2	50	23	15	*$^{13}C_2$Erythromycin*						
Roxithromycin	837.5	158.2/679.6	158.2	60	32	18	*Oleandomycin-phosphat-dihydrat*	688.7	158.1/544.3	158.1	50	27	15
Carbamazepine	237.1	194.2/192.2	194.2	38/46	29/32	18/11	Dihydro-carbamazepin	239.2	194.2/222.2	222.2	56	30	13
Primidone	219.0	91.2/162.2	162.2	27	20	15	$^{13}C_1^{15}N_1$ Carbamazepine	240.2	195.2/193.2	195.2	54/54	27/35	17/13
Benzotriazole	120.5	64.9/91.8	64.9	48	29	12	Benzotrizole-d4	123.8	69.1/96.0	69.1	40	35	12
Methyl-Benzotriazole	134.3	76.9/79.2	76.9	50	31	13	*Dimethyl-Benzotriazol*	147.8	91.3/93.1	93.1	54	32	15
Bisphenol-A	226.9	133.0/211.9	133	–87	–40	–10	Bisphenol-A-d16	241.1	142.1/222.9	222.9	–79	–34	–10
Method 1 (TUB)													
Sulfamethoxazole	253.9	108.0/155.9	155.9	–	20	–	d4-Sulfamethoxazole	258.3	112.1/160.0	160.0	–	20	–
Sulfamethazine	278.9	124.0/185.5	185.5	–	16	–	Sulfamethazine-N4-acetyl	321.0	124.4/185.5	185.5	–	23	–
Trimethoprim	291.0	122.9/230.0	230.0	–	24	–	d4-Sulfamethoxazole	258.3	112.1/160.0	160.0	–	20	–
Clarithromycin	749.4	157.8/591.3	157.8	–	20	–	Clarithromycin-N-methyl-d3	752.4	160.9/592.5	592.5	–	20	–
Erythromycin-H_2O	716.5	158.2/558.2	558.2	–	20	–	Clarithromycin-N-methyl-d3	752.4	160.9/592.5	592.5	–	20	–
Roxithromycin	837.5	158.1/679.4	158.1	–	25	–	Clarithromycin-N-methyl-d3	752.4	160.9/592.5	592.5	–	20	–

(*Continued*)

Table 12.4 Substance specific MS/MS settings for analytes and internal standards from the analytical lab at Eawag and BfG and TU Berlin (DP: declustering potential, CE: collision energy, CXP: cell exit potential). If no labelled internal standard was available, a surrogate was used (*italic*) (*Continued*).

Analyte	Pre-cursor m/z	Product ion I/II m/z	Quantifier m/z	DP (V)	CE (V)	CXP (V)	Internal Standard	Pre-cursor m/z	Product ion I/II m/z	Quantifier m/z	DP (V)	CE (V)	CXP (V)
Method 2													
Ibuprofen	205.0	161.2/158.8	161.2	–44	–10	–8	Ibuprofen-d3	208.1	161.3/164.3	164.3	–44	–10	–12
Diclofenac	294.0	214.0/249.8	249.8	–34	–16	–16	Diclofenac-d4	300.7	218.0/256.9	256.9	–35	–17	–15
Clofibric acid	212.8	85.0/126.8	126.8	–28	–13	–7	Clofibric acid-d4	219.0	85.1/132.9	132.9	–33/–30	–14/–20	–3/–9
Naproxen	229.0	169.0/169.9	169.9	–30	–27	–13	Naproxen-C13/d3	233.3	168.9/169.9	169.9	–30	–26	–11
Bezafibrate	360.2	154.0/274.0	274.0	–30	–27	–8	Bezafibrate d4	366.1	159.9/280.0	280.0	–47/–44	–39/–24	–12/–15
Mephenamic acid	240.9	197.0/181.1	197	–52	–24/–36	–13	Mephenamic acid d3	243.4	200/181.9	200.0	–56	–24	–14
Iopamidol	777.9	386.9/558.9	558.9	64	32	33	*Desmethoxy-iopromid (DMI)*	761.9	528.8/743.7	528.8	64	32	33
Iomeprol	777.9	531.7/687.0	531.7	62	40	18	*Desmethoxy-iopromid (DMI)*						
Iopromide	792.0	558.7/572.9	572.9	60	35	16	*Desmethoxy-iopromid (DMI)*						
Iohexol	822.1	602.7/803.9	803.9	56	28	21	*Desmethoxy-iopromid (DMI)*						
Ioxitalamic acid	644.8	408.0/583.6	583.6	58	25	26	*Desmethoxy-iopromid (DMI)*						
Ioxitalamic sodium	666.9	666.9	666.9	58	25	26	*Desmethoxy-iopromid (DMI)*						
Diatrizoate	614.7	337.1/361.0	361.0	72	25	21	*Desmethoxy-iopromid (DMI)*						
Paracetamol	152.2	64.8/110.2	110.2	39	23	5	*3-Acetamidophenol*	152.2	64.8/110.2	110.2	39	23	5
Method 4													
NDMA	75.0553	–	75.0553	–	–	–	NDMA-D_6	81.0930	–	81.0930	–	–	–
NDEA	103.0866	75.0553	103.0866	–	60 (CID)	–	NPYR-D_8	109.1212	92.0982	109.1212	–	80 (HCD)	–
NDPA	131.1179	89.0709	131.1179	–	50 (HCD)	–	NDPA-D_{14}	145.2058	97.1212	145.2058	–	50 (CID)	–
NDBA	159.1492	103.0866	159.1492	–	50 (CID)	–	NDPA-D_{14}	145.2058	97.1212	145.2058	–	50 (CID)	–
NPYR	101.0709	55.0542	101.0709	–	80 (HCD)	–	NPYR-D_8	109.1212	92.0982	109.1212	–	80 (HCD)	–
NPIP	115.0866	69.0699	115.0866	–	70 (HCD)	–	NPYR-D_8	109.1212	92.0982	109.1212	–	80 (HCD)	–
NMOR	117.0659	87.0679/ 86.0600	87.0679	–	50 (CID)	–	NMOR-D_8	125.1161	95.1181	125.1161	–	50 (CID)	–

Nitrosamine analysis: CE for collision-induced dissociation (CID) and higher energy C-trap dissociation (HCD) are given in arbitrary units.
[a] '–' denotes that ion was not monitored in this mode in the final method.

12.3.1 Nardo

(i) Site description

The Nardo aquifer is located in the Salento Peninsula in southern Italy, approximately 8 km from the coast of the Ionian Sea (Masciopinto & Carrieri, 2002). Water recharge acts as a salt intrusion barrier, due to reduction of the groundwater table consequent to excessive groundwater pumping in this arid climate. A lot of wells in the area are used for both agricultural and domestic purposes. In the catchment area there are three wastewater treatment plants (WWTPs), Galatone, Galatina and Copertino, with water flows of approximately 1,300, 3,000 and 9,000 m^3d^{-1}, respectively; there are no pharmaceutical industries connected to these WWTPs. After secondary activated sludge treatment, the effluents of all three WWTPs are transported in the 8 km long open Asso channel to a sinkhole (a natural cavity) where the water recharges in the aquifer. The channel discharges 140 Ls^{-1} during dry weather, and up to 450 Ls^{-1} in the winter due to additional rainfall. The aquifer consists mainly of fractured sandstone (5–7 m thick), limestone (30 m thick) and dolomite deposits, and has a hydraulic conductivity of 7.9 10^{-3} ms^{-1}. The water table is approximately 32 m below the ground surface, and the groundwater flows along preferential, horizontal pathways.

The five sampling points for the Nardo case study site are shown in Figure 12.1. Sampling point S1 was the secondary effluent from WWTP Galatone, point S2 was the aquifer recharge point at the sinkhole, points S3 and S4 were wells 320 m and 500 m down gradient to the groundwater flow, (retention time between sinkhole and well was two days for S3 and five days for S4); point S5 was outside the recharge area and provided background values for groundwater quality. Samples were collected once in November 2006 and three times in 2007 (February, May, September).

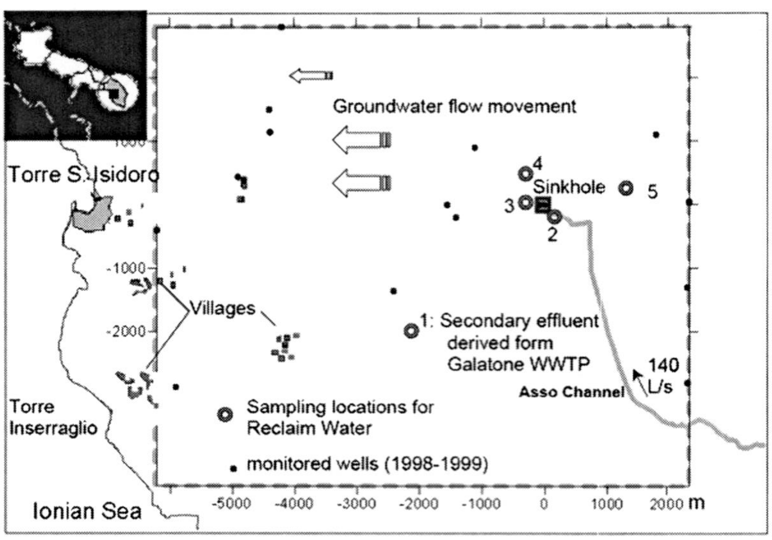

Figure 12.1 Location of Nardò case study site with the sampling points (S1–S5); S5 was the local background concentration (provided by C. Masciopinto, IRSA/CNR, 2006)

(ii) Nardo, trace organic pollutants

In the WWTP effluent (S1), the micropollutants were detected at commonly found concentrations (Ternes, 2001; Joss *et al.* 2005; Khetan & Collins, 2007), and no significant elimination occurred during the channel passage to the sinkhole (in some cases S2 values were even higher than S1 concentrations). The background groundwater did not show any trace organic contamination, therefore Figure 12.2(a–c) show only four of the five samples. Not found in the effluent samples were roxithromycin, sulfadiazine, sulfadimethoxine, sulfamethazine, trimethoprim, ibuprofen, clofibric acid, bezafibrate, paracetamol, iohexol, ioxitalamic acid, estrone, tolyltriazole, N-nitrosopyrrolidine (NPYR), N-nitrosopiperidine (NPIP). Estrogens and nitrosamines were only analyzed in two measurement campaigns. Carbamazepine (mean concentration $C_{av} = 720$ ng/L), sulfamethoxazole ($C_{av} = 180$ ng/L) and clarithromycin ($C_{av} = 160$ ng/L) were regularly found in WWTP effluent. The concentrations of contrast media and ibuprofen markedly fluctuated over time (for example iomeprol from <LOQ up to 7,000 ng/L).

In the wells (S3 and S4), the concentrations of carbamazepine and sulfamethoxazole were decreased by 36–45% compared to the aquifer recharge site (S2). Since these compounds, especially carbamazepine, are persistent in aquifer passage (Drewes *et al.* 2002; Ternes *et al.* 2007), the decrease of carbamazepine in concentration was probably only due to dilution. This statement is supported by conductivity measurements over the 4 campaigns with mean values ($\pm$ standard deviation) of: S1: 1552 $\pm$ 189 $\mu S/cm^{-1}$, S2: 1198 $\pm$ 224 $\mu S/cm^{-1}$, S3: 803 $\pm$ 440 μS/cm^{-1}, S4: 1004 ± 154 μS/cm^{-1}, S5: 1080 ± 161 μS/cm^{-1}, which result in removal rates of 33–16% from S2 to S3 and S4, respectively.

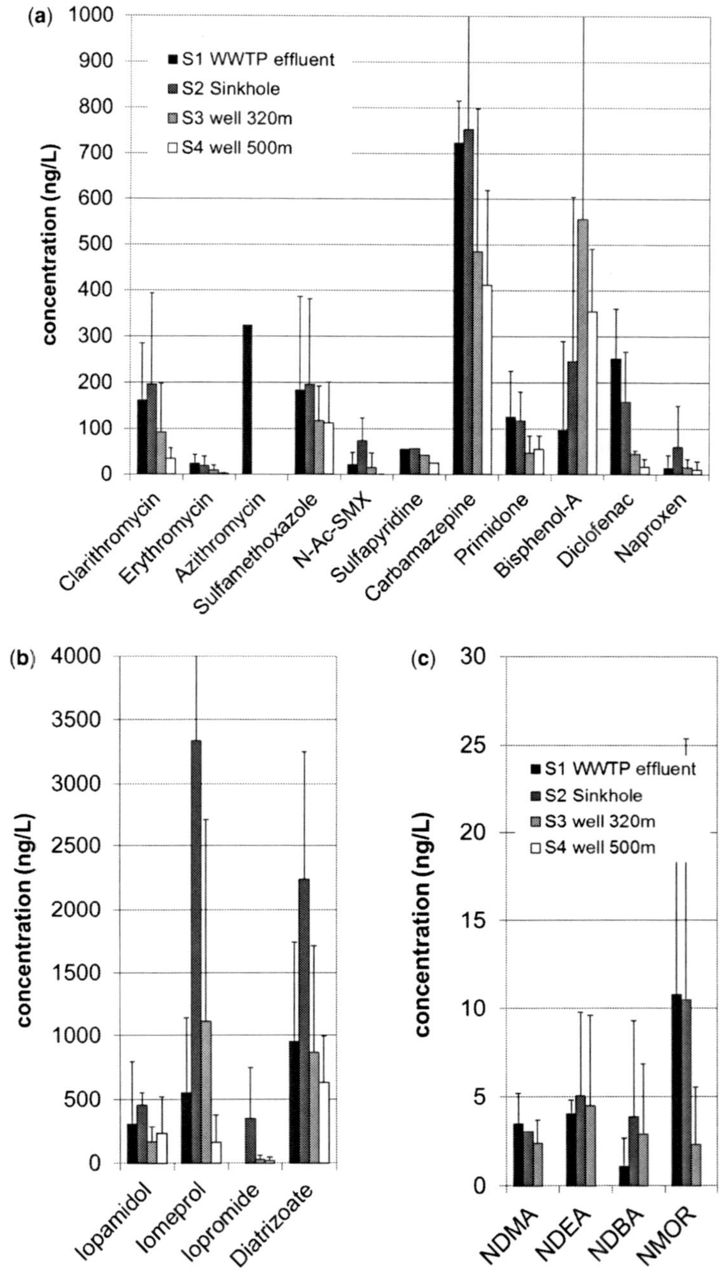

Figure 12.2 Mean concentration and standard deviation of four sampling campaigns at Nardo site. The background groundwater sample (S5) does not show any trace organic contamination and is therefore not depicted

In contrast, concentrations of the macrolide antibiotic clarithromycin in the recovery wells, was more decreased compared to that of carbamazepine and SMX (on average 82% removal from S2 to S4), indicating that an additional elimination process was taking place (Feitosa-Felizzola *et al.* 2009). Azithromycin (another macrolide antibiotic) was eliminated more quickly than clarithromycin due to sorption processes. Also diclofenac was more decreased in the aquifer than carbamazepine (90% S2 to S4).

Concerning the presence of nitrosamines four out of nine analysed compounds were detected in the treated wastewater samples. In the aquifer (S3, as S4 and S5 were not analysed), the nitrosamine concentrations were lower than in the wastewater, mainly for NMOR. Bisphenol A, and the contrast media diatrizoate, iopamidol and iomeprol were observed in the wells in the highest concentrations of all single compound concentration. Most relevant in the well at 500 m (S4) was diatrizoate with a maximum concentration of 1 μg/L. Due to fluctuating input it was not possible to determine elimination rates of ICM compounds in aquifer passage, but other findings in groundwater wells confirm the persistence of contrast media (Putschew *et al.* 2000; Sacher *et al.* 2001).

At the Nardò site, the aquifer is fractured and very permeable, and due to the high flow velocity (or the low residence time from the injection to the recovery well), only a negligible reduction of fairly persistent compounds along the groundwater pathways can be expected.

12.3.2 Sabadell

(i) Site description

The city of Sabadell lies 20 km north of Barcelona, has approximately 200,000 inhabitants, and occupies an area of 38 km^2. The mean rainfall is 600–700 mm per year. The water reuse system is based on the 40 km long Ripoll river, 7 km of which crosses Sabadell city (Levantesi *et al.* 2010). The river originates in the 'Sant Llorenç del Munt i l'Obac' nature park, near the village of Sant Llorenç Savall. There are three discharge locations for the secondary effluents of Sabadell's WWTP into the Ripoll river (see Figure 12.3), namely: (i) Colobrers Stream: up to 8,000 m^3day^{-1} (upstream of the reuse area), (ii) Torrella Mill: up to 10,000 m^3day^{-1}, and (iii) Sant Oleguer Mill: up to 12,000 m^3day^{-1} (downstream of the reuse area). Once in the river, the treated wastewater infiltrates and reaches the aquifer, which is mainly formed by sand and gravel. The groundwater is then recovered in a mine 7 m below the riverbed. The water is disinfected with UV, and chlorinated (1.0–1.2 ppm). Subsequently the pumped water is used to irrigate the Taulí Park and for urban reuse applications such as street cleaning.

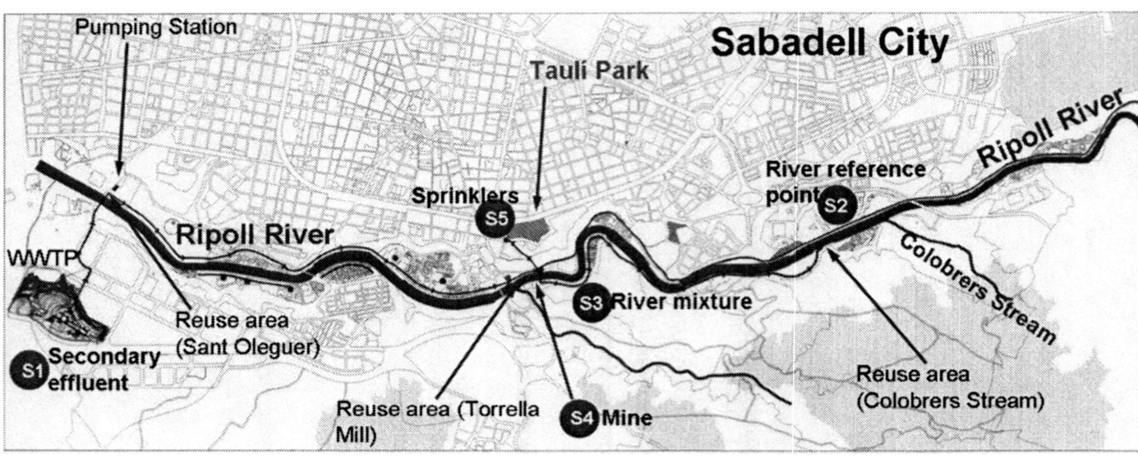

Figure 12.3 Location of the sampling points S1–S5 at the case study site in Sabadell (provided by Sabadell City Hall, 2006)

There are two more WWTPs upstream of the study area: Castellar del Vallès and Sant Llorenç Savall. Castellar del Vallès has 20,000 inhabitants and Sant Llorenç Savall 2,200. The WWTPs have only rudimentary primary treatment, and both discharge their effluents into the Ripoll river.

Samples were taken from five sampling locations: S1: secondary effluent from Sabadell's WWTP; S2: Ripoll river upstream of the first discharge point (representative of the water entering the study area, including water from the WWTPs upstream with only rudimentary primary treatment), S3: river water after mixing of Sabadell's WWTP effluent with the river water upstream, S4: groundwater recovered from the aquifer at the mine, S5: water from the sprinklers of Taulí Park after UV irradiation and chlorination (no details on doses given). Samples were collected three times in 2007 (March, July, November).

(ii) Sabadell, trace organic pollutants

It was difficult to appropriately determine the mean concentrations of the three sampling campaigns because of variations in discharge and technical problems. The concentrations in the first campaign were markedly higher than in the second campaign, particularly in the Ripoll river reference sample; this variation is explained by high variations in discharge from the WWTPs upstream. In the third sampling campaign, the amount of internal standard added to the samples for method 1 was too low, and therefore the neutral compounds and antibiotics could not be quantified. The mean concentration shown in Figure 12.4 therefore shows a high variation, since it is partly based on only two measurements with high differences. Nitrosamines were only determined once in the second sampling program. Analyzed but not found (or rarely found in the WTTP effluent) were roxithromycin, sulfamethazine, ibuprofen, N-nitrosopyrrolidine (NPYR), N-nitrosopiperidine (NPIP), N-nitrosodiethylamine (NDEA), and N-nitrosodibutylamine (NDBA).

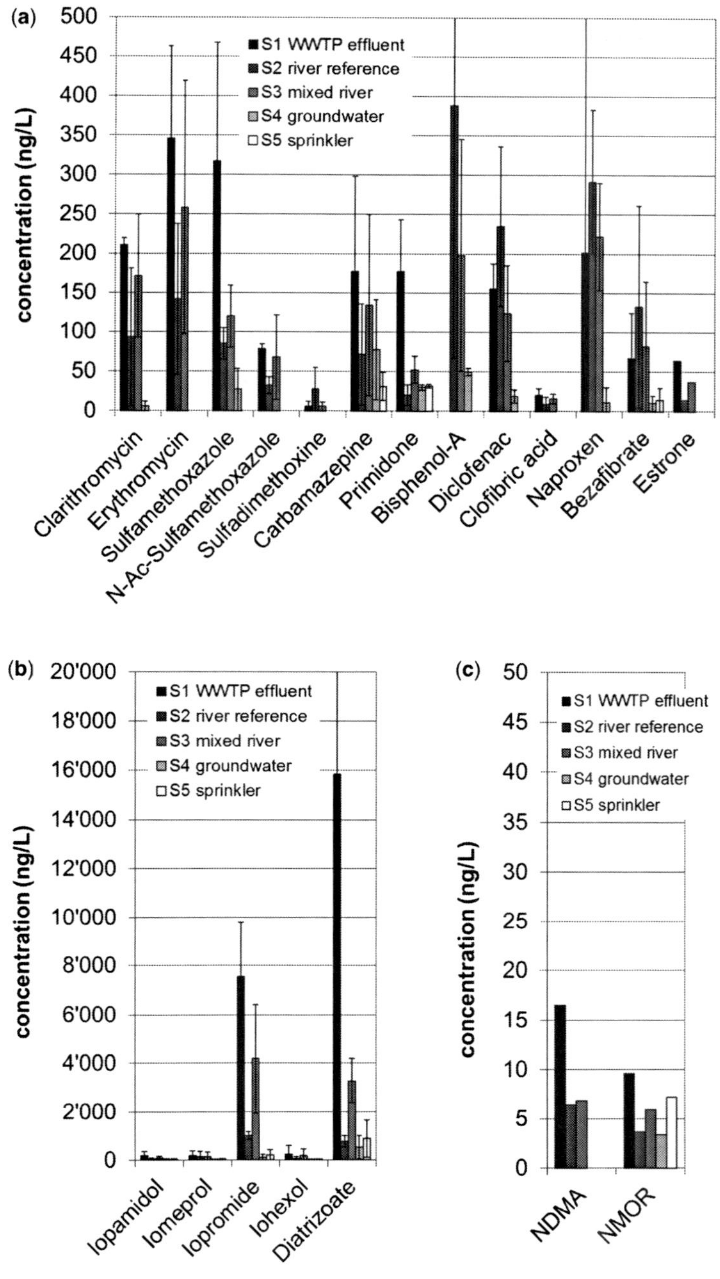

Figure 12.4 Mean concentration and standard deviation of measured compounds during three sampling campaigns at Sabadell sites (only two measurements for compounds from method 1)

In the WWTP effluent (S1), the concentrations of iopromide and diatrizoate were rather high, in the µg/L range. In the secondary effluent the concentrations of erythromycin, sulfamethoxazole, naproxen, clarithromycin, carbamazemine, primidone and diclofenac varied from 100–500 ng/L. In the Ripoll river at the reference point upstream from the disposal site (S2), micropollutants were also found in high concentrations because there is another WWTP with only rudimentary primary treatment discharging further upstream. Concentrations at the reference point were only slightly lower than in the WWTP effluents. At the river site with mixed water (S3), the concentrations were between those at S1 and S2. After aquifer passage (S4), there was a significant reduction in concentration for nearly all compounds. The persistent carbamazepine, primidone, sulfamethoxazole and contrast media were still detected, with the concentrations of carbamazepine and primidone being in a similar range to that in the infiltrating river water. In the sprinkler water (S5), carbamazepine, primidone (both at 10–100 ng/L) and diatrizoate (250–1,800 ng/L) were still present. Treatment with UV and chlorination, which are intended for disinfection, is not effective for removing these compounds. The measurements of chloride, potassium and electrical conductivity as a tracer during the measuring campaigns gave evidence for a reduction of concentration of 18–35% between S3 and S4. The mean chloride concentration in the river reference is 30% larger than the concentration in the WWTP effluent (S1), suggesting a higher load of salts coming from upstream of the river (data see book chapter 2). Concerning nitrosamines, N-nitrosodimethylamine (NDMA) and N-nitrosomorpholine (NMOR) were present in the WWTP effluent at low concentrations up to 10 and 17 ng/L, respectively. While NDMA was removed during aquifer passage to levels below LOQ, NMOR was still present even after UV treatment.

12.3.3 Shafdan

(i) Site description

The Shafdan water recycling facilities are close to Tel Aviv in Israel. Secondary municipal effluent (activated sludge process, approx. 1.5 Mio. population equivalent) is treated by the soil aquifer treatment (SAT) for artificial groundwater recharge (conventional SAT, Aharoni & Cikurel, 2006). The Shafdan SAT system, the largest in Israel and one of the largest in the world, saves approximately 135 million m^3/year of drinking water by supplying reclaimed water, of almost drinking water quality, to agricultural irrigation for the south of the country. The system has been operating for more than 30 years, on a one day flooding (1 m/d infiltration velocity) and 2 days drying mode. The SAT recharge capacity is decreasing due to lack of space for construction of new artificial recharge fields and gradual clogging of the older fields. Further problems include fouling of effluent pipelines by organic matter, and manganese precipitation caused by clogging of the irrigation pipes due to anaerobic aquifer conditions, resulting in high manganese concentrations (Oren *et al.* 2007). The contact time of the reclaimed water in the Shafdan aquifer is approximately one year.

As a potential alternative to the conventional SAT system, and a possible solution to increase recharge capacities in Shafdan, a hybrid SAT system was investigated in the RECLAIM WATER project (Figure 12.5). After the conventional activated sludge process (S1), the water is filtered by a membrane ultrafiltration unit (S2, molecular weight cut off (MWCO) = 30–50,000 g/mol, manufacturer = Trihigh, pressure = 0.3–1.0 bar, 4 units of 25 m^2 each), and subsequently delivered to a dug well 3.6 m in diameter. The dug well infiltrates the water at a filtration rate of 10–12 m/d.

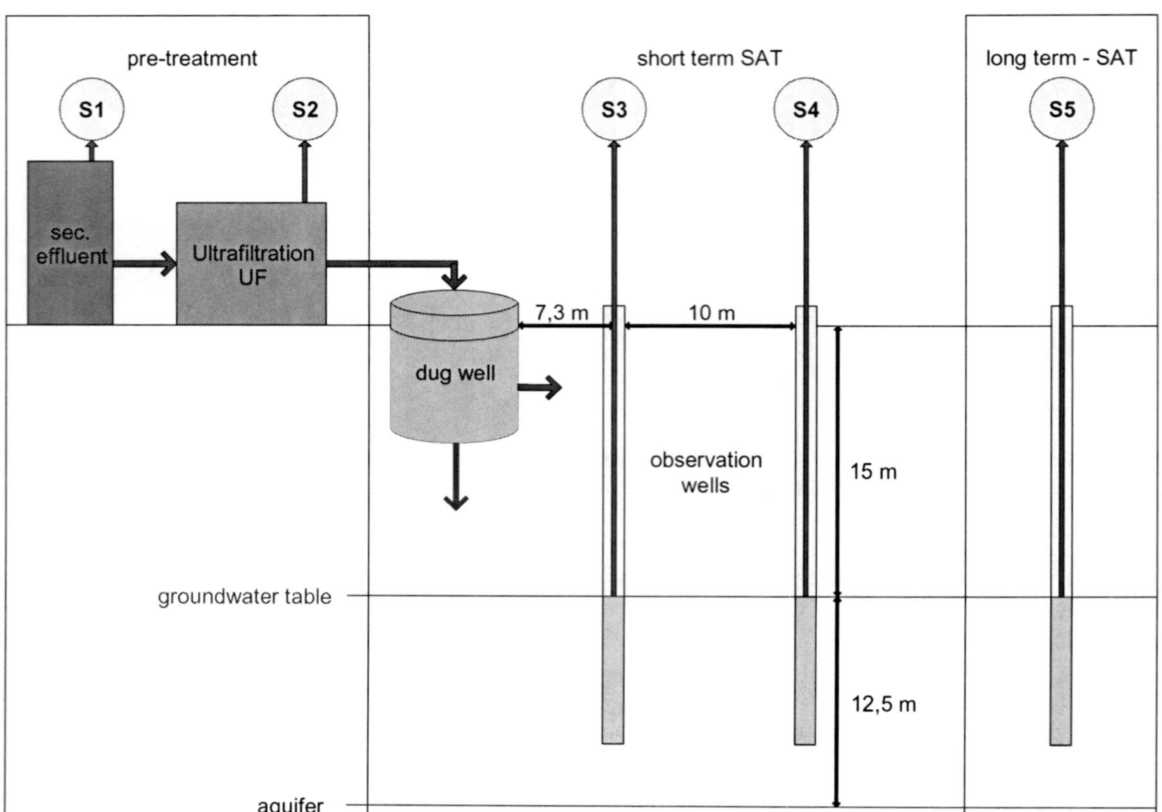

Figure 12.5 Sampling positions of the hybrid SAT demonstration site at Shafdan, Israel

Observation wells provided sampling options at 7.3 m (S3) and 17.3 m (S4) from the infiltration point. The hydraulic retention of the hybrid system was 20–60 days (short SAT, S3 and S4), whereas the hydraulic retention time of the conventional SAT system is 6–12 months (long term SAT, S5). Samples from the long-term SAT were obtained via observation well S5. Three sampling programs were carried out in 2007 (June, October, December).

(ii) Shafdan, trace organic pollutants

High rates of organic matter removal were seen for both short and long term SAT; about 80% of DOC was removed with short term SAT (S4) and more than 90% with long term SAT (S5), (see also data in book chapter 2, 6 and 15). The composition of effluent-derived bulk organic carbon may have a strong effect on the biological removal of the target trace organic compounds, which is especially important for biomass that has adapted to aerobic aquifer conditions (Rauch-Williams *et al.* 2010).

The Shafdan effluent was only partially nitrified, but values of <1 mg/L NH_4^+ could be achieved for both post-treatments, as denitrification takes place in the soil body using the remaining bioavailable DOC. For other nutrients such as phosphorous, values of <30 µg Phosphate/L could be achieved in the long SAT system (Chapter 6), however for the short SAT system, the phosphorous contents remained much higher.

There were multiple pharmaceutical products in the effluent of the Shafdan WWTP. The succession of measured macrolides, sulfonamide antibiotics, naproxen, bezafibrate, and carbamazepine, in the effluent and subsequent soil passage, are shown in Figure 12.6. The macrolide antibiotics such as roxithromycin (ROX), clarithromycin (CLA), and anhydro-erythromyin (ERY-H_2O), were rapidly removed after only a few days of soil contact (S2, S3). The ultrafiltration did not reduce the concentration of polar micropollutants, as it excludes only particles and colloids, not dissolved compounds.

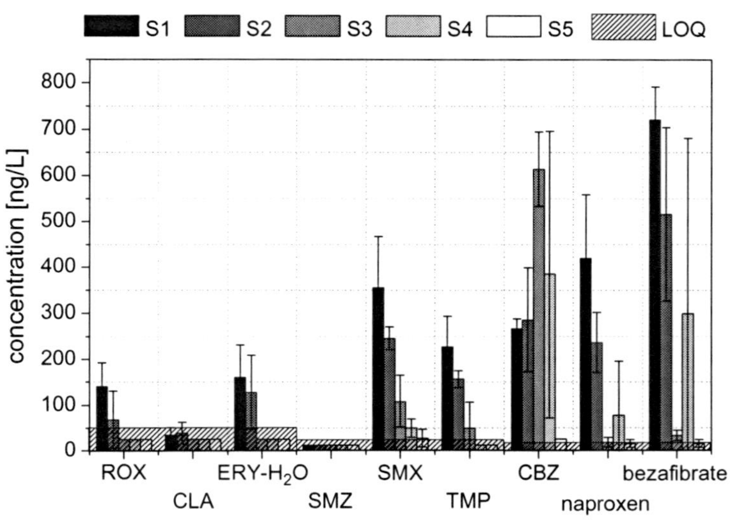

Figure 12.6 Mean concentration and standard deviation of macrolides, sulfonamides (n = 4), carbamazapie, naproxen and bezafibrate (n = 3) concentration in process succession at Shafdan. The marked area represents the limit of quantification. (S1: sec. effluent, S2: UF, S3: well 1, S4: well 2, S5: long SAT)

The SAT removal rates were remarkable high, especially for macrolides, as even at high dug well filtration rates, ERY-H_2O, CLA and ROX were not detected (S3-S4, S5). The sulfonamide; sulfamethazine (SMZ) was not found in any sample, obviously it is not applied in Israel. The highest concentration of antibiotics in the effluent were for sulfamethoxazole (SMX) and trimethoprim (TMP), with mean values of 350 ng/L and 230 ng/L, respectively. Although the concentrations of both compounds were greatly reduced during SAT, SMX could still be detected after long SAT (S5). For naproxen, bezafibrate, and carbamazepine, there were marked differences between their fate in long and short SAT regimes. During long SAT, the concentration of all three compounds dropped below the LOQ. This is especially notable for carbamazepine, which is considered to be a stable anthropogenic tracer compound (Benotti & Snyder, 2009; Gaser *et al.* 2011). A possible explanation could be that its concentration in the effluent was low and not stable over the LOQ vs. time, so in the groundwater it cannot be found above the LOQ.

Three typical refractory polar trace pollutants were found, namely the corrosion inhibitors 1H-benzo-1,2,3-triazole (benzotriazole, BTri) and its two methylated analogues (tolyltriazole, 4-TTri and 5-TTri). BTri was found at 0.5–2.5 µg/L, 4-TT at 0.2–1.5 µg/L, and 5-TTri at 0.2–0.4 µg/L. Although BTri showed higher removal potential for long term SAT than did 4-TT and 5-TT, all three chemicals were detected at >200 ng/L in S5 (Figure 12.7).

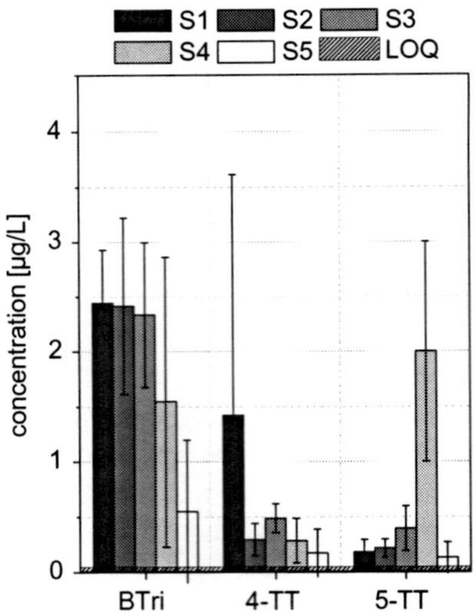

Figure 12.7 Mean concentration and standard deviation of benzotriazole, 5-tolyltriazole and 4-tolyltriazole

For the nitrosamines, the compounds with concentrations above LOQ are shown in Figure 12.8. Large variations in concentration led to high standard deviations for both effluent and UF samples. NDMA concentrations decreased considerably during SAT, to levels <5 ng/L from S3 on, however there were no further significant decreases in concentration in the aquifer between S3 and S4. Unexpectedly the concentration of detected nitroamines in S5 in general were higher than those of S4. From these data we conclude that neither the long SAT (S5) or the short SAT treatment, is suitable for fully removing nitrosamines.

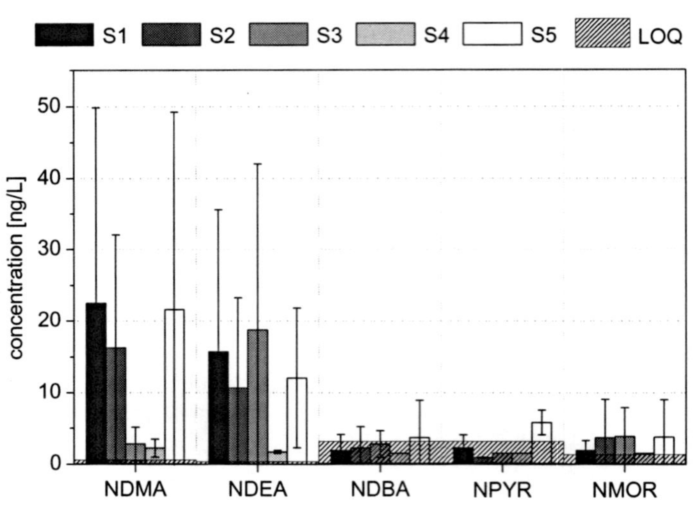

Figure 12.8 Mean concentration and standard deviation of different Nitrosamines during the treatment path (S1: sec. effluent, S2: UF, S3: well 1, S4: well 2, S5: long SAT)

Three iodinated contrast media (ICM), iopamidol, iopromide and iohexol, were single substance compounds with the highest effluent concentration, up to 5 µg/L and more. Iopamidol was quite stable during short term SAT, and was removed to <LOQ in long term SAT (Figure 12.9), just like Iopromide and Iohexol. Iopromide was also completely removed during short term SAT and Iohexol was considerably reduced here. Iomeprol was present in the effluent at low concentrations of some 100 ng/L, and was decreased by 50% within the short term SAT (S4); no iomeprol >LOQ was found in the long-term SAT (S5). Among the ICM's, the behaviour of diatrizoate was an exception, and it's stability throughout the treatment path was confirmed, with concentrations ranging from 200–700 ng/L. Diatriozate was only reduced little further by long-term SAT. The bulk parameter adsorbable organic iodine (AOI), includes as a bulk organic parameter, the sum of all ICM besides other organic compounds containing iodine. The relatively stable behaviour of AOI (Figure 12.9) in short term SAT, suggests that the diagnostic ICM compounds iohexol and iopromide are almost removed as parent compounds but are not oxidized completely. About 33% of the effluent AOI (mean 22.5 µg/L) can be tracked back to measured ICM single compounds (50% of each compound concentration = AOI). In long

SAT, a significant decreased in concentration of AOI, from 18 to 12 μg/L, confirms that longer soil contact times (probably combined with anaerobic aquifer conditions) are favourable for further AOI decrease. However only a partial removal of AOI can be stated in long term SAT.

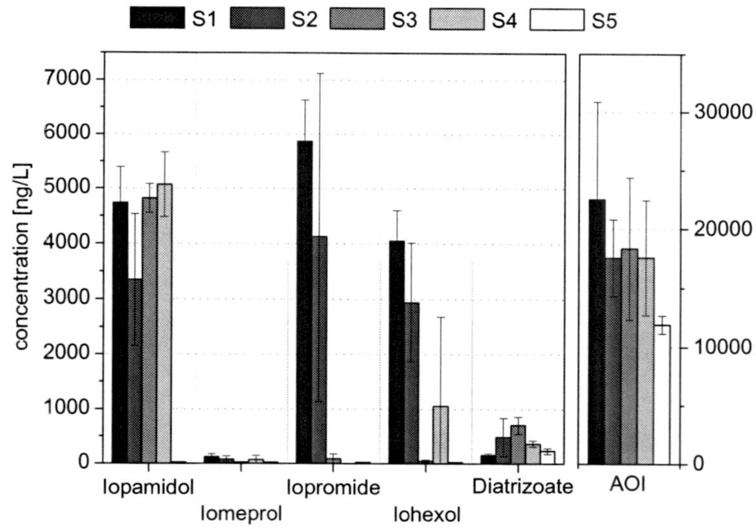

Figure 12.9 Mean concentration and standard deviation of ICM and AOI during the treatment path of short SAT (S4) and long SAT (S5). S1: sec. effluent, S2: UF, S3: after dug well infiltration

12.3.4 Gaobeidian

(i) Site description

Gaobeidian is a district in the east of Beijing, the capital of China. Here an artificial groundwater recharge demonstration plant is located along the west side of the Gaobeidian Wastewater Treatment Plant (GWWTP), which applies coagulation, ozonation, slow sand filtration, well injection, and recovery from a separate well (aquifer storage transfer and recovery, ASTR) (Figure 12.10). Samples were taken from the secondary effluent of GWWTP (S1), the coagulated effluents (S2) and the ozonated effluents (S3). The post-wastewater treatments are: (i) coagulation by polyaluminium chloride (PACL) at a dosage of 30 mg/L, and subsequent rapid sand filtration (10 m/h); and (ii) ozonation by 10–15 mg O_3/L, which is equivalent to 2–3 mg O_3/mg DOC_0. After ozonation the effluent percolates through a slow sand filter, and is subsequently injected into the saturated aquifer zone (S4). Sample S5 was taken from the extraction well after a soil aquifer passage of 34 m, equivalent to 2–3 months of hydraulic retention time. The Gaobeidian demonstration site has been sampled four times (Dec. 2006, June, July and Oct. 2007).

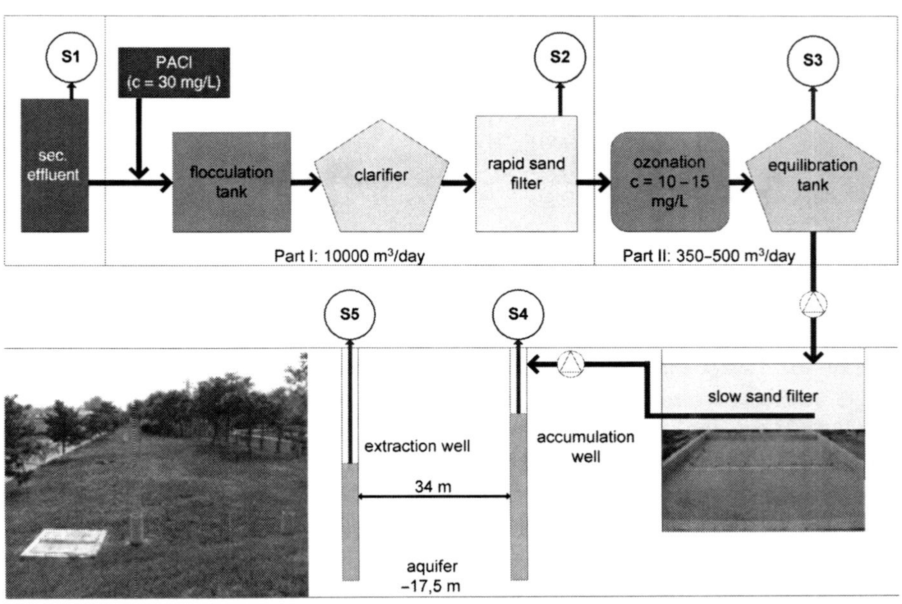

Figure 12.10 Schematic diagram of the Gabeidian water recycling demonstration plant in Beijing; (provided by Zhao Xuan, Tsinhua University, 2010)

(ii) Gaobeidian, trace organic pollutants

In the secondary effluent (S1), there was great variation in the range of concentrations of antibiotics among analytes and sampling events (Figure 12.11). Clarithromycin and anhydro-erythromycin were close to their limit of quantification (LOQ), and roxithromycin was found at concentrations of approximately 150 ng/L, only SMX was present at higher concentration of 620 ng/L. No significant reduction in concentration of micropollutants was observed as a result of coagulation (Figures 12.11 and 12.12).

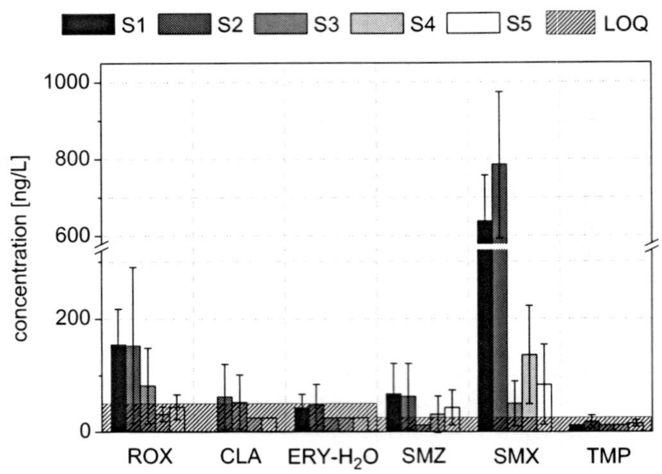

Figure 12.11 Mean concentration and standard deviation of macrolides and sulfonamides at STP Gaobeidian, Beijing. [S1: sec. effluent, S2: coagulated effl., S3: ozonated effl., S4: injection well, S5: production (extraction) well]

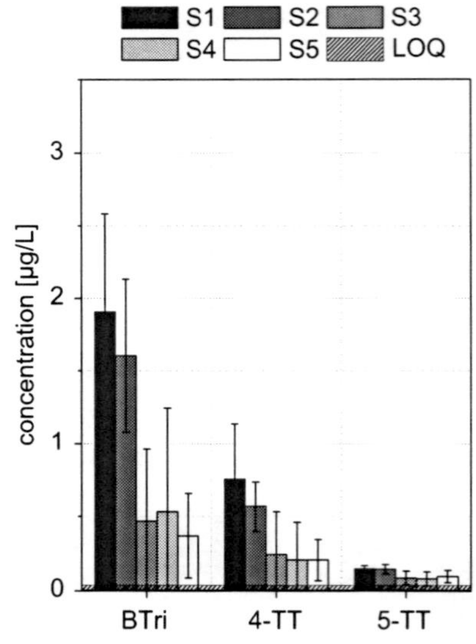

Figure 12.12 Mean concentration of benzotriazole, 4-Tolytriazole and 5-Tolytriazole at STP Gaobeidian

Ozonation was effective at reducing the concentrations of sulfamethoxazole which was reduced from several hundred ng/L to values close to the LOQ. The short aquifer residence time did not significantly influence the antibiotic concentration; SMX had the highest concentration (83 ng/L) of antibiotics in the production well (S5).

In the treated municipal effluent, the concentration of benzotriazole (BTri) was one of the highest for single compounds, with highest values of 2.5 µg/L. The impact of ozone treatment on BTri and its methylated analogues (tolyltriazole, 4-TTri and 5-TTri) was very marked (Figure 12.12), in contrast to coagulation. The subsequent short term ASTR did not significantly change the concentration of BTri, 4-TT and 5-TT, and all three compounds were still at >LOQ in the final extraction well.

The highest concentration of a single ICM compound was found for iopamidol, with a maximum value of almost 6 µg/L (Figure 12.13). In the treatment train there was about 40% reduction of iopamidol during the passage from S1 to the production well. Iomeprol could not be quantified (<LOQ). Iopromide and iohexol showed comparable behaviour,

with effluent concentration of 1–2 µg/L, and a steady decline of mean values from S1 to the production well of 150–200 ng/L. The average concentration of diatrizoat was relatively stable over the applied processes, and varied between 1.3 and 1.5 µg/L. So diatrizoate can be considered as an anthropogenic tracer for wastewater, so the recharged aquifer was fully filled with reclaimed water. From the diagram of adsorbable organic iodine (AOI) compounds (Figure 12.13), we learn that very few ICM were completely removed. The bulk parameter remained more or less stable over the treatment train, however the sum of ICM only accounted for 32% of the measured AOI value in the municipal effluent [(4 µg/L iopamidol +1 µg/L Iopromide + 1.5 µg/L Iohexol + 1.5 µg/L diatrizoate)/2 = 4 µg/L AOI vs. 12.5 µg/L AOI measured)] and for about 25% of AOI in the final product water (rule of thumb, 50% of ICM concentration = AOI concentration, Jekel and Putschew 2006).

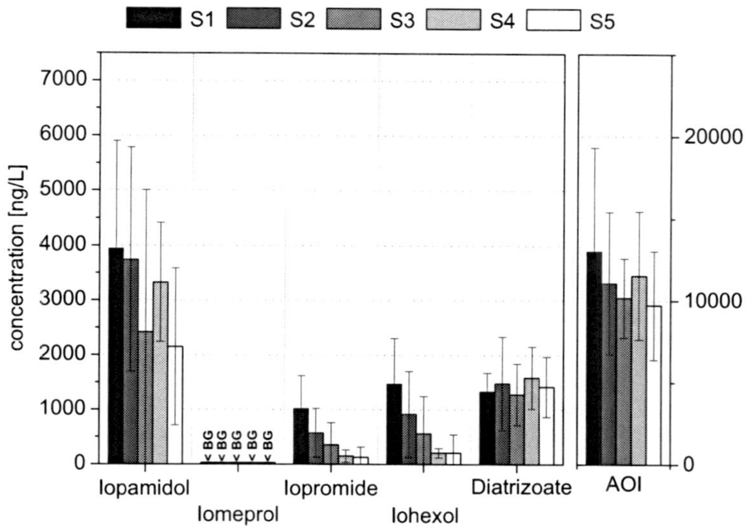

Figure 12.13 Mean concentration, with standard deviation, of iodinated contrast media and adsorble organic iodine (AOI). S1: secondary effluent, S2: coagulated effluent, S3: ozonated effluent, S4: observation well, S5: production well

12.3.5 Wulpen/Torrele

(i) Site description

The demonstration site in Wulpen/Torrele, Belgium, applies the maximum of advanced water treatment, as displayed schematically in Figure 12.14. After a conventional wastewater treatment with nutrient removal (activated sludge process, pre-denitrification, and phosphorous precipitation), the treated effluent (S1) passes to the Torrele reclamation plant, where treatments include an integrated membrane process. After chlorination to control bio-growth, the effluent flows to five, parallel, ZeeWeed® ultrafiltration (UF, ZW500C) trains. The concentration of HOCl in the inlet of UF is 1.5 mg/ in the winter and 2.75 mg/L HOCl in the summer season. The UF filtrate (S2) enters a buffer reservoir and is chloraminated to control biofouling on the subsequent reverse osmosis (RO) membranes; monochloramine is dosed to 0.5 mg/L in winter and 1.5 mg/L in summer. From here the water is pumped to the RO system, which is equipped with low energy membranes from DOW (30LE-440) with a mean recovery of 77% and a flux maximum of 20 LMH (1 LMH = 10^{-3} m^3 h^{-1} m^{-2}). The RO filtrate (S3), to which sodium hydroxide is dosed to increase the pH to about 6.5, is then piped approximately 2.5 km to the recharge/extraction site of St-André, in the municipality of Koksijde. The recharge capacity is approximately 2,500,000 m^3/year, and occurs in an unconfined sandy aquifer. From the top of the infiltration basin to the pumping depth, the sands are fine to medium size and contain shells; there are also some small irregular clayey and peaty layers. After recharging, the water is recaptured using 112 new wells with filters between 8 and 12 m deep (Van Houtte & Verbauwhede, 2008). The horizontal separation between the infiltration pond and the abstraction wells varies between 33 m and 153 m, with an average of 59 m. The minimum retention time is 30 to 35 days, and the mean retention time is 55 days. The recovered water is conveyed to the potable water production facility at St. André, which consists of an aeration step, rapid sand filtration, a reservoir, and UV disinfection, prior to distribution (S4). For mass balance reasons (i.e. to identify rejected organic compounds in the concentrate) the RO brine was also sampled (S5).

(ii) Wulpen/Torrelle, trace organic pollutants

The Wulpen site was sampled three times in 2007 (Jan., July, Oct. 2007) for the determination of the full list of trace pollutants, and for disinfection by-products (N-nitrosamines).

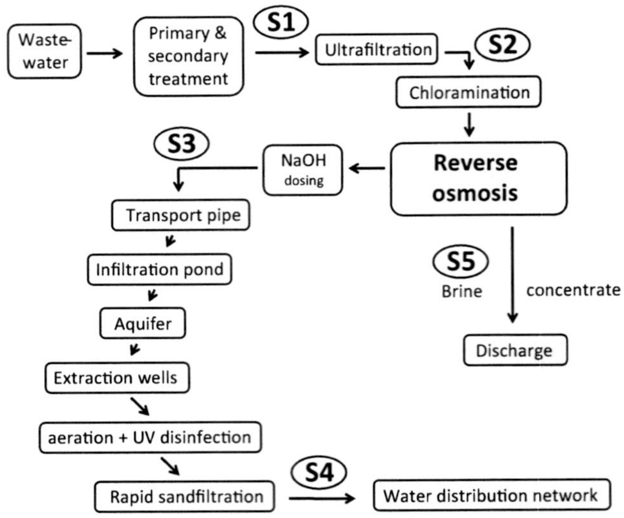

Figure 12.14 Flow diagram of the water reuse scheme Torrelle/St-André showing sampling points S1–S5

Pharmaceuticals from the investigated classes (for example antibiotics, contrast media, antiepileptics, and corrosion inhibitors) were detected in the raw sewage influent and in the secondary effluent (S1) of the WWTP in Wulpen (Wulpen was the only WWTP plant, where raw sewage water was measured, data not shown here). For some compounds (such as carbamazepine, primidone, clarythromycin, and benzotriazole), there was no or only a very limited removal during conventional treatment, indicating why a more advanced treatment, such as reverse osmosis (RO), was selected for the production of water later recovered and treated for use as drinking water. The series of UF prior to RO membranes has often occurred in the past years for the treatment of municipal effluents for water reuse (Markus 2009; Schnoor 2009). Due to the physical disinfection of secondary effluents by UF, and a complete exclusion of particulate and suspended solids, the colloidal fouling is well controlled. A low concentration of bacteria in the feed water of the RO is needed to control bio-fouling on the surface of the homogenous solution diffusion membrane.

Figures 12.15 and 12.16 show that the remaining concentrations of pharmaceuticals are only present up to the UF permeate (S2). The concentrations of all trace organic compounds were well below the LOQ for the RO permeate (S3) and for final drinking water (S4). The effluent concentrations (S1) for macrolides and sulfonamides were comparable to the other demonstrations sites, however for diatrizoate, iopromide, benztriazoles, and carbamazipine, the effluent concentrations were well above 1 μg/L, even exceeding >5 μg/L for benzotriazole. The integrated membrane approach proved to be an efficient technique for removal of organic micropollutants from the water phase; however a critical point is the disposal of the RO-concentrate. For high concentration effluent contaminants (such as carbamazepine, benzotriazole, and diatriozat), the brine concentration (S5) reached 10 to 20 μg/L, and were generally 3–4 fold higher than the effluent concentrations (S1).

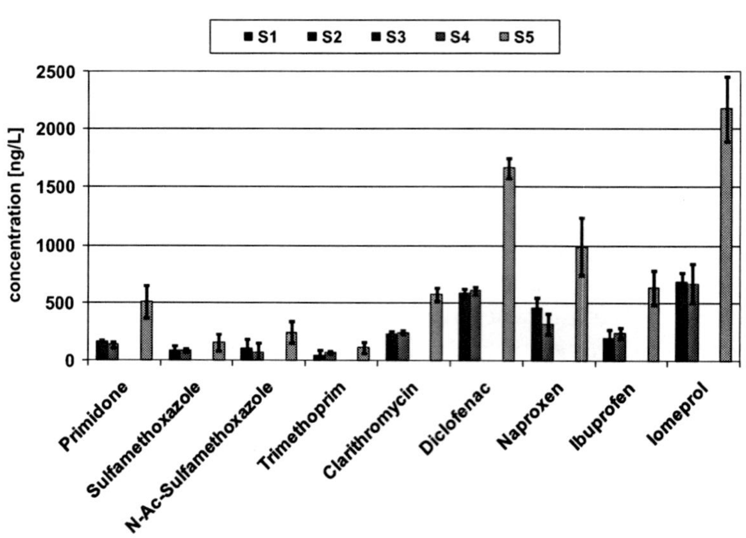

Figure 12.15 Concentration in ng/L for 9 selected pharmaceutical compounds at Wulpen/Torrelle. Mean from three sampling programs (with standard deviation)

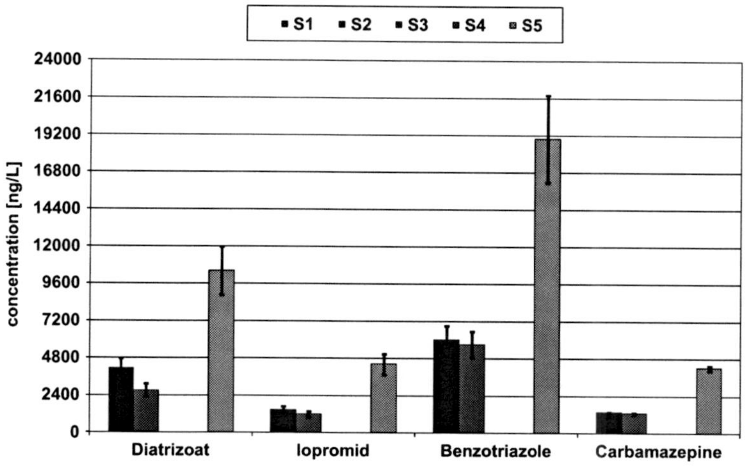

Figure 12.16 Concentration in ng/L for four selected organic compounds at Wulpen/Torrelle. Mean of three sampling programs (with standard deviation)

For the nitrosamines, the concentrations in the secondary effluent (S1) were in general low, not exceeding 10 ng/L for NDMA and 5 ng/L for NMOR, and were below the LOQ for the other nitrosamines. In March and October 2007, no NDMA was formed during chlorination/chloramination before the UF/RO units at the low doses were employed; however, in August 2007 about 5 ng/L NDMA were formed during chlorination at the larger dose of 2.75 mg/L free chlorine. About 50% of NDMA was excluded by reverse osmosis, and NMOR was excluded to a larger extent, resulting in levels below the LOD after RO (Figure 12.17). No nitrosamines were detected in the product water (S4) suggesting a further degradation of NDMA traces in the aquifer, although photo-degradation is also possible in the infiltration ponds (Plumlee and Reinhard, 2007).

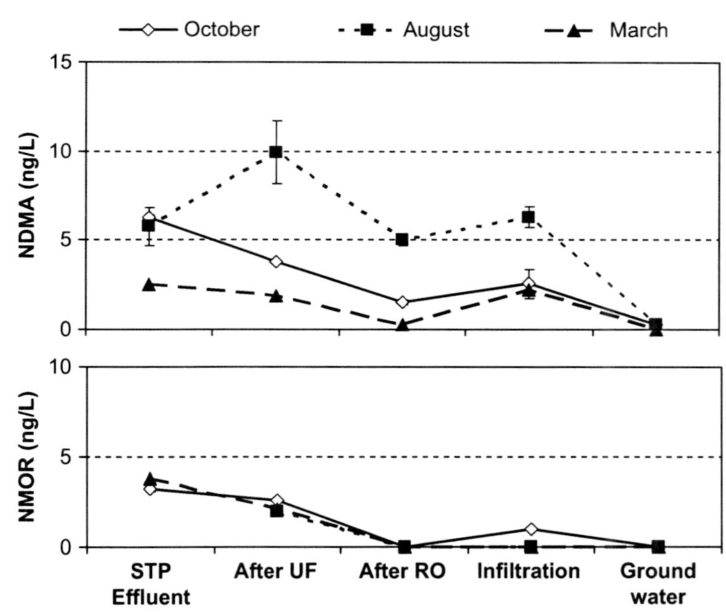

Figure 12.17 Concentrations of NDMA and NMOR along the treatment train at the Wulpen/Torrelle plant during sampling programs in 2007 (data is mean and standard deviation of two replicates). Values below the limit of quantification (LOQ) of 1 ng/L were set to half of the LOQ, values below the limit of detection were set to zero. Reprinted from Krauss *et al.* (2010) with permission

12.4 CROSS SITE ANALYSIS

There are challenges in making direct comparison of water quality at different managed aquifer recharge (MAR) sites, such as the water reuse demonstration sites in the RECLAIM WATER project, because of differences in local conditions, raw water quality, applied treatment technologies, and water residence times. However, we can distinguish between two main types of aquifers: (i) those that are confined by a low permeability layer, and (ii) those that are unconfined and allow water to infiltrate through permeable soils and vadose aquifer zones. The confined aquifer requires injecting water via a well, whereas in unconfined aquifers, recharge is via basins and galleries (Figures 12.18 and 12.19, Dillon *et al.* 2009).

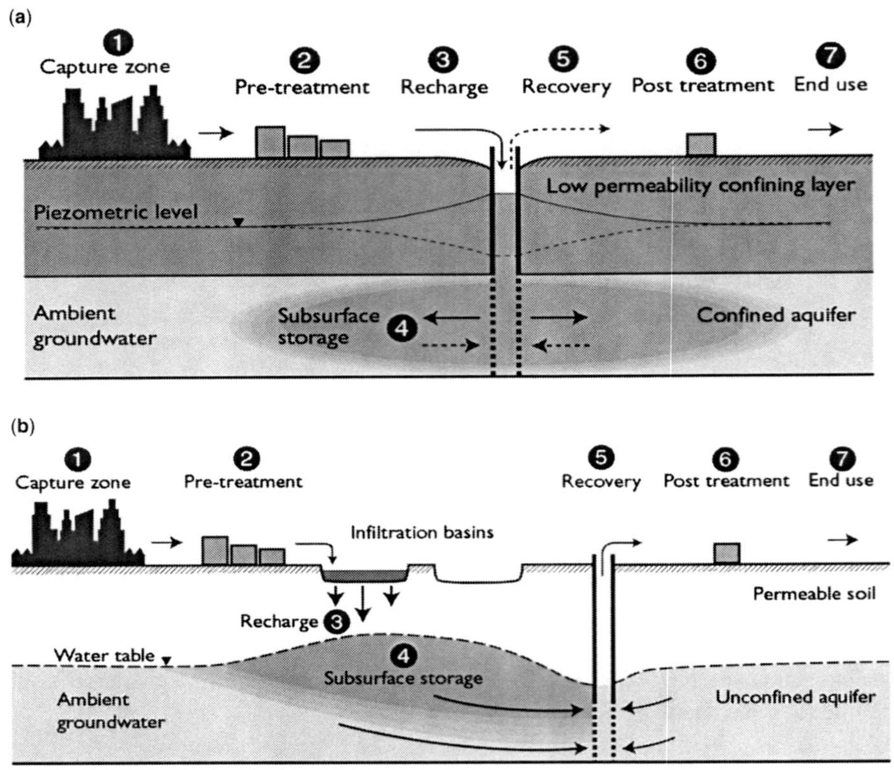

Figure 12.18 a, b Two examples of managed aquifer recharge, (a) confined aquifer, with aquifer storage recovery (ASR), and (b) unconfined aquifer with soil aquifer treatment (SAT), showing the seven elements common to each system (Dillon *et al.* 2009, reproduced with permission)

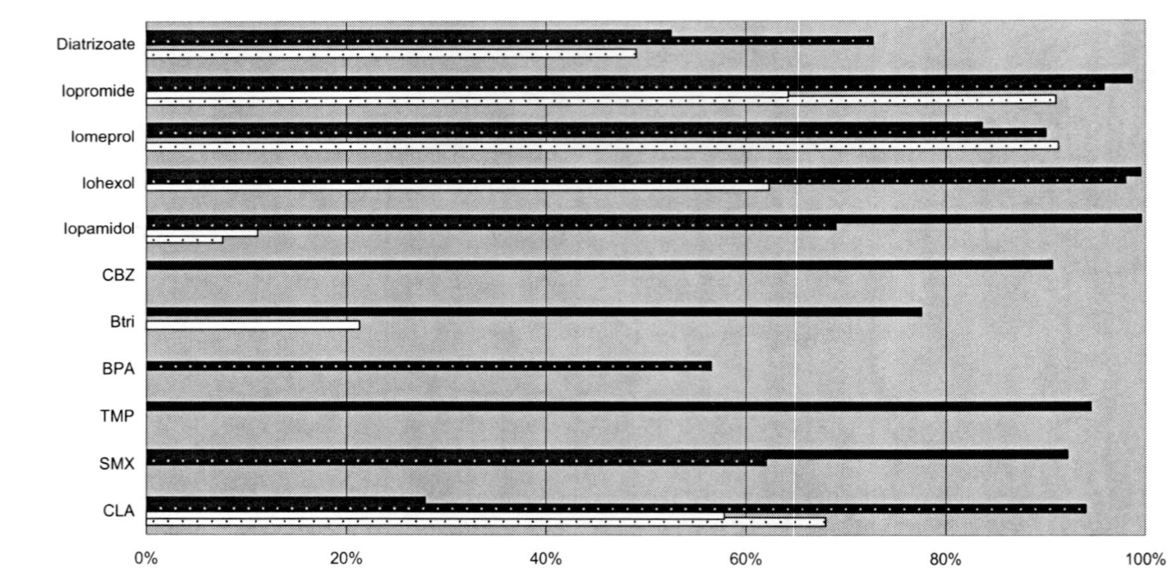

	CLA	SMX	TMP	BPA	Btri	CBZ	Iopamidol	Iohexol	Iomeprol	Iopromide	Diatrizoate
■ R (Shafdan, l.t.)	28%	92%	95%		78%	91%	100%	100%	84%	99%	53%
▣ R (Sabadell)	94%	62%		57%	-7%	0%	69%	98%	90%	96%	73%
□ R (Gaobeidian)	58%	-67%	-13%		21%		11%	62%		64%	-11%
⊡ R (Nardo)	68%	-4%		-162%	-652%	0%	8%		91%	91%	49%

Figure 12.19 Mean removal rates over all sampling campaigns for "infiltration through soil aquifer sites" (Shafdan, long term SAT and Sabadell) as well as "direct injection aquifer sites" (Gaobeidian and Nardo), considering dilution in the aquifer. The table below the diagram shows the numbers used for the figure (compounds <LOQ if empty)

The five demonstration sites of the present study were grouped into two categories as follows: (i) direct injection aquifers (presumably confined aquifers): Gaobaidien (China), Nardo (Italy); (ii) infiltration through soil aquifer (presumably unconfined aquifers): Sabadell (Spain), Shafdan (Israel), Wulpen/Torrele (Belgium). This classification reflects the aquifer redox conditions to some degree, as the infiltration through soil aquifers usually have more contact surface to the unsaturated soil body and thus contains more oxygen than do the direct infiltration aquifers.

The seven key elements of the five investigated demonstration sites are summarised in Table 12.5.

Table 12.5 Comparison of 7 key elements of the 5 demonstration sites, grouped according to whether their managed aquifer recharge systems were "direct injection aquifers" (likely confined aquifers) or "infiltration through soil aquifers" (likely unconfined aquifers).

	Direct injection aquifers		Infiltration through soil aquifers		
	Gaobeidian	**Nardo***	**Shafdan****	**Sabadell**	**Wulpen/Torrele**
1. Capture zone	WWTP	WWTP + river	WWTP	WWTP + river	WWTP
2. Pre-treatment	Coagulation, RSF, ozonation, SSF	None	UF	None	Chlorination UF, chlorination RO
3. Recharge	Injection well	Natural sinkhole	Percolation basins, dug well	Bank filtration	Pond infiltration
4. Sub-surface storage	Shallow aquifer	Karst aquifer	Sandy aquifer	Alluvial aquifer	Shallow aquifer
5. Recovery	Observation well	Production well	Production wells	Mine well	Production wells
6. Post-treatment	None	None	Aeration	UV	Aeration, RSF, UV
7. End use	Washing	Agriculture	Agriculture	Irrigation	Drinking water

*The Nardò site is an aquifer storage transfer and recovery site (ASTR) but due to high flow velocities there is a low residence time in the aquifer between the injection and recovery wells. RSF: rapid sand filtration, SSF: slow sand filtration, UF: ultrafiltration, RO: reverse osmosis, UV disinfection.
**Shafdan, long term soil aquifer system.

For the iodinated contrast media, benzotriazole, bisphenol A and some antibiotics the mean elimination in four of the managed aquifer recharge sites was calculated and are presented in Figure 12.19. The data given are mean removal rates of measured compounds from the aquifer infiltration or injection site to the final well in the aquifer over all applied sampling campaigns (see earlier chapters). The focus of this analysis is on the removal capacity of the two aquifer systems "direct injection" and "infiltration through soil". The removal rates were calculated by considering dilution factors at each of the analysed sites derived in general by the ratio of the carbamazepine concentration prior to infiltration with the concentration measured in the product well. Table 12.6 summarizes the background for the calculation of removal rates and accounted dilution factors.

Table 12.6 Basis for removal rate calculation in Figure 12.19 and accounted dilution factor at four MAR demonstration sites.

	Gaobeidian	**Nardo**	**Shafdan, long term SAT**	**Sabadell**
Removal rate calculated by	S5 over S3 (final product well vs. injected water)	S4 over S2 (final product well vs. injected water)	S5 over S1 (final product well vs. sec. effluent)	S4 over S3 (product well vs. mixed river water)
Dilution factor calculated by	The site relies on secondary effluent only	Ratio of carbamazepine, S4 vs. S2	the site relies on secondary effluent only	Ratio of carbamazepine S4 vs. S3
Considered dilution factor	None	55%	None	58%

The Wulpen/Torrele reclamation site was excluded from the cross site analysis because after RO treatment the infiltration water in general is below the limited of quantification for all measured trace organic compounds (with the exception of NDMA in two campaigns, Figure 12.17) and so the calculation of removal rates were not possible. Moreover, due to visibility reasons for the data analysis, the Shafdan short term SAT data are not considered. Here the removal rates are little lower than in long term SAT but are not in contrast to long term SAT (see 12.3.3).

As shown in Figure 12.19, the "filtration through soil aquifers" sites Shafdan (long term SAT) and Sabadell exhibit in general higher removal rates than the "injection aquifer" sites. This is especially true for SMX which is removed significantly only at Shafdan.

The high removal rates found for X-ray contrast media have to be evaluated with care. ICM are known as very refractory compounds, especially diatrizoate. Since their concentrations in the WWTP effluents and at the aquifer injection sites vary substantially over time due to irregular usage patterns, elimination can only be evaluated if samples are taken over a considerable long time frame. This was not the case at the investigated sites. Nevertheless, a trend can be observed for better elimination at the sites with "infiltration through soil aquifers" than at the "direct injection aquifers".

As a summary of the cross site analyse, we assume that due to different aquifer systems, the redox conditions in "infiltration through soil" aquifers are more favourable for the removal of most trace organic compounds. For the "direct infiltration aquifers" Gaobeidian and Nardo (in book Chapter 2, Table 1 identified as confined aquifers) the redox potentials are likely lower than for the infiltration through soil aquifers (Shafdan, short term SAT and Sabadell). The higher removal potentials for trace compounds under elevated redox conditions were confirmed by other studies also (Grünheid *et al.* 2005; Rauch-Williams *et al.* 2010), suggesting that the bioavailable dissolved organic carbon (BDOC) stimulates soil biomass growth, especially under aerobic conditions, and induces secondary substrate utilization for the removal of trace organic compounds.

Risk estimation

This preliminary risk assessment provides information about which chemical compounds need to be particularly considered in future monitoring. The maximum observed concentrations of organic compounds in the treated effluents and product waters (production wells) of the demonstration sites (all sites and samples pooled), are given in Table 12.7. The principle of this risk estimate is to compare predicted concentrations in the environment (predicted environmental concentration, PEC) to the lowest concentration that does not have an effect on exposed organisms (predicted no-effect concentration, PNEC). The PEC can be estimated by models or measured in the field. The latter approach is used here by employing field measurement from the demonstration sites (measured environmental concentration, MEC). The PNEC values are derived from the lowest PNEC of the most sensitive species found in studies where organisms such as fish, algae, crustaceans, or bacteria are exposed to different concentration of the test compounds. This MEC to PNEC approach represents a measure for eco-toxicological risk. Human health risks are more difficult to derive and little or no data is available on human health risk studies on the considered compounds. Therefore human health risk is not considered in this risk estimation.

Table 12.7 Highest measured effluent (MEC_{effl}) and product water concentrations (MEC_{Pro}) and relevant no effect concentrations (PNEC) given in the literature.

Compound	MEC_{effl} (μg/L)	MEC_{Pro} (μg/L)	PNEC (μg/L)	Species	MEC_{effl}/PNEC	MEC_{Pro}/PNEC	Reference
Bisphenol A	0.76	0.5	1.6		0.48	0.31	European Union Risk Assess. Report, 2008
Benzotriazole	6	9.5	30		0.20	0.32	Steber J. & W. Hater 1997
Carbamazepine	1.38	0.63	2.5	*Ceriodaphnia dubia*	0.55	0.25	Ferrari *et al.* 2003
Clarithromycin	0.3	0.05	0.2	*Pinnularia subcapitata*	1.5	0.30	Isidori *et al.* 2005
Iopamidol	7.68	5	10000		0.0	0.0	Steger-Hartmann *et al.* 1999
Iohexol	4.69	1	1000	*Daphnia magna*	0.00	0.00	Steger-Hartmann *et al.* 2002
Sulfamethoxazole	0.72	0.2	0.59	Cyanobacteria	1.22	0.34	Ferrari *et al.* 2004

As the highest observed concentrations for MEC_{effl} or MEC_{Pro} were used for calculating the risk, a "worst case scenario" was applied. The MEC/PNEC ratios for bisphenol A, benzotriazole and carbamazepine were between of 0.20–0.55, which means that these compounds in worst case could reach half of its "no effect concentration". For the worst case scenario, the antibiotics clarithromycin and sulfamethoxazole each had MEC/PNEC ratios of >1 for effluent, and in the product water the MEC/PNEC ratio was below 1 for sulfamethoxazole (0.34) and for clarithromycin (0.30). This means that some of the exposed bacteria or higher organisms may be inhibited, especially in the effluent discharges. For iopamidol and iohexole the MEC/PNEC ratios are very small as the ICM are not of toxicological concern and have very high PNEC values. We conclude that the environmental risk of using product well waters from the artificial RECLAIM WATER recharge sites after SAT is acceptable. Nevertheless, actions will be necessary to control the risk in the long term, as for example when usage patterns change.

12.5 CONCLUSIONS

Organic micropollutants in product water are of specific concern especially if the water will be used for (indirect) potable reuse. During the present Reclaim Water project, a wide range of analytical methods were developed, validated, optimized and applied in laboratory experiments as well as on five international demonstration sites for managed aquifer recharge. Although treatment efficiencies varied widely (depending on the applied technology, local wastewater matrix, hydro-geological conditions, and hydraulic retention times), it was possible to derive general outcomes from the composite data at the five international demonstration sites.

In the secondary effluents, carbamazepine was found at all sites except in Beijing (China), where the antiepileptic is not prescribed. The mean effluent concentration of carbamazepine was about 0.5 μg/L, with single concentrations varying between 60–>1,000 ng/L depending on sampling date and location. During infiltration the concentration of carbamazepine remains stable (or even increases) at Shafdan (short term SAT). In the case of Nardo and Sabadell, carbamazepine declines partially. We made dilution by carbamazipine free groundwater responsible for the decline of the average concentration. Only in Shafdan, long term SAT, the final carbamazepine concentration after one year of SAT was below the limit of quantification and much lower than the measured effluent values over 4 sampling campaigns (30 ng/L vs. 300 ng/L). Assuming that carbamazepine is a conservative tracer, these results can be explained either by (i) varying input concentration, or (ii) simply by dilution, as it is the case in Nardo and Sabadell.

According to the demonstration site results, we can consider diatrizoate also as an anthropogenic tracer as relatively stable concentrations were measured in Shafdan for long as well as for short term SAT. The same situation was monitored in Nardo and Sabadell if dilution factors were accounted for (according to carbamazipine changes).

In general, iodinated contrast media contributed to one of the highest load of analysed organic single substances in domestic effluents and in recharged groundwater. Diatrizoate, iopamidol and iomeprol were detected at μg/L levels over the whole treatment train, with concentrations varying from some hundreds of ng/L up to 4–5 μg/L even in product waters. For the Shafdan demonstration plant, about 32% of the AOI could be traced back to the single iodinated contrast media concentrations in the secondary effluents. This fraction decreased with SAT at longer hydraulic retention times to <25%, which is consistent with the findings of Putschew and Jekel, 2006.

The mean WWTP effluent concentrations of the sulfonamide and macrolide antibiotics (sulfamethoxazole, sulfamethazine, clarithromycin, roxithromycin, erythromycin) varied from 100–600 ng/L, depending on the degree of medical application and intensity of wastewater treatment. Prior to infiltration (during wastewater treatment), the mean concentration did not decrease by more than 50% if no advanced treatment like ozonation or RO membrane filtration were applied. However in SAT, the removal efficiency increased and led to lower concentrations in the abstracted product waters, where maximum values of 100 ng/L antibiotics were rare. Concentrations of macrolides and trimethoprim were below those of the sulfonamides.

Nine different N-nitrosamines were detected in demo site effluents where chlorine disinfection was applied; concentrations ranged from 0.5–55 ng/L, with large fluctuations between the sampling events. These results underline that optimization potential for disinfection practices exist at some sites. Only two nitrosamines compounds were found in groundwater at one site at very low concentration. Other trace organic compounds were detected mainly in effluents and first treatment steps, namely bisphenol A, estrone, ibuprofen, diclofenac, naproxen, benzafibrate, benzotriazole and tolyltriazoles. Endocrine compounds (method III) were not found in any investigated product waters, as they were removed to values below LOQ at all sites.

The cross site analysis revealed better elimination of micropollutants at "infiltration through soil aquifer" sites (unconfined sites) than at "direct injection aquifers". Unconfined sites exhibit vadose zones with oxygen above the groundwater level, for which higher redox potentials can be assumed. The relevance of elevated redox potentials for a more effective micropollutant removal in the subsurface was also confirmed by previous studies (Water Research Foundation, 2010; WateReuse report, 2008; WateReuse report 2007; Maeng *et al.* 2011). In a brief environmental risk assessment, only the antibiotics clarithromycin and sulfamethoxazole showed MEC/PNEC ratios of >1 in worst case scenarios (highest effluent concentration). However the MEC/PNEC ratios for all product waters were <1 for all considered compounds. Accordingly the risk of using product waters is considered as acceptable for all RECLAIM WATER demonstration sites.

ACKNOWLEDGEMENTS

The partners from the different demonstration sites are acknowledged for their logistical support, and sample collection and preparation using the SPE cartridges. We greatly appreciate the help of M.N. Ayuso-Gabella and M. Salgot from the

University of Barcelona, Spain (Sabadell site), L. Balest, G. Mascolo and C. Masciopint from the National Research Council (CNR) in Bari, Italy (Nardo site), H. Cikurel and A. Aharoni from Mekorot, Israel (Shafdan site), X. Cheng, L. Yu and X. Zhao, Beijing, China (Gaobeidian site), and J. Cauwenberghs and E. van Houtte, Belgium (Wulpen/Torrele site). The European Commission is acknowledged for co-funding the RECLAIM WATER Project under Contract No. 018309 in the Global Change and Eco-system sub-priority of the 6th Framework Programme for Research and Technological Development.

REFERENCES

Aharoni A. and Cikurel H. (2006). Mekorot's research activity in technological improvements for the production of unrestricted irrigation quality effluents. *Desalination*, **187**(1–3), 347–360.

Arye G., Dror I. and Berkowitz B. (2011). Fate and transport of carbamazepine in soil aquifer treatment (SAT) infiltration basin soils. *Chemosphere*, **82**, 244–252.

Benotti M. and Snyder S. A. (2009). Pharmaceuticals and endocrine disruption compounds: implications for groundwater replenishment with recycled water. *Ground Water*, **47**(4), July-Aug., 499–502.

Dillon P., Pavelic P., Declan P., Beringen H. and Ward J. (2009). Managed Aquifer Recharge: An Introduction, Waterlines Report Series No. 13, Australian Government/National Water Commission.

Drewes J. E., Heberer T. and Reddersen K. (2002). Fate of pharmaceuticals during indirect potable reuse. *Wat. Sci. Technol.*, **46**(3), 73–80.

European Union Risk Assessment Report: 4,4′-Isopropylidenediphenol (BPA) (2008). (to be read in conjunction with published EU RAR of BPA, 2003), EUR 24588 EN – Joint Research Centre – Institute for Health and Consumer Protection. K. Aschberger, S. Munn, H. Olsson, S. Pakalin, G. Pellegrini, S. Vegro and A. B. Paya Perez (ed.), ISBN: 978-92-79-17541-1.

Feitosa-Felizzola J., Hanna K. and Chiron S. (2009). Adsorption and transformation of selected human-used macrolide antibacterial agents with iron(III) and manganese(IV) oxides Environ. *Pollution*, **157**(4), 1317–1322.

Ferrari B., Mons R., Vollat B., Fraysse B., Paxeus N., Lo Giudice R., Pollio A. and Garric J. (2004). Environmental risk assessment of six human pharmaceuticals: are the current risk assessment procedures sufficient for the protection of the aquatic environment? *Environmental Toxicology and Chemistry*, **23**, 1344–1354.

Ferrari B., Paxeus N., Lo Giudice R., Pollio A. and Garric J. (2003). Ecotoxicological impact of pharmaceuticals found in treated wastewaters: study of carbamazepine, clofibric acid, and Diclofenac. *Ecotoxicology and Environmental Safety*, **55**, 359–370.

Gaser G., Rona M., Voloshenko A., Shelkov R., Lev O., Elhanany S., Lange F. T., Scheurer M. and Pankratov I. (2011). Evaluation of micropollutant tracers. II. Carbamazepine tracer for wastewater contamination from a nearby water recharge system and from non-specific sources. *Desalination*, **273**(2–3), 398–404.

Giger W., Schaffner C. and Kohler H. P. E. (2006). Benzotriazole and tolyltriazole as aquatic contaminants. 1. input and occurrence in rivers and lakes. *Environ. Sci. Technol.*, **40**, 7186–7192.

Göbel A., Thomsen A., McArdell C. S., Joss A. and Giger W. (2005). Occurrence and sorption behavior of sulfonamides, macrolides, and trimethoprim in activated sludge *Treatment. Environ. Sci. Technol.*, **39**(10), 3981–3989.

Grünheid S., Amy G. and Jekel M. (2005). Removal of bulk dissolved organic carbon (DOC) and trace organic compounds by bank filtration and artificial recharge. *Water Research*, **39**(14), 3219–3228.

Heberer T., Reddersen K. and Mechlinski A. (2002). From municipal sewage to drinking water: fate and removal of pharmaceutical residues in the aquatic environment in urban areas. *Wat. Sci. Techn.*, **46**(3), 81–88.

Isidori M., Lavorgna M., Nardelli A., Pascarella L. and Parrella A. (2005). Toxic and genotoxic evaluation of six antibiotics on non-target organisms. *Science of the Total Environment*, **346**, 87–98.

Joss A., Keller E., Alder A. C., Göbel A., McArdell C. S., Ternes T. and Siegrist H. (2005). Removal of pharmaceuticals and fragrances in biological wastewater treatment. *Wat. Res.*, **39**, 3139–3152.

Khetan S. K. and Collins T. J. (2007). Human pharmaceuticals in the aquatic environment: A challenge to green chemistry. *Chem. Rev.*, **107**, 2319–2364.

Krauss M. and Hollender J. (2008). Analysis of nitrosamines in wastewater: exploring the trace level quantification capabilities of a hybrid linear ion trap/orbitrap mass spectrometer. *Anal. Chem.*, **80**, 834–842.

Krauss M., Longrée P., van Houtte E., Cauwenberghs J. and Hollender J. (2010). Assessing the fate of nitrosamine precursors in wastewater treatment by physicochemical fractionation. *Environ. Sci. Technol.*, **44**, 7871–7877.

Kümmerer K. (ed.) (2008). Pharmaceuticals in the Environment, Sources, Fate, Effects, and Risks. 2nd edn, Springer. ISBN 9783540746638.

Levantesi C., La Mantia R., Masciopinto C., Bockelmann U., Ayuso-Gabella M., Salgot M., Tandoi V., Van Houtte E., Wintgens T. and Grohmann E. (2010). Quantification of pathogenic microorganisms and microbial indicators in three wastewater reclamation and managed aquifer recharge facilities in europe. *Science of the total Environment*, **408**(21), 4923–4930.

Maeng S. K., Sharma S. K., Lekkerkerker-Teunissen K. and Amy G. (2011). Occurence and fate of bulk organic matter and pharmaceutically active compounds in managed aquifer recharge: a review. *Water Research*, **45**, 3012–3033.

Markus M. R. (2009). The groundwater replenishment system. *Journal American Water Works Association*, **101**(2), 49–51.

Masciopinto C. and Carrieri C. (2002). Assessment of water quality after 10 years of reclaimed water injection: the nardò fractured aquifer (southern italy). *Ground Water Monitoring & Remediation, Winter*, **22**(1), 88–97.

Massmann G., Greskowiak J., Dünnbier U., Zuehlke S., Knapped A. and Pekdeger A. (2006). The impact of variable temperatures on the redox conditions and the behaviour of pharmaceutical residues during artificial recharge. *Journal of Hydrology*, **328**(1–2), 141–156.

Massmann G., Sultenfuss J., Dünnbier U., Knappe A., Taute T. and Pekdeger A. (2008). Investigation of groundwater a residence times during bank filtration in Berlin: multi-tracer approach. *Hydrological processes*, **22**(6), 788–801.

McArdell C. S., Alder A. C., Göbel A., Löffler D., Suter M. J. F. and Ternes T. A. (2006). Analytical methods. In: Human Pharmaceuticals, Hormones and Fragrances: The Challenge of Micropollutants in Urban Water Management, T. A. Ternes and A. Joss (eds), IWA Publishing, London, UK. ISBN: 1843390930.

MUNLV-NRW – Ministerium für Umwelt und Naturschutz, Landwirtschaft und Verbraucherschutzdes Landes Nordrhein Westfalen (2007). Eintrag von Arzneimitteln und deren Verhalten und Verbleib in der Umwelt. Literaturstudie – Fachbericht 2. Fachberichte des Ministeriums für Umwelt und Naturschutz, Landwirtschaft und Verbraucherschutz des Landes Nordrhein Westfalen, Landesamt für Natur, Umwelt und Verbraucherschutz (LANUV), in German.

Nielsen U., Pedersen B. M. and Halling-Sørensen B. (2007). Begrænsning af humane lægemiddelrester og antibiotikaresistens i spildevand med fokus på reduktion ved kilden. (limitation of pharmaceuticals and antibiotic resistance in sewage with focus on source reduction) miljøprojekt Nr. 1189, Danish EPA.

Oleksy-Frenzel J., Wischnack S. and Jekel M. (2000). Application of ion-chromatography for the determination of the organic-group parameters AOCl, AOBr and AOI in water. *Fresenius J. Anal. Chem.*, **366**, 89–94.

Oren O., Gavrieli I., Burg A., Guttman J. and Lazar B. (2007). Manganese mobilization and enrichment during soil aquifer treatment (SAT) of effluents, the dan region sewage reclamation project (shafdan), israel. *Environ. Sci. Technol.*, **41**, 766–772.

Patoli A. A., Patoli B. B. and Mehraj V. (2010). High prevalence of multi-drug resistant Escherichia Coli in drinking water samples from Hyderabad. *Gomal Journal of Medical Sciences*, **8**(1).

Plumlee M. H. and Reinhard M. (2007). Photochemical attenuation of N-nitrosodimethylamine (NDMA) and other nitrosamines in surface water. *Environmental Science & Technology*, **41**(17), 6170–6176.

Putschew A. and Jekel M. (2006). Iodinated X ray contrast media in "organic pollutants in the water cycle". In: T. Reemtsma and M. Jekel (eds), Wiley-VCH, ISBN 3-527-31297-8.

Putschew A., Wischnack S. and Jekel M. (2000). Occurrence of triiodinated X-ray contrast agents in the aquatic environment. *Science of the Total Environment*, **255**(1–3), 129–134.

Rauch-Williams T., Hoppe-Jones C. and Drewes J. E. (2010). The role of organic matter in the removal of emerging trace organic chemicals during managed aquifer recharge. *Water Research*, **44**, 449–460.

Reemtsma T., Miehe U., Duennbier U. and Jekel M. (2010). Polar pollutants in municipal wastewater and the water cycle: occurrence and removal of benzotriazoles. *Water Res.*, **44**(2), 596–604.

Reemtsma T., Weiss S., Müller J., Petrovic M., Gonzalez S., Barcelo D., Ventura F. and Knepper T. P. (2006). Polar pollutants entry into the water cycle by municipal wastewater: A european perspective. *Environ. Sci. Technol.*, **40**, 5451–5458.

Sacher F., Lange F. T., Brauch H. J. and Blankenhorn I. (2001). Pharmaceuticals in groundwaters: analytical methods and results of a monitoring program in Baden-Württemberg. *Germany J. Chromatogr.*, **A938**, 199–210.

Scheytt T. J., Mersmann P., Rejman-Rasinski E. and These A. (2007). Tracing pharmaceuticals in the unsaturated zone. *Journal of Soils and Sediments*, **7**(2), 75–84.

Schnoor J. L. (2009). NEWater future? *Environmental Science & Technology*, **43**(17), 6441–6442.

Steber J. and Hater W. (1997). Ökologische Beurteilung von Kühlwasserbehandlungsmitteln. *VGB Kraftwerkstechnik*, **77**, 976.

Steger-Hartmann T., Länge R. and Schweinfurth H. (1999). Environmental risk assessment for the widely used iodinated X-ray contrast agent iopromide (ultravist). *Ecotoxicology and Environmental Safety*, **42**(3), 274–281.

Steger-Hartmann T., Länge R., Schweinfurth H., Tschampel M. and Rehmann I. (2002). Investigations into the environmental fate and effects of iopromide (ultravist), a widely used iodinated X-ray contrast medium. *Wat Res*, **36**, 266–274.

Tao R., Ying G. G., Su H. C., Zhou H. W. and Sidhu J. P. S. (2010). Detection of antibiotic resistance and tetracycline resistance genes in enterobacteriaceae isolated from the pearl rivers in south China. *Environmental Pollution*, **158**(6), 2101–2109.

Ternes T. A. (2001). Analytical methods for the determination of pharmaceuticals in aqueous environmental samples. *Trends Anal. Chem.*, **20**(8), 419–434.

Ternes T. A., Bonerz M., Herrmann N., Teiser B. and Andersen H. R. (2007). Irrigation of treated wastewater in Braunschweig, Germany: An option to remove pharmaceuticals and musk fragrances. *Chemosphere*, **66**, 894–899.

Van Houtte E. and Verbauwhede J. (2008). Operational experience with indirect potable reuse at the flemish coast. *Desalination*, **218**(1–3), 198–207.

Water Research Foundation (2010). Removal and Fate of EDCs and PPCPs in Bank Filtration Systems, Order #3136, TZW.

WateReuse Report (2007). Reclaimed water aquifer storage and recovery. In: T. Clinton (ed.) 157, ISBN: 978-1-934183-03-8.

WateReuse Report (2008). Development of indicators and surrogates for chemical contaminant removal during wastewater treatment and reclamation. In: J. E. Drewes (ed.), 169 pages, 03-014-01.

Chapter 13

Water quality analysis – bulk organic compounds

Saroj K. Sharma, Mathias Ernst, Martin Jekel and Gary Amy

13.1 OVERVIEW OF SELECTED WATER QUALITY PARAMETERS

Removal of bulk organic compounds is one of the main concerns in treatment of wastewater treatment plant effluents for water reuse applications as it impacts the quality of the water for intended applications. The refractory organic compounds remaining after advanced water treatment are of special concern. The organic matter present in wastewater treatment plant effluents (commonly known as effluent organic matter, EfOM) is a mixture of (i) natural organic matter (NOM), (ii) soluble microbial products (SMPs), (iii) easily biodegradable organic matter and (iv) synthetic organic compounds (SOCs). The concentrations and proportions of each of these components in the effluent depends on the source of drinking water, type of drinking water treatment, water use in the service area and type of wastewater treatment. The EfOM that remains in reclaimed water is hypothesized to be comprised of sources such as NOM, SMPs and remaining anthropogenic trace organics (Drewes *et al.*, 2006).

EfOM is a disinfection by-product (DBP) precursor, exerts higher coagulant and oxidant demands, influences nitrification and denitrification processes as well as the removal of micropollutants by biodegradation. Some components of EfOM, for example, protein-like organic matter, are also responsible for the fouling of membranes (Lee *et al.* 2004; Sharma *et al.* 2009) and adsorbents. Furthermore, some aesthetic (color, taste, and odor) and operational problems (corrosion, regrowth) associated with NOM in drinking water also affect wastewater reuse applications. In general, EfOM affects essentially all chemical and biological processes in aquatic environments (Shon *et al.* 2006).

The fate of EfOM during different water treatment processes targeting water reclamation and reuse is not fully understood due to its complexity, heterogeneous composition, and varied sources. A better understanding of EfOM enables a water treatment plant receiving EfOM-impacted or -dominated waters to improve and/or select other treatment approaches to efficiently remove EfOM. Moreover, an understanding of EfOM characteristics contributes to better wastewater and drinking water treatment, proper effluent disposal and sustainable water management practices.

While both NOM and EfOM are complex, heterogenous mixtures, they can be characterised according to size, structure, functionality and reactivity. The characteristics of EfOM can be differentiated from drinking water NOM using several analytical methods: (i) specific UV absorbance (SUVA); (ii) size-exclusion chromatography in combination with a dissolved organic carbon detector (SEC-DOC), which is also known as LC-OCD (liquid chromatography with on-line organic carbon detection); (iii) fluorescence spectroscopy (excitation-emission matrix [F-EEM] and fluorescence index [FI]); and (iv) total nitrogen (TN) and dissolved organic nitrogen (DON). Characterization of organic matter in soil aquifer treatment (SAT) systems has revealed that the distinct properties of EfOM, as compared to NOM, are a lower SUVA, a higher FI value, a higher polysaccharides peak (PS) in SEC-DOC/LC-OCD, a clear protein-like peak in F-EEM, and a higher level of DON (Drewes *et al.* 2006).

Protocols for characterization of bulk organic matter fractions present in effluents from wastewater treatment plants and SAT effluents or reclaimed water were further developed in the EU project RECLAIM WATER. The protocols developed aimed at methods and steps to distinguish between EfOM and "background" NOM. The intermediate level protocols included measurement of SUVA, biodegradable dissolved organic carbon (BDOC) and mean oxidation number of carbon (MOC), whereas the advanced level protocol included F-EEM, SEC-DOC/LC-OCD and TN/DON. These protocols were applied to samples from laboratory-scale batch and column experiments (using primary and secondary effluents, with or without pre-ozonation) and to samples from case study sites by measuring DOC and TN concentrations, UVA_{254}, SUVA, F-EEM and SEC-DOC. The DOC, SUVA, TN and F-EEM/FI measurements were performed at UNESCO-IHE laboratories (The Netherlands) while the DOC and LC-OCD measurements were performed at TU Berlin (Germany). Furthermore, other project partners also measured some basic parameters like DOC, UVA and COD in their respective studies.

13.2 SAMPLING, STORAGE AND PROCESSING

All water samples for EfOM analysis should be handled properly and should be analysed within a short period of time. Except for few newly developed spectrophotometry-based (e.g., UV-Vis) technology for on-line monitoring, generally it is not possible to conduct detailed in-situ analyses of organic matter. Therefore, the representative samples must be taken carefully and transported to a qualified laboratory for analysis. Besides external contamination during handling, some components of EfOM are subject to slow biodegradation and hydrolysis. Rapid analysis should be envisaged specifically for samples rich in biodegradable organic matter. Prior to shipment, the samples should be kept at temperature of 4°C or below after sampling.

Specially cleaned glassware (preferably "soak cleaned") should be used with hard plastic screw cups and Teflon, polypropylene (PP) or polyethylene (PE) inlays. When dealing with samples with low TOC content, higher precautionary measures are required. For low-TOC waters, use of glass or Teflon sampling bottles are recommended, as some perfluoroalkoxy (PFA) products may leach monomers. Other precautionary measures include (i) proper labeling of the sampling bottles, (ii) rinsing the sampling bottle with the sample, where feasible, and (iii) not freezing the samples or treating with any additives or preservatives. Samples dedicated to dissolved organic analysis are generally filtered through 0.45 μm filter immediately after sampling.

It is necessary to use cooling boxes for the shipment of samples. The shipment method should consider the arrival time for the sample and the time between sampling and analysis. Non-cooled samples should be analysed within 24 hours after sampling (cooled: up to 72 hours). Another alternative is pasteurisation of closed sampling bottles at 70°C for 30 minutes.

These requirements for sampling and sample handling were strictly followed during the RECLAIM WATER project. The samples from different case study sites were collected during three to four different sampling campaigns and analysed for consistency and reproducibility of the results.

13.3 ANALYTICAL METHODS

The following parameters or methods can be applied for characterisation of EfOM.

(i) Total organic carbon (TOC) and Dissolved Organic Carbon (DOC)
(ii) Chemical Oxygen Demand (COD) (generally instead of or in absence of, TOC/DOC) and mean oxidation number of carbon (MOC)
(iii) UV Absorbance @ 254 nm (UV_{A254})
(iv) Specific UVA (SUVA = UVA_{254}/DOC)
(v) Differential UVA (ΔUVA)
(vi) XAD Resin Fractionation
(vii) Dissolved Organic Nitrogen (DON)
(viii) Liquid Chromatography with On-Line Organic Carbon Detection (LC-OCD)
(ix) Fluorescence Excitation Emission Matrix (F-EEM)

A brief description of the selected analytical methods applied during the RECLAIM WATER project is presented in the following paragraphs.

***TOC and DOC*:** TOC is defined as the sum of all organically bound carbon in dissolved and non-dissolved organic species. The first characterisation of NOM and EfOM is the measurement of non-purgeable total (unfiltered) and dissolved (0.45 μm) organic carbon by employing TOC analyzers, which provide a bulk measure of the amount of organic matter present. DOC can be measured using two categories of methods: combustion and wet oxidation. Generally, combustion methods are more accurate in DOC measurement than wet oxidation methods, specifically for seawater with high chlorine concentration. The most successful DOC analysis methods are persulfate oxidation, ultraviolet irradiation, and a combination of the two (Sharp, 1993), specifically for the determination of low concentrations (Dafner and Wangersky, 2002).

***COD and MOC*:** Sometimes COD is measured as a surrogate of TOC, when TOC analyzers are not available. The COD is used as a measure of the oxygen equivalent of the organic matter content of a sample that is susceptible to oxidation by the strong chemical oxidant potassium dichromate ($K_2Cr_2O_7$). The mean oxidation number of carbon

(MOC) is given by:

$$MOC = 4 - 1.5 \times \frac{COD_o}{TOC}$$

Where COD_O is the organic COD. The MOC value must always lie in the range (−4 to + 4), the possible range of oxidation numbers/states of C, for biochemical processes. Values outside this range are chemically impossible and indicate a problem during the analytical determination of COD or TOC. The MOC concept allows one to discriminate between oxidative/reductive chemical, and physical processes (Vogel *et al.* 2000).

***UV_{254} and SUVA*:** Measurement of UV absorbance of the filtered (0.45 μm) sample at 254 nm with a spectrophotometer provides a measure of the aromatic character of the sample. With these results, specific UV absorbance (SUVA = $UVA_{254} \times 100/DOC$) is calculated with a classification scheme of SUVA >4 L/mg-m for sources dominated by humic substances and SUVA sources <2 L/mg-m for sources dominated by non-humic material (Edzwald and Tobiason, 1999). Humic substances are more dominant in NOM while non-humic material is more dominant in EfOM.

***F-EEM*:** A fluorescence excitation-emission matrix (EEM) is developed by scanning over an excitation range of 240 to 450 nm by 10-nm increments and an emission range of 290 to 530 nm by 2-nm increments using a FluoroMax-3 spectrofluorometer (HORIBA Jobin Yvon, Inc., USA). The result is a three-dimension spectrum in which fluorescence intensity (arbitrary units) is represented as a function of excitation and emission wavelengths. There are three dominant peak areas observed in EEM: (i) humic/fulvic-like organic matter peak (at excitation = 330–350 nm and emission = 420–480 nm), (ii) humic-like organic matter peak (at excitation = 250–260 nm and emission = 380–480 nm) and (iii) protein-like organic matter peak (at excitation = 250–280 nm and emission = 280–350 nm) (Baker and Lamont-Black, 2001; Leenheer and Croué, 2003). EfOM, compared to NOM, typically shows more protein-like and less humic-like organic matter. Based on an EEM, a fluorescence index (FI) can be calculated by the ratio of fluorescence intensities at emissions of 450 and 500 nm at an excitation 370 nm. A higher FI (∼1.7 − ∼2.0) reflects organic matter of an autochthonous (microbial) origin while a lower FI (∼1.3 − ∼1.4) reflects organic matter of an allochthonous (terrestrial) origin. EfOM typically shows an autochthonous signature (higher FI) while NOM typically shows an allochthonous signature (lower FI) (Donahue *et al.* 1998; McKnight *et al.* 2001). Fluorescence is sensitive to factors such as pH, solvent polarity, temperature, redox potential of the medium, and interactions with metal ions and organic substances (Westerhoff *et al.* 2001; Leenheer and Croué, 2003).

During F-EEM analysis of the samples, a uniform method of sample preparation was adopted. Filtered water samples were diluted to 1 mg/L DOC with 0.01 N KCl, and adjusted to a pH of 3.0. The blank sample (0.01 N KCl) fluorescence was subtracted from all spectra.

***TN and DON*:** Measurement of dissolved organic nitrogen (DON) is accomplished by a two-step process, first involving elimination of dissolved inorganic nitrogen (DIN) by continuous dialysis through dialysis bags with a molecular weight cutoff (MWCO) of 100 Daltons, followed by direct measurement of DON as total nitrogen (TN) with a TN analyzer. Because the concentration of the inorganic nitrogen species (DIN) is often considerably higher than that of the organic nitrogen species, DON measurements are subject to substantial error. Pre-treatment such as dialysis can improve the accuracy of DON measurement, separating DIN species (nitrite, nitrate, and ammonia) from DON (Lee, 2005); however this method may not be very useful for DON measurement in wastewater due to matrix effects.

***LC-OCD*:** The DOC and the DON of the samples are analysed by LC-OCD-OND (Liquid Chromatography with Organic Carbon Detection and Organic Nitrogen Detection). This analyser is manufactured by DOC-laboratory Dr. Huber (Karlsruhe, Germany) (Huber, 2007). The LC-OCD-OND uses a packing column based on the principle of size exclusion chromatography (SEC). Large molecules cannot diffuse into the pores of the stationary phase and in the chromatogram appear before the smaller molecules. The stationary phase of the column consists of a porous gel based on polymers (Toyopearl®). It is a synthesis of ethylenglycol and methacrylate produced by Tosoh Bioscience. Because of its chemical structure the column material has hydrophilic as well as hydrophobic properties. The gel has a specific size exclusion characteristic for different molecule sizes. The manufacturer has investigated this size exclusion with polyethylenglycol, dextrane and globular proteins which serve as reference substances of different sizes. The LC-OCD-OND separates natural organic substances which are hydrophilic and only slightly acid, but of very different molecular sizes into different size fractions including biopolymers, humic substances (HS), building blocks (BB), and low molecular weight acids (LMWA). These individual peaks can be integrated and expressed as mg/L as DOC.

13.4 SELECTED RESULTS FROM LABORATORY AND CASE STUDIES

Details of application of various analytical methods and protocols developed for bulk organic compounds and applied to the samples from different case study sites are presented elsewhere in this book and in RECLAIM WATER (2008a&b). This section presents some selected examples of application of the different bulk organic matter analysis methods to illustrate how these methods can provide insight into the fate of bulk organic compounds during water reclamation.

13.4.1 Laboratory studies

DOC and SUVA

Soil column studies were conducted at UNESCO-IHE using 5-m long columns filled with silica sand (0.8–1.25 mm) and operated at a hydraulic loading rate of 1.25 m/day under different conditions. Different types of wastewater treatment plant effluents were applied to the soil columns and the removal of bulk organic compounds were monitored (after acclimation of the soil columns for about 60 days) using different analytical protocols. Table 13.1 summarises the average DOC and SUVA of primary, secondary, ozonated secondary and MBR effluents before and after soil columns simulating soil aquifer treatment (SAT) (Katukiza, 2006; Kenneth, 2007; Kahawita, 2008).

Table 13.1 DOC and SUVA values of different effluents before and after simulated SAT.

	Before SAT		After SAT	
Effluent applied	**DOC (mg/L)**	**SUVA (L/mg·m)**	**DOC (mg/L)**	**SUVA (L/mg·m)**
Primary Effluent	36.0	2.30	15.5	2.95
Secondary Effluent	11.6	2.90	9.7	3.40
Secondary Effluent + Ozonation (1 mg O_3: 1 mg DOC)	16.0	2.35[a] 1.05[b]	7.8	2.36
MBR Effluent	11.4	2.72	10.3	2.91

[a]before ozonation, [b]after ozonation.

Table 13.1 shows that for all the effluents tested, DOC levels decreased and SUVA values increased with soil passage, showing the preferential removal of non-humics over humics in the soil columns. The amount of DOC removed in the column is also an indicator of biodegradable dissolved organic carbon (BDOC), which was confirmed by a strong correlation between the DOC removal profile and biomass profile along the depth of the column. Furthermore, as expected, ozonation decreased the SUVA values of secondary effluent which increased again after the soil passage. The average SUVA value of the effluent from a soil column fed with ozonated secondary effluent was 2.36 L/m-mg, which is less than the SUVA value for effluent of soil column fed with un-ozonated secondary effluent. This implies that the ozonated water has a lower aromaticity and consists of lower molecular weight organic compounds.

COD and MOC

Figure 13.1 presents the COD profile along the soil column (operating with primary effluent) for filtered (using 0.45 μm filter) and unfiltered samples. The non-filtered COD removal was 86% while filtered (soluble) COD removal was 55%. The filtered and non filtered COD levels of the soil column effluents were 38 mg/L and 42 mg/L, respectively, implying that both physical straining and degradation processes are significant with respect to COD removal and the difference in effluent DOCs is small due to the absence of suspended solids. This is in agreement with Idelovitch *et al.* (2003) who found that there is virtually no difference between $COD_{non\text{-}filtered}$ and $COD_{filtered}$ after SAT with partially treated effluent from a wastewater treatment plant.

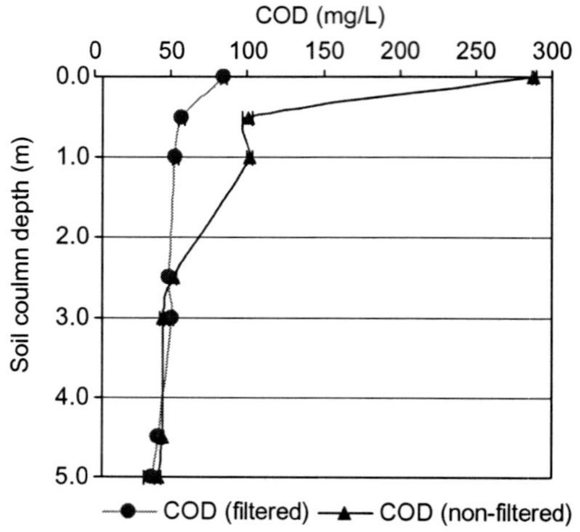

Figure 13.1 COD profile along the soil column when operated with settled primary effluent at HLR = 1.25 m/day under aerobic conditions

The mean oxidation number of carbon (MOC) profiles along the depth of the soil column at different HLRs for experiments with primary effluents are presented in Figure 13.2. These results show that MOC is in the range −3 to 1, which indicates that DOC removal during SAT is biochemical (oxidation) process (Sharma *et al.* 2007; Vogel *et al.* 2000).

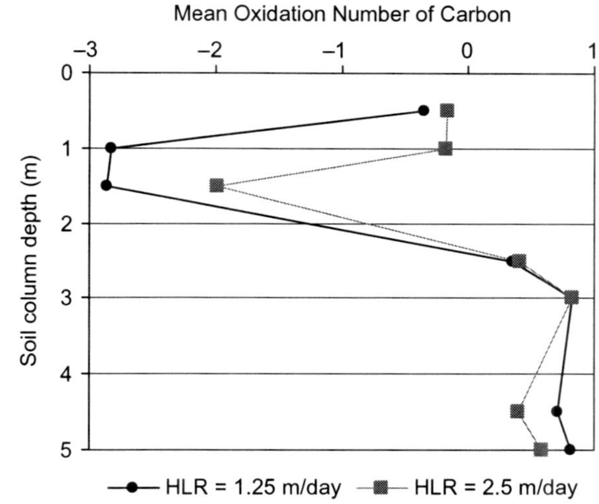

Figure 13.2 MOC profile along the soil column when operated with settled primary effluent at different HLR under aerobic conditions

F-EEM

Figure 13.3 shows typical F-EEM spectra of a secondary effluent before and after soil column passage. Both influent and effluent clearly show two characteristic peaks of humic-like and humic/fulvic-like organic matter fractions at excitation wavelengths of around 250 nm and 330 nm, respectively. The F-EEM did not clearly show any substantial protein-like peaks in the secondary effluent from the WWTP. This was attributed to the fact that either the protein content in the effluent was very low or due to that fact that this plant treats both domestic and industrial wastewater and the composition of the industrial waste could have limited the detection of the protein-like peak. Some protein-like peaks were, however, observed in some of the effluents from soil column, possibly due to the formation of soluble microbial products (SMPs) during the soil passage (in soil column which was operating for more than a year) (Westerhoff & Pinney, 2000).

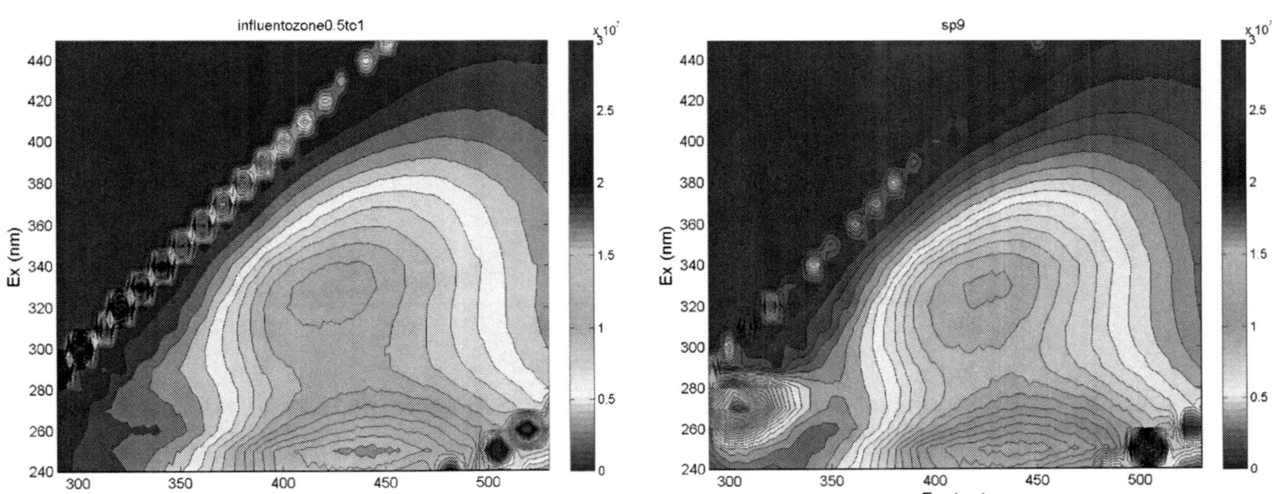

Figure 13.3 F-EEM of secondary effluent before and after SAT using soil column (5 m depth, HLR = 1.25 mg/L)

Table 13.2 presents the change in intensity of the humic-like peaks of the secondary effluent during soil passage under different process conditions. It shows that there is slight decrease in both humic peaks in all cases, likely due to some removal of humic-like fractions in soil columns. It was also observed that pre-ozonation decreased the peaks of humic like materials and the decrease was higher for a higher dosage of ozonation. These results clearly show that F-EEM can be used to fingerprint the organic matter characteristics and to analyse their fate during soil passage.

Table 13.2 Comparison of intensity of peaks of humic-like organic materials in secondary effluent before and after soil column experiments under different process conditions.

Water type	Influent Humic-like peak intensity ($\times 10^6$)		Effluent Humic-like peak intensity ($\times 10^6$)	
	Ex~250 nm	**Ex~330 nm**	**Ex~250 nm**	**Ex~330 nm**
Secondary effluent	29.20	19.50	25.10	16.80
Ozonated secondary effluent (0.5 mg O_3 : 1 mg DOC)	16.84	11.40	13.59	9.52
Ozonated secondary effluent (1 mg O_3 : 1 mg DOC)	4.07	2.77	4.74	3.39

LC-OCD

Soil column studies were conducted with secondary effluent under different redox conditions at TU Berlin using 1 m long columns filled with quartz sand (0.7–1.2 mm). Figure 13.4 shows the three chromatograms of (i) secondary effluent from the WWTP Berlin-Ruhleben, (ii) effluent of an aerobic column and (iii) effluent of an anaerobic column. It was observed that DOC removal was higher under aerobic condition than under anaerobic condition. Secondary effluent DOC consisted of large bio-available organic compounds (biopolymers) followed by humic substances (HS), precursor compounds of HS followed by low molecular weight acids (LMWA). About 19% of the total secondary effluent DOC was reduced during soil passage under anaerobic conditions, most of the organics in the range of biopolymers but also some HS and LMWA. For the same contact time, under aerobic conditions the DOC removal was about 37% with a complete removal of biopolymers and increased reduction for other smaller molecules.

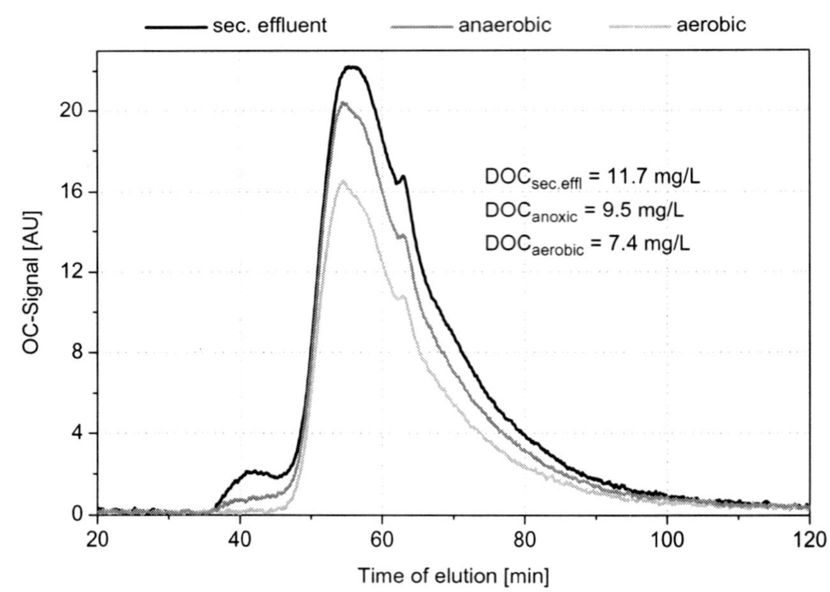

Figure 13.4 LC-OCD chromatograms of secondary effluent (WWTP Berlin-Ruhleben) and after an aerobic and anaerobic soil column (column length = 1 m; contact time = 6 days)

13.4.2 Case studies

The bulk organic compounds present in different samples from six case study sites were characterized by organic matter measurements as a bulk parameter and further by LC-OCD, TN and F-EEM measurements. As an example, Table 13.3 compares the DOC and TN concentrations of some selected samples of a case study site in Gaobeidian, China. The DOC removal after infiltration is about 60%. Analysis showed that background groundwater has much higher DOC than that of secondary effluent. Furthermore, the total nitrogen concentrations of secondary effluent and background groundwater at this site are comparable.

Table 13.3 DOC and TN measurement of samples from a case study site (Gaobeidian, China).

Parameter	Secondary effluent	After infiltration (SAT)	Background groundwater (local)
DOC (mg/L)	5.4	2.1	7.3
TN (mg/L)	32.2	28.2	29.6

Figure 13.5 presents the F-EEM spectra for these samples and the intensities of characteristic peaks in the spectra are presented in Table 13.4.

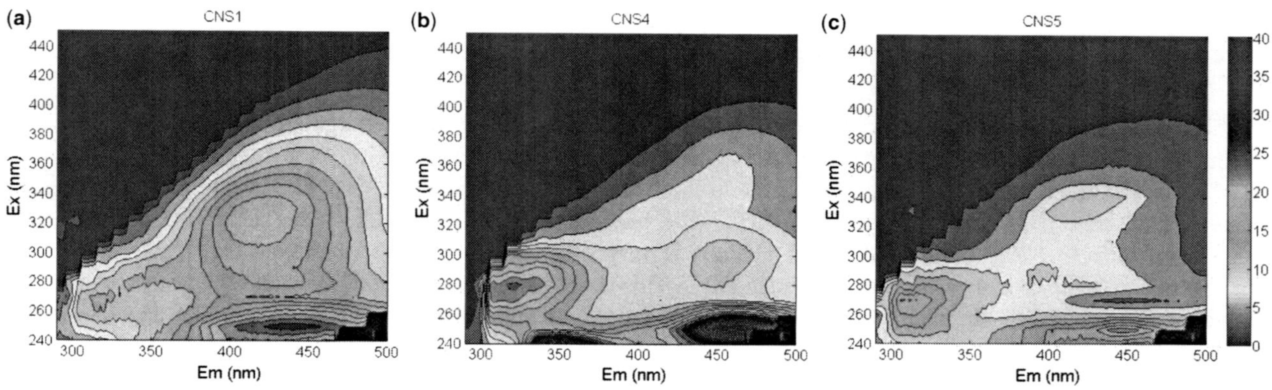

Figure 13.5 F-EEM spectra of samples from a case study site: (a) secondary effluent (b) after infiltration and (c) background groundwater

Table 13.4 Intensities of characteristic peaks in F-EEM spectra of samples from a case study site.

Sample	Humic/fulvic-like	Humic-like	Protein-like	FI
Secondary effluent	19.2	25.9	12.2	1.57
After infiltration	12.0	43.5	24.8	1.45
Background groundwater	11.9	23.4	19.7	1.55

Three characteristic organic matter fraction peaks were visible, in the F-EEM spectra of all the three samples (Figure 13.5). The humic/fulvic-like peak by decreased by about 30% whereas the humic-like peak (2nd) increased after SAT indicating a breakdown of larger humic/fulvic-like organic matter fractions or a peak shift. There was an increase in the protein-like peak after infiltration indicating some groundwater pollution, as background groundwater also has a high protein-like peak and TN.

SEC-DOC analysis of the secondary effluent and infiltrated water showed that while the biopolymers and humic substances (HS) are largely removed during soil aquifer treatment, the low molecular weight (LMW) acids appear to be persistent.

Summary of sample analyses from case study sites

The DOC concentrations of secondary effluents from the case study sites ranged from 5–15 mg/L, and in the final reclaimed waters (ranged from <1–5 mg/L. Average DOC removal rates for secondary effluent after soil passage under different treatment process conditions in the series Sabadell, Nardo, Shafdan (short-term SAT), Shafdan (long-term SAT), Beijing, Wulpen are 51%; 65%; 75%, 90%, 75%, 15%, respectively (Figure 13.6). These differences in organic carbon removal are mainly due to differences in the water reclamation treatment trains, but also differences in organic water matrix, managed aquifer systems and geo-chemical conditions.

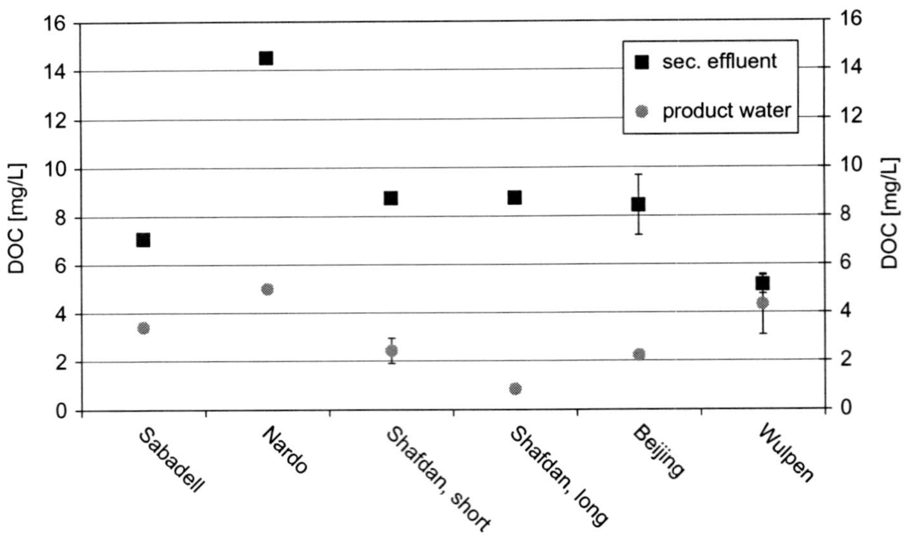

Figure 13.6 DOC of effluent and final groundwater at six different demonstration sites for groundwater recharge

F-EEM analysis of the influent (secondary effluents) from case study sites showed that all of the samples had some protein-like peaks, confirming the contribution of organic matter of wastewater origin in the samples. It was observed that in all of the cases more than 80% of the protein-like materials (peak intensity) were removed during soil passage (although DOC removals ranged from 15 to 90%). The peak removal of humic-like materials at the different sites ranged from 60% to 80%, showing that soil passage is very effective in removal of protein-like material and also removes other organic matter fractions substantially. This implies that soil passage would be very appropriate pre-treatment or complete treatment for water reuse applications as improved removal of protein-like and humic-like material means reduction in the fouling potential of membranes and distribution system as well as reduced DBP formation potential. In all cases the F-EEM plots of the final water were comparable to those of the native groundwater.

A decrease in FI was also observed during the treatment processes at all sites investigated. This shows that during SAT, the preferential removal of autochthonous organic matter (microbially derived, with higher FI) over allochthonous organic matter (from terrestrial plant origin, with low FI) takes place.

Figure 13.7 shows typical LC-OCD chromatograms of groundwater samples extracted from recovery wells at the six case study sites. The succession of DOC fractions versus the elution time of the size exclusion chromatography column shows, that even if the bulk parameter DOC is close to each other (like for Nardo, Beijing and Wulpen), the size exclusion "fingerprint" of each sample may vary largely. Due to matrix effects and total resolution, the specific identification of single organic compounds is not possible by LC-OCD. Nevertheless organic compound groups and their distributions in product waters can be identified. Next to organic carbon detection the parallel detection of UV 254 nm and total organic nitrogen offers further insights and allows tracking of unsaturated organic constituent as well as of protein like compounds.

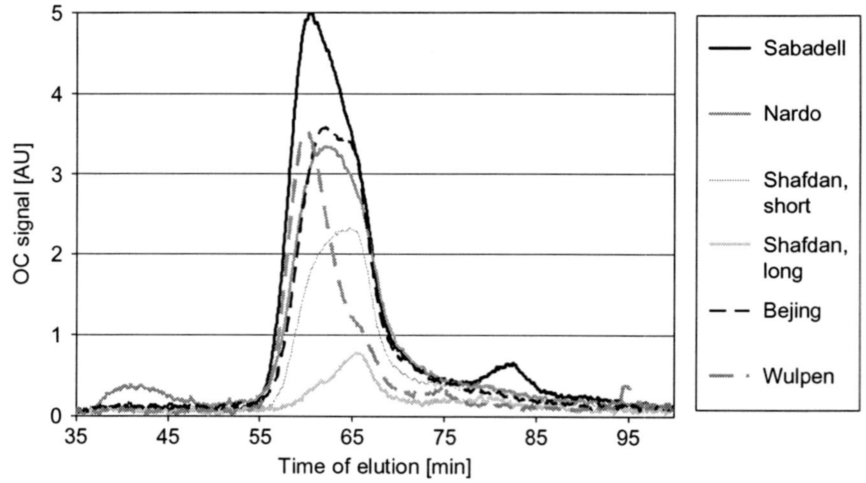

Figure 13.7 DOC characterisation by LC-OCD of abstracted groundwater from six SAT case study sites: Shafdan (short term SAT) and Shafdan (long term SAT)

In general, SEC-DOC/LC-OCD and F-EEM offer the potential to identify specific fractions originating from wastewater or natural water (effluent organic matter vs. natural organic matter) and describe their fate and behaviour during water treatment. Besides single compounds that could be identified as wastewater tracers such as carbamazepine and diatrizoate (see book Chapter 12), the analytical combination of SEC and F-EEM provides evidence of specific organic fractions originating from domestic wastewaters.

13.5 SUMMARY AND CONCLUSIONS

Analysis of bulk organic compounds in the samples of laboratory-scale batch and column studies and case study sites, by applying different analytical protocols, revealed the following:

(i) Soil passage preferentially removes non-humic over humic organic matter fractions, as shown by increased SUVA after soil passage.
(ii) The MOC concept can be used to verify that organic matter removal during soil passage is a biochemical process rather than adsorption.
(iii) Protein-like organic matter fractions and non-humic organic matter are preferably removed during soil passage. Humic-like organic matter fractions are also removed during soil passage to some extent.
(iv) EfOM exhibits lower SUVA values and higher FI values compared to NOM, indicating that the EfOM source is derived from autochthonous (microbial) sources.
(v) A distinct difference in F-EEMs between NOM and EfOM is the presence of a clear protein-like peak, which is present in EfOM, but not in (terrestrial) NOM.
(vi) EfOM contains a higher DOC content of biopolymers (or polysaccharide) constituents that exhibit non-aromatic and hydrophilic characters.
(vii) Compared to NOM, EfOM is enriched in DON. NOM and EfOM can be clearly discriminated in the chromatograms by content of organic carbon and nitrogen organic compounds.
(viii) The developed approach of using several state-of-the-art analytical techniques shows promise to reveal new insights into the removal mechanisms of organic matter and to serve as a tool to predict organic matter removal in SAT systems in an attempt to improve their design and operation. This was also proven from the analysis of different samples from case study sites.

ACKNOWLEDGEMENTS

This chapter summarizes the main activities carried out in EU RECLAIM WATER project under Work Package 3 "Analysis of Organic Compounds" for the development and implementation of bulk organic matter analysis protocols. We are thankful to all the project partners who contributed to this activity.

REFERENCES

Baker A. and Lamont-Black J. (2001). Fluorescence of dissolved organic matter as natural tracer of ground water. *Ground Water*, **39**(5), 745–750.

Dafner E. V. and Wangersky P. J. (2002). A brief overview of modern directions in marine DOC studies Part I—Methodological aspects. *J. Environ. Monit.*, **4**, 48–54.

Donahue W. F., Schindler D. W., Page S. J. and Stainton M. P. (1998). Acid-induced changes in DOC Quality in an experimental whole-lake manipulation. *Environ. Sci. & Technol.*, **32**(19), 2954–2960.

Drewes J. E., Quanrud D. M., Amy G. L. and Westerhoff P. K. (2006). Character of organic matter in soil aquifer treatment systems. *Journal of Environmental Engineering*, **132**(110), 1447–1458.

Edzwald J. K. and Tobiason J. E. (1999). Enhanced Coagulation: USA Requirements and a Broader View. Removal of Humic Substances from Water. IAWQ/IWSA Joint Specialist Group on Particle Separation, Trondheim, Norway.

Huber S. (2007). LC-OCD applications. DOC-Labor Dr. Huber, Germany (online). http://www.doc-labor.de/english_pages/What_is_LC-OCD_About/What_is_LC-OCD_about_2007-2.pdf

Idelovitch E., Icekson-Tal N., Avraham O. and Michail M. (2003). The long-term performance of Soil Aquifer Treatment (SAT) for effluent reuse. *Water Science and Technology: Water Supply*, **3**(4), 239–246.

Kahawita K. I. (2008). Effect of Advanced Oxidation Pre-treatment and MBR Pretreatment on Performance of Soil Aquifer Treatment Systems. MSc Thesis, MWI 2008–30, UNESCO-IHE, Delft, The Netherlands.

Katukiza A. Y. (2006). Effect of Water Quality and Process Parameters on Removal of Organic Matter During Soil Aquifer Treatment. MSc Thesis, MWI 2006-15, UNESCO-IHE, Delft, The Netherlands.

Kenneth M. (2007). Characterisation of Organic Matter Removal During Soil Aquifer Treatment and Analysis of the Effect of Pre-ozonation on Removal Efficiency. MSc Thesis, MWI 2007-19, UNESCO-IHE, Delft, The Netherlands.

Lee N., Amy G., Croue J.-P. and Buisson H. (2004). Identification and understanding of fouling in low-pressure membrane (MF/UF) filtration by natural organic matter (NOM). *Water Research*, **38**, 4511–4523.

Lee W. (2005). Occurrence, molecular weight and treatability of dissolved organic nitrogen. Doctoral Dissertation, Arizona State University, USA.

Leenheer J. A. and Croué J.-P. (2003). Characterizing Dissolved Aquatic Organic *Matter. Environ. Sci & Technol.*, **37**(1), 18A–26A.

McKnight D. M., Boyer E. W., Doran P. T., Westerhoff P. K., Kulbe T. and Anderson D. T. (2001). Spectrofluorometric characterization of aquatic fulvic acid for determination of precursor organic material and general structural properties. *Limnol. Oceanorg.*, **46**(1), 38–48.

Reclaim Water (2006). Analytical Procedure to Identify and Distinguish between EfOM and NOM. EU Reclaim Water Project, Deliverable no. D3.2.

Reclaim Water (2008a). Monitoring of Developed and Adapted Parameters. EU Reclaim Water Project, Milestone Report M 3.2.

Reclaim Water (2008b). Water Reclamation Technologies. EU Reclaim Water Project, Deliverable no. D1.2/D5.2.

Sharma S. K., Katukiza A. and Amy G. (2007). Effect of water quality and process parameters on the removal of effluent organic matter (EfOM) during soil aquifer treatment. *Proceedings of the ISMAR6 Conference (28 October – 2 November 2007)*, Phoenix, USA, 272–284.

Sharma S. K., Al-Sakkaf R. and Amy G. (2009). Effect of soil aquifer treatment (SAT) pretreatment of wastewater effluent on the performance of MF/UF/NF membranes. *Proceedings of 7th IWA World Congress in Water Reuse and Reclamation (20–25 September 2009)*, Brisbane, Australia.

Sharp J. H. (1993). The dissolved organic carbon controversy: An update. *Oceanography*, **6**(2), 45–50.

Shon H. K., Vigneswaran S. and Snyder S. A. (2006). Effluent organic matter (EfOM) in wastewater: constituents, effects, and treatment. *Critical Reviews in Environmental Science and Technology*, **36**(4), 327–374.

Vogel F., Harf J., Hug A. and Rohr P. R. V. (2000). The mean oxidation number of carbon (MOC)- a useful concept for describing oxidation processes. *Water Research*, **34**(10), 2689–2702.

Westerhoff P., Chen W. and Esparza M. (2001). Organic compounds in the environment fluorescence analysis of a standard fulvic acid and tertiary treated wastewater. *J. Environ. Qual.*, **30**, 2037–2046.

Westerhoff P. and Pinney M. (2000). Dissolved organic carbon transformations during laboratory-scale groundwater recharge using lagoon-treated wastewater. *Waste Management*, **20**(1), 75–83.

Part C

Water Reclamation Technologies in MAR

Chapter 14

Treatment trains utilising natural and hybrid processes

Saroj K. Sharma, Mathias Ernst, Arne Hein, Martin Jekel, Bruce Jefferson and Gary Amy

14.1 NATURAL SYSTEMS FOR WATER RECLAMATION

Natural wastewater treatment systems (soil/aquifer- or vegetation-based) rely on natural processes comprising different physical, chemical and biological removal mechanisms and combinations thereof, for improvement in water quality. Natural treatment systems like soil aquifer treatment (SAT) and constructed wetlands (CW) are robust barriers, can remove multiple contaminants, minimise the use of chemicals, use relatively low energy and have a small carbon footprint. These systems have been applied for wastewater treatment (as pre-treatment, main treatment or post treatment) and reuse in different parts of the world. These systems are very appropriate for developing countries and countries in transition and at the same time equally applicable in developed countries (Amy & Drewes, 2007; Sharma *et al.* 2008).

SAT is a sustainable and robust "engineered" water treatment as well as groundwater recharge and recovery technology in which treated wastewater effluents (secondary or primary) applied to the ground are infiltrated and further purified through physical, chemical and biological mechanisms taking place in unsaturated (vadose) and saturated (aquifer) zones during soil passage. It is a low cost and appropriate managed aquifer recharge (MAR) technology which can produce effluent of acceptable quality for indirect potable reuse (Sharma & Amy, 2010). SAT has been proven to provide efficient removal of contaminants at different scales (Nema *et al.* 2001; Idelovitch *et al.* 2003; Fox *et al.* 2006) and can be implemented as either a complete treatment (Jimenez & Chávez, 2004) or as a polishing step (Fox *et al.* 2006).

CWs are man-made analogs of natural wetlands that optimally exploit the biogeochemical cycles that normally occur in these systems for the purpose of wastewater treatment. Systems can be surface flow or subsurface flow (the latter one to avoid odour problems and mosquito proliferation) and are most commonly planted with emergent macrophytes such as reed, cattails and papyrus (Kaldec & Wallace, 2008). CWs can efficiently remove organic carbon, particulates and pathogens. CWs are attractive for wastewater treatment at a household or community level and at a larger scale also for the recovery of nutrients to minimize the eutrophication potential of the receiving water bodies. The technology can be operated with either vertical or horizontal flow which provides different treatment pathways. CWs are an established technology in many water reuse schemes, and in addition also offer ancillary benefits such as biomass production, habitat provision and landscape improvement (Rousseau *et al.* 2008).

Natural systems have a great potential for contaminant removal from wastewater effluents by sorption and biological activity for water reclamation and reuse. The suitability and performance of such natural treatment systems, however, depend on source water quality, process conditions applied, hydrogeological conditions and water quality goals to be achieved by treatment. It is expected that with proper design (including pre-treatment and post treatment) and further improvement of these systems, and especially by synthesizing suitable hybrids or integrates of SAT and CW, a comprehensive system for wastewater treatment and reuse can be developed which can be applied for treatment of different types of wastewater, at different-scales and in different regions of the world. Therefore, in the context of RECLAIM WATER project, several laboratory-scale batch and soil column studies as well as field studies were conducted by different project partners under different process conditions to obtain more insight of the removal capacities of these systems for different contaminants, their reliability as well as operational and economic aspects.

14.2 OVERVIEW AND METHODS OF NATURAL TREATMENT SYSTEMS RELATED STUDIES UNDER RECLAIM WATER

Table 14.1 gives an overview of the type of treatment trains utilising natural systems investigated in the context of the RECLAIM WATER project and the partners involved. An overview of the full-scale case studies and the applied scope of treatment is presented in Table 14.2.

Table 14.1 Natural treatment system trains investigated at laboratory and pilot-scale.

	Involved Reclaim Water partner			
Treatment Train	**CNR-IRSA**	**Cranfield University**	**UNESCO- IHE**	**TU Berlin**
Primary Effluent + SAT			✓	
Primary Effluent + Reed bed + SAT		✓		
Primary Effluent + Coagulation + SAT			✓	
Secondary Effluent + SAT	✓	✓	✓	✓
Secondary Effluent + Reed bed + SAT		✓		
Secondary Effluent + Ultrafiltration + SAT				✓*
MBR Effluent + SAT	✓	✓	✓	
MBR Effluent + Reed bed + SAT		✓		

*Data from the pilot site in Shafdan, Israel.

Table 14.2 Natural system based treatment trains investigated at different case studies.

Case study site Treatment Train	**Tula Valley (Mexico)**	**Nardò (Italy)**	**Sabadell (Spain)**	**Shafdan (Israel)**	**Atlantis (RZA)**	**Parafield (Australia)**
Primary effluent blended with freshwater + irrigation/infiltration (unintended SAT)	✓					
Secondary effluent blended with freshwater + SAT		✓	✓			
Secondary effluent + SAT				✓		
Secondary effluent + UF + SAT				✓		
Secondary effluent blended with stormwater + reed bed + SAT					✓	
Stormwater + constructed wetland + ASTR						✓

A summary of the experimental setup and methods used in by different partners in various studies are presented in the following paragraphs.

Column studies with fractured aquifer material conducted by Istituto di Ricerca sulle Acque (IRSA), Italy: The IRSA activities were aimed at exploring effective technologies to mitigate possible risks posed by chemicals and pathogens in reclaimed wastewater streams used for groundwater recharge into a fractured aquifer. An experimental investigation was carried out to evaluate the feasibility of fractured aquifer recharge adopting effluents from conventional activated sludge (CAS) in comparison with the effluent form a membrane bioreactor (MBR).

The local case study site (Nardò, Italy) refers to a fractured aquifer where treated municipal secondary effluents are collected and injected through a natural sinkhole, with an average flow rate of about 140–200 L/s. Water quality for the artificial aquifer recharge is particularly relevant since the overall equilibrium of the local water table is heavily affected by withdrawals for drinking and irrigation uses. In order to simulate the local fractured aquifer, laboratory tests were performed on two horizontal columns containing layers of limestone slabs and local soil. The number of slabs and the volume of soil spread on their surfaces were defined to ensure that in every cross-section of the filter the porosity was similar to the real fractured subsoil (5%).

Filter 1 was fed with the secondary effluent from the municipal wastewater treatment plant of Bari, Italy before the chlorination step; the city of Bari has two treatment plants and the effluent for this investigation was taken from the facility located in the west side of city and serving half of the city population plus the industrial area. Filter 2 was fed with effluent from a lab-scale membrane bioreactor. The latter consisted of a 6 L bioreactor equipped with a submerged Zenon hollow fiber microfiltration module, with a membrane surface of 0.047 m^2. The system was mechanically stirred and aerated in order to maintain residual dissolved oxygen concentrations above 2 mg/L and to favor COD removal and complete nitrification. The system was continuously fed on pre-settled municipal wastewater taken at the same municipal plant of Bari, screened at 1 mm, stored at 4°C and mixed.

Investigation of different pre-treatment options by Cranfield University, UK: Research activities at Cranfield University aimed to compare the effect of different pre-treatment options on the performance of constructed wetland and SAT system. For this purpose, a conventional activated sludge (CAS) process, a membrane bio-reactor (MBR) and vertical flow reed beds (VFRB) were operated as single pre-treatment units and also as part of treatment trains. The effluent from each unit or train was fed to SAT soil columns. Vertical flow reed beds (VFRB) are of particular interest as they are soil based treatment and hence represent a potential buffering step for SAT in regards to soil clogging and pollutants accumulation. Initially both MBR and CAS systems were operated at a sludge retention time (SRT) of 6 days and hydraulic retention time (HRT) of 8 hours. After some months, these systems are operated at a 20 days SRT and 8 hours HRT. The VFRBs are fed eight times per day for ten minutes at 125 mL/min for the ones following the CAS and the MBR systems, and 25 mL/min for the one fed with primary treated wastewater due to the higher TSS and BOD loadings in this type of effluent. The different sludge retention times tested in the CAS and the MBR enables the impact of upstream operating conditions to be assessed.

The SAT columns (1 m long, filled with aquifer materials from Shafdan, Israel) were operated under unsaturated conditions, at constant head of 5 cm with a cycle of 7 days drying and 7 days wetting periods, hence simulating realistic full- scale SAT operation.

Sand column studies conducted by UNESCO-IHE, Netherlands: The soil column setup at UNESCO-IHE consisted of 2 PVC columns of 54 mm internal diameter and 2.5 m length each connected in series, with 11 sampling points at intervals of 0.5 m. Silica sand of grain size 0.8–1.25 mm was used as filter media. The hydraulic loading rate (HLR) was controlled by adjusting the valves. For soil column studies, HLR varied from 1.25 m/day to 2.5 m/day. DOC, UVA_{254}, SUVA, oxygen and nitrogen profiles along the depth of the column were monitored. Each time after changing the process conditions, the soil column system was allowed to attain steady state with respect to DOC removal before taking samples to study the effect of a particular parameter.

In order to conduct batch experiments in the laboratory using filter media with biofilm (simulating soil aquifer treatment), bio-sand was prepared in the laboratory using 0.8–1.25 mm silica sand and secondary effluent from the wastewater treatment plant. This was done by putting 3 kg of silica sand in a 10-litre container and applying secondary effluent to this reactor. The secondary effluent in the reactor was continuously stirred without mixing the filter media. The secondary effluent was renewed by withdrawing the supernatant and feeding the reactor with a new batch of secondary effluent after every 4 days for 30 days. This was done to allow growth of heterotrophic bacteria (biofilm) on the media to produce bio-sand. Batch experiments were conducted by transferring 25 g of the bio-sand (prepared earlier) to each of triplicate 0.5-litre glass bottles that were later fed with primary or secondary effluent and aerated to obtain a dissolved oxygen concentration up to ~8 mg/L. The triplicate batch reactors were operated under similar conditions, anoxic and aerobic, as desired. Samples were taken periodically and DOC and UVA were measured

Sand column studies conducted by TU Berlin, Germany: Laboratory soil columns operated: length 1 m, diameter = 0.14 m, volume = 15 L. Material: industrial sand (SiO) with a particle size: 0.7–1.2 mm, for redox controlled conditions: hydraulic loading rate (flow, filtration rate, residence time: 2.4 L/day, 0.05 m/day, 20 days; 0.8 L/day, 0.17 m/day, 6 days; 0.24 L/day, 0.50 m/day, 2 days. For temperature controlled conditions: aerobic, 0.45 L/day, 0.08 m/day, 6 days; inlet water: continuously delivered secondary effluent of wastewater treatment plant (WWTP) Berlin Ruhleben (annual mean values: NH_4-N = 0.9 mg/L, NO_2-N = 0.2 mg/L, NO_3-N = 6.3 mg/L, DOC = 11.4 mg/L, TOC = 12.2 mg/L, pH 7.4, conductivity = 1159 μS/cm).

The results of case studies are presented in detail in Chapters 2 through 9 of this book. The parameters monitored in these investigations are shown in Table 14.3. This chapter mainly focuses on the results of the laboratory-based studies and summarizes the results obtained in terms of performance of different treatment trains for removal of different contaminants, and operational as well as economic aspects.

Table 14.3 Parameters monitored in laboratory and pilot-scale as well as in the case studies.

Parameter	CNR-IRSA	Cranfield University	UNESCO- IHE	TU Berlin	Tsinghua University	Case studies
TSS	✓	✓				✓
COD	✓	✓			✓	✓
DOC			✓	✓	✓	✓
Nitrogen	✓	✓	✓	✓	✓	✓
Phosphorus		✓		✓	✓	✓
Micropollutants		✓		✓	✓	✓
Heavy Metals						✓
Microorganisms	✓	✓				✓
Viruses	✓*					✓

*Data from the field-site in Nardò, Italy.

14.3 PERFORMANCE OF TREATMENT TRAINS

14.3.1 Bulk organics removal

In several studies, it was observed that natural systems required some ripening or acclimation time (ranging from 30 to 60 days) to achieve steady state with respect to DOC removal. This can be attributed to the development of a biofilm on the filter media. In general, natural treatment systems (SAT and CWs) were found to be effective in removing a considerable fraction (20 to 65%) of the bulk organic matter present in wastewater treatment plant effluents (primary and secondary), specifically the readily and slowly biodegradable organic matter fractions. The degree of bulk organics removal was influenced by hydraulic retention time and type of the effluent applied. With proper pre-treatment and by providing adequate HRT, comparable effluent DOC concentrations can be achieved for primary effluent and secondary effluent after soil passage. The coagulation of the primary effluent further improved bulk organic matter removal during SAT and consequently this pre-treatment can be applied to reduce the residence time if the other water quality parameters are comparable and cost considerations are favorable compared to SAT of secondary effluent. No significant DOC removal (about 10% only) was observed when MBR effluent was applied to soil columns or reed bed systems, indicating that MBR effluent is low in biodegradable DOC.

In general, biopolymers are completely removed by soil passage. DOC was only partially removed (biodegradable DOC only) by soil passage and under anoxic conditions the removal is relatively lower. DOC removal was dependent on the previous biological treatment (type of organic matter left in the effluent), being highest in primary effluent, comparable in secondary effluent and coagulated primary effluent and slightly lower for MBR effluent.

Primary effluent + SAT

Soil column studies were conducted at UNESCO-IHE (The Netherlands) to study the effectiveness of SAT under different water quality and process conditions. The primary and secondary effluents from a wastewater treatment plant located in Hoek van Holland, The Netherlands were used as influent for SAT columns. The DOC concentrations of primary and secondary effluents from the wastewater treatment plants were 35–37 mg/L and 12–15 mg/L, respectively.

SAT was effective in treatment of primary effluent from wastewater treatment plants (with respect to DOC removal), when adequate depth (travel time) is provided. It was also observed that both nitrification and denitrification can be achieved during SAT, if proper redox conditions and adequate depth is provided in soil columns (Katukiza, 2006; Hailu, 2008; Sharma *et al.* 2007). Figure 14.1 shows the DOC and UVA profile along the depth of the soil column fed with primary effluent at different loading rates after ripening. The effluent DOC concentration of the primary effluent increased from 15.2 mg/L to 18.4 mg/L when the hydraulic loading rate (HLR) was increased from 1.25 m/day to 2.5 m/day. The corresponding DOC removal rates were 54% and 46%, respectively. Thus, there was about an 8% reduction in DOC removal when the HLR was increased from 1.25 m/day to 2.5 m/day. However, when calculated in terms of total DOC loading on the column, DOC removal increased considerably. Furthermore, it was observed that effluent DOC at a HLR of 1.25 m/day was about 15 mg/L which is close to that obtained from conventional activated sludge systems (Sharma *et al.* 2007).

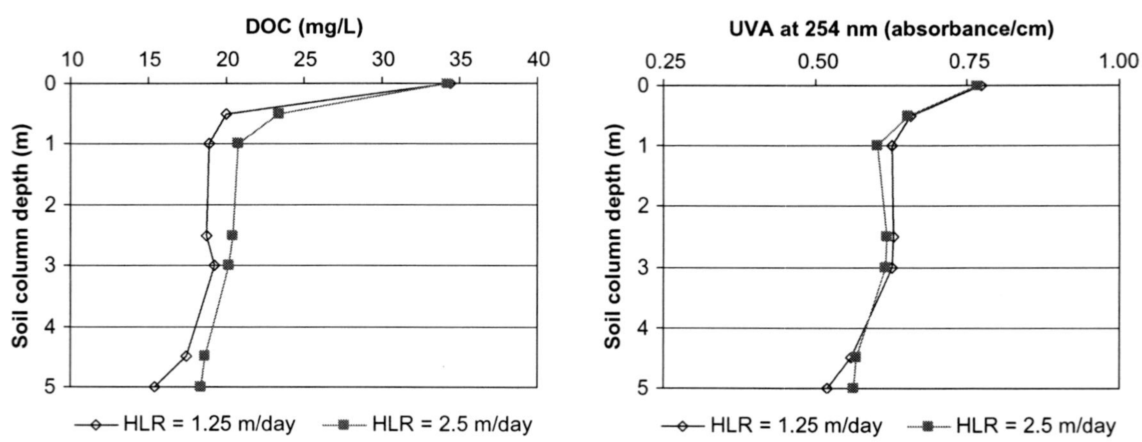

Figure 14.1 DOC and UVA profile along the depth of the soil column at different hydraulic loading rates with primary effluent under aerobic conditions (column depth = 5 m, filter media = sand 0.8–1.25 mm, temp = 20°C)

The overall trend regarding UVA is the same for both HLR values showing some decrease in the first 0.5 m of the column due to high removal of non-humic organic matter in the top layers. The SUVA values of the primary effluent increased from less than 2.3 L/mg-m to greater than 2.9 L/mg-m in first 1 m of the soil column. This indicates the preferential removal of non-humics in the top part of the soil column.

Figure 14.2 presents nitrate profiles along the depth of the column at the two HLRs. In the top one meter of the column there is nitrification of ammonium present, leading to an increase in nitrate concentration. As there is depletion of oxygen in the subsequent depth of the column, the denitrification of the nitrate starts. As expected, at higher HLR, nitrification and denitrification activities shifted deeper down in the soil column.

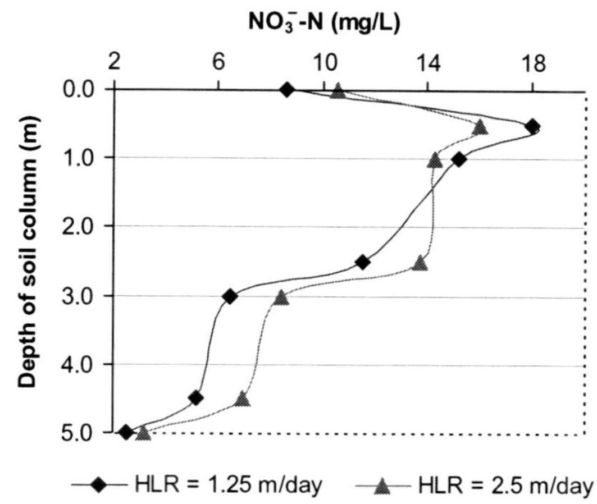

Figure 14.2 NO_3^--N profile along the soil column at different HLR when operated under aerobic conditions with settled primary effluent (column depth = 5 m, filter media = sand 0.8–1.25 mm, temp = 20°C)

Advanced primary effluent + SAT

In order to improve the performance of SAT systems using primary effluent, the effect of pre-coagulation of primary effluent on subsequent DOC removal during SAT was analyzed. Using the optimum dose of coagulant established with jar test (40 mg Fe^{3+}/L), the batch and soil column experiments conducted earlier were conducted with primary effluent and removal of DOC was monitored. It was observed that in batch studies the overall DOC removals were was 29%, 46% and 56% for secondary effluent (SE), primary effluent (PE) and coagulated primary effluent (C-PE), respectively. The DOC of the primary effluent decreased from 35.9 mg/L to 28.7 mg/L by coagulation which ultimately reduced to 15.4 mg/L at the end of batch test with acclimated sand (Hirpa, 2008).

Figure 14.3 presents the removal of DOC along the depth of the laboratory-scale soil columns for different types of wastewater treatment plant effluents. It was found that coagulation of primary effluent further improves bulk organic matter removal during SAT and consequently this pre-treatment can be applied to reduce the residence time if the other water quality parameters are comparable and cost considerations are favorable compared to SAT of secondary effluent. Furthermore, it was also observed that SAT is equally effective in treatment of coagulated primary effluent from wastewater treatment plants (with respect to DOC removal) as in the case of treatment of secondary effluent or MBR effluent, when adequate depth (travel time) is provided.

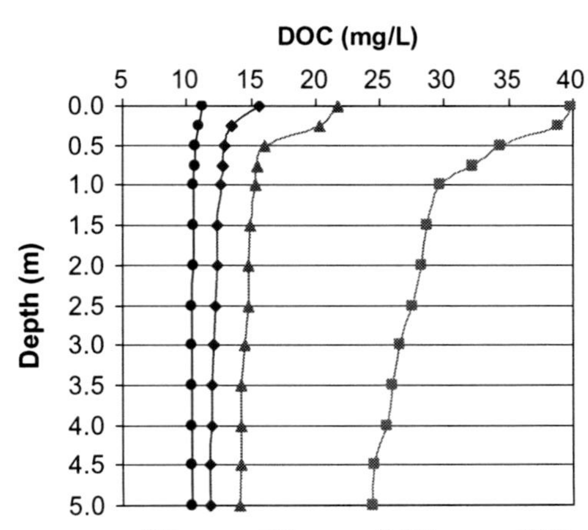

Figure 14.3 DOC removal in laboratory-scale soil column with different types of wastewater treatment plant effluent under oxic conditions (HLR = 1.25 m/day, depth of soil column = 5 m, sand media size = 0.8–1.25 mm, SE = secondary effluent, PE = primary effluent, C-PE = coagulated primary effluent and MBR = MBR effluent)

Secondary effluent + SAT

The use of secondary effluent from conventional activated sludge (CAS) treatments, possibly blended with freshwaters, for the artificial groundwater recharge is one of the most practiced options, as also extensively investigated within this project by several partners (IRSA, CRAN, UNESCO-IHE, TUB), with reference to the case studies at the sites of Nardò (Italy), Sabadell (Spain), Shafdan (Israel), Gaobeidian (China) and Atlantis (South Africa).

When monitoring the behavior of a horizontal column simulating a fractured aquifer (as found at the Nardò site), with reference to a static batch reference, it was observed that there was a positive effect of the fractured soil passage for the water quality improvements (Figure 14.4). As shown, in case of non-nitrified CAS effluent, the passage through a fractured aquifer SAT column with high HRT (20–90 days) led to the complete removal of the ammonium with no nitrate formation observed. It implies that both nitrification and denitrification process was taking place in the column and the low concentration of the nitrate observed during the test indicates that the rate of denitrification was higher than that of nitrification. The lower nitrification rate observed during wastewater filtration was likely caused by a reduction of dissolved oxygen observed in the column (from 1.1 at the inlet to 0.2–0.3 mg/L at the outlet), with respect to that measured in the batch tank with static (i.e., non-filtered) wastewater (1.1–1.3 mg/L).

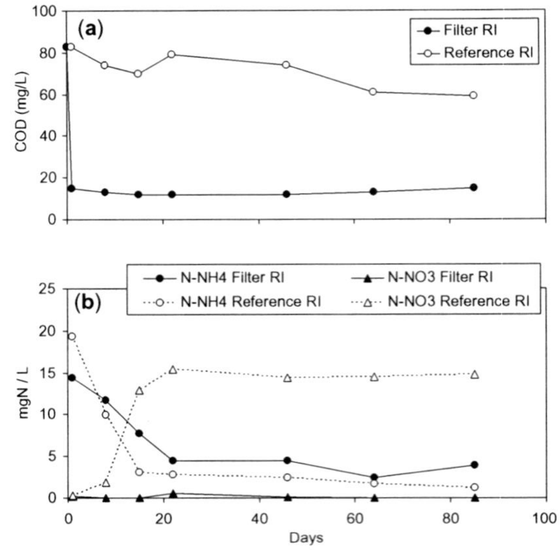

Figure 14.4 COD and nitrogen concentrations in filter (soil column) and reference batch fed with CAS effluent

Soil column studies at UNESCO-IHE showed that DOC removal from secondary effluent was about 30% at a HLR of 1.25 m/day (effluent DOC concentration = 10.3 ± 0.7 mg/L). When operated at a HLR of 2.5 m/day with the average influent concentration of 11.6 ± 0.3 mg/L, the soil column effluent DOC concentration was 9.7 ± 0.1 mg/L. In general, there was no significant change in % DOC removal and UVA along the depth of the soil column with secondary effluent when the hydraulic loading rate was increased from 1.25 m/day to 2.5 m/day (Figure 14.5) (Kenneth, 2007; Sharma *et al.* 2007; Sharma *et al.* 2008b). Fluorescence EEM analysis showed that SAT of secondary effluent can partially remove the bulk organic matter, preferentially removing non-humic fractions of the organic matter.

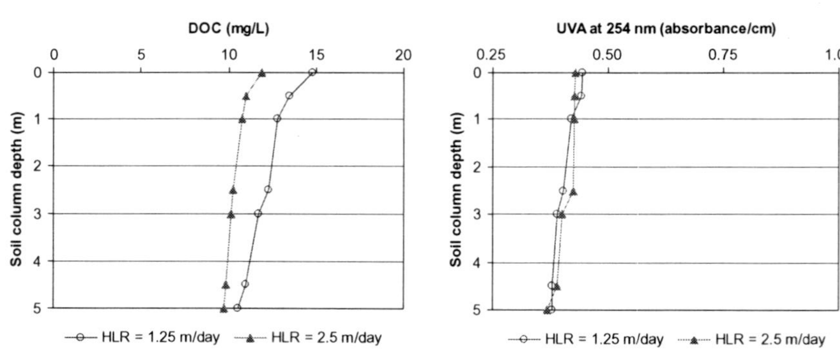

Figure 14.5 DOC and UVA profile along the depth of the soil column at different hydraulic loading rates with secondary effluent, under aerobic conditions (filter media = sand 0.8–1.25 mm)

Soil column studies at TU Berlin (using 1 m long soil column containing sand media of 0.7–1.2 mm size and HRT = 10 days) showed that DOC removal efficiency from the secondary effluent can be ranked in a series of aerobic column (R_{DOC} = 33%) > anoxic (R_{DOC} = 31%) > anaerobic (R_{DOC} = 16%). As given in Table 14.4, for temperature controlled soil columns (depth = 0.5 m, HRT = 6 days) best removal occurs at 15 °C (R_{DOC} = 28%), followed by the 25°C (R_{DOC} = 25%) and the 5°C (R_{DOC} = 19%) column as it was already observed for most of the antibiotics (see Figures 14.12 and 14.13).

Table 14.4 Mean DOC concentrations and removal (n = 28). SD: standard deviation.

	Influent mg/L	*SD*	Effluent mg/L	*SD*	Removal %
Aerobic			8.0	*0.7*	33.4
Anoxic	12.1	*1.5*	8.4	*0.9*	30.8
Anaerobic			10.1	*1.2*	16.3
5°C			9.8	*0.7*	18.5
15°C	12.1	*1.5*	8.7	*0.6*	28.2
25°C			9.1	*1.0*	24.8

Further characterization of DOC

DOC characterization by liquid chromatography organic (Haberkamp *et al.* 2007) carbon detection (LC-OCD) showed highest removal rates for the fraction of biopolymers (>90%) in the aerobic column but also relatively high removals for other redox conditions (Figure 14.6a). The fractions of humic substances and low molecular weight acids are removed partially (median removal between 30–42%) under aerobic conditions, but with lower efficiency at anoxic and anaerobic conditions (median removal between 12–22%). For temperature controlled columns best removal of DOC vs. molecular weight was observed at higher temperatures with a median maximum at 10°C (Figure 14.6b). The 5°C column reduced biopolymers, smaller molecules and humic substances less efficiently. Like the anoxic and anaerobic columns, the UV chromatograms of temperature controlled columns showed formation of UV active compounds (data not shown).

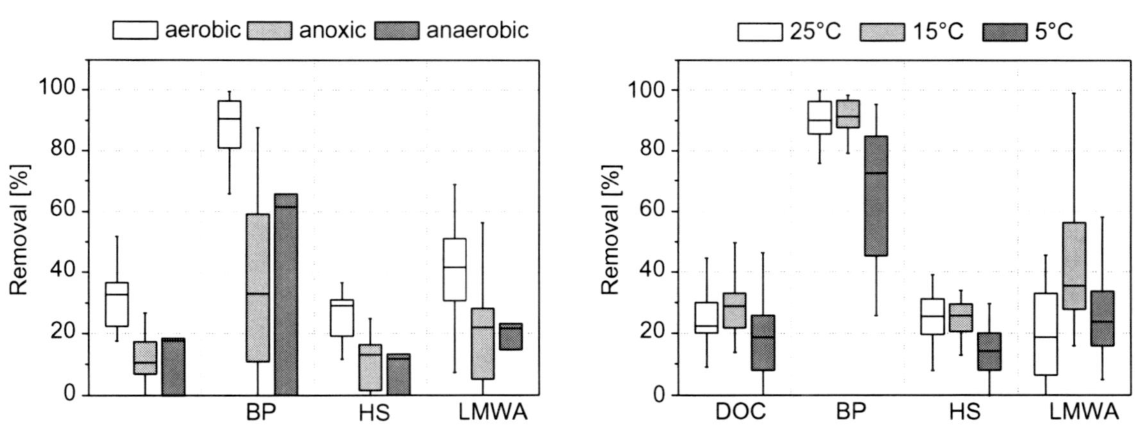

Figure 14.6 Removal of DOC fractions measured by LC-OCD (DOC = By-pass peak, BP = Biopolymer, HS = humic substances, LMWA = low molecular weight acids) in redox controlled soil columns (n = 19, box plots: Percentile 25–75, Outlier, Median); temperature: 20 ± 5°C as well as in temperature controlled soil columns (n = 19)

Comparison of laboratory column tests vs. field site Shafdan, Israel

The influence of DOC matrix, hydraulic retention time and groundwater recharge conditions can be identified by comparing laboratory soil columns experiments at TU Berlin (secondary effluent, DOC_0 = 11.4 mg/L, hydraulic retention time of 20 days) with the effluent of the WWTP Shafdan, Israel (DOC_0 = 9.3 mg/L, hydraulic retention time of 365 days). In the later case, effluent DOC can be removed by the unconfined aquifer by a median value of 93 % compared to 33% in Berlin. The highest removal rates can be seen in the biopolymer fraction, both for demo site (median = 98%) and lab site (90%). However humic substances are still removed to a large extent in Israel (median = 94%) compared to only 30% in the Berlin column plant. Also the differences of LMWA are high as for Shafdan median values of 93% are reached, compared to just 42% in Berlin (Fig. 14.7).

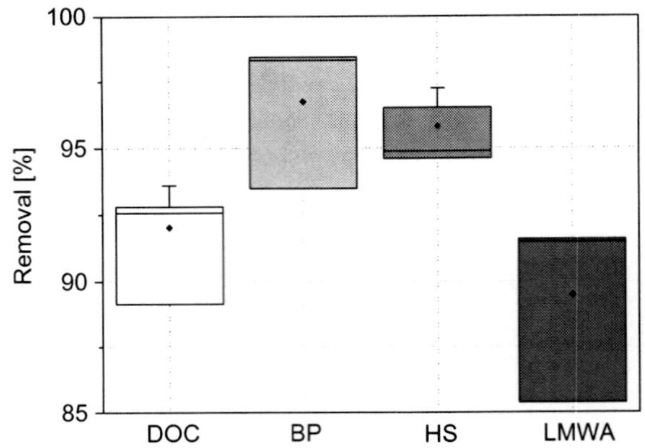

Figure 14.7 Removal of DOC fractions ($DOC_{effl.}$ *vs.* $DOC_{SAT,12M}$) measured by LC-OCD (DOC = bypass peak, BP = biopolymer, HS = humic substances, LMWA = low molecular weight acids, n = 3–4) at the demonstration site Shafdan, Israel [Box plots: Percentile 25–75, Outlier, Median, Mean (diamond)]

Studies at Cranfield University comparing the removal efficiencies of the different treatment trains showed that the best COD removal was achieved by the treatment train including CAS, reed bed and SAT column with 92% removal. The reed bed treating primary effluent followed by SAT column provided a slightly better COD removal (90%) than the treatment train composed of MBR followed by reed bed and SAT (89%). The COD removal for soil column fed with CAS and MBR effluent were <80% and 89% respectively. This shows that provision of reed bed before soil column increased the COD removal from CAS effluent.

Ozonation/Advanced oxidation + Secondary effluent + SAT

In order to further optimize the performance of SAT, the effect of pre-ozonation and advanced oxidation pre-treatment on SAT of secondary effluent was analysed by conducting lab-scale soil column and batch experiments at UNESCO-IHE (Kenneth, 2007; Kahawita, 2008). It was also observed that pre-ozonation (O_3:DOC = 1:1) of the secondary effluent transformed the refractory OM into readily (and slowly) biodegradable OM resulting in an increased DOC removal during SAT by 10 to 20% in 5 m long ripened soil columns operating at a HLR of 1.25 m/day. DOC removal efficiency increased with increased ozone dosage. Pre-ozonation can be used to oxidize target organic micropollutants (MPs). Pre-ozonation can provide additional carbon substrate and oxygen to drive heterotrophic micropollutant biodegradation (co-metabolism). It was found that the effect of advanced oxidation (peroxone) pre-treatment depends on the alkalinity of the water. For water with high bicarbonate alkalinity, there was no significant difference in DOC removal between pre-ozonation (O_3:DOC = 1 mg:1 mg) and advanced oxidation (peroxone, O_3:DOC:H_2O_2 = 1 mg: 1 mg: 0.5 mg). When the alkalinity of the secondary effluent was reduced, the DOC removal efficiency was higher with advanced oxidation pre-treatment as compared to pre-ozonation only (Sharma *et al.* 2008b).

Laboratory-scale studies, simulating the treatment technologies applied at the Gaobeidian (China) case study site, including coagulation, ozonation, slow sand infiltration and soil aquifer treatment were carried out at INET (China) to analyse the performance of the ozonation and subsequent soil columns treatment to optimize the operation strategy of the demonstration site. There were two sets of columns to simulate the artificial slow sand infiltration basin and soil aquifer treatment respectively. Each soil column system consisted of two filter columns (0.25 m in diameter with bed height of 1.5 m) and the HRT of columns was 20 days. The soil columns were filled respectively with quartz sand (0.5–1 mm) and aquifer material (<0.25 mm size), taken from 9–17 m deep underground at Gaobeidian site. The ozone dose applied was 10–15 mg/L. The influent to the system was secondary effluent from Gaobeidian WWTP polished with polyaluminium chloride (10–30 mg/L) coagulation. The parameters analysed included bulk organics (DOC, COD, AOC, BDOC, F-EEM, UV_{254}), N and P.

It was observed that ozonation can improve the biodegradability of effluent, by increasing AOC and changing the MW distribution. The simulated SAT showed efficient removal for the organic fraction of MW<1 kD, especially the <500 D fraction. Both soil and quartz sand columns demonstrated the favourable removal of organics (DOC, UV_{254}), NH_4-N and P. However, soil from Gaobeidian site showed higher removal efficiencies than the quartz sand. Ozonation followed by aquifer material column system removed 62% DOC, with the final effluent DOC of 1.80 mg/L. Ozonation followed by quartz sand column system removed 48% DOC, with the final effluent DOC of 2.46 mg/L. Removal efficiencies of the system with aquifer material column were 74% UV_{254}, 99% PO_4-P and 99% total phosphorus (TP), while the system with quartz sand columns removed 57% UV_{254}, 84% PO_4-P and 84% TP. The humic-like content in the effluent was effectively reduced by ozonation and soil treatment as demonstrated by EEM analysis. The humic-like content in the effluent from system with aquifer material column was close to that of local groundwater at Gaobeidian site.

MBR + SAT

The effectiveness of SAT of MBR effluent for water reclamation with respect to pathogen and organic micropollutant removals has already been discussed in previous sections. Batch-scale and soil column studies at UNESCO-IHE showed that MBR permeate was very attractive for aquifer recharge during SAT, with product water containing less refractory DOC than secondary effluent of conventional wastewater treatment plant. The average DOC concentration of the MBR permeates was 11.4 mg/L which decreased to 10.3 mg/L after SAT. With an increase of residence time in a 5 m soil column, from 96 hours to 160 hours (changing HLR from 1.25 m/day to 0.75 m/day) DOC removal was increased from 7.5% to 13.5% (Kahawita, 2008). As expected, DOC removal was higher under aerobic conditions as compared to anoxic conditions (Figure 14.8). A comparison of DOC removal in soil columns with different types of influent including MBR effluent is presented in Figure 14.3. It shows that of all the combinations tested, MBR effluent + SAT produces the effluent with lowest DOC concentration. Studies at IRSA also showed that MBR effluent is very suitable for recharge in fractured aquifer and has a very low suspended solids content and hence likely causes less particulate clogging.

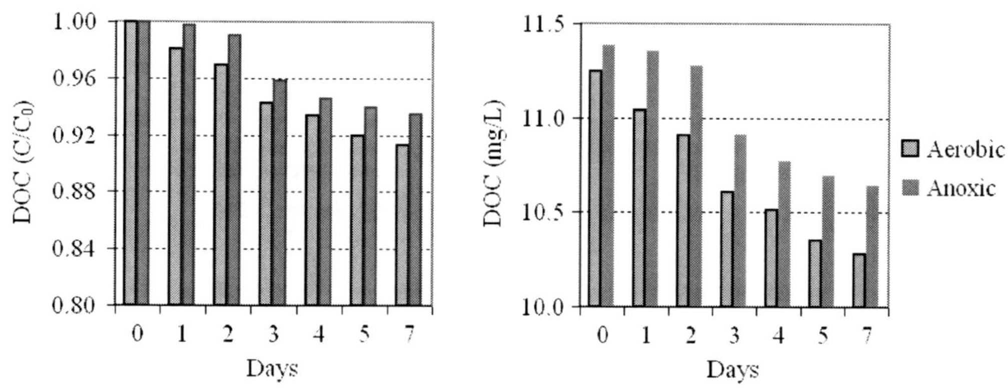

Figure 14.8 DOC removal from MBR permeate during batch tests under aerobic and anoxic conditions (biologically acclimated silica sand 0.8–1.25 mm) (Kahawita, 2008)

14.3.2 Nutrient removal

Both nitrification and denitrification was shown to occur in lab-scale soil columns, in the case of availability of oxygen and organic substrate respectively. Only limited phosphorus removal was achieved upon soil passage. Furthermore it was observed that the phosphorus content in water after soil passage is independent of the type of pretreatment (only primary, secondary or MBR).

Laboratory studies at UNESCO-IHE showed that both nitrification and denitrification can be achieved during SAT when proper process conditions are provided. When adequate depth and appropriate process conditions are provided, SAT is equally effective in removal of DOC and nitrogen species from primary effluent as in the case of treatment of secondary effluent (Sharma *et al.* 2007).

With regard to the analysis of nitrogen compounds, studies at Cranfield University showed that nitrification was similar in all trains. However when looking at the single treatment units, nitrification in the reed bed fed with CAS effluent was higher than in the reed bed fed with MBR effluent. Ammonia and total nitrogen (TN) were less removed and nitrification was lower in the reed bed fed with primary effluent than in the other reed bed trains, which is consistent with the fact that this train is operated as a secondary treatment unit (Figure 14.9).

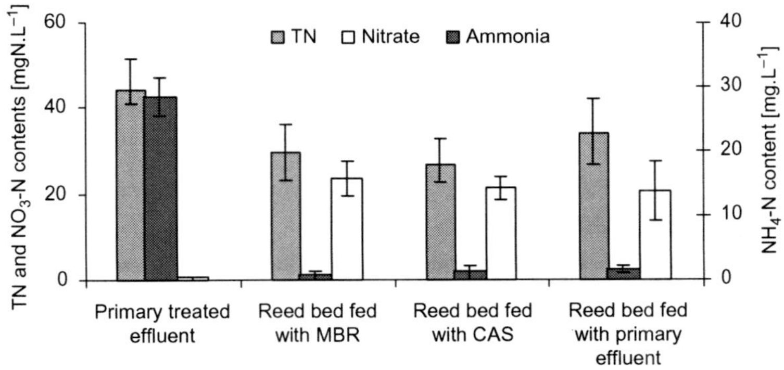

Figure 14.9 Concentrations of nitrogen compounds after reed beds (Besançon, 2010)

For treatments trains including SAT column, the treatment train including MBR resulted in the lowest effluent ammonia contents compared to the treatment train including CAS and reed bed treating primary effluent with average ammonia concentrations of 0.17, 0.35 and 0.84 mg N/L respectively (Figure 14.10). However, when comparing the nitrification of the different SAT columns, the one fed with primary treatment followed by reed bed demonstrated the highest nitrification with 25% while nitrification in the column fed with effluent from the reed bed treating CAS or MBR was 12 and 8%, respectively.

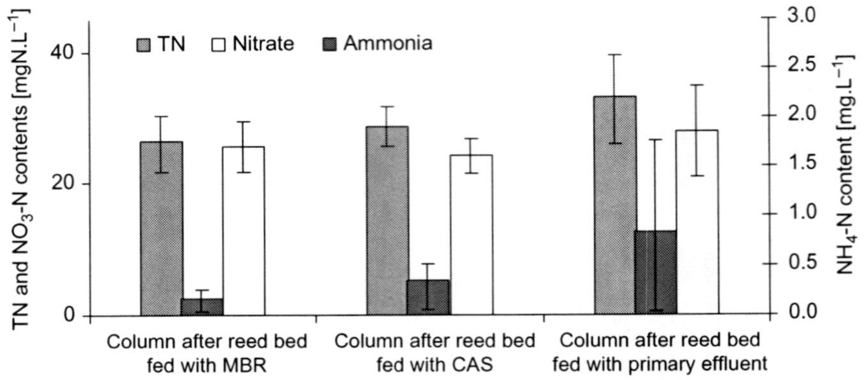

Figure 14.10 Concentrations of nitrogen compounds after SAT columns (Besançon, 2010)

Regarding phosphorus concentrations, no significant variations were found between the reed beds effluents. Similarly, phosphate contents did not vary significantly (Figure 14.11). However, the column fed with primary effluent followed by reed bed showed the highest average TP removal efficiency of 49%.

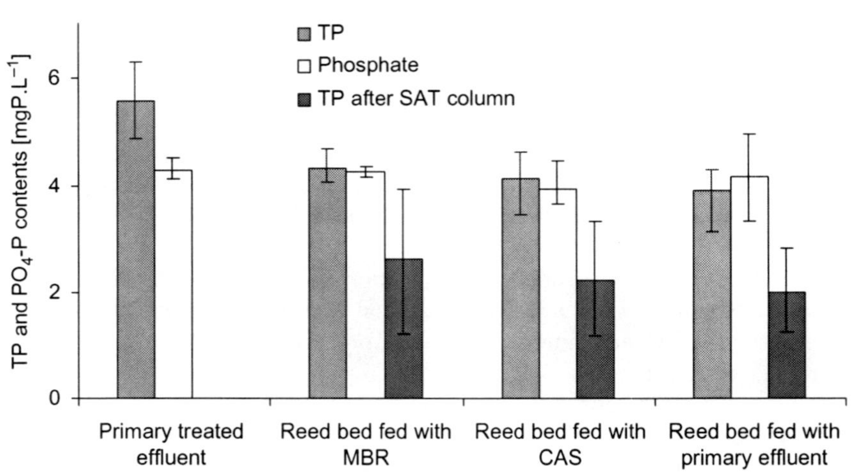

Figure 14.11 Concentrations of phosphorus compounds after reed beds and SAT columns (Besançon, 2010)

14.3.3 Organic micropollutant removal

There was a wide variation among different treatment trains in terms of organic micropollutant removal efficiencies, which was mainly influenced by the characteristics of the organic micropollutant, redox conditions applied as well as the temperature. It was observed that adsorption and bio-degradation were the two main mechanisms responsible for removal of organic micropollutant in natural treatment systems.

Fate of antibiotics during soil passage

Removal rates for macrolide and sulfonamide antibiotics as well as alterations in the size distribution of the dissolved organic matter were investigated in laboratory-scale sand column tests at the TU Berlin (Germany) using secondary effluent at different redox conditions (aerobic, anoxic, anaerobic) and temperature conditions (5°C, 15°C, 25°C). The macrolides roxithromycin (ROX), clarithromycin (CLA), dehydro-erythromycin (ERY-H_2O) and sulfonamides, sulfamethazine (SMZ) and sulfamethoxazole (SMX), plus trimethoprim (TMP) were chosen due to their refractory character and frequent occurrence in mainly effluent impaired surface waters in concentrations ranging from 0.05 to 1.7 μg/L (Hirsch *et al.* 1999; McArdell *et al.* 2003).

Six columns were filled with quartz sand (particle size: 0.7–1.2 mm) and were operated in parallel under different redox conditions (aerobic, anoxic and anaerobic) and at different temperatures (5, 15, 25°C). Temperature controlled columns were running under aerobic conditions and were supplied by an aerated tank (stored at 5°C). The HRTs for the redox controlled columns (1 m length) and the temperature controlled columns (0.50 m length) were 10 days and 6 days, respectively. Secondary effluent from the WWTP Ruhleben (Berlin) was spiked with six antibiotics by a peristaltic pump into the influent to the columns at concentrations of 2–5 μg/L each. Before the spiking tests, sand columns were already running with secondary effluent for 6 months to establish a biomass in the column which is well adapted to the feed water. The adaptation process was monitored by DOC, NO_3 and ortho-PO_4 measurements. After the establishment of stable water quality conditions in the sand column effluent the feed water was spiked with antibiotics. Results of sand column studies are presented in Figures 14.12 and 14.13 as box plots (boxes: percentile 25–75, whisker: min and max, line: median). The relative high error bars for removal in redox controlled columns can be explained by seasonal variations of the feed water, which was taken monthly from WWTP Ruhleben and by unstable temperature in the lab (15–25°C).

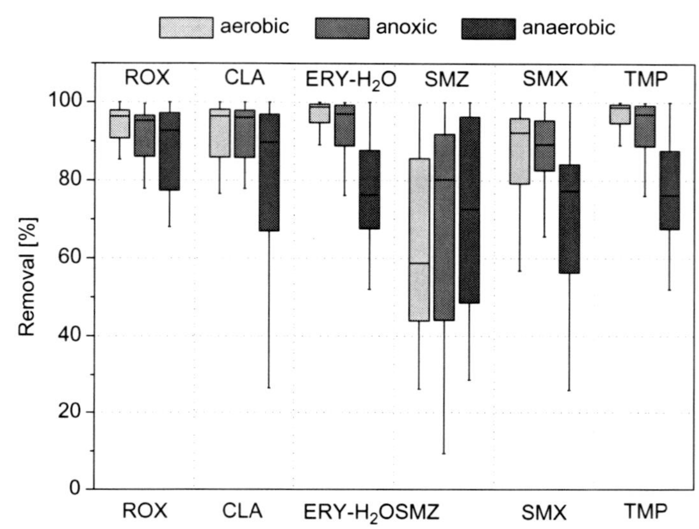

Figure 14.12 Box plots of removal of macrolides, sulfonamides and TMP in aerobic sand columns tests (secondary effluent + SAT) at different redox conditions (n = 18) (Column depth = 1 m; filter media = 0.7–1.2 mm sand, HRT = 10 days)

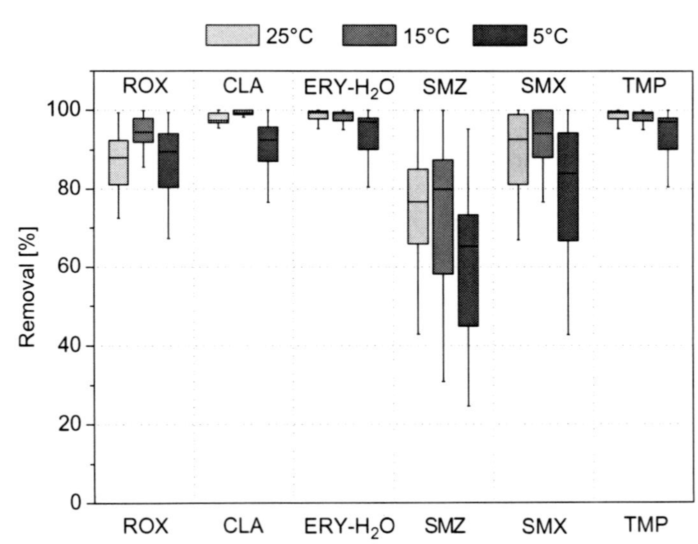

Figure 14.13 Box plots of removal of macrolides, sulfonamides and TMP in sand columns tests conducted at TU Berlin (secondary effluent + SAT) at different temperature conditions (n = 18) (Column depth = 1 m; filter media = 0.7–1.2 mm sand, HRT = 6 days)

For different redox conditions in the sand columns variations between the two classes of antibiotics were observed (see Figure 14.12). All macrolides and TMP are nearly completely eliminated (median removal rates $R > 90\%$), with marginal differences between aerobic and anoxic conditions. Under anaerobic conditions the median removal is limited to approx. 75% for ERY-H_2O and TMP; the two sulfonamides SMZ and SMX are removed with lower removal rates (Figure 14.12) compared to the macrolides. Generally, SMX is more effectively removed (85–99%) than SMZ (55–80%) and best under aerobic conditions. In contrast to the other investigated antibiotics SMZ is removed clearly better in the anoxic and anaerobic columns than in the aerobic column. But the removal of SMZ is highly variable under all conditions. In general the removal in the anaerobic column varies widely due to unstable anaerobic conditions during the testing period, as shown by NO_3- measurements.

The observed removal rates for macrolides and SMX in redox controlled columns do not correspond with other tests conducted with soil columns within the RECLAIM project by DHI and RWTH due to differences in setup and used sand material. However, Heberer *et al.* (2008) described also good elimination for macrolides by bank filtration and Jekel & Grünheid (2005) observed a better removal of around 30% for SMX under aerobic than under anoxic conditions in soil columns operating with surface water. The limited removal of SMZ was also observed by Hamscher *et al.* (2005) who showed that that SMZ leached into groundwater (>240 ng/L) from sandy soils where liquid manure was applied before.

Analysis with the aerobic temperature controlled columns (HRT = 6 days) showed that best removal of CLA, ERY-H_2O and TMP occurred at elevated temperatures (15, 25°C). At 5°C their removal rate decreased and was less stable. Clearly best removal could be observed for ROX and SMX at 15°C, compared to less removal at 5°C. With respect to results from redox controlled columns, SMZ is poorly eliminated compared to other antibiotics but with the same preference for 15°C. Preferred removal at 15°C indicates that the compounds are removed by both biodegradation and adsorption. Biodegradation increases with temperature while sorption decreases with temperature. Good removal at 15°C indicates that both processes are effective. Massmann *et al.* (2006) also showed that temperature influences the microbial activity in their effectiveness in degradation of some pharmaceuticals because the removal at an artificial recharge site was an order of magnitude higher in summer (T = 14–25°C) compared to winter (T = 0–14°C). Furthermore, Grünheid *et al.* (2008) also demonstrated comparable removal (>99% in 3 days) at 15°C and 25°C and slower degradation at 5°C (40 % in 3 days) for SMX in sand column studies with surface water.

In conclusion:

- Removal effectiveness via SAT depends on redox and temperature conditions. In general antibiotics can be removed via soil passage under all conditions but with different kinetics and efficiencies.
- Degradation is comparable under aerobic and anoxic conditions but lower under anaerobic conditions.
- Antibiotics are best removed at high temperatures; in the study best removal occurred at 15°C.
- Elimination rates observed via SAT are comparable to bank filtration field studies.

Micropollutants removal in reed beds

Studies at Cranfield University showed that micropollutants removal by reed beds depended on micropollutant concentration and on the amount of available organic carbon source. The removal efficiencies of diclofenac were coherent with the literature (Kimura *et al.* 2007). This can be explained by its polarity and hence the negligible sorption. Micropollutants like ibuprofen, naproxen and diclofenac were not detected in the solid phase after any of the studied treatment steps (Figure 14.15). On the contrary, tonalide and galaxolide were mainly transferred to the solid phase after treatment, except for CAS where the compounds were well distributed between the solid and the liquid phase. The best removals of the studied micropollutants were obtained with the CAS followed by reed bed with 83% removal for carbamazepine, 82% for galaxolide, 91% for tonalide and 91% for diclofenac (Figures 14.14 and 14.15). The reed bed fed with MBR effluent also provided good performances just after the reed bed with CAS effluent, and even better removal efficiencies for ibuprofen and naproxen, respectively with 99 and 93%.

14.3.4 Pathogen removal

In general, virus removal between 0 and 2 log units during soil passage was found in laboratory-scale experiments over the distance of 2 m. Observed bacteria removal varied between 1 and 5 log units depending upon the effluent water quality and process conditions applied. In case of soil columns fed with membrane-filtered effluents, heterotrophic plate counts increased during soil passage.

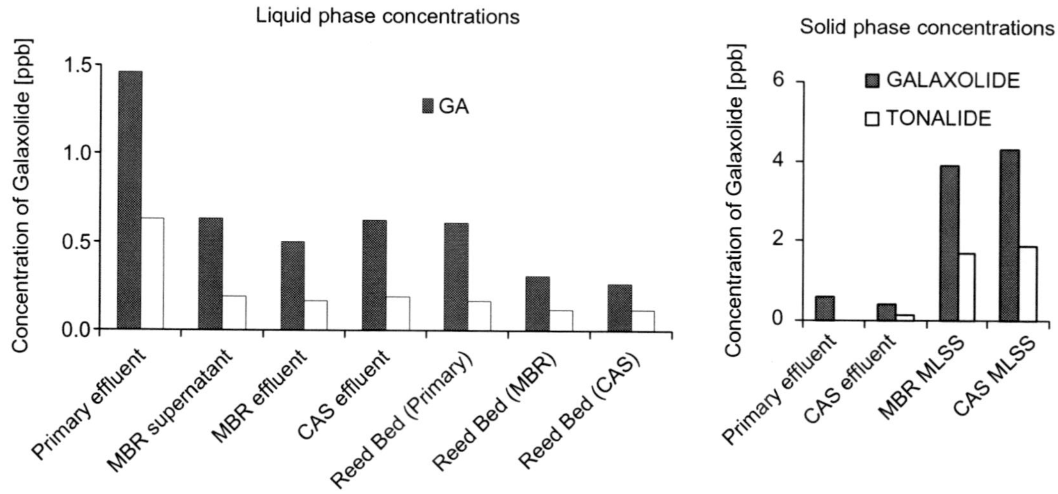

Figure 14.14 Galaxodine and tonalide concentrations in liquid and solid phases

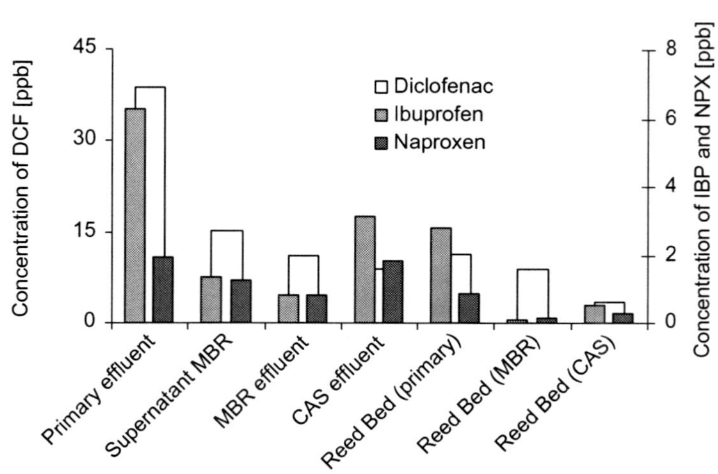

Figure 14.15 Concentrations of diclofenac, ibuprofen and naproxen in the effluents along the treatment trains

Pathogen removal in reed bed and during SAT

The effect of different pre-treatments on the performance of a constructed wetland in the form of vertical flow reed bed (VFRB) and SAT were studied at Cranfield University (UK). Some results of the studies conducted at Cranfield University on comparison of pathogen removal in natural treatment system are presented in Figure 14.16. The reed bed fed with primary effluent showed the lowest removals of *E. coli* and total coliforms compared to the other reed beds with log reduction typically below 2 (Figure 14.16a). As illustrated in Figure 14.16b, the column fed with MBR effluent followed by reed bed showed the best removal of *E. coli.* After all trains including SAT columns, *E. coli* and total coliforms were reduced, respectively, by 4.8 and 5 log units except for the column fed with CAS followed by reed bed which showed a 4.4 log reduction for both *E. coli* and total coliforms. It is to be noted, however, the reed bed degraded the MBR effluent quality for *E. coli* and total coliforms increasing the bacterial count by a factor 6.

Pathogen removal in fractured aquifers

Soil column experiments were carried out at IRSA (Italy) to evaluate the feasibility of fractured aquifer recharge, contrasting effluents from a CAS process in comparison with the effluent from a MBR. In order to simulate the local fractured aquifer (the Nardò site, Italy), laboratory tests were performed on two horizontal columns (2 m × 0.2 m) containing layers of limestone slabs and local soil. SAT tests with the CAS effluent showed that the pathogen indicators decreased rapidly and reached final concentration in the same range observed for the reference groundwater at the Nardò site (Table 14.5).

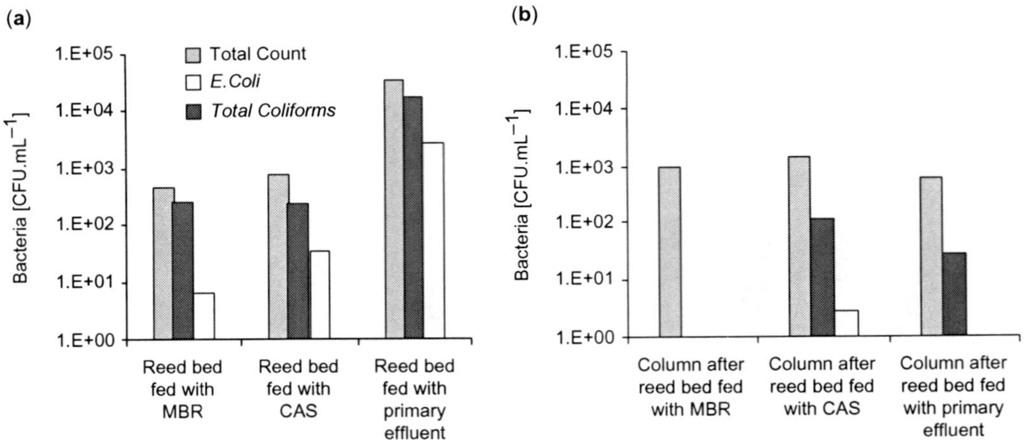

Figure 14.16 Bacteria content in reed bed (a) and SAT (b) effluents

Table 14.5 CAS effluent quality and SAT removal efficiencies.

Parameter	Unit	CAS effluent	CAS + SAT effluent	SAT removal	Nardò, GW background
TSS	mg/L	24	–	–	<2
COD	mg/L	65	12	82%	5
Ammonium	mg N/L	25	4.4	82%	<0.3
Nitrate	mg N/L	2	0.5	75%	40
Nitrite	mg N/L	<0.05	<0.05	–	0
Total bacterial count	CFU/mL	210000	390	3 Log	718
Faecal coliforms	MPN/100 mL	160900	5	5 Log	374
Enterococci	CFU/100 mL	260000	4	5 Log	161
Clostridium spores	CFU/100 mL	45000	190	2 Log	173
Somatic coliphages	PFU/100 mL	1200	1	3 Log	131

In terms of microbiological indicators, the MBR effluent quality was found suitable for direct injection without specific disinfection treatments (Table 14.6).

Table 14.6 Effluent quality after the biological treatments in comparison with the case study municipal effluent and groundwater reference.

Parameter	Unit	MBR effluent	MBR+SAT effluent	SAT removal	Nardò, GW background
TSS	mg/L	<2	–	–	<2
COD	mg/L	28	16	43%	5
Ammonium	mg N/L	<0.3	<0.3	–	<0.3
Nitrate	mg N/L	36	24	35%	40
Nitrite	mg N/L	<0.05	<0.05	–	0
Total bacterial count	CFU/mL	4100	1000	76%	718
Faecal coliforms	MPN/100 mL	4	0	100%	374
Enterococci	CFU/100 mL	0	0	–	161
Clostridium spores	CFU/100 mL	20	95	−375%	173
Somatic coliphages	PFU/100 mL	1.5	0.03	98%	131

14.3.5 Soil clogging potential

Clogging of the infiltration basin and the aquifer is one of the main concerns with respect to application of SAT for water reclamation and reuse. When primary effluent is used for SAT, some pre-treatment of the effluent would be necessary to reduce the clogging and rapid decrease of the infiltration rate. Even when secondary effluent is used, due to removal of organic polymers and suspended solids during soil passage, some biological or particulate clogging of infiltration basin and consequent decrease in infiltration rate is likely (Rice, 1974), which should be managed by properly maintaining wetting and drying cycle and well as periodic maintenance of the infiltration basin. No conclusive results or evidence regarding clogging potential was obtained during the study as most of the experiments were short-term. It is expected that some granular media filtration or membrane- based pre-treatment helps, avoiding soil colmation due to removal of particulate matter as well as precipitants.

Studies at Cranfield University showed that, in terms of TSS and turbidity removals, the reed bed fed with primary effluent proved to be less efficient than the other trains, leaving the effluents with 16.3 mg TSS/L and a 7.1 NTU turbidity compared to reed bed fed with MBR and CAS for which values of 7 mg TSS/L and 0.6 NTU were observed. Despite the variations in quality of the reed beds effluents, all SAT columns proved efficient in buffering these variations as for example solids contents in all columns effluents were below 1 mg TSS/L and turbidity below 0.9 NTU. The MBR effluent contained extremely low level of TSS (below the detection limit) and it was not been possible to estimate the clogging potential based on relatively short lab-scale tests. The column fed with CAS effluent showed the lowest infiltration rates from the early stages of operation. The infiltration rates in some soil columns decreased significantly over time, resulting in 42 to 58% reduction between the beginning and the end of the monitoring period. In case of columns fed with effluent from MBR and CAS after reed bed treatment, the reductions were relatively low.

Laboratory-scale investigations at IRSA showed that clogging of carstic aquifer sediments mainly due to filtration of inert and organic fractions which can be measured as total suspended solids, chemical precipitation playing a minor or irrelevant role. With CAS effluent, a clogging event (i.e., less than 1% of the original permeability) was predicted to occur after 10 years of operation, in good agreement with the trend observed at the real site, where clogging of the well next to the injection site occurred after about 16 years.

14.4 OPERATIONAL ASPECTS

14.4.1 Soil aquifer treatment

Soil aquifer treatment (SAT) of wastewater effluent is accomplished through infiltration via spreading wastewater on a basin, vadose zone injection well and direct injection methods. A widely used recharge method is rapid infiltration through spreading recharge basins; direct injection of reclaimed water is used when suitable land is not available (Fox *et al.* 2005) and/or where aquifers are confined. The advantages of surface spreading include: (a) groundwater supply may be replenished in the vicinity of metropolitan and agricultural areas where groundwater over-drafting is severe and (b) surface spreading provide added benefits of the treatment effect of soils and transporting facilities of aquifers (Asano and Cotruvo, 2004). The disadvantages include: (a) large land requirement and (b) clogging at the infiltration interface caused by biological, chemical and physical processes. A chemical process caused by algal photosynthesis and a change in the soil pH leads to precipitation of carbonate, gypsum, phosphorus and other chemicals in the soil cause clogging below the infiltration interface (Bouwer, 2002). The advantages of direct injection include: (a) small land requirement and (b) a confined aquifer can be recharged. The disadvantages include (a) high wastewater quality pretreatment requirement compared to surface spreading method making SAT system indirectly expensive and (b) requirement for frequent backwashing of the clogging layer which requires a dedicated pump in the well making the system more expensive.

The operation of a SAT system involves wetting and drying cycles. This is particularly important to avoid clogging of the system and at the same time influencing the redox conditions in the soil\layer and transforming nitrogen species (Amy & Drewes, 2007). At the basin-soil interface, the combined effect of sedimentation, filtration, aeration, and microbial growth leads to the formation of a biologically active zone that may become impermeable. The wetting and drying cycles which takes several days help to prevent clogging and therefore the basin-soil interface does not become impermeable. When the recharge basin dries, the clogging layer dries, decomposes, shrinks, cracks, and curls up, restoring its permeability (Fox *et al.* 2005; Bouwer, 2002). During the drying cycle re-aeration takes place and the redox potential increases. Wetting and drying cycles typically vary from 8 hours dry-16 hours flooding to 2 weeks dry-2 weeks flooding (Pescod, 1992). Ideally, SAT systems should have a number of basins so that some basins can be flooded while others are drying.

SAT technology makes use of soil to treat the reclaimed wastewater, the treatment process occurs through infiltration, soil percolation, and transport through the groundwater aquifer. During groundwater recharge through the vadose zone and transport through the groundwater aquifer, water quality improvements occur. Sorption and biodegradation processes in the soil are mechanisms that can reduce or remove microbial and other contaminants in wastewater. In case sufficient travel time or travel distance is provided, SAT is effective for removal of organic matter, trace organics, salts, nitrogen species, heavy metals as well as microbes (bacteria and viruses). Total suspended solids are typically 1 to 2 mg/L in the percolate from SAT systems as a result of filtration through soil layers. DOC removal efficiencies during SAT of primary, secondary and tertiary wastewater effluents were found to be in the range of 10 to 60%, 12 to 94% and 19 to 80%, respectively (Sharma *et al.* 2008). Both nitrification and denitrification can be achieved during SAT. However, nitrogen removal is a function of detention time, DOC to nitrogen ratio and redox conditions. Nitrogen removals of 38 to 83%, phosphorus removals of 29 to 97% and heavy metal removals of 50 to 90% have been observed in different field SAT sites. Pathogens are filtered out by the soil and adsorbed onto clay particles and organic matter. Bacteria and virus removals of 2 to 4 log have been reported (Crites *et al.* 2006). Furthermore, it has been observed that SAT is equally effective in treatment of primary effluent as in the case of treatment of secondary effluent. The implication is that, rather than just being a secondary effluent polishing step, SAT may also provide the equivalent of secondary biological treatment (Sharma *et al.* 2007).

It is to be noted that the range of removal efficiencies quoted above for DOC, nitrogen and microbes from different types of wastewater effluents are based on the field studies worldwide. The upper limit of the removal efficiencies can be achieved only under the optimum conditions with respect to travel time/travel distance, hydraulic loading rate and redox conditions. A travel distance of at least 10 m or a travel time of at least 10 days together with the optimal redox conditions is required to achieve average to optimal removal. Additionally, SAT is more effective for removal of bio-degradable DOC. Therefore, DOC removal from given wastewater also depends on the fractions of biodegradable and non-biodegradable DOC present in the effluent and process conditions applied. Furthermore, it should be clear that although some heavy metal and phosphorus are removed during SAT, it is not sustainable for long-term operation of SAT systems as the removal mechanism relies on adsorption.

Reliability, process complexity and flexibility

SAT is highly reliable and relatively simple to operate compared to conventional above-ground wastewater treatment and reclamation systems. The main requirements of the system are availability of land (or provision of a well) and a suitable aquifer. Furthermore, energy is required to pump the treated water out of the aquifer. SAT is a natural treatment system with least maintenance requirements. It does not require highly skilled technical personnel for operation. The frequency of wetting and drying cycles may be controlled according to the field situations. The main operational problems are related to clogging of the infiltration basin and surface cleaning.

SAT is flexible in terms of required raw water quality and pretreatment requirements. SAT can be used for treatment of primary, secondary or tertiary effluents. SAT is a robust system and is capable of coping with momentary and seasonal flow and quality variations. It can be adapted for changes in quantity or flow rates of the influent by changing the loading rate. Several pre-treatments such as sedimentation, pre-oxidation and filtration can be used to improve the quality of the SAT influent and to enhance the removals of different contaminants. The type and extent of pre-treatment and post treatment, however, depend on the intended application of the SAT effluent.

14.4.2 Constructed wetlands

Different types can be distinguished depending on the plant species (submerged, emergent or floating). Within subsurface treatment systems, further classification can be made with respect to horizontal, vertical flow and hybrid configurations, a combination of both aerobic and anoxic conditions in stages (Kadlec & Knight, 1996).

In horizontal-flow constructed wetlands, the wastewater flows slowly through the bed in a horizontal path until it reaches the outlet zone where it is collected. Within the bed, different conditions are present, aerobic zones can be found near reeds' roots and rhizomes (reeds allow oxygen transfer from leaves to roots), further in and away from the rhizosphere anaerobic bacteria are also present .Vertical flow constructed wetlands typically consist of a bed of gravels topped with sand where reeds are growing. The wastewater is dosed intermittently onto the bed flooding the surface. Then, water gradually drains down through the bed and is collected by a drainage network at the base. The bed drains completely free and this allows air to refill the bed. On the next flooding, air get trapped providing, with the help of aeration pipes and reeds roots, a good oxygen transfer and hence nitrifying conditions (Sundaravadivel & Vigneswaran, 2001).

The operational parameters of constructed wetlands are based on feeding regime, harvesting and resting time. But the critical operating parameter is the flow rate going through the bed as it controls the hydraulic retention time, the nutrients loading rate and the bed oxygenation levels. The determination of the constructed wetland area is based on the pollutant concentration which is the most difficult or important to remove. The wetland should then provide satisfactory treatment for all other parameters of concern.

A reed bed used as tertiary treatment showed benefits in terms of SAT soil clogging prevention. It also proved to enhance CAS effluent quality for nutrients, bacteria and micropollutants giving comparable results to a treatment train including MBR and reed bed. However, the reed beds had detrimental effects on MBR effluent for some parameters such as solids and bacteria. When looking at the SAT columns, only turbidity and ammonia removals were enhanced in the pre-treatment including reed beds compared to SAT infiltration of MBR and CAS effluents. As expected, the reed bed fed with primary effluent showed the lowest effluent quality for COD, nitrogen compounds, solids and bacteria as it is only a secondary treatment. However, phosphorous removal was enhanced in this reed bed because of a better nutrients balance. The reed beds showed a good buffering for nitrification, leveling the nitrate content between CAS and MBR effluents. Nevertheless, the reed bed reduced the MBR effluent quality for solids and bacteria, so that the reed bed fed with MBR had similar bacteria content to the reed bed fed with CAS. On the other hand, bacteria removal efficiency was promoted in the SAT column after the reed bed fed with MBR.

During 6 months of experiments at Cranfield University, no specific trend regarding clogging propensity of the columns was observed for the two trains involving MBR and CAS followed by reed bed when compared to infiltration of secondary effluents. However, the treatment train composed by MBR followed by reed bed and SAT would be a safer option for indirect infiltration as it was found that bacteria removal was promoted after the SAT soil columns. Considering reuse standards for indirect infiltration, TN, ammonia, micropollutants and coliforms levels in effluents after reed beds are higher than the required values. A longer SRT for the CAS and the MBR units or the addition of a denitrification step should solve these problems as the values are not much higher than the reuse standards.

Reliability, process complexity and flexibility

Constructed wetlands have a high buffering capacity providing a fairly stable effluent quality. However, reliability is medium and controllability low. Indeed, risks of failure may increase at low temperature, during peak flows, toxic discharge and if the matrix clogs. The complexity is low and does not require highly skilled operators. The degress flexibility is medium and depends on the initial design. Indeed, beds lay-out in parallel or series allows diverting flow and provides a multi-barrier system (Bixio & Wintgens, 2006). Furthermore, hybrid systems combining CW and SAT or other treatment technologies offer high potential for increasing removal efficiencies for different contaminants and improving reliability and flexibility

14.5 ECONOMIC ASPECTS

14.5.1 Soil aquifer treatment

Natural treatment systems are relatively cheaper than conventional wastewater treatment system provided the land is available at reasonable costs. The cost of a SAT system is site specific and depends on the cost of (i) pipelines from the wastewater treatment plant to SAT site, (ii) available land, and (iii) energy requirements for O&M systems (mainly for pumping water out of the extraction well). In general, the cost of SAT is relatively lower than that of conventional above the ground treatment system and its operation is simple and no chemical or expensive treatment units and equipment are required. Table 14.7 shows the indicative ratios of total annual costs of wastewater treatment technologies in comparison to SAT.

Table 14.7 Comparison of total annual costs for SAT and other technologies (Nema *et al.* 2001).

Treatment system	Indicative cost ratio (Based on total annual costs)
Soil Aquifer Treatment	1.00
Activated Sludge	1.65
Trickling Filter	1.70
Anaerobic Filter	1.35
Upflow Anaerobic Sludge Blanket (UASB)	1.17

A cost savings of 30% can be achieved by direct SAT treatment of primary effluent by bypassing the secondary treatment. SAT treatment is also expected to be less than 40% of the cost of treatment above ground with renovated effluent quality equal to or better than that of a conventional tertiary wastewater treatment plant depending on the characteristics of wastewater and also climatic conditions to facilitate primary clarification (Bouwer, 1991). Based on the experience in Dan region, Israel, typical capital costs of SAT (excluding storage and conveyance) are estimated to be 0.23 to 0.25 €/m^3 of water treated. The corresponding O&M costs are 0.10 to 0.15 €/m^3. The total capital costs of SAT system including pumping systems, storage systems, and pipelines to the infiltration fields is estimated to be 0.45 to 0.50 €/m^3 (Bixio and Wintgens, 2006). Furthermore, additional costs will be incurred in post-treatment of the SAT effluent, depending upon the intended application (irrigation. industrial or potable/non-potable municipal) and prevailing water quality guidelines and standards for such water reuse.

14.5.2 Constructed wetlands

The major investment cost is the land acquisition and the construction (digging, water proof layer to avoid groundwater contamination, sands matrix) (Bixio & Wintgens, 2006). Consequently, it depends highly on the location and the design (climate, square meter cost, terrain characteristics). For instance, horizontal flow reed beds require a higher footprint than a vertical flow configuration, but on the other hand they necessitate less matrix material and digging. Operation and maintenance costs are relatively cheaper for larger installations; it is also cheaper for free-water-surface than subsurface flow configuration. Maintenance cost is mainly due to labour cost for inspection, effluent quality control, pipe-network and pump maintenance and harvesting. The energy consumption is almost restricted to the pumping, and can be reduced using gravitational flow. The sludge production is highly depending on the pre-treatment; indeed wetlands used as tertiary treatment produce very little sludge. For an area between 0.5 and 20 ha, corresponding to a flow of 1000 m^3/d, CAPEX and annual OPEX of subsurface flow reed beds have been estimated respectively between 10 and 4000 k€, and 1 and 75 k€ (or 0.3 to 20 cents Euros per m^3) (Kadlec & Knight, 1996). The wetland matrix cost for a flow of 1000 m^3/d ranges between 70 and 3000 k€, considering a mix of gravel and sand. Compared to conventional activated sludge or MBR, constructed wetlands are cost competitive for installation from 50 p.e. (Reicherter, 1999). It is to be noted that total cost of treatment can be reduced with the additional benefits from valuable plants production.

14.6 CONCLUSIONS

Natural treatment systems like SAT and CWs are robust, environment friendly and have high potential for removal of bulk organic matter, nitrogen and microorganisms (pathogens) from wastewater treatment plant effluents (primary or secondary). These natural systems can also remove many organic micropollutants to some extent. The efficiency of these systems, however, depends mainly up on the effluent water quality and process conditions applied. The performance of these natural treatment systems can be further improved by providing proper pre-treatment like coagulation, sedimentation or ozonation.

It was observed that SAT is equally effective in treatment of coagulated primary effluent from wastewater treatment plants (with respect to DOC removal) as in the case of treatment of secondary or MBR effluent, when adequate depth (travel time) is provided. Furthermore, ozonation pre-treatment increased the DOC removal efficiency of SAT systems and redox condition was found to have significant effect on the removal of different types of organic micropollutants during soil passage. This study clearly showed that combination of CW and SAT improved removal of bulk organics, nutrients and organic micropollutants and reduced clogging potential. It is expected that hybrids or integrates of SAT and CW systems could be attractive and cost-effective technological options for wastewater treatment and reuse applications in developed as well as developing countries.

ACKNOWLEDGEMENTS

This chapter is based on the results of the soil column and batch studies and related activities carried out in EU RECLAIM WATER project under Work Package 1 "Assessment and development of water reclamation technologies" and Work Package 5 "Specific studies on removal and fate of microorganisms and organic compounds", which has been presented in detail in project Deliverable report no. D1.2/D5.2. We are thankful to all the project partners who contributed to this activity.

REFERENCES

Amy G. and Drewes J. (2007). Soil aquifer treatment (SAT) as a natural and sustainable wastewater reclamation/reuse technology: Fate of wastewater effluent organic matter (EfOM) and trace organic compounds. *Environ Monit Assess*, **129**, 19–26.

Asano T. and Cotruvo J. A. (2004). Groundwater recharge with reclaimed municipal wastewater: health and regulatory considerations. *Water Research*, **38**, 1941–1951.

Besançon A. (2010). Technologies and Multi-barrier Systems for Sustainable. Groundwater Recharge and Irrigation, PhD thesis Cranfield University, https://dspace.lib.cranfield.ac.uk/bitstream/1826/5726/1/Axelle_Besancon_Thesis_2010.pdf.

Bouwer H. (1996). Issues in artificial recharge. *Water Science & Technology*, **33**(10–11), 381–390.

Bouwer H. (2002). Artificial recharge of ground water: hydrogeology and engineering. *Hydrogeology Journal*, **10**, 121–142.

Bixio D. and Wintgens T. (2006). Water Reuse System Management Manual. *AQUAREC*, European Commission, Direcotrate-General for Research, Brussels.

Crites R. W., Joe Middlebrooks E. and Reed S. C. (2006). Natural Wastewater Treatment Systems. CRC – Taylor and Francis Publications.

Fox P., Houston S. and Westerhoff P. (2006). Advances in Soil Aquifer Treatment Research for Sustainable Water Reuse. American Water Works Association, USA.

Fox P., Aboshanp W. and Alsamadi B. (2005). Analysis of soils to demonstrate sustained organic carbon removal during soil aquifer treatment. *Journal of Environmental Quality*, **34**(1), 156–163.

Grünheid S., Hubner U. and Jekel M. (2007). Impact of temperature on biodegradation of bulk and trace organics during soil passage in an indirect reuse system. *Proceedings of the 6th IWA Conference on Wastewater Reclamation and Reuse for Sustainability*, Antwerpen, Begium.

Haberkamp J., Ruhl A. S., Ernst M. and Jekel M. (2007). Impact of coagulation and adsorption on DOC fractions of secondary effluent and resulting fouling behaviour in ultrafiltration. *Water Research*, **41**, 3794–3802.

Hamscher G., Pawelzick H. T., Höper H. and Nau H. (2005). Different behaviour of tetracyclines and sulfonamides in sandy soils after repeated fertilization with liquid manure. *Environmental Toxicology and Chemistry*, **24**, 861–868.

Heberer T., Massmann G., Fanck B., Taute T. and Dunnbier U. (2008). Behaviour and redox sensitivity of antimicrobial residues during bank filtration. *Chemosphere*, **73**(4), 451–460.

Hirsch R., Ternes T., Haberer K. and Kratz K. L. (1999). Occurrence of antibiotics in the aquatic environment. *Science of the Total Environment*, **225**(1–2), 109–118.

Hirpa H. S. (2008). Soil Column Studies for Removal of Bulk Organic Matter and Trace Organics from Water during Soil Passage. MSc Thesis MWI 2008-33, UNESCO-IHE, Delft, The Netherlands.

Idelovitch E., Icekson-Tal N., Avraham O. and Michail M. (2003). The long-term performance of soil aquifer treatment (SAT) for effluent reuse. *Water Science and Technology: Water Supply*, **3**(4), 239–246.

Jimenez B. and Chávez A. (2004). Quality assessment of an aquifer recharged with wastewater for its potential use as drinking source: "El mezquital valley" case. *Water Science and Technology*, **50**(2), 269–276.

Jekel M. and Gruenheid S. (2005). Bank filtration and groundwater recharge for treatment of polluted surface waters. *Water Supply*, **5**(5), 57–66.

Kahawita K. I. (2008). Effect of Advanced Oxidation Pre-treatment and MBR Pretreatment on Performance of Soil Aquifer Treatment Systems. MSc Thesis MWI 2008-30, UNESCO-IHE, Delft, The Netherlands.

Kadlec R. H. and Wallace S. B. (2008). Treatment Wetlands. CRC press.

Katukiza A. Y. (2006). Effect of Water Quality and Process Parameters on Removal of Organic Matter During Soil Aquifer Treatment. MSc Thesis MWI 2006-15, UNESCO-IHE, Delft, The Netherlands.

Kadlec R. H. and Knight R. L. (1996). Treatment Wetlands. Lewis Publishers, Boca Raton, FL, 893. ISBN 0-87371-930-1.

Kenneth M. (2007). Characterisation of Organic Matter Removal During Soil Aquifer Treatment and Analysis of the Effect of Pre-ozonation on Removal Efficiency. MSc Thesis MWI 2007-19, UNESCO-IHE, Delft, The Netherlands.

Kimura K., Hara H. and Watanabe Y. (2007). Elimination of Selected Acidic Pharmaceuticals from Municipal Wastewater by an Activated Sludge System and Membrane Bioreactors. *Environmental Science and Technology*, **41**(10), 3708–3714.

Massmann G., Greskowiak J., Dunnbier U., Zuehlke S., Knappe A. and Pekdeger A. (2006). The impact of variable temperatures on the redox conditions and the behaviour of pharmaceutical residues during artificial recharge. *Journal of Hydrology*, **328**(1–2), 141–156.

McArdell C. S., Molnar E., Suter M. J. F. and Giger W. (2003). Occurrence and fate of macrolide antibiotics in wastewater treatment plants and in the glatt valley watershed, Switzerland. *Environmental Science & Technology*, **37**(24), 5479–5486.

Nema P., Ojha C. S. P., Kumar A. and Khanna P. (2001). Techno-economic evaluation of soil-aquifer treatment using primary effluent at Ahmedabad, India. *Water Research*, **35**(9), 2179–2190.

Pescod M. B. (1992). Waste Water Treatment and Use in Agriculture. *FAO*, [Online] http://www.fao.org/docrep/T0551E/t0551e06.htm.

Reicherter (1999). Kosten und betriebsdaten von kleinkläranlagen. Seminar Des Bayerischen Industrieverbands Steine Und Erden e.V., 6 October 1999, Hirschaid, Germany.

Rice R. C. (1974). Soil clogging during infiltration of secondary effluent. *Journal of Water Pollution Control Federation*, **46**, 708–716.

Rousseau D. P. L., Lesage E., Story A., Vanrolleghem P. A. and De Pauw N. (2008). Constructed wetlands for water reclamation. *Desalination*, **218**, 181–189.

Sharma S. K., Katukiza A. and Amy G. (2007). Effect of Water Quality and Process Parameters on the Removal of Effluent Organic Matter (EfOM) During Soil Aquifer Treatment. *Proceedings of the ISMAR6 Conference (28 October – 2 November 2007)*, Phoenix, USA, 272–284.

Sharma S. K., Harun C. M. and Amy G. (2008). Framework for assessment of performance of soil aquifer treatment systems. *Water Science and Technology*, **57**(6), 941–946.

Sharma S. K., Musabe K. and Amy G. (2008b). Effects of pre-ozonation and advanced oxidation on removal of effluent organic matter (EfOM) during soil aquifer treatment. *Proceedings of 12th Annual Water Reuse and Desalination Research Conference*, (5–6 May 2008), Denver, USA.

Sharma S. K. and Amy G. (2010). Chapter 15: Natural treatment systems. In: Water Quality and Treatment: A Handbook on Drinking Water, K. E. James (ed.) 6th edn, American Water Works Association and McGraw Hill Inc., USA.

Sundaravadivel M. and Vigneswaran S. (2001). Constructed wetlands for wastewater treatment. *Critical Reviews in Environmental Science and Technology*, **31**(4), 351–409.

Chapter 15

Membrane based treatment trains for managed aquifer recharge

Christian Kazner, Adriano Joss, Thomas Ternes, Emmanuel van Houtte and Thomas Wintgens

15.1 MEMBRANES IN WATER RECLAMATION

Membranes play a key role in many state-of-the-art water reclamation schemes (Wintgens *et al.* 2005, 2008) as they provide a barrier to main effluent constituents such as particulates, pathogens, dissolved compounds and trace pollutants. They can be employed as pre-treatment prior to managed aquifer recharge with reclaimed water to:

- protect the hydraulic capacity of the aquifer,
- protect the treatment ability of the soil-aquifer system while preventing adverse geochemical reactions,
- protect the recovered water quality,
- prevent any negative trend in groundwater quality to comply with legislation.

The pressure-driven membrane filtration processes used in water and wastewater treatment can be classified according to the particle size of material components which they separate. One distinguishes microfiltration (MF), ultrafiltration (UF), nanofiltration (NF), and reverse osmosis (RO) (Mulder, 2004). Pressure driven membrane processes are divided in two main groups: i) Porous membranes (used in MF and UF) for the retention of macroscopic particles and molecules with a molar mass >2000 g/mol, and ii) Dense membranes (used in NF and RO) for the separation of smaller molecules (<2000 g/mol), atoms and ionic species (Melin and Rautenbach, 2007). Porous MF and UF membranes are used for separating suspended particles. UF has the additional advantage of removing the whole range of pathogenic microorganisms from protozoa to bacteria and viruses. Nanofiltration membranes are dense enough to retain even macromolecules (including many trace pollutants) down to a molecular size of about 200 g/mol. Reverse osmosis eliminating nearly all dissolved solids is preferably used in large-scale desalination either for the production of drinking water from sea and brackish water or for state-of-the-art reclamation schemes mainly leading to potable reuse.

As shown in the investigated case studies (Chapters 3 to 10) and explained in the chapter about natural treatment methods (chapter 14) soil and aquifer treatment can provide substantial removal of microbiological contamination with removal rates up to 5 log units. As an alternative to the extensive natural treatment typically involving long residence time and large areas, membrane systems can be tailored to the specific requirements regarding the quality of the injectant and product water at a small footprint. The broad range of membrane based treatment trains utilizes all types of pressure driven membranes at different stages of the treatment often in combination with other engineered or natural processes to remove certain compounds which are not sufficiently rejected by a certain type of membrane or to provide an additional barrier to certain key constituents such as viruses or micropollutants (Asano *et al.* 2007).

Combinations of adsorption of granular or powdered activated carbon (GAC or PAC) and nanofiltration such as GAC/NF, NF/GAC and PAC/NF are examples of novel hybrid processes which can provide stable high quality product water and allow fast and simple adaptation to changing feed water qualities (Kazner *et al.* 2009; Meier & Melin, 2005; Roorda *et al.* 2005; Verliefde *et al.* 2008a; 2008b). A general overview of typical membrane based treatment trains, types and configurations typically applied in water reclamation and reuse is given in Asano *et al.* (2007).

While membrane bioreactor (MBR) and RO technology are mature technologies and applied for water reclamation in many cases, numerous emerging hybrid processes and process combinations are currently under development. Thus the work in the RECLAIM WATER project focused on laboratory-scale and pilot scale studies in parallel to field studies as full scale reference to gain new insight of their removal capacities, particularly for organic micropollutants, as well as operational and economic aspects.

Quality goals for advanced water treatment

According to the pre-treatment objectives the main quality aspects to be considered for municipal wastewater after centralized biological treatment are given in Table 15.1 (adapted from AQUAREC, Bixio and Wintgens, 2006).

Table 15.1 Comparison of typical concentration of pollutants after biological treatment (Metcalf & Eddy, 1979; Gujer, 1999; Henze *et al.* 2002; Tchobanoglous, 2003), requirement for indirect groundwater recharge (according to Bixio and Wintgens, 2006), proposed quality for direct groundwater recharge and drinking water quality requirement (the Council of the European Commission, 1998).

Parameter	Unit	Secondary effluent	Polishing goal for indirect infiltration	Polishing goal for direct infiltration	Drinking water quality
pH		6.7–7.5	7–9	>6.5 and <9.5	>6.5 and <9.5
Conductivity	mS/cm	0.7–1.2	0.7	<0.7	<2.5
Alkalinity	$mgCaCO_3/L$				
TSS	mg/L	5–25			
COD	mg/L	5–50	70–100	<5	<5
TOC	mgC/L	15–40			
Total N	mgN/L	5–40	<25	<25	
NH_4^+	mgN/L	0–2	<0.2	<0.5	<0.5
NO_2^-	mgN/L	0		<0.5	<0.5
NO_3^-	mgN/L	5–40	<25	<25	<11.3
Total P	mgP/L	0–1			
PO_4^{--}	mgP/L	0–1			
SO_4^{--}	mgS/L	20–50	30	<30	<250
As	μg/L		5	<10	<10
B	mg/L		0.2	<1.0	<1.0
Cd	μg/L	0.3–1.5	3	<5.0	<5.0
Cl^-	mg/L	30–90	100	<250	<250
Cr	μg/L	20	25	<50	<50
Hg	μg/L		0.5	<1.0	<1.0
Pb	μg/L	10–30	5	<25 <10 in 2013	<25 <10 in 2013
Zn	μg/L	40–300			
Micropollutants	μg/L	0.001–100	<0.1	depending on compound	depending on compound
Faecal coliforms	CFU/100 mL	10^4–10^6	<10^4	0	0
Faecal streptoc.	CFU/100 mL	10^4		0	0
Helminth ova, Giardia	1/L	200	<1	0	0

15.2 OVERVIEW AND METHODS OF MEMBRANE TREATMENT RELATED STUDIES UNDER RECLAIM WATER

In high-grade water reclamation applications dense membranes build the key barrier against numerous contaminants, particularly pathogens, micropollutants and bulk organics. Dual membrane processes combining porous membranes with reverse osmosis followed by advanced oxidation (see Figure 15.1: treatment train No. 1) represent the state of the art in advanced water treatment for high-quality water reuse (Bixio and Wintgens, 2006; Asano *et al.* 2007; Alturki *et al.* 2010). Most of the large scale systems for planned indirect potable reuse rely on this treatment train. Some prominent examples are the Groundwater Replenishment System in Orange County, CA, the NEWater reclamation plants in Singapore, and the Western Corridor Recycled Water Project in Brisbane/Southeast Queensland, Australia. The quality of the product water allows direct injection as applied in Orange County, but in some cases pond technology was chosen for infiltration, e.g. in Torreele, Belgium. Although very effective for the elimination of contaminants, the dual membrane process is associated with rather high capital and operating cost, and involves the use of chlorinated chemicals to suppress membrane fouling. Furthermore cost-efficient methods for the treatment of RO concentrate containing salts, nutrients, disinfection by-products and micropollutants are not yet available.

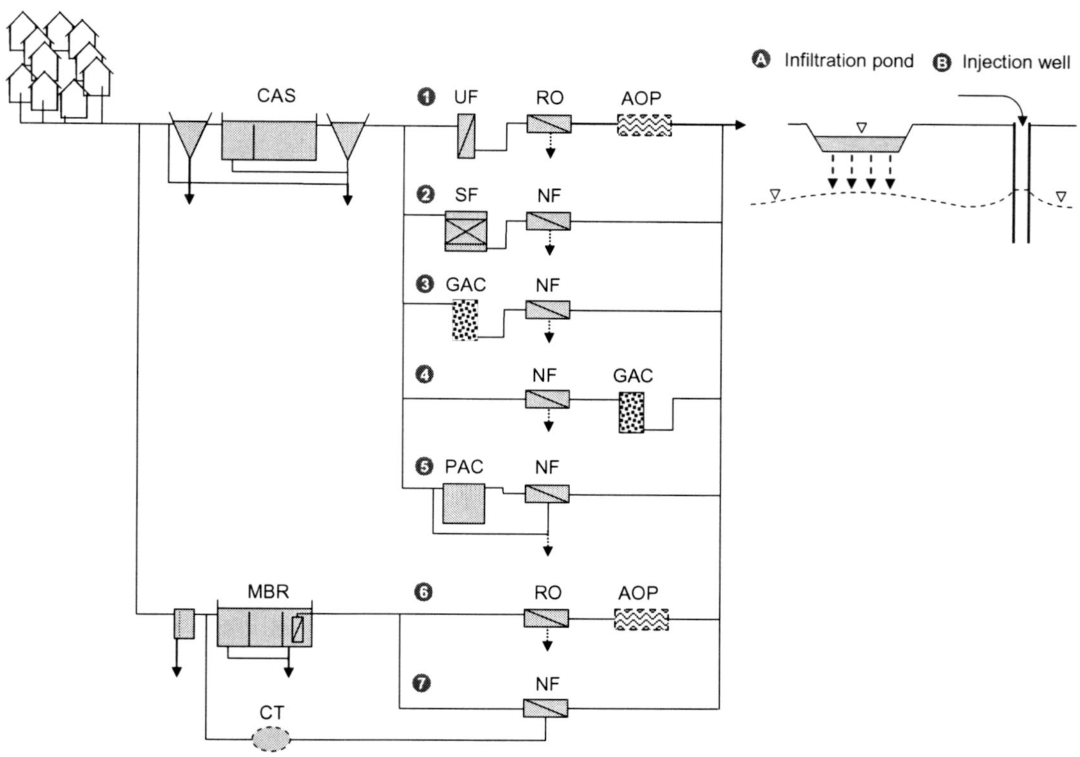

Figure 15.1 Treatment trains for high-quality water reuse via MAR with NF or RO based pre-treatment (CT = concentrate treatment)

Against this backdrop, nanofiltration provides an interesting alternative to reverse osmosis with a number of advantages in particular lower rejection of monovalent salts and thus less problematic membrane concentrates (Bellona and Drewes, 2007; Yangali-Quintanilla, 2010). The lower operating pressure leads to reduced energy demands of 0.6–1.2 kWh/m^3 compared to 1.5–2.5 kWh/m^3 with RO (Asano *et al.* 2007). While nanofiltration builds an excellent barrier against microbial contaminants including viruses on the same level as reverse osmosis (Trussell *et al.* 2000; Asano *et al.* 2007) the rejection of certain small-size organic contaminants is often lower than in reverse osmosis systems (Bellona *et al.* 2004, Yoon and Lueptow, 2005). To achieve a comparable removal of micropollutants, nanofiltration has to be combined with additional separation processes such as adsorption to activated carbon (Verliefde, 2008a).

As standard method for organics removal, several projects (Snyder *et al.* 2007; Metzger, 2010; Zwickenpflug *et al.* 2010) investigated the behaviour and optimisation of activated carbon in effluent treatment. Likewise nanofiltration (NF) was covered in a number of research projects during the last decade, particularly for the removal of micropollutants (*e.g.* Schäfer *et al.* 2005; Nghiem *et al.* 2004 and 2005; Drewes *et al.* 2008; Verliefde, 2008b; Yangali Quintanilla, 2010). Nanofiltration for the removal of organic micropollutants has also been already employed in large scale projects such as at the Méry-sur-Oise water treatment plant (Beros *et al.* 2003) where NF ensures the safe removal of pesticides from surface water for drinking water production.

In a comprehensive literature review Bellona *et al.* (2004) identified the main factors affecting the rejection of organic solutes in nanofiltration and reverse osmosis. According to his findings, micropollutant properties characterised by molecular weight, molecular size, acid dissociation constant, octanol-water partition coefficient and diffusion coefficient, as well as membrane properties defined by molecular weight cut-off, membrane pore size, surface charge, hydrophilicity and membrane surface roughness, determine the degree of solute rejection. Verliefde (2008a) highlighted the importance of the solute charge. He also proposed to add granular activated carbon after nanofiltration in drinking water treatment to adsorb the compounds poorly retained in nanofiltration.

Yangali Quintanilla (2010) further studied the previous work by developing a quantitative structure-activity relationship (QASR) model combined with an artificial neural network (ANN) model to predict the rejection of organic micropollutants in nanofiltration and reverse osmosis in drinking water treatment and water reuse based on the molecular structure of the target solute. The applicability of the QASR-ANN model was proven in lab-scale tests. Meanwhile the quantitative structure property relationship (QSPR) methodology was also successfully transferred to activated carbon (Dickenson and Drewes, 2010). Schrader (2006) investigated direct nanofiltration of effluent from conventional wastewater treatment plants and proved the successful application using spiral-wound modules. Besides a small study on concentrate treatment, his work focused on fouling and anti-fouling strategies due to the high fouling potential of effluent organic matter.

15.2.1 Emerging membrane based treatment trains

Ernst (2000) was the first to report on treatment trains combining nanofiltration and activated carbon for effluent polishing prior to managed aquifer recharge. Using mainly tests at laboratory scale, he identified powdered activated carbon (PAC) and granular activated carbon (GAC) following an 'open' nanofiltration membrane as most efficient and best operable combination. Further biofouling was detected as the main fouling mechanism in direct nanofiltration and it was concluded that porous membranes should serve as a proper pre-treatment, as long as no fouling resistant or backflushable nanofiltration membranes were available. Since then, several treatment trains combining nanofiltration and activated carbon as shown in Figure 15.1 have been developed in search of optimised hybrid processes and proposed for advanced water treatment.

The combination of powdered activated carbon followed by nanofiltration was first applied in landfill leachate treatment (Eltner, 1998; Eilers, 2000; Meier *et al.* 2002) and later adapted to effluent treatment (Eilers, 2000; Meier & Melin, 2005; Meier, 2008) using capillary nanofiltration membranes (Frank *et al.* 2001; Futselaar *et al.* 2002). Investigating the effect of powdered carbon on nanofiltration, Meier (2008) showed that activated carbon particles have a significant influence on the operability, particularly in terms of membrane abrasion, which occurred at high shear rates above 18,000 s^{-1}. At lower shear rates particle deposition, mainly the small particle fraction, lead to the formation of a cake layer, which reduced the permeate flux and compromised the solute rejection.

Roorda (2005) and Verliefde (2008a; 2008b) touched upon micropollutant removal in activated carbon nanofiltration hybrid processes. A comparative study of the micropollutant removal during the possible process combinations for advanced water treatment has not been performed. Open questions further exist with regard to the long-term behaviour and operability of nanofiltration activated carbon hybrid processes as the previous tests were conducted either in lab- to bench-scale or in pilot scale for shorter periods.

15.2.2 Membrane studies under Reclaim Water

Table 15.2 presents the treatment trains investigated within Reclaim Water which employ membrane processes and the involved project partners. Ultrafiltration treating secondary effluent was investigated in pilot scale at Shafdan. The permeate was infiltrated through a dug well and then treated with a short time soil aquifer treatment (SAT).

Table 15.2 Membrane based treatment train investigated in the scope of Reclaim Water.

	Involved partner						
Treatment Train	**EAWAG**	**RWTH**	**IRSA**	**CRAN**	**IHE**	**Torreele/ Wulpen**	**Shafdan**
CAS + UF + RO + MAR*						case study	
CAS + Direct NF		lab/pilot tests					
CAS + PAC/NF		lab/pilot tests					
CAS + NF/GAC		lab/pilot tests					
CAS + UF + SAT**							case study
MBR + NF (LPRO)	lab/pilot tests						
MBR + SAT***			lab/pilot tests	lab/pilot tests	lab/pilot tests		
MBR + Reed bed + SAT***				lab/pilot tests			

*cf. chapter 3, **cf. chapter 4, ***cf. chapter 14.

Besides the RO treatment train based on the dual membrane process (UF + RO), several combinations of nanofiltration and activated carbon were investigated. Additionally MBR/NF was selected using a dense NF membrane which allows high recoveries and minimized waste streams.

Membrane bioreactor treatment studies. The MBR studies conducted within RECLAIM WATER are presented in Chapter 14 (natural systems) and Chapter 4. The studies of MBR/NF are presented in Chapter 15.3.6. For an improved removal of trace organics and foulants activated carbon can be added to the activated sludge tank. Baumgarten *et al.* (2007) and Li *et al.* (2011)

have demonstrated that relatively small amounts of PAC can increase the removal of micropollutants significantly. It can also be employed as an emergency measure to cope with hazardous events due spills within the catchment area.

Torreele/Wulpen case study, Belgium. Besides some plants in Spain, the Wulpen-Torreele reclamation scheme represents a prominent European example where this treatment train is applied. It is operational since July 2002 after pilot testing from 1997–99 (Van Houtte, 1998; Van Houtte and Vanlerbeghe, 2004). Within RECLAIM WATER this plant was chosen to investigate the actual performance and reliability of such a system including the removal efficacy of the subsequent soil aquifer treatment (details are given in Chapter 3).

NF-activated carbon hybrid processes: As adsorption on membranes and spiking of micropollutants often lead to unclear results of lab scale experiments, the two investigated process combinations are operated in parallel on a long-term basis in pilot scale with detection of the real concentrations of selected micropollutants such as antibiotics, contrast media, and other substances (primidone, diclofenac, carbamazepine and benzotriazole) by SPE enrichment prior to LC-MS-MS measurement according to Ternes and Joss (2006). The tests by RWTH Aachen University, Germany employed the set-ups and experimental conditions as described below.

Direct nanofiltration. Direct NF with a 1.5 mm capillary nanofiltration module NF50 M10 from Norit X-Flow (Duin *et al.* 2000; Futselaar *et al.* 2002; 2003) was investigated in pilot scale at the Aachen-Soers WWTP treating the effluent from the sand filter. The molecular weight cut-off (MWCO) of the membrane is about 200 g/mol. The capillary NF was operated with a trans-membrane pressure between 1.5 and 4.0 bar and a permeate flux ranging from 20 to 25 $L/m^2 \cdot h$. The membrane was operated in cross flow mode with a cross flow velocity of 1.0 to 1.2 m/s and a recovery between 75 and 85%. A backwash was applied every 45 minutes for 1 minute for additional cake layer control.

PAC/NF treatment. The experiments (see Figure 15.2) were run with the capillary nanofiltration membrane NF50 M10 under the same conditions as the direct NF tests and PAC Norit SAE Super at 10 to 100 mg/L. The PAC/NF process cannot employ standard spiral wound NF modules as high particulate loads from PAC *quickly* block the module. PAC adsorbs the organic and inorganic substances in the retentate and thus provides combined concentrate treatment (Kazner *et al.* 2009). Main disadvantage is the limited applicability of oxidative chemicals for membrane cleaning and the requirement of a cross-flow operation mode with high velocities to avoid capillary blockage and control the cake layer.

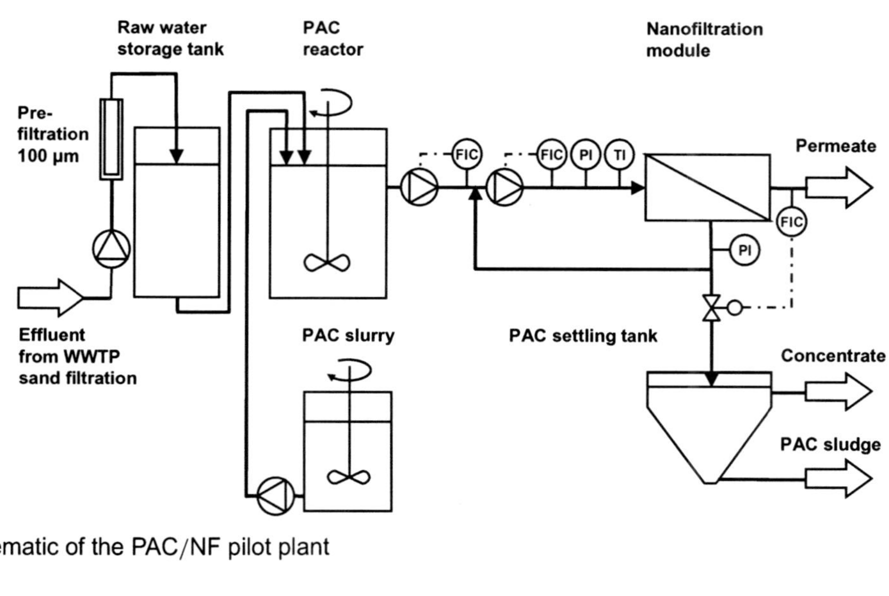

Figure 15.2 Schematic of the PAC/NF pilot plant

GAC/NF treatment. Granular activated carbon filtration can be used as pre-treatment prior to nanofiltration to reduce the content of bulk and trace organics, suspended solids and bacteria. Depending on the effluent quality GAC adsorbers require a frequent backwash to prevent the development of a biofilm and the transition to biologically active GAC treatment which reduces the ability to remove trace organics as observed by Snyder *et al.* (2007). The GAC filter (Norit NRS Carbon GA 0.5–2.5) at the Aachen-Soers WWTP was operated with an empty bed contact time of 30 minutes. The tests were performed using a Norit X-Flow capillary NF membrane as described above. Earlier pilot tests by Roorda *et al.* (2005) were conducted at the Apeldoorn WWTP. Roorda *et al.* highlight the following advantages: High fluxes (30 $L/m^2 \cdot h$ instead of 10–20 $L/m^2 \cdot h$) at low transmembrane pressures, high recovery above 94% without scaling, low brine production (<6% instead of 10–20%), reduced chemical cleanings (<1 per month), double barrier for pathogens as well as EDCs, and cost similar to direct NF.

NF/GAC treatment. Direct nanofiltration followed by granular activated carbon (GAC = Norit NRS Carbon EA 0.5 – 2.5) filtration is another promising treatment train providing a high-quality effluent with a partial desalination depending on the rejection characteristics of the selected NF membrane. While for advanced removal of organic micropollutants a membrane with a low MWCO would be preferable, it is beneficial to use a more open NF type to reduce the concentrate and increase the recovery (Kazner *et al.* 2008b).

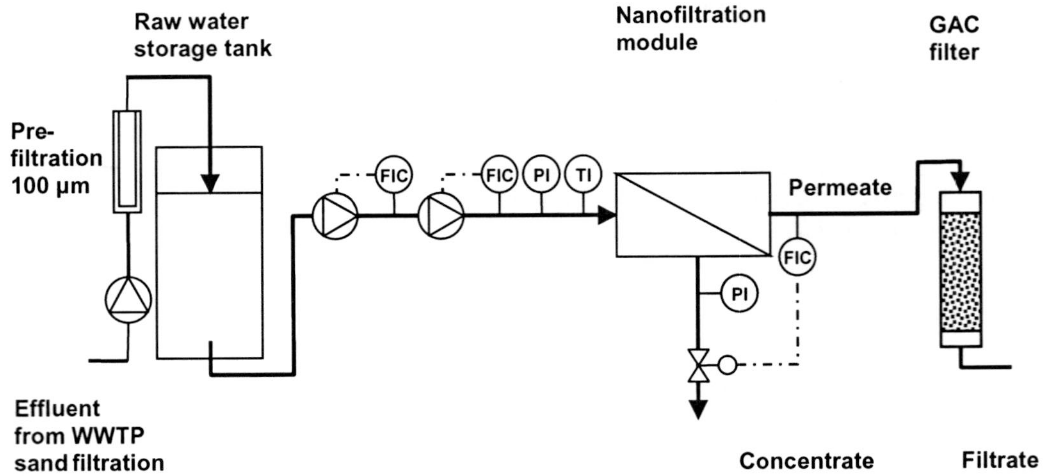

Figure 15.3 Schematic of the NF/GAC plant.

The process combination was tested in pilot scale at the Aachen WWTP treating the effluent after secondary treatment followed by sand filtration for one year (Figure 15.3). Similar to PAC/NF, nanofiltration and activated carbon filtration are expected to act complementary in removing organic micropollutants as compounds not well rejected by nanofiltration should be more readily adsorbed by the GAC filtration (Heijman *et al.* 2007). The EfOM is almost completely removed from the GAC feed water, if the GAC follows the nanofiltration. Thus one major advantage of NF/GAC is the significantly extended bed volumes of the GAC filter (Ernst, 2000). Additionally it provides a double barrier minimizing the potential break-through of trace contaminants (Snyder *et al.* 2007). The experiments employed also the capillary nanofiltration membrane NF50 M10 supplied by Norit X-Flow (Futselaar *et al.* 2002) in a 4″ module. The NF was operated in cross-flow mode with a permeate flux of 20 $L/m^2 \cdot h$ and a recovery of 75%.

MBR plus NF with concentrate recirculation. Dense membranes have great potential for retaining most of the relevant pollutants: salts, ions, organic micropollutants, pathogens. Besides the rather high technical and energy requirement, the main disadvantage of this separation process is the production of significant amounts of brines (up to 25% of the treated water). This drawback is mostly overcome by disposal (e.g. into the sea) or by separate treatment. The option to recycle the brines to the biological first treatment step was studied within RECLAIM WATER (Figure 15.4). At EAWAG in Switzerland MBR/NF was investigated in pilot scale at the experimental hall with a spiral wound NF90 module from DOW Filmtec treating MBR permeate from a ZENON pilot plant.

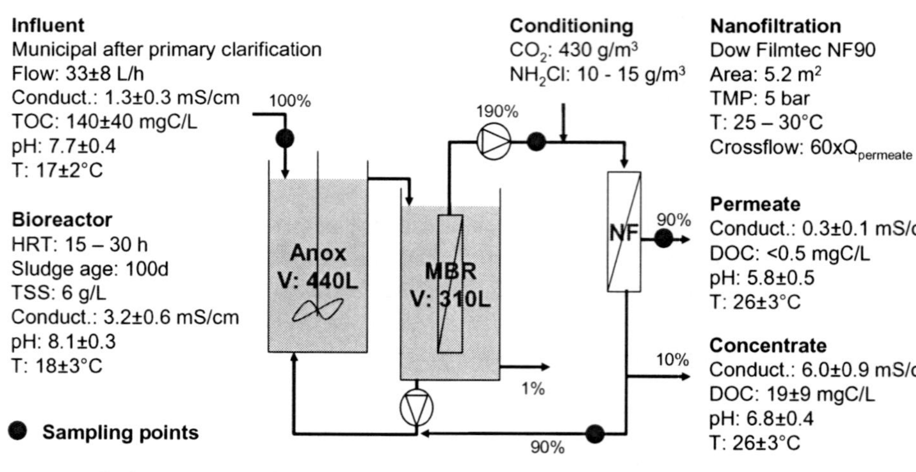

Figure 15.4 Summary of chosen operating parameters: Percentage indicates water flow relative to the influent. The concentrate was treated by ozone during a second test phase (not shown)

More than one year of continuous operation confirmed that proper conditioning before the nanofiltration allows avoiding significant precipitation of salts on the nanofiltration but let them occur in the membrane bioreactor and therefore to increase the water yield to 92%: the salts are being discarded either as precipitates with the excess sludge or as solutes of the concentrate (Joss *et al.* 2011).

15.3 PERFORMANCE OF TREATMENT TRAINS

15.3.1 Dual membrane treatment by UF/RO

The removal efficacy regarding the main relevant parameters is shown in Table 15.3. The RO product water complies with all requirements of drinking water guidelines and applied standards for infiltration (cf. Chapter 3). The infiltration water DOC is around 0.5 mg/L. The total volume of recovered water amounts to 3.5 million m^3/year to safeguard full recovery of the infiltrated water and mixing with ambient groundwater (mixing rate 70%/30%). Conductivity is reduced from 1200 μS/cm in the effluent to 300 μS/cm in the recovered water. All pathogens are fully removed. Organic micropollutants were also almost completely rejected by RO. Only NDMA, NMOR and bisphenol A were still detectable. All compounds were removed below the LOQ during MAR. During the investigation period the UV post-treatment was not in operation. Additional removal is therefore attributed to subsoil processes.

Table 15.3 Average removal rates of contaminants from WWTP effluent and concentrations of the extracted water of the Torreele reclamation plant and St-André aquifer recharge scheme.

	UF removal %	UF/RO removal %	UF/RO/MAR removal %	Extracted water
DOC	26.5	>90.7	76.3	2 mg/L
Ammonia (NH_4-N)	44.6	88.4	70.5	0.20 mg/L
Nitrate (NO_3-N)	9.6	90.3	96.8	0.37 mg/L
Total nitrogen	17.5	>82.0	>82.0	<2 mg/L
Total phosphorus	27.3	86.4	72.7	0.06 mg/L
Electrical conductivity	14.6	95.4	82.9	247 μS/cm
Microbiological parameters				
Total bacteria count (22 °C)	99.992	99.999	99.994	2.9 CFU/mL
Total bacteria count (37 °C)	99.592	100	99.999	0.17 CFU/m/L
Total coliforms	100	100	100	0 CFU/100mL
E.Coli	100	100	100	0 CFU/100mL
Enterococci	99.999	100	100	0 CFU/100mL
Clostridium spores	100	100	99.996	0.1 CFU/100mL
Bacteriophages	99.6	100	97.8	450 PFU/mL
Organic micropollutants				
Benzotriazole	−1.0	66.1	>95.8	<10 ng/L
Carbamazepine	2.7	>99.8	>99.8	<2 ng/L
Clarithromycin	−48.1	>98.1	>98.1	<2 ng/L
Diclofenac	−3.5	>99.2	>99.2	<2 ng/L
Ibuprofen	−19.5	>99.0	>99.0	<2 ng/L
Naproxen	30.2	>99.6	>99.6	<2 ng/L
Sulfamethoxazole	3.6	>97.8	>97.8	<2 ng/L
N-Ac-SMX	−5.9	>96.9	>96.9	<10 ng/L
Trimethoprim	30.4	>98.2	>98.2	<2 ng/L
Primidone	15.4	>96.9	>96.9	<2 ng/L
Iomeprol	7.7	>99.7	>99.7	<2 ng/L
Iopromide	18.9	>99.9	>99.9	<2 ng/L
Diatriozate	34.4	>99.95	>99.95	<10 ng/L
AOI	−25.3	94.5	91.8	<2 ng/L
Estrone E1	24.8	>90.8	>90.8	<1 ng/L
Bisphenol-A	50.2	68.5	>89.5	<5 ng/L
N-Nitrosodimethylamine (NDMA)	−7.2	8.5	>85	<0.7 ng/L

15.3.2 Direct NF treatment

Direct NF treatment provides a high removal capacity regarding bulk organics producing high quality water from tertiary effluent. Absence of particulates and DOC removal rates above 90% (DOC < 0.5 mg/L) confirm the excellent permeate quality suitable for direct injection. Permeate DOC values were between 0.3 to 0.5 mg/L at DOC removal rates between 85 and 95%. Similar rates were observed in other studies (Schrader, 2006).

The applied NF membrane reduced salinity measured as electrical conductivity by 15 to 30% removing mainly bivalent ions, such as sulfate, while chloride was removed to minor extent. The Donnan effect may even lead to elevated concentrations of monovalent ions in the permeate.

Nanofiltration is expected to remove organic micropollutants to a high degree. As described by Bellona *et al.* (2004) the properties of the compounds, such as molecular weight, polarity and p*Ka*, as well as the membrane properties, such as MWCO and surface charge, influence the rejection performance. There is no general model predicting the rejection performance regarding a specific substance and membrane without testing. For the determination of the performance of the capillary nanofiltration, the pilot test at the Aachen-Soers WWTP investigated the removal of selected organic micropollutants as given in Table 15.4. The LC-MS measurement was performed by Federal Institute of Hydrology, Germany.

Table 15.4 Selected organic micropollutants.

Analytes	Short name	Use	MW g/mol	pK_a	Log K_{OW}	Charge at pH 7
Benzotriazole	BTA	Complexing agent	119.13	8.37	1.44	+
Carbamazepine	CBZ	Anti-epileptic	236.27	0.37	2.45	0
Clarithromycin	CLA	Antibiotic (macrolide)	747.95	8.90	3.16	+
Diatrizoic acid	DIATR	X-ray contrast medium	613.90	3.4	1.37	−
Diclofenac	DCF	Analgesic/anti-inflammatory	296.15	4.15	4.50	−
Erythromycin	ERY	Antibiotic (macrolide)	733.93	8.88	3.06	+
Iohexol	IOHEX	X-ray contrast medium	821.14		−3.05	0
Iomeprol	IOMEP	X-ray contrast medium	777.00			0
Iopamidol	IOPAM	X-ray contrast medium	777.09		−2.42	0
Iopromide	IPM	X-ray contrast medium	791.00	10.2	−2.05	0
Primidone	PRIM	Anticonvulsant	218.25		0.90	0
Sulfamethoxazole	SMX	Antibiotic (sulfonamide)	253.28	5.60	0.89	−

Direct nanofiltration showed the expected incomplete removal of small size molecules in the range of 40 to 70% while larger ones such as contrast media are retained almost completely (Figure 15.5). The selected organic micropollutants were rejected by the investigated capillary nanofiltration to a varying degree as follows (Figures 15.5 and 15.6):

- Very high rejection of neutral and negatively charged molecules with a molecular weight significantly larger than the MWCO due to electrostatic repulsion and/or steric hinderance, prominent examples are contrast media,
- Lower but still high rejection of positively charged molecules with a molecular weight significantly larger than the MWCO due to a combination of electrostatic attraction and size exclusion, prominent examples are macrolides,
- Rejection of molecules with a molecular weight close to the MWCO in good accordance to the PEG rejection curve, if compounds are hydrophilic (log $K_{OW} < 2$) irrespective of their charge due to size exclusion, examples are benzotriazole, primidone and sulfamethoxazole,
- Rejection of molecules with a molecular weight close to the MWCO significantly below the PEG rejection curve if compounds are more hydrophobic (log $K_{OW} > 2$) irrespective of their charge, examples are carbamazepine, diclofenac and clofibric acid,
- Poor rejection of strong hydrophobic compounds (log $K_{OW} > 3$), persistent adsorption to the membrane and plant can lead to breakthrough of compounds and permeate concentrations above feed concentrations, examples are bisphenol A and 17α-ethinylestradiole (Kazner *et al.* 2008a).

The investigated capillary nanofiltration must be regarded as relatively loose NF rejecting many organic micropollutants typically found in secondary effluent only partially. In comparison to dense nanofiltration membranes, it becomes obvious that the investigated open NF requires post- or pre-treatment, *e.g.* by adsorptive or oxidative processes to reduce the target compounds to the levels below 100 ng/L or even below the LOQ (around 1 to 10 ng/L) as mostly applied in indirect potable reuse schemes.

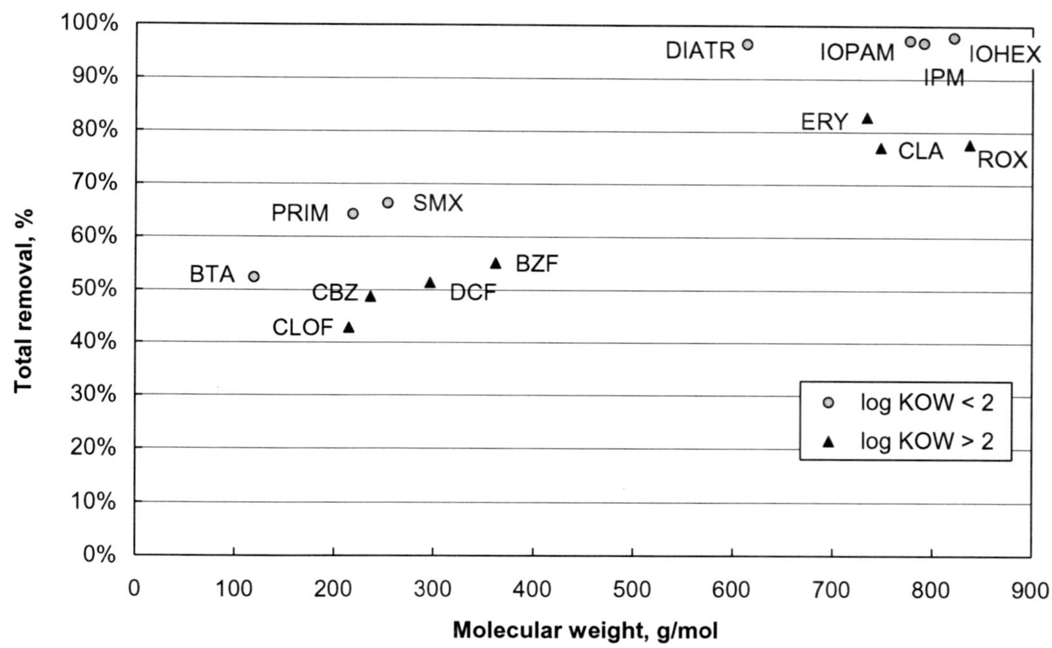

Figure 15.5 Total removal of selected pharmaceuticals by nanofiltration in pilot scale short-term tests depending on molecular weight and log K_{OW} (75% recovery, J = 20 LMH, v_{CF} = 1.25 m/s)

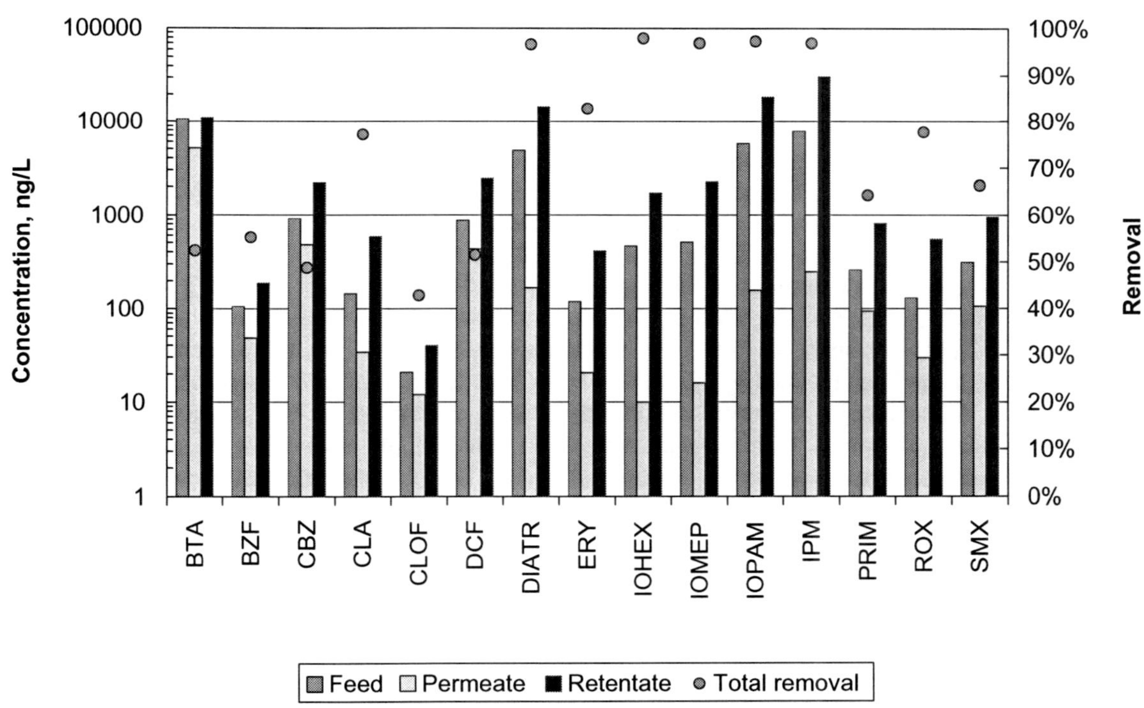

Figure 15.6 Concentrations and removal of selected pharmaceuticals for the direct NF pilot test with the capillary NF membrane (Flux = 20 LMH, Recovery = 75%)

15.3.3 PAC/NF treatment

The PAC/NF process has a very high removal efficiency regarding bulk organics. With relatively low PAC doses of about 25 to 50 mg/L most of the effluent organic matter can be removed. Table 15.5 gives the average removal of DOC and UVA as well as concentrations at the different sampling points. Direct nanofiltration as well as powdered activated carbon combined with nanofiltration can produce high quality water from tertiary effluent. DOC values below 0.5 mg/L and removal rates above 90% confirm the excellent quality suitable for direct injection (Kazner *et al.* 2009).

Table 15.5 DOC and UVA average concentrations, SUVA and removal rates.

Parameter	Raw water	Adsorption		Nanofiltration	
	$C_{influent}$	$R_{Ads.}$	C_{Feed}	R_{total}	$C_{Permeate}$
Without PAC					
DOC	4.6 mg/L	–	–	93.4%	0.3 mg/L
UVA 254	0.126 cm^{-1}	–	–	90.0%	0.0126 cm^{-1}
SUVA	2.7 L/mg · m	–	–	–	4.2 L/mg · m
With 25 mg/L PAC					
DOC	5.7 mg/L	29.8%	4.0 mg/L	94.7%	0.3 mg/L
UVA 254	0.136 cm^{-1}	34.6 %	0.089 cm^{-1}	90.4%	0.0130 cm^{-1}
SUVA	2.4 L/mg · m	–	2.2 L/mg · m	–	4.3 L/mg · m

Already direct NF (i.e. without PAC) provided high quality permeate with 0.3 mg/L DOC and a total removal of 93.4%. Addition of low adsorbent doses removed about 30% of the raw water DOC while the total removal increased only slightly. The different removal rates for DOC and UVA indicated that UV relevant molecules such as humic substances possess a higher affinity to adsorb. The average SUVA in the range of 2.4 to 2.7 L/mg · m was slightly above the typical effluent value of 2 L/mg · m. While adsorption reduced the SUVA to 2.2 L/mg · m, it increased again after NF to around 4 L/mg · m indicating a higher aromaticity on a very low DOC level. Fluorescence EEM spectroscopy has been investigated in Kazner *et al.* (2008c).

A substantial removal of organic micropollutants is one of the major advantages of PAC/NF (see Table 15.6). Based upon the required permeate quality and the given raw water concentrations, the carbon dose rate can be adapted to the required removal rate. For the determination of the performance of the PAC adsorption and the nanofiltration, the pilot test at the Aachen-Soers WWTP investigated the removal of selected organic micropollutants (Kazner *et al.* 2007) as given in Table 15.4. The permeate produced by the PAC/NF process proved to be of very high quality with concentrations <100 ng/L for most of the micropollutants (PAC dose 50 mg/L, Figure 15.7). Without PAC dosage the permeate concentrations of some substances of concern such as benzotriazole or carbamazepine are significantly higher (BTA: 5.1 μg/L, CBZ: 0.5 μg/L, Figure 15.7). Direct nanofiltration without PAC dosage showed the expected incomplete removal of small size molecules in the range of 50 to 80% while larger ones such as contrast media are retained almost completely (Figure 15.7). With a PAC dose of 50 mg/L the total removal increased considerably reaching 88% for SMX (before 66%) to 99.9% for carbamazepine (before 49%). It was demonstrated that poorly adsorbable contrast media are well retained by the NF while the total removal of well adsorbable small size molecules such as diclofenac or carbamazepine improved significantly due to adsorption to PAC already at low to moderate dosage. For poorly adsorbable solutes such as sulfamethoxazole, low PAC doses (~10 mg/L) as typically applied in drinking water treatment could not improve the removal efficiency. Due to cake enhanced concentration polarization removal rates may even deteriorate at low PAC dose (compare Figure 15.8). For example, the high rejection of X-ray contrast media was slightly reduced from >97% (0 mg/L PAC) to 82–90% (10 mg/L) due to increased concentration polarization.

Table 15.6 Average rejection rates for direct NF and PAC/NF (NF: X-Flow NF50M10, 75% recovery, 20 L/m^2 · h, PAC: Norit SAE Super $C_{DOC,0}$ = 5–6 mg/L).

	Unit	NF	PAC adsorption		PAC/NF	
PAC dose	mg/L	0	10	50	10	50
Bulk organics						
DOC	%	90–95	20	50–60	90–95	90–95
Microbiological parameters[(1)]						
Protozoa	log	6–7 (>6)	0	0	6–7 (>6)	6–7 (>6)
Bacteria	log	6–7 (3–6)	0	0	6–7 (3–6)	6–7 (3–6)
Viruses	log	6–7 (3–5)	0	0	6–7 (3–5)	6–7 (3–5)

(Continued)

Table 15.6 Average rejection rates for direct NF and PAC/NF (NF: X-Flow NF50M10, 75% recovery, 20 L/m$^2 \cdot$ h, PAC: Norit SAE Super $C_{DOC,0} = 5-6$ mg/L) (*Continued*).

	Unit	NF	PAC adsorption		PAC/NF	
Salinity						
Conductivity	%	15–20	0	0	12	12
Sulfate	%	70–90	0	0	60–70	60–70
Chloride	%	5–20	0	0	3–5	−5–0
Organic micropollutants[(2)]						
Benzotriazole	%	50	80	95	85	98
Primidone	%	65	50	85	55	95
Carbamazepine	%	50	75	98	85	99.9
Diclofenac	%	50	60	85	80	99
Clofibric acid	%	40	10	ø	35	ø
Bezafibrate	%	50	55	ø	80	ø
Antibiotics						
Sulfamethoxazole	%	65	30	70	50	90
Macrolides (CLA, ERY, ROX)	%	80	80	>99	95	>99
Iodinated contrast media						
Iopamidol	%	99	20	75	90	99
Iomeprol	%	99	20	75	90	97
Iopromid	%	99	30	80	90	99
Iohexol	%	99	20	75	90	99
Diatrizoate	%	95	0	50*	85	95*
Endocrine disruptors						
17a-Ethinylestradiol	%	30	70	95	88	94
Bisphenol A	%	10	80	90	98	98

[(1)] Estimation based on literature (Trussell, 2000; Toze, 2006), values in brackets are reported in Asano *et al.* (2007) reflecting membrane integrity concerns. PAC/NF is anticipated to perform similar to NF.
[(2)] Measurement at BfG (Ternes and Joss, 2006).
ø Concentrations below LOQ.
*Value for 50 mg/L estimated based on removal rates at 25 and 100 mg/L.

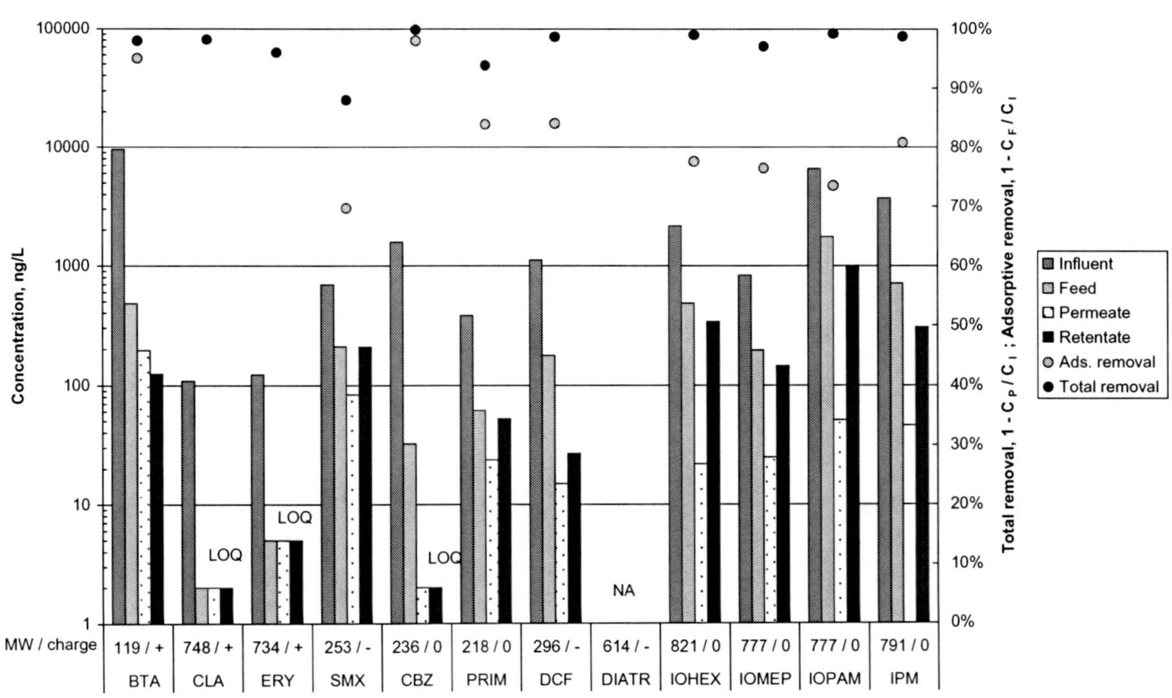

Figure 15.7 Concentration and total removal rates by PAC/NF for selected micropollutants (50 mg/L PAC)

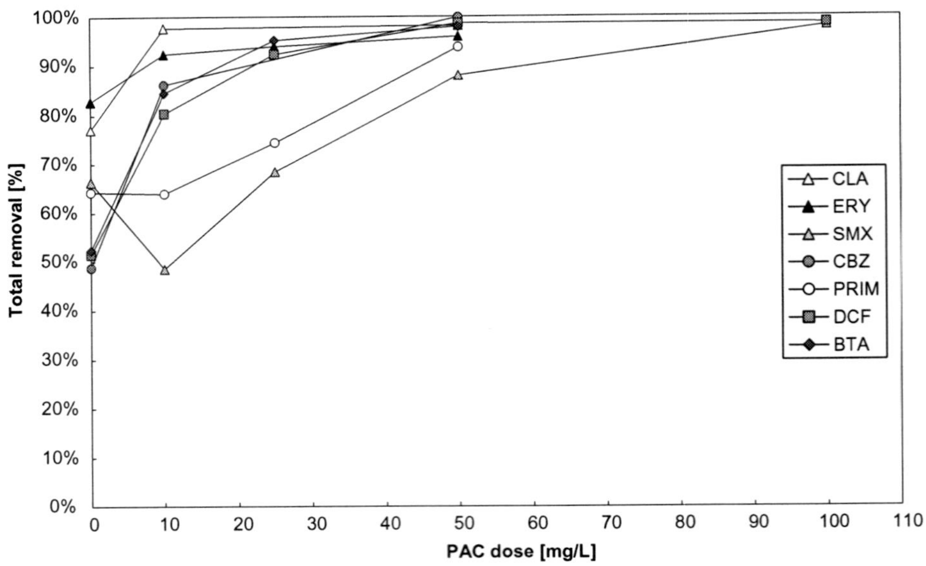

Figure 15.8 Total removal rates for selected micropollutants at varying PAC dose (Norit SAE Super, total contact time 1.2 h)

The capillary NF membrane provides a high rejection of bivalent ions, e.g. sulfate in the range of $94 \pm 2\%$ while chloride rejection of the relatively open membrane is about $35 \pm 5\%$ (Norit, 2002). During the PAC/NF pilot tests, the total removal for sulfate was found to be in the range of 60 to 70%. Due to Donnan effects and cake enhanced concentration polarization from PAC cake layer formation, there was no significant chloride removal (Herr, 2007). Removal of calcium and magnesium is in the range of 50% (Salzbrenner, 2005).

15.3.4 GAC/NF treatment

The results of the pilot tests are given in Table 15.7. Bulk organics removal reached levels up to 99%. With increased loading of the GAC filter, the total DOC removal by GAC/NF slowly decreased to 95%. The permeate DOC concentration was below the LOQ.

Table 15.7 Average concentrations and removal rates for DOC and selected pharmaceuticals for GAC/NF (runtime 1700 hours, GAC NRS Carbon GA).

Parameter	GAC influent mg/L	GAC effluent mg/L	NF permeate mg/L	GAC removal %	GAC/NF removal %
Bulk organics					
DOC	5.24	1.54	<0.17	70.6	>96.8
Nutrients					
Ammonium (NH_4-N)	0.09	0.06	0.03	33.3	66.7
Nitrate (NO_3-N)	5.2	5.6	5.9	−7.7	−13.5
P total	0.05	0.04	0.001	20.0	98.0
Salinity					
EC	882	861	778	2.4	11.8
Chloride	79	77	77	2.5	2.5
Sulphate	59	59	16	0	72.9
Micropollutants	µg/L	µg/L	µg/L		
Benzotriazole	2.990	<0.0052	<0.0059	>99.8	>99.8
Carbamazepine	1.046	<0.0009	<0.0006	>99.9	>99.9
Primidone	0.373	<0.0014	<0.0009	>99.6	>99.8
Sulfamethoxazole	0.515	<0.0051	<0.0012	>99.0	>99.8

The removal of the selected micropollutants was stable below the LOQ. Roorda *et al.* (2005) investigated the micropollutant removal. He observed relatively low removal efficiencies for the GAC stage. While caffeine was adsorbed to >95%, most of the other micropollutants were only removed by 36 to 83%.

The removal with GAC filters as effluent treatment is mainly governed by the adsorbability of the specific compound. In rapid small scale column tests (Snyder *et al.* 2007), the compounds with the highest hydrophility such as sulfamethoxazole (SMX) exhibited the fastest breakthrough. Snyder *et al.* (2007) observed a 10% breakthrough of SMX after 5,000 bed volumes whereas estrone was still significantly removed after more than 50,000 bed volumes. Reduced adsorption rates and earlier breakthrough can be expected at high DOC effluent concentrations due to pore blockage. It can be concluded that the overall organic micropollutant removal of a GAC/NF system is quite complex and case specific. It depends mainly on:

- Effluent DOC and micropollutant concentrations
- Design and operation of GAC adsorber (type of GAC, contact time, backwash, regeneration of carbon)
- Rejection characteristics of the NF membrane (MWCO of membrane, operation mode, e.g. recovery, flux)

The salinity removal is mainly influenced by the performance of the NF and thus the type and operation of the selected nanofiltration membrane.

15.3.5 NF/GAC treatment

Pathogen removal and salinity removal are comparable to direct NF. The removal of bulk organics was stable at around 95% with DOC concentrations in the NF/GAC filtrate of around 0.2 mg/L.

The micropollutant removal measured at ambient effluent concentration without spiking is given below in Table 15.8. NF/GAC showed a very stable removal below the LOQ even after more than one year of continuous operation. Also poorly adsorbable and rejected solutes such as SMX were very well removed.

Table 15.8 Average concentrations and removal rates for DOC and selected pharmaceuticals for NF/GAC (runtime 9500 hours, GAC NRS Carbon EA).

Parameter	NF feed mg/L	NF permeate mg/L	GAC filtrate mg/L	NF removal %	NF/GAC removal %
Bulk organics					
DOC	4.98	0.46	0.24	90.8	95.2
Nutrients					
Ammonium (NH_4-N)	0.09	0.03	0.02	66.7	77.8
Nitrate (NO_3-N)	5.7	5.6	6.0	1.8	−5.3
P total	0.09	0.03	0.03	66.7	66.7
Salinity					
EC	978	836	826	14.5	15.5
Chloride	160	164	159	−2.5	0.6
Sulphate	65	15	14	76.9	78.5
Micropollutants ($n = 11$)	µg/L	µg/L	µg/L		
Benzotriazole	2.090	1.660	<0.030	20.6	>99.8
Carbamazepine	0.831	0.490	<0.004	41.0	>99.9
Primidone	0.263	0.135	<0.004	48.7	>99.8
Sulfamethoxazole	0.449	0.175	<0.010	61.0	>99.8
Micropollutants ($n = 2$)	µg/L	µg/L	µg/L		
Bezafibrate	0.079	0.017	<0.010	78.5	>87.3
Clarithromycin	0.116	<0.005	<0.005	>95.7	>95.7
Erythromycin	0.145	<0.010	<0.010	>93.1	>93.1
Sulfamethoxazole	0.718	0.332	<0.010	53.8	>98.6
Trimethoprim	0.101	0.048	<0.005	52.5	>95.0
Diatrizoate	1.416	0.030	<0.010	97.9	>99.3
Iohexol	0.138	<0.010	<0.010	>92.8	>92.8
Iomeprol	0.427	0.005	<0.010	98.8	>97.7
Iopamidol	3.566	0.064	<0.010	98.2	>99.7
Iopromide	1.850	0.029	<0.025	98.4	>98.6

NF-GAC hybrid processes provide water of highest quality with DOC concentrations below 0.5 mg/L and concentrations of organic micropollutant below the limit of quantification. Although nanofiltration as such is already an efficient water recycling technology, the addition of granular activated carbon features some advantages, in particular with regard to an improved removal of small size pharmaceuticals and industrial chemicals as shown earlier by Verliefde *et al.* (2007) and Roorda *et al.* (2005). The present study revealed the following key insights in the long-term behaviour of the investigated processes:

- GAC adsorption for advanced treatment of WWTP effluent can provide a very good removal of most of the refractory micropollutants. However, high EfOM concentrations can limit the carbon efficiency. With an empty bed contact time of 15 min it can be expected that micropollutants break through after 7,000 to 10,000 bed volumes depending on the carbon characteristics. From the economic and operational point of view it appears advantageous to employ a reactivated carbon with a large degree of meso-pores and high inner surface as this seems to be more resilient against competitive effects from bulk organics. The appropriate carbon type should be selected based on tests using a representative effluent.
- Coupling GAC with NF post-treatment can upgrade the water quality in particular regarding pathogens, multivalent ions and large micropollutants, e.g. iodinated contrast media.
- GAC adsorption following nanofiltration showed very stable removal results over a period of more than a year. Removal of bulk organics prior to GAC filtration produced an influent water quality comparable to drinking water treatment. For these applications a long runtime of the adsorber can be accomplished, particularly if high-grade drinking water carbons are employed. This could also be required to obtain authorisation for managed aquifer recharge according to drinking water standards.
- The combination of NF/GAC exceeds the performance of GAC/NF in terms of the achievable bed volumes which were well above 20,000 for NF/GAC compared to about 7,000 for GAC/NF.

Both GAC-NF hybrid processes feature also advantages in managed aquifer applications as the salt content of the product water is closer to natural conditions and NF concentrate could be treated more easily on-site.

15.3.6 MBR/NF treatment

Pathogen removal has not been tested, but since MBR/NF represents a double membrane system it is expected that the pathogen removal will be >6 log units (double barrier). Only a small fraction of the organics is still present in the permeate as shown by the low value of only 0.6 ± 0.3 mg DOC/L as well as the total colour removal. Out of the 23 different micropollutants only benzotriazole and to a lower degree propranolol and diclofenac were found in concentrations significantly above the limit of quantification, proving that most are retained between 90% and 100% (Figure 15.9). The mass balance shown in Figure 15.10 confirms that biological degradation and disposal with the discarded concentrate are the main removal routes.

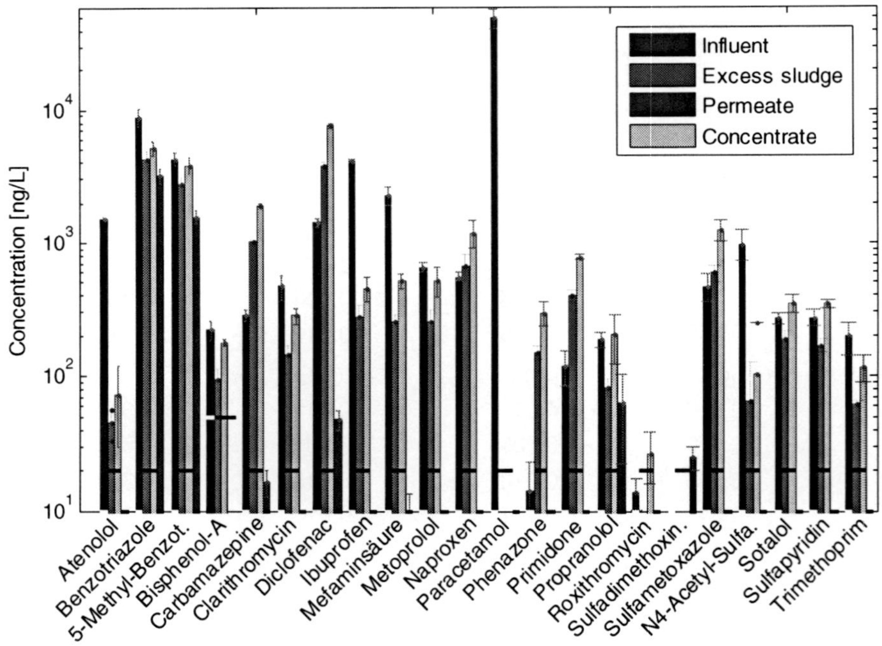

Figure 15.9 Micropollutant sampling campaign of five consecutive days sampled each flow proportionally over 24 h. Only the benzotriazoles are found in significant amounts in the permeate. Propranolol and diclofenac slightly above, all the others below the limit of quantification (horizontal black line)

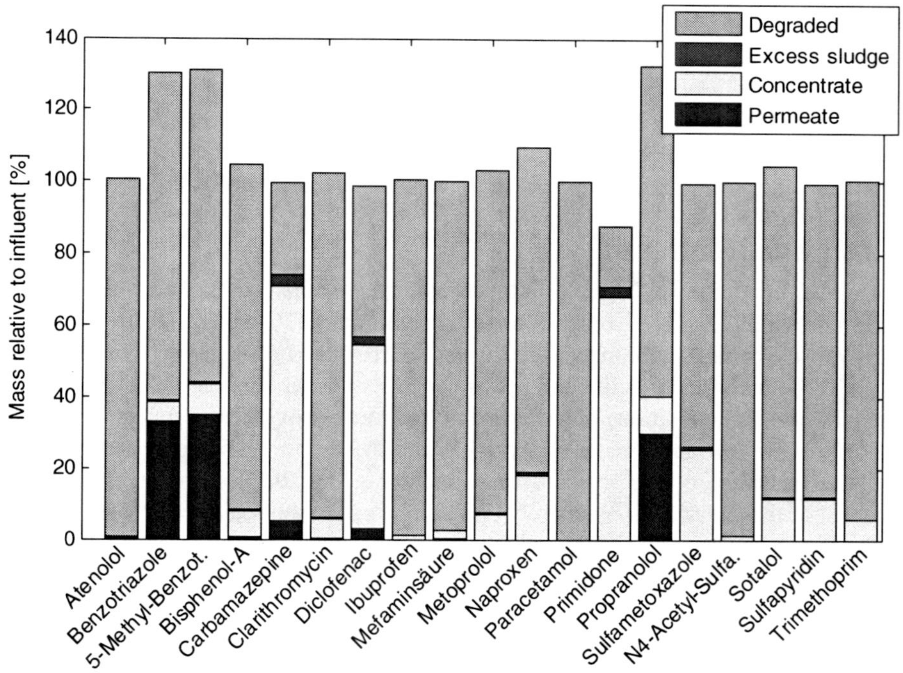

Figure 15.10 Mass balance of the micropollutants. The influent load is normalized to 100% for each compound. Biological degradation and concentrate discharge are confirmed as primary output path.

The ozonation of the recycled concentrate allows further reducing the concentration in the permeate (Figure 15.11): all measured compounds but the two benzotriazoles are found at concentrations $<10\ \text{ng} \cdot \text{L}^{-1}$. Further also the discharged concentrate is significantly depleted from micropollutants. The ozonation was achieved by a simple bubble column inserted in the recycle line (i.e. between the NF and the MBR). The results confirm that ozonation is suitable for reducing the micropollutant content in the concentrate: ozonating the recycle stream or only the discharged concentrate are both seen as options. The rationale of treating the recycle is to allow biological degradation of the formed ozonation products by taking them back to the MBR; this feasibility study does not yet allow a fully-fledged cost/efficiency evaluation. Details are discussed in Joss *et al.* (2011).

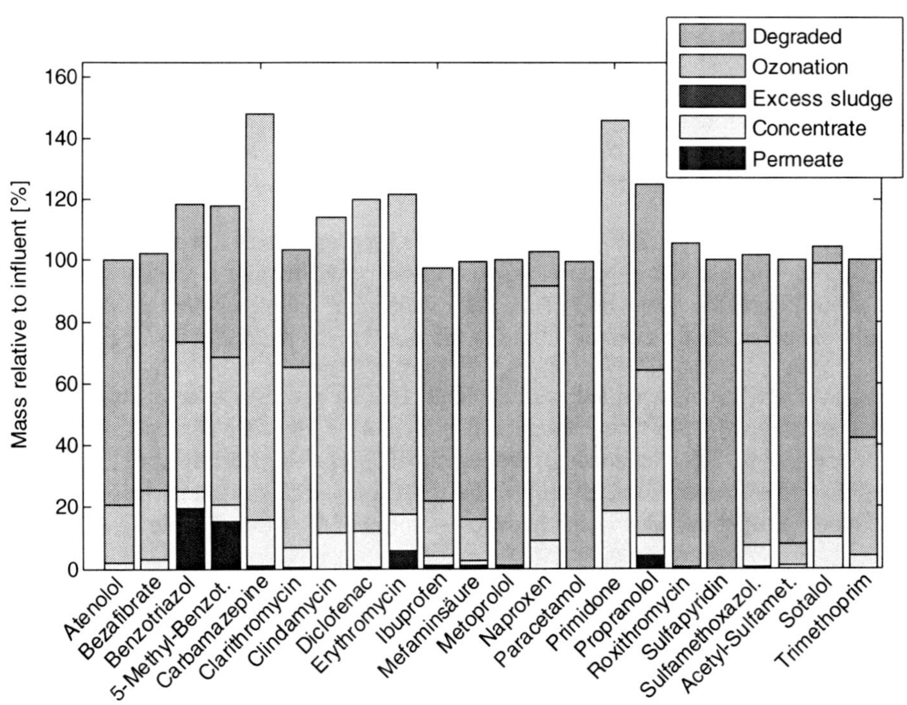

Figure 15.11 Mass balance of the micropollutants in the sampling campaign with ozonation of the NF to MBR recycle. The influent load is normalized to 100% for each compound. Applied ozone dose: $16\ \text{gO}_3 \cdot \text{m}^{-3}$ corresponding to $0.85\ \text{gO}_3 \cdot \text{g DOC}^{-1}$

15.4 OPERATIONAL PERFORMANCE

Besides the tests which were conducted at full scale plants, all processes were studied at pilot scale with supporting lab studies. All processes provide permeate compliant with stringent water quality requirements and suitable for managed aquifer recharge by well injection. Due to the limitation of the scale and the duration of the studies, the remarks regarding the economic and operational performance are somewhat limited and would require additional investigation prior to full scale application to proof the findings and their advantages under site specific conditions.

15.4.1 Dual membrane treatment

The 10 years of successful operation at Torreele/Wulpen and many other cases where the dual membrane process was chosen proved the high stability and reliability of the treatment train (cf. Chapter 3). Currently it provides the highest water quality which is mostly better than conventional water sources used for drinking water production. Automation and online control units verify a constant quality and performance also under changing raw water qualities and quantities as well as during irregular process conditions (e.g. failure of treatment components). Integrity monitoring is still under further development to provide fast and simple methods. While the dual membrane systems show a very stable and reliable performance, the treatment of the reject streams including the RO concentrates can still be regarded as a drawback. Currently most of the full scale systems rely upon discharge to the sea as economic solution to dispose of the waste. Many approaches for concentrate treatment and management are currently under investigation (cf. Chapter. 16).

15.4.2 PAC/NF treatment

From the operational point of view the PAC/NF process is very complex and challenging. Several factors influence the performance of the PAC/NF process (*cf.* Table 15.9). The most challenging part is the optimisation of the nanofiltration especially regarding the control of the cake layer formation and fouling. According to Meier (2008) the formation of a cake layer can hardly be avoided completely and it is advantageous to avoid abrasion. The PAC/NF process should be operated at slightly carbon depositing process conditions with shear rates between 6,000 and 18,000 s^{-1} and use hydraulic flushes and chemical cleaning to limit the formation of a cake layer. During previous pilot tests and the experiments presented here, backflush as available in capillary nanofiltration was the main method for cake layer control. Meier and Melin (2005) demonstrated that non-backflushable membranes, such as spiral wound module also with wide spacers, were quickly blocked and therefore cannot be employed in PAC/NF. Theoretical considerations and practical experiments confirmed the importance of the carbon particle size for the formation of the cake layer (Meier, 2008) with fine particles below 0.4 µm depositing preferentially. Scanning electronic microscopy revealed that the membrane was covered by a 30 to 50 µm thick cake layer of particles and agglomerates sized from 0.5 to 3 µm ($v_{CF} =$ 1.25 m/s, $C_{PAC} = 0.2–0.5$ g/L). A reduced retention of the sulphate from around 96% to 87% was attributed to cake enhanced concentration polarisation as described by Hoek *et al.* (2002) and Hoek and Elimelech (2003).

General operational settings. PAC is added prior to the NF with a contact time of about 1 h which was found to be sufficient for good adsorption efficiency. The PAC remained in contact with the feed water for about another 15 minutes during the cross-flow recirculation. The nanofiltration was operated with a transmembrane pressure ranging between 1.5 and 4.0 bar. The membrane was operated in cross flow mode with a cross flow velocity of 1.0 to 1.2 m/s and a recovery of about 75 to 85%. For additional cake layer control a back-wash was applied every 45 min for 1 min.

Operational flexibility. PAC/NF can be designed to cope with varying amounts and qualities of raw water. However adjustment of the PAC dosage to *changing* water flows and dosages will require a precise and reliable dosage system. Due to a modular set-up in blocks the adaptation of the NF to changing quantities occurs through switching NF blocks on and off. Additionally the permeate flux can be varied in the range of 10 to 25 $L/(m^2 \cdot h)$ equivalent to 40 to 100%.

Operational reliability. PAC/NF can be regarded as a technically feasible process for large scale centralized applications. It cannot be applied in decentralized systems where a regular control and maintenance of all parts are not ensured. Chemicals and activated *carbon* pose a moderate potential risk for the personnel but can be reduced through proper handling and design of the plant. The operational risk is expected to be comparable to RO plants. Nevertheless our operational experiences with the pilot plant showed a couple of challenges which have to be mastered prior to full scale application:

- A challenging control of the cake layer formation.
- A constant demand for chemical cleaning every 10 to 14 days (chemical cleaning could be replaced by chemical enhanced back-wash, e.g. 4 times a day).
- A fast increase of the TMP when applying higher PAC doses whereas extended backwash or reduced filtration cycles should be able to tackle this problem.
- Reduced lifetime of module while a more stable operation avoiding extreme conditions with a pressure above the limits should extend the lifetime significantly.

Table 15.9 Factors affecting the performance of the PAC/NF process with regard to rejection of organic solutes and economic efficiency (Bellona *et al.* 2004; Crittenden *et al.* 2005; Snyder *et al.* 2007; Schäfer *et al.* 2005; Yangali Quintanilla, 2010).

	Powdered activated carbon	Nanofiltration
Material properties	Inner surface Pore size distribution Particle size distribution Raw material and origin Shape	MWCO and porosity Surface charge Surface roughness Applicable pH range Resistance against oxidants
Water characteristics	Temperature pH value Water matrix (EfOM) Salinity	Temperature pH value Raw water matrix (foulants) Ion concentration (e.g. Ca^{2+})
Solute characteristics	log K_{OW} and log D Charge (pKa) Diffusivity	Molecular weight log KOW Charge (pKa) Molecular size
Operational conditions	Carbon concentration Reactor type Mixing Contact time (PAC reactor and crossflow recirculation loop)	Permeate flux Recovery Transmembrane pressure Crossflow velocity Backflush duration and cycle Membrane cleaning method Dosage of chemicals (e.g. antiscalants)

15.4.3 Direct NF

Operational flexibility. NF plants can be designed to cope easily with changing amounts and qualities of raw water. Future extensions are easy to implement if already foreseen in the design of the piping and automation system. The process provides high quality treatment on a very small footprint and thus should be easily integrated in existing WWTPs which often lack sufficient space for large extensions.

Operational reliability. Fouling and biofouling pose major obstacles for long term operation of NF systems. Thus fouling control and in some cases scaling control are the main *issues* from an operational perspective. Recent studies investigated the preferential operational settings of direct NF systems for fouling control, such as the influence of an air flush and forward flush (Bonné *et al.* 2003; Schrader, 2006), of coagulation as pre-treatment (Schrader, 2006), of an electric field (Schrader, 2006) and the influence of the raw water conditions (Schrader, 2006).

15.4.4 NF/GAC treatment

Operational performance in terms of flexibility and reliability is comparable to a direct NF system. The fully automated pilot plant at the Aachen Soers WWTP was operated for more than a year without major interruption. Still fouling required a chemical cleaning about every three to four weeks. Bed volumes are expected to be above 30,000 for less adsorble critical compounds such as sulfamethoxazole. The process complexity consisting of a complex NF unit and a GAC filter is relatively high. At a flux of 20 $L/m^2 \cdot h$ the capillary NF plant was operated at a transmembrane pressure of about 2.5 to 3.5 bar. Transfer of the tests in full scale applications appear to be straightforward.

15.4.5 GAC/NF treatment

According to Roorda *et al.* (2005) the optimum process conditions were found at a constant flux of 30 $L/m^2 \cdot h$ and a TMP of 4.5 bar at a cross-flow velocity of 1.0 m/s and an anti-scalant dosage of 3 mg/l (4 Aqua OSM 92). The membrane was cleaned every 30–45 min with a combined air flush and a forward flush 4–5 times a day with a chemical flush with acid (HCl at pH 2). Intensive chemical cleaning was only necessary every 3 weeks. Compared to direct NF which showed a rapid flux decrease, the operation of the ANF system was stable at 25 to 30 LMH.

15.4.6 MBR/NF treatment

Long-term tests with the MBR/NF unit at a minimzed liquid discharge mode confirmed the high risk of scaling of the dense membrane. As preventive measure operational conditions such as permeate recovery, pH and anti-scalants have to be adjusted to the ion composition to avoid any scaling. Details are given in Joss *et al.* (2011).

15.5 ECONOMIC ASPECTS

15.5.1 Dual membrane treatment

Lazarova *et al.* (2003) reports total treatment costs for dual membrane systems to be in the range of 0.22–0.44 US$/m^3. Due to meanwhile significantly increased energy costs these specific costs might have risen slightly.

The costs for the new Orange County GWRS plant are estimated at 0.28 US$/m^3 for operation and maintenance and 0.25 US$/m^3 for capital costs based on a 20 years depreciation (Deshmukh, 2004 and GWRS, 2005). Energy consumption contributes with 0.12 US$/m^3 to the O&M costs, equivalent to 45%. The power consumption for the US installations was estimated to be in the range between 1.1 and 1.2 kWh/m^3 which is significantly below current sea water RO desalination systems. The dual membrane system of the Torreele plant has a specific consumption of 0.82 kWh/m^3. The actual value for the New Water plant in Singapore is around 1.0 kWh/m^3 while the estimated consumption was 1.2 kWh/m^3. Typical energy demand values are given in Table 15.10.

Table 15.10 Typical energy use required by various RO applications (Orchard, 2007).

RO applications	Energy use (kWh/m^3)
Conventional treatment of surface water	0.2–0.4
Reclamation of municipal wastewater to reuse	0.5–1.0
Brackish water desalination	0.7–1.2
Seawater desalination	2.5–3.2

15.5.2 NF-AC hybrid systems

From the economic point of view the presented pilot studies do not indicate strong differences in costs for the three hybrid processes as the overall costs are mainly driven by the operational and capital costs for the nanofiltration and to a minor degree by the carbon price and consumption. However, the specific costs for the activated carbon treatment are significantly lower in NF/GAC than in PAC/NF and GAC/NF. Due to a number of constraints any detailed estimation at this stage is hampered by significant uncertainties, such as lack of experience in full scale application, variation of costs due to site specific aspects (*e.g.* plant size, quality upgrading requirement) and future development (*e.g.* optimised membrane characteristics and increased competition lowering specific NF membrane prices).

Furthermore the main question for an optimised implementation of a process with dense membranes remains the disposal of the concentrates and the achievement of high recoveries requiring an optimum embedding of the NF and AC process in the overall wastewater treatment concept. Thus cost comparisons for NF in combination with AC treatment options have to be done on the basis of the case specific boundary conditions.

The applied capillary NF membrane implies investment costs in the range of 2000–2500 € per 20 m^2 module for medium to large full scale plants. Thus the specific price ranges between 100 and 125 €/m^2. With a Lang factor of 3 to 5, a specific design permeate flux of 25 L/(m^2 · h), and an average recovery of 70 %, the specific investment costs for a full scale plant varies between 20,000 and 42,000 €/(m^3/h). Based upon an average depreciation of 10 %/year the specific capital costs range between 0.23 and 0.48 €/m^3. The operational costs largely depend on the local boundary conditions in terms of labour and energy costs. The operational costs can be estimated to 0.20 to 0.25 €/m^3 (Meier, 2008) including PAC dosage of 50 mg/L (= 0.06 €/m^3). GAC costs can be slightly lower but higher investment costs are required for the GAC installation. The total costs can therefore be estimated to range between 0.43 and 0.73 €/m^3.

Energy demand for NF depends on several factors, such as plant size, raw water salinity, plant configuration, type of NF membrane and module, and the raw water temperatureThe specific energy requirement for AC-NF hybrid processes can be estimated based upon full scale RO and NF water reclamation references. According to this the energy requirement should range between 0.5 and 1.0 kWh/m^3.

15.5.3 MBR/NF treatment

Economic viability: The costs for NF are expected to be in the same range as for reverse osmosis. The specific price might vary around 0.30 to 0.50 €/m^3.

Energy consumption for NF depends on several factors, such as plant size, raw water salinity, plant configuration, type of NF membrane and module, raw water temperature, etc. In Mery-sur-Oise (drinking water production), the NF required a mean of 0.35 kWh/m^3 and varied between 0.27 in summer and 0.50 kWh/m^3 in winter (Cyna *et al.* 2002). Direct NF for effluent reclamation is expected to consume less than 1 kWh/m^3.

15.6 SUMMARY AND CONCLUSIONS

Emerging hybrid processes combining activated carbon and nanofiltration as MBR/NF have been studied to identify the optimum solution for advanced water treatment in water reclamation and reuse. With a focus on the removal of bulk and trace organic compounds several promising process combinations have been identified, namely powdered activated carbon followed by nanofiltration (PAC/NF), granular activated carbon followed by nanofiltration (GAC/NF) and nanofiltration followed by granular activated carbon (NF/GAC).

All investigated activated carbon nanofiltration hybrid processes as well as MBR/NF provide very high quality product water with dissolved organic carbon concentrations clearly below 0.5 mg/L and organic micropollutant concentrations close or below the limit of quantification. This complies with the strict Californian standard. If a high degree of salinity removal is not required, nanofiltration-activated carbon hybrid processes can be regarded as a reliable and economic alternative to dual membrane processes (UF/RO). Due to lower filtration pressures the hybrid processes feature lower energy consumption and produce less problematic concentrates mainly consisting of organics and multivalent ions which can be precipitated. They feature also some advantages in indirect potable reuse applications such as managed aquifer recharge since the salt content of the product water is closer to natural conditions.

15.6.1 Removal rates

In comparison to a single stage NF or a single adsorption stage, all three hybrid processes provide a superior product quality. In NF/GAC and GAC/NF the removal of the selected micropollutants was slightly better than in PAC/NF (Table 15.11). Overall each of the investigated hybrid processes can treat WWTP effluent to a quality suitable for high quality water reuse applications. If higher rejection rates for salts are required, NF/GAC and GAC/NF can employ tighter NF membranes than the investigated capillary nanofiltration membrane. The PAC/NF process requires the use of a backwashable nanofiltration system such as the investigated capillary nanofiltration. The optimum carbon dose was determined at 50 mg/L using the PAC type SAE Super. In the light of the overall excellent product water quality with only slight differences between the respective hybrid processes, the process selection could be made on the basis of the cost effectiveness and operational performance. Rapid small scale column tests showed that GAC types with a high percentage of mesopores are favourable for applications with high DOC concentrations whereas waters with low DOC such as NF permeate allow the use of high-grade drinking water GAC to maximise the runtime of the filter.

Table 15.11 Removal rates of investigated activated carbon, nanofiltration and hybrid processes for key contaminants in water reuse (values may vary according to type of PAC and NF).

	PAC***	PAC	Cap NF	PAC/NF	PAC/NF	GAC/NF	NF/GAC
PAC dose/BV	10 mg/L	50 mg/L	0 mg/L	10 mg/L	50 mg/L	<7000	20000
Bulk organics							
DOC	20%	50–60%	90–95%	90–95%	90–95%	90–95%	>95%
Micropollutants							
Benzotriazole	80%	95%	50%	85%	98%	>99.8%	>99.8%
Primidone	50%	85%	65%	55%	95%	>99.8%	>99.8%
Carbamazepine	75%	98%	50%	85%	99.9%	>99.9%	>99.9%
Sulfamethoxazole	30%	70%	65%	50%	90%	>99.8%	>99.8%
Macrolides	80%	>99%	80%	95%	>99%	>99%*	>98%
Contrast media	20–30%	75–80%	99%	90%	97–99%	>99%*	>99%
Diatrizoate	0%	50%	95%	85%	94%	>99%*	>99%
Salinity							
Conductivity	–	–	15–20%	12%	12%	12%	15–20%
Sulfate	–	–	>75%	60–70%	60–70%	>70%	>75%
Chloride	–	–	5–20%	3–5%	–5–0%	3%	5–20%
Pathogens**	–	–	>6 log	>6 log	>6 log	>6 log	>6 log

*qualified estimation, not investigated in detail.
**Asano *et al.* 2007, NRMMC–EPHC–NHMRC, 2008.
***SAE Super, other carbon types might deviate in rejection rates.

The advanced analytical methods for characterisation of the bulk organics (Kazner *et al.* 2008c) revealed that the product water from AC-NF processes is lower in organic carbon than typical drinking water since most of the soluble microbial products are removed alongside the typical drinking water humic substances. Fluorescence spectrometry proved to be a very powerful tool allowing to detect wastewater borne bulk organic substances in samples with DOC concentrations below the limit of quantification (<140 μg/L).

15.6.2 Comparison with other alternative processes

AC-NF hybrid processes compete with PAC/UF and PAC/MF as well as with double membrane processes, *i.e.* UF/RO, which is the benchmark process in high-grade water reuse (Table 15.12). The removal of all key contaminants occurs in AC-NF hybrid processes on a level comparable to UF/RO. The more open membrane however retains less salts and monovalent ions such as nitrate, so that AC-NF hybrid processes cannot be employed, if severe salinity problems prevail in the catchment area. Similar to UF/RO, AC-NF processes present a double barrier for key contaminants. While UF/RO provides a double barrier for pathogens, AC-NF processes feature a double barrier for organic compounds. If a natural post-treatment such as managed aquifer recharge is foreseen, remaining contamination in particular NDMA can be well removed during the soil passage. Typical European wastewater also has lower nitrosamine concentrations close to the strict Californian recommendation for recharge so that an AOP post-treatment will not be necessary in most cases. Powdered activated carbon combined with porous membranes is a good alternate process for effluent polishing, if the advantages of nanofiltration such as removal of organics and multivalent ions are not required. However PAC/UF and PAC/MF are no multi-barrier systems.

Table 15.12 Technical performance of activated carbon, nanofiltration and hybrid processes.

Critical parameter	UF	NF	PAC	GAC	PAC/UF	PAC/NF	GAC/NF	NF/GAC	UF/RO MBR/RO MBR/NF
Removal									
Bulk organics	0	++	+	++	++	+++	+++	+++	+++
Organic micropollutants	–	+(+)	+(+)	++	++	++	+++	+++	+++
Microbial parameters	+++	+++	–	–	+++	+++	+++	+++	+++
Salts	–	++	–	–	–	+	++	++	+++
Fouling	0	0	+	–	+	+	++	0	++

Legend: +++ excellent, ++ very good, + good, 0 moderate, – poor

15.6.3 Treatment of NF concentrate

One main advantage of the PAC/NF and GAC/NF process is the significant reduction of trace organics in the NF concentrate. Direct NF generates a concentrate with about fourfold elevated micropollutant concentrations reaching levels up to 30 μg/L at a recovery of 75%. A PAC dosage of 50 mg/L reduced the micropollutant concentrations in the liquid phase of the concentrate below the feed concentrations.

In contrast to activated carbon pre-treatment in PAC/NF and GAC/NF, NF/GAC does not contribute to concentrate treatment. In cases where a high recovery is desired or the discharge of the NF concentrates is not possible due to local conditions, additional treatment of the NF concentrate should be foreseen, *e.g.* with a second GAC filter, which enables to return the concentrates to the WWTP. Low grade GAC with a high percentage of meso-pores was found to be advantageous for these applications.

MBR/NF with concentrate recycle to the bioreactor proved to be feasible but required strict control of the inorganic and organic solutes to limit fouling and scaling processes.

15.6.4 Operational aspects

The capillary NF showed a very stable performance without any unintentional interruptions in all process combinations. The operational strategy of the NF can be adapted to the respective raw water quality and hybrid process type. Pre-treatment with GAC provided the best feed water quality and reduced the fouling rates. When aiming at an optimised performance of the GAC filter, NF/GAC is superior to the pre-treatment with activated carbon. After almost 20,000 bed volumes bulk and trace organics were still removed stably. It was not necessary to backflush the filter. The operation of the PAC/NF process was the most challenging. The cake layer control was identified as the operational key issue to limit increased concentration polarisation and reduce the transmembrane pressure. Frequent backflushing at the expense of reduced net fluxes proved to have the best potential for a stable operation of PAC/NF.

From the operational perspective NF/GAC was the most favourable combination (compare Table 15.13), but requires a good feed water quality to allow for direct NF. For the treatment of poorer raw water qualities GAC/NF and PAC/NF are good alternatives. Further PAC/NF has still a high optimisation potential to reduce the carbon usage through recycling within the PAC/NF unit and the surplus carbon back to the activated sludge system, as developed by Metzger (2010) for PAC treatment of WWTP effluent.

Table 15.13 Advantages and disadvantages of the activated carbon-nanofiltration processes.

	PAC/NF process	GAC/NF process	NF/GAC process
Advantages	Adjustable carbon dose	Combined concentrate treatment	Reduced cake layer formation
	Adsorption on increased concentration level	Adsorption on influent concentration level	Employment of various NF modules
	Combined concentrate treatment	Reduced fouling	Separate optimisation of NF and AC
		Employment of various NF modules	No fouling of GAC filter
		Separate optimisation of NF and AC	Reactivation of GAC
Disadvantages	Carbon cake layer formation	High fouling of GAC filter	No treatment of concentrate
	Increased waste production	Frequent replacement of GAC	Design of GAC filter for maximum load
	No carbon reactivation	Limited reactivation of GAC	

15.6.5 Conclusion

Activated carbon-nanofiltration as well as MBR/NF hybrid processes are promising alternatives to existing treatment trains. They are well suited for high-grade water reuse applications. The present study demonstrated the feasibility of all processes and identified advantages and disadvantages of the processes. The optimum process combination depends on local boundary conditions such as size of plant, raw water characteristics and plant location. In the light of growing water scarcity and increasing concerns about organic micropollutants activated carbon treatment in combination with nanofiltration could be applied in an increasing number of cases with appropriate boundary conditions.

Future applications should be embedded in integrated wastewater treatment concepts including concentrate treatment. Particularly combinations with membrane bioreactors are interesting options for innovative sustainable water recycling solutions.

ABBREVIATIONS AND NOMENCLATURE

AC	Activated carbon
ANF	Adsorption filtration followed by nanofiltration
AOI	Adsorbable organic iodine
AOP	Advanced oxidation process
BTA	Benzotriazole
Cap NF	Capillary nanofiltration
CAS	Conventional activated sludge
CBZ	Carbamazepine
CLA	Clarithromycin
COD	Chemical oxygen demand
DIATR	Diatrizoic acid
DCF	Diclofenac
DNF	Direct nanofiltration
DOC	Dissolved organic carbon
EEM	Excitation emission matrix
EfOM	Effluent organic matter
ERY	Erythromycin
GAC	Granular activated carbon
ICM	Iodinated contrast media

IOHEX	Iohexol
IOMEP	Iomeprol
IOPAM	Iopamidol
IPM	Iopromide
LC-MS	Liquid chromatography-mass spectrometry
LMH	Permeate flux rate expressed in $L \cdot m^{-2} \cdot h^{-1}$.
LOQ	Limit of quantification
LPRO	Low pressure reverse osmosis
MAR	Managed aquifer recharge
MBR	Membrane bio-reactor
MF	Microfiltration
MW	Molecular weight
MWCO	Molecular weight cut off
MS	Mass spectrometry
NF	Nanofiltration
PAC	Powdered activated carbon
PRIM	Primidone
RO	Reverse osmosis
SAT	Soil aquifer treatment
SMX	Sulfamethoxazole
SUVA	Specific ultraviolet absorption
TMP	Transmembrane pressure
TOC	Total organic carbon
UF	Ultrafiltration
UV	Ultraviolet radiation
UVA	Ultraviolet absorption
WWTP	Wastewater treatment plant

REFERENCES

Alturki A. A., Tadkaewa N., McDonald J. A., Khan S. J., Price W. E. and Nghiem L. D. (2010). Combining MBR and NF/RO Membrane Filtration for the Removal of Trace organics in Indirect Potable Water Reuse Applications. *Journal Of Membrane Science*, **365**(1-2), 206–215, doi:10.1016/j.memsci.2010.09.008.

Asano T., Burton F. L., Leverenz H. L., Tsuchihashi R. and Tchobanoglous G. (2007). Water Reuse – Issues, Technologies, and Applications. McGraw Hill, NY, ISBN 978-0-07-145927-3.

Baumgarten S., Schröder H. F., Charwarth C., Lange M., Beier S. and Pinnekamp J. (2007). Evaluation of Advanced Treatment Technologies for the Elimination of Pharmaceutical Compounds. Gewässerschutz-Wasser-Abwasser GWA 206, 1/1 – 1/10, ISBN 978-3-938996-12-6.

Bellona C., Drewes J. E., Xu P. and Amy G. (2004). Factors affecting the rejection of organic solutes during NF/RO treatment – A literature review. *Water Research*, **38**(12), 2795–2809.

Bellona C. and Drewes J. E. (2007). Viability of a low-pressure nanofilter in treating recycled water for water reuse applications: A pilot-scale study. *Water Research*, **41**, 3948–3958.

Beros M., Ventresque C., Bablon G., Tanghe N. and Chagneau G. (2003). Betriebserfahrung Mit Einer NF-GroßAnlage Zur Trinkwasseraufbereitung. Proceedings of the 4th ATSV Conference "Membrantechnik in der Wasseraufbereitung und Abwasserbehandlung", ISBN 3-921955-25-4.

Bixio D. and Wintgens T. (2006). Water Reuse System Management Manual – AQUAREC, Project Report, Office for Official Publications of the European Communities, Luxembourg, 2006, ISBN.

Bonné P. A. C., Hiemstra P., Van der Hoek J. P. and Hofman J. A. M. H. (2002). Is direct nanofiltration with air flush an alternative for household water production for Amsterdam? *Desalination*, **152**(2002), 263–269.

Crittenden J. C., Trussell R. R., Hand D. W., Howe K. J. and Tchobanoglous G. (2005). Water Treatment: Principles and Design. 2nd edn, John Wiley & Sons, Hoboken, NJ, ISBN 0-471-11018-3.

Cyna B., Chagneau G., Bablon G. and Tanghe N. (2002). Two years of nanofiltration at the Mery-sur-Oise plant, France. *Desalination*, **147**, 69–75.

Deshmukh S. (2004). Operational Experience with Membrane Filtration Systems for Groundwater Recharge. Proc. Aquarec Workshop, Thessaloniki, Greece, 11–12, March 2004.

Dickenson E. R. V. and Drewes J. E. (2010). Quantitative structure property relationships for the adsorption of pharmaceuticals onto activated carbon. *Water Science and Technology*, **62**, 2270–2276.

Drewes J., Bellona C., Xu P., Amy G., Filteau G. and Oelker G. (2008). Comparing Nanofiltration and Reverse Osmosis for Treating Recycled Water, AwwaRF Report 91212, IWA Publishing, London, UK, ISBN: 9781843392163.

Duin O., Wessels P., van der Roest H., Uijterlinde C. and Schoonewille H. (2000). Direct nanofiltration or ultrafiltration of WWTP effluent? *Desalination*, **132**, 65–72.

Eilers L. (2000). Verfahrenskombination von Nanofiltration und Adsorption an Pulverkohle zur Kontinuierlichen Abwasserreinigung (Process Combination of Nanofiltration and Adsorption on Activated Carbon for Contious Wastewater Treatment). PhD thesis, Department of Chemical Engineering, RWTH Aachen University, http://sylvester.bth.rwth-aachen.de/dissertationen/2001/027/ (accessed 10 June 2011).

Eltner A. (1998). Behandlung hoch belasteter AbwäSser durch Aktivkohleadsorption und Aktivkohle/Nanofiltration – Verfahrensvergleich und Bewertung (Treatment of Highly Polluted Wastewater by Activated Carbon and Activated Carbon/Nanofiltration – Comparison and Evaluation of Technologies). PhD thesis, Department of Chemical Engineering, RWTH Aachen University.

Ernst M. (2000). Abtrennung organischer Verbindungen aus kommunalen Kläranlagenabläufen mittels Nano- und Ultrafiltrationsmembranen mit dem Ziel der künstlichen Grundwasseranreicherung (Separation of organic compounds from municipal wastewater treatment plant effluents by means of ultra- and nanofiltration membranes for artificial groundwater recharge), VDI series 15, Environmental Engineering 221, VDI-Verlag, Dusseldorf.

Frank M., Bargeman G., Zwijnenburg A. and Wessling M. (2001). Capillary hollow fiber nanofiltration membranes. *Separation and Purification Technology*, **22–23**, 499–506.

Futselaar H., Schonewille H. and van der Meer W. (2002). Direct capillary nanofiltration – a new high-grade purification concept. *Desalination*, **145**, 75–80.

Futselaar H., Schonewille H. and van der Meer W. (2003). Direct capillary nanofiltration for surface water. *Desalination*, **157**, 135–136.

Heijman S. G. J., Verliefde A. R. D., Cornelissen E. R., Amy G. and van Dijk J. C. (2007). Influence of natural organic matter (NOM) fouling on the removal of pharmaceuticals by nanofiltration and activated carbon filtration. *Water Science & Technology - Water Supply*, **7**(4), 17–23.

Herr J. (2007). Entfernung von organischen Stoffen aus Kläranlagenabläufen mit der Verfahrenskombination Pulveraktivkohle-Nanofiltration. Diploma Thesis, RWTH Aachen University.

Hoek E. M. V., Kim A. S. and Elimelech M. (2002). Influence of crossflow membrane filter geometry and shear rate on colloidal fouling in reverse osmosis and nanofiltration separations. *Environmental Engineering Science*, **19**, 357–372.

Hoek E. V. and Elimelech M. (2003). Cake-enhanced concentration polarisation: a new fouling mechanism for salt-rejecting membranes. *Environmental Science and Technology*, **37**, 5581–5588.

Joss A., Baenninger C., Foa P., Koepke S., Krauss M., McArdell C. S., Rottermann K., Wei Y., Zapata A. and Siegrist H. (2011). Water reuse: >90% water yield in MBR/RO through concentrate recycling and co_2 addition as scaling control. *Water Research*, **45**(18), 6141–51.

Kazner C., Fink G., Ternes T., Wintgens T. and Melin T. (2007). Removal of Organic Micropollutants by Nanofiltration in Combination with Adsorption on Powdered Activated Carbon for Artificial Groundwater Recharge with Reclaimed Wastewater, Proceedings of the 5th IWA Micropol & Ecohazard 2007 conference, 17 – 20 June 2007, DECHEMA e.V., Frankfurt/Main, Germany, 259–265.

Kazner C., Lehnberg K., Kovalova L., Wintgens T., Melin T., Hollender J. and Dott W. (2008a). Removal of endocrine disruptors and cytostatics from effluent by nanofiltration in combination with adsorption on powdered activated carbon. *Water Science and Technology*, **58**(8), 1699–1706.

Kazner C., Wintgens T. and Melin T. (2008b). Factors Affecting the Performance and Optimization of the PAC/NF Process for Water Reclamation. Proc. of the MDIW conference 2008, Toulouse.

Kazner C., Baghoth S., Sharma S., Amy G., Wintgens T. and Melin T. (2008c). Comparing the effluent organic matter removal of direct NF and powdered activated carbon/NF as high quality pretreatment options for artificial groundwater recharge. *Water Science & Technology*, **57**(6), 821–827.

Kazner C., Meier J., Wintgens T. and Melin T. (2009). Capillary Nanofiltration Coupled with Powdered Activated Carbon Adsorption for High Quality Water Reuse. *Water Science & Technology*, **60.1**, 251–259.

Lazarova V., Hills S. and Birks R. (2003). Using recycled water for non-potable, urban uses: a review with particular reference to toilet flushing. *Water Supply*, **3**(4), 69–77.

Li X., Hai F. I. and Nghiem L. D. (2011). Simultaneous activated carbon adsorption within a membrane bioreactor for an enhanced micropollutant removal. *Bioresource Technology*, **102**(9), 5319–5324.

Meier J., Melin T. and Eilers L. (2002). Nanofiltration and adsorption on powdered adsorbent as process combination for the treatment of severly contaminated waste water. *Desalination*, **146**, 361–366.

Meier J. and Melin T. (2005). Wastewater reclamation by the PAC/NF process. *Desalination*, **178**, 27–40.

Meier J. (2008). Auswirkungen von Pulveraktivkohle auf die Nanofiltration von Kläranlagenablauf (Effect of Powdered Activated Carbon on the Nanofiltration of Sewage Plant Effluent). PhD thesis, Department of Chemical Engineering, RWTH Aachen University, http://darwin.bth.rwth-aachen.de/opus3/volltexte/2008/2556/ (accessed 10 June 2011).

Melin T. and Rautenbach R. (2007). Membranverfahren, Grundlagen der Modul- und Anlagenauslegung, 3rd edn, Springer, Berlin, Heidelberg, New York, ISBN 978-3-540-34327-1.

Metzger S. (2010). Einsatz von Pulveraktivkohle zur weitergehenden Reinigung von kommunalem Abwasser (Application of Activated Carbon for the Advanced Treatment of Municipal Wastewater). PhD thesis, Technical University of Berlin, Oldenbourg Industrieverlag, ISBN: 978-3835632318.

Mulder M. (2004). Basic Principles of Membrane Technology, 2nd edn, Springer, Netherlands, ISBN 978-0-7923-4247-2.

Norit (2002). Technical Data Sheet NF50 M10, Norit X-Flow BV, Enschede, NL.

Nghiem L. D., Schäfer A. I. and Elimelech M. (2004). Removal of natural hormones by nanofiltration membranes: measurement, modeling, and mechanisms. *Environmental Science & Technology*, **38**(6), 1888–1896.

Nghiem L. D., Schäfer A. I. and Elimelech M. (2005). Nanofiltration of hormone mimicking trace organic contaminants. *Separation Science and Technology*, **40**, 2633–2649.

Roorda J. H., Wortel N. C. and van Dalen R. (2005). New process for treatment of organically fouled water: experiences with WWTP effluent. *Desalination*, **178**, 141–148.

Salzbrenner J. (2005). Betrieb einer Pilotanlage zur Aufbereitung von KläRanlagenablauf zu Industriellem Brauchwasser. Diploma Thesis, RWTH Aachen/TU Cottbus.

Sayed S., Tarek S., Dijkstra I. and Moerman C. (2007). Optimum operation conditions of direct capillary nanofiltration for wastewater treatment. *Desalination*, **214**, 215–226.

Schäfer A. I., Fane A. G. and Waite T. D. (2005). Nanofiltration: Principles and Applications. Elsevier Advanced Technology, Oxford, UK.

Schrader G. A. (2006). Direct Nanofiltration of Waste Water Treatment Plant Effluent, PhD Thesis, University of Twente, ISBN 90-365-2332-X, available online: http://www.utwente.nl/tnw/mtg/publications/mstpublications/pdf/2006_Schrader.pdf

Snyder S., Adham S., Redding A. M., Cannon F. S., DeCarolis J., Oppenheimer J., Wert E. C. and Yoon Y. (2007). Role of membranes and activated carbon in the removal of endocrine disruptors and pharmaceuticals. *Desalination*, **202**, 156–181.

Ternes T. and Joss A. (Eds) (2006). Human Pharmaceuticals, Hormones and Fragrances – The Challenge of Micropollutants in Urban Water Management, IWA Publishing, London, ISBN 978-1-8433-9093-0.

Tchobanoglous G., Burton F. L. and Stensel H. D. (2003). Wastewater Engineering, Treatment and Reuse, Metcalf & Eddy, 4th edn, McGraw Hill, Boston.

Toze S. (2006). Water reuse and health risks – real vs. perceived. *Desalination*, **187**, 41–51.

Trussell R., Adam S. and Jackangelo J. (2000). Potable Water Reuse and the Role of Membranes, Proceedings of the 3rd ATSV Conference "Membrantechnik in der Wasseraufbereitung und Abwasserbehandlung" 2000, paper A22, Aachen.

Van Houtte E., Verbauwhede J., Vanlerbege F., Demunter S. and Cabooter J. (1998). Treating different types of wastewater with micro- and ultrafiltration for further desalination using reverse osmosis. *Desalination*, **117**, 49–60.

Van Houtte E. and Verbauwhede J. (2004). Closing the water cycle: twenty months of operational experiences in Torreele (Flanders). Proc. Intl Workshop on Implementation and Operation of Municipal Wastewater Reuse Plants, Thessaloniki, Greece, 11–12, March 2004.

Verliefde A. R. D., Heijman S. G. J., Cornelissen E. R., Amy G. L., van der Bruggen B. and van Dijk J. C. (2007). Influence of electrostatic interactions on the rejection with NF and assessment of the removal efficiency during NF/GAC treatment of pharmaceutically active compounds in surface water. *Water Research*, **41**(15), 3227–3240.

Verliefde A. R. D., Cornelissen E. R., Heijman S. G. J., Verberk J. Q. J. C., Amy G. L., Van der Bruggen B. and van Dijk J. C. (2008a). The role of electrostatic interactions on the rejection of organic solutes in aqueous solutions with nanofiltration. *Journal of Membrane Science*, **322**, 52–66.

Verliefde A. R. D. (2008b). Rejection of Organic Micropollutants by High Pressure Membranes (NF/RO), PhD thesis, TU Delft, NL, Water Management Academic Press, ISBN 978-9-0895-7005-5.

Wintgens T., Melin T., Schafer A., Khan S., Muston M., Bixio D. and Thoeye C. (2005). The role of membrane processes in municipal wastewater reclamation and reuse. *Desalination*, **178**(1-3), 1–11.

Wintgens T., Melin T., Salehi F. and Hochstrat R. (2008). Emerging contaminants and treatment options in water recycling for indirect potable use. *Water Science and Technology*, **57**, 99–107.

Yangali-Quintanilla V. A. (2010). Rejection of Emerging Organic Contaminants by Nanofiltration and Reverse Osmosis Membranes: Effects of Fouling, Modelling and Water Reuse, CRC Press/Balkema, Leiden, NL, ISBN 978-0-415-58277-3.

Yoon Y. and Lueptow R. M. (2005). Removal of organic contaminants by RO and NF membranes. *Journal of Membrane Science*, **261**(1-2), 76–86 doi:10.1016/j.memsci.2005.03.038.

Zwickenpflug B., Böhler M., Sterkele B., Joss A., Siegrist H., Traber J., Gujer W., Behl M., Dorusch F., Hollender J., Ternes T. and Fink G. (2010). Einsatz von Pulveraktivkohle zur Elimination von Mikroverunreinigungen aus Kommunalem Abwasser (Use of powdered activated carbon for the removal of micropollutants from municipal wastewater). Final report of the MicroPoll project, EAWAG on behalf of the Swiss Federal Office for the Environment, Dübendorf, CH, http://www.eawag.ch/forschung/eng/schwerpunkte/abwasser/strategie_micropoll/pak_eawag/Abschlussbericht_MicroPoll_PAK.pdf (accessed 10 June 2011).

Chapter 16

Treatment of reject streams from dense membrane processes

How Yong Ng, Winson Lay, Lai Yoke Lee, Guihe Tao, Harry Seah, Jessica Benner, Thomas Ternes and Chris Thoeye

16.1 TREATMENT OF RO CONCENTRATE

The reverse osmosis (RO) process has been widely adopted as the technology of choice for modern water reclamation application. RO is essentially a water desalting process whereby its end products, within the context of water reclamation, consist of a stream of high quality permeate suitable for indirect potable or direct non-potable use, and a stream of RO concentrate or brine.

The consideration of brine disposal is critical in the planning of a RO water reclamation facility. The generated brine contains both moderate to high concentration of organics and inorganic compounds, which necessitate careful consideration in its management and disposal. The method of brine disposal is site specific, though a number of basic brine disposal methods are known (AWWA, 2004). State-of-the-art treatment relies on evaporation ponds, falling film evaporators, crystallization, spray dryers, membrane distillation, discharge to the wastewater collection system, land application, deep well injection or disposal to surface waters (Asano *et al.* 2007; Khan *et al.* 2009). All available methods have serious shortcomings either from the environmental or the economic perspective. Promising emerging technologies for concentrate desalination are capacitive deionisation (CDI) (this study), electrodialysis (ED) (Adham *et al.* 2006) and forward osmosis (FO) (Adham *et al.* 2007).

At water reclamation facilities where the brine is to be discharged into inland or sensitive water bodies, proper brine treatment will be needed prior to discharge. The concern is on the potentially adverse environmental impacts of the brine. This concern tends to be manifested in stringent regulation and high disposal cost imposed on the discharged brine. Meeting regulatory requirements and reducing disposal cost, in this instance, form the driver to putting in place a dedicated brine treatment process.

However, where there is a suitable saline sink as the brine receiving body, for example, well-mixed sea or ocean, the impacts of brine from a water reclamation facility on the environment tend to be mitigated. It is then usually possible to fulfill regulatory requirements by mixing and co-disposing the brine from a water reclamation facility together with the secondary effluent from a nearby sewage treatment plant. The driver for brine treatment, in this instance, is to view the brine as an untapped water resource, whereby the objective is to increase water resource by treating and recovering brine from a water reclamation facility.

Depending on the background and driver for different situation, the approach taken for the treatment of brine from water reclamation facilities may differ. Two approaches were undertaken within the RECLAIM WATER project on the treatment of the RO brine stream:

(a) Treatment of brine to meet regulatory requirements and to reduce disposal cost; and
(b) Treatment of brine to increase water recovery of the water reclamation process

The goal of the brine treatment studies was to provide cost-effective technology solutions to water utilities for the management of their RO brine streams.

16.2 CAPACITIVE DEIONISATION (CDI)

In Singapore, NEWater is the product from a **multiple barrier** water reclamation process using advanced dual-membrane (microfiltration/ultrafiltration and reverse osmosis) and ultraviolet technologies. It is a high-grade water of drinking water quality reclaimed from domestic used water. By closing the water loop, NEWater is a key pillar of PUB's integrated water resource management. The four water resources of Singapore are water from local catchment, imported water, desalinated water and NEWater. Together, they make up a robust and diversified water supply strategy known as the Four National Taps.

In the current NEWater production, brine generated from the RO process is more than 20% of the RO feed flow. The brine therefore presents an additional source of water that is untapped, as it contains moderate to high concentration of ions and organics, which prevents its direct reuse.

Within the framework of the RECLAIM WATER project, a study by the National University of Singapore and Public Utilities Board, Singapore was undertaken to develop a brine treatment process that is not constrained by water chemistry, and capable of recovering water from the currently untapped source of RO brine from the NEWater factories.

The objective is to increase water recovery of the NEWater factories to more than 95% at the required water quality by treating and recovering RO brine. Correspondingly, the volume of brine to be disposed will reduce to less than 5%, and thereby achieving waste minimization. If successful, the positive implication derived from this project is the multiplying effect of water reuse, whereby water resource is increased by increasing recovery of the water reclamation loop. The integration of brine treatment within the NEWater process is schematically illustrated in Figure 16.1.

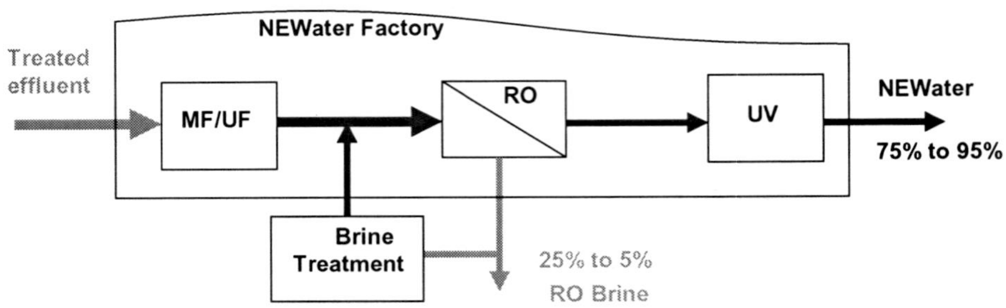

Figure 16.1 Objective of increasing water recovery with RO brine treatment

16.2.1 Concept

The brine treatment process comprises the CDI technology and an integrated pre-treatment. CDI is the core of the brine treatment process. CDI targets at the removal of ions including divalent and trivalent ions in the RO brine that cause scaling in RO. The removal of these inorganic compounds is critical, in order to recycle the treated brine as feed water to the RO and thereby increase water recovery. The pretreatment comprises biological activated carbon (BAC) with an optional second step of microfiltration (MF) or ultrafiltration (UF) prior to the CDI. The purpose of the pretreatment is to remove organics, microorganisms and particulates that would otherwise foul the CDI.

The CDI process is a low-pressure electrochemical process that removes dissolved ions from an electrolyte stream (brine) in the presence of an electrical field. The CDI process has been explored for removal of inorganic compounds from different types of water, namely, seawater, groundwater, industrial process water and wastewater (Weldemoed and Schutte, 2005; Garrido *et al.* 2006; ENPAR Technologies Inc., 2007).

The CDI process cycle consists of purification phase, regeneration phase and purge phase. During the purification phase, an electrical field with a potential difference of about 1.2 – 1.5 volts (direct current) between the two electrodes will remove the dissolved ions from the water as its passes through the electric field. The anions and cations will be attracted to the opposite charge and directed to the respective electrode until saturation occurs. During the purification phase, product with lower conductivity will be generated. Regeneration then takes place by reversing the potential. Hence, the ions are expelled into the rinse water and eventually purged out from the cell into a concentrate stream. The concentrate stream is purged out during the purging phase. In practice, more than 80% of water can be recovered with CDI process (Dietz, 2004). The working principle and picture of a CDI pilot unit are shown in Figure 16.2.

16.2.2 Plant set-up

RO brine

The RO brine from the second stage RO process of a water reclamation plant reclaiming treated domestic (secondary) effluent was used.

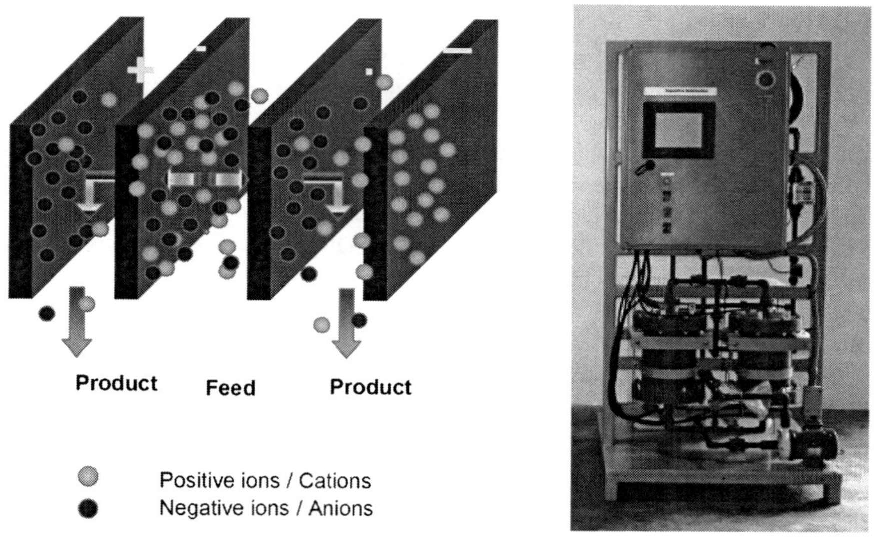

Figure 16.2 Working principle and a pilot unit of the CDI

Pretreatment

Pretreatment methods consisting of different combinations using BAC with or without UF membrane on the RO brine were used prior to desalting using the CDI process. The empty bed contact time (EBCT) in the BAC column was maintained at 60 min. Packing material used was granular activated carbon (Filtrasorb® F400D, Calgon Carbon Co.). The activated carbon had an effective size and density of 0.65 mm and 0.47 g cm^{-3}, respectively. In Pretreatment 1 (P1), RO brine was treated using BAC followed by filtration through a 5-μm cartridge filter before feeding to the CDI process. A submerged polyvinylidene fluoride (PVDF) flat sheet UF membrane with an effective surface area of 0.1 m^2 was placed after the BAC column in Pretreatment 2 (P2). The UF membrane has a nominal pore size of 0.08 μm and was operated at a flux of 10 L m^{-2} h^{-1}. Continuous aeration with compressed air at a flow rate of 6 L min^{-1} was used to provide cross-sectional shearing effect on the membrane surface. The pH was adjusted to 6.5 ± 0.2 in the submerged UF membrane tank using 0.1N HCl.

CDI process

The bench-scale CDI process (Model: DesEL 400, ENPAR Technologies Inc., Canada), with a flow rate of up to 250 mL min^{-1} and an electrode surface area of 0.7 m^2, was used for desalting the pretreated RO brine. Fouling of the CDI unit was assessed based on the pressure build-up monitored using a pressure gauge located between the peristaltic pump and the CDI cell, as well as the decrease in permeate flow rate during the purification cycle. The process cycle and product quality were controlled using a programmable logistic controller. A water recovery of 85% was maintained throughout the study for the different pretreatment methods.

16.2.3 Analytical methods

The water quality of samples from each treatment stage, namely RO brine, BAC effluent, UF effluent, and CDI product and concentrate were characterized for pH, conductivity, TDS, TOC and ion concentrations. The pH was measured using a pH meter (Model: F-54 BW, Horiba Ltd, Japan) while conductivity was quantified using a conductivity meter (Model: LF 538, WTW, Germany). TDS was quantified in accordance with the Standard Methods (APHA, 1998) and TOC was analyzed using a total organic carbon analyzer (TOC-VCSH, Shimadzu, Japan). The anion and cation concentrations were determined using ion chromatography (Model: Dionex LC 20 Chromatography, Dionex Corporation, USA). The anions analyzed were chloride (Cl^-), nitrate (NO_3^-), phosphate (PO_4^{3-}) and sulphate (SO_4^{2-}) and cations were sodium (Na^+), potassium (K^+), magnesium (Mg^{2+}) and calcium (Ca^{2+}). Chemical oxygen demand (COD) was quantified only for RO brine and BAC effluent in accordance with the Standard Methods (APHA, 1998).

16.2.4 Water quality

The typical water qualities of the CDI product pretreated using BAC, with or without submerged UF, are summarized in Table 16.1. In general, the CDI product water is comparable or of better quality than the RO feed water.

16.2.5 Operational issues

A pilot-scale study of the brine treatment process with a CDI unit of 5 m^3/d capacity was conducted at the Kranji NEWater factory. The pilot-scale studies investigated the sustainable operation and energy consumption of the brine treatment process.

An operational issue was the pressure buildup caused by fouling within the CDI unit. The typical increase of the CDI cell pressure with operation time is shown in Figure 16.3, and is to be minimized for sustainable operation. The effectiveness of the pretreatment was important. Generally, the fouling rate on CDI was lower using UF filtrate as compared to without UF. In this instance, the combined BAC-UF-CDI process was able to provide a consistent product quality with more than 85% TDS removal. The TOC concentration of the CDI product water was also comparable to the RO feed water while lower concentrations of cationic and anionic species were achieved. The CDI product water was subsequently treated by a dedicated RO system. The batch test results are shown in Table 16.2. RO permeate was of good quality including low SiO_2 concentration, indicating the advantage of using a dedicated RO operated at optimum conditions with better performance in recovering high quality water from the CDI product water. The results showed that the brine treatment process could potentially achieve more than 95% total water recovery for the NEWater factory.

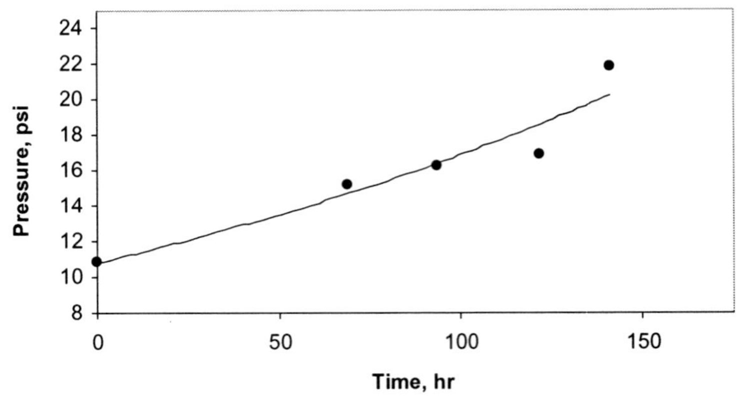

Figure 16.3 Typical change of CDI cell pressure with operation time

Table 16.1 Water quality of RO feed water, RO brine and CDI product.

Parameters	Units	RO Feed Water	RO Brine (75% Recovery)	CDI Product (85% recovery)	
				Without UF	With UF
Conductivity	μS/cm	722	1895	221.5	201
TDS	mg/L	483.7	1215	142	129
TOC	mg/L	6.7	31	7.0	6.6
Chloride	mg/L	176.0	337.8	43.8	43.2
Nitrate	mg/L	27.0	83.2	3.1	4.6
Fluoride	mg/L	3.2	7.2	N.D.	N.D.
Sulphate	mg/L	69.0	207.6	10.4	12.2
Phosphate	mg/L	17.0	28.7	6.7	5.6
Calcium	mg/L	21.0	91.8	3.8	4.5
Magnesium	mg/L	9.0	14.7	1.6	1.3
Potassium	mg/L	22.0	55.7	3.5	3.2
Sodium	mg/L	109.0	300.5	25.6	19.1

N.D. – not detectable.

The power consumption of the treatment process is estimated at 0.85 kWh/m^3, lower than the target of 1 kWh/m^3 computed based on a pressure-driven membrane process. This may be attributed to the reason that the CDI process does not require high pressure pumps and is thus able to achieve the treatment efficiency at lower energy consumption (Weldemoed & Schutte, 2005).

Table 16.2 Water quality of dedicated RO to treat CDI product water

Parameters	Units	RO Permeate
Conductivity	μS/cm	9.97
TDS	mg/L	6.15
TOC	mg/L	<0.1
Chloride	mg/L	1.47
Nitrate	mg/L	0.9
Fluoride	mg/L	0.06
Sulphate	mg/L	0.14
Phosphate	mg/L	<0.08
Calcium	mg/L	0.037
Magnesium	mg/L	<0.027
Potassium	mg/L	0.112
Sodium	mg/L	1.23
SiO_2	mg/l	0.25

16.2.6 Cost estimation

Table 16.3 gives a cost estimation that is computed for a plant of 100 m^3/d capacity based on the CDI pilot-scale study of the brine treatment process at an exchange rate of 1 EUR:2 SGD. A significant portion of the unit cost is attributed to the current high cost of the CDI technology. Estimated costs associated with the CDI technology include both capital cost of the CDI unit and the maintenance cost of electrode replacement, which together amount to more than 50% of the total unit cost at 21 Euro cents per m^3 product.

Table 16.3 Cost estimation of a 100 m^3/d plant based on the CDI pilot scale study.

Cost Item	Unit cost (€/m^3 product)
Capital Costs	
CDI	0.15
Pretreatment	0.05
Onsite cost–infrastructure and construction	0.02
Total Capital Costs	**0.22**
Operations and Maintenance(O&M) Costs	
Energy	0.05
Manpower	0.02
Electrode replacement	0.06
Chemicals & Maintenance	0.01
Total O&M Costs	**0.14**
Contingency allowance	
Contingency allowance factor @ 10%	**0.03**
Total	**0.39**

In view of the relatively early development stage of the CDI technology, it should be noted that the above cost estimation based on existing data will not likely reflect the true economic potential of the technology. With wider usage and technology acceptance, it is anticipated that the unit cost of treatment would reduce further with increasing capacity and advancement in the CDI technology.

16.2.7 Conclusions

The integrated pre-treatment with CDI technology was demonstrated to be a feasible option to treat and recovery at least 85% of the RO brine from the water reclamation facility. A lower pH and TOC concentration in the pretreated RO brine was also able to reduce fouling on the CDI process. At least 2 times longer CDI processing time, with minimal changes to the CDI pressure build-up and permeate flow rate were attained with RO brine pretreated with BAC-UF (having pH of 6.4 – 6.5

and 13.4 mg L^{-1} TOC) compared with BAC effluent. More than 80% ion removals were achieved using CDI process from the pretreated RO brine, except for PO_4^{3-} and TOC, which removals were only ranged from 50–80% and 48–65%, respectively. The product water from the integrated pre-treatment with CDI process was comparable or of better quality than the RO feed water to the water reclamation plant. The RO brine treatment and recovery process using CDI process had an estimated 15% lower power consumption compared with a pressure-driven membrane process.

16.3 OZONATION

16.3.1 Removal of emerging pollutants

Concentrate resulting from a reverse osmosis (RO) reactor processing secondary WWTP effluent exhibited a 3–4 times higher contamination of pharmaceuticals such as antibiotics, beta blockers, acidic pharmaceuticals and contrast media which is consistent with TOC enrichment in the brine (e.g. Wulpen: TOC of WWTP effluent: 11.6 mg/l; of the brine: 45.8 mg/l; Benner *et al.* 2008).

Ozonation might be a potential measure to remove most of the emerging compounds (Figure 16.4), as well as pathogens due to its disinfection properties. At all ozone concentrations applied (0.33, 0.76, 1.6 mg O_3/mg TOC) a significant oxidation of the target compounds were observed. However, it can be expected that a large number of oxidation products are formed as identified for the beta blockers metoprolol and propranolol in raw and treated wastewater (Benner *et al.* 2009a; Benner *et al.* 2009b). Several emerging contaminants were recalcitrant with ozone doses up to 1.6 mgO_3/mg TOC such as the iodinated contract media, benzotriazole, bezafibrate and clofibric acid. Similar results were reported by Ternes *et al.* (2003) and Huber *et al.* (2005) for the ozonation of WWTP effluents using ozone doses of 0.2–0.65 mgO_3/mg TOC.

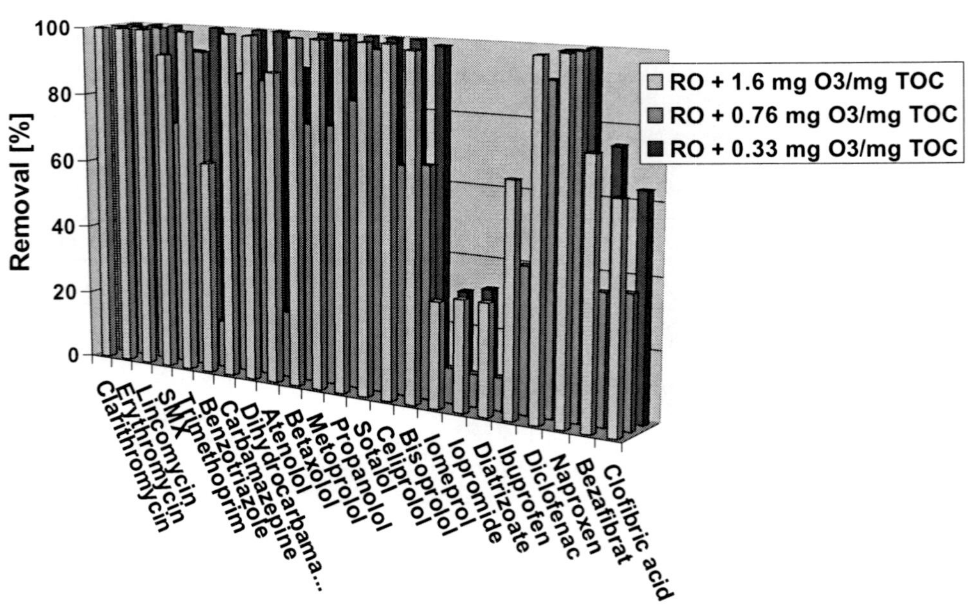

Figure 16.4 Removal of organic micropollutants after ozonation of brine using different ozone doses

16.3.2 Increase of biodegradability by ozonation

Ozonation increases the biodegradability only at elevated ozone doses (Figure 16.5). However, even at the two highest ozone doses applied, COD removal was not significant, whereas the BOD increased from 2 to only ~7–8 mg/L. Hence, a subsequent biological treatment step (or as a recycle in the WWTP) might reduce also the COD to a limited extent. This conclusion can be supported by the production of biodegradable dissolved organic carbon (BDOC) during ozonation. Ozonated and non ozonated RO concentrate inoculated with activated sludge from a wastewater treatment plant indicated only a minor increase of BDOC in the ozonated samples over a time period of 10 days.

16.3.3 Bromate formation

Bromate, a potential human carcinogen, is formed during ozonation of water containing an appreciable concentration of bromide (von Gunten, 2003). Therefore, there is a high potential of elevated bromate concentrations by the ozonation of brine, due to very high bromide concentrations present in the brine (e.g. 1200 μg/L Benner *et al.* 2008). Benner *et al.*

(2008) investigated the bromate formation by applying different ozone doses to a brine sample spiked with propranolol and metoprolol (1 µM). The ozone dose needed for a 90% oxidation of propranolol and metoprolol (~8 mg O_3/L and ~11 mg O_3/L, respectively) caused bromate concentrations of ca. 24 µg/L and 35 µg/L, respectively. These are relatively low bromate concentrations for a discharged effluent. Hutchinson *et al.* (1997) proposed not to exceed the bromate concentration beyond 3.0 mg/L (100 times higher than measured before) for the protection of aquatic organisms based on long term adverse effects. However, data on long term and chronic influence of bromate to the aquatic ecosystems is still very limited and further studies are needed.

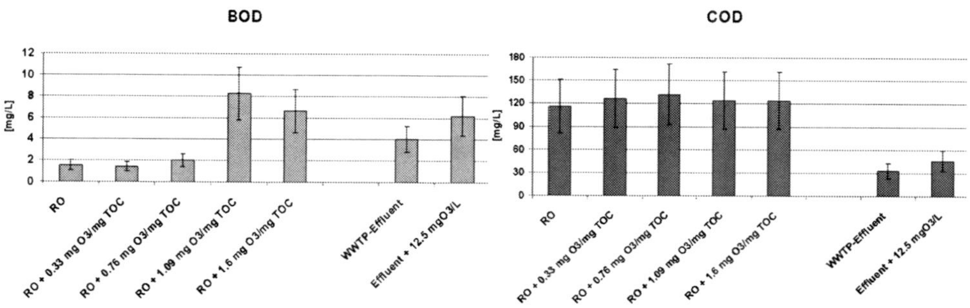

Figure 16.5 Influence of ozone on COD and BOD concentrations in brine

16.3.4 Toxicity of RO concentrate

The performed tests for general toxicity of the RO concentrate are based on the so called pT-method, which evaluate the toxic effects of water samples (Krebs 2005). The reduction of bioluminescence (Vibrio fischeri) of 35 % for the effluent and 33 % for the brine sample as well as the Lemna minor test indicated a pT-value of 2 for both samples. Neither the toxicity test with algae nor the Ames-test indicating mutagenicity nor daphnia magna showed any significant effect for both samples. The estrogenic effects were monitored with two tests, the E-screen and YES-test. The E-Screen test for effluent and brine revealed a 17-β-estradiol equivalent concentration of about 15 ng/L and 80 ng/L and the YES assay equivalent concentration of 54 ng/L and 115 ng/L, respectively.

Positive effects could be detected in the comet assay (indicating genotoxicity), because the WWTP effluent as well as the brine sample showed a significant DNA-damaging potential. However, the ozonated samples exhibited also elevated effects in the Comet assay. Especially the ozonated concentrate sample indicated a positive correlation between the increase of ozone dosage and the extent of the effect in the assay: this specific toxicity parameter is increased by ozonation.

16.3.5 Oxidation product formation

Several studies (Ternes *et al.* 2003; Huber *et al.* 2005; McDowell *et al.* 2005; Andreozzi *et al.* 2005; Ikehata *et al.* 2006; Huber *et al.* 2003; Vieno *et al.* 2007) have shown that ozone treatment can be very efficient in the oxidation of a wide range of micropollutants (e.g. beta blockers, antibiotics, estrogens). Estrogenic activity (Huber *et al.* 2004) as well as antibiotic effects (Dodd *et al.* 2006) are significantly reduced during ozonation. However, economically feasible ozone doses will not mineralize micropollutants but form oxidation products (OP) instead.

Major oxidation products (OPs) of the beta blockers propranolol and metoprolol formed during ozonation in aqueous solution were identified. The OPs of the investigated beta blockers mainly contained aldehyde moieties, several of them with known genotoxic and carcinogenic activity (Richardson *et al.* 2007). Additionally, hydroxylated aromatic rings were formed and many compounds containing phenolic moieties are known to show estrogenic activity. A translation to other compounds reveals that ozonation of a compound containing activated aromatic moieties might lead to the formation of a high number of product with unknown toxicological and ecotoxicological properties.

16.4 GRANULAR ACTIVATED CARBON WITH MICROFILTRATION (BIO MAC)

Taking into account the expected increase of wastewater reuse and the projected predominance of membrane technology, treatment of the produced membrane concentrates becomes an emerging issue. Membrane processes are highly efficient as a separation technique, but pollutants are retained and concentrated in the reject brine. For NF and RO, most organic constituents as well as a significant amount of the dissolved solids and emerging pollutants, such as pharmaceuticals and endocrine disruptors are found back in the concentrate. Therefore, disposal of these concentrates or discharge into the surface water should, within the framework of sustainable development, not be considered.

The patented BioMAC (Biological membrane assisted carbon filtration) technology uses a combination of GAC (granular activated carbon) and micro- or ultrafiltration for the direct production of reclaimed water from WWTP effluent or for the treatment of concentrate streams. The combination of activated carbon sorption and membrane filtration has been extensively documented in literature. The novelties of the BioMAC are:

(1) Membranes and recirculation are used to concentrate developed specialists. The inclusion of membranes and activated carbon as carrier material in the BioMAC is essential to be able to retain and concentrate specialist organisms that can biodegrade so-called recalcitrant COD (chemical oxygen demand).
(2) The carrier material is partially non-submerged in order to obtain high specialist activity. Thin and fully aerated biofilms show higher activity, resulting higher reactor efficiency. Experience shows this higher activity can best be obtained by partially not submerging the carrier material and hereby keeping the developed biofilms relatively dry. The applied recirculation ratio can be used to optimize this effect.
(3) Secondary effects of recirculation over the BioMAC carrier material. The recirculation over the BioMAC carrier material (i) dilutes the incoming concentrations, (ii) enhances mineralization in the biofilms and therefore decreases (iii) biofilm sludge production and (iv) fouling of the BioMAC.

16.4.1 Plant set-up

Preliminary lab-scale BioMAC tests (permeate flow 0.5–2.5 $l \cdot h^{-1}$) were carried out at Ghent University in 2001. Results are reported in Van Hege *et al.* (2002). Within the framework of this workpackage the BioMAC concept was scaled up (Figure 16.6 and 16.7): a 0.5–2.0 $m^3 \cdot h^{-1}$ BioMAC pilot plant was designed, constructed and started up at the site of the IWVA (Intermunicipal Water Company of the Veurne region) reuse plant. The pilot plant is directly fed with the mixed UF-RO concentrate originating from this full-scale reuse plant (Figure 16.7).

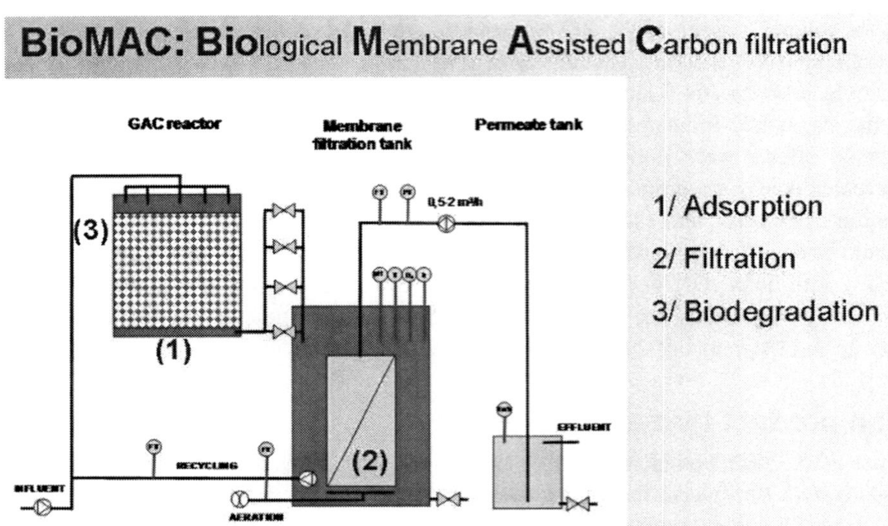

Figure 16.6 Schematic overview of the BioMAC plant

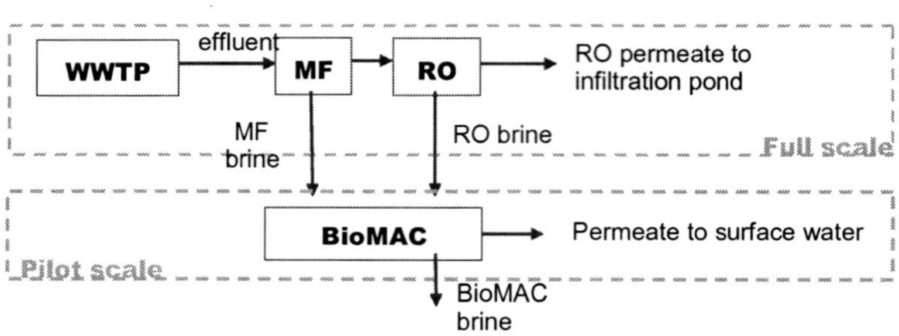

Figure 16.7 Schematic overview of the BioMAC pilot plant receiving concentrate from the full-scale IWVA reuse plant

16.4.2 Removal capacity

Numerous pollutants were analyzed during a 9 month sampling campaign, in which 24 h composite samples were taken in the BioMAC influent (mixed UF-RO concentrate) and effluent. The number of samples, average influent-effluent concentrations, standard deviations and removal percentages can be found in Table 16.4. A summary of these results can be found in Table 16.5. Furthermore, the BioMAC influent and effluent was grab sampled once (early in the research period: 29/01/07) and analyzed for antibiotics, contrast media, acidics and estrogens (Figures 16.8 to 16.11). Removal rates were around 40 to 50%.

Table 16.4 Average influent and effluent concentration and removal rates of the BioMAC tests treating the UF/RO reject streams of the Torreele plant.

Parameter	Unit	# samples in/effluent[1]	Average influent	Average effluent	Removal (%)
BOD	mg/l	40	5.9±5.7	1.4±4.2	76
COD	mg/l	42	110.8±23.2	64.1±28.9	42
CODfilt	mg/l	42	93.9±21.1	57.9±15.2	38
TOC	mg/l	24	31.4±8.0	22.6±6.0	28
DOC	mg/l	24	22.6±6.9	20.0±5.4	25
SS	mg/l	41	14.3±6.9	10.9±13.0[2]	24
Dry Solids	mg/l	30	2293±514	2177±385	5
Ash Content	mg/l	30	1994±424	1874±338	22
NH_4-N	mg/l	42	3,3±3,7	0.4±0.9	87
Org-N	mg/l	42	4.2±1.2	2.6±0.6	38
KjN	mg/l	42	7,5±4,3	3.0±1.2	60
NO_2-N	mg/l	42	0,9±0,9	0.5±0.5	50
NO_3-N	mg/l	42	14.2±6.3	165.8±17.5	−19
TN	mg/l	42	22.6±8.6	20.3±7.9	10
PO_4-P	mg/l	42	1.3±0.9	0.9±0.4	30
Org-P	mg/l	39	0.68±0.22	0.33±0.32	52
TP	mg/l	42	2.1±1.0	1.4±0.8	34
Cl	mg/l	14	602±113	600±114	0
Conductivity	μS/cm	29	3486±652	3255±576	7
Anionic detergents	mg/l	12	0.188±0.047	0.000±0.000	100
FOG (Fats, Oils, Grease)	mg/l	6	0.28±0.18	0.33±0.82	−18
Hardness	°F	30	90.1±15.2	75.5±9.0	−5
Ca	mg/l	30	305±50	248±29	19
Mg	mg/l	30	34.1±7.0	33.1±6.3	3
Fe-total	mg/l	27	0.6±0.2	0.1±0.1	77
Fe-soluble	mg/l	26	0.2±0.1	0.1±0.0	75
Heavy Metals					
As	mg/l	3	4.12±3.57	2.72±2.50	34
Cr	mg/l	8	0.025±0.049	0.009±0.009	64
Cu	mg/l	8	0.013±0.007	0.014±0.012	−11
Mn	mg/l	30	0.164±0.045	0.044±0.056	73
Ni	mg/l	8	0.031±0.025	0.028±0.017	9
Zn	mg/l	8	0.085±0.020	0.094±0.041	−10
Herbicides					
Atrazine	μg/l	7	0.054±0.076	0.010±0.013	82
Simazine	μg/l	7	0.191±0.206	0.029±0.028	85
Diuron	μg/l	7	1.203±1.435	0.096±0.134	92
Halogenated Hydrocarbons – Trihalomethanes (THMs)					
Br-dichlorometh. $CHBrCl_2$	μg/l	7	4.19±4.18	6.87±14.08	−64

(*Continued*)

Table 16.4 Average influent and effluent concentration and removal rates of the BioMAC tests treating the UF/RO reject streams of the Torreele plant (*Continued*).

Parameter	Unit	# samples in/effluent[1]	Average influent	Average effluent	Removal (%)
Di-Br-chlorometh. $CHBr_2Cl$	μg/l	7	0.543 ± 1.157	2.01 ± 4.16	−271
Trichloromethane $CHCl_3$	μg/l	7	9.1 ± 7.3	11.2 ± 20.1	−23
Microbiological[3]					
Total coliforms	CFU/100ml	6	5333 ± 5248	807 ± 1097	85
E. Coli	CFU/100ml	6	123 ± 160	72 ± 156	42
Enterococcus	CFU/100ml	6	20.3 ± 44.0	1.3 ± 2.4	93
Total germ count	CFU/ml	6	46300 ± 75756	18350 ± 35164	60

[1]All 24th composite samples
[2]Due to serious and subsequent membrane integrity problems and because of scaling in the permeate buffer tank, SS contamination was detected in the permeate
[3]Due to serious and subsequent membrane integrity problems, microbiological contamination was detected in the permeate

Table 16.5 Removal efficiencies for standard parameters and selected micropollutants.

Removal efficiency	Basic parameters	Micropollutants
70–100%	BOD, NH_4-N, Fe-sol, Fe-tot, Mn	Anionic detergents, Fe-total, Fe-soluble, Mn, Atrazine, Simazine, Diuron, Total coliforms, Enterococcus
40–70%	COD, KjN, NO_2-N, Org-P	As, Cr, E. Coli, Total germ count
20–40%	COD filtrated, TOC, DOC, SS, Ash content, Org-N, PO_4-P, TP	–
$\leq$20%	Dry solids, NO_3-N, TN, Cl, Conductivity, FOG, Hardness, Ca, Mg	Cu, Ni, Zn, $CHBrCl_2$, $CHBr_2Cl$, chloroform

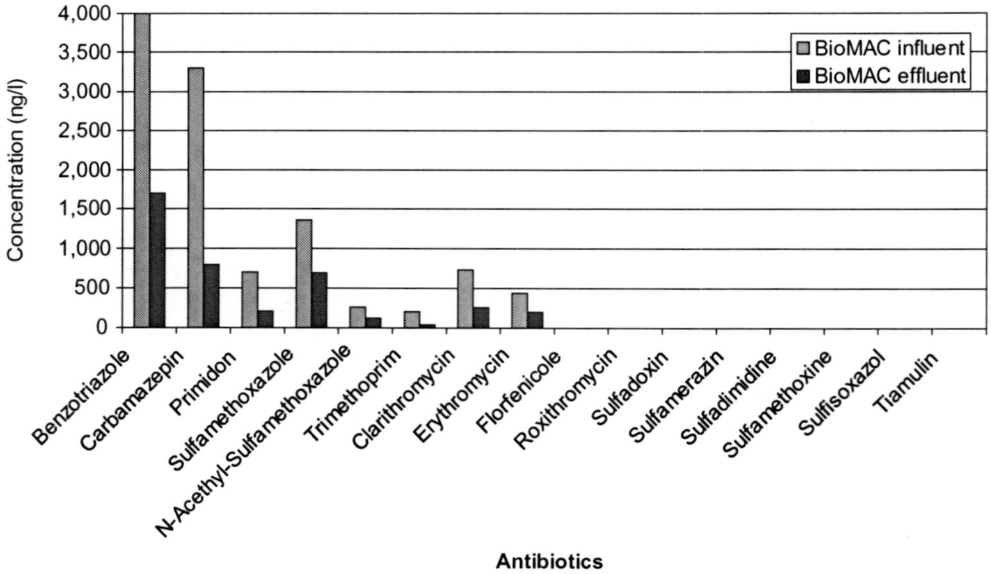

Figure 16.8 Results for antibiotics

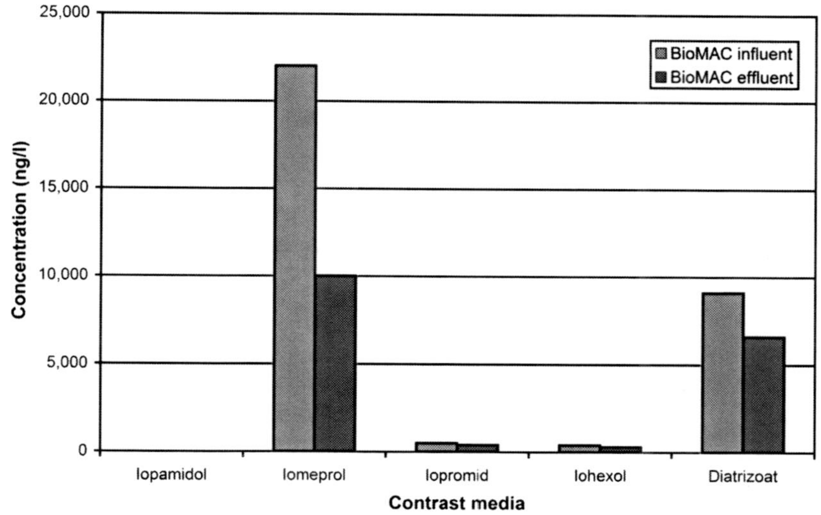

Figure 16.9 Results for contrast media

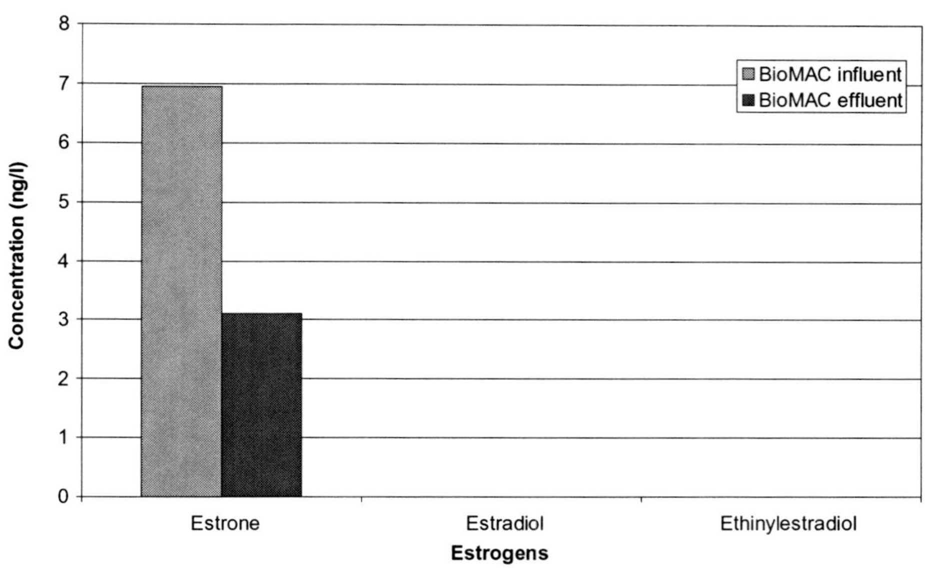

Figure 16.10 Results for estrogens

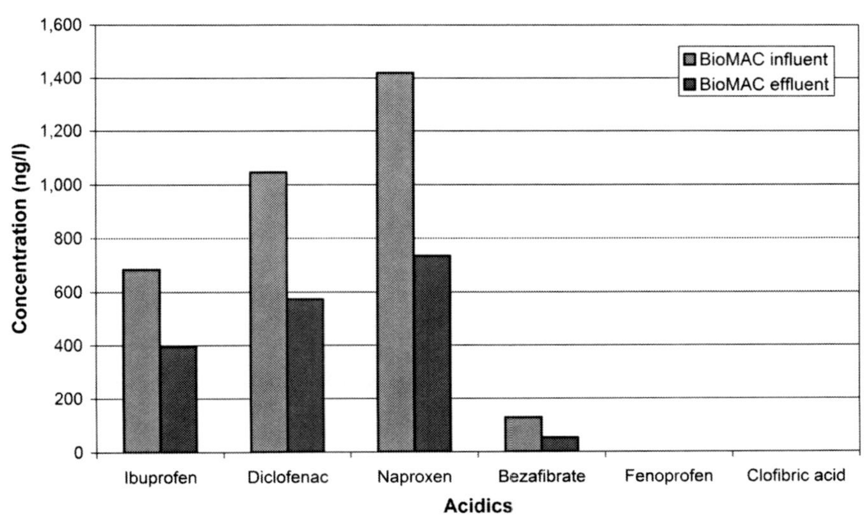

Figure 16.11 Results for acidics

16.4.3 Operational regime and conclusions

Without addition of antiscalant and when applying low fluxes, chemical membrane cleaning is required on average once every week. Adaptation of recycling rates and depth of the submerged GAC filter are insufficient tools to optimize the denitrification process. Additional carbon source dosing is necessary for this, but this will make the concept more complex. Optimal PID control of the filtration cycles is necessary to minimize membrane fouling A diverse and abundant biology could be detected on the GAC particles. Biodegradation does play an important role in the BioMAC process. However, quantifying this mechanism is difficult. The whole BioMAC process can be automated. A full-scale BioMAC plant will require rather low manpower. Vertically mounted vent pipes are required in the GAC for a stable BioMAC process. Even with a pre-treatment of 400 µm, short (2 minutes) but frequent (every 2 hours) backwashing of the GAC is necessary to prevent it from clogging with influent SS and fine AC particles. This backwash procedure also serves as sludge waste.

Following conclusions can be made regarding process flexibility: It is possible to treat mixed UF-RO concentrate in the BioMAC. However, even when dosing antiscalant, gross flux is limited to 15–20 LMH. Operating the BioMAC in different hydraulic regimes (e.g. treating the double of an incoming concentrate flow, and applying gross fluxes until 45 lmh) seems not possible. This is one of the two critical factors which lead to a high payback period of the BioMAC concept. BioMAC is a robust process. A plant shut-down of several days does not have an influence on future filtration stability or removal efficiencies. Cost is estimated at 0.22 €/m^3 product water (Table 16.6), equivalent to 0.055 €/m^3 RO permeate at a recovery of 80%.

Table 16.6 Cost estimation.

Cost item	**Unit price (EUR/ m^3 Product)**
Capital Costs	
Equipment (BioMac or Ozone generator etc ...)	BioMAC
Pretreatment (if required)	not necessary
Onsite cost – infrastructure and construction	0.127
Operations and Maintenance Costs	
Energy	0.044
Manpower	0.027
Chemicals	0.002
Maintenance	0.024
TOTAL O&M	0.097
TOTAL COSTS	**0.224**

16.5 SUBSURFACE FLOW REED BED

Both, the UF and the RO process produce reject streams which are currently discharged into the adjacent canal with the part of effluent that has not been treated. The canal leading to the open sea after some kilometres is brackish, so the salinity does not have a major negative effect on the quality. The treatment concept is shown in Figure 16.12.

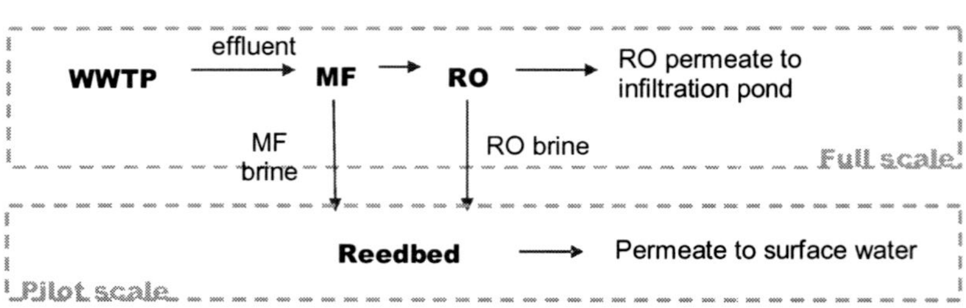

Figure 16.12 Schematic overview of the Reed bed pilot receiving concentrate from the full-scale IWVA reuse plant

16.5.1 Plant set-up

For 5 years IWVA performed tests with the UF/RO concentrate using a subsurface flow reed bed (constructed wetland). The characteristics of the reedbed are given in Table 16.7.

Table 16.7 Characteristics of experimental subsurface flow reed bed (Torreele).

Construction	October 2003
Dimensions	3 m on 3 m
Vegetation	Phragmites australis
Substratum	70 cm of sand 3–8 mm
Source water	Concentrate of RO mixed with UF backwash water
Flux	400 l/h

16.5.2 Removal capacity

The results of the monthly sampling showed that suspended solids and nitrogen are removed substantially (Table 16.8). The organic content (e.g. COD) is removed to a minor extent and the phosphorous content remains as it was. Values of TDS are added to show that the quality of both waters matched. The feed flow to the reed bed was 400 l/h. This flow is higher than for conventional use of reed beds but the flow was adjusted to the space available for future construction with the given amount of concentrate.

Table 16.8 Comparison between the concentrate before and after treatment with an experimental reed bed at IWVA's Torreele facility.

Parameter	Concentrate before reed bed	Effluent of the reed bed	Removal efficiency
Total Dissolved Solids (mg/l)	2.266 (1.677–3.195)	2.280 (1.834–3.193)	0%
Suspended solids (mg/l)	13 (5–25)	1.2 (0.2–2.8)	91%
Total nitrogen (mg N/l)	22.3 (11.5–38.9)	13.6 (3.6–30.0)	39%
Total phosphorous (mg P/l)	1.7 (1.1–2.5)	1.5 (1.0–2.0)	<20%
COD (mg/L)	88 (72–125)	67 (50–91)	24%
BOD (mg/L)	<4–11	<4–6	<50%

16.6 CONCLUSIONS

Several promising RO concentrate treatment methods have been investigated. While the CDI tests proved that the general feasibility of this desalination, proper pretreatment to remove the potential foulants was shown to be crucial. Ozonation of the concentrate is an effective method for many of the organic compounds. However some micropollutants such as iodinated contrast media were oxidized only at a low percentage. Adsorptive processes such as the tested BioMAC process are able to remove a broad range of compounds but require more optimization, particularly regarding GAC runtime, backwash and the interplay between the GAC and the membrane unit. Simple natural methods such as the tested reedbeds are simple and robust in operation but serve mainly for the removal of suspended solids and reduce nitrogen partially.

REFERENCES

Acero J., Haderlein S. B., Schmidt T. C., Suter M. J. F. and von Gunten U. (2001). MTBE oxidation by conventional ozonation and the combination ozone/hydrogen peroxide: efficiency of the processes and bromate formation. *Environ. Sci. Technol*, **35**, 4252–4259.

Adham S., Oppenheimer J. and Kumar M. (2006). Innovative approaches to RO concentrate management: beneficial reuse and concentrate minimization. *Proceedings of the Water Environment Federation*, **11**, 4334–4344.

Adham S., Oppenheimer J., Liu L. and Kumar M. (2007). Dewatering Reverse Osmosis Concentrate from Water Reuse Applications using Forward Osmosis, 05-009-01. WateReuse Foundation, Alexandria, VA, USA, ISBN: 978-1-934183-02-1.

American Water Works Association (2004). Water Desalting Planning Guide for Water Utilities. John Wileys & Sons.

Andreozzi R., Canterino M., Marotta R. and Paxeus N. (2005). Antibiotic removal from wastewaters: the ozonation of amoxicillin. *Journal of Hazardous Materials*, **122**(3), 243–250.

Asano T., Burton F. L., Leverenz H. L., Tsuchihashi R. and Tchobanoglous G. (2007). Water Reuse – Issues, Technologies, and Applications. Mc Graw Hill, NY, ISBN: 978-0-07-145927-3.

Benner J., Salhi E., Ternes T. A. and von Gunten U. (2008). Ozonation of reverse osmosis concentrate: kinetics and efficiency of beta blocker oxidation. *Water Research*, **42**(12), 3003–3012.

Benner J. and Ternes T. A. (2009a). Ozonation of metoprolol: elucidation of oxidation pathways and major oxidation products. *Environmental Science & Technology*, **43**(14), 5472–5480.

Benner J. and Ternes T. A. (2009b). Ozonation of propranolol: formation of oxidation products. *Environmental Science & Technology*, **43**(13), 5086–5093.

Khan S. J., Murchland D., Rhodes M. and Waite T. D. (2009). Management of concentrated waste streams from high-pressure membrane water treatment systems. *Critical Reviews in Environmental Science and Technology*, **39**(5), 367–415.

Dietz S. (2004). Improved electrodes for capacitive deionization. *Proceedings of the 2004 NSF Design, Service and Manufacturing Grantees and Research Conference*, Birmingham, AL, January 2004.

Dodd M. C., Buffle M. O. and Von Gunten U. (2006). Oxidation of antibacterial molecules by aqueous ozone: moiety-specific reaction kinetics and application to ozone-based wastewater treatment. *Environmental Science & Technology*, **40**(6), 1969–1977.

Elovitz M. and von Gunten U. (1999). Hydroxyl radical/ozone ratios during the ozonation processes. *I. The RCT Concept, Ozone Sci. Eng*, **21**, 239–260.

ENPAR Technologies Inc. (2007). ENPAR's DesEl system: case histories. http://www.enpar-tech.com/pdf/DesEl%20Case%20Histories-4-5.pdf.

Garrido S., Aviles M., Ramirez A., Calderon C., Ramirez-Orozco A., Nieto A., Shelp G., Seed L., Cebrian M. and Vera E. (2006). Arsenic Removal from Water of Huautla, Morelos, Mexico using Capacitive Deionization. WEFTEC 2006.

Huber M. M., Canonica S., Park G. Y. and Von Gunten U. (2003). Oxidation of pharmaceuticals during ozonation and advanced oxidation processes. *Environmental Science & Technology*, **37**(5), 1016–1024.

Huber M. M., Ternes T. A. and von Gunten U. (2004). Removal of estrogenic activity and formation of oxidation products during ozonation of 17 alpha-ethinylestradiol. *Environmental Science & Technology*, **38**(19), 5177–5186.

Huber M. M., Gobel A., Joss A., Hermann N., Loffler D., McArdell C. S., Ried A., Siegrist H., Ternes T. A. and von Gunten U. (2005). Oxidation of pharmaceuticals during ozonation of municipal wastewater effluents: a pilot study. *Environmental Science & Technology*, **39**(11), 4290–4299.

Hutchinson T. H., Hutchings H. J. and Moore K. W. (1997). A review of the effects of bromate on aquatic organisms and toxicity of bromate to oyster (crassostrea gigas) embryos. *Ecotoxicology and Environmental Safety*, **38**(3), 238–243.

Ikehata K., Naghashkar N. J. and Ei-Din M. G. (2006). Degradation of aqueous pharmaceuticals by ozonation and advanced oxidation processes: a review. *Ozone-Science & Engineering*, **28**(6), 353–414.

Krebs F. (2005). The pT-method as a hazard assessment scheme for wastewaters. In: Small-Scale Freshwater Toxicity Test Methods, C. B. a. J. F. FéRard (ed.), Springer, Berlin, pp. 1–23.

McDowell D. C., Huber M. M., Wagner M., Von Gunten U. and Ternes T. A. (2005). Ozonation of carbamazepine in drinking water: identification and kinetic study of major oxidation products. *Environmental Science & Technology*, **39**(20), 8014–8022.

Richardson S. D., Plewa M. J., Wagner E. D., Schoeny R. and DeMarini D. M. (2007). Occurrence, genotoxicity, and carcinogenicity of regulated and emerging disinfection by-products in drinking water: a review and roadmap for research. *Mutation Research-Reviews in Mutation Research*, **636**(1–3), 178–242.

Ternes T. A., Stuber J., Herrmann N., McDowell D., Ried A., Kampmann M. and Teiser B. (2003). Ozonation: a Tool for removal of pharmaceuticals, contrast media and musk fragrances from wastewater? *Water Research*, **37**(8), 1976–1982.

Vieno N. M., Harkki H., Tuhkanen T. and Kronberg L. (2007). Occurrence of pharmaceuticals in river water and their elimination a pilot-scale drinking water treatment plant. *Environmental Science & Technology*, **41**(14), 5077–5084.

von Gunten U. (2003). Ozonation of drinking water: part II. disinfection and by-product formation in presence of bromide, iodide or chlorine. *Water Research*, **37**(7), 1469–1487.

Welgemoed T. J. and Schutte C. F. (2005). Capacitive Deionization Technology^{TM}: an alternative desalination solution. *Desalination*, **183**, 237–340.

Part D

Design and Management of MAR Systems

Chapter 17

General design considerations

Peter J. Dillon

17.1 INTRODUCTION

Managed aquifer recharge (MAR) is a collection of techniques in water resources management and water supply that integrates above and below ground components of the water cycle to meet human needs and environmental objectives. Hence in theory the design of MAR projects could be complicated. However in reality understanding the opportunities and constraints presented by climate, geomorphology, hydrology, hydrogeology and geochemistry in any locale allows options worthy of pursuit to be relatively quickly identified. 'No regret' decisions can be made easily if those aspects of landscape function are sufficiently well understood. In areas without MAR projects, further effort on characterising aquifers will be needed to gain clarity about what types of projects are likely to be successful and where they should be sited. Projects with recycled water as the source present small additional challenges that can easily be addressed within the systematic framework presented here.

For water engineers who can read the specifications for a treatment technology to determine whether it is appropriate for a given situation, MAR can at first sight appear inconvenient. The specifications for the rate at which an aquifer can store, treat and yield water must be learned by observation, testing and inference from relevant experience. That is the aquifer seems like a 'black box' technology where we find out how a system performs by testing it. However, there is good news. A growing scientific understanding of how aquifers perform is gradually being documented and the general principles and processes are known but the rate constants and tolerance limits for perturbations are not. Combinations of laboratory and in-situ testing in aquifers can provide the information that represents the 'specifications' for aquifer storage and treatment functions and allow it to be considered alongside traditional options.

This chapter overviews the steps in establishing MAR projects, including identifying objectives, siting, design, operation and monitoring as a prelude to the following chapters addressing hydrogeological investigations, water quality risk assessment and management, public perception and integrated water resources management.

Firstly, however, it is necessary to start from a bigger picture, concerning the range of options beside MAR that are capable of achieving the same objectives, and also in relation to spatial scale where the whole of the water catchment and groundwater system needs to be considered.

17.2 IDENTIFYING PROJECT OBJECTIVES – WHAT ARE THE OPTIONS?

MAR can be used to store water from various sources, such as reclaimed water, stormwater, mains water, desalinated seawater, rainwater or even groundwater from other aquifers. With appropriate pre-treatment before recharge and sometimes post-treatment on recovery of the water, it may be used for drinking water supplies, industrial water, irrigation, sanitation and to sustain ecosystems. Where the population is increasing, such as in urbanising areas, and where the climate is becoming more variable, increasing the storage in aquifers will be increasingly important as a buffer to secure supplies (Dillon *et al.* 2009a; van Steenbergen & Tuinhof, 2010; McCartney & Smakhtin, 2010).

Common objectives for MAR projects include:

- securing and enhancing water supplies by increasing storage and reducing evaporation,

- improving groundwater quality, and
- maintaining groundwater levels to protect groundwater quality and the biodiversity of groundwater-dependent ecosystems.

Consequential benefits may also include:

- reducing costs of water supply in comparison with alternative supplies,
- reducing greenhouse gas emissions in comparison with alternative supplies,
- improving health of communities through provision of safer water supplies,
- maintaining agricultural and industrial production in comparison with demand management alone,
- preventing salt water from intruding into coastal aquifers,
- improving coastal water quality by reducing urban discharges,
- mitigating floods and flood damage, and
- facilitating urban green space and amenity improvements that increase land value.

Given the importance of these objectives it would be rational to expect monitoring of the performance of MAR projects to assess the degree to which they achieved their stated objectives with reporting on successes and failures so that future projects are better informed at inception. This may require baseline monitoring for several years before a project is implemented. This is a luxury that may not be affordable for all projects, but at national level a selection of demonstration projects should be documented so that project impacts are quantified credibly, and embrace all key performance indicators.

17.2.1 Options for storage increase

Other receptacles for holding increased storage include reservoirs, ponds and tanks, soil moisture and natural wetlands (McCartney & Smakhtin, 2010). An advantage of aquifers is that they are already the dominant storage of freshwater on earth accounting for more than 90% of all freshwater, are relatively ubiquitous in comparison with dam sites. Also evaporation is eliminated or at least is at much lower rates than for other forms of storage. With a warming climate it has been estimated for every degree Celcius increase in temperature that evaporation rates from temperate surface water bodies will increase by 4% (Cai & Cowan, 2008). Furthermore establishment of large dams has become problematic due to increased public awareness over recent decades of their impacts on riverine ecosystems. (WCD, 2000).

A comparison of capital costs of alternative storage systems for projects in Australia in 2008 showed that capital costs, neglecting land costs, for large dams are cost efficient compared with other surface storages, and are similar to those for aquifer storage and recovery projects. However infiltration basins were found to be almost an order of magnitude cheaper under favourable conditions (Table 17.1) (Dillon *et al.* 2009b).

Table 17.1 Indicative costs (2008) and land area requirements of alternative storages in Australia including managed aquifer recharge projects (from Dillon *et al.* 2009b).

Type of storage	Storage size range costed (m^3)	Unit capital cost of storage† (Euro/m^3)	Land surface area required (m^2/m^3 water)
Rainwater tank – polyethylene	2–10	140	0.5
Concrete tank – trafficable	1,000–4,000	700	0.2
Pre-cast concrete panel tank	4,000–8,000	180	0.25
Lined earthen dam impoundment	4,000–8,000	8	0.6
Large dam – gravity or concrete	350,000–200,000,000	3–7	0.1–0.2
Aquifer storage and recovery#	75,000–2,000,000	3–7	0.001*
Pond infiltration/soil aquifer treatment#	200,000–600,000	0.7–1.4	0.02–0.06**

†excluding land cost.
#storage size used here for MAR is the mean annual recharge volume. Actual storage volume of recoverable water may be many times this amount, however in brackish aquifers recoverable volume from earlier years will depreciate due to mixing.
*if detention storage is required to capture stormwater, the area required may be 0.02 to 0.1 m^2/m^3 water depending on runoff from catchment and capture efficiency.
**for hydraulic loading rates of 17 to 50 m/yr.

Taking account of land area, especially where land value is high, such as in or near urban areas (>Euro 700/m^2 is common in major cities), ASR has potential to have the least total capital costs per unit volume of water storage of alternatives in Table 17.1. Infiltration systems where hydraulic loading rates are high also increase their cost advantage over dams. It has long been known that the conjunctive use of surface and aquifer storages can reduce the costs of operating either system on its own (Johnston *et al.* 1973). This broadens our perspective of Table 17.1, which only compares cost of storage for individual storages, not cost of supply. Generally the annual supply from a large dam will be less than its storage capacity, whereas for a rainwater tank the annual supply may be many times its capacity.

17.2.2 Options for water quality improvement

In cases where MAR is considered primarily as part of a treatment train to improve water quality and make it more uniform (for example bank filtration systems, Grischek and Ray 2009; Massmann *et al.* 2006, 2008) the costs of engineered treatment processes that give the equivalent level of treatment can be compared. In some locations where operation and maintenance expertise is lacking or intermittency of power supplies can constrain engineered treatment effectiveness, these factors should be taken into account. Verifiable water treatment functions where there is adequate residence time within an aquifer may include filtration of colloidal material, disinfection (pathogen removal) and biodegradation of some trace organics. The costs of equivalent engineered treatment processes can be compared using levelised cost. (The levelised cost is the stream of capital and maintenance and operating costs over the life of a project converted using a discount rate to the equivalent annualised cost and dividing this by the annualised volume supplied, converted in the same way as annualised cost.)

Levelised costs of equivalent engineered treatment processes were; slow sand filters or biofilters (Euro $1.75/m^3$), rapid sand filtration plus chlorination (Euro $0.11/m^3$), GAC plus chlorination (Euro $0.13/m^3$) compared with stormwater aquifer storage transfer and recovery (ASTR) (Euro $0.78/m^3$). That is if aquifer storage is not required, it would be much cheaper to provide GAC plus chlorination or ultraviolet disinfection than to develop ASTR. Details of cost parameters are given in Dillon *et al.* (2009b).

That evaluation did not include a bank filtration project. For such a project the costs are simply the difference in cost between pumping directly from the river and pumping from a well a sufficient distance from the stream to provide adequate treatment of induced recharge plus the cost of protecting the quality of shallow groundwater in that vicinity, and of assessing likely contaminants in the stream. The benefits of such systems are likely to be expressed in terms of reduced disease burden on the supplied community with metrics such as reduced incidence of; reported diarrhoea cases, hospital bed days, infant deaths, school absenteeism and lost days of work. Obviously this type of intervention also goes hand in hand with improving hygiene practices and sanitation, which combined can have very significant impacts where pre-existing supplies are exposed to faecal contamination.

17.2.3 Options for sustaining groundwater levels and dependent ecosystems

In general where irrigation supplies exceed long-term rates of natural replenishment groundwater levels decline. In unconfined aquifers, the first impacts are generally in degradation of groundwater-dependent ecosystems, and subsequently in reduced yields and/or increased salinity of groundwater in irrigator's wells. In confined aquifers the impacts may be first evidenced by reduced yields of irrigator's wells, increased groundwater salinity and, in cases where compressive clays are depressurised, land subsidence. Improving irrigation practices, selection of higher valued and more water-efficient crops, capping demand at a sustainable level and alternative supplies are all parts of the solution (e.g. Dillon *et al.* 2009a). If MAR is included in the solution package, the cap can be higher than it otherwise would be and the value of irrigation not having to be foregone or the cost of bringing in alternative independent supplies (e.g. recycled water) are surrogates for the value of the MAR project. This may be compared against the costs of MAR to determine a benefit/cost ratio and assess attractiveness. Efficient outcomes are unlikely to be achieved without the engagement of all stakeholders, and the final solution is likely to include a mix of options. For example the source of water for recharge may be used directly by irrigators when required and stored in an aquifer when demand is less than supply, such as at Bolivar SA (Dillon *et al.* 2003).

Currently some jurisdictions do not separate land and water rights and especially in these cases, but also more generally, the benefits of MAR outlined above when explicitly quantified in the situation can be used to galvanise groundwater users groups to invest in MAR and other projects to protect the aquifer and their own interests.

17.2.4 Whole catchment and groundwater system context

Before progressing to the steps in development of a MAR project, it is necessary to take a broader look at the catchment that will supply the source water, whether it is a natural watershed or an urban stormwater or sewage collection system, to understand its characteristics that could impinge or even inhibit formation of a MAR project. Similarly for the groundwater system. Considering projects one at a time within a single catchment-groundwater system is pragmatic for the first project, especially demonstration projects, but is not the model to be promoted for all projects.

Thinking ahead as a water resources management agency rather than as a utility with a need to supply a particular area with a certain volume of water at particular levels of reliability and quality, it is possible that exercising some options for MAR may preclude others occurring that would have had higher value to the region. In most jurisdictions there will be a water resources master plan that is updated periodically. That serves as a good basis for determining the water needs and supply options over the long term and allowing MAR to be considered strategically.

Such planning needs to involve surface water and groundwater and address both quantity and quality. That is, effective implementation of MAR requires integrated water resources management and coordination is required among agencies responsible for component sectors. Table 17.2 shows the way governments typically partition management issues that must be addressed in concert.

Table 17.2 Water resources management and environmental protection issues to be addressed in establishing MAR projects (after Dillon *et al.* 2009b).

Attribute	Quantity	Quality
Management issue	Water and storage entitlements and allocations	Human health and environment protection
Resource		
Surface water	• Environmental flow requirements • Water allocation plans and surface water entitlements • Inter-jurisdictional agreements	• Catchment pollution control plan • Water quality requirements for intended uses of recovered water • Risk management plan for water quality assurance
Groundwater	• Groundwater allocation plan and groundwater entitlements • Resource assessment accounting for groundwater-dependent ecosystems • Demand management • Allocatable capacity and entitlement for additional storage in the aquifer • Inter-jurisdictional agreements	• Groundwater quality protection plan for recharged aquifer • Water quality requirements for intended uses of groundwater • Risk management plan for water quality assurance beyond attenuation zone, accounting for aquifer biogeochemical processes

The water resources management issues are generally more efficient to address first to assess the viability of MAR projects. These therefore assume precedence over water quality evaluations that generally require more exact localised information on aquifer properties and source water quality, some of which is likely to require site-specific investigations.

17.3 STEPS IN ESTABLISHING A MAR PROJECT

An approach adopted in Australia for the development of MAR projects is given below. This causes a proponent to undertake a desktop 'entry level assessment' based on existing information to establish the viability then degree of difficulty for a project. Following that are a sequence of risk assessments based on identified hazards to ensure that through the identified preventive measures all risks to public health and the environment are acceptably low. If the cost of required investigations and preventive measures is within the budget available the project can proceed.

17.3.1 Viability assessment

There are five essential elements for every successful MAR project:

- a sufficient demand for recovered water
- an adequate source of water for recharge
- a suitable aquifer in which to store and recover the water
- sufficient land to harvest and treat water
- capability to effectively manage a project

Australian Guidelines for Managed Aquifer Recharge (NRMMC-EPHC-NHMRC, 2009a) require a desktop (entry level) assessment, essentially the above checklist, to assess the viability of a proposed MAR project together with subsequent questions to assess its likely degree of difficulty.

Viability assessment focuses first on demand for recovered water because demand is the driver for investment in a sustainable project that protects the quality of native groundwater and of recovered water. Generally this will produce the revenue stream to pay for the capital and operating costs of the project. Additional non-monetary benefits may be accounted for in other ways, for example by government co-investment or establishment of market trading mechanisms (e.g. reduced nitrogen discharge to sea or carbon credits awarded for greenhouse gas savings). If the demand is stable, seasonal or occurs only during droughts and emergencies the mix of investment may differ, with governments playing a larger role in security of supply.

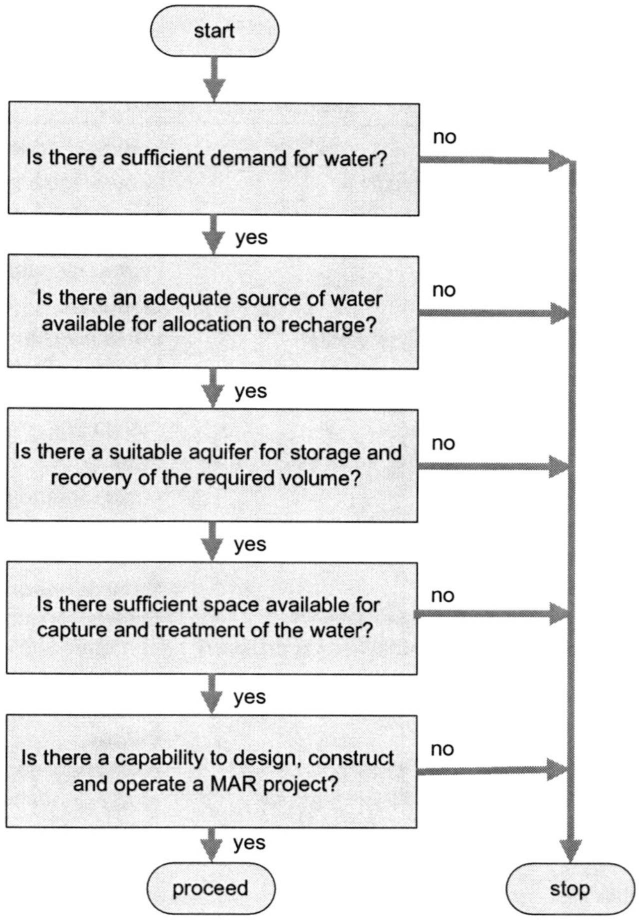

Figure 17.1 A checklist for considering whether to undertake a managed aquifer recharge project (from NRMMC, EPHC and NHMRC, 2009a)

The next question addresses entitlements to source water quantity (the upper left side of Table 17.2), which may have various temporal patterns and levels of security. A project is non-viable if surface water resources are already over-allocated because further extraction for recharge will impact on rights of water users downstream and on surface water dependent ecosystems. The MAR operator would need to buy the rights of other water users according to catchment-wide water allocation plans in order for the project to proceed.

Without a suitable aquifer MAR is not possible. Permeability and storage capacity need to be sufficient that an adequate volume of water can be stored and recovered. In future, groundwater allocation plans could take into account of the physical capability of the aquifer to store additional water. Even artesian aquifers may be used to store additional water under some circumstances. Likely to be a tighter constraint is the ability to recover stored water so as not to adversely impact on other groundwater users. This would be assessed in subsequent stages if the project is deemed viable.

Aquifer characteristics affecting suitability as a storage target are shown in Table 17.3. Mapping of key attributes provides a way to help with site selection within a groundwater system. For example, the organisation Geoscience Australia has used well yield and salinity of groundwater to map large areas and multiple aquifers in assessing the most prospective targets for large scale MAR projects. Mapping and prioritising MAR targets in other parts of Australia has also been undertaken using readily available hydrogeological data and in South Africa a recharge master plan has been prepared based on mapping (Murray *et al.* 2007). For infiltration type projects, the hydraulic conductivity of the least permeable layer between the surface and the aquifer needs to be known. This often is at a depth deeper than soil surveys and unsaturated zone formation properties are rarely recorded with enough information from water well or mineral exploration drilling to detect these layers. Therefore although infiltration sites can be the most economical form of MAR (as shown in Table 17.1) their investigations costs can be high to assess likely infiltration rates with acceptable groundwater mound heights and recoverability of recharged water.

Generally rates of surface runoff are higher than rates at which an aquifer can be recharged, so some form of detention storage may be required. In the case of recycled water that comes from a sewage treatment plant, the plant itself may already have adequate buffering storage. In established urban areas this can be the tightest constraint on the volume of stormwater that may be harvested and recharged. In some cases, supplemental treatment may be required prior to recharge (to be revealed in subsequent investigations) and this will require space.

Table 17.3 Characteristics of aquifers and their influence on potential for managed aquifer recharge (adapted from Dillon and Jimenez 2008).

Characteristic	Feature and influence on managed aquifer recharge	
Permeability	**Moderate to high** • High rates of recharge possible • Recharged water can be dispersed • Lower capital and energy costs per unit of water recovered	**Low to moderate** • Lower rates of recharge possible • Recharged water more localised • Greater capital and energy costs per unit of water recovered
Confinement	**Unconfined** • Surface infiltration methods viable • Unprotected from surface contamination • Storage capacity depends on depth to watertable	**Confined** • Well injection methods only • Protected from surface contamination • Storage capacity depends on aquifer thickness
Thickness	**Thick** • High storage potential • More sensitive to salinity stratification if native groundwater is brackish	**Thin** • Low storage potential • May limit rate of recovery by wells
Uniformity of hydraulic properties	**Homogeneous** • Minimal mixing and higher recovery efficiencies if native groundwater is brackish	**Heterogeneous** • Lower recovery efficiencies if native groundwater is brackish • In karstic and fractured rock systems, limited ability to contain recharged water
Salinity of groundwater	**Fresh** • Recovery efficiency not limiting • Requirement to protect wider range of beneficial uses of aquifer (higher treatment costs)	**Saline** • Recovery efficiency can limit effectiveness • Less beneficial uses to protect, so treatment requirements less onerous
Lateral hydraulic gradient	**Gentle** Recharged water contained closer to point of recharge	**Moderate to steep** • Recharged water dispersed downgradient and lower recovery efficiencies if native groundwater is brackish
Consolidation	**Consolidated** • Easier to complete wells • Easier to maintain recharge wells to prevent irrecoverable clogging	**Unconsolidated** • Screens required for injection and recovery wells • Land subsidence a consideration
Aquifer mineralogy	**Unreactive with recharge water** • Recovered water quality unaffected by geochemical reactions with aquifer matrix • Likelihood of clogging of injection wells is sometimes increased	**Reactive with recharge water** • Consider metal (eg As) mobilisation, Fe and H_2S effects on recovered water and groundwater • In carbonate aquifers, less onerous treatment required to avoid clogging of injection wells
Redox state of native groundwater	**Aerobic** • Higher rates of inactivation of pathogens and biodegradation of some endocrine disrupting chemicals	**Anaerobic** • Higher biodegradation rates for trihalomethanes

As = arsenic, Fe = iron; H_2S = hydrogen sulfide.

Finally, the capability to manage a MAR project needs to be established. Disciplines required include hydrogeology, biogeochemistry, geotechnology, water treatment, and risk management. Undertaking demonstration projects with inputs from those experienced in MAR elsewhere is a way to build the local capability, among proponents, regulators and researchers. Networking through organisations such as IAH Commission on MAR (www.iah.org/recharge) and the MAR-NET network can facilitate this process.

These basic water access entitlement and planning regulation issues are addressed before considering water quality which generally requires more detailed knowledge. Water allocation decisions will generally be based on existing information set within the context of the whole water catchment and aquifer system, and on the proposed volumes of recharge and recovery. These either result in a decision to proceed to the next step or that the proposed project is unviable.

Positive answers to the questions in Figure 17.2 suggest that a project is potentially viable. This leads to an assessment of the degree of difficulty. Note that concurrence or close proximity of locations where there is demand for water, surface water available, space for capture and/or treatment and access to a suitable target aquifer define the most prospective sites for MAR projects. If water trading is allowed within the jurisdiction and water can be transferred in the aquifer or by pipeline to end users, the demand for water need not be co-located. This allows more degrees of freedom in site selection.

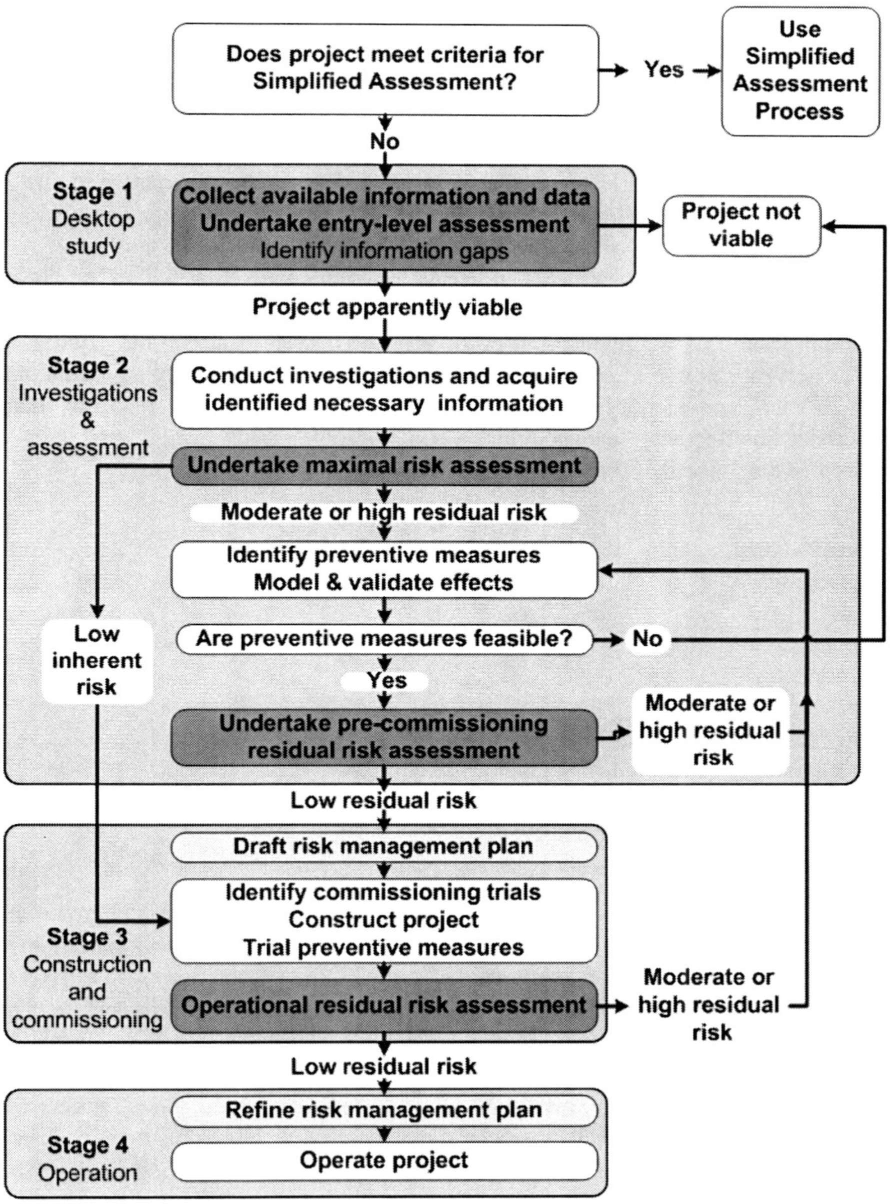

Figure 17.2 Risk assessment stages in MAR project development (from NRMMC-EPHC-NHMRC, 2009a)

17.3.2 Degree of difficulty assessment

This is a further simple checklist of 14 questions that help inform proponents of the degree of difficulty of their conceived project (NRMMC, EPHC and NHMRC, 2009a). This serves as a guide to the amount of effort required in project investigations and commissioning trials in order to manage human health and environmental risks. For example storing water in an aquifer used as drinking water supply will require considerably more management effort than storing water in a similar aquifer that is brackish because projects will need to demonstrate that they protect all existing beneficial uses of the aquifer. Similarly projects undertaken near groundwater-dependent ecosystems will require more monitoring and possibly a higher level of pre-treatment than those in the same aquifer but further from such ecosystems.

Costs of MAR investigations and trials are not trivial and, having completed this checklist, the proponent should know whether their proposed project has a low or high degree of difficulty and the types of information which will be of most value in the investigation stage. Because of the costs of these investigations it is normal to first seek assurance that at least the core approvals for MAR are likely to be obtained from the regulator in the jurisdiction, before undertaking such investigations, noting that some approvals will not be possible until after the investigation stage. Thus the work to this point is a screening stage using the available information to help the proponent estimate the likely cost of investigations, the likely constraints under which the project would operate and thereby to assist in deciding whether investment in MAR investigations is warranted.

17.3.3 Investigations and risk assessment

If the project looks potentially viable, having taken account of the likely degree of difficulty, at this the first stage, the Australian MAR Guidelines (NRMMC-EPHC-NHMRC, 2009a) lead proponents through the investigations (Stage 2) and commissioning trials (Stage 3) to an operational project (Stage 4). This process takes account that some information on which to base risk assessments will not be available without first having characterised the aquifer, the native groundwater and the surface water intended for recharge. Figure 17.2 is a process diagram showing the stages leading to an operational project.

The MAR Guidelines are part of the Australian Guidelines for Water Recycling within the National Water Quality Management Strategy (NRMMC-EPHC-NHMRC, 2006, 2008, 2009a, 2009b). The MAR guidelines specifically aim to protect the environmental values of all intended uses of recovered water and of the aquifer beyond a transient attenuation zone, and to prevent adverse impacts. This is done in stages 2, 3 and 4 by assessing potential hazards and the risks associated with each, and identifying preventive measures to manage the risks. In Stage 3 data becomes available to verify that risks are effectively managed during commissioning trials.

The hazards addressed in the guidelines are:

(1) Pathogens
(2) Inorganic chemicals
(3) Salinity and sodicity
(4) Nutrients
(5) Organic chemicals
(6) Turbidity/particulates
(7) Radionuclides
(8) Pressure, flow rates, volumes and levels
(9) Contaminant migration in fractured rock and karstic aquifers
(10) Aquifer dissolution and aquitard and well stability
(11) Impacts on groundwater dependent ecosystems
(12) Greenhouse gas emissions

For each hazard the guidelines document sources or causes, the effect on public health and environment, how it can be managed, including preventive measures, the proposed validation, verification and operational monitoring, and list the acceptance criteria for the various stages of risk assessment (Figure 17.2).

A simplistic view that treating water to near drinking standards before recharge will protect the aquifer and recovered water is incorrect. Aquifers are not inert, they may behave like bioreactors causing significant changes in quality of stored water (Dillon & Toze, 2005; Dillon *et al.* 2008; Vanderzalm *et al.* 2009). For example chlorination, to remove pathogens that would be removed in the aquifer anyway, can result in water recovered from some aquifers containing excessive chloroform. In some locations, drinking water injected into potable aquifers has resulted in excessive arsenic concentrations on recovery due to reactions between injected water and pyrite containing arsenic. Source water that has been desalinated to a high purity dissolves more minerals within the aquifer than water that has been less treated. Consequently the MAR guidelines adopt a scientific approach accounting for three ways that aquifers interact with recharged water:

(1) Sustainable hazard removal – the guidelines allow for pathogen inactivation, and biodegradation of some organic contaminants during the residence time of recharged water in the soil and/or aquifer under specified conditions within an attenuation zone of finite size,
(2) Ineffective hazard removal – these hazards need to be removed prior to recharge because they are either not removed (e.g. salinity) or removal is unsustainable (e.g. adsorption of any metals and organics that are not subsequently biodegraded, or excessive nutrients or suspended solids),
(3) New hazards introduced by aquifer interaction (e.g. metal mobilization, hydrogen sulphide, salinity, sodicity, hardness, or radionuclides) – there is a need to change the quality of recharge water to avoid these (e.g. change acidity-alkalinity, reduction-oxidation status or reduce nutrients).

The response of an aquifer to any water quality hazard depends on specific conditions within the aquifer, including temperature, presence of oxygen, nitrate, organic carbon and other nutrients and minerals, and prior exposure to the hazard. The guidelines indicate the state of current knowledge on attenuation rates of pathogens and organic compounds under a range of conditions, and provides for new local knowledge to be taken into account in assessing risks and determining sizes of attenuation zones and the siting of monitoring wells.

17.4 SITE SELECTION AND AQUIFER CHARACTERISATION

These were discussed under viability assessment and would be assisted by mapping of aquifers for their potential as targets for MAR in the vicinity of unmet future water demands and sources of recharge water. This can indicate the strategic priority sites within a catchment and groundwater system. This information should then be fed into land planning processes to reserve sites that are eminently suited to MAR. In areas with unconfined aquifers the adjacent lands should be zoned for groundwater quality protection and data on the unsaturated zone, such as geophysical logs, should be recorded as part of well-construction permitting in order for the lack of information to be addressed over time and facilitate simpler investigations for infiltration-type MAR sites. Gale (2005) has reported on site selection considerations in semi-arid catchments for a range of types of MAR projects (Dillon, 2005).

In general some form of groundwater model will be developed to assist with interpretation of data, depending on the complexity of the site and issues encountered. This could range from a one dimensional analytical model of groundwater flow to a three-dimensional reactive solute and heat transport model. MAR operations are very helpful perturbations of the groundwater system to enable calibration and validation of models (e.g. Pavelic *et al.* 2006) which conversely are helpful for interpreting data on changes in pressure, temperature and water quality analytes within the aquifer. A calibrated model can be used to guide operations at a MAR site so that it can achieve water quantity and quality objectives in recovered water and meet environmental requirements within the aquifer and connected groundwater-dependent ecosystems.

17.5 OPERATION AND MAINTENANCE

The location, design, operation and monitoring of a managed aquifer recharge project must take account of potential operational issues, including:

- clogging of injection wells or basins
- poor recoverability of recharge water (allowing for the reliability and continuity of source-water supply)
- interactions with other groundwater users
- managing purge water, basin scrapings and water treatment byproducts.

Chapter 6 of the Australian MAR Guidelines (NRMMC-EPHC-NHMRC, 2009a) discusses each of these issues, providing information on their causes, consequences, assessment methods and preventive measures. These issues are of primary importance to project proponents and operators, but are of less concern to regulators responsible for human health and environmental protection. The greatest emphasis is given to clogging and recovery efficiency, because these are vital to the technical and economic viability of managed aquifer recharge projects, and hence to the perceived affordability of investigations and trials.

Further information on design, operation and maintenance of surface infiltration and injection well systems, based on experience at managed aquifer recharge projects, predominantly from the United States is given by NRC (1994, 2008), EWRI/ASCE (2001), Bouwer (2002), Pyne (2006) and Bouwer *et al.* (2008). The European Commission (2001) also summarised knowledge generated in a number of European projects.

In overview, monitoring and evaluation of the data during operations reveals potential problems before they grow to become too complex to address efficiently. This is especially so for clogging which for ASR wells can lead to chronic decline in injection rates if not carefully managed. If recovered water quality differs markedly from expectations the conceptual model of the aquifer storage zone needs to be revised. This could lead to different ways of operating the site in order to regain the efficiency expected or revised expectations.

17.6 MONITORING

Monitoring is an integral part of the risk assessment of managed aquifer recharge. It forms a central component of the risk management framework. In particular, the following elements require monitoring:

- Operational procedures and process control
- Verification of recycled water quality and environmental performance
- Validation, and research and development.

Managed aquifer recharge monitoring programs should:

- Have clear objectives,
- Aim to maximise information content and the value of measurements in relation to the objectives,
- Be commensurate with the complexity and risk of the proposed managed aquifer recharge scheme,
- Be integrated with the risk assessment and management processes being adopted.

Guidance on four types of monitoring for MAR projects is given in Chapter 7 of the Australian MAR Guidelines (NRMMC-EPHC-NHMRC, 2009a); baseline, validation, operational and verification monitoring.

Baseline monitoring provides information for the maximal risk assessment. It is used to define the state of the system before commissioning a MAR project.

Validation monitoring is essential when there is a reliance on the treatment capacity of the aquifer. It quantifies the treatment efficacy of any new or uncharacterised treatment steps, such as subsurface treatment at a new locality. It may also include pre-treatment and post-treatment technologies, and exploration of water-quality deterioration (e.g. arsenic release, or formation of disinfection byproducts). Validation monitoring usually includes monitoring surrogate parameters and the use of piezometers. Much of the validation monitoring will occur during investigations and commissioning stages of project development.

Operational monitoring is fundamental to the risk management of operational MAR projects. It provides timely information for use as critical control points in the risk management plan and to evaluate the systems operation for example, volumes and pressures. It often includes supervisory control and data acquisition (SCADA) and web-based reporting systems that provide near real-time data. Operational monitoring is fundamental to setting appropriate critical limits and management responses such as:

- Shutting down recharge due to poor quality source water,
- Shutting down recovery due to salinity limits being reached in recovered water,
- Backflushing ASR wells at the onset of clogging,
- Changing dose concentrations in pretreatment processes.

Verification monitoring is required for operational projects to confirm that the MAR system and its components are performing as anticipated. Verification monitoring can be performed on a compliance basis, and is necessary for regulatory agencies assessing the whole of the managed aquifer recharge system. Where there is reliance on an attenuation zone to meet the aquifer's pre-existing environmental values, then verification would require one or more observation wells at the perimeter of that zone. On decommissioning of a managed aquifer recharge operation, the verification monitoring specified in the site management plan needs to be undertaken until the aquifer has been restored to its ambient environmental values.

17.7 CONCLUSIONS

Managed aquifer recharge projects, like any projects dependent on natural systems perform best if appropriate effort goes into characterising the aquifer and the source of water for recharge. In this way the system can be sited, designed and operated to demonstrably meet its intended objectives, including all water resources management and environmental requirements. Systems are designed around an aquifer's capabilities and additional preventive measures such as excluding source water of unacceptable quality, inclusion of pre-treatment and/or post-treatment processes and restricting uses of recovered water can be used where necessary to achieve the projects objectives.

Appropriate monitoring and alerts and knowledgeable operators who respond to warning signs, and can revise operations to address contingencies are keys to sustaining MAR operations. Even tightly constrained systems can be run effectively over very long time periods. Awareness of changes in catchments that may present new hazards and ongoing research to validate aquifer processes under changed conditions are part of the duty of care of water supply operators.

Aquifers are abundant natural assets that can be better used as buffer storages, to improve water quality and to protect vulnerable groundwater resources and dependent ecosystems, through augmentation with 'natural' and recycled waters. It is important that MAR projects are operated to demonstrably meet the standard of care that a community expects relevant to the end uses of recovered water from the MAR system and by others drawing on the same aquifer. Guidelines now exist that establish the duty of care, and further guidelines or derivations may be needed to address situations in countries without the resources to commit to the design, operation and monitoring of projects, especially where drinking water is the end use.

ACKNOWLEDGEMENTS

Work reported here is a synthesis of numerous projects supported by the Australian Government through Department of Industry Innovation Science and Research under the International Collaboration Program with the European

Commission's 7th Framework Program, the National Water Commission through the Raising National Standards Program, the Environment Protection and Heritage Council in development of Guidelines and the CSIRO Water for a Healthy Country National Research Flagship. The work would not have been possible without numerous partners on MAR projects in Australia who have generously supported the research upon which this synthesis is based.

REFERENCES

Bouwer H. (2002). Artificial recharge of groundwater: hydrogeology and engineering. *Hydrogeology Journal*, **10**(1), 121–142.

Bouwer H., Pyne R. D. G., Brown J., St Germain D., Morris T. M., Brown C. J., Dillon P. and Rycus M. J. (2008). Design, Operation and Maintenance for Sustainable Underground Storage Facilities. American Water Works Association Research Foundation Report. Denver, USA.

Cai W. and Cowan T. (2008). Evidence of impacts from rising temperature on inflows to the Murray – Darling Basin. *Geophysical Research Letters*, **35**, L07701, doi: 10.1029/2008GL033390.

Dillon P. and Toze S. (eds) (2005). Water Quality Improvements During Aquifer Storage and Recovery. American Water Works Assoc., Research Foundation Report 91056F, 286p + 2CDs.

Dillon P., Martin R., Rinck-Pfeiffer S., Pavelic P., Barry K., Vanderzalm J., Toze S., Hanna J., Skjemstad J., Nicholson B. and Le Gal La Salle C. (2003). Aquifer storage and recovery with reclaimed water at Bolivar, South Australia. *Proc. Aust. Water Assoc.*, 2nd Australian Water Recycling Symposium, Brisbane 1–2 Sept 2003.

Dillon P., Gale I., Contreras S., Pavelic P., Evans R. and Ward J. (2009a). Managing aquifer recharge and discharge to sustain irrigation livelihoods under water scarcity and climate change. IAHS Publ, **330**, 1–12.

Dillon P., Page D., Vanderzalm J., Pavelic P., Toze S., Bekele E., Sidhu J., Prommer H., Higginson S., Regel R., Rinck-Pfeiffer S., Purdie M., Pitman C. and Wintgens T. (2008). A critical evaluation of combined engineered and aquifer treatment systems in water recycling. *Water Science And Technology*, **57**(5), 753–762.

Dillon P., Pavelic P., Page D., Beringen H. and Ward J. (2009b). Managed Aquifer Recharge: An Introduction. National Water Commission Waterlines, Report No 13, Feb 2009, 65p. http://www.nwc.gov.au/www/html/996-mar–an-introduction—report-no-13-feb-2009.asp (accessed 11 Feb 2010).

Dillon P. J. (2005). Future management of aquifer recharge. *Hydrogeology Journal*, **13**(1), 313–316.

European Commission. (2001). Artificial recharge of groundwater. *Final Report Of EC Project ENV4-CT95-0071*, Directorate General for Research, EUR 19400.

EWRI/ASCE. (2001). Standard Guidelines for Artificial Recharge of Ground Water. EWRI/ASCE 34-01. Artificial Recharge Of Ground Water Committee, Standards Development Council, Environmental and Water Resources Institute, American Society of Civil Engineers.

Gale I. (2005). Strategies for Managed Aquifer Recharge (MAR) in semi-arid areas. UNESCO IHP. 34p. www.iah.org/recharge.

Gale I. (2005). Strategies for Managed Aquifer Recharge (MAR) in semi-arid areas. UNESCO IHP. 34p. www.iah.org/recharge (accessed 11 Feb 2010).

Grischek T. (200X) BF.

Grischek, T. and Ray, C. (2009). Bank filtration as managed surface-groundwater interaction. *Intl. Journal of Water*, **5**(2), 125–139.

Johnston P. R., Laurenson E. M. and Howell D. T. (1973). A design procedure for the conjunctive use of surface and groundwater storages. *Australian Water Resources Council Technical Paper No 3*, Aust Govt Publishing Service.

Massmann G., Greskowiak J., Dünnbier U. and Zuehlke S. (2006). The impact of variable temeratures on the redox conditions and the behavior of pharmaceutical residues during artifical recharge. *J. Hydrol*, **328**(1–2), 141–156.

Massmann G., Dunnbier U., Heberer T. and Taute T. (2008). Behaviour and redox sensitivity of pharmaceutical residues during bank filtration – investigation of residues of phenazone-type analgesics. *Chemosphere*, **71**(8), 1476–1485.

McCartney M. and Smakhtin V. (2010). Water Storage in an Era of Climate Change: Addressing the Challenge of Increasing Rainfall Variability. Intl. Water Management Institute Blue Paper, 14p.

Murray R., Tredoux G., Ravenscroft P. and Botha F. (2007). Artificial Recharge Strategy. – A national approach to implement artificial recharge as part of water resource planning, (accessed 11 Feb 2010) http://www.dwaf.gov.za/Documents/Other/Water%20Resources/ARStrategyforSAJun07SecA.pdf.

NRC (National Research Council). (1994). Ground Water Recharge Using Waters of Impaired Quality. National Academies Press, Washington DC.

NRC (National Research Council). (2008). Prospects for Managed Underground Storage of Recoverable Water. National Academies Press, Washington DC.

NRMMC-EPHC-AHMC. (2006). Australian Guidelines for Water Recycling (Phase 1): Managing Health and Environmental Risks. (Natural Resources Ministerial Management Council, Environment Protection and Heritage Council and Australian Health Ministers' Conference), Canberra.

NRMMC-EPHC-NHMRC. (2008). Australian Guidelines for Water Recycling (Phase 2): Augmentation of Drinking Water Supplies. (Natural Resource Ministerial Management Council, Environment Protection and Heritage Council and National Health and Medical Research Council), Canberra, www.ephc.gov.au/taxonomy/term/39.

NRMMC-EPHC-NHMRC. (2009a). Australian Guidelines for Water Recycling (Phase 2): Managed aquifer recharge. (Natural Resource Management Ministerial Council, Environment Protection and Heritage Council and National Health and Medical Research Council), Canberra, www.ephc.gov.au/taxonomy/term/39.

NRMMC-EPHC-NHMRC. (2009b). Australian Guidelines for Water Recycling (Phase 2): Stormwater Harvesting and Reuse. (Natural Resource Management Ministerial Council, Environment Protection and Heritage Council and National Health and Medical Research Council), Canberra, www.ephc.gov.au/taxonomy/term/39.

WCD (World Commission on Dams) 2000. Dams and Development: A New Framework for Decision-Making. London: Earthscan.

Pavelic P., Dillon P. J. and Simmons C. T. (2006). Multi-Scale characterization of a heterogeneous aquifer using an ASR operation. *Ground Water*, **44**(2), 155–164.

Pyne R. D. G. (2006). Groundwater recharge and wells. 2nd edition, ASR Systems, Florida.

Vanderzalm J., Sidhu J., Bekele G-G., Pavelic P., Toze S., Dillon P., Kookana R., Hanna J., Barry K., Yu X., Nicholson B., Morran J., Tanner S. and Short S. (2009). Water Quality Changes During Aquifer Storage and Recovery. Water Research Foundation. Denver, USA.

Van Steenbergen F. and Tuinhof A. (2010). Managing the Water Buffer for Development and Climate Change Adaption: Groundwater recharge, retention, reuse and rainwater storage. Published by BGR, CPWC, Netherlands National Committee IHP-HWRP, 92p. www.iah.org/recharge.

Chapter 18

Use of groundwater models for prediction and optimisation of the behaviour of MAR sites

Wolfram Kloppmann, Avi Aharoni, Haim Chikurel, Peter Dillon, Irina Gaus, Joseph Guttman, Tom Kraitzer, Sarah Kremer, Costantino Masciopinto, Konrad Miotlinski, Paul Pavelic, Marie Pettenati and Géraldine Picot-Colbeaux

18.1 GROUNDWATER MODELLING AND ARTIFICIAL RECHARGE: WHAT MODEL FOR WHAT PROBLEM

18.1.1 Models as tool to design and operate a MAR system in a given legal context

The principal benefits of the concept of Managed Aquifer Recharge (MAR) for water management are twofold:

(1) It allows storage of large quantities of surface water (including surface runoff, stormwater, reclaimed water, and also freshwater from desalination) at those periods of the hydrological year when availability exceeds demand and to restore them when demand exceeds availability.
(2) The underground passage (unsaturated and saturated zone) can constitute a complementary treatment step, due to physical, chemical and biological processes that will affect water quality.

The major challenge to meet is a potential conflict of the MAR system with other uses, mainly in terms of water quality (risk of degradation of chemical or biological background quality for one or more parameters due to infiltration or infiltration-induced chemical processes in the storage medium). In order to address those benefits and risks, the designer or operator of any MAR system will need to address the following key questions:

(1) How efficient is my system in terms of recovery of the recharged water?
(2) How long will the water and solutes reside in the system and be in contact with reactive minerals and biofilms?
(3) In which way will the recharge-recovery cycles affect the quality of the recovered water and the background water, on short term and long term?

These questions will be asked from the very beginning of the planning phase and over the whole lifetime of the MAR project and **groundwater models** provide the unique possibility to preview the feasibility of the MAR system in the regional context, to optimize the choice of the site, the configuration of an appropriate recharge-recovery system, to optimize operating conditions in a way to meet fixed quantity and quality targets. Those targets are most frequently quantified through the key parameters recovery efficiency, residence time, and recovered water quality compared to target quality. **Recovery efficiency** (Lowry & Anderson, 2006) as a measure of success of a MAR system, is defined either by the percentage of injected water that can be recovered or by the percentage of usable water (meeting a defined target quality, e.g. drinking or irrigation water standards) compared to the injected volume per cycle of injection. **Residence time** or transfer time (in the case of Aquifer Storage, Transfer and Recovery, ASTR) will be the average duration of water and solutes in the reservoir determining the time available for water-rock interactions and bio-geochemical reactions. Residence time will largely influence the potential for degradation of pollutants and thus the efficiency of Soil Aquifer Treatment (SAT) systems and to estimate this parameter may therefore be legally compulsory. A typical example are the Californian draft regulations applying to new recharge projects (CDPH, 2011).

They define the minimum retention time to allow identification of treatment failures and implement remediation actions and to guarantee the overall treatment efficiency. **Target quality** for the groundwater quality is generally defined by law and water reuse guidelines. **Potential use** of recovered water is also defined in these guidelines. . The **quality of the recovered water** has to be constantly compared to this target quality and is determined by the quality of recharged water, the attenuation potential of the unsaturated and saturated aquifer and the background water, frequently but not necessarily of lower quality than the recharged water (e.g. freshwater injection into saline aquifers). The underlying processes controlling water quality are (1) mixing by advection and hydrodynamic dispersion (2) density dependent stratification (3) ambient groundwater displacement and (4) reactions within the aqueous phase induced by mixing and reactions with aquifer material and subsurface microbial communities.

Those key parameters are determined by *intrinsic, physical factors* (aquifer properties like permeability, effective porosity, dispersivity, preferential flow zones and the natural hydraulic gradients, aquifer mineralogy and background water quality) and by *operational variables* (e.g. storage period, volumes recharged, and recharge/recovery rates). The intrinsic properties are generally constant but may evolve over the lifetime of a MAR system (e.g. through clogging, Perèz-Paricio & Carrera, 1999; Pavelic *et al.* 2007. Intrinsic factors will depend on the selection and design of the system, operational variables on the strategy for MAR operation. Groundwater models can help for both purposes. Currently available groundwater models allow quantifying recovery efficiency, residence time and quality of recovered water. However, a complete response to the questions listed above may need the use of state of the art models (reactive transport models) or go beyond the current capacities of the state of the art (bio-geochemical reaction modelling). Even simple models (analytical models) can provide sufficient information at least for preliminary design or evaluation of MAR systems but, most frequently, numerical models will be used. Standard numerical models will nowadays be able to simulate up to full 3D advective (e.g. through particle tracking) and dispersive flow and transport of water and solutes. Supplementary features may be needed as, in the order of increasing complexity:

- Density driven flow (in the context of highly saline waters, like in coastal aquifers)
- Sorption and (bio-)degradation of solutes (e.g. through sorption isotherms, degradation factors)
- Variable saturation flow (in the case of a significant thickness of the unsaturated zone, in particular if the latter plays an important role for water quality improvements in SAT systems)
- Geochemical reactions through the combined use of flow-transport models and thermodynamic equilibrium models or thermo-kinetic models taking into account the reaction kinetics
- Biologically mediated geochemical reactions (specific models available)
- Aquifer ecosystem simulation accounting for changing microbial communities and their assimilation of nutrients, competencies for metabolism of organic contaminants, enzymatic attack on pathogens and their influences directly and through microbial products (such as polysaccharides) on aquifer porosity and hydraulic conductivity.

In this chapter we will outline the use of groundwater models for MAR focusing on the strategic phases of a MAR system (site selection, MAR design, MAR operation), on the data needs to build and run such models as well as on the model selection (types of models). Applications of different types of models within the case studies of Reclaim Water will illustrate this synthesis.

Site selection, feasibility of MAR, predimensioning of the MAR system and design of the associated monitoring system

Decision-making processes on MAR require an initial feasibility study in the regional context. A first step will be the selection of the target reservoir (aquifer) and of the potential sites for such a MAR system. Initial lack of data is a rather general drawback of feasibility or pre-dimensioning modelling. In some cases, regional groundwater models are available which offer the possibility to simulate the impact of MAR on an aquifer system and to select potential sites for recharge systems. Such regional models generally do not offer the spatial resolution required for designing the MAR system in its technical details and only the construction of a pilot will provide sufficient data on the relevant hydrogeological parameters to evaluate feasibility of a larger scale Aquifer Recharge. In an early study, Merritt (1985) investigated the feasibility of freshwater storage and recovery in saline aquifers in South Florida using a numerical solute-transport model (INTERA model) including buoyancy effects. This regional study considers a hypothetical reservoir, representative of the regional aquifers in South Florida taking into account the variability of hydraulic parameters in this area. It shows the impact of hydrodynamic dispersion (mixing), density effects, background hydraulic gradients, MAR system configuration and operating conditions on the recovery efficiency, thus providing a conceptual framework for site selection in a regional context. Gau *et al.* (2006) used a variable saturated flow model (FEMWATER[1], Lin *et al.* 1997) to investigate infiltration conditions through the unsaturated zone at regional scale in

[1]http://www.epa.gov/ceampubl/gwater/femwater

an alluvial, coastal plain in Taiwan. Through the use of the model they established a hierarchy of hydrogeological factors that determine arrival times of infiltrated water at the groundwater table through a weighting factor (in the order of importance: hydraulic conductivity, water content, shape of the water release curve, depth of water in infiltration pond, depth of groundwater table) and derived an index of infiltration efficiency. Suitable areas for a MAR site were then determined on the basis of a regional survey of soil parameters and mapping of the infiltration efficiency index. This study is an example for the use of models for the selection of a site with optimal recharge conditions within a target aquifer system. An important output from modelling can be a hierarchy of relevance of soil parameters which can, in turn, guide measurement campaigns in the field.

Other studies investigate, through numerical modelling, the response of a regional aquifer system to MAR and evaluate if fixed water management targets can be reached by MAR. Shammas *et al.* (2008) established a regional flow and transport model to evaluate efficiency of MAR to combat saline intrusion into a coastal aquifer in Oman and investigate water quality improvement through salinity reduction. They made use of 3D-MODFLOW (McDonald & Harbaugh, 1988) as saturated groundwater flow model combined with MT3DMS and PMPATH for multi-species transport modelling (advection). They determined the possibility to stabilise water levels at aquifer scale through MAR and its efficiency for saline intrusion stabilisation and even displacement of the saline-fresh interface. Yet, MAR would only re-establish the water balance when accompanied by complementary measures (pumping reduction in specific parts of the aquifer). The model also allowed testing recharge scenarios and helped selecting well fields for injection. In this study, the model was calibrated with data from a pilot ASTR study site. Preliminary injection tests and pilot studies are frequently conducted at shallow depths and cannot predict the behaviour of a large-scale multi aquifer system. In such cases, as stressed by Giao *et al.* (1998), only modelling can assess the response of deeper aquifers and the system as a whole. Their study investigates a 500 m thick graben filling below Bangkok city and the response of the strongly overexploited regional multi-aquifer system to MAR. Modelling provided answers to the following questions: (1) impact of different patterns and number of wells as well as of depths of pumping/recharge on piezometric heads, (2) changes in land subsidence induced by overpumping, (3) soil parameters required for subsequent studies. The study used a quasi-3D model approach (2D horizontal flow in the aquifers, 1D vertical exchange with adjacent aquitards), taking also into account clogging induced by injection and combines it with a consolidation model to calculate rebound of soil layers concerned by injection.

The subsequent step, after regional investigation of the feasibility and impact of a MAR system on a basin, a multi-aquifer system or an individual aquifer and after site selection, will be the design of the infiltration/injection-recovery system. The recharge system will be chosen from a huge variety of techniques of artificial aquifer recharge (Dillon, 2005). Type and dimension will depend, among others, on the available water and the storage capacity of the aquifer reservoir as well as on the target pumping rates and the need for complementary treatment. The second important choice is the geometry of the recovery well array. In the simplest case, injection and recovery well are identical (Aquifer Storage and Recovery, ASR system). When looking upon ASTR or SAT in particular, depths, number and spacing of the recovery wells with respect to the injection point(s) become crucial. Here, again, groundwater modelling is frequently used to plan the system configuration. Regional models will rarely be directly used for site scale planning as their spatial resolution will not be sufficient but they can provide the boundary conditions for local models or "nest" them through a local increase of grid density. A typical example is shown below (Adelaide case study) where an array of 4 injection and 2 recovery wells was optimised through a local model (Pavelic *et al.* 2004, 2005; Kremer *et al.* 2008) to achieve the required residence time and recovery efficiency in a brackish aquifer with a constrained volume of injectant.

Two particular problems concerning MAR need special features in the used hydrodynamic models: density driven flow and flow/transport in the unsaturated zone. The first becomes relevant when dealing with injection of freshwater into a brackish or saline aquifer, background salinity being linked to seawater intrusion into a coastal aquifer or to other natural salinity sources. The second aspect cannot be neglected when investigating infiltration pond-like MAR system and when the unsaturated zone plays an important role for pollutant removal (SAT) and/or if the residence time of the infiltrated water in the USZ is significant. Ward *et al.* (2007, 2008, 2009) investigated in detail the effects of MAR-induced density flow in a hypothetical multi-layer aquifer using FEFLOW, a finite element code for up to 3D transport in porous media (Diersch, 2005). They come to the conclusion that even for relatively fast pumping rates and a low density contrast between recharged and ambient water, there is still the potential for significant reduction in recovery efficiency due to density driven flow after 10 cycles. Ward *et al.* (2009) further generalised this in aquifers with lateral flow to describe and test the use of non-dimensional parameters to characterise expected recovery efficiency. Izbicki *et al.* (2008) studied the possibility to recharge an aquifer through a very thick (117m) unsaturated zone with several perching clay layers and modelled water and solute flow with the TOUGH2 software package (Pruess *et al.* 1999). The model succeed in simulating the progress of the wetting front (over 3.5 years), of salt and nutrient transport and the study concluded that MAR is feasible even in zones with thick USZ. This is confirmed by the study of Nichols *et al.* (2007) on combined USZ-SZ modelling of a large scale operational MAR scheme where Columbia river water is used for nuclear material production and then disposed of through infiltration into a complex, vertically heterogeneous porous aquifer with a 80–90m thick USZ.

Optimisation of the operating scheme

Even once the geometry of the MAR system fixed, a large variety of options for the operating of the system remains open. Most important choices are the injection and pumping rates: the choice of a negative injection or pumping rate will even transform an injection well into a pumping well and *vice versa* thus leaving the choice of reversing the flow direction in the system. Other operational choices concern the rhythm of injection, e.g. alternating between wet and dry cycles for spreading basins in SAT. In this case, flow and transport models are usefully integrated with mathematical optimization techniques (Eusuff & Lansey, 2004) thus allowing for the simulation of the complex behaviour of the aquifer system and the identification of optimum management strategies considering the management objectives and the associated constraints, including minimisation of operating costs. Eusuff & Lansey (2004) provide an overview of simulation-optimisation management models since the early 1980s progressively taking into account flow field and later solute transport through a variety of linear and nonlinear programming techniques. Eusuff & Lansey (2004) themselves integrate 3D flow (via MODFLOW), solute transport (MT3D), considering biodegradation, and particle tracking (MODPATH) through an optimisation algorithm (shuffled complex evolution, SCE). They test maximisation of recharge and, independently, of extraction taking into account legal constraints on the water level, nutrient concentrations, and water particle residence times. Such integrated management models bear a much higher potential than planning tools which do not integrate a physical model of the MAR reservoir (e.g. Uddameri, 2006, presenting a model deriving optimal monthly storage and extraction rates meeting water supply needs at minimum costs).

Groundwater models are also a valuable tool for the handling and analysis of data produced by an operating MAR scheme. In this configuration, models are linked to a project data base and continuously fed by operational data and groundwater data (wellheads, quality). They investigate causal links between changes of the operating conditions and the system responses at local scale. Experiments, like artificial tracer tests, on the aquifer response can be easily interpreted if a model is available. New data, on the other hand, provide complementary constraints to the model calibration and can be used for model validation. On the whole, the used models will lead to a better understanding of the effects of an operating MAR system on the receiving reservoir.

Artimo *et al.* (2008) show an integrated tool for data storage, visualisation and analysis for a projected large-scale MAR project in south-western Finland. It combines a 3D geological and hydrogeological model (geodatabase) with a groundwater flow model that has also the capacity to integrate solute transport models in the aim to reach a thorough quantitative understanding of the aquifer system. This tool allows for planning from the predimensioning to operating conditions. It is further designed as a means for communication among scientists, engineers and decision-makers. Integrated modelling tools like the described can be used to meet the legal constraints on groundwater management, defined in the EU Water Framework Directive (Directive 2000/60/EC)[2] and its daughter directive on groundwater (Directive 2006/118/EC), with an obligation to define and measure criteria of good chemical status and quality trends of groundwater bodies.

Modeling of MAR-induced bio-geochemical processes and their impact on pollutant degradation, fixation or release

The introduction of any water into a groundwater body with a chemistry (or even temperature) different from the background conditions will trigger a re-equilibration of aqueous species and of water and solutes with the aquifer minerals they are in contact with. Reequilibration of aqueous species will mainly occur in zones where recharged water and background water mix. Reequilibration with the aquifer material will take place in the whole "bubble" or "bottlebrush"-shape volume where the recharged water proportion is >0%. Each injection-recovery cycle will induce new displacements of the chemical equilibria in the MAR system. This picture is further complicated by the presence of indigenous microbial communities in the aquifer and the introduction of new species via the reclaimed water. Certain microbial strains are able to catalyse geochemical reactions; in particular redox reactions in aquifers frequently involve microbiota which depend on the energy provided by those reactions. Such bio-geochemical reactions are non-equilibrium reactions with their kinetics determined by microbial growth and metabolism. Where reactions are slow this may limit the use of codes relying solely on the concept of thermodynamic equilibrium. The practical consequences of the reaction chains triggered by managed aquifer recharge are crucial for the functioning of the MAR system and the quality of the recovered water:

- Sorption or biodegradation of organic pollutants will change water quality, at least temporarily in the case of reversible sorption. In most cases, these processes induce an improvement of water quality, which is made use of through the concept of SAT. If the metabolites of biodegradation are more toxic than the original compounds, biodegradation can lead to complementary risks.
- Sorption may fix inorganic pollutants like heavy metals more or less temporarily on solid phases like clay minerals, hydroxides or organic matter. Any change of the chemical boundary conditions may lead to a remobilisation of those compounds (e.g. through injection-recovery cycles)

[2]http://ec.europa.eu/environment/water/water-framework/index_en.html

- Dissolution-precipitation processes may lead to clogging (e.g. by carbonates or hydroxides), co-precipitation of inorganic compounds (including pollutants like heavy metals) and also to their mobilisation.
- Redox reactions will determine the aqueous speciation of elements with several redox states (e.g. N, S...), degradation of organic matter (including organic pollutants) and involve also directly the aquifer minerals (mainly oxides and hydroxides).

Modelling of those processes occurring in MAR systems ranged from the simulation of relatively simple problems (carbonate dissolution-precipitation, fluoride release...) to state of the art bio-geochemical models attempting to fusion biological and geochemical concepts. First attempts of applying coupled reactive transport models on MAR go back to the 1980s (Van Breukelen *et al.* 1998 and references therein) but only recently thermo-kinetic equilibrium models, taking into account a large variety of inorganic/organic aqueous species and water-mineral reactions under various salinity conditions, have become widely available, harnessed to flow models, and have been applied to MAR systems (e.g. PHT3D, Prommer *et al.* 2003, Greskowiak *et al.* 2005). As the occurrence of geochemical reactions is triggered by groundwater and vadose zone flow, in most cases, geochemical modelling was linked to flow and transport modelling. Mixing processes and flow velocity, together with reaction kinetics, will determine where and when certain reactions occur. Under steady state conditions, a geochemical zoning may develop with a distinct reaction sequence occurring along the flowpath (Stumm & Morgan, 1981). This chemical zoning has been shown to be intimately linked to microbial zoning, i.e. distinct segments of the aquifer where a specific functional group of microorganisms is predominant (Bethke *et al.* 2008).

Gaus *et al.* (2001, 2002) investigated a typical problem of trace contaminant release from the chalk aquifer matrix induced by ASR in Dorset, UK. This study combines physical modelling of flow and conservative transport of chloride with a geochemical model of fluorite dissolution. The 2D radial groundwater flow model used, SWIFT/486 (Williams, 2000), allows for dual porosity flow and was calibrated to match chloride concentrations measured during the recovery phases. Fluoride contents are systematically around 10% higher than expected from simple mixing with native groundwater. PHREEQC (Parkhurst & Appello, 1999) in its column (1D) flow and transport mode was first calibrated to match chloride as conservative tracer; the results were comparable to those obtained by SWIFT. Among the several hypotheses tested – high F content in matrix porosity, immediate dissolution of a small amount of fluorite (CaF_2), and kinetically hindered fluorite dissolution – the latter was found to explain best the observed F-excess.

Dissolution/precipitation of carbonates was investigated and modelled by several authors mainly with the aim to determine geochemical controls of clogging of MAR wells. Yobbi (1996) used the PHREEQC code, without coupling to a flow-transport model, simply to calculate the saturation status of recharged effluents and receiving saline background waters in Florida and concluded that injection would enhance rather than diminish porosity through calcite and dolomite dissolution. Dissolution-precipitation reactions are also included in the coupled CLOG code (Pérez-Paricio, 2001, Pérez-Paricio & Carrera, 1999, Pérez-Paricio *et al.* 1998) integrating models of physical, biological and chemical clogging of MAR systems.

Most recent applications of reactive transport models to MAR are not limited to a single process but integrate a large variety of interdependent reactions. Van Breukelen *et al.* (1998) show the first example PHREEQC modelling of large scale, long-term (25 years) MAR system where Rhine water was infiltrated in a coastal dune sand aquifer in the Netherlands. This study includes carbonate dissolution-precipitation, redox reactions (oxygen consumption and denitrification by oxidation of organic matter) and cation-exchange and could successfully model the reactive transport of the variable input signal measured in the infiltrated river water. It shows that parameters like Ca^{2+} and alkalinity cannot be satisfyingly reproduced without considering redox processes (CO_2-production by oxidation of organic matter induces dissolution of calcite). Considering reaction kinetics is of prime importance when dealing with reactions which are slow non-equilibrium reactions mediated by bacteria (Hunter *et al.* 1998). Van Breukelen *et al.* considered a first order rate law for organic matter oxidation. A more realistic and popular model linking bacterial growth to nutrient concentrations is the Monod model (Lobry *et al.* 1992, Kargi, 2009 and references therein). Monod kinetics were used as reaction rates of organic matter breakdown to simulate the deep well recharge of oxygenated water into a deep pyrite bearing aquifer (Saaltink *et al.* 2003).

Generally, biologically catalysed redox reactions, are modelled through a single-step or a two-step approach (Greskowiak *et al.* 2005). In the single-step approach, redox reactions include simultaneously organic matter oxidation and the reduction of a terminal electron acceptor (TEAP) like dissolved oxygen, nitrate, sulphate, Fe^{III}, Mn^{IV}, ... The two-step approach or partial equilibrium approach (PEA, Brun & Engesgaard, 2002) separates the oxidation step from the electron accepting step. The first step is assumed to be rate-limiting, the second step is then modelled as equilibrium reaction. The PEA approach was used by Greskowiak *et al.* (2006) to model redox zoning and degradation of the pharmaceutical residue phenzone in glacial sediments artificially recharged by surface water through infiltration ponds in Berlin. This model, using the reactive transport code PHT3D (Prommer *et al.* 2003) takes into account organic matter mineralisation (following Monod kinetics), denitrification, manganese reduction, calcite and rhodochrosite (Mn-carbonate) dissolution-precipitation as well as the temperature dependence of organic matter breakdown.

It simulated successfully all redox sensitive parameters, except, to some extent, manganese, and of phenazone and showed the importance of seasonal temperature variations for the redox zoning.

A similar approach was used by Greskowiak *et al.* (2005) to investigate an ASR system in a confined brackish aquifer. The objective was to model carbon cycling between dissolved and particulate organic carbon, biomass, and inorganic carbon, considering the different TEAPs for a redox sequence reaching sulphate reduction and methanogenesis. PHT3D, the combined transport (MT3DMS) and geochemical (PHREEQC) model, incorporates, as kinetic reactions, DOC mineralisation and bacterial growth/decay, using Modod kinetics and the PEA concept. To estimate the reaction rates within the calibration process, a parameter estimation program (PEST, Doherty, 2002) was used. The model manages to reproduce satisfyingly all geochemical observations over an injection duration of 800 days. The authors conclude on the necessity to develop further alternative conceptual models tested in a large variety of settings before reaching a stage where reactive multicomponent transport models can be used as predictive tools.

The Shafdan case study (see below) illustrates that conceptual obstacles are still to be overcome, in particular for specific redox sensitive elements (Mn), and that a generalised coherent conceptual framework for integrating non-equilibrium bacterially mediated reactions and equilibrium reactions is still to be developed.

Modelling spreading of pathogens in fractured/karstic and porous aquifers with reference to legal constraints/thresholds

Securing the microbial safety of drinking after supplies is usually based on the use of multiple barriers, from catchment to consumer, to prevent the contamination of drinking-water or to reduce contamination to levels not injurious to the health. Drinking water production history suggests that raw water filtration is a long-known and very effective process for eliminating pathogens from the water (Schoenen, 2002) by increasing the efficacy of successive water treatments. Several authors developed this important topic (Adelman *et al.* 1998; Reinoso *et al.* 2008) also in fractured aquifers (Abbaszadegan *et al.* 1998). The former study concluded that the EPA (1997) has proposed a setback distance of 160 m (or 525 ft) for municipal wells in Nebraska, which represents a significant cost for utilities, as more than 2000 wells failing under the groundwater disinfection rule are used for public water supplies. Abbaszadegan (2001) study reports that, though the mean setback distance (3400 ft or 1036 m) from contaminated surface water far exceeded the mean EPA's (1997) distance (525 ft) in United States, the majority of wells used for drinking supplies are subject to full conventional water treatments (coagulation, sedimentation, filtration and disinfection).

Masciopinto *et al.* (2008) at the Nardò karstic and fractured aquifer established a setback distance over 8000 m in order to have a complete pathogen inactivation, though the distance of 500 m is currently respected by local communities for drinking water distribution.

The above descriptions reinforce that the uncertainty of mathematical model parameters and approximations in model assumptions together with the poorly understood processes governing virus transport and fate in groundwater, suggest the need for research to refine the credit given for "natural disinfection" and modify the groundwater disinfection rule.

18.1.2 Model data requirements and hydrogeological characterisation

Data requirements

The choice of the model will depend on the complexity of the system and the problem to be treated, the data situation and the available financial resources. Data requirements and the associated costs increase considerably with the complexity of the model (Figure 18.1). All combinations of spatial type (1D-3D) with saturation state and (reactive) transport modelling are in principle possible. The complexity and data requirements are nevertheless increasing steeply when combining full 3D with unsaturated flow and/or reactive transport. The present limits of models are reached for 3D unsaturated flow and reactive transport that takes into account bacterially controlled redox reactions. Each type of model has its own minimum data requirements, a synthesis is provided in Table 18.1. It may be necessary to increase the complexity of the chosen model if the observed reality (e.g. piezometric heads, tracer breakthrough…) cannot be satisfyingly described. Typical example is the consideration of aquifer heterogeneity which may need a shift from a 2D to a quasi-3D or 3D model. Historical hydrogeological (wellheads, recharge, discharge) or geochemical data are necessary to calibrate and validate transient state models both for flow and geochemical observations. In the other cases, where no historic data are available, pre-dimensioning modelling could be considered. Pre-dimensioning modelling can give insight in potential reactions that might occur at the site. A major issue is the scale of investigation that has to be adapted to the scale of the studied system. Pavelic *et al.* (2006) tested several independent methods to determine permeability from lab to field scale (plus regional scale). Core permeability tests provide data at the smallest scale (0.1 m vertical length scale). Downhole EM flowmeter tests sample at the intermediate scale (1–10 m). Pumping tests yield a bulk measure of K over the full aquifer thickness (tens of metres). Core data may be biased by sampling of more consolidated horizons and underestimate the overall permeability. In the cited study, the K values were strongly heterogeneous and by a factor of

50 lower than the effective K determined by pumping tests. Flowmeter tests provided insight in the vertical heterogeneity of the aquifer, to be taken into account when setting up a stratified quasi-3D model, whereas pumping tests yielded values of the effective K at a scale most relevant for the MAR operations.

Complexity and data requirements	Spatial type	Saturation state	Transport/reactive transport
	1D	Only saturated flow	Dispersive-diffusive transport of non-sorbed solutes
			Transport taking into account Kd values
			Transport taking into account sorption and (bio-)degradation
	2D	1D unsaturated 2-3D satured	Transport taking into account aqueous equilibria and liquid-solid-gas equilibria
			Transport taking into account aqueous equilibria and liquid-solid-gas equilibria and kinetics
----- state of the art	Pseudo 3D		Transport taking into account patho en behaviour
	Full 3D	Full 3D unsaturated saturated	Transport taking into account bacterial growth, bacterially mediated reactions (redox) and liquid-solid-gas equilibria

Figure 18.1 Types of numerical models that can be used for management of aquifer recharge.

Table 18.1 Data requirements per type of model and methods for data acquisition

Data category/type	Type of model	Methods
Geometry of the aquifer system (topographic surface, impermeable top and bottom layers, intermediate semi-or impermeable layers), lithology	Overall geometry: Analytical solutions, 1D-full 3D numerical modelsIntermediate semi- or impermeable layers: Multilayer-3D numerical model	Digital elevation model, surface geological mapping, borehole data (lithological description of cuttings, geophysical logging…), surface geophysical methods
Piezometric well heads Time series of well heads	Steady state 1D-3D models Transient state 1D-3D models	Wellhead measurements (pressure capt. Manual…) over time, geophysical measurements (resistivity, EM…)
Natural recharge rates	Steady state 1D-3D models Transient state 1D-3D models	Rainfall, effective rainfall estimated on the basis of a balance rain-runoff-evaporation or the chloride method
Artificial recharge rates	Steady state 1D-3D models Transient state 1D-3D models	
Pumping/recovery rates (flow, chronology)	Steady state 1D-3D models Transient state 1D-3D models	Pumping/discharge measurements: Flows and timing of pumping are determined as routine data in ASR/ASTR systems. Wellheads and tracer behavior will also be influenced by pumping outside the MAR site, from local to regional scale. Historical pumping data may be difficult to obtain but are needed to calibrate transient state hydrodynamic models
Surface waters: Geometry of streams, lakes, sea shore, infiltration ponds or other AR	2D-3D model taking into account exchange with surface water bodies	Mapping: Topography and geology may be available in analogous or numerical form, in the first case, they need to be digitized. If the scale of available maps is not adapted to the site scale, complementary mapping might be needed (detailed outcrop of the aquifer boundaries…). GIS (Geographical information system) techniques will be used in most cases). Measurements of river and lake levels will determine the hydraulic gradient with respect to the connected groundwater body

(Continued)

Table 18.1 Data requirements per type of model and methods for data acquisition (*Continued*).

Data category/type	Type of model	Methods
Hydraulic parameters: Permeability, transmissivity	All hydrodynamic or transport models	Small scale: permeametry (core permeability laboratory tests) Medium scale: Downhole flow meter measurements: determine vertical permeability distribution in a well before pumping and after the establishment of steady state conditions during pumping (Molz *et al.* 1994, application on MAR in Pavelic *et al.* 2006). Site scale: pumping tests, tracer breakthrough data, Forced Gradient Tracer Tests (FGTT), may be performed during ASR/ASTR cycles. Pumping tests are extensively used to determine the average bulk permeability/transmissivity of the investigated aquifer. Drawdowns in the pumping and observation wells are generally interpreted through analytical solutions of the flow equation.
Hydrodispersive parameters: Porosity, dispersivity	1D-3D transport models including reactive transport models	Forced gradient or natural gradient Tracer Tests (FGTT, NGTT).
Mineralogy of the aquifer material	Reactive transport models taking into account water-rock interactions	Mineralogical and chemical analyses of the aquifer material: There is a large variety of analytical techniques to determine the mineral phases present in an aquifer. Most frequently, X-ray diffractometry (XRD) is used that provides semi-quantitative data on the major, and to some extent, minor phases. Oriented thin sections can be used for detailed analysis of clay minerals. Combined with bulk chemical analysis (e.g. after acid or alkaline attack), mineralogy of minor and trace minerals can be determined through modal analysis. Such data are required for chemical equilibrium calculations.
Chemistry of the solid phase	Reactive transport models taking into account water-rock interactions	Large choice of analytical methods for soil and rock analysis. Semi-quantitative mineralogical and quantitative whole rock chemical analysis can be combined in order to determine a theoretical modal mineralogy (major and minor mineral species)
Water chemistry (background waters, injectates, injection-recovery cycle monitoring), including physico-chemical parameters (pH, temperature, conductivity, redox potential, alkalinity…) Time series of water chemistry	Transport models (tracer concentrations are needed) Reactive transport models, complete water chemistry is needed to calculate speciation of the considered elements	Appropriate sampling, transport and analytical techniques depending on the required parameters. In situ measurements for non-conservative parameters (e.g. pH, Eh, T, O_2)

Like for every hydraulic groundwater model, MAR models need to fix boundary conditions (limits of the aquifer system, recharge-discharge conditions), here the geometry of the receiving groundwater body, its relationship with underlying or neighboring aquifers, with surface water bodies are of importance. Natural variations of recharge/discharge have to be known, in addition to injection/infiltration and recovery rates induced by the MAR system.

Tracers

Features relevant to MAR operations that can be introduced by tracers studies are (1) determination of mixing fractions of recharged water and ambient groundwater and (2) the residence time of recharged water before recovery as influenced by groundwater velocity and hydrodynamic dispersion.

The use of tracers offers a unique insight into subsurface flow patterns and provides direct access to relevant hydrodynamic parameters, to both the advective and dispersive properties of the aquifer (dispersivity, porosity…), even in media with preferential flow paths like fissured aquifers and karsts (e.g. Maloszewski & Zuber, 1985, Seiler *et al.* 1989). Tracer experiments are therefore currently used to design and dimension MAR systems, in the aim to calibrate hydrogeological models of flow and transport in a MAR system (Pavelic *et al.* 2006).

Two main categories of tracers can be distinguished:

(1) **Artificial or human-applied tracers** which are voluntarily introduced into a groundwater system, to study its response and
(2) **Intrinsic or environmental tracers** which are naturally present in the system or were released inadvertently in the past.

A large variety of tracers exist, including salts or, generally spoken, inorganic ions, organic compounds including dyes, isotopes of the water molecule and of solutes, noble gases, particles, micro-organisms, and temperature. Summaries are given by Flury & Wai (2003) and Quast *et al.* (2006) (for artificial tracers), Cook & Herczeg, 2000, Clark & Fritz, 1997, Pearson *et al.* 1991 (for environmental tracers). The latter provide a comprehensive overview on the advantages and disadvantages of the different types of groundwater tracers. Artificial tracer tests belong to two main categories, natural or forced gradient tracer tests (NGTT and FGTT respectively, Ptak *et al.* 2004). In the first case the tracer is injected into the non-influenced flow-field and tracer flow velocities will depend on natural gradients. In the second case, flow will be accelerated through pumping, which leads to conditions closer to the operating conditions of a MAR system. MAR infiltration tests can be considered and used as large-scale long-term tracer tests, with durations and injected quantities commonly not reached by classical tracer tests (Williams, 2000; Pavelic *et al.* 2006)

Use of intrinsic/environmental tracers

Intrinsic geochemical tracers (the simplest intrinsic tracers to use are salinity and temperature) including isotope signatures can only be used when a significant contrast exists between the injectate and the background water and that the concentrations/isotope ratios are relatively homogeneous in space and time. Nevertheless, the choice of intrinsic tracers is large so that usable individual tracers or combinations should be found in most circumstances. Environmental isotopes have been used extensively to investigate or monitor MAR systems (Kloppmann *et al.* 2009 and references therein): they serve (1) as dating tool (^{3}H, ^{3}H/^{3}He, ^{14}C), (2) indicators of groundwater dynamics: the isotopes of the water molecule (δ^{18}O, δ^2H) are ideal tracer of dispersion, mixing proportions…., boron is usually contained in domestic wastewaters at high concentrations, (3) tracers of reactive transport, indicating degradation of organic carbon, denitrification, sulphate reduction (C, N, O, S isotopes). When comparing the use of intrinsic (environmental) tracers to extrinsic (artificial) tracers, the following advantages of intrinsic tracers can be stated (Kloppmann *et al.* 2009):

- No need to add an artificial substance to the injectate. This simplifies the logistics of tracer tests and avoids legal constraints on the introduction of toxic, radioactive or undesirable substances into groundwater resources.
- The chemical quality of secondary or tertiary treated wastewater is generally well constrained by the standard quality monitoring program of the treatment plant and potentially strong chemical variations of raw wastewaters are attenuated due to the treatment process. This allows controlling the input function of the intrinsic tracers, especially for inorganic components like B.
- Environmental tracers can be used over much longer time scales than an introduced artificial tracer. Injection can be followed over months to years, provided that the input concentrations or isotopic compositions are monitored or can reasonably be considered as constant. Retardation effects, as seen here for B, may not be significant over long time scales of a MAR operation.
- Environmental tracers can even be used *a posteriori* after years of functioning of a MAR system to investigate the extent of penetration of the injectate plume into the natural groundwater system and calculate mixing ratios.
- Several tracers can be combined that show different behaviour (e.g. stable isotopes of the water molecule, salinity, specifically chosen solutes or isotopes) to assess multi-component mixing (Phillips & Gregg, 2003). Using several artificial tracers simultaneously is possible but costly.

One obstacle to the use of environmental tracers is the potential lack of information on the input function. Artificial tracing allows, in theory, perfect control of the input flow so that uncertainties for modelling are minimised. Nevertheless, monitoring the input concentrations or isotope ratios with a reasonable frequency over the duration of the injection test can overcome this drawback.

18.1.3 Groundwater modelling and artificial recharge: Model selection

Analytical solutions of flow and transport equations

Analytical solutions of flow and transport equations were the first tools to quantify groundwater and solute flow (Bruggemann, 1999; Wexler, 1992) and they are still widely applied for small scale short-term projects, (Janssen & Hemke, 2004) e.g. to interpret pumping tests and tracer tests (Sauty & Kinzelbach, 1988). The latter are, in general, performed in order to determine aquifers parameters for the subsequent development of numerical models. The main advantage of analytical solutions is that they are straightforward, much less (computing-) time-consuming than numerical models and that they can be applied when the data situation is too poor to develop a numerical model. Their main restriction is the limitation to situations where the subsoil conditions can be reduced to a simple configuration (e.g. homogeneous and isotropic aquifers with simple geometry under steady state conditions). Nevertheless, more sophisticated analytical models have been elaborated, adapted to multiwell configurations in anisotropic aquifers, in particular for pump-and-treat remediation of contaminated groundwater, a problem with some similarities to MAR (Chang & Yeh, 2007; Yeo & Lee, 2003).

Sedighi *et al.* (2006) give an overview of analytical and semi-analytical approaches so far used for the modelling of MAR systems. They present a semi-analytical model applicable on the steady state flow and conservative transport modelling of ASR system allowing for a horizontal stratification of the aquifer. This model was able to predict successfully chloride concentrations in an ASR system in Florida and results were compared with numerical models (MODFLOW/MT3DMS) and produced comparable results in significantly shorter time. Pavelic *et al.* (2006) come to a more differentiated conclusion when comparing analytical solutions developed for single-well testing to a numerical model solute breakthrough induced by injection in a multi-layer aquifer in S Australia. The analytical model, neglecting aquifer heterogeneity, was able to predict satisfyingly recovered water quality OR individual piezometer breakthrough but not both simultaneously whereas the numerical model represented both reasonably. Analytical models have also been developed for pathogen transport in fractured MAR reservoirs (Masciopinto, 2008). More details on these case studies are provided in Chapter 19.2.

Numerical flow models

A large number of numerical modelling tools are available nowadays and have been applied to MAR problems in porous aquifers. General features of those models, whether they use finite-difference or finite-element techniques, are their ability to simulate 3D advective, diffusive and dispersive flow and transport of solutes under steady-state or transient state conditions in multilayer and/or anisotropic aquifers. Additional requirements for many MAR systems may be the possibility to take into account density dependent flow and coupled saturated-unsaturated flow, as well as simple fixation-degradation models like sorption isotherms and degradation factors. Here we will shortly present, *pars pro toto*, the MARTHE software (Thiery, 1993), as an example for a fairly complete package well adapted to MAR problems, like the modelling of the Wulpen site in Belgium (case study 4) and the Shafdan site in Israel (case study 1). In addition to basic groundwater hydraulics, it allows for modelling of discontinuities, such as open bodies of water with free surface (infiltration basins …), and a continuous treatment of the saturated and unsaturated zones (Vanderborght *et al.* 2005, Herbst *et al.* 2005) which is particularly important for modelling SAT systems. The domain to be modelled is discretized as a rectangular "tartan" model: each row and each column of the grid has a constant width, but the user is free to adapt the width of each of these rows and columns so as to fit local heterogeneities, the amount of available data and the accuracy sought. It is also possible to create locally finer cells by nesting sub-grids. Density effects caused by heterogeneous salinity and/or temperature fields are taken into account so that this software is well adapted to saline intrusion problems and to MAR in saline aquifers. For the calculation of transport, the user can choose between different calculation techniques, according to the type of problem encountered (predominance of convection or dispersion): (1) TVD method (Total Variation Diminishing) with flux limiter; finite volumes method; (2) method of characteristics (MOC) using particles; (3) Random Walk method. Transport in the aquifer considers advective, diffusive and dispersive components. Several options are available: (1) exponential decrease of the effluent source as a function of time; (2) chain degradation; (3) retardation factor or partition coefficient, Kd (adsorption-desorption); (4) double porosity (equilibrium or with kinetic); (5) Freundlich and Langmuir isotherms. A particularly seducing feature in the context of MAR is the possibility to simulate chemical interactions between water, solutes and matrix can through the coupling of MARTHE with a chemical model dedicated to the geochemical context. Up to now, reactive transport modelling has been performed through coupling of MARTHE with Specific Chemical Simulators (SCS, Kloppmann *et al.* 2001); coupling with PHREEQC (Parkhurst & Appelo, 1999) and TOUGHREACT (Xu *et al.* 2004) is underway but the choice is, a priori, not limited. Models such as PHT3D (Prommer *et al.* 2003) have been used to simulate combined flow and multi-species and heat transport together with geochemistry and biodegradation kinetics to interpret processes in many varied MAR operations (e.g. Greskowiak *et al.* 2005, 2006; Prommer & Stuyfzand, 2005; Wallis *et al.* 2010)

Geochemical models

Several thermodynamic and thermo-kinetic codes are in use, based on the same fundamental concept of thermodynamic equilibrium among aqueous species and between aqueous species, gases and minerals and several of them have been applied on MAR problems. Dillon *et al.* (2005) summarise three different tools, their complexity ranging from a simple prescreening tool to a state of the art flow and reactive transport model, applied to MAR sites in Australia. ASRRI as a robust and simple risk-assessment tool (Miller *et al.* 2002) uses degradation rates to determine storage or transfer times necessary for contaminant reduction below a given threshold for one- or two-well systems respectively. The simple analytical 2D-spreadsheet model EASY-LEACHER 4.6 (Stuyfzand, 1998) simulates oxic water injection into an anoxic aquifer considering cation exchange, labile and more stable organic matter breakdown, together with pyrite oxidation. PH3T combines a widespread flow model (MODFLOW) with a transport module (MT3DMS) and the geochemical model most frequently used for MAR reactive transport modelling: PHREEQC (Parkhurst & Appelo, 1999). The latter is a program for aqueous speciation, batch-reaction of water, solutes, gases and minerals, one-dimensional transport, and inverse geochemical calculations. All are based on the same concepts as PHREEQC and comprise EQ3/6 (Wolery, 1992) and MINTEQA2 (Allison *et al.* 1991).

Other examples include GEOCHEMIST'S WORKBENCH® (Bethke 2002), GEMS (Karpov *et al.* 1997), and CHESS (Van der Lee, 1998). All geochemical codes are associated with a base of thermodynamic and kinetic data. Generally, each code has a standard database that can be enlarged or adapted by the user; nowadays, databases common to different codes and applications are in use. In any case, the user has to know in some detail the characteristics of the database he chooses, to check its coherency and adaptation to the problem to be studied and to cite the database version indicating any eventual changes. The choice of the geochemical code will depend on the necessity to take into account high salinities (which may require the use of a particular model of ion activity like the Pitzer model) and kinetics, in particular for redox reactions. A synthesis of the characteristics of common geochemical models is provided by IAEA (2006).

Pathogen transport models

Traditionally, mathematical models treating pathogens either as finite particles or as dissolved molecules, are used for a preliminary estimation of pathogen movement in aquifers (Chrysikopoulos & Sim, 1996; Schijven & Hassanizadeh, 2000), and subsequent evaluation of setback distances (Faulkner *et al.* 2003). Colloids present in subsurface formations are mainly mineral micro particles in the form of metal oxides, humic macromolecules, bacteria and viruses (Stumm & Morgan, 1981). Colloids generally range in size from 1 nm to 10 μm; although, biocolloids such as *bacteria* range in size from 0.2 to 5 μm and they are much larger than dissolved contaminant molecules (Chrysikopoulos & Sim, 1996). Consequently, the sorption of bacteria onto solids is described by colloid filtration theory (Yao *et al.* 1971). *Viruses* are at the lower end of the colloid size distribution with size ranging from 0.02 to 0.3 μm (Brock & Madigan, 1991). Traditionally, the sorption of viruses behaving as solutes is represented by a non-equilibrium, first-order kinetic process (Bales *et al.* 1991); whereas, the sorption of viruses behaving as colloids can be described by colloid filtration theory (Sim & Chrysikopoulos, 1995, 1996; Chrysikopoulos & James, 2002). Pathogen count variations caused by particle desorption/sorption phenomena on rock/soil surfaces have been also described by Wang *et al.* (1981), Meschke & Sobsey (1998), Hendry *et al.* (1999) and Mallén *et al.* (2005).

The theory of the kinetics and dynamics of particles deposition has been also developed by Johnson and Elimelech (1996), considering the effects of *geochemical heterogeneity* (Bhattacharjee *et al.* 2002) on the surface charge of soil particles. In the rock fractures favorable and unfavorable patches may also occur on the surface of the fractures. Thus, as suggested by Johnson & Elimelech (1996), the whole surface of fracture can be partitioned in several patches which are favorable or unfavorable for colloids deposition by having different charge characteristics. Anyway, at a regional scale, continuous recharge operations may saturate soil and rock surfaces for both attachment/detachment and sorption/desorption effects and, consequently, the effects during the non-stationary phase of pathogen transport in groundwater caused by geochemical heterogeneity can be assumed negligible. On the contrary, at a laboratory scale, the attachment/detachment and sorption/desorption effects may produce non negligible effects on particle count and specific model equations able to account for the non-stationary phase of the pathogens transport are required.

18.2 CASE STUDIES

18.2.1 Case study 1: Shafdan

Context

Different types of models were applied to design a novel MAR system in the framework of the Dan Region Reclamation Project (Israel), one of the world's largest (135 Mm^3 per year) and oldest soil aquifer treatment (SAT) systems using secondary treated wastewater from the Tel Aviv area which is recharged into an aquifer via infiltration basins and recovered by pumping well galleries with a retention time of the system of 6–12 months. The reclaimed water is used for agricultural irrigation in the South of Israel. Water quality issues at the site are of great concern, especially redox related geochemical interactions since manganese mobilisation has been observed recently on several occasions (Oren *et al.* 2007). The infiltration capacity of the existing facilities reaching its limits, partly due to clogging, dug wells are tested as a new injection system in the framework of the Reclaim Water project. The aim is to evaluate their potential to complete the existing large scale infiltration ponds and to overcome some of the operating problems inherent to ponds (e.g. loss of permeability due to clogging, limited possibility of extension). A system of dug wells around the infiltration basins closer to the extraction wells, leading to a shorter residence time, is considered. The water infiltrated through the dug well would undergo ultrafiltration as an extra treatment to guarantee the quality of recovered water and to compensate for the shorter residence time.

Modelling has accompanied the successive phases of design, installation and testing of the prototype, using different type of hydrodynamic and geochemical models during (1) pre-dimensioning, (2) injection test and optimisation of operating conditions, (3) assessment of hydrochemical conditions over long-term injection, and (4) scenarios of pollutant transfer.

The following scheme summarises the phases of the prototype design and testing and the accompanying modelling activities (Figure 18.2).

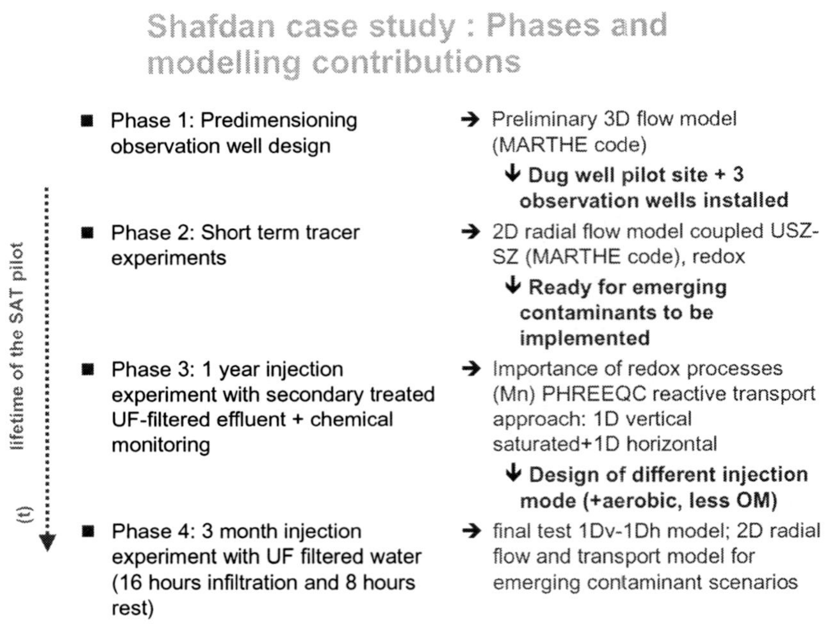

Figure 18.2 Shafdan (Israel) SAT site: project phases of dug well infiltration pilot and parallel modelling activities

Pre-dimensioning of the dug-well injection system and monitoring array

Pre-dimensioning modelling was carried out to assess the impact of the dug well infiltration on the groundwater flow in the direct surroundings and to advise on lay-out of the pilot site (more specifically on the location of Observation wells 1 and 3). This modelling was executed with the data available at the onset of the project and was thus prone to uncertainties. Dug well and observation wells were oriented along a flowpath, based on the piezometrical data and the derived groundwater flow gradient which is not natural but influenced by the groundwater extraction wells in the area and the infiltration of tertiary treated wastewater in the infiltration basins of the Shafdan site.

A preliminary 3D flow and transport model was developed using the finite difference code MARTHE (Thiery, 1995; Herbst *et al.* 2005) to estimate the ideal location of the observation wells and the displacement of the infiltrated water in the dug well during the planned 1-year infiltration experiment.

Results (Figure 18.3) show that due to the very low gradient the infiltrated water in the dug well will form a lens of infiltrated water below the dug well that is almost circular. This lens is very little displaced during the 1 year infiltration experiment that started in March 2007.

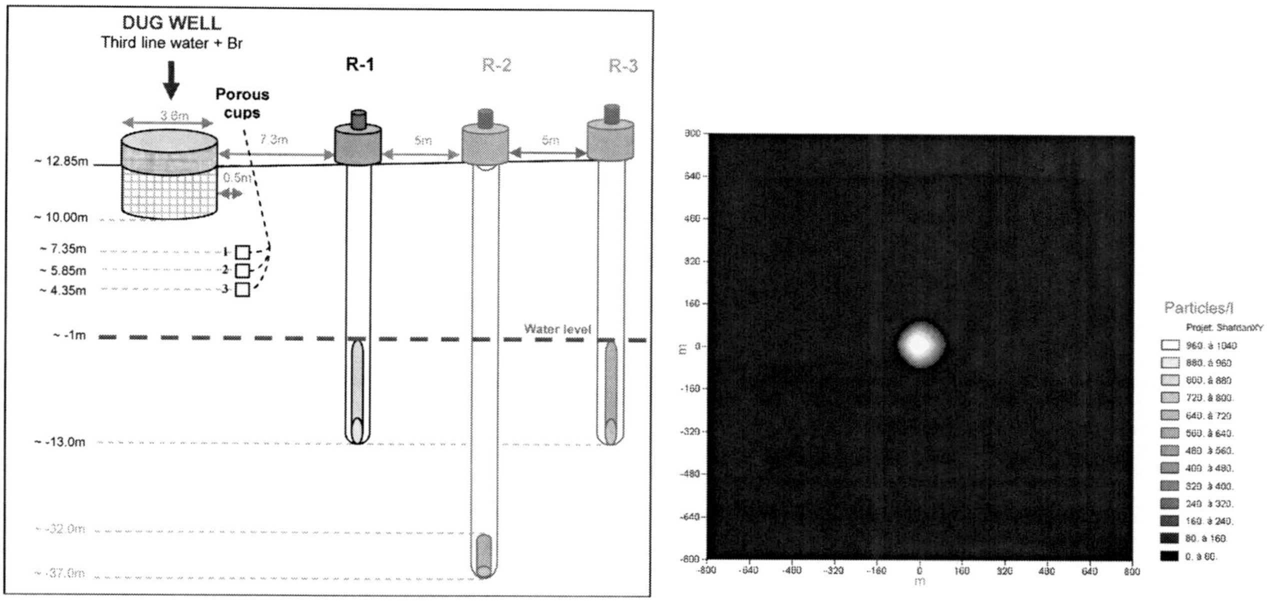

Figure 18.3 Horizontal representation of the calculated tracer concentrations after 365 days infiltration in the dug well (injected concentration 1000 particles/l) based on an aquifer permeability of 10^{-5} m/s (distances from the injection point in metres)

The results of this preliminary 3D flow modelling had important implications for the conceptualisation of the coupled model that was subsequently developed. Because the existing gradient is so low, the problem could be modelled radial symmetrical problem. This reduced the initial 3D model setup to a 2D radial symmetrical setup for unsaturated flow and solute transport modelling and a 1D vertical 1D horizontal model including geochemical reactions.

Optimising injection conditions through 2D radial dispersive–diffusive transport modelling

2D radial hydrogeological modelling with MARTHE was performed to characterise the flow regime in detail using two tracer tests with KBr as an artificial tracer for calibration and validation (Gaus *et al.* 2007a, b). Furthermore isotope monitoring (B, Li) was performed during the first tracer-test in order to assess the potential of those tracers as intrinsic tracers (Kloppmann *et al.* 2009).

The unsaturated flow was modelled using Richards's (1931) equation and Van Genuchten's retention model (1980), using a radial axi-symmetric two dimensional grid, capable of representing the extension of the infiltrating tracer bubble. The simulation was performed in transient conditions during 45 days (second tracer test). The porous medium (top: +12.85 m, bottom −47.15 m) is modelled as two homogeneous sand layers separated by a clay layer (top at −17.5 m, bottom −32.15 m).

Hydraulic conductivity of the model was calibrated with the 2nd tracer test data, and its optimum situated in the range 9.10^{-5} m/s –5.10^{-6} m/s (distinctly for sand and shale layers) with porosity of 18% for sand and shale layers and 25% for the unsaturated zone. While the fit for R-1 is very good (Figure 18.4), for R-3, the peak concentration is matched, but the maximum observed concentrations before the breakthrough are higher than the modelled concentrations. This suggests winter rain infiltration remobilised trapped pore water of the first tracer test in the unsaturated zone towards the saturated zone and lead to a Br increase in R-3 before the Br of the second tracer-test arrived. Modelling confirmed this and showed that after the first tracer test Br tracer remains trapped in the unsaturated zone long after the tracer has passed. The 2D model was, in the following, used for simulations of pollutant breakthrough, taking into account biodegradation and sorption.

Modelling the redox zoning and Mn release with 1Dv-1Dh Monod-kinetics-reactive transport models

In a first step, a simple kinetic geochemical batch modelling was developed, using the PHREEQC2 code and the *phreeqc.dat* database (Parkhurst & Appelo, 1999), to identify dominant redox reactions that can be attributed to the presence of dissolved organic matter (DOM represented by CH_2O in the database) in the infiltrating water (Gaus *et al.* 2007b). Assuming that the DOM degradation is determined by microbial activity, a multiplicative Monod rate expression, was used to describe the rate of the DOM degradation.

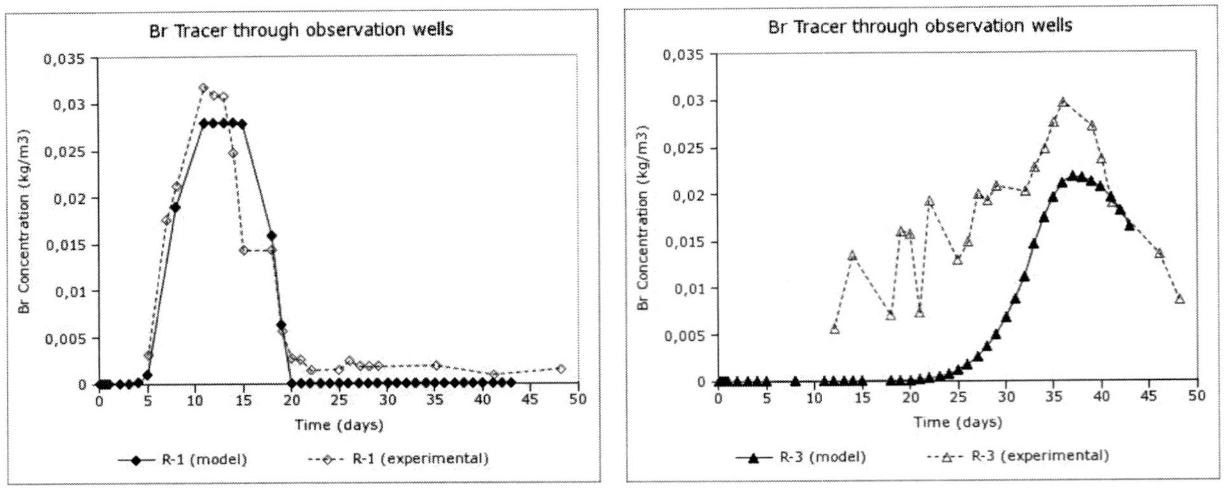

Figure 18.4 Comparison between the simulated and measured tracer concentration (Br) in observation wells R-1 and R-3 for the second tracer test

Parameters measured in Shafdan aquifer before injection and during the one year infiltration pilot started on April 10th 2007, such as NO_3^-, Mn^{2+}, Fe^{2+} concentrations show the high impact of redox chemistry in the aquifer (Figure 18.6.). A typical redox sequence was established: During the 10–12 days of travel time to R-1 and 36–37 days of travel time to R-3 gradually the oxygen concentration diminished, dissolved organic carbon was oxidized and while nitrate is still present in R1, it has disappeared completely in R-3. Mn and Fe increase in the saturated zone compared to the injected water, the first in R1 and R3, the latter only in R3 (maximum of 1 mg/l). Over the whole period of measurements (long-term injection), a tendency of Mn decrease is observed for the R1 well. This might indicate a a slow displacement of the redox zonation within the aquifer, eventually due to changes in the injection conditions (intermittent 8 h infiltration-16h relaxation) at the end of the injection experiment. Another possibility is the progressive depletion of the solid phase. A significant decrease in DOC is also observed from 5.8 mg/l in the dug well to 1.3 mg/L in the aquifer (May 2007, well R-3). DOC concentrations in the dug well change due to changing mixing proportions of ultrafiltrated water and "3rd line" water. The redox sequence of Shafdan site was modelled by adapted Monod kinetics, including inhibition factors such as limited TEAP concentrations (Van Cappellen & Wang, 1996). Kinetic modelling of dissolution/precipitation and redox processes have been included in the PHREEQC code and the underlying database has been adapted in order to control kinetically the redox transformation for each principal electron acceptor. DOM is considered to be the ultimate source of chemical energy for microbial and inorganic redox reaction and bacterial growth is implicitly limited by the DOM limiting concentration. Dissolution-precipitation reactions of Pyrolusite, Quartz, Ferrihydrite, Kaolinite, Calcite, Albite, Microcline, observed in the aquifer matrix, were considered. The biodegradation of organic substrates is represented by a series of reactions (aerobic respiration, denitrification, manganese reduction, iron reduction) and the degradation kinetic is assumed to be of Monod type. In order to model microbially mediated reactions coupled to an inorganic reaction system, redox is decoupled in the database (Sengör *et al.* 2007). Elements with different oxidation states are treated as separate components. Thus PHREEQC database has to be entirely rewritten to introduce redox species whose concentrations are controlled by microorganisms. A fundamental difficulty is to reproduce the MnO_2 reduction by microorganisms because in the redox couple Mn(IV)/Mn(II), the oxidized state Mn^{4+} is not stable in solution in opposition to the couple Fe(III)/Fe(II) with stable Fe^{3+}. To ensure the continuity of the redox sequence and express MnO_2 reduction as the same way as $Fe(OH)_3$ reduction by microorganisms, it was decided to create a transition phase called $Mn^{4+}_{virtual}$. Pyrolusite is chosen to represent MnO_2 and the equilibrium of this mineral phase with the aqueous phase was imposed as a constraint in PHREEQC. Saturation Index (SI) of MnO_2 dissolution does not enter in the kinetic expression of MnO_2 reduction by micro-organisms and therefore the $Mn^{4+}_{virtual}$ does not affect the chemical system.

Batch interaction of Dug Well water and soil mineralogy

For the first batch modelling, DW-water is used (after secondary treatment with UF) to test the DOM degradation model. In order to take into account the DOM initially present in the soil solution, a concentration of DOC_CH_2O of 18 $mg \cdot L^{-1}$ was chosen. This value is voluntarily elevated, compared to injected wastewater, so that DOM will not be a limiting factor for bacterial growth and the subsequent redox reactions. The batch modeling results are in good agreement with biodegradation processes controlled by different population of bacteria. Figure 18.5 shows the important decrease of DOM caused by the rapid consumption of EA by biomass during 1 day. Biodegradation of organic matter is progressively coupled to the reduction of the terminal electron acceptors (TEAPs) such as O_2, NO_3, Mn(IV) and Fe(III). N_2, Mn^{2+} and Fe^{2+} are produced from DOM degradation (Figure 18.6).

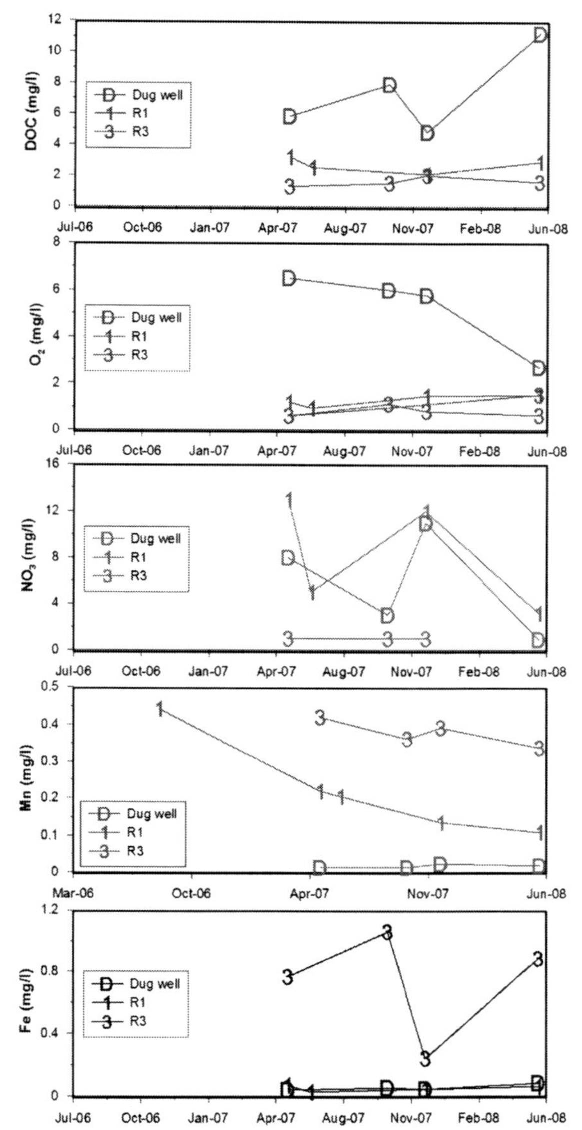

Figure 18.5 Measured redox zoning in the aquifer downstream the dug well (observation wells R1 and R3)

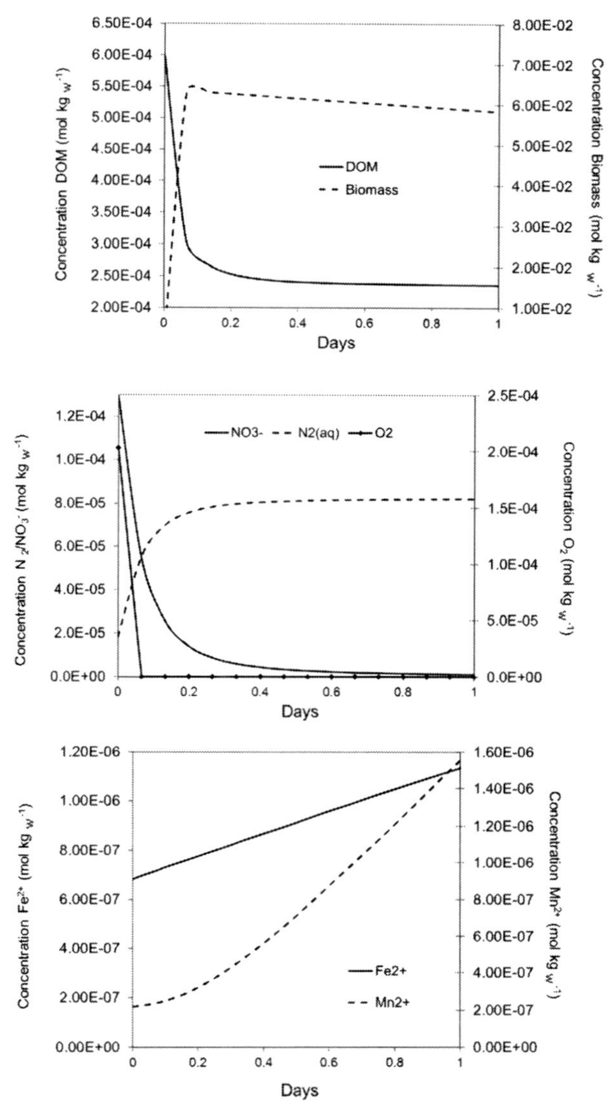

Figure 18.6 Simulated DOM degradation, bacterial growth, denitrification, O_2 reduction and Mn^{2+} production by MnO_2 reduction versus time

The second modelling step consisted in simulating the percolation of DW-water in contact with Shafdan soil mineralogy through the unsaturated zone (UZ) to the water table with a vertical 1D transport PHREEQC column. Considering UZ to be oxygenated, only aerobic bacteria are considered. Surface complexation with ferrihydrite is also considered following the Dzombak & Morel (1987) model.

The solution at the outlet of the vertical column is then mixed with initial R1 water (background before the construction of the dug well) and the solution obtained after mixing is injected in a horizontal column to simulate the evolution of groundwater quality (saturated zone) again using an 1D transport geochemical model (PHREEQC) considering all microbially driven redox sequences. The simulated water composition is compared to R-1 measured water composition. The best fit for major aqueous species observed in R1 is obtained when 80% of outlet column solution is mixed with 20% of initial groundwater composition. The transport velocity used in the geochemical model is deduced from the 2D radial flow simulation. Parameters controlling degradation of DOM by micro-organisms have been calibrated in order to rapidly consume O_2 to attain reducing conditions in groundwater. In order to better understand the problem of iron and manganese at the Shafdan site, we have focused our attention on the evolution of R1 water composition in the case of the total consumption of O_2 during the first steps of modelling. Figure 18.7 shows the evolution of R1 water composition versus time. Results are similar to the first batch test except for Fe(III) reduction. Biomass decreases rapidly because TEAs are not available anymore. Moreover, DOM remains constant after complete reduction of O_2 and NO_3^- while Mn(IV) continues to be reduced. This is due to the strong need of DOM by bacteria to reduce O_2 and NO_3^- opposed to the much lesser need of DOM for MnO_2 reduction.

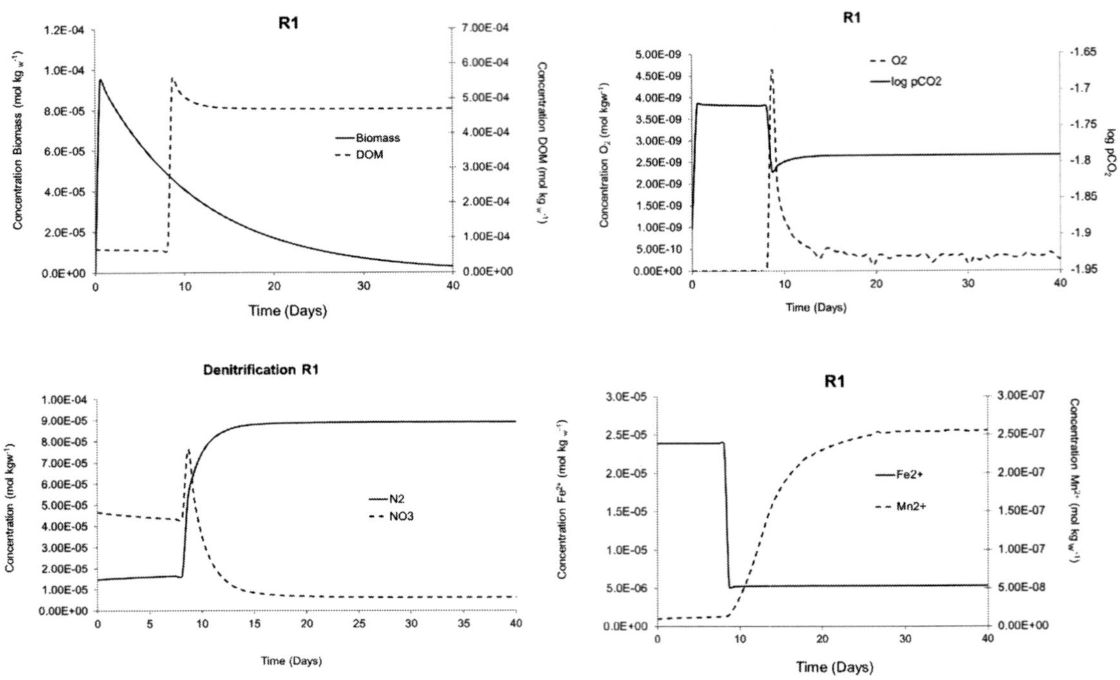

Figure 18.7 Simulated evolution of TEAPs and reduction reaction product concentrations at well R1, versus time

The geochemical column model 1Dv-1Dh (PHREEQC) demonstrates that groundwater chemistry is principally controlled by geochemical signature of dug well water passing through UZ. The arrival of these oxygenated waters at the water table in case of incomplete consumption of O_2 in UZ by bacteria, can cause $Fe(OH)_3$ and MnO_2 precipitation near the water table. Microbial control of the redox state in the aquifer explains the increase of Mn^{2+} and Fe^{2+}: O_2 is rapidly consumed in UZ by aerobic bacteria. Nevertheless, the potential reoxidation of pyrolusite and ferrihydrite is not currently allowed in the model.

Future investigation have therefore to consider, in a first time, the results of mineralogical analysis of Shafdan soil after injection in order to confirm $Fe(OH)_3$ and MnO_2 precipitation near R1. A decisive step will be to find a modelling concept that allows including the reversibility of redox reactions in PHREEQC. This would allow simulation of recurring oxidative conditions in the aquifer.

Scenarios of pollutant transfer modelled by non-conservative 2D radial transport modelling

Based on the 2D radial transport model, and using the parameters of the best fit of KBr tracer experiments, the hypothetical case of NDMA and SMX spill in the injection dug well was simulated and the modelled concentrations in the observation wells evaluated. Hypotheses were chosen to define scenarios of compound spill and simulate failure of the treatment system, assuming "worst case" concentrations, leak duration, sorption and degradation behaviour deduced from experimental results of fate studies within Reclaim Water project. Several scenarios were performed to identify the possibility of contamination of the pumping well.

Sorption and degradation parameters were deduced from column experiments and 1D column modelling using MARTHE and are compared to literature studies (Table 18.2). According to Ternes *et al.* (2008), sorption under real conditions, such as slow sand filtration or subsoil passage, compared to laboratory experiments, can be expected to be less important and persistence should be comparatively higher. The removal of some compounds is facilitated by redox conditions. According to Jekel & Grünheid (2005) Sulfamethoxazole shows a better removal under anoxic/anaerobic conditions whereas Iopromide was removed under both anaerobic and aerobic conditions.

Table 18.2 Synthesis of degradation and sorption parameters

Parameter	Unit	Compounds	Set of values from experiments on Wulpen sand	Set of values from literature studies	Global behaviour
Linear sorption		NDMA	nd	0.00–0.40	Sorption neglected
KD	(–)	CBM	0.20	0.30–8.40	High sorption
		SMX	0.15	0.10–1.50	Low sorption
		BT	0.15–0.20		Low sorption
		IOP	No sorption-0.02		Sorption neglected
Half-life time		NDMA	nd	1.3d–22.5d	Moderate persistence
T_{50}	day	CBM	7	328d	High persistence
		SMX	No degradation	Few months	Moderate to high persistence
		BT	No degradation	29 ± 4d	No degradation
		IOP	1–3		Moderate persistence

References: Yang *et al.* 2005; Gan *et al.* 2006; Drewes *et al.* 2006; Haruta *et al.* 2008; Hart *et al.* 2004; Löffler *et al.* 2005; Scheytt *et al.* 2005; Reclaim Water, 2008; Heberer *et al.* 2008; Williams & Adamsen 2006; Ternes *et al.* 2002; Cordy *et al.* 2004; Jekel & Grünheid, 2005; Godfrey *et al.* 2007; nd: not determined.

For the NDMA contamination scenario, concentrations of 50 ng/L, which is comparable to maximum values measured in the effluents before SAT treatment, and 100 ng/L are injected continuously into the dug well. Sorption is judged to be negligible and several levels of degradation are tested ($T_{50} = 20$; 15; 10; 5 days) and compared to a simulation without degradation. The simulation runs for 90 days of injection and the results are shown for the observation well R3 and for a distance of 30 m from the dug well, both at 20 m of depth.

Figure 18.8 and Figure 18.9 show breakthrough curves of NDMA at different distances of the dug well (17 m (R3) and 30 m) and different T_{50} degradation half life values as well as the contaminant distribution around the well after 35 days for the case of no degradation (maximum concentrations). Only the hydraulic gradient induced by injection was considered (no pumping). Measured concentrations in R1 and R3 lie in the order of magnitude of the concentrations obtained by simulations with $T_{50} = 5$ d to 10 d.

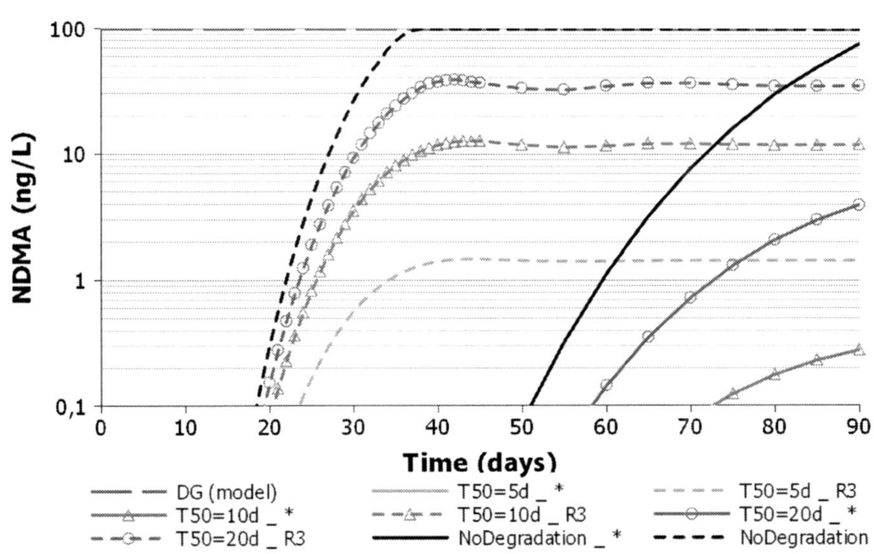

Figure 18.8 Simulated NDMA breakthrough in the R-3 observation well and at a distance of 30 m downstream from the dug well (*, full lines) for 100 ng/l injected concentration, different degradation half life times, considering the hydraulic gradient induced by injection only (no pumping)

A Dirac-like pulse of 100 ng/L NDMA (1 h of injection) without degradation would still induce a breakthrough maximum with measurable concentrations in R3 (0.2 ng/L).

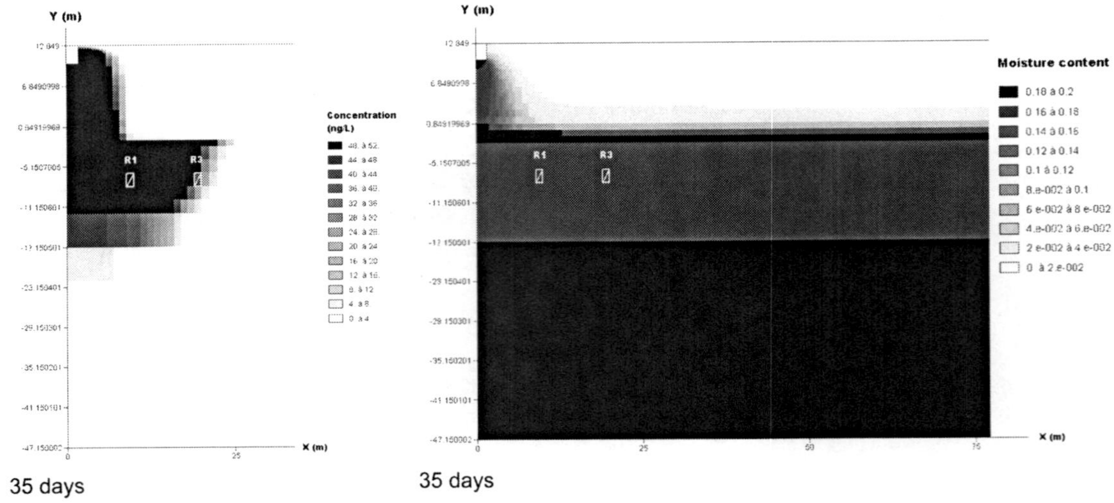

Figure 18.9 Simulated moisture contents (right) and NDMA breakthrough (left) in the in the unsaturated and saturated zone in the vicinity of the dug well for 100 ng/l injected concentration and non degradation (worst case), considering the hydraulic gradient induced by injection only (no pumping)

For the SMX contamination scenario, two cases of continuous injection are run: one with 1000 ng/L and a second with 1500 ng/L, compared to maximum measured concentrations in the order of 1000 ng/L in the effluent. No degradation processes were taken into account and several levels of sorption are tested and compared to a simulation without sorption. The simulation running for 90 days of injection are shown for the observation well R3 (17 m from the dug well) and for a distance of 30 m from the dug well, both at 20 m of depth (Figure 18.10).

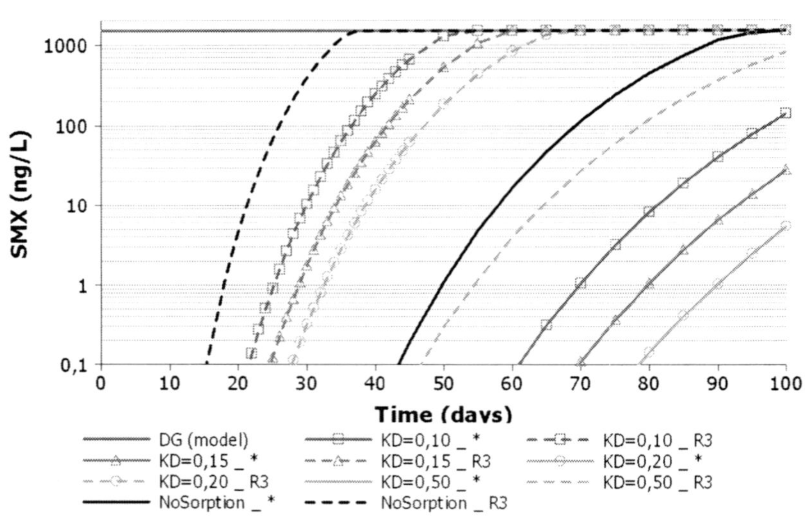

Figure 18.10 Simulation results at the observation well R3 and for a distance of 30m from the dug well (*) of SMX contamination scenarios by leakage into the dug well (1500 ng/L injected). Sorption process is taken into account by volumic partition coefficient (KD)

Simulation results show a retardation of SMX arrival at the observation well. Obviously, this retardation is more important when the sorption is high (i.e. the volumetric partition coefficient is high). For all scenarios, SMX concentrations in the observation well attain the injected concentration; the pollutant is only delayed so that the concentration plateau is reached after different travel times. This is coherent with the measured concentrations that are in the same order of magnitude throughout the system (from treated effluent over SAT to R3 well).

18.2.2 Case study 2: Adelaide

Context

The Aquifer Storage Transfer and Recovery (ASTR) research project located at Salisbury, South Australia seeks to demonstrate that wetland-treated urban stormwater, when injected into an initially brackish aquifer and recovered from dedicated recovery wells, can create a safe and reliable drinking water supply (Rinck-Pfeiffer *et al.* 2005; Swierc *et al.* 2005; Regel *et al.* 2006). Unlike ASR, which utilizes the same well for injection and recovery, the groundwater flow system established through ASTR by having separate injection and recovery wells offers more uniform residence times in the aquifer, which is expected to lead to more predictable levels of chemical and microbial attenuation of contaminants necessary for the provision of water of potable quality.

Two distinct phases of modelling have been conducted at Salisbury site. The first phase aims to optimise the number and arrangement of injection and recovery wells through a range of ASTR scenarios taking into account the extent of mixing with the brackish ambient groundwater and the minimum allowable residence time for biodegradation to occur. The second phase goes beyond the planning aspects to simulate actual and proposed ASTR operations using a revised conceptual model of the groundwater system based on aquifer characterization and solute breakthrough data.

Hydrogeology

The target aquifer for ASTR is known as the 'T2' aquifer and its composition ranges from fossiliferous and marly limestone through to siliceous calcarenite. Locally, the T2 is encountered at depths of between 160 and 220 m below ground surface and contains three distinct hydro-facies (sub-aquifers): T2a,b,c (Gerges, 2005; 2006a; 2006b). Downhole EM flowmeter surveys consistently indicate the deepest sub-aquifer (T2c) to be the most transmissive. In order to contain the distribution of the injected water and prevent short-circuiting between injection and recovery wells that would otherwise reduce the effectiveness of the aquifer to attenuate contaminants before recovery, only the upper two sub-aquifers T2a,b are being targeted for ASTR operations. Aquifer transmissivity of this upper interval is around 50 m^2/day. Available evidence suggests that the aquifer is heterogeneous in nature, with flow characteristics that are typical of porous media rather than of fissures or karst features. The ambient groundwater quality is brackish, with a TDS of around 2000 mg/L. The regional groundwater flow direction is towards the west at an average hydraulic gradient of 0.0015. Much stronger local gradients also occur due to two ASR wells situated around 500 m NNE of the ASTR site.

Basic concepts of applied modelling approach

Methods of analysis

Two methods of analysis were undertaken to predict the movement and mixing of the injected water within the aquifer. The first method used a conservative solute as tracer to quantify the mixing between the two-end members of waters. Assuming that the solute concentration in the injected water differs sufficiently from the concentration in the ambient groundwater, a relative concentration, or so called mixing fraction *f*, can be estimated at any time from a simple mass balance given by Equation 18.1:

$$f(t) = \frac{C(t)_{mix} - C_{amb}}{C_{inj} - C_{amb}} \qquad (18.1)$$

where C_{inj} is the conservative solute concentration in injected water, C_{amb} is the conservative solute concentration in ambient groundwater and $C(t)_{mix}$ is the conservative solute concentration in the mixture at any particular time.

The second analytical method involves the calculation of the residence time of the injected water. Due to the difficulties resulting from the aquifer heterogeneity which leads to unstable kinematic age computation from a particle tracking method, the age of the water was calculated using a direct simulation of groundwater age (Goode 1996; Varni & Carrera 1998). Nevertheless, for an easier computational process in transient flow, the method was adapted by assuming a tracer present at some defined concentration in the injected water but absent in the ambient groundwater, and subject to simple exponential decay, as described by the following Equation 18.2:

$$\frac{C(t)}{C_o} = e^{-\lambda t_{eff}} \qquad (18.2)$$

where t_{eff} is the effective residence time of the groundwater in the aquifer (d), λ is the decay rate constant (1/d), $C(t)$ is the tracer concentration at some time t (mg/L) and C_o is the tracer concentration in the injectant (mg/L).

Numerical model for the design of the well field

The first stage of the modelling involved simulations of a range of possible ASTR scenarios performed with a 2D model to define the optimal number and arrangement of injection and recovery wells for the full-scale ASTR trial (Pavelic *et al.* 2004, 2005). A semi-analytical method for predicting the distribution of injected water "fronts" was also developed in parallel which was shown to compare favourably with the simulated results, thus providing a verification of the numerical approach. The numerical analysis took account of the quantity and sequence of flushing, orientation and magnitude of local and regional hydraulic gradients, uncertainties in aquifer properties (particularly aquifer heterogeneity), the proportion of water available for recovery, time-lag between injection and recovery, to assess the relative efficiency of two-, four- and six-well systems and of the combinations of well spacing and injection rates. Simulations were constrained by salinity (determined by the proportion of mixing between injected and background waters) and travel time of the injected water within the aquifer.

Development of a 3D numerical model

A 3D flow and conservative solute transport model was developed for the ASTR site as detailed in Kremer *et al.* (2008). The study involved two conceptual models, herein referred to as conceptual model 1 (CM1) and conceptual model 2 (CM2), that were based upon the aquifer characterisation work and used a layered-cake structure to describe the heterogeneity within the aquifer. The two alternative conceptual models differ by either including or excluding the volume of the T2 aquifer not intersected by the open interval of the injection/recovery wells (Figure 18.11). A triangular finite-element mesh was used on a 10,000 m × 10,000 m domain in the planar view with the ASTR system positioned at its centre. The mesh was most highly refined around each well and gradually diminished with increasing distance from the areas of interest. Boundary conditions and parameters were set from field data. CM1 and CM2 were calibrated using a trial-and-error approach constrained by EC data from breakthrough curves observed at four wells during flushing operations at the ASTR site between 2006 and 2007 involving the injection of water into two central wells. The models that best fit the parameters during the calibration process was then used as a predictive tool to simulate actual current and proposed future ASTR operations.

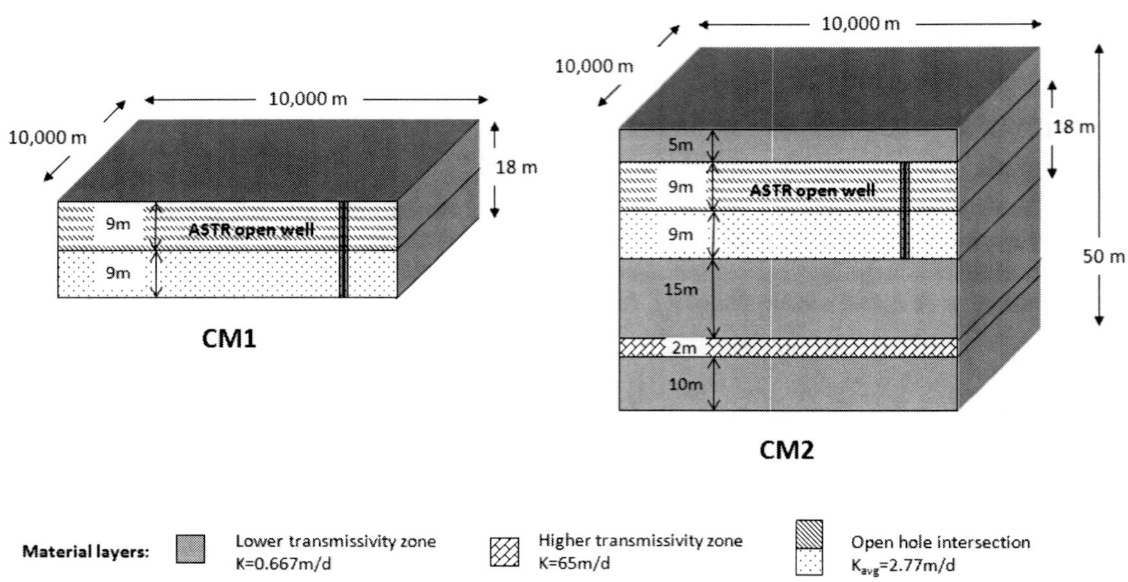

Figure 18.11 Schematic 3D view of conceptual models CM1 and CM2 showing material layers and open interval of ASTR wells (from Kremer *et al.* 2008)

Codes used and specific adaptation

Flow and solute transport processes were simulated using the commercial code FEFLOW Version 5.1 (Diersch, 2005) which is a three dimensional finite-element package capable of simulating contaminant and heat flow and transport. The simulations were done under saturated porous media conditions in a confined aquifer. Due to the different injection and recovery phases in the trial, all the simulations were transient both for flow and solute transport. Density effects and temperature changes were not considered since the salinity and the temperature contrasts between the two end-members waters were assumed insufficient to warrant inclusion.

The "mixing fraction simulations" were implemented in FEFLOW using a relative concentration ranging from 0 to 1, set at the value 0 for the ambient groundwater and the value 1 for the injected water. The "residence time simulations" were undertaken using a known decay rate then used to calculate the residence time from the concentration response (Equation 18.2) after first accounting for mixing (Equation 18.1).

Model-assisted design of ASR and ASTR well arrays

Combinations of well spacing and injection rates were tested with respect to the constraints on salinity and travel time using semi-analytical and 2D modelling methods (Pavelic *et al.* 2004, 2005). The initial results for homogeneous simulations during the first phase of the study revealed that both the salinity and travel time constraints could be met with a six-well system arranged within a quadrilateral domain with a uniform inter-well separation distance of 75 m (Figure 18.12).

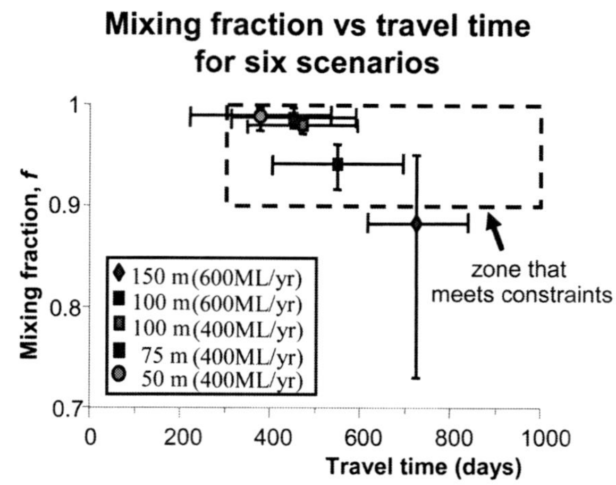

Figure 18.12 Combinations of well spacing and injection rates in a 6-well rhombic configuration, tested for their ability to meet the salinity (mixing fraction) and travel time constraints

Subsequent collection of local aquifer hydraulic property data indicated that partially-penetrating wells needed to be constructed to preclude the preferential flow that would invariably occur if the T2c sub-aquifer was also intercepted. Following this new condition, additional numerical results suggested that a reduction in the inter-well spacing from 75 m to 50 m was needed. Note that the anticipated scale of the ASTR trial was limited to approximately 200,000 m^3/year as a trade-off for the partial penetration of wells. The layout of the 4 injection wells (IW1-IW4) and 2 recovery wells (RW1-RW2) was established in 2006–07 at an inter-well spacing of 50 m (Figure 18.13).

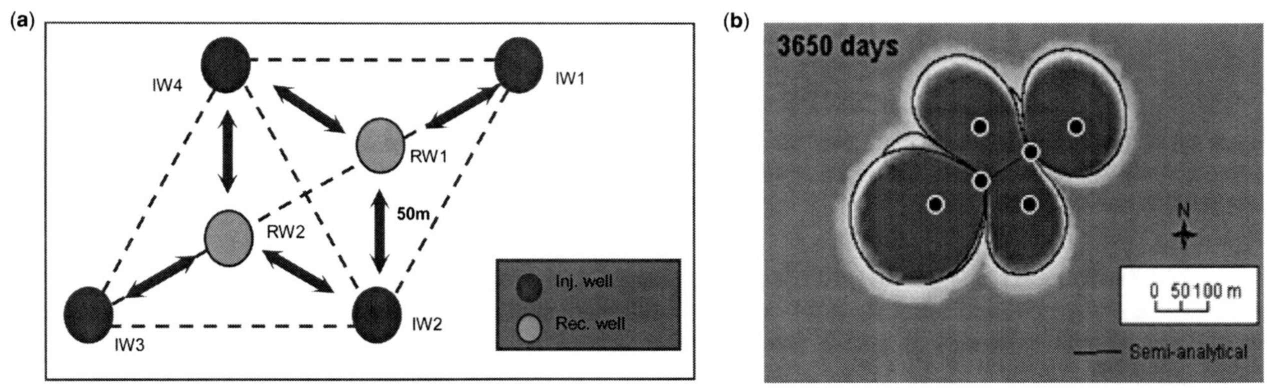

Figure 18.13 Retained 6-well rhombic configuration (a) and solute distribution after 10 years of injection comparing the results of semianalytical and 2D numerical modelling (b)

Optimising operating conditions with 3D dispersive–diffusive transport model

Many of input parameters of 3D conceptual models CM1 and CM2 were considered sufficiently well-known to be directly set from field data (e.g. regional groundwater flow, local gradient from adjacent Parafield ASR operations), from the aquifer characterization work (e.g. storage coefficient, aquifer transmissivity) or else from the literature (e.g. molecular diffusion). Porosity, anisotropy, hydraulic conductivities, and dispersivity were calibrated using a trial-and-error approach constrained by transport field data during a transient simulation. The calibration process evaluated using the root-mean-square error (RMSE) of the mixing fraction (Zheng & Bennett, 2002) showed consistently better performances for CM2 than for CM1. This reveals that some degree of vertical leakage into the adjacent strata, and most notably into T2c needed to be considered. Two alternative sets of parameterization of CM2 are referred to as Model A and Model B and differ in terms of the degree of vertical leakage of recharge water into T2c, which is significantly higher in model B, and in the porosity values with 0.7 for model A while 0.25 for model B. Porosity set on model A is out of the realistic range of values. Both models A and B were found to fit the field data equally well (RMSE of 0.81 and 0.88 respectively) but result in very different solute distribution profiles (Figure 18.14).

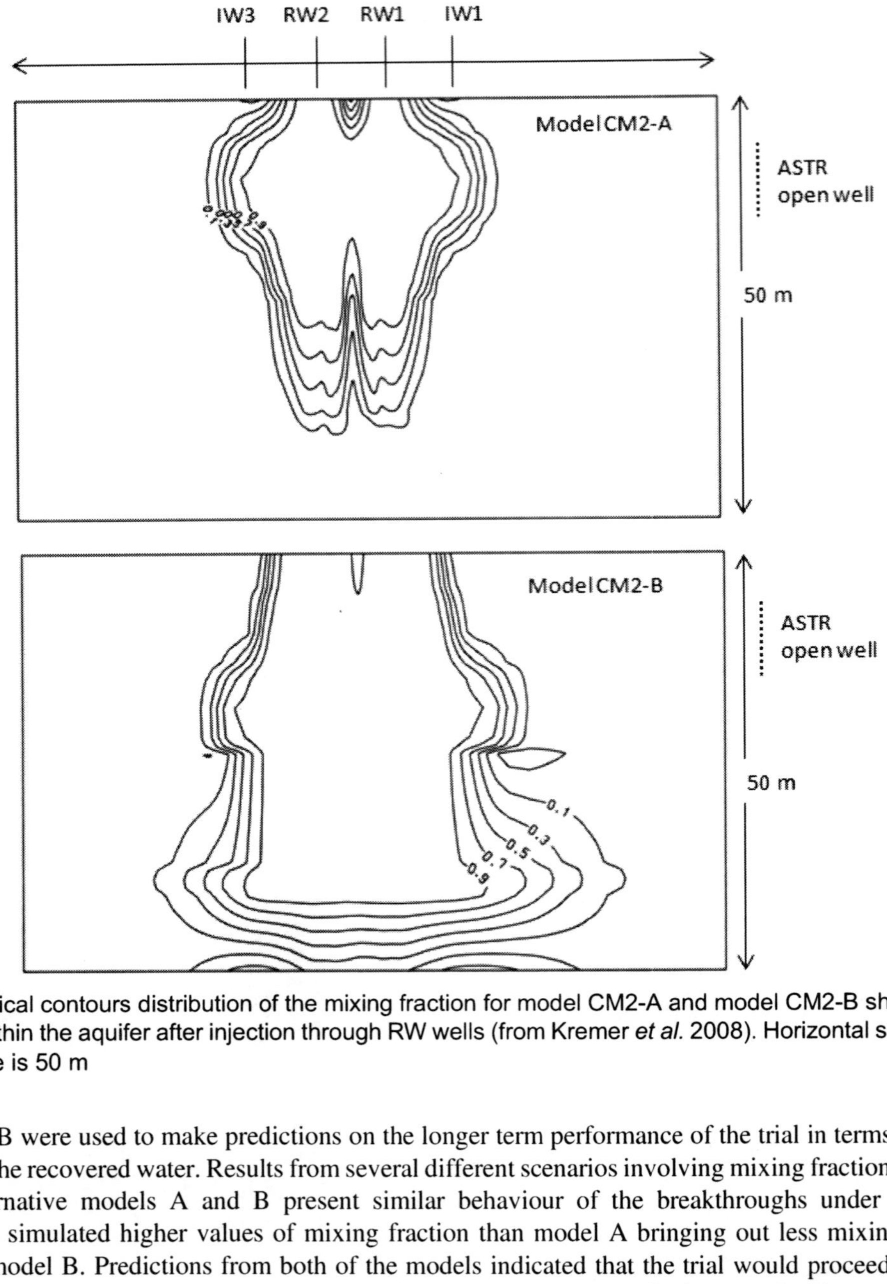

Figure 18.14 Vertical contours distribution of the mixing fraction for model CM2-A and model CM2-B showing the extent of the fresh plume within the aquifer after injection through RW wells (from Kremer *et al.* 2008). Horizontal scale is about 500 m while vertical scale is 50 m

Models A and B were used to make predictions on the longer term performance of the trial in terms of the salinity and residence time of the recovered water. Results from several different scenarios involving mixing fraction prediction showed that the two alternative models A and B present similar behaviour of the breakthroughs under the same induced stresses. Model B simulated higher values of mixing fraction than model A bringing out less mixing between the two end-members in model B. Predictions from both of the models indicated that the trial would proceed satisfactorily with the salinity of the recovered water suitable to ensure drinking water as shown for instance in Figure 18.15 which shows the predicted recovered water quality remains below 500 mg/L TDS over a 6-year period.. Consistency in the simulated breakthroughs from the two models shows the robustness of the models used as predictive tools and allowed a limited sensitivity analysis focused only on key parameters.

The residence time of the recovered water is an important variable to ensure sufficient time for inactivation attenuation of microbial pathogens and biodegradation of trace organics. Results from a simulation of an annual injection/recovery cycle involving 172.8 ML stored in the ground showed a minimum residence time of around 240 days and a difference of about 60 days between the two conceptual models (Figure 18.16). Both models suggest there is sufficient time to provide the required number of log removals to produce water of potable quality if inactivation rates are similar to those at a nearby reclaimed water ASR site in the same aquifer (Dillon & Toze, 2005).

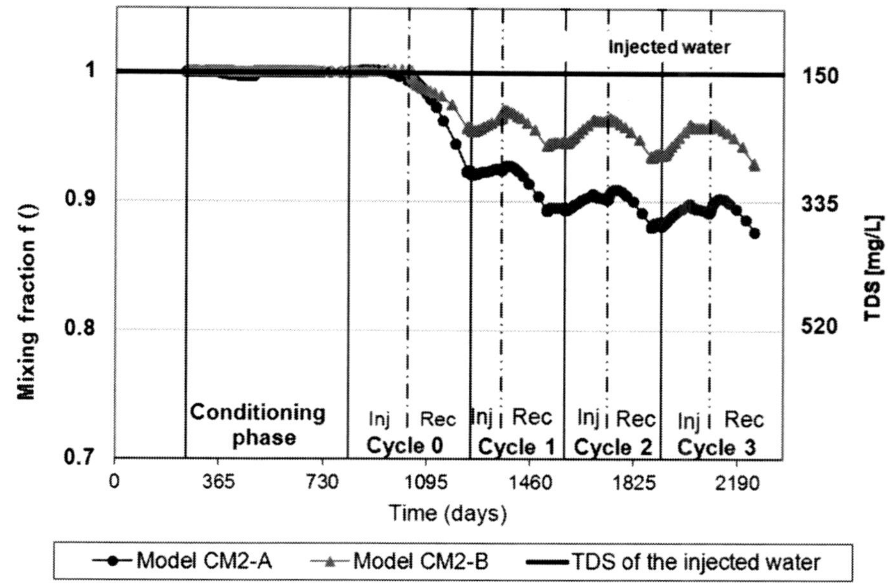

Figure 18.15 Simulated mixing fraction of the water at the RW wells during 6 years of ASTR operations taking into account a conditioning phase (from 250 to 820 days) and four cycles of injection (Inj) via IW wells, and extraction (Rec) via RW wells, for model CM2-A and CM2-B

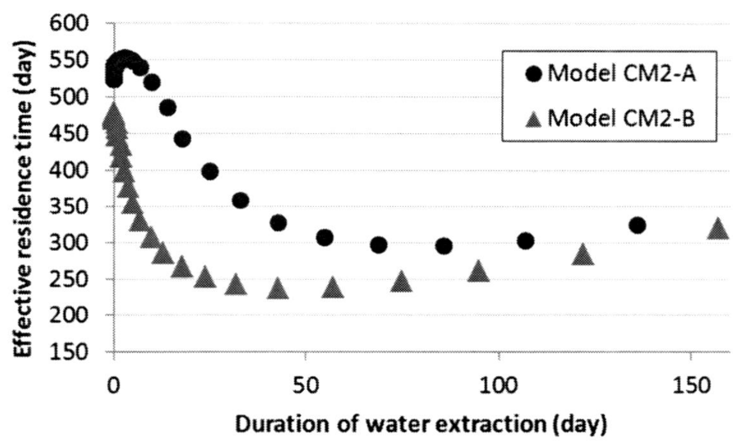

Figure 18.16 Comparison of simulated residence time in the aquifer of the injected water recovered from inner recovery wells RW versus duration of water extraction (absolute time), for model CM2-A and CM2-B, after a previous injection of 172.8 ML during 100 days through outer injection wells IW1 to IW4.

Geochemical modelling using PHREEQC was also performed at this site to compare geochemical processes between injection and recovery (Vanderzalm *et al.* 2011).

18.2.3 Case study 3: Nardò

Context

At the Nardò site in Southern Italy secondary treated municipal effluent is injected into a natural sinkhole in a coastal area. The sinkhole has been loaded in the past 17 years, by the Asso Channel which outflows municipal effluent from treatment plants and local surface drainages at an average constant rate of about 140 L/s. In the coastal aquifer, the groundwater was sampled in monitoring wells located at different distances from the sinkhole. At Nardò site, the experimental investigations have shown that groundwater is subjected to a faecal contamination and enteric viruses can be responsible for severe and, in some cases, fatal illnesses. Treatment technologies have to be combined with wellhead protection measures to reduce the risk of waterborne outbreaks. Health Canada (2004) guidelines established that treatment technologies should achieve at least a 4-log (i.e. 99.99%) reduction credits and/or inactivation of pathogens. The experimental study carried out at Nardò, suggests that 4-log reduction credits in a well positioned downgradient from injection can be achieved only after a filtration time of $\tau + 85$d (τ is the travel time required by a fluid particle to get to the observation well at current time t, see Figure 18.17).

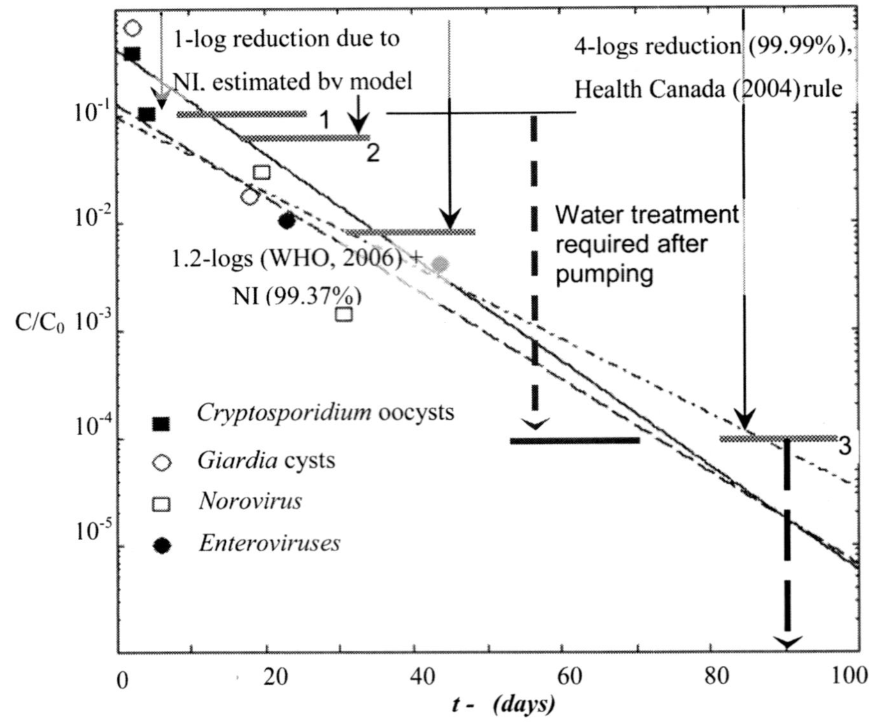

Figure 18.17 Observed and fitted (model outcomes) pathogen counts in relation to travel time for three wells positioned downgradient with respect to the groundwater flow at distances of 320m (solid line), 500m (dashed line), 3000m (dot and line), respectively, from the injection. The measured concentrations (circles and squares) are referred to the sampling period (February 2007). The arrows indicate: 1) Natural inactivation (NI) after 500 m in ground water; 2) Minimum target for water treatments by considering the reference level of risk of 10^{-6} DALYs per person (WHO, 2006) derived from *Cryptosporidium* count in groundwater pumped 500 m from the injection; and 3) 4-log disinfection credits required by Health Canada (2004) rule to have safe drinking ground water

The mathematical model determined a mean filtration time of 5.5 d for pathogen particles arriving at into the Well SP4 (500m from injection), the natural inactivation (NI) at Nardò site can contribute to the reduction of the treatment technology up to 1-log (90%) reduction credits, and a disinfection water treatment with UV light, or ozone, chlorine or chlorine dioxide after pumping, it is required in order to provide high-quality pathogen-free drinking water. Lab hypochlorination tests (Masciopinto *et al.* 2007) suggested that Italian drinking water requirements (D. Lgs. 31/2001 or EU Regulation (EC) no. 1882/2003) can be achieved in pumped groundwater after 5–15 min of contact time, with 3.3/3.7-log reductions, on average, in the pumped ground water. Moreover, the simulated breakthrough curves determined by model (Figure 18.17) suggests the setback distance of 3000 m to meet "natural disinfection credits", to achieve 4-logs virus reduction at Nardò groundwater. This distance can also be achieved by assuming an apparent velocity of water filtration into the fractures of 18.2 m/d and the required time for pathogen inactivation equal to 165 d, i.e. ($\tau = 70d + 95d$) derived from the model solution (see Figure 18.17).

Experimental recharge tests showed that the quality of the water recovered in a well is a function of the injectant water quality and, in particular, of the aquifer transmissivity and soil/rock porosity. As a consequence the latter hydrogeological parameters are very important in ASTR studies.

In fractured carbonate aquifers, such as in the Salento (South Italy) region, experimental pumping-tracer tests have shown that the groundwater flow can be addressed in a 3D set of extensive horizontal and parallel fractures with variable apertures. The transmissivity of the 3D set of fracture at an assigned location (x, y) can be estimated by a pumping test results, assuming that all fractures of the set have the same mean aperture. Hence, the latter can be considered as the fundamental parameter for modelling groundwater flow in fractured aquifers. A regional study of the covariance of the mean fracture aperture of the 3D set requires information from several pumping tests.

These estimations were carried out in 38 boreholes distributed throughout the study area (8.4 × 7.6 km). The values were obtained directly by solving the steady radial flow equation around every borehole under pumping. Every aperture estimation $\overline{b}$ was considered representative of flow resistances (which include wall ruggedness, fracture interconnections and tortuosity effects) over the whole saturated thickness of the aquifer surrounding the pumping well. In particular the $\overline{b}$ value was determined for each borehole, by inverting the solution of De Glee's formula (Bear, 1979: 347). The summation of the fracture apertures in the vertical section is defined by means of the effective aquifer porosity n which, as it is well known, is the uniform ratio of the void-space per unit volume of aquifer (Bear & Verruijt, 1987: 19–30; Borgia *et al.* 2002) and it follows that in the idealized set we have:

$$n = \frac{\sum_{i=1}^{N_f} b_i \times \Delta x \Delta y}{B \times \Delta x \Delta y} \tag{18.3}$$

where $\sum b_i$ is the sum of the all local (or nodal) apertures in the vertical aquifer column with horizontal area $\Delta x \Delta y$ (L^2) and thickness B. If $N_f(-)$ is the total number of the fractures of the parallel set and all fractures of the set have been assumed as having the same mean aperture $\bar{b}$, it follows that:

$$n = \frac{N_f\,\bar{b}}{B}; \quad \longrightarrow \quad N_f = \frac{n_{\overline{1pt}}\,B}{\bar{b}}$$

The apertures given by pumping tests allowed a spatial indicator variogram of the study area to be drawn. Though the De Glee formula may provide the hydraulic conductivity (or transmissivity) values directly, the stochastic generation of apertures b_i where i, ranging from 1 to the number of grid nodes in each fracture plane of the idealization (see Figure 18.18b), requires the covariance model of the fracture apertures.

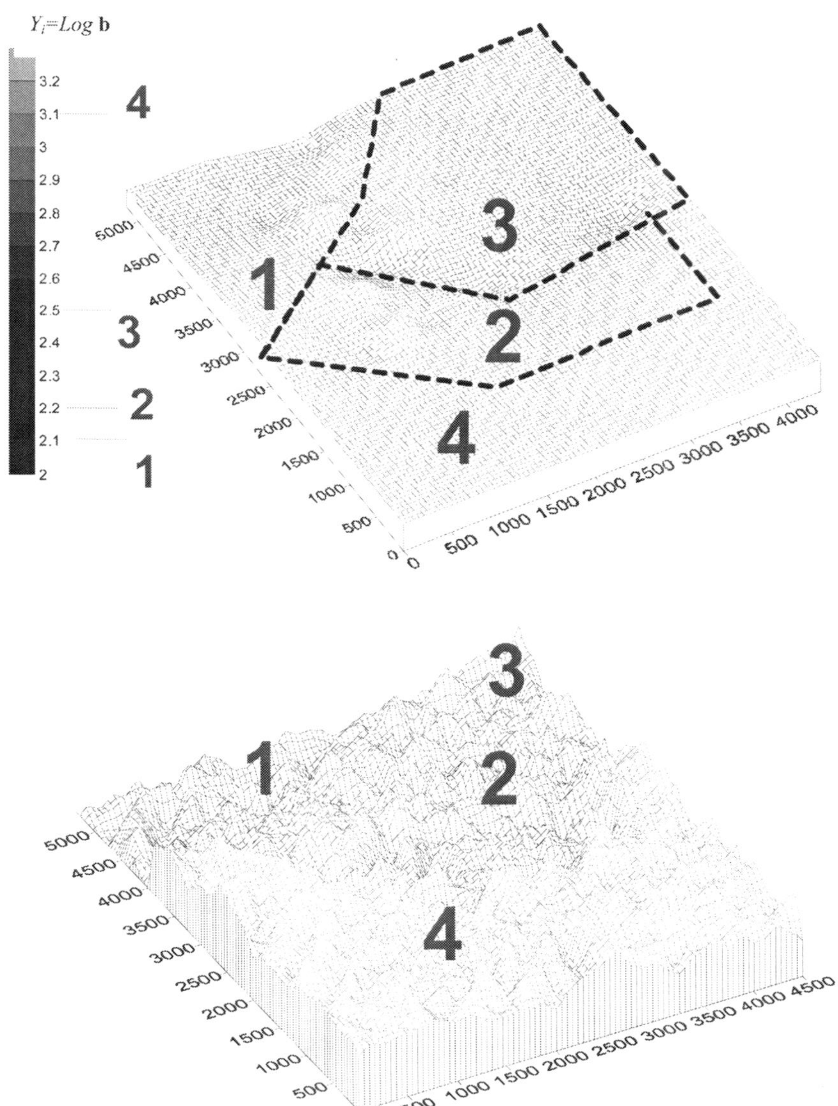

Figure 18.18 Above: Kriging of mean apertures $\bar{b}$ estimated on boreholes and selection of four sub domains of a square area of 4500 × 5500 m^2 characterized by different mean values of Y_i. Below: Fracture apertures generated by composing four stochastic generations based on the above four Y_i values

The empirical variogram model of apertures was fitted by the software packages as SURFER (1999) common to both the exponential and (fractional Gaussian noise) fGn models. The statistical information given by variogram models can be used to fit the profile of the rock surface in the fractal procedure or to evaluate the apertures of the generic single fracture of the 3D parallel set.

To improve the fit of the fracture aperture representation to field data, the whole domain can be divided into sub regions by using a spatial interpolation (kriging) of Log-apertures estimation in the 38 locations. In every sub domain a more appropriate mobile average Y_i can be evaluated and a smoothing of discrete values Y_{is} can be defined as follows:

$$Y_1 = \alpha Y_{is} + (1 - \alpha)\overline{Y}; \quad i = 1, 2, .., m; \tag{18.4}$$

where $\overline{Y}$ is the mean of Log-apertures in Table 18.1, m is the number of the of sub domains selected and α is the smoothing constant with values ranging from 0 to 1. For $\alpha = 0$, there is only one mean value $Y_i = \overline{Y}$ in the whole domain. For $\alpha = 1$, there are m values of Y_i equal to sub domain spatial averages Y_{is}.

This procedure produced a fracture aperture distribution properly conditioned by the spatial interpolation of available data.

Flow model equations and numerical formulation

For each realization of the aperture field every fracture of the parallel set can be discretized in a two-dimensional grid. The water flow in each grid element (or channel) of the fracture, with constant aperture b_i and bounded by two parallel plates, is governed by solution of the Navier-Stokes' equation (Brown, 1987):

$$Q_x = -\frac{1}{12} b_i^3 \nabla\varphi \frac{\gamma_w}{\mu} \Delta y \tag{18.5}$$

where γ_w/μ is about 10^7 $m^{-1}s^{-1}$ for water at 20 °C; $\nabla\phi$ (m) is a function of the piezometric head at the two extremities of the channel with constant length Δx and constant cross-section $\Delta y \times b$.

The cubic law (18.5) could be generalized to account for non laminar regimes. These regimes may occur in fractures where high water velocity around pumping wells together with large fracture apertures (above 2 mm) lead to Reynolds (Re) up to 40–50. By using the Darcy-Weisbach (Bear, 1972: 126) friction factor f, the flow discharge in the rectangular channel $\Delta x \times b_i$ is then defined as (Bear, 1972: 126):

$$Q_x = \left(\frac{4g}{f}\frac{\Delta\varphi}{\Delta x} b_i\right)^{1/2} \times b_i \Delta y \tag{18.6}$$

To guarantee mass-conservation at each grid node (without sinks and sources), the following equation can be used:

$$\sum_J Q_{ij} = \sum_J \left(\frac{\varphi_i - \varphi_j}{\Omega_{ij}}\right)^{1/2} = 0 \tag{18.7}$$

where the number J of the converging channels in each grid node is three, instead of four, along the borders of the studied area. This means that no-flux boundary conditions are fixed in the borders where piezometric heads not are imposed, and that water can enter or leave the aquifer system from the nodes at prescribed heads. As the fracture thickness varies abruptly between adjacent nodes, a linear interpolation of the gradient gives a channel conductance, $1/\Omega_{ij}$, that is proportional to the harmonic mean of the nodal values of apertures b_i. By imposing this equation for all grid nodes, a system of equations is obtained; its solution can be found after imposing suitable boundary conditions. In addition the piezometric heads in grid nodes of domain can be used also as boundary (Dirichlet) conditions. The Successive Over Relaxation (SOR) method can be easily implemented in a code to solve the equation system, following linearization by means of the Mc Laurin series (Abramovitz & Stegun, 1972). System solutions, i.e. the piezometric heads in each grid node, can be obtained using a tolerance equal to 10^{-6} m.

The domain studied of the Nardò was a square of side 8400×8600 m^2, whereas both Δx and Δy were selected equal to 200 m (1892 nodes). The total outflow of the parallel set of fractures is then obtained multiplying the single fracture discharge by the number of the fractures (N_f) in the set (see Figure 18.19).

Pathogen transport in a fractured/karstified aquifer: Analytical solutions combined with finite difference numerical approach

Following the colloid filtration theory, the governing partial differential equation for the stationary phase of the pathogen transport in a two-dimensional fracture with spatially variable apertures, assuming that the pathogens are stable of equal size (monodisperse), may deposit onto fracture surfaces but do not penetrate the rock matrix surrounding the fracture and that may undergo first-order inactivation of liquid phase and deposited pathogens with different inactivation rate coefficients, is given by (Abdel-Salam & Chrysikopoulos, 1995):

$$\frac{\partial C(t,x,y)}{\partial t} + \frac{2}{b(x,y)}\frac{\partial C^*(t,x,y)}{\partial t} = \nabla \cdot [\mathrm{D} \cdot \nabla C(t,x,y) - \mathrm{U}C(t,x,y)] - \lambda C(t,x,y) - \frac{2}{\mathrm{b}(x,y)}\lambda^* C^*(t,x,y) \tag{18.8}$$

where C^* is the concentration of pathogens deposited on the fracture surfaces, expressed as mass of colloids per unit area of the fracture surface, ∇ is the two-dimensional vector gradient operator (del); $\nabla \cdot$ denotes divergence ($\nabla \cdot \mathrm{G} = \partial G_x/\partial x + \partial G_x/\partial y$ where G is an arbitrary two-dimensional vector); D is the hydrodynamic dispersion coefficient; U is the interstitial ground water flow velocity vector, which is determined from the numerical solution of flow equation (Masciopinto *et al.* 2008); and λ and λ^* are the pathogen inactivation constants for suspended and adsorbed virus particles, respectively.

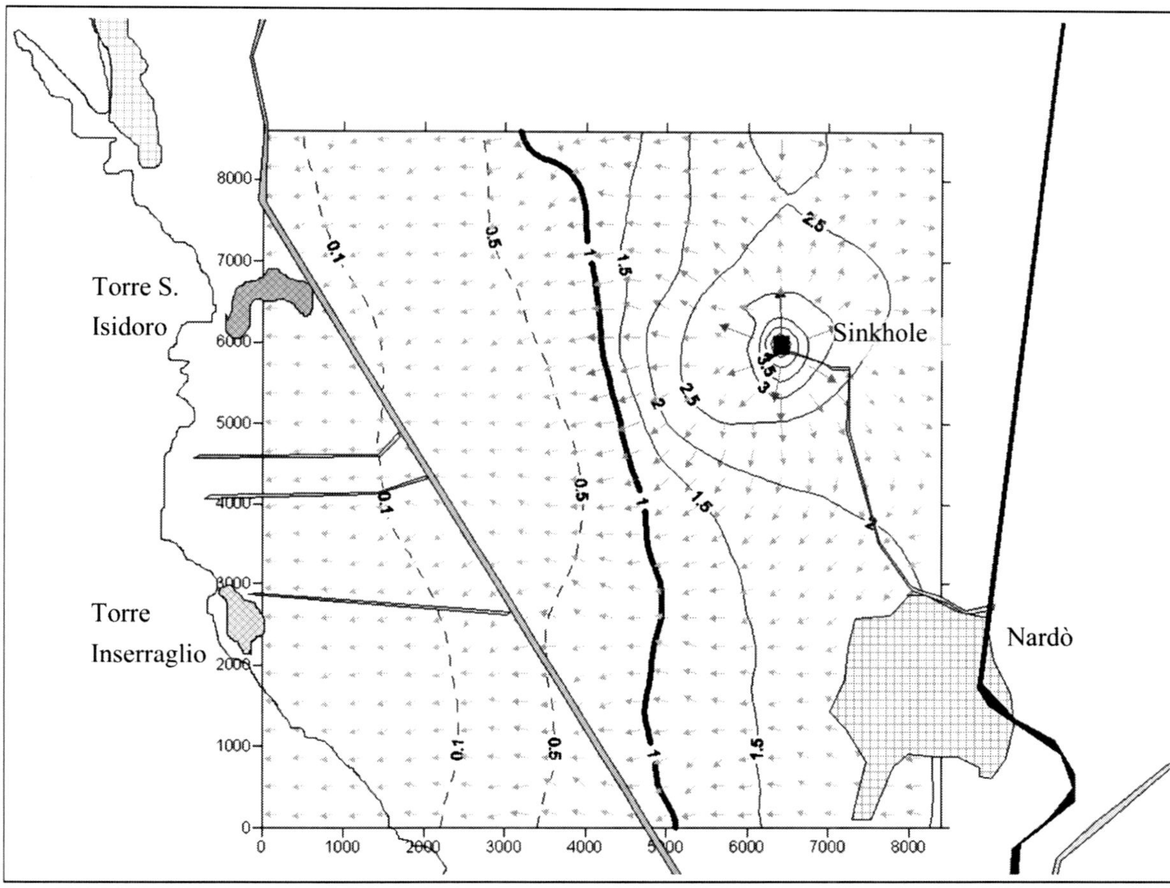

Figure 18.19 Groundwater flow solution at the Nardò site (Winter, 2007): heads (contour lines) and vectors proportional to the water velocity during the continuous artificial injection of 170 L/s into the sinkhole

For pathogen transport in fractured media such as the Nardò aquifer, where the water velocities in the various fractures are high, the molecular diffusion and lateral Taylor dispersion can be considered negligible and, consequently, the Lagrangian solution approach is applicable. The Lagrangian approach was introduced by Taylor (1921) in his fundamental work on turbulent diffusion, and since then it has been employed in numerous contaminant transport investigations (Dagan, 1989; Cvetkovic & Dagan, 1994; Wood, 1998). For an observer that follows a pathogen particle in the flow field of a fracture, the particle appears to be stationary (null velocity). Employing the Lagrangian approach Equation (18.8) reduces to:

$$\frac{dC[t,x(t)]}{dt} = -\frac{2}{b(x,y)}\frac{\partial C^*(t,x,y)}{\partial t} - \lambda C(t,x,y) - \frac{2}{b(x,y)}\lambda^* \mathrm{C}^*(t,x,y) \tag{18.9}$$

In the Nardò sinkhole the wastewater injection has been operated under uniform and almost continuous flowrates since 1991. Moreover, during the injection period, sorption phenomena affected the transport of pathogens in fractures, especially during the transient periods and until equilibrium conditions inside the fractures were achieved. The accumulation of pathogens onto the fracture surfaces can be described by the following generalized mass balance expression

(Chrysikopoulos & Abdel-Salam, 1997; Sim & Chrysikopoulos, 1998), as a function of the reverse and forward pathogen deposition rate coefficients, r_r and r_f. The latter is defined as

$$r_f = \kappa[U_x^2(x,y) + U_y^2(x,y)]^{1/2}F(C^*) \tag{18.10}$$

where U_x and U_y are the velocity components in the x- and y-direction, respectively, κ(L) is the colloids (i.e., pathogens) deposition coefficient, depending on several physicochemical properties of particles and collectors as well as on microscopic colloid deposition mechanisms (i.e., Brownian diffusion, van der Waals and electrostatic forces); and $F(C^*)$ is the dynamic blocking function (DBF), which takes into account the effect of previously deposited particles on subsequent colloid deposition by specifying the portion of the fracture surface area available for deposition (surface exclusion effect) during stationary phase. The DBF depends on the shape and size of colloids, geometry of sorbent, flow characteristics, and physicochemical properties of the interstitial fluid. The value of the DBF ranges between one for a fracture surface free of colloids ($C^* = 0$) and zero for a fracture surface completely covered (monolayer) by deposited colloids.

The integration of governing equation leads to the following analytical solution:

$$C[t, \mathbf{x}(t)]/C_0 = \exp[(-A_1 + A_2)t] \tag{18.11}$$

where the coefficients $A_2(t^{-1})$ and $A_1(t^{-1})$ have to be estimated at the prefixed time $t = \tau$, which is required by numerical particles released at $t = 0$ at the injection section (or sinkhole) to arrive into observation section (or well). In particular it can be set:

$$A_1 = \left[\frac{2}{b^2(x,y)} r_f + \lambda\right] \tag{18.12}$$

where r_f is given by Equation (18.12) when at the assigned time $t = \tau$ the average water velocity $\overline{\mathbf{U}} = \mathbf{x}(\tau)/\tau$ is evaluated, whereas

$$A_2 = \frac{2}{b^2(x,y)} r_r r_f \frac{1}{r_r + \lambda^*} \exp[-(r_r + \lambda^*)\,\tau] \tag{18.13}$$

The solution given by Equation (4) (see Figure 18.20) can be found only after that both groundwater flow and the numerical particle tracking models have been applied in order to determine the time $t = \tau$ and pathogen particle counts collected at the observation wells. Moreover A_1 and A_2 can be determined after that also the reverse and forward pathogen deposition rate (r_r, r_f), DBF, κ and inactivation constants (λ and λ*) have been estimated.

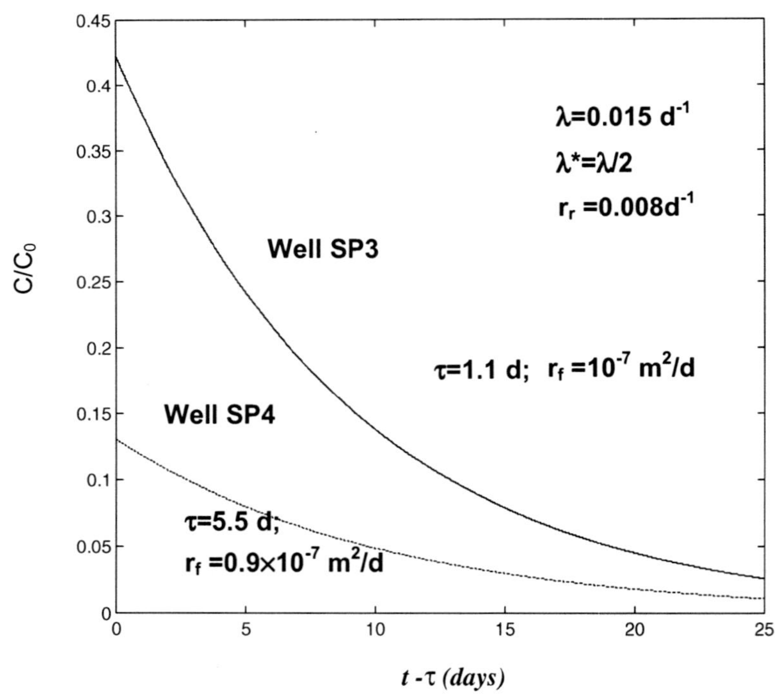

Figure 18.20 Normalized counts given by analytical solution (4) for pathogen transport in horizontal fractures with mean aperture of 1 mm, at assigned distances (320 m and 500 m) from the injection and subject to the pathogen deposition/detachment and natural inactivation. The assigned values of modal time τ required by numerical particles to arrive into Well SP3 ($\tau = 1.1$ d) and Well SP4 ($\tau = 5.5$ d), respectively, together with maximum counts have been determined by applying the particle tracking numerical code. The details of the numerical particle tracking procedure are reported by Masciopinto (2007)

These numerical estimates were performed using MATLAB (The MathWork, Inc. - version 4.2c.1, 1994), by imposing $r_r = 3\ y^{-1}$ (Chrysikopoulos and Abdel-Salam, 1997), $\lambda = 0.015\ d^{-1}$, $\lambda^* = \lambda/2$, $b = 1$ mm (i.e., mean value) and, $r_f = 10^{-7}$, 0.9×10^{-7} and $0.4 \times 10^{-7} m^2/d$ with average water velocities of 50, 45 and 20 m/d, for well #SP3, #SP4, and at 3000 m respectively, deposition coefficient (κ) of 10^{-8} m and DBF equal to 0.2. On the basis of the above mentioned values were determined A2 equal to 0.103, 0.096 and 0.015 d^{-1} at distances of 320 m ($\tau = 1.1$ d), 500 m ($\tau = 5.5$ d) and 3000 m ($\tau = 70$ d) from the injection site, respectively, whereas A1 had the correspondent values of 0.295, 0.195 and 0.095 d^{-1}.

Other aspects: How rainfall, pH and organic compounds interfere with pathogens transport

Of the 37 Nardò total samples analyzed, only 8 (22%) samples resulted positive to the enteric virus with PCR and, among the total 37 × 4 (= 148) determinations, it was possible to measure the quantitative estimation of genomic copies only on six samples. Similarly, a low (9%) recovery in quantitative RT-PCR analyses of the environmental samples was also investigated by Laverick *et al.* (2004). A previous study (Masciopinto *et al.* 2007) at the same Nardò site detected only 3 (17%) positive samples of enteric viruses among 18 water samples analyzed. Moreover, as in the present work, it was also noted that sampled water in wells was positive to the *Enteroviruses*, whereas neither in samples of the WWTP effluent nor of the injectant water *Enteroviruses* were not detected.

Obviously, if the presence of pathogens in municipal effluents could be questionable when a minimum rate of illness that may cause virus diffusion is absent (even if there are high counts of faecal pathogenic indicators), the absence of virus in samples of injectant water which produce the contamination in wells positioned downgradient, it could be due to the presence of both dissolved organic compound and salt ions in water samples, which interfere with the filtration process by causing increased virus adsorption on loamy sand a calcspar particles in subsurface. Indeed, in the same water samples (i.e. where the enteric viruses were missing), it was detected the presence of *Giardia* and *Crypto* oocysts at high counts and accordingly, high counts of the all faecal pathogenic indicators. Following Schmoll *et al.* (2006, p. 75), a decrease in salt concentration of the ground water, such as would occur after a rainfall event, can cause disaggregation-detachment of viruses and bacteria from soil particles and dissolved ions. Redman *et al.* (1997) suggests that small changes in salt concentrations can dramatically affect the mobilization of some organisms in groundwater systems. The implication of this discussion is that salt concentration in ground water may be of greater significance for pathogen transport than pH, although it is important to consider that no factor will act alone. The effect of calcium, magnesium, iron(II) and sodium ions, which are usually dissolved in the studied ground water and WWTP effluents, on the *Poliovirus 1* adhesion-aggregation in groundwater samples was also investigated by Gassilloud and Gantzer (2005). The above reduction of salt concentration from 0.7–0.9 g/L in injectant water and WWTP effluent to 0.4–0.5 g/L in the wells, could also explain the reason why the *Enteroviruses* were detected on the wells and not in both the injectant water and WWTP effluent during February 2007. Moreover the maximum *Norovirus* count (7600 gc/L) was detected in the well (SP3) at 320 m from the injection, during a water sampling carried out only two days after a rainfall event (11.4 mm), which was 3 day long (from 17 to 19 May 2007). During the same water sampling it was also noted the highest counts of *Giardia* and *Cryptosporidium* oocysts, into the same well.

18.2.4 Case study 4: Wulpen

Context and problem to be investigated

At St-André tertiary treated municipal effluent is infiltrated through an infiltration pond into the dunes of a coastal area to enable sustainable groundwater management. This scheme intends to restore the groundwater quality and enhance the ecological values of the dune areas as the extraction of natural groundwater is reduced (Van Houtte & Verbauwhede, 2005). The effluent comes from a municipal wastewater treatment plant at Wulpen which operates primary treatment and activated sludge secondary treatment. Tertiary treatment (UF/RO) then follows at the adjacent Torreele plant. The tertiary treated water is then transported by a pipeline to the recharge/extraction site of St-André (Figure 18.21). Tertiary treated wastewater is artificially recharged through two infiltration ponds then form a lens of artificially recharged water in the dunes' fresh water lens. The recharged water is recovered by 112 extraction wells located around the ponds. Observation wells (WP21-WP22 and WP24-WP23) are located between the pond and the extraction wells on groundwater flowpath. The monitoring of these observation wells allows understanding of the recharge water behaviour (water level, physico-chemical state, compound concentrations, …). The large amount of hydrogeological data available for the Wulpen site and the 3D regional model developed by Vanderbohede *et al.* (2008) allows studying in more detail the subsurface transport of the emerging contaminants along the flowpath between infiltration ponds and recovery wells.

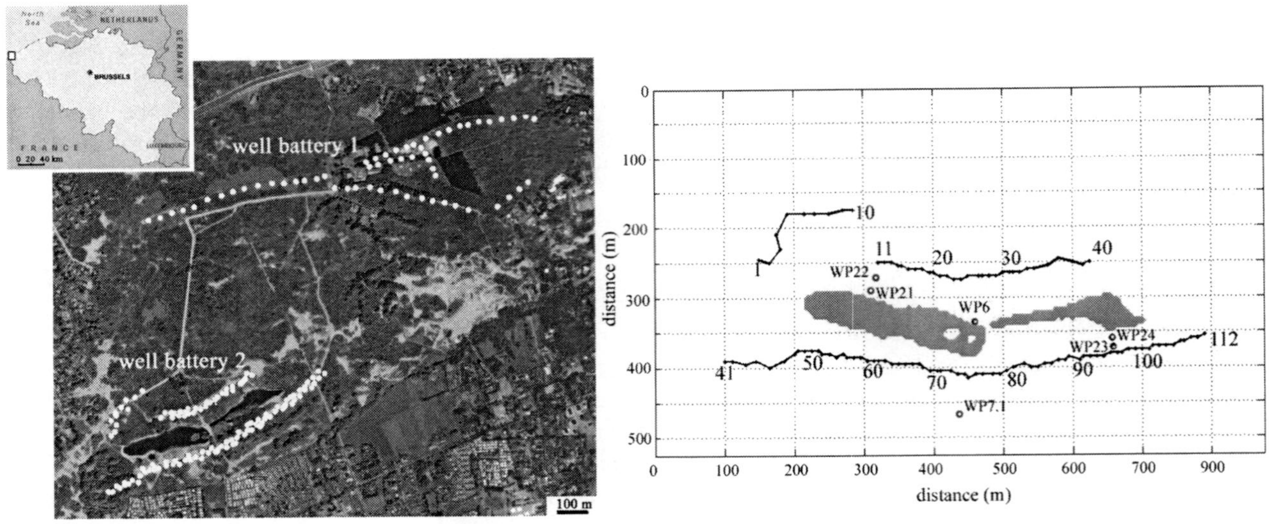

Figure 18.21 Photo and scheme of the artificial recharge pond (in grey color) and 112 extraction wells located around (Vandenbohede *et al.* 2008)

This case study shows how upscaling of modelling can be performed from 1D simulation of column laboratory experiments to 2D modelling of breakthrough scenarios embedded in the framework of a regional 3D model. Furthermore, a regional study based on environmental isotopes (B, Li, O, H) was conducted to provide further constraints to the local to regional models in terms of horizontal and vertical confinement of the infiltration-recovery zone (Kloppmann *et al.* 2008).

Parameters describing the behaviour of the selected contaminants in saturated and unsaturated conditions and under various redox regimes were first modelled on the basis of fate study column experiments from Reclaim Water. Through the introduction of the sorption and degradation behaviour in a 2D flow and transport model, contaminant-specific simulations were performed for NDMA and SMX. This allows predicting the concentrations in the recovered water in case of failure as input for subsequent risk assessment.

Scenarios of non-conservative pollutant transport within the ASTR system

In a first step, a 1D column unsaturated conservative flow model (Marthe Code) was established on the basis of Li tracer breakthrough through soil columns of Wulpen dune sands (saturated and unsaturated column studies performed by RWTH laboratory). In a second step, PhACs experimental behaviours were modelled: Carbamazepine, Sulfamethoxazole, Benzotriazole and Iopromide. Only sorption and (bio-)degradation processes are taken into account. The other parameters remain identical to the conservative Li tracer transport model. Linear sorption partition coefficients and half-life times T_{50} were derived and compared with literature values. Figure 18.22 shows measured and modelled SMX breakthrough in Wulpen soil column experiments. No degradation is observed in the results of column experiment. In order to fit the model to experimental results, a linear sorption process is taken into account with a $KD = 0.15$. Carballa *et al* (2008) measured distribution coefficient (Kd) on digested sludge and obtain values between 5.8 and 61.5 L/kg. Heberer *et al.* (2008) found SMX to be almost fully removed in the permanently anoxic wells in a sandy aquifer after bank filtration while seasonally oxic wells contained significant residues. Ternes *et al.* (2008) observe no significant removal in batch experiments with sand under natural aerobic and anoxic conditions.

On the basis of regional scale 3D modelling (Vandenbohede *et al.* 2008) and a tracer test conducted in August 2006 by the University of Ghent in cooperation with IWVA whereby a NaF solution was added to the infiltration pond, a local 2D plan model was developed and calibrated using the MARTHE code (Thiery *et al.* 1993). Hydraulic conductivities used in the 2D plan model are presented in Figure 18.23. The flow is considered at steady state whereas the transport can be varied with time. Unsaturated flow is approached through the concept of 'pseudo-unsaturated' flow considering the existence of a minimal water quantity (minimal saturation of a few percent) into the soil whatever the hydraulic head. Spillover is allowed for when the hydraulic head is higher than the topography.

Coupled modelling of breakthrough at the pumping wells after accidental spill of NDMA and SMX combines the results of conservative flow and transport 2D modelling of Wulpen site and of the sorption/degradation experiments (unsaturated and saturated column). Several scenarios, including "worst case" scenarios were assumed to simulate a leakage of NDMA and SMX into the groundwater due to a failure in the treatment system. For the NDMA contamination scenario, a concentration of 10 ng/L is injected continuously into the pond (worst case as all measured values were lower). No sorption process are taken into account and several level of degradation are tested ($T_{50} = 20$; 15; 10; 5 days) and compared to a simulation without degradation. The simulation running for 40 days of injection and the concentrations of NDMA and SMX are shown for the pumping well (Figure 18.24). Even for low degradation compared to experimental observation ($T_{50} = 20$ and 15 days), the simulations show a low concentration below 3–4 ng/L of NDMA in the pumping well.

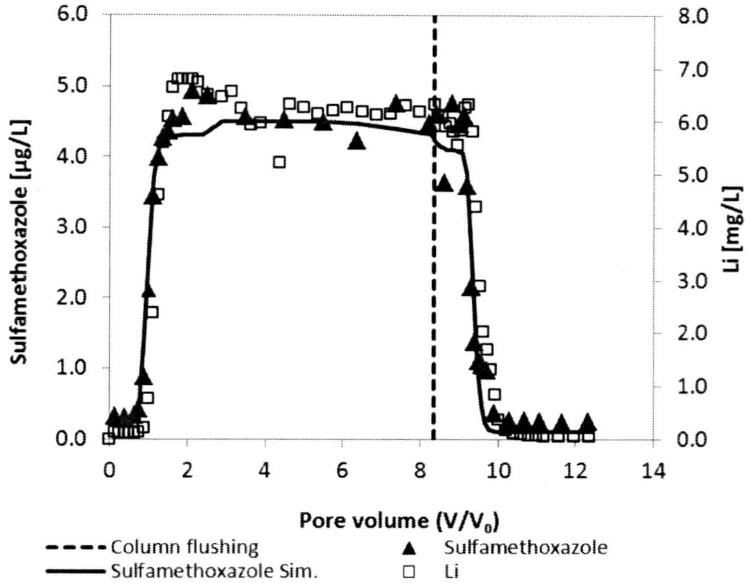

Figure 18.22 SMX breakthrough measured in Wulpen saturated sand column and 1D column modelling results (Sulfamethoxazole Sim.)

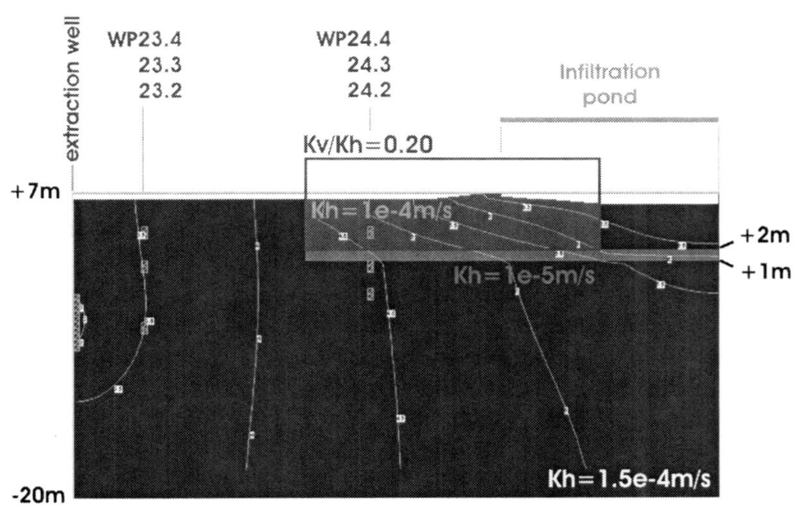

Figure 18.23 Hydraulic parameters take into account in the best case and hydraulic head simulated

The SMX contamination scenario considers two types of injection: one with 36 ng/L corresponding to 20% of maximal concentrations and a second with 180 ng/L corresponding to maximal concentrations (worst case). The leakage SMX compound is considered to be active for 30 minutes (Dirac-like signal). No degradation process is taken into account and several level of sorption are tested and compared to a simulation without sorption.

Simulation results show a retardation of SMX arrival into the pumping well. Obviously, this retardation positively correlated with the volumic partition coefficient. For all scenarios, SMX concentration in the pumping well is much lower than what is injected (1–3 ng/L for 36 ng/L injected and 7–10 ng/L for 180 ng/L injected). This fact is not due to sorption effects but to the dispersion occurring into the Wulpen aquifer and due to the Dirac peak of the injection. Sorption will essentially increase the residence time of the compound in the groundwater and so on the arrival time in the pumping well with retardation around 10 days compared to a conservative tracer.

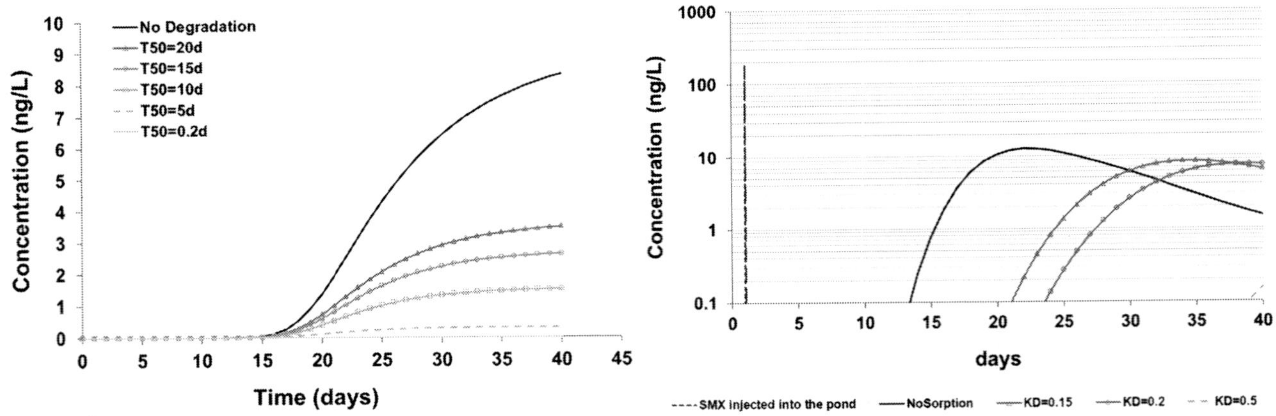

Figure 18.24 Simulation results at pumping well of contamination scenarios by leakage of NDMA (left, 10 ng/L continuous injection) and SMX (right, 180 ng/L Dirac-like pulse injection, 30 min) into the pond. Sorption process is taken into account for SMX by volumic partition coefficient (KD), biodegradation for NDMA by T_{50} half life

18.3 MODELLING OF MAR SYSTEM: LEARNING FROM THE RECLAIM WATER CASE STUDIES

The four described case studies cover various MAR systems in different hydrogeological and hydroclimatic contexts. The problems addressed through modelling for each of the investigated Reclaim Water sites are as diverse as the settings (synthesis in Table 18.3). The range of applied tools varies accordingly, from analytical solutions over 1D to 3D numerical models, taking into account conservative transport of dissolved species (salinity in particular) and pathogens as well as pathogen and contaminant attenuation, which requires combined/integrated models. Organic contaminant attenuation was addressed through the modelling of laboratory column studies of emerging pollutants (Wulpen case study), deriving biodegradation and sorption parameters, and then upscaled to field scale by 2D profile and radial numerical flow and transport models for the Wulpen and Shafdan sites. Inorganic contaminant release from the aquifer matrix was investigated for the Shafdan site through a reactive transport model taking into account the flow in the unsaturated and saturated zone (1D vertical and 1D horizontal reactive transport), the kinetics of organic matter breakdown induced by bacterial activity in the aquifer, the resulting redox zoning in the aquifer and the specific problem of manganese release. The injection of reclaimed freshwater into a multilayer saline aquifer system was modelled with a 3D conservative transport model for the Adelaide case study, aiming to configure and optimise the ASTR system and the operating conditions so to meet the constraints on recovery salinities and residence time. Pathogen transport and attenuation in a fractured/karstified carbonate aquifer in Nardò (Italy) was simulated for a 2D set of parallel fractures allowing for pathogen deposition on fracture surfaces using a combination of an analytical approach with numerical flow modelling and particle tracking.

The **Adelaide** site (Australia) shows how modelling activities accompany all planning phases of an ASTR system in a brackish groundwater body after the initial choice (Pavelic *et al.* 2002) and evaluation (Pavelic *et al.* 2001) of the target aquifer. Semi-analytical and numerical (FEFLOW) finite element 2D models were used to investigate the impact of well configuration (2–6 well patterns were tested) and of well distance and injection rate on the two key regulatory constraints, maximum acceptable salinity (mixing fraction) and residence time. This allowed the design and construction of the well field that was subsequently completed by observation wells and tested through the monitoring of flushing conducted over 2 years. The acquired data served for calibrating a 3D multilayer model, a 6-layer model was found to represent best the observations (Kremer *et al.* 2008). Scenarios of the evolution of mixing fractions were then calculated till 2012. Refinement of model after new data collected over the complete conditioning phase is underway and it is projected to apply ASRRI (analytical) and/or PHT3D (numerical) models to assess the fate of pathogens, trace organics and biogeochemical processes when sufficient data becomes available.

At **Nardò** (Italy) the main concern is faecal contamination of a fractured/karstified carbonate aquifer. The main goal of the conducted modelling and experimental study is to estimate natural inactivation within the aquifer as a function of travel time, and to evaluate the need for additional disinfection treatment of the recovered water. The analytical solution for pathogen transport in parallel and horizontal fractures was developed and combined with the particle tracking numerical solution. The natural inactivation at Nardò can contribute to a single log reduction of pathogen disinfection credits for the 5.5 d travel time between sinkhole and a 500 m distant pumping well. Consequently additional treatments are needed to achieve 4-log water disinfection credits for virus inactivation. Also, the minimum distance of recovery of ground water from the injection point to achieve full 4-log of natural disinfection was estimated to be over 3000 m.

Further laboratory investigations were conducted on the impact of pH, salinity, and organic compounds on virus/bacteria adsorption and disaggregation in ground water.

Table 18.3 Overview of the RECLAIM WATER sites investigated through hydrodynamic (flow, transport, reactive transport) modelling

Case study	Type of reservoir	MAR system	Problems/questions addressed	Modelling approach
Adelaide	Mainly porous, confined multilayer brackish aquifer	ASTR field with 2 injection and 4 recovery wells	Optimising well field design and operating conditions considering salinity and residence times	Semianalytical and 2-3D numerical multilayer model with conservative transport
Nardò	Fractured, karstified carbonate aquifer	ASTR with injection in a karst sinkhole and well recovery	Pathogen transport and natural attenuation in fracture system with limited filtration capacity	Analytical solution for 2D flow and transport in a series of parallel fractures combined with numerical particle tracking and flow mode
Wulpen	Coastal unconfined dune sand aquifer underlain by saline groundwater			1D modelling of organic pollutant column experiments, 2D transport modelling on a profile from injection ponds to recovery wells considering sorption and biodegradation
Shafdan	Sandy unconfined aquifer with important unsaturated zone	SAT with basin infiltration, and well gallery for recovery. Pilot of dug well injection	Design and performance evaluation of dug well injection system.Problem of redox zoning and associated Mn-release	Simple 3D model for initial screening and pilot design. 2D radial unsaturated flow and transport modelling, including sorption and biodegradation. 1Dv and 1Dh unsaturated-saturated reactive transport model including kinetic laws of bacterial growth and organic matter degradation

In **Wulpen** (Belgium), reclaimed water is treated to drinking water quality through advanced tertiary treatment before injection through infiltration ponds into a coastal dune sand aquifer and recovery by two well galleries. An important point to address with respect to regulatory constraints is the confinement of the SAT system with respect to the host aquifer. Secondly, the attenuation of trace organic compounds investigated through WP3 of Reclaim Water, was to be simulated for this system to evaluate effects of dilution-sorption-biodegradation for accidental failure scenarios. The conclusions from a regional 3D model (using MOCDENS3D, Vandenbohede *et al.* 2008) on the confinement of the reclaimed water within the injection-recovery zone were confirmed through a geochemical study using environmental isotopes of boron, lithium, oxygen and hydrogen (Kloppmann *et al.* 2008). The behaviour of emerging pollutants was investigated on soil columns from Wulpen (WP5) and transport/degradation parameters modelled with an 1D numerical model. In a second step, a small scale 2D model was set up to investigate tracer breakthrough on a profile from the infiltration ponds to the recovery wells and attenuation evaluated for NDMA and SMX compounds. It can be concluded that, even for the maximum observed concentrations for both compounds, concentrations at the recovery wells are not of concern.

The Shafdan case study (Israel) offers the opportunity to investigate the long-term behaviour of a large scale Soil Aquifer Treatment system over a timescale of several decades. Specific problems occurred only after 20 years of successful operation, like the reduction of infiltration capacity, partly due to physical and chemical clogging, and the sudden appearance of significant concentrations of dissolved manganese in several wells. A new injection system through dug wells was developed and tested during the project, destined to complete the existing infiltration basins, and the design and testing of the dug-well prototype was accompanied by different modelling activities. A preliminary 3D flow and conservative transport model was set up a the onset of the project to plan and design the dug well prototype and the monitoring wells. The radial pattern of solute transport obtained by the 3D model showed that a 2D radial model could satisfyingly describe the system so that a 2D unsaturated flow and transport model was set up to model tracer tests and transfer scenarios of accidental peaks of organic contaminants. This model was used to simulate NDMA and SMX transfer from the injection well to the observation wells. Redox conditions were addressed through a 1D vertical unsaturated and 1D horizontal saturated reactive transport model. This model combines kinetics of microbiological degradation of organic matter (Monod kinetics), kinetic control of electron acceptors, with water-mineral interactions (dissolution-precipitation reactions). It was possible to reproduce the observed redox sequence in the Shafdan system even if this example demonstrates the limits of current bio-geochemical concepts, in particular for an element like Mn for which no stable oxidised (Mn^{IV}) aqueous species exists. Clearly, the consideration of bacterially catalysed, non-equilibrium reactions, and, simultaneously, of water-mineral equilibria through coherent bio-geochemical models is currently one of the major challenges in modelling such complex systems as are MAR systems using reclaimed water. Such models are a prerequisite for risk assessment in the field of MAR, if we want to go beyond the use of bulk parameters like sorption isotherms and biodegradation half lifes for when looking on the behaviour of organic contaminants. A thorough understanding of biogeochemical reactions occurring both in the unsaturated and the saturated zone is needed to develop predictive tools that are able to address and foresee potential drawbacks of MAR systems like pollutant breakthrough, release of trace contaminants from the aquifer material and mineral precipitations leading to a degradation of the hydraulic properties of the receiving reservoir.

REFERENCES

Abdel-Salam A. and Chrysikopoulos C. V. (1995). Modeling of colloid and colloid-facilitated contaminant transport in a two-dimensional fracture with spatially variable aperture. *Transport in Porous Media*, **20**(3), 197–221.

Abramovitz M. and Stegun I. A. (eds) (1972). Handbook of mathematical functions. 10th edn, NBS.

Adelman D. D., Stansbury J. and Tabidian M. A. (1998). A risk/cost analysis to manage viral contamination of groundwater19th Biennial Conference of the International-Association-on-Water-Quality – Water Quality International 98, Vancouver, Canada.

Allison J. D., Brown D. S. and Novo-Gradac K. J. (1991). MINTEQA2/PRODEFA2: A Geochemical Assessment Model for Environmental Systems, Version 3.0 User's Manual. USEPA, Washington, DC.

Appelo C. A. J. and Postma D. (2005). Geochemistry, Groundwater and Pollution, 2nd edn. A. A. Balkema Publishers, Leiden, The Netherlands, p xviii + 649 pages.

Artimo A., Sarapera S. and Ylander I. (2008). Methods for integrating an extensive Geodatabase with 3D modeling and data management tools for the Virttaankangas artificial recharge project, Southwestern Finland. *Water Resources Management*, **22**, 1723–1739.

Bales R. C., Hinkle S. R., Kroeger T. W., Stocking K. and Gerba C. P. (1991). Bacteriophage adsorption during transport through porous media: chemical perturbations and reversibility. *Environ. Sci. Technol.*, **25**, 2088–2095.

Barry K. (2008). Flow and solute transport observations and modelling from the first phase of flushing operations at the Salisbury ASTR site. CSIRO: Water for a Healthy Country National Research Flagship.

Bear J. (1979). Hydraulics of Groundwater. McGraw-Hill Book Company, INC. New York, pp. 320–347.

Bear J. and Verruijt A. (1987). Modeling Groundwater Flow and Pollution. D. Reidel Publishing Company, 414 P.

Bethke C. M., Ding D., Jin Q. S. and Sanford R. A. (2008). Origin of microbiological zoning in groundwater flows. *Geology*, **36**, 739–742.

Bethke C. M. (2002). The Geochemist's Workbench®: A User's Guide to Rxn, Act2, Tact, React, and Gplot. Release 4.0, http://www.geology.uiuc.edu/~bethke/

Bhattacharjee S., Ryan J. N. and Elimelech M. (2002). Virus transport in physically and geochemically heterogeneous subsurface porous media. *Journal Of Contaminant Hydrology*, **57**, 161–187.

Borgia G. C., Bortolotti V. and Masciopinto C. (2002). Valutazione del contributo della porosità effettiva alla trasmissività di acquiferi fratturati con tecniche di laboratorio e di campo (in Italian, with English and French abstracts). *IGEA, Groundwater Geoengineering*, **17**, 31–43.

Brock T. D. and Madigan M. T. (1991). Biology of Microorganisms. 6th edn, Prentice Hall, pp. 874.

Brown S. R. (1987). Fluid flow through rock joints: the effect of surface roughness. *Journal of Geophysical Research*, **92**(B2), 1337–1347.

Bruggeman G. A. (1999). Analytical Solutions of Geohydrological Problems. Elsevier, Amsterdam, Netherlands.

Carballa M., Fink G., Omil F., Lema M. J. and Ternes T. (2008). Determination of the solid-water distribution coefficient (Kd) for pharmaceuticals, estrogens and musk fragrances in digested sludge. *Water Research*, **42**, 287–295.

CDPH (2011). Groundwater Replenishment Reuse Draft Regulation. California Department of Public Health. http://www.cdph.ca.gov/certlic/drinkingwater/Documents/Recharge/DraftRechargeReg-2011-11-21.pdf. (accessed 21 NOV 2011)

Chang Y. C. and Yeh H. D. (2007). Analytical solution for groundwater flow in an anisotropic sloping aquifer with arbitrarily located multiwells. *J. Hydrol.*, **347**, 143–152.

Chrysikopoulos C. V. and Sim Y. (1996). One-dimensional virus transport in homogeneous porous media with time dependent distribution coefficient. *J. Hydrol.*, **185**, 199–219.

Chrysikopoulos C. V. and Abdel-Salam A. (1997). Modeling colloid transport and deposition in saturated fractures. *Colloids and Surfaces A – Physicochemical and Engineering Aspects*, **121**, 189–202.

Chrysikopoulos C. V. and James S. C. (2002). Transport of neutrally buoyant and dense variably sized colloids in a two-dimensional fracture with anisotropic aperture. *Transport in Porous Media*, **51**, 191–210.

Clark I. D. and Fritz P. (1997). Environmental Isotopes in Hydrogeology. CRC Press/Lewis Publishers, Boca Raton, FL.

Cook P. G. and Herczeg A. L. (2000). Environmental Tracers in Subsurface Hydrology. Kluwer Academic Publishers, Boston.

Cordy G. E., Duran N. L., Bouwer H., Rice R. C., Furlong E. T., Zaugg S. D., Meyer M. T., Barber L. B. and Koplin D. W. (2004). Do pharmaceuticals, pathogens, and other organic waste water compounds persist when waste water is used for recharge? *Groundwater Monitoring and Remediation*, **24**(2), 58–69.

Cvetkovic V. and Dagan G. (1994). Transport of kinetically sorbing solute by steady random velocity in heterogeneous porous formations. *J. Fluid Mech.*, **265**, 189–215.

Dagan G. (1989). Flow and Transport in Porous Formations. Springer-Verlag, Berlin Heildelberg, Germany.

Diersch H.-J. G. (2005). FEFLOW: Finite Element Subsurface Flow & Transport Simulation System. WASY GmbH Institute for Water Resources Planning and Systems Research.

Dillon P. (2005). Future management of aquifer recharge. *Hydrogeol. J.*, **13**, 313–316.

Dillon P., Toze R., Pavelic P., Vanderzalm J., Barry K. E., Ying G.-G., Kookana R. S., Skjemstad J., Nicholson B., Miller R., Correll R. L., Prommer H., Greskowiak J. and Stuyfzand P. (2005). Water Quality Improvements During Aquifer Storage and Recovery at Ten Sites. Proceedings ISMAR5, Berlin.

Doherty J. (2002). PEST – model-independent parameter estimation: user's manual. 5th edn, Watermark Numer. Comput., Brisbane, Australia.

Drewes J. E., Hoppe C. and Jennings T. (2006). Fate and transport of N-nitrosamines under conditions simulating fullscale groundwater recharge operations. *Water Environment Research*, **78**(13), 2466–2473.

Dzombak D. A. and Morel F. M. M. (1987). Development of a data base for modelling adsorption of inorganics on Iron and Aluminum oxides. *Environmental Progress*, **6**, 133–137.

Eusuff M. M. and Lansey K. E. (2004). Optimal operation of artificial groundwater recharge systems considering water quality transformations. *Water Resources Management*, **18**, 379–405.

Faulkner B. R., Lyon W. G., Khan F. A. and Chattopadhyay S. (2003). Modeling leaching of viruses by the Monte Carlo method. *Water Res.*, **37**, 4719–4729.

Flury M. and Wai N. N. (2003). Dyes as tracers for vadose zone hydrology. *Rev. Geophys.*, **41**, 37.

Gan J., Bondarenko S., Ernst F., Yang W., Ries S. B. and Sedlak (2006). Leaching of N-nitrosodimethylamine (NDMA) in Turfgrass soils during wastewater irrigation. *J. Environ. Qual.*, **35**, 277–284.

Gassilloud B. and Gantzer C. (2005). Adhesion-aggregation and inactivation of Poliovirus 1 in groundwater stored in hydrophobic container. *Applied and Environmental Microbiology*, **71**(2), 922–920.

Gau H. S., Hsieh C. Y. and Liu C. W. (2006). Application of grey correlation method to evaluate potential groundwater recharge sites. *Stochastic Environmental Research and Risk Assessment*, **20**, 407–421.

Gaus I., Cikurel H., Picot G., Aharoni A., Guttman Y. and Kloppmann W. (2007a). Alternative SAT treatment of secondary effluents using a combination of UF and short term SAT for increasing infiltration rates at the Shafdan site (Israel), in ISMAR6 – 6th International Symposium on Managed Aquifer Recharge – Phoenix – USA – 28/10-02/11/2007.

Gaus I., Cikurel H., Picot G., Aharoni A., Guttman Y., Kloppmann W. and Azaroual M. (2007b). Alternative SAT treatment of secondary effluents using a combination of UF and short term SAT for increasing infiltration rates at the Shafdan site (Israel), in IWA WRR2007 – 09-12/10/2007 – Anvers – Belgique.

Gaus I., Shand P., Gale I. N. and Eastwood J. (2001). Geochemical modelling of groundwater quality changes during aquifer storage and recovery (ASR) in the dual porosity chalk aquifer, england. In: 10th International Symposium on Water-Rock Interaction. R. Cidu (ed.), A. A. Balkema Publishers, Villasimius, Italy.

Gaus I., Shand P., Gale I. N., Williams A. T. and Eastwood J. C. (2002). Geochemical modelling of fluoride concentration changes during Aquifer Storage and Recovery (ASR) in the Chalk Aquifer in Wessex, England. *Quarterly Journal of Engineering Geology and Hydrogeology*, **35**, 203–208.

Gerges G. (2005). Greenfield Railway Station ASTR Project–Briefing on Aquifer Tests, Report prepared for ASTR Technical Committee by NZG Groundwater Consultant.

Gerges G. (2006a). Technical specification for drilling and well completion ASTR project at Greenfield's Railway Station (to be read as Parafield Gardens oval) 16th March 2006.

Gerges G. (2006b). Parafield oval ASTR wells completion and aquifer Tests results. *Preliminary draft for Technical Meeting August 2006.*

Giao P. H., Phien-Wej N. and Honjo Y. (1998). FEM quasi-3D modelling of responses to artificial recharge in the Bangkok multiaquifer system. *Environmental Modelling and Software*, **14**, 141–151.

Godfrey E., Woessner W. W. and Benotti M. J. (2007). Pharmaceuticals in on-site sewage effluent and groundwater, Western Montana. *GroundWater*, **45**(3), 263–271.

Goode D. J. (1996). Direct simulation of groundwater age. *Water Resources Research*, **32**(2), 289–296.

Greskowiak J., Prommer H., Massmann G. and Nutzmann G. (2006). Modeling seasonal redox dynamics and the corresponding fate of the pharmaceutical residue phenazone during artificial recharge of groundwater. *Environ. Sci. Technol.*, **40**, 6615–6621.

Greskowiak J., Prommer H., Vanderzalm J., Pavelic P. and Dillon P. (2005). Modeling of carbon cycling and biogeochemical changes during injection and recovery of reclaimed water at Bolivar, South Australia. *Water Resour. Res.*, **41**.

Hart D. S., Davis L. C., Erickson L. E. and Callender T. M. (2004). Sorption and partitioning parameters of benzotriazole compounds. *Microchemical Journal*, **77**, 9–17.

Haruta S., Chen W., Gan J., Simunek J, Chang A. C. and Wu L. (2008). Leaching risk of n-nitrosodimethylamine (NDMA) in soil receiveing reclaimed wastewater. *Ecotoxicology and Environmental Safety*, **69**, 374–380.

Health Canada (2004). Guidelines for Canadian Drinking Water Quality: Supporting Document–Enteric Viruses. (www.hc-sc.gc.ca/ewh-semt/pubs/water-eau/enteric-enterovirus/index-eng.php).

Heberer T., Massmann G., Franck B., Taute T. and Dünnbier U. (2008). Behaviour and redox sensitivity of antimicrobial residues during banck filtration. *Chemosphere*, **73**, 451–460.

Hendry M. J., Lawrence J. R. and Maloszewski P. (1999). Effects of velocity on the transport of two bacteria through saturated sand. *Ground Water*, **37**(1), 103–112.

Herbst M., Fialkiewicz W., Chen T., Putz T., Thiery D., Mouvet C., Vachaud G. and Vereecken H. (2005). Intercomparison of flow and transport models applied to vertical drainage in cropped lysimeters. *Vadose Zone J.*, **4**, 240–254.

Hunter K. S., Wang Y. F. and Van Cappellen P. (1998). Kinetic modeling of microbially-driven redox chemistry of subsurface environments: coupling transport, microbial metabolism and geochemistry. *Journal of Hydrology*, **209**, 53–80.

IAEA (2006). Applicability of monitored natural attenuation at radioactively contaminated sites. Technical Reports Series No. 445, http://www-pub.iaea.org/MTCD/publications/PDF/TRS445_web.pdf

Izbicki J. A., Flint A. L. and Stamos C. L. (2008). Artificial recharge through a thick, heterogeneous unsaturated zone. *Ground Water*, **46**, 475–488.

Janssen G. and Hemke K. (2004). Groundwater models as civil engineering tools. In: FEM_MODFLOW, Kovar, Hrkal and Bruthans (eds.), Karlovy Vary, Czech Republic.

Jekel M. and Grünheid S. (2005,). Bank filtration and groudwater recharge for treatment of polluted surface waters. *Water Science and texhnology: Water Supply*, **5**(5), 57–66.

Johnson P. N. and Elimelech M. (1996). Colloid transport in geochemically heterogeneous porous media: modeling and measurements. *Env. Sci. and Techn.*, **30**, 3284–3293.

Kargi F. (2009). Re-interpretation of the logistic equation for batch microbial growth in relation to monod kinetics. *Letters in Applied Microbiology*, **48**, 398–401.

Karpov I. K., Chudenko K. V. and Kulik D. A. (1997). Modeling chemical mass transfer in geochemical processes: thermodynamic relations, conditions of equilibria, and numerical algorithms. *Am. J. Sci.*, **297**, 767–806.

Kloppmann W., Van Houtte E., Gaus I., Picot G., Vandenbohede A., Lebbe L., Guerrot C., Millot R. and Wintgens T. (2008). Monitoring reverse osmosis treated wastewater recharge into a coastal aquifer by environmental isotopes (B, Li, O, H). *Environ. Sci. Technol.*, **42**(23), 8759–8765.

Kloppmann W., Chikurel H., Picot G., Guttman J., Pettenati M., Aharoni A., Guerrot C., Millot R., Gaus I. and Wintgens T. (2009). B and Li isotopes as intrinsic tracers for injection tests in aquifer storage and recovery systems. *Applied Geochemistry (2009)*, doi: 10.1016/j.apgeochem.2009.03.006.

Kloppmann W., Thiery D., Kervevan C., Bourguignon A., Negral P. and Casanova J. (2001). Chemistry-transport coupled modelling of the aspo groundwater system (sweden) since the last glaciation. In: 10th International Symposium on Water-Rock Interaction, R. Cidu (ed.), A. Balkema Publishers, Villasimius, Italy.

Kremer S., Pavelic P., Dillon P. J. and Barry K. (2008). Flow and solute transport observations and modelling from the first phase of flushing operations at the salisbury astr site. CSIRO: Water for a Healthy Country National Research Flagship.

Laverick M. A., Wyn-Jones A. P., Kremer S., Pavelic P., Dillon P. J. and Carter M. J. (2004). Quantitative RT-PCR for the enumeration of norovirus (norwalk-like viruses) in water and sewage. *Letters in Applied Microbiology*, **39**, 127–136.

Lin H. C., Richards D. R., Yeh G. T., Cheng J. R., Cheng H. P. and Jones N. L. (1997). FEMWATER: a three-dimensional finite element computer model for simulating density-dependent flow and transport in variably saturated media. Technical Report CHL-97-12. Waterways Experiment station. U. S. Army Corps of Engineers, Vicksburg, MS 39180–6199, July 1997.

Lobry J., Flandrois J., Carret G. and Pave A. (1992). Monod's bacterial growth model revisited. *Bulletin of Mathematical Biology*, **54**, 117–122.

Löffler D., Römbke J., Meller M. and Ternes T. (2005). Environmental fate of pharmaceuticals in water/sediment systems. *Environ. Sci. Technol.*, **39**, 5209–5218.

Lowry C. S. and Anderson M. P. (2006). An assessment of aquifer storage recovery using ground water flow models. *Ground Water*, **44**, 661–667.

Mallén G., Maloszewski P., Flynn R., Rossi P., Engel M. and Seiler K. P. (2005). Determination of bacterial and viral transport parameters in a gravel aquifer assuming linear kinetic sorption and desorption. *Journal of Hydrology*, **306**, 21–36.

Maloszewski P. and Zuber A. (1985). On the theory of tracer experiments in fissured rocks with a porous matrix. *J. Hydrol.*, **79**, 333–358.

Masciopinto C. (2007). Biodegradation of wastewater nitrogen compounds in fractures: laboratory tests and fields observations. *Journal of Contaminant Hydrology*, **92**(1–4), 230–254.

Masciopinto C., La Mantia R. and Chrysikopoulos C. V. (2008). Fate and transport of pathogens in a fractured aquifer in the Salento area, Italy. *Water Resources Research*, **44**, W01404, doi: 10.1029/2006WR005643, 1–18.

Masciopinto C., La Mantia R., Carducci A., Casini B., Calvario A. and Jatta E. (2007). Unsafe tap water in households supplied from groundwater in the Salento region of Southern Italy. *Journal of Water and Health*, **5**(1), 129–148.

Masciopinto C., Benedini M., Troisi S. and Straface S. (2000). Conceptual models and field test results in porous and fractured media. In: Groundwater Pollution Control, K. L. Katsifarakis (ed.), WIT Press, pp. 245–292.

McDonald M. G. and Harbaugh A. W. (1988). A modular three-dimensional finite-difference ground-water flow model: U. S. geological survey techniques of water-resources investigations, Book 6, Chap. A1, 586.

Merritt M. L. (1985). Subsurface storage of freshwater in South Florida: a digital model analysis of recoverability. *U. S. Geological Survey Water-Supply Paper*, **2261**, 44pp.

Meschke J. S. and Sobsey M. D. (1998). Comparative adsorption of norwalk virus, poliovirus 1 and F+ RNA coliphage MS2 to soil suspended in treated wastewater, water. *Sci. Technol.*, **38**(12), 187–189.

Miller R., Corell R., Dillon P. and Kookana R. (2002). ASRRI: a predictive index of contaminant attenuation during aquifer storage and recovery. In: Management of Aquifer Recharge for Sustainability, P. J. Dillon (ed.), Proceedings of the 4th International Symposium on Artificial Recharge (ISAR4), Adelaide, Sept. 22–26, 2002, Swets & Zeitlinger, Lisse, ISBN. 90 5809 527 4, pp. 69–74.

Molz F. J., Boman G. K., Young S. C. and Waldrop W. R. (1994). Borehole flowmeters: field application and data analysis. *Journal of Hydrology*, **163**, no. 3–4, 347–371.

Oren O., Gavrieli I., Burg A., Guttman J. and Lazar B. (2007). Manganese mobilization and enrichment during soil aquifer treatment (SAT) of effluents, the dan region sewage reclamation project (Shafdan). *Israel. Env. Sci. & Tech.*, **41**, 766–772.

Parkhurst D. L. and Appelo C. A. J. (1999). User's Guide to PHREEQC (Version 2): A Computer Program for Speciation, Batch-reaction, One-dimensional Transport, and Inverse Geochemical Calculations.

Pavelic P., Dillon P. and Robinson N. (2004). Groundwater modelling to assist well-field design and operation for the ASTR trial at Salisbury, South Australia. CSIRO Land and Water Technical Report 27/04.

Pavelic P., Dillon P. and Robinson N. (2005). Modelling of well-field design and operation for an aquifer storage transfer and recovery (ASTR) trial. In: Recharge Systems for Protecting and Enhancing Groundwater Resources. Proceedings of the 5th International Symposium on Management of Aquifer Recharge (ISMAR5), Berlin, Germany, 11–16 June 2005. IHP-VI Series on Groundwater No. 13.

Pavelic P., Dillon P. J. and Simmons C. T. (2006). Multiscale characterization of a heterogeneous aquifer using an ASR operation. *Ground Water*, **44**, 155–164.

Pavelic P., Dillon P. J., Barry K. E., Vanderzalm J. L., Correll R. L. and Rinck-Pfeiffer S. M. (2007). Water quality effects on clogging rates during reclaimed water ASR in a carbonate aquifer. *J. Hydrol.*, **334**, 1–16.

Pavelic P., Dillon P., Martin R., Traegar B. and Simmons C. (2001). Multi-scale permeability characterisation of a confined carbonate aquifer targeted for aquifer storage and recovery. Proc IAH XXXI Congress: "New Approaches to Characterising Groundwater Flow". K.-P. Seiler and S. Wohnlich, (eds), Swets, Zeitlinger Lisse ISBN 902 651 848 X, pp. 859–862.

Pavelic P., Dillon P. J. and Simmons C. T. (2002). Lumped parameter estimation of initial recovery efficiency during aquifer storage and recovery. In: Management of Aquifer Recharge for Sustainability, P. J. Dillon (ed.), Proceedings of the 4th International Symposium on Artificial Recharge (ISAR4), Adelaide, Sept. 22–26, 2002, Swets & Zeitlinger, Lisse, ISBN. 90 5809 527 4, pp. 285–290.

Pearson F. J., Balderer J. W., Loosli H. H., Lehmann B. E., Matter A., Peters T., Schmassmann H. and Gautschi A. (eds) (1991). Applied isotope hydrology – a case study in Northern Switzerland, studies in environmental. *Sciences*, **43**, Elsevier Science Publishers B. V., Amsterdam-London-New York-Tokyo.

Perez-Paricio A. and Carrera J. (1999). Validity and sensitivity analysis of a new comprehensive clogging model. In: F. Stauffer, W. Kinzelbach, K. Kovar and E. Hoehn (eds) ModelCARE 1999 Conference. Int Assoc Hydrological Sciences, Zurich, Switzerland.

Perez-Paricio A. (2001). Integrated Modelling of Clogging Processes in Artificial Groundwater Recharge. PhD thesis, Technical University of Catalonia (UPC).

Perez-Paricio A., Benet I., Saaltink M. W., Ayora C. and Carrera J. (1998). CLOG: A code to address the clogging of artificial recharge systems. In: International Symposium on Computer Methods in Porous Media Engineering, J. M. Crolet (ed.), Giens, France.

Prommer H., Barry D. A. and Zheng C. (2003). MODFLOW/MT3DMS-based reactive multicomponent transport modeling. *Ground Water*, **41**, 247–257.

Pruess K., Oldenburg C. and Moridis G. (1999). TOUGH2 User's Guide, Version 2.0 Lawrence Berkeley National Laboratory. Report LBNL-43134, Berkeley, CA, USA, 197 p.

Ptak T., Piepenbrink M. and Martac E. (2004). Tracer tests for the investigation of heterogeneous porous media and stochastic modelling of flow and transport – a review of some recent developments. *J. Hydrol.*, **294**, 122–163.

Quast K. W., Lansey K., Arnold R., Bassett R. L. and Rincon M. (2006). Boron isotopes as an artificial tracer. *Ground Water*, **44**, 453–466.

Redman J. A., Grant S. B., Olson T. M., Hardy M. E. and Estes M. K. (1997). Filtration of recombinant Norwalk virus particles and bacteriophage MS2 in quartz sand: importance of electrostatic interactions. *Environ. Sci. Technol.*, **31**, 3378–3383.

Regel R., Rinck-Pfeiffer S., Dillon P., Pavelic P., Page D., Purdie M., Pitman C., Naumann B. and Gerges N. (2006). The ASTR project – turning stormwater into water of a potable quality. Australian Water Association South Australian Branch Regional Conference, 21 July 2006, Victor Harbour.

Regli S., Rose J. B., Haas C. N. and Gerba C. P. (1991). Modelling the risk from Giardia and Viruses in drinking water. *Journal American Water Works Association*, **83**, 76–84.

Reinoso R. and Becares E. (2008). Environmental inactivation of Cryptosporidium parvum oocysts in waste stabilization ponds. *Microbial Ecology*, **56**, 585–592.

Richards L. A. (1931). Capillary conduction of liquids through porous mediums. *Physics*, **1**, 318–333.

Rinck-Pfeiffer S., Pitman C. and Dillon P. (2005). Stormwater ASR And ASTR (aquifer storage transfer and recovery) in practice and under investigation in South Australia. In: Recharge systems for protecting and enhancing groundwater resources. Proceedings of the 5th International Symposium on Management of Aquifer Recharge (ISMAR5), Berlin, Germany, 11–16 June 2005. IHP-VI Series on Groundwater No. 13.

Saaltink M. W., Ayora C., Stuyfzand P. J. and Timmer H. (2003). Analysis of a deep well recharge experiment by calibrating a reactive transport model with field data. *J. Contam. Hydrol.*, **65**, 1–18.

Sauty J. P. and Kinzelbach W. (1988). On the identification of the parameters of groundwater mass transport. In: *Groundwater Flow And Quality Modelling*, Custodio Emilio, Gurgui Antonio and Lobo Ferreira João Paulo (eds.), North Atlantic Treaty Organization. Scientific Affairs Division. Springer, 1988, pp. 843.

Scheytt T., Mersmann P., Lindstädt R. and Heberer T. (2005). Determination of sorption coefficients of pharmaceutically active substances carbamazepine, diclofenac, and ibuprofen, in sandy sediments, chemosphere. **60**(2005), 245–253.

Schijven J. F. and Hassanizadeh S. M. (2000). Removal of viruses by soil passage: overview of modeling, processes, and parameters. *Critical Reviews in Environmental Science and Technology*, **30**, 49–127.

Schoenen D. (2002). Role of disinfection in suppressing the spread of pathogens with drinking water: possibilities and limitations. *Water Research*, **36**, 3874–3888.

Schmoll O., Howard G., Chilton J. and Chorus I. (2006). Protecting groundwater for health/managing the quality of drinking-water sources. IWA Publishing, UK.

Sedighi A., Klammler H., Brown C. and Hatfield K. (2006). A semi-analytical model for predicting water quality from an aquifer storage and recovery system. *J. Hydrol.*, **329**, 403–412.

Seiler K.-P., Maloszewski P. and Behrens H. (1989). Hydrodynamic dispersion in karstified limestone and dolomites in Upper Jura of Franconian Alb, FRG. *J. Hydrol.*, **108**, 235–247.

Shammas M. I. (2008). The effectiveness of artificial recharge in combating seawater intrusion in Salalah coastal aquifer. *Oman. Environmental Geology*, **55**, 191–204.

Sim Y. and Chrysikopoulos C. V. (1995). Analytical models for one-dimensional virus transport in saturated porous media. *Water Resour. Res.*, **31**, 1429–1437. (Correction, Water Resour. Res, **32**(5), pp. 1473, (1996).

Sim Y. and Chrysikopoulos C. V. (1998). Three-dimensional analytical models for virus transport in saturated porous media. *Transport in Porous Media*, **30**, 87–112.

Stumm W. and Morgan J. J. (1981). Aquatic chemistry. Wiley/Interscience, New York, 780p.

Stuyfzand P. J. (1998). Quality changes upon injection into anoxic aquifers in the netherlands: evaluation of 11 experiments. In: Artificial Recharge of Groundwater, Proc. 3rd Intern. Symp. on Artificial Recharge, J. H. Peters (ed.), Amsterdam, The Netherlands, Balkema, pp. 283–291.

Prommer H. and Stuyfzand P. J. (2005). Identification of temperature-dependent water quality changes during a deep well injection experiment in a pyritic. *Aquifer. Environ. Sci. Technol.*, **39**, 2200–2209.

Swierc J., Page D., van Leeuwen J. and Dillon P. (2005). Preliminary Hazard Analysis and Critical Control Points Plan (HACCP) – Salisbury Stormwater to Drinking Water Aquifer Storage Transfer and Recovery (ASTR) Project, CSIRO Land and Water Technical Report 20/05.

Taylor G. I. (1921). Diffusion by continuous movements. *Proc. London Math. Soc.*, **2**, 196–214.

Ternes T. A., Meisenheimer M., McDowell D., Sacher F., Brauch H.-J., Haist-Gulde B., Preuss G., Wilme U. and Zulei-Seibert N. (2008). Removal of pharmaceuticals during drinking water treatment. *Environ. Sci. Technol.*, **36**(17), 3855–3863.

Ternes T. A., Meisenheimer M., McDowell D., Sacher F., Brauch H. J., Gulde B. H., Preuss G., Wilme U. and Seibert N. Z. (2002). Removal of pharmaceuticals during drinking water treatment. *Environ. Sci. Technol.*, **36**, 3855–3863.

Thiéry D. (1993). Modélisation des aquifères complexes – prise en compte de la zone non saturée et de la salinité. *Calcul Des Intervalles De confiance. (In French.) Rev. Hydrogéol*, **1993**(4), 325–336.

Uddameri V. (2007). A dynamic programming model for optimal planning of aquifer storage and recovery facility operations. *Environmental Geology*, **51**, 953–962.

Van Breukelen B. M., Appelo C. A. J. and Olsthoorn T. N. (1998). Hydrogeochemical transport modeling of 24 years of Rhine water infiltration in the dunes of the Amsterdam water. *Supply. J. Hydrol.*, **209**, 281–296.

Van Cappellen P. and Wang Y. F. (1996). Cycling of iron and manganese in surface sediments: a general theory for the coupled transport and reaction of carbon, oxygen, nitrogen, sulfur, iron, and manganese. *American Journal of Science*, **296**(3), 197–243.

Van der Lee J. (1998). Thermodynamic, mathematical concepts of CHESS. Tech. Rep. LHM/RD/98/39, CIG-ÉCole Des Mines De Paris, Fontainebleau.

Van Houtte E. and Verbauwhede J. (2005). Artificial recharge of treated wastewater effluent enables sustainable groundwater management of a dune aquifer in Flanders, Belgium. ISMAR 5 conference, Berlin.

Vandenbohede A., Houtte E. and Lebbe L. (2008). Groundwater flow in the vicinity of two artificial recharge ponds in the Belgian coastal dunes. *Hydrogeol. J.*, **16**, 1669–1681.

Vanderborght J., Kasteel R., Herbst M., Javaux M., Thiery D., Vanclooster M., Mouvet C. and Vereecken H. (2005). A test of numerical models for simulating flow and transport in soils using analytical solutions. *Vadose Zone J.*, **4**, 206–221.

Varni M. and Carrera J. (1998). Simulation of groundwater age distribution. *Water Resources Research*, **34**(12), 3271–3281.

Wallis I., Prommer H., Simmons C. T., Post V. and Stuyfzand P. J. (2010). Evaluation of conceptual and numerical models for arsenic mobilization and attenuation during managed aquifer. *Recharge. Environ. Sci. Technol.*, **44**, 5035–5041.

Wang D., Gerba C. P. and Lance (1981). Effect of soil permeability on virus removal through soil columns. *Appl. Environ. Microb.*, **42**, 83–88.

Ward J. D., Simmons C. T. and Dillon P. J. (2007). A theoretical analysis of mixed convection in aquifer storage and recovery: How important are density effects? *J. Hydrol.*, **343**, 169–186.

Ward J. D., Simmons C. T. and Dillon P. J. (2008). Variable-density modelling of multiple-cycle aquifer storage and recovery (ASR): importance of anisotropy and layered heterogeneity in brackish aquifers. *J. Hydrol.*, **356**, 93–105.

Ward J. D., Simmons C. T., Dillon P. J. and Pavelic P. (2009). Integrated assessment of lateral flow, density effects and dispersion in aquifer storage and recovery. *J. Hydrol*, **370**, 83–99.

Wexler E. J. (1992) Analytical solutions for one-, two-, and three-dimensional solute transport in ground-water flow systems with uniform flow, techniques of water-resources investigations of the united states geological survey, book 3: applications of hydraulics. Chapter B7. United States Geological Survey, Denver, Colorado.

Williams C. F. and Adamsen F. J. (2006,). Sorption-esorption of carbamazepine from irrigated soils. *J. Environ. Qual.*, **35**, 1779–1783.

Williams A. (2000). Using an aquifer storage and recovery (ASR) trial as a large scale tracer test. In: International Conference on Tracers and Modelling in Hydrogeology (TraM 2000), A. Dassargues (ed.), Int Assoc Hydrological Sciences, Liege, Belgium.

Wolery T. (1992). EQ3NR: a computer program for geochemical aqueous speciation–solubility calculations. Rep. UCRL-MA-110662, Lawrence Livermore National Lab., CA.

Wood B. D. (1998). A connection between the lagrangian stochastic-convective and cumulant expansion approaches for describing solute transport in heterogeneous porous media. *Advances in Water Resources*, **22**, 319–332.

Xu T., Sonnenthal E., Spycher N. and Pruess K. (2004). TOUGHREACT user's guide: a simulation program for nonisothermal multiphase reactive geochemical transport in variably saturated geologic media. Lawrence Berkeley National Laboratory. Berkeley, CA. (LBNL-55460).

Yang W. C., Liu W. P. and Green R. (2005). Degradation of N-nitrosodimethylamine (NDMA) in landscape soils. *J. Environ. Qual.*, **34**, 336–341.

Yao K.-M., Habibian M. T. and O'Melia C. R. (1971). Water and waste water filtration: concepts and applications. *Environmental Science and Technology*, **5**, 1105–1112.

Yates M. V. and Jury W. A. (1995). On the use of virus transport modelling for determining regulatory compliance. *J. Environ. Qual.*, **24**, 1051–1055.

Yeo I. W. and Lee K. K. (2003). Analytical solution for arbitrarily located multiwells in an anisotropic homogeneous confined aquifer. *Water Resour. Res.*, **39**(5), 1133.

Yobbi D. K. (1996). Simulation of subsurface storage and recovery of treated effluent injected in a saline aquifer, St. Petersburg, Florida. In: Water-Resources investigations report 95–4271, USGS (ed.), U. S. GEOLOGICAL SURVEY, Tallahassee, Florida.

Zheng C. and Bennett G. D. (2002). Applied contaminant transport modeling. 2nd edn, Wiley-Interscience, New York, New York.

Chapter 19

Risk assessment and risk management in Managed Aquifer Recharge

Declan Page, Maria Neus Ayuso-Gabella, Irena Kopač, Davide Bixio, Peter Dillon, Miquel Salgot de Marçay and Bettine Genthe

This chapter presents the methodologies used for risk assessment (RA) and risk management (RM) in MAR in Australia and within the European Union, qualitative and quantitative approaches adopted within the RECLAIM Water project and case studies where the outcomes of these approaches are presented.

19.1 METHODOLOGIES FOR RISK ASSESSMENT AND MANAGEMENT

Risks to human health and the environment are dealt with in each country based on the particular legislative environment. General methodologies for risk assessment in Australia and in the European Union associated with MAR are summarised in this section.

19.1.1 European Union

While there are no provisions in European Union (EU) wide legislation focussed explicitly on RA and RM for MAR, there are many pieces of legislation and policy affecting them. Key Union wide regulatory elements having a strong impact on RA/RM for MAR include the Water Framework Directive 2000/60/EC (WFD), the Ground Water Daughter Directive 2006/18/EC (GWD), the Environmental Impact Assessment Directive 85/337/EEC (EIAD), the Nitrate Directive 91/676/EEC (ND) and for water to be intended for human consumption, the Drinking Water Directive 98/83/EC (DWD).

The WFD reinforces the well established requirement of taking into account the precautionary principle, relying in particular on the determination of any potentially adverse effects of the reclaimed water and on a scientific assessment of the risk. It recognises that preventative measures or treatment shall have to be employed in each case consistent with the perceived level of risk. Whilst specifying preventative measures to be employed, the WFD consents also case-by-case assessment where field data or model ecosystems would allow more precise preventative measures to be calculated and applied.

Article 16 (2) of the WFD prescribes that risk assessment for chemical contaminants should be performed under Council Regulation (EEC) No 793/93, Council Directive 91/414/EEC (both repealed by Regulation (EC) 1907/2006), and Directive 98/8/EC, or targeted risk-based assessment focusing solely on aquatic eco-toxicity and on human toxicity via the aquatic environment. This simplified risk-based assessment procedure based on scientific principles should take particular account of:

- Evidence regarding the intrinsic hazard of the substance concerned, and in particular its aquatic eco-toxicity and human toxicity via aquatic exposure routes, and
- Evidence from monitoring of widespread environmental contamination, and
- Other proven factors which may indicate the possibility of widespread environmental contamination, such as production or use volume of the substance concerned, and use patterns.

Other specific actions for the protection of the groundwater resources connected to the WFD are included in the ND and in the GWD. In particular, Annex I of the GWD prescribes that the results of the application of the quality standards for pesticides in the manner specified for the purposes of the GWD will be without prejudice to the results of the risk assessment procedures required by the above mentioned Directive 91/414/EEC (repealed by Regulation (EC) 1907/2006) or Directive 98/8/EC.

If recovered water is intended for use as a drinking water supply, then the main legal instruments for the control of drinking water quality is the DWD. The existing DWD focuses on compliance monitoring of the final product. In light of the precautionary principle, however, parametric indicators set out in the DWD are not sufficient to say that recovered water from MAR sites is safe. The water supplier is requested to perform analyses on the chemicals and microorganisms not covered by the limits but that can be present in the water and that can pose a risk to public health. Reclaimed water can possibly contain microbiological and chemical contamination of a much larger number of compounds than the minimum set included in the DWD, which refers to pristine catchments. The starting point for risk assessment is provided by the Decision 2455/2001/CE of the European Parliament and of the Council establishing a first list of priority substances ubiquitously present in wastewater, based on a procedure for the definition of the priorities based on the principles of monitoring and modelling.

The European Commission is considering further development of preventive risk management to update/revise DWD. Areas of intervention should include the introduction of the principles of RA/RM, and of the basis of those principles, revision of microbiological and chemical parameters with an increasing decision making role of the single water utility. Several EU member states and federal regions have already adopted their own standards and regulations in this sense, while others are reportedly waiting for revision of the DWD to introduce such concepts. For example, in France, Article 18 of the Decree 2001-1220 shifts emphasis in national drinking water legislation towards a preventative quality assurance approach that encompasses RA and RM strategies. RA/RM principles and methods used in drinking water draw on other methods, particularly on Hazard Analysis and Critical Control Points (HACCP). As the HACCP approach is compulsory in the food sector, in EU member states where water utilities fall under the provisions of Food Safety Acts, such as for instance in Lithuania and in Austria, development and implementation of a HACCP is already an obligation.

Similarly, in Spain, the central water authority is evaluating the implementation of WHO style water safety plans (WSPs), see also the section below specific to Spain. The WSP can be defined as the adaptation of HACCP to the specific requirements of the drinking water sector. Although some MAR projects are under way in Spain, with experience being gained through pilot studies, a specific regulation is still lacking in the country. Every MAR project is regarded as a different case, and different authorisations are required, depending on the environmental matrices involved.

In the new hydrological plan developed by the Spanish Environment Department, MAR is included in only those using the riverbed of main rivers to recharge, and only two, the Llobregat River and Pla de Palma, use a recycled water source. However, there are many more pilot studies using reclaimed water as a source. In most of the cases, MAR is considered as a treated wastewater reuse case, thus it has to attain the Royal Decree concerning water reuse (R. D. 1620/2007 of 7th September 2007). This Royal Decree defines the quality that the reclaimed water must have depending on its final reuse purpose, considering MAR as one of them. Besides, and in view of the lack of regulations, a groundwater law is being prepared by the Spanish Environmental Department, which will also consider MAR.

The Spanish case study involved in the RECLAIM Water project, Sabadell, uses a riverbed as a means of recharging the underlying alluvial aquifer. To date, no authorisation procedure for the recharge of the aquifer is required, as the system is not interpreted as a "pure" artificial recharge scheme by the administration. It is considered that the WWTP discharges the effluent in the Ripoll River, hence contributing to the preservation of its ecological flow, and as such is not constrained to discharge regulations. It is not considered as an indirect artificial recharge site but specifically as an "ecological river flow maintenance" site. For this type of reuse application, the water quality or sampling routine required have not been specifically defined yet, the decree only indicates that "the minimum quality required will be studied case by case".

In the UK, impetus for WSP implementation was given by the regulator stating that drinking-water improvement schemes for the next five year investment programme would only receive regulator support if they were identified through WSP methodology. Compliance monitoring was initially viewed as the main WSP verification stage. However, additionally from the beginning of 2008, the hazard identification and risk assessment elements of the WSP framework were made regulatory requirements and WSPs began to feature in the regulator's audit programme. Drinking-water improvement programmes from 2009 onward must be identified through the water safety plan approach.

Furthermore, other incentives for implementation of preventive risk management are the provisions of the European Council Directive 2004/35/CE on Environmental Liability. This Directive prescribes that it is incumbent on the operator to demonstrate that he is not negligent/at fault. Another strong incentive to operate with state of the art risk management procedures is the requirement for the development of financial security instruments and markets.

HACCP methodology

HACCP is a tool which evaluates the hazards and establishes control systems which focuses on prevention rather than on the final product. The tool can be applied to the whole water treatment train and to the environment where water is disposed.

Initially, seven basic principles of the HACCP system were stated, namely (Codex Alimentarius, 1997):

(1) Hazard analysis and determination of preventive measures – conduct a hazard analysis by indentifying each hazard, assess the likelihood of occurrence and severity and identify the preventive control measures in place to control that hazard. Hazards are usually described as any biological/microbiological, chemical or physical agent that may contaminate the product (in our case groundwater).
(2) Identification of the Critical Control Points (CCP) – determine the Critical Control Points (CCPs) in the process and operational procedures that can be controlled to minimise the likelihood of each hazard occurring.
(3) Determination of the critical limits for every CCP – establish critical limit(s) for each CCP which must be met to ensure that the CCP is under control.
(4) Monitoring of the CCP – establish a system to monitor the control of the CCP, which is a planned sequence of events, observations, measurements and records needed for assessment whether a CCP is under control.
(5) Corrective measures – implements the corrective action to be taken in case of monitoring indicating that a particular CCP is not under control.
(6) Verification/validation – establish procedures for verification, including supplementary tests to ensure that the HACCP system is working effectively.
(7) Registers – establish documentation concerning all relevant procedures and records to meet these principles and their application.

The seven principles are necessary elements for developing a self-controlled procedure based on the HACCP system in order to ensure the control of the risks which are significant for the safety of the reclaimed water. The water-related authorities – like other authorities – adopt the concept that self-control is a necessity in order to not oversize the bodies of the administration and to incorporate the stakeholders in the important task to ensure public health and success of any environmental or agrofood project. Additional examples of adaption of the HACCP methodology are given the following section on Australia.

19.1.2 Australia

The Australian Guidelines for Water Recycling (Phase 2): Managed Aquifer Recharge (called the MAR guidelines) are the foundational document which describe the approach adopted to risk assessment and management for MAR in Australia. The MAR guidelines form an integral part of the National Water Quality Management Strategy (NWQMS), upon which contemporary water resource management in Australia is based. The MAR guidelines build on the policies and principles of the strategy, and on other key NWQMS guidelines, such as drinking water. The MAR guidelines include assessment of the aquifer which is what sets them apart from the other water recycling examples in the Phase 1 guidelines (NHMRC–EPHC–AHMC, 2006).

Although the subsurface component provides water storage and treatment functions, it may add hazards to stored water and create other environmental problems. The Australian MAR guidelines aim to provide a sound and consistent basis for protecting human health and the environment for MAR operations. The guidelines focus on the protection of aquifers and the quality of recovered water in managed aquifer recharge projects. Where managed aquifer recharge is part of water recycling projects, the MAR guidelines are used in conjunction with the Australian Guidelines for Water Recycling (Phase 1) (NHMRC– EPHC–AHMC, 2006). If stormwater is the source of the water to be recharged, then Australian Guidelines for Water Recycling: Stormwater Harvesting and Reuse (NRMMC-EPHC–NHMRC, 2009a) should be used. If recovered water is intended for use as a drinking water supply, then Australian Guidelines for Water Recycling: Augmentation of Drinking Water should also be used.

Australian approach to risk management in MAR

In Australia, the risks to human health and the environment, including the receiving aquifer, are managed through the development of a risk management plan for MAR. This plan involves the 12 fundamental elements (developed on the 7 HACCP principals) adopted in Phase 1 of Australia's guidelines for water recycling (NRMMC–EPHC–AHMC 2006); these elements are shown in Figure 19.1.

The 12 elements of the framework presented in the Phase 1 Australian guidelines for Water Recycling apply as much to MAR as to other applications of recycled or drinking water management. The 12 elements are not sequential; they should all be followed to ensure that the risk management plan is comprehensive. The resulting 'managed aquifer recharge risk management plan' is a documented system for the management of aquifer recharge. The central philosophy of the MAR guidelines is that it is better to prevent hazardous events from occurring than to clean up their effects afterwards. The multiple barrier approach is a key concept in the management of risks in aquifer recharge. This same approach is well established as a means of protecting drinking water quality in Australia (NHMRC–NRMMC, 2004), and internationally by the WSP approach described above (WHO, 2006). The application of the multiple barrier approach through the MAR risk management plan should encompass every component of the MAR system.

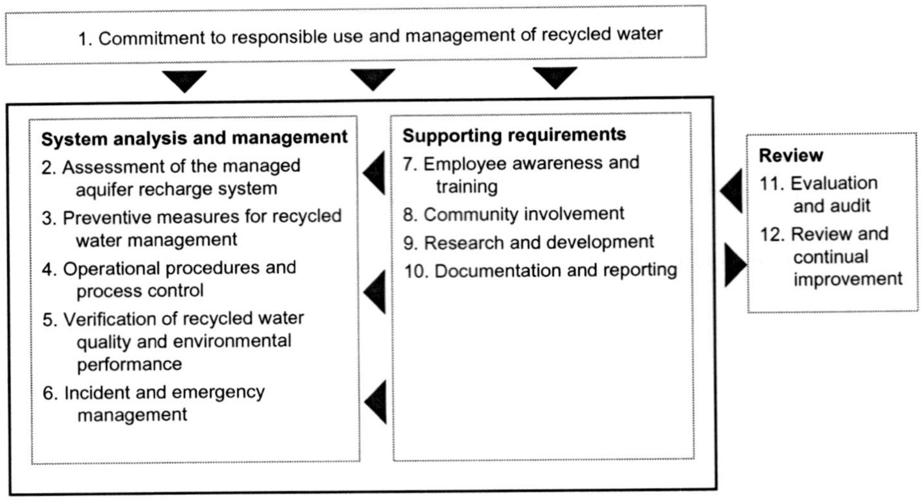

Figure 19.1. Elements of the framework used in Australia for managing water quality and use

Australian approach to stage project development in Managed Aquifer Recharge

MAR risk assessments are interspersed between the various project development stages to address catchment and groundwater plans as well as local government requirements. Figure 19.2 shows a series of risk assessments that are designed to ensure protection of human health and the environment, as in the Phase 1 guidelines (NRMMC–EPHC–AHMC 2006). These risk assessments allow staged decision points for investment, based on an informed understanding of the next required level of investigations.

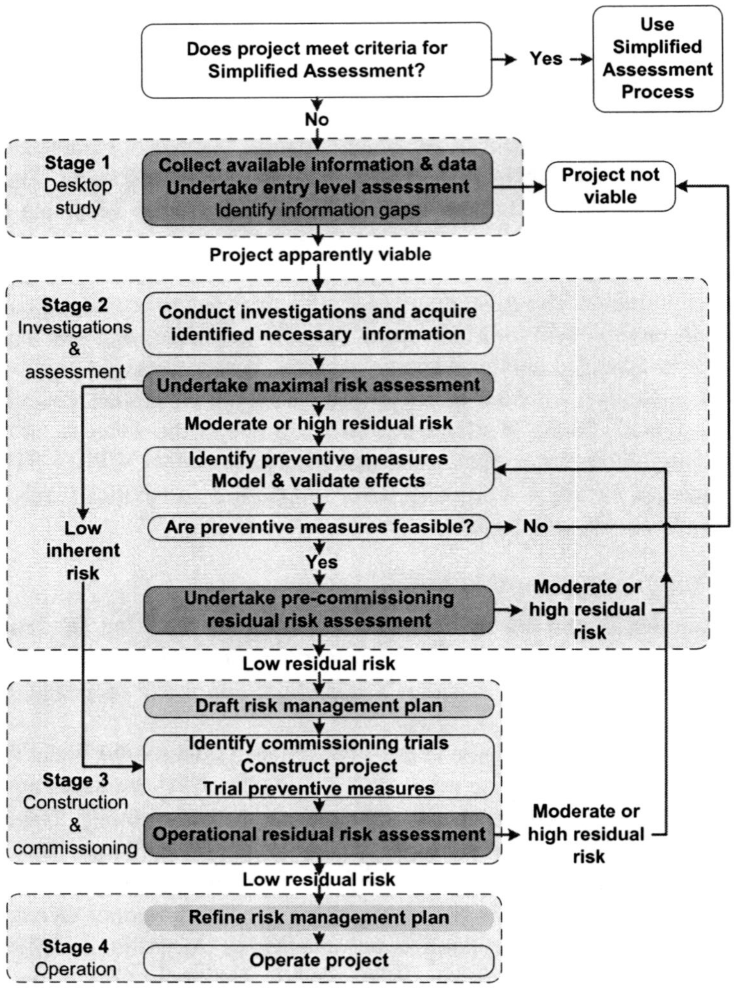

Figure 19.2. Risk assessment stages in managed aquifer recharge project development

New MAR projects begin with the Stage 1 entry level risk assessment which largely address water allocation issues that are usually adequately determined without detailed site-specific information. Governance of these issues will generally be in the hands of a state or regional water resources management agency. The entry level assessment gives an indication of the effort likely to be required to demonstrate low risks to human health and the environment. The Stage 1 entry level assessment is intended to inform on the likely degree of difficulty of the MAR project, and hence inform proponents of the extent of field investigations needed in Stage 2.

Stage 1 is the most cost-effective stage at which to abandon MAR projects for which the potential rewards do not justify the high degree of difficulty. If the potential value of recycled water generated is large, an investment in Stage 2 investigations can focus on the highest risk issues. Causes of the high degree of difficulty may be resolved with determination of viable preventive measures, such as pre- or post-MAR treatment.

At Stage 2, risk is assessed at two levels – maximal risk and residual risk. The maximal risk (also referred to as unmitigated or inherent risk) is risk in the absence of any preventive measures. Whereas, the residual risk is the risk after consideration of preventive measures, including potential aquifer treatment. A residual risk assessment provides an indication of the safety and sustainability of the MAR scheme and must be less than the upper limits of tolerable risk.

Following investigations in Stage 2 maximal risk is determined for each hazard. If the responsible authority in the jurisdiction assesses the maximal risk to be low for all hazards the MAR project may then proceed directly to construction. The more usual case will require preventive measures related to some hazards to reduce risk and reassessment of residual risk at pre-commissioning stage. This step estimates the residual risk of commissioning the project. Preventive measures, operational procedures and incident and emergency management plans are intended to give confidence that the project will be safe during commissioning trials (Stage 3). If residual risks fail to reach acceptance criteria, preventive measures are added and residual risks reassessed until residual risks are determined to be low, or the project proponent determines that the expense of these measures makes the project unviable.

The risks for each MAR project will depend on the quality of the source of water, the intended uses of recovered water and the environmental values of the aquifer. While all projects follow the same risk assessment pathway, the level of effort required in risk assessment and management can vary markedly between projects based on the specific risk profile of the project. For example, MAR projects intending to produce drinking water supplies (e.g. the Parafield case study site) will in general require substantially more effort than those producing irrigation supplies. For many MAR projects, the level of some risks can only be approximated before full-scale implementation and validation monitoring.

Following construction of the MAR scheme, commissioning trials are run to enable validation of processes that could not be measured until recharge occurs, and to allow verification of the efficacy of the preventive measures. At this stage, Stage 3 (Figure 19.2), an accurate calculation of residual risk, an operational residual risk assessment, can be made. A low residual risk assessed at Stage 3 provides a basis for ongoing operation of the site and development of risk management plans (including verification and operational monitoring and reporting) (Stage 4). The risk management plans may be subject to periodic review subject to monitoring results. In the event that the forecast low risks are not achieved, the proponent needs to identify and adopt additional preventive measures and perform further commissioning trials to proceed with the project.

Examples of the application of this risk assessment methodology to Australian MAR sites has been documented in a recent studies by Page *et al.* (2010a; 2010d).

19.2 CHEMICAL RISK ASSESSMENT METHODOLOGY

Many chemical hazards (both inorganic and organic) found in wastewater have human health guideline values which indicate the acceptable level if found in a drinking water supply. The term ‘trace organics’ more specifically refers to a range of emerging chemicals, such as: pharmaceuticals, endocrine disrupters, disinfection by-products and flame retardants. These emerging contaminants pose a challenge to public health regulation as in many cases there is no toxicological data or guideline value from which to derive the potential human health risk.

A chemical risk assessment was conducted on the case study sites to assess which organic chemicals had potential to be a human health risk. The approach selected in the framework of the RECLAIM WATER project was based on the USEPA guidelines for chemical risk assessment (USEPA, 1987, 1998, 2002), also adopted by the Australian Guidelines for Water Recycling (NRMMC EPHC & NHMRC, 2008). A preliminary work applying these concepts appears in Ayuso-Gabella *et al.* (2007). The methodology is based on comparing the amount of a certain chemical compound with a reference value or daily dose intake reported within the USEPA guidelines (USEPA, 1987, 1998, 2002), WHO drinking water guidelines (WHO, 2004) or other sources of data. Lower exposure to recycled water through irrigation (e.g. at the Nardò, Sabadell and Shafdan case study sites) is unlikely to result in any significant risk to human health from chemicals; however, the risk quotient approach using drinking water guidelines allows comparison of risk associated with different chemicals.

In the present section, a systematic desktop assessment of potential adverse health effects of organic chemicals at the concentrations observed at the case study sites assuming that recovered water were available for drinking is presented. The assessment requires the quantification of trace organic chemical. Full details on the methodology are shown in Ayuso-Gabella *et al.* (2009).

The risk quotients (RQ) method is the most widely used method of assessing risk from trace organic chemicals by comparing the measured chemical concentrations as a ratio for guideline values. Health values are concentrations below which no adverse health effects are expected if the water is consumed over a lifetime. The health values are calculated assuming an average daily intake of 2 litres of water for an individual with a 70 kg body weight over 70 years of water consumption. All values were calculated using the equations used in the Australian Guidelines for Water Recycling phase 2: Augmentation of Drinking Water Supplies (NRMMC EPHC & NHMRC, 2008).

19.3 CHEMICAL RISK ASSESSMENT OF THE CASE STUDY SITES

19.3.1 Source waters

A selection of organic chemicals were analysed for at the case study site source waters. The list of chemicals tested for is detailed in the risk assessment work package (Ayuso-Gabella *et al.* 2009) and also in the work package devoted to the detection of organic chemicals (Ernst *et al.* 2009). The Australian case study site analysed a larger set of organic chemicals that were more likely to be found in stormwater such as herbicides and petroleum products. The results of the chemicals assessed and the risks to human health are detailed below.

Of the organic chemicals tested for in the source waters, most were detected in at least one of the seven case study sites. The total number of chemicals analysed for each case study site source waters and the percentage of detections at each site is given in Table 19.1. The list of chemical groups analysed for includes a broad range of chemical groups with different physico-chemical characteristics and toxic effects (see Table 19.2). Table 19.2 summarizes the number of samples, median and max results for each compound analyzed in source waters at each site, as well as the RQ value for each compound (using the median value to calculate it). The compounds that were not detected in any of the case studies do not appear in Table 19.2 and include the following: NDPhA (N-Nitrosodiphenylamine), NMEA (N-Nitrosomethylethylamine), florfenicole, sulfadiazine, sulfadimethoxine, sulfadoxine, sulfamerazine, sulfisoxazole, tiamulin, ioxitalamic acid, ethinylestradiol EE2, mefenamic acid, methadon, morphine, nordiazepam, oxazepam and paracetamol.

Table 19.1 Total number of chemicals analysed for each case study site in the source waters and the percentage of detections.

Compound detection in source water at each site	Atlantis		Gaobeidian		Nardò		Parafield		Sabadell		Shafdan		Torreele	
	n	Detected (%)	n	Detected (%)	n	Detected (%)	n	Detected (%)	n	Detected (%)	n	Detected (%)	n	Detected (%)
DBPs														
THMS							11	9						
HAAs							*							
N-nitrosamines					9	50	*		5	40	9	72	7	62
Pharmaceuticals														
Antibiotics	5	80	6	73	11	55	2	0	9	50	6	79	13	29
Contrast media	5	30	5	80	6	56	2	0	5	73	5	95	5	60
Hormones					1	0	2	0	3	33	3	33	3	33
Other	14	50	4	47	7	50	2	0	7	69	4	100	6	86
VOC's							11	0						
PAH's							11	0						
Dioxin and Furans							11	0						
PCB's							11	0						
Complexing agents			3	100	3	33	11	0	3	100	3	100	3	56
Pesticides and herbicides	1	0	1	75	1	0	11	63	1	100	1	0	1	0
Other chemicals					1	50	11	45	1	0			1	100

*monitored as disinfection by product formation potential.

Table 19.2 Endocrine disruptor and personal care products chemical risk quotients in the case study source waters.

	Units	LOQ	Atlantis				Gaobeidian				Nardò				Parafield				Sabadell				Shafdan				Torreele				Hth. v.	Reference
			n	med	max	RQ med	n	med	max	RQ med	n	med	max	RQ med	n	med	max	RQ med	n	med	max	RQ med	n	med	max	RQ med	n	med	max	RQ med		
DBPs																																
N-nitrosamines																																
N-Nitrosodibutylamine (NDBA)	ng/L	3									2	<LOQ	<LOQ	–					1	<LOQ	–	–	4	8.2	45	**1.4**	1	<LOQ	–	–	6	USEPA
N-Nitrosodiethylamine (NDEA)	ng/L	0.3									2	4.1	4.6	0.41					1	<LOQ	–	–	4	13	32	**1.3**	2	0.81	1.3	0.08	10	AGADWS
N-Nitrosodimethylamine (NDMA)	ng/L	0.7									2	3.5	4.7	0.35					1	17	–	**1.7**	4	9.8	53	0.98	3	5.8	6.3	0.58	10	AGADWS
N-Nitrosodipropylamine (NDPA)	ng/L	1.1									1	<LOQ	–	–									4	1.2	2.4	0.24	1	<LOQ	–	–	5	USEPA
N-Nitrosomorpholine (NMOR)	ng/L	1.3									2	11	20	**11**					1	9.6	–	**9.6**	4	11	53	**11**	2	3.5	3.8	**3.5**	1	AGADWS
N-Nitrosopiperidine (NPIP)	ng/L	3.2									2	<LOQ	<LOQ	–					1	<LOQ	–	–	4	3.2	6.9	0.86	2	1.1	1.4	0.29	3.7	OEHHA
N-Nitrosopyrrolidine (NPYR)	ng/L	3									2	<LOQ	<LOQ	–									3	2.4	3.7	0.12	2	<LOQ	<LOQ	–	20	USEPA
Pharmaceuticals																																
Azithromycin	µg/L	0.01									1	0.32	–	0.08																	3.9	AGADWS
Clarithromycin	µg/L	0.02	2	0.03	0.04	0.00	5	0.02	0.06	0.00	4	0.15	0.31	0.00					2	0.21	0.22	0.00	4	0.03	0.13	0.00	3	0.05	0.23	0.00	250	AGADWS
Erythromycin	µg/L	0.01	2	0.18	0.23	0.01	5	0.02	0.06	0.00	4	0.02	0.05	0.00					2	0.35	0.46	0.02	4	0.16	0.22	0.01	3	<LOQ	<LOQ	–	17.5	AGADWS
N-Acetilsulfamethoxazole	µg/L	0.01									4	0.07	0.09	0.00					2	0.08	0.08	0.00					1	0.06	–	0.00	35	AGADWS
Roxythromycin	µg/L	0.02	2	<LOQ	<LOQ	–	5	0.05	0.19	0.00	4	0.02	0.03	0.00					2	<LOQ	<LOQ	–	4	0.11	0.23	0.00	3	<LOQ	<LOQ	–	150	AGADWS
Sulfamethoxazole	µg/L	0.01	2	3.4	3.7	0.10	5	0.45	0.65	0.01	4	0.10	0.48	0.00					2	0.32	0.47	0.01	4	0.15	0.73	0.00	3	0.05	0.18	0.00	35	AGADWS
Sulfamethazine	µg/L	0.01					5	0.04	0.14	0.00	4	<LOQ	<LOQ	–					2	<LOQ	<LOQ	–	4	0.01	0.01	0.00	3	<LOQ	<LOQ	–	35	AGADWS
Sulfapyridine	µg/L	0.01									1	0.06	–	0.00																	35	AGADWS
Trimethoprim	µg/L	0.01	2	0.76	0.94	0.01	5	0.01	0.03	0.00	2	<LOQ	<LOQ	–					3	0.01	0.26	0.00	4	0.05	0.07	0.00	3	0.11	0.15	0.00	70	AGADWS
Contrast media																																
Diatrizoate	µg/L	0.01	2	<LOQ	<LOQ	–	4	1.2	1.8	0.00	4	1.0	1.7	0.00					3	11	28	0.02	4	0.31	0.84	0.00	3	3.0	6.9	0.00	600	LTD
Iohexol	µg/L	0.02	2	<LOQ	<LOQ	–	4	1.7	2.2	0.00	2	<LOQ	<LOQ	–					3	0.02	0.64	0.00	4	4.0	4.7	0.01	3	<LOQ	<LOQ	–	720	AGADWS
Iomeprol	µg/L	0.02	2	<LOQ	<LOQ	–	4	<LOQ	<LOQ	–	4	0.56	1.1	0.00					3	0.22	0.34	0.00	4	0.15	0.30	0.00	3	0.60	1.4	0.00	375	LTD
Iopamidol	µg/L	0.02	2	0.84	1.7	0.00	4	3.7	6.5	0.01	3	0.05	0.87	0.00					3	0.25	0.29	0.00	4	5.1	7.7	0.01	3	<LOQ	<LOQ	–	400	AGADWS
Iopromide	µg/L	0.02	2	0.09	0.15	0.00	4	1.0	1.6	0.00	4	<LOQ	<LOQ	–					3	6.6	10.1	0.01	4	6.8	7.8	0.01	3	0.38	3.8	0.00	750	AGADWS
Hormones																																
Estradiol E2	ng/L	1																	1	<LOQ	–	–	4	<LOQ	<LOQ	–	3	1.0	1.9	0.01	175	AGADWS
Estrone E1	ng/L	0.5									2	<LOQ	<LOQ	–					1	64	–	**2.1**	4	<LOQ	<LOQ	–	3	4.5	6.4	0.15	30	AGADWS
Other																																
Benzoylegconine	µg/L	0.02	2	0.10	0.16	0.01																									20	LTD
Bezafibrate	µg/L	0.01	2	<LOQ	<LOQ	–	4	<LOQ	<LOQ	–	4	<LOQ	<LOQ	–					3	0.06	0.13	0.00	4	0.68	0.80	0.00	2	<LOQ	<LOQ	–	300	AGADWS

(Continued)

Table 19.2 Endocrine disruptor and personal care products chemical risk quotients in the case study source waters (*Continued*).

	Units	LOQ	Atlantis				Gaobeidian				Nardò				Parafield				Sabadell				Shafdan				Torreele				Hth. v.	Reference
			n	med	max	RQ med	n	med	max	RQ med	n	med	max	RQ med	n	med	max	RQ med	n	med	max	RQ med	n	med	max	RQ med	n	med	max	RQ med		
Carbamazepine	µg/L	0.01	2	2.7	3.5	0.03					4	0.71	0.85	0.01					2	0.18	0.30	0.00					3	1.4	1.4	0.01	100	AGADWS
Clofibric acid	µg/L	0.01	2	<LOQ	<LOQ	–	4	0.04	0.05	0.00	4	<LOQ	<LOQ	–					3	0.02	0.03	0.00	4	<LOQ	<LOQ	–	3	<LOQ	<LOQ	–	750	AGADWS
Codein	µg/L	0.02	2	1.4	1.8	0.03																									50	AGADWS
Diazepam	µg/L	0.02	2	0.03	0.04	0.01																									2.5	AGADWS
Diclofenac	µg/L	0.01	2	0.90	0.92	0.50	3	0.20	0.22	0.11	4	0.28	0.35	0.16					2	0.16	0.19	0.09	4	0.37	0.80	0.21	2	0.59	0.79	0.33	1.8	AGADWS
Ibuprofen	µg/L	0.02	2	9.6	11	0.02	4	0.04	0.21	0.00	4	<LOQ	<LOQ	–					3	<LOQ	<LOQ	–	3	0.31	0.41	0.00	2	0.21	0.26	0.00	400	AGADWS
Naproxen	µg/L	0.01	2	<LOQ	<LOQ	–	4	0.07	0.09	0.00	4	0.04	0.06	0.00					3	0.04	0.57	0.00	4	0.5	0.66	0.00	2	0.46	0.70	0.00	220	AGADWS
Primidone	µg/L	0.01	2	<LOQ	<LOQ	–					3	0.16	0.20	0.00					2	0.18	0.24	0.00					3	0.14	0.21	0.00	125	LTD
Temazepam	µg/L	0.02	2	0.07	0.08	0.01																									5	AGADWS
Complexing Agents																																
4-Tolyltriazole	µg/L	0.1					5	0.73	1.3	0.10	1	<LOQ	–	–					1	0.30	–	0.04	4	0.39	4.0	0.06	3	0.1	0.27	0.01	7	AGADWS
5-Tolyltriazole	µg/L	0.1					5	0.16	1.7	0.02	1	<LOQ	–	–					1	0.24	–	0.03	4	0.25	0.45	0.04	3	0.1	0.24	0.01	7	AGADWS
Benzotriazole	µg/L	0.1					5	2.3	3.2	0.33	1	0.24	–	0.03					1	2.5	–	0.36	4	2.2	3.3	0.32	3	5.4	6.0	0.77	7	AGADWS
Other chemicals																																
Adsorbable organic iodine	µg/L	0.5					5	11	39	0.18	1	11	–	0.18									4	21	32	0.36	3	21	30	0.35	60	ADWG
Bisphenol-A	µg/L	0.01									4	0.04	0.39	0.00					2	<LOQ	<LOQ	–					3	0.08	0.17	0.00	200	AGADWS

<LOQ: value below the limit of quantification; AGADWS: Australian Guidelines for Water Recycling: Augmentation of Drinking Water Supplies; LTD: Lowest Terapeutic Dose; USEPA: United States Environmental Protection Agency (IRIS website); OEHHA: Office of Environmental Health Hazard Assessment (California Environmental Protection Agency).

Note: For those compounds were data included measured values and values <LOQ, the values <LOQ were substituted for the actual LOQ for the compound in order to calculate the median value.

Atlantis case study has two different source waters: the secondary effluent of Wesfleur WWTP and stormwater. Both source waters were monitored for herbicides, antibiotics, contrast media and other pharmaceuticals in two sampling campaigns. The results given in Table 19.2 correspond to the source water (secondary effluent or stormwater) with the highest values for the selected compound used to perform the risk assessment. Among the five antibiotics monitored, four could be detected, with sulfamethoxazole being detected in the highest concentration (3.7 μg/L). For contrast media, two out of five compounds could be detected, in 30% of the samples analyzed. Among the other pharmaceuticals, eight out of fifteen could be detected. Ibuprofen was the pharmaceutical detected in the highest concentration (10.6 μg/L). Carbamazepine could be detected with a maximum concentration of 3.5 μg/L. For the herbicides, only clofibric acid was monitored and it could not be detected in any of the sampling campaigns.

Gaobeidian uses secondary effluent of Gaobeidian WWTP as source water. Five samplings were performed on source waters at Gaobeidian, six antibiotics were monitored and all of them were detected, but in very low concentrations. The highest value detected corresponded to sulfamethoxazole, 0.65 μg/L. Four out of five contrast media compounds could be detected. The highest values corresponded to iopamidol (6.5 μg/L) and iohexol (2.2 μg/L). Other pharmaceuticals were also monitored, and three out of four could be detected: diclofenac, ibuprofen and naproxen. Three complexing agents were monitored at this site, and all of them could be detected. Benzotriazole was the complexing agent detected at the highest concentration, 3.2 μg/L. Clofibric acid was the only pesticide monitored, and it was detected at a low concentration. Absorbable organic iodine was also detected with a median value of 11.0 μg/L and the maximum value 39.0 μg/L.

At the Nardò site, the effluents of three different WWTPs are gathered and transported in an open channel to the sinkhole, where they are infiltrated to the subsurface. The effluent from Galatone WWTP accounts for the highest volume of source water, and it was sampled in the four sampling campaigns. Nine nitrosamines were monitored but only three were detected. NMOR was the nitrosamine found at the highest concentrations, with a median value of 10.8 ng/L and a maximum value of 20 ng/L. Seven out of eleven antibiotics could be detected at Nardò. All of the antibiotics were detected in low concentrations. In this site, six contrast media were monitored, and half of them were detected. The highest concentration detected was for diatrizoate, 1.7 μg/L. Only one hormone was investigated at this site, estrone E1, but was not detected. Other pharmaceuticals were monitored included carbamazepine, diclofenac, naproxen and primidone were detected. One out of the three complexing agents monitored could be detected; benzotriazole. Clofibric acid was not detected. Adsorbable organic iodine was monitored in only one sample, and the concentration was 10.8 μg/L. Bisphenol A was also monitored, and the highest concentration detected was 0.39 μg/L.

For Parafield, the organic chemicals in the influent stormwater were monitored on 11 campaigns. The Parafield case study had a separate monitoring program of over 300 organic chemicals as described by Page *et al.* (2009; 2010d). There were no pharmaceuticals detected in the influent water, and a single THM (dichloromethane 1.4 μg/L) was detected. The most frequently detected pesticide was simazine with a maximum concentration of 0.86 μg/L. In addition, methyl blue active substances (a general class for detergents and surfactants) were detected with a maximum concentration of 120 μg/L. Additional sampling using passive samplers detected ng/L concentrations of other herbicides and chemicals as reported by Page *et al.* (2010d).

For Sabadell case study, the secondary treated effluent of the Ripoll River WWTP was sampled in 3 sampling campaigns.

Five N-nitrosamines were monitored in only one of the sampling campaigns performed. NDMA was detected at a concentration of 16.5 ng/L and NMOR 9.6 ng/L. Presence of pharmaceuticals was expected as the Taulí Hospital sends its wastewater to the Ripoll River WWTP. Five out of nine antibiotics monitored could be detected. Contrast media was detected in medium to high concentrations in all sampling campaigns, with peak values of 27.7 μg/L of diatrizoate and 10.1 μg/L of iopromide. Hormones were only monitored in one sampling campaign, and only estrone E1 was detected, at a concentration of 64 ng/L. Other pharmaceuticals detected were bezafibrate, carbamazepine, diclofenac, primidone and naproxen. All complexing agents monitored in the secondary effluent of the Ripoll River WWTP were detected in the only sampling campaign they were tested for, benzotriazole being the one detected at the highest concentration (2.5 μg/L). For the herbicides, only clofibric acid was monitored and it could be detected in all sampling campaigns at low concentrations. Bisphenol A could not be detected.

The Shafdan site also uses the secondary effluent of a WWTP as source water. Four sampling campaigns were performed. Nine nitrosamines were monitored and seven could be detected. NDBA, NDEA NDMA and NMOR were found at high concentrations, with median values of 8.2, 12.8, 9.8 and 10.9 ng/L respectively, and reaching peak values of 45.2, 32.4, 53.0 and 53.5 ng/L respectively. Six antibiotics were monitored at this site, and detected in 79% of the samples. Contrast media was detected at concentrations ranging from 0.15 μg/L (median value for iomeprol) to 6.8 μg/L (median value for iopromide). Hormones were not detected in any of the samplings. Other pharmaceuticals monitored were bezafibrate, diclofenac, ibuprofen and naproxen. Complexing agents were detected in all sampling campaigns, and benzotriazole was detected at the highest concentrations (median value of 2.2 μg/L and maximum value of 3.3 μg/L). Clofibric acid could not be detected, and absorbable organic ionic median value was 21.4 μg/L.

In Torreele case study the secondary effluent of the Wulpen WWTP was monitored in 3 sampling campaigns. Different N-nitrosamines were detected in all sampling campaigns, with NDMA being the most frequently detected and also in the highest concentration (maximum of 6.3 ng/L). NMOR concentrations were also high, with a median value of 3.5 ng/L and a maximum value of 3.8 ng/L. In Torreele, 13 antibiotics were monitored and only clarithromycin, sulfamethoxazole, N-Acetil-sulfamethoxazole and trimethoprim could be detected, at very low concentrations. Contrast media was detected in high concentrations in 60% of the samples, with peak values of 6.9 μg/L for diatrizoate and 3.8 μg/L for iopromide. Concentrations of hormones varied more than the antibiotics, contrast media or other pharmaceuticals, and varied between campaigns. Of the other pharmaceuticals analysed, carbamazepine was detected in the highest concentration (1.4 μg/L), followed by diclofenac and naproxen at trace levels. Other pharmaceuticals detected were primidone and ibuprofen at the limit of detection. All complexing agents monitored could be detected, with benzotrizole being the most abundant (6.0 μg/L, maximum concentration). For the herbicides, only clofibric acid was monitored and it could not be detected in any of the sampling campaigns. Bisphenol A had a maximum concentration was of 0.17 μg/L.

Table 19.2 summarises the organic chemical risk assessment performed for the detected chemicals in the source waters at the case study sites. All chemical groups except N-nitrosamines and hormones have screening RQs (median) below one. For the N-nitrosamines, NMOR (N-nitrosomorpholine) was monitored in four sites and in all of them the RQ (calculated with the median value) was much higher than one: 11 at Nardò and Shafdan, 9.6 at Sabadell and 3.5 at Torreele. NDMA (N-Nitrosodimethylamine) RQ value was 1.7 in Sabadell. NDBA and NDEA RQ values in Shafdan were 1.4 and 1.3, respectively. These results indicate that nitrosamines could pose a risk for the human health if the water was used for drinking. For the hormones, estrone E1 had a median RQ value of 2.1 in Sabadell.

Most of the monitored compounds that could be detected had median RQ values close to zero, generally well below one. The compounds whose RQs median values were higher than one pose a risk for the human health, and their possible presence in final waters must be monitored. There were no detections of these compounds at the Parafield site, where the untreated stormwater is used. Due to restrictions of the analytical methods at the time of this study, it was determined the concentrations of the majority of the organic chemicals detected in the source waters (Table 19.2) were too low in the recovered water within the case studies to be accurately determined (Table 19.4).

19.3.2 Recovered waters

Despite the low detection limits, many of the compounds were still detected in the recovered waters of the case study sites. The percentage of compounds detected in every group decreased after the treatment train. Table 19.3 summarizes the number of chemicals analysed for each case study site in the recovered waters and their percentages of detection. Table 19.4 summarizes the number of samples, median and max results for each compound analyzed in recovered waters at each site, as well as the RQ value for each compound (using the median value to calculate it). The compounds that were not detected in any of the case studies do not appear in Table 19.2. These compounds are the following: NDPhA (N-Nitrosodiphenylamine), NDPA (N-Nitrosodipropylamine), NMEA (N-Nitrosomethylethylamine), NPIP (N-Nitrosopiperidine), NPYR (N-Nitrosopyrrolidine), azithromycin, florfenicole, N-acetilsulfamethoxazole, sulfadiazine, sulfadimethoxine, sulfadoxine, sulfamerazine, sulfisoxazole, tiamulin, ioxitalamic acid, estradiol E2, ethinylestradiol EE2, benzoylegconine, codein, diazepam, ibuprofen, mefenamic acid, methadon, morphine, nordiazepam, oxazepam, paracetamol, temazepam and clofibric acid.

At the Atlantis case study site only carbamazepine, with a maximum value of 0.03 μg/L out of 24 could be detected in the recovered water; two orders of magnitude lower than in the source water.

The Gaobeidian case study site also had pharmaceuticals detected in the recovered water. All six antibiotics were detected in the recovered water in some of the campaigns, and at similar concentrations to the source water, thus showing no removal or poor removal. Four contrast media compounds were also detected. Lower concentrations were observed for iohexol and iopromide, with median values in the recovered water of 0.07 and 0.05 μg/L, respectively. However, no removal was measured for diatrizoate. Other pharmaceuticals tested (three) could not be detected in the recovered water, only diclofenac was detected. Complexing agents were removed along the treatment, and much lower values were obtained in the recovered water. Adsorbable organic iodine was also reduced along the treatment train, but still high values up to 13 μg/L.

Nitrosamines and estrone E1 could not be detected in the recovered water for Nardò. Some antibiotics could still be detected, in 21% of the samples, namely clarithromycin, sulfamethoxazole and sulfapyridine, but at lower concentrations than in source water. Contrast media was also removed, and the three compounds detected (diatrizoate, iomeprol and iopamidol) had lower concentrations than in source water. Other pharmaceuticals detected in the recovered water were carbamazepine, diclofenac and primidione, all of them at lower concentrations than in source water. Benzotriazole was detected at a higher concentration than in source water, 2.4 μg/L. Adsorbable organic iodine was present at levels similar to those in source water and bisphenol A maximum concentration was lower (0.05 μg/L).

Table 19.3 Total number of chemicals analysed for each case study site in the recovered waters and the percentage of detections.

Compound detection in source water at each site	Atlantis		Gaobeidian		Nardò		Parafield		Sabadell		Shafdan		Torreele	
	n	Detected (%)	n	Detected (%)	n	Detected (%)	n	Detected (%)	n	Detected (%)	n	Detected (%)	n	Detected (%)
DBPs														
THMS														
HAAs														
N-nitrosamines					9	0			5	20	5	67	6	0
Pharmaceuticals														
Antibiotics	5	0	6	43	11	21			8	11	6	29	13	0
Contrast media	5	0	5	55	6	68			5	40	5	25	5	0
Hormones					1	0			3	0	3	8	3	0
Other	14	4	4	7	7	54			7	56	4	7	6	0
VOC's														
PAH's							1	0						
Dioxin and Furans							1	0						
PCB's							1	0						
Complexing agents			3	73	3	33	1	0	3	100	3	58	3	33
Pesticides and herbicides	1	0	1	0	1	0	1	0	1	0	1	0	1	0
Other chemicals					1	100	1	0	1	100			1	33

For Sabadell, one nitrosamine was still detected in the recovered water, NMOR, at a concentration of 3.3 ng/L. Two out of eight antibiotics could also still be detected, at very low concentrations in only 11% of the samples. Contrast media concentrations were substantially reduced, and the highest value measured was for diatrizoate, 1.1 μg/L. Hormones could not be detected in the recovered water, and other pharmaceuticals detected at low concentrations were bezafibrate, carbamazepine, diclofenac, naproxen and primidone. Complexing agents were still present in the recovered water, and 4-tolyltriazole was detected at a higher concentration (2.0 μg/L) than in the source water. Bisphenol A was not detected in source water but it was detected in the recovered water, at a maximum concentration of 0.05 μg/L.

At Shafdan four nitrosamines were detected in the recovered water: NDBA, NDEA, NDEMA and NMOR. The maximum value measured was for NDEA, 9.9 ng/L, but NMOR was the only one that could pose a risk for the human health, with a median value of 2.8 ng/L. Antibiotics concentrations were reduced after the treatment train, and could be detected in 29% of the samples. Two out of five contrast media were still present in the recovered water, diatrizoate and iopamidol, at much lower concentrations than in source water. Estrone E2 was not detected in source water but it was in the recovered water, at a maximum concentration of 1.2 μg/L. Another pharmaceutical detected was diclofenac. Complexing agents were still present in the recovered water, but at lower concentrations than in source water. Benzotriazole median value was 0.39 μg/L. Adsorbable organic iodine was still high in the recovered water, with a maximum value of 20 μg/L, but lower than in the source water.

In Torreele, the only compounds detected among the 38 compounds monitored in the recovered water were benzotriazole (0.24 μg/L, maximum value), adsorbable organic iodine (2.6 μg/L, maximum value) and bisphenol A (0.23 μg/L, maximum value).

Table 19.4 summarises the organic chemical risk assessment performed for the detected chemicals at the case study sites. All chemical groups except nitrosamines (NMOR) have screening RQs (median) below 1. The median RQ value for NMOR in Sabadell case study was 3.3, and 2.8 in Shafdan. This RQ values indicate a risk for the human health in case the final water is used for drinking water purposes. However, in the case of Sabadell and Shafdan, the recovered water is not used as drinking water, but as irrigation and streets cleaning water, thus decreasing considerably the risk. The amount of water that can be inhaled when parks irrigation or streets cleaning are performed is between 3 and 4 orders of magnitude inferior to the amount ingested if used as drinking water.

Most of the compounds detected in final waters have median RQ values well below one, close to zero, thus not posing a risk for the human health.

Table 19.4 Organic chemical risk quotients in the case study recovered waters.

	units	LOQ	Atlantis				Gaobeidian				Nardò #500m				Parafield				Sabadell				Shafdan				Torreele				Hth. v.	Reference
			n	med	max	RQ med	n	med	max	RQ med	n	med	max	RQ med	n	med	max	RQ med	n	med	max	RQ med	n	med	max	RQ med	n	med	max	RQ med		
DBPs																																
N-nitrosamines																																
N-Nitrosodibutylamine (NDBA)	ng/L	3									1	<LOQ	–	–					1	<LOQ	–	–	2	4.4	8.6	0.73	1	<LOQ	–	–	6	USEPA
N-Nitrosodiethylamine (NDEA)	ng/L	0.3									1	<LOQ	–	–					1	<LOQ	–	–	3	1.0	9.9	0.10	1	<LOQ	–	–	10	AGADWS
N-Nitrosodimethylamine (NDMA)	ng/L	0.7									1	<LOQ	–	–					1	<LOQ	–	–	3	2.3	3.0	0.23	1	<LOQ	–	–	10	AGADWS
N-Nitrosomorpholine (NMOR)	ng/L	1.3									1	<LOQ	–	–					1	3.3	–	**3.3**	3	2.8	9.7	**2.8**	1	<LOQ	–	–	1	AGADWS
Pharmaceuticals																																
Antibiotics																																
Clarithromycin	μg/L	0.02	2	<LOQ	<LOQ	–	4	0.03	0.05	0.00	4	0.04	0.05	0.00					2	<LOQ	<LOQ	–	4	0.03	0.05	0.00	3	<LOQ	<LOQ	–	250	AGADWS
Erythromycin	μg/L	0.01	2	<LOQ	<LOQ	–	4	0.03	0.03	0.00	4	<LOQ	<LOQ	–					2	0.01	0.01	0.00	4	0.01	0.04	0.00	3	<LOQ	<LOQ	–	17.5	AGADWS
Roxythromycin	μg/L	0.02	2	<LOQ	<LOQ	–	5	0.04	0.11	0.00	4	<LOQ	<LOQ	–					2	<LOQ	<LOQ	–	4	0.05	0.11	0.00	3	<LOQ	<LOQ	–	150	AGADWS
Sulfamethoxazole	μg/L	0.01	2	<LOQ	<LOQ	–	5	0.07	0.86	0.00	4	0.11	0.22	0.00					2	0.03	0.05	0.00	4	0.04	0.17	0.00	3	<LOQ	<LOQ	–	35	AGADWS
Sulfamethazine	μg/L	0.01					5	0.04	0.11	0.00	4	<LOQ	<LOQ	–					2	<LOQ	<LOQ	–	4	<LOQ	<LOQ	–	3	<LOQ	<LOQ	–	35	AGADWS
Sulfapyridine	μg/L	0.01									1	0.03	–	0.00																	35	AGADWS
Trimethoprim	μg/L	0.01	2	<LOQ	<LOQ	–	5	0.01	0.02	0.00	2	<LOQ	<LOQ	–					3	<LOQ	<LOQ	–	4	0.01	0.01	0.00	3	<LOQ	<LOQ	–	70	AGADWS
Contrast media																																
Diatrizoate	μg/L	0.01	2	<LOQ	<LOQ	–	4	1.4	2.1	0.00	4	0.63	0.99	0.00					3	0.28	1.1	0.00	4	0.20	0.29	0.00	3	<LOQ	<LOQ	–	600	LTD
Iohexol	μg/L	0.02	2	<LOQ	<LOQ	–	4	0.07	0.7	0.00	2	<LOQ	<LOQ	–					3	<LOQ	<LOQ	–	4	<LOQ	<LOQ	–	3	<LOQ	<LOQ	–	720	AGADWS
Iomeprol	μg/L	0.02	2	<LOQ	<LOQ	–	4	<LOQ	<LOQ	–	4	0.08	0.48	0.00					3	<LOQ	<LOQ	–	4	<LOQ	<LOQ	–	3	<LOQ	<LOQ	–	375	LTD

Iopamidol	µg/L	0.02	2	<LOQ	<LOQ	–	4	1.7	4.1	0.00	3	0.09	0.56	0.00	3	0.02	0.05	0.00	4	0.02	0.18	0.00	3	<LOQ	<LOQ	–	400	AGADWS
Iopromide	µg/L	0.02	2	<LOQ	<LOQ	–	4	0.05	0.41	0.00	4	<LOQ	<LOQ	–	3	0.06	0.23	0.00	4	<LOQ	<LOQ	–	3	<LOQ	<LOQ	–	750	AGADWS
Hormones																												
Estrone E1	ng/L	0.5									2	<LOQ	<LOQ	–	1	<LOQ	–	–	4	0.50	1.2	0.02	3	<LOQ	<LOQ	–	30	AGADWS
Other																												
Bezafibrate	µg/L	0.01	2	<LOQ	<LOQ	–	4	<LOQ	<LOQ	–	4	<LOQ	<LOQ	–	3	0.01	0.02	0.00	4	<LOQ	<LOQ	–	2	<LOQ	<LOQ	–	300	AGADWS
Carbamazepine	µg/L	0.01	2	0.02	0.03	0.00					4	0.44	0.63	0.00	2	0.08	0.14	0.00					3	<LOQ	<LOQ	–	100	AGADWS
Diclofenac	µg/L	0.01	2	<LOQ	<LOQ	–	3	0.02	0.18	0.01	4	0.01	0.04	0.01	2	0.02	0.03	0.01	4	0.02	0.35	0.01	2	<LOQ	<LOQ	–	1.8	AGADWS
Naproxen	µg/L	0.01	2	<LOQ	<LOQ	–	4	<LOQ	<LOQ	–	4	<LOQ	<LOQ	–	3	0.01	0.03	0.00	4	<LOQ	<LOQ	–	2	<LOQ	<LOQ	–	220	AGADWS
Primidone	µg/L	0.01	2	<LOQ	<LOQ	–					3	0.07	0.08	0.00	2	0.03	0.03	0.00					3	<LOQ	<LOQ	–	125	LTD
Complexing Agents																												
4-Tolyltriazole	µg/L	0.1					5	0.21	0.39	0.03	1	<LOQ	–	–	1	2.0	–	0.28	4	0.05	0.42	0.01	3	<LOQ	<LOQ	–	7	AGADWS
5-Tolyltriazole	µg/L	0.1					5	0.10	0.22	0.01	1	<LOQ	–	–	1	0.20	–	0.03	4	0.05	0.29	0.01	3	<LOQ	<LOQ	–	7	AGADWS
Benzotriazole	µg/L	0.1					5	0.44	0.71	0.06	1	2.4	–	–	1	1.9	–	0.27	4	0.39	1.3	0.06	3	0.21	0.24	0.03	7	AGADWS
Other chemicals																												
Adsorbable organic iodine	µg/L	0.5					4	7.6	13	0.13	1	12	–	0.20					4	12	20	0.21	3	0.80	2.60	0.01	60	ADWG
Bisphenol-A	µg/L	0.01									4	0.35	0.51	0.00	2	0.05	0.05	0.00					3	0.02	0.23	0.00	200	AGADWS

<LOQ: value below the limit of quantification; AGADWS: Australian Guidelines for Water Recycling: Augmentation of Drinking Water Supplies; LTD: Lowest Terapeutic Dose; USEPA: United States Environmental Protection Agency (IRIS website).

Note: for those compounds were data included measured values and values <LOQ, the values <LOQ were substituted for the actual LOQ for the compound in order to calculate the median value.

A chemical characterisation of case study sites source waters was also conducted including metals and other inorganic compounds, nutrients and organic chemicals. Trace organic chemical contaminants tested for were found in low concentrations (Table 19.2), which indicate that the reclaimed effluent source waters are similar. The presence of toxic contaminants is often a function of the industrial effluent component of the source water and therefore stormwater and wastewater quality is primarily determined by the presence and nature of industries discharging waste and the treatment processes applied. For example, the Parafield stormwater catchment area is mainly residential and commercial although there is also some light manufacturing and automotive industries which may contribute hydrocarbons to the stormwater quality.

Of the organic chemicals analysed most were detected in at least one of seven study sites taken from the source waters (Table 19.2). Pharmaceuticals were commonly detected in low concentrations (ng/L to low µg/L) in the wastewater sources and pesticides and herbicides were more commonly detected in the stormwater sources consistent with the reviewed international studies (Ayuso-Gabella *et al.* 2009). Despite the detection of pharmaceuticals in the wastewater derived source waters, the human health risk assessment indicates a low health significance.

A reduction of screening RQ (mean) values was observed for most sites. These results are consistent with other studies reporting natural filtering and attenuation of pharmaceuticals and other pollutants within aquifers. However, the removal efficiencies differ for each case study site. When biodegradation occurs, redox (e.g. the presence or absence of oxygen) can have a significant impact (Ying *et al.* 2008). Some organic chemicals are most effectively removed under aerobic conditions while others are only removed under anoxic conditions (Dillon *et al.* 2009). Given the strong influence of redox conditions on removal rates of many organic chemicals, it is optimal to have different redox zones in the MAR system such that water is exposed to differing conditions to maximise opportunity for organic chemical degradation (Dillon *et al.* 2009).

The dominant human health risk of water recycling via aquifers is related to pathogens and not to chemicals. A quantitative assessment of human health risks from pathogens is given in the following section.

19.4 QUANTITATIVE MICROBIAL RISK ASSESSMENT METHODOLOGY

The quantitative microbial risk assessment (QMRA) methodology adopted in RECLAIM WATER follows the approach outlined in WHO (2004) and NRMMC-EPHC-AHMC (2006). The traditional approach to identifying tolerable risk has been to define maximum levels of infection or disease. However, this approach fails to consider the varying severity of outcomes associated with different hazards. This shortcoming can be overcome by measuring severity in terms of disability adjusted life years (DALYs). DALYs have been used extensively by agencies such as the World Health Organization (WHO) to assess disease burdens (WHO, 2004). For microbial risk assessment three representative pathogens; rotavirus, *Cryptosporidium* and *Campylobacter*, were used to assess the risk of viruses, protozoa and bacteria as described in WHO (2004) and NRMMC-EPHC-AHMC (2006). As the risk estimates are probability distribution functions (PDFs), the mean, median and 95th percentile were routinely calculated for each pathogen risk. The tolerable mean risk adopted is 10^{-6} DALYs per person per year (WHO 2004).

For the risk assessment case study sites discussed in this Chapter, a qualitative residual risk assessments were initially performed as part of the RECLAIM WATER project (Ayuso-Gabella *et al.* 2007, 2009). In furthering the qualitative understanding of the pathogenic hazards at each case study site, a QMRA was performed to determine the residual risk for each case study site and the value of the aquifer treatment. The residual risks are risk probability estimates assuming normal operating conditions i.e. where source waters are not exposed to unusual hazard inputs and treatment processes are operating according to specifications. The QMRA reported below is derived from the work of Page *et al.* (2010b; 2010d and Ayuso-Gabella *et al.* (in press).

19.5 QMRA OF THE CASE STUDIES

This section considers QMRA case studies that form part of the larger RECLAIM WATER project. Each risk case study site utilises a non-traditional water source and an engineered water treatment train coupled to an aquifer recharge system for augmenting urban drinking water supplies or irrigation. The three case studies using the recovered water for augmenting drinking water supplies are: Atlantis (South Africa), Parafield (Australia) and Torreele/St. André (Belgium). The three case studies using the recoverd water for crops irrigation, green space irrigation or street cleaning are: Nardò (Italy), Sabadell (Spain) and Shafdan (Israel). Each case study was assessed using a QMRA approach and the aquifer treatment contribution compared across the four case study sites. Special attention has been given to the contribution of the aquifer barrier within the broader treatment train and its importance in managing human health risks.

The treatment trains and important attributes of the case studies have been previously summarised in Part A of this book. They range from almost total reliance on the subsurface passage and residence time for water quality improvement (e.g. Shafdan, Israel) to advanced tertiary treatment (e.g. Torreele/St-André, Belgium) where there is almost no reliance on the aquifer for water quality improvements. At Atlantis, South Africa, Parafield, Australia and Shafdan, Israel the aquifer plays an important complementary role to the engineered treatment systems.

The risk models for simulating hazard reduction, consumption, infection and disease burden were constructed using MS Excel program [2003] enhanced with @ Risk Industrial v. 4.5 and v. 5.5 (Palisade Corp, USA). The minimum value calculated was 1.0×10^{-10} DALYs per person per year.

A triangular probability distribution function (PDF) describing each engineered treatment barrier was adopted from literature for each pathogen (Smeets *et al.* 2006; NRMMC-EPHC-NHMRC, 2006; Cleary, 2005). The triangular distribution was defined by a minimum, most likely and maximum $\log_{10}$ removal value (Smeets *et al.* 2006; NRMMC-EPHC-NHMRC, 2006) and is shown in Table 19.5 and Table 19.6.

The risks associated with recontamination at each treatment barrier were not specifically assessed. For the aquifer treatment barrier, the product of two PDFs; the aquifer residence time and a daily pathogen decay rate (expressed in $\log_{10}$/day) were used to calculate the $\log_{10}$ removal value. Initial pathogen numbers in the stormwater and wastewater were derived from literature (Kocwa-Haluch & Zalewska, 2002; NRMMC-EPHC-NHMRC, 2006; NRMMC-EPHC-NHMRC, 2009a; Robertson *et al.* 2006; Sedmark *et al.* 2005; Montemayor *et al.* 2005; Westrell, 2004). For the Atlantis case study, the two source waters were mixed to derive a final PDF of pathogen concentration for the injectant. Each of the PDFs (including pathogen numbers and aquifer and non-aquifer treatments) was subsequently used in the Monte Carlo simulations to calculate the residual risk.

Once the residual risks were calculated for each MAR scheme a sensitivity analysis was performed which standardises the factors which affect risk and is termed the factor sensitivity (FS) (Zwietering and van Gerwen, 2000). For each MAR scheme the residual risk was then recalculated in the absence of each barrier in turn (such as the aquifer treatment barrier). The FS is a ratio calculated by dividing the revised residual risk estimate (in DALYs) when a factor (e.g. a treatment step) is removed from the treatment train [denoted *N*(*Barrier*)], by the baseline mean risk, *N*(*Mean*) also in DALYs from the residual risk assessment and then $\log_{10}$ transforming the ratio (Equation 1):

$$FS = \log_{10}[N(Barrier)/N(Mean)] \tag{19.1}$$

Higher FS values means the factor has a larger effect on risk. Following assessment of FS a risk-based approach for determining suitable aquifer residence times for MAR schemes is proposed. Aquifer treatment uses the surrogate parameter, aquifer residence time to estimate the value of the aquifer treatment as part of the multiple barrier system. Simulations of changes in the aquifer residence time allow the aquifer barrier to be quantified and compared to the acceptable risk, 1.0×10^{-6} DALYs (WHO 2004). This allows the determination of a required average residence time and associated monitoring can be utilised to manage this barrier within the treatment system.

19.5.1 Aquifer barrier treatment characterisation

Aquifer treatment characteristics were derived from the probability distribution functions (PDFs) of the residence time in the aquifer and the reported $\log_{10}$ decay rates for pathogens (see Tables 19.5 and 19.6). The aquifer and engineered (non-aquifer) treatment barrier characteristics are reported as $\log_{10}$-removals (Table 19.7) which conveys the order of magnitude of the removal for each of the reference pathogens.

Removal $\log_{10}$ values for each treatment barrier were considered additive and thus the Torreele/St-André scheme with multiple engineered barriers resulted in the highest $\log_{10}$ removal for the 'non-aquifer' treatment component. All $\log_{10}$ removal values accredited to aquifers were capped at a maximum of 6.0 $\log_{10}$ consistent with the reported values for engineered treatments in NRMMC-EPHC-NHMRC (2008). Each of the MAR sites placed a different value on the aquifer removal characteristics compared to the engineered treatments. Some relied heavily on the aquifer, like Shafdan and Atlantis, whereas others like Torreele/St-André had extensive redundancy in their system due to a long treatment train of engineered barriers and as such relied little on the aquifer. Torreele/St-André and Sabadell had a low calculated $\log_{10}$ removal capability for the aquifer whereas Parafield, Shafdan and Atlantis had greater calculated treatment capacities due to the longer residence times of water in the subsurface at these sites. Pathogen decay rates applied impacted considerably the effectiveness of the aquifer in removing pathogens.

In order to provide safe drinking water with MAR an integrated approach to managing risks needs to be adopted which includes characterisation of the aquifer treatment barrier. To date there have been no reported case studies where the aquifer treatment barrier of a MAR scheme is accredited with $\log_{10}$ removals for pathogens much like in conventional drinking water treatment. In valuing the treatment capacity, integrity and independence of aquifers, MAR can be utilised in the same way as conventional engineered water treatment in an integrated water supply system.

The value of the aquifer barrier was determined by the relative $\log_{10}$ removal characteristics with respect to the reference pathogens (Table 19.7). The $\log_{10}$ removals for *Campylobacter* are potentially >6.0 $\log_{10}$, if the decay rate measured at Parafield is used for the sites, as is in Atlantis and Torreele/St André. For instance, in Sabadell and Shafdan the decay rates used come from a review of different studies published on pathogens in groundwater (John and Rose, 2005), and in Nardò a much lower decay rate measured in the site (La Mantia *et al.* 2008) is used. A similar value of >6.0 $\log_{10}$ units removal is attributed to other water treatment technologies such as reverse osmosis (NRMMC-EPHC-AHMC, 2006). For *Cryptosporidium* the value of the aquifer treatment is dependant on the residence time. *Cryptosporidium*

Table 19.5 Probability distribution functions used for the quantitative risk assessment in case studies using the recovered water for crops irrigation, green space irrigation or streets cleaning purposes.

Site	Nardò			Sabadell			Shafdan		
Pathogen	**Campylobacter**	**Cryptosporidium**	**Rotavirus**	**Campylobacter**	**Cryptosporidium**	**Rotavirus**	**Campylobacter**	**Cryptosporidium**	**Rotavirus**
Pathogen source water number (n/L)	U(10^3, 10^5)[a]	LN(225.7, 83.7)[b]	LN(1342.6; 6329.9)[c]	U(10^3, 10^5)[a]	LN(225.7, 83.7)[b]	LN(1342.6; 6329.9)[c]	U(10^3, 10^5)[a]	LN(225.7, 83.7)[b]	LN(1342.6; 6329.9)[c]
WWTP (log_{10})	T(1, 2, 3.5)[d]	T(0.5, 1, 1.5)[d]	T(0.5, 1, 2.1)[d]	T(1, 2, 3.5)[d]	T(0.5, 1, 1.5)[d]	T(0.5, 1, 2.1)[d]	T(1, 2, 3.5)[d]	T(0.5, 1, 1.5)[d]	T(0.5, 1, 2.1)[d]
Channel/river mixture (log_{10})	T(0, 0.7, 1.6)[e]	T(0.1, 1, 2)[e]	T(0, 0.5, 1.4)[e]	LL(−1.6, 1.9, 4.4)[h]	E(0.63)[h]	T(−1, 0, 2)[h]			
Subsurface storage (days)	T(13, 23, 65)	T(13, 23, 65)	T(13, 23, 65)				T(180, 270, 365)	T(180, 270, 365)	T(180, 270, 365)
Subsurface pathogen decay rate (log_{10}/day)	0.038[g]	0.057[g]	0.12[g]	T(0.02, 0.08, 1.5)[i]	N(0.012, 0.0030)[f]	T(0.012, 0.16, 0.83)[j]	T(0.007, 0.2, 1.4)[i]	N(0.012, 0.0030)[f]	T(0.03, 0.33, 1.3)[j]
UV (log_{10})				T(2.0, 3.0, 4.0)[a]	T(2.0, 3.0, 3.5)[a]	T(1.0, 2.0, 3.5)[a]			
Chlorination (log_{10})				T(2.0, 3.0, 4.0) [d]	T(0.0, 0.0, 0.5)[d]	T(1.0, 1.5, 3.0)[d]	T(2.0, 3.0, 4.0) [d]	T(0.0, 0.0, 0.5)[d]	T(1.0, 1.5, 3.0)[d]
Rapid sand filtration (log_{10})				T(0.0, 0.0, 0.5) [d]	T(0.0, 0.0, 0.5) [d]	T(0.0, 0.0, 0.5)[d]			
Volume ingested (L)[d,k]	U(10^{-3}, 10^{-4})	U(10^{-3}, 10^{-4})	U(10^{-3}, 10^{-4})	U(10^{-3}, 10^{-4})	U(10^{-3}, 10^{-4})	U(10^{-3}, 10^{-4})	U(10^{-3}, 10^{-4})	U(10^{-3}, 10^{-4})	U(10^{-3}, 10^{-4})
Exposures per year[k]	U(183, 365)	U(183, 365)	U(183, 365)	U(183, 365)	U(183, 365)	U(183, 365)	U(183, 365)	U(183, 365)	U(183, 365)

T: Triangular distribution; LL: Loglogistic; LN: Lognormal distribution; N: Normal distribution; U: Uniform distribution; E: Exponential distribution
[a] Adapted from NRMMC–EPHC–AHMC, 2006; Westrell, 2004.
[b] Adapted from Montemayor *et al.* (2005).
[c] Adapted from Sedmark *et al.* (2005).
[d] Adapted from NRMMC–EPHC–AHMC (2006).
[e] Data from Nardò site on indicators was used to construct the PDFs (Böckelmann *et al.* 2009; Levantesi *et al.* 2010; Aharoni *et al.* 2008).
[f] From Sidhu *et al.* (2010).
[g] Single value only, using indicators data, from La Mantia *et al.* (2008).
[h] Data from Sabadell site on indicators was used to construct the PDFs (Böckelmann *et al.* 2009; Levantesi *et al.* 2010; Aharoni *et al.* 2008).
[i] Adapted from John and Rose (2005).
[j] Adapted from Pedley *et al.* (2006).
[k] The scenario considered is an accidental aerosols inhalation during irrigation by growers/irrigators. This scenario was the worst case as per Ayuso-Gabella *et al.* submitted).

Table 19.6 Probability distribution functions used for the quantitative risk assessment in case studies using the recovered water for augmenting urban drinking water supplies.

Site	Atlantis			Parafield			Torreele/St-André		
Pathogen	**Campylobacter**	**Cryptosporidium**	**Rotavirus**	**Campylobacter**	**Cryptosporidium**	**Rotavirus**	**Campylobacter**	**Cryptosporidium**	**Rotavirus**
Pathogen source water number (n/L)	LN(3.9, 9.8)[a]/ U(10^{1}–10^{4})[d]	LN(0.5, 1.2)[a]/ LN(200, 100)[c]	LN(0.3, 0.6)[a]/ LN(443, 220)[b]	LN(3.9, 9.8)[a]	LN(0.5, 1.2)[a]	LN(0.3, 0.6)[a]	U(10^{1}–10^{4}) [d]	LN(200, 100)[a]	LN(443, 220)[c]
Artificial wetland (log_{10})	T(1.5, 2, 2.5)[e]	T(0.5, 0.5, 1.0)[e]	T(0.0, 0.0, 0.0)[e]	T(1.5, 2.0, 2.5)[e]	T(0.5, 0.5, 1.0)[e]	T(0.0, 0.0, 0.0)[e]			
WWTP (log_{10})	T(0.6, 1.4, 3.7)[e]	T(0.4, 1.8, 3.8)[e]	T(0.2, 1.7, 2.3)[e]				T(0.6, 1.4, 3.7)[f]	T(0.4, 1.8, 3.8)[f]	T(0.2, 1.7, 2.3)[f]
UF (log_{10})							T(4.0, 4.0, 7.0)[f]	T(3.0, 3.0, 7.0)[f]	T(4.0, 4.0, 6.5)[f]
RO (log_{10})							T(4.0, 4.0, 7.0)[f]	T(3.0, 3.0, 7.0)[f]	T(2.7, 3.0, 6.5)[f]
UV (log_{10})							T(2.0, 3.0, 4.0)[f]	T(2.0, 3.0, 4.0)[f]	T(2.0, 2.0, 3.0)[f]
Subsurface storage (days)		T(182, 365, 730)			LN(241, 58)			T(30, 35, 40)	
Subsurface pathogen decay rate (log_{10}/day)[g]	5.6	N(0.012, 0.0030)	N(0.0055, 0.0036)	5.6	N(0.012, 0.0030)	N(0.0055, 0.0036)	5.6	N(0.012, 0.0030)	N(0.0055, 0.0036)
Rapid sand filtration (log_{10})							T(0.8, 1.5, 3.3)[f]	T(0.8, 2.9, 5.4)[f]	T(0.1, 0.5, 3.9)[f]
UV (log_{10})				T(2.0, 3.0, 4.0)[e]	T(2.0, 3.0, 4.0)[e]	T(2.0, 2.0, 3.0)[e]	T(2.0, 3.0, 4.0)[f]	T(2.0, 3.0, 4.0)[f]	T(2.0, 2.0, 3.0)[f]
Chlorination (log_{10})	T(2.0, 4.0, 6.0)[e]	T(0.0, 0.0, 0.5)[e]	T(1.0, 2.0, 3.0)[e]	T(2.0, 4.0, 6.0)[e]	T(0.0, 0.0, 0.5)[e]	T(1.0, 2.0, 3.0)[e]			
Volume ingested (L)[h]	2	2	2	2	2	2	2	2	2
Exposures per year[h]	365	365	365	365	365	365	365	365	365

T: Triangular distribution; LN: Lognormal distribution; N: Normal distribution; U: Uniform distribution
Engineered treatment efficacy log_{10} removal efficiencies come from; except Torreele/St-André from Ayuso-Gabella *et al.* (2007)
[a] 95th Percentile as per TableA3.1 of the Guidelines for Stormwater Harvesting and Reuse: *Campylobacter* 15 n/L; *Cryptosporidium* 1.8 n/L; rotavirus 1 n/L (NRMMC-EPHC-NHMRC 2009a).
[b] Robertson *et al* (2006).
[c] Cited in Kocwa-Haluch and Zalewska (2002).
[d] Adapted from NRMMC–EPHC–AHMC (2006).
[e] Adapted from Smeets *et al.* (2006); NRMMC–EPHC–NHMRC (2006).
[f] From Ayuso-Gabella *et al.* (2007).
[g] From Sidhu *et al.* (2010). Single value only for *Campylobacter*.

decay rate has only been measured at Parafield (Sidhu *et al.* 2010) and to date, there are no other published studies that measure this decay rate in aquifers. Then, the highest $\log_{10}$ unit removals by aquifer treatment for *Cryptosporidium* are obtained at Atlantis, Shafdan and Parafield, that have longer residence times. In Sabadell, a very low residence time is considered, and the triangular function created for the residence time could be adjusted according to hydrogeological modelling and tracer tests studies, that are still required at the site. Rotavirus had different $\log_{10}$ removals in the aquifer (Table 19.7) depending on the decay rates used (Tables 19.5 and 19.6). For Atlantis, Parafield and Torreele/St. André the decay rate used was the one measured at Parafield aquifer, which was low, thus rendering low removals. For Nardò, the decay rate for viruses was measured in the site, and it was higher than for bacteria (La Mantia *et al.* 2008). For Shafdan and Sabadell, decay rates from the literature were used (Pedley *et al.* 2006), and removal for Shafdan was very high.

Table 19.7 Calculated aquifer barrier and engineered treatment (non-aquifer) removal efficiency in $\log_{10}$ units.

Case study		**Atlantis**		**Nardò**		**Parafield**		**Sabadell**		**Shafdan**		**Torreele/St-André**	
Treatment removal	**$\log_{10}$**	**Aquifer**	**Non-aquifer**	**Aquifer**	**Non-aquifer**	**Aquifer**	**Non-aquifer**	**Aquifer**	**Non-aquifer**	**Aquifer**	**Non-aquifer**	**Aquifer**	**Non-aquifer**
Campylobacter	Min	≥6.0	4.1	0.5	1	≥6.0	5.5	≤0.1	3.6	2.7	3	≥6.0	13.4
	Most likely	≥6.0	7.4	1.2	2.7	≥6.0	9.0	0.5	8.2	≥6.0	5	≥6.0	16.9
	Max	≥6.0	12.2	2.5	5.1	≥6.0	12.5	1.5	15	≥6.0	7.5	≥6.0	29.0
Cryptosporidium	Min	0.3	0.9	0.8	0.6	0.1	2.5	≤0.1	2.7	0.2	0.5	≤0.1	11.2
	Most likely	5.0	2.3	1.8	2	2.8	3.5	≤0.1	4.6	3.1	1	0.4	16.7
	Max	≥6.0	5.3	3.7	3.5	≥6.0	5.5	0.3	8	≥6.0	2	0.9	31.2
Rotavirus	Min	≤0.1	1.2	1.6	0.5	≤0.1	3.0	≤0.1	1.5	≥6.0	1.5	≤0.1	8.3
	Most likely	2.5	3.7	3.8	1.5	1.4	4.0	2.1	4.5	≥6.0	2.5	0.2	17.2
	Max	≥6.0	5.3	≥6.0	3.5	≥6.0	≥6.0	≥6.0	11.1	≥6.0	5.1	0.7	25.2

Note: In those case studies where the recovered water is used for irrigation purposes (Nardò, Sabadell and Shafdan) the non-aquifer treatment does not include the postharvest decay and washing treatments.

Knowledge of both the aquifer residence time and the rate of decay are essential for enabling the treatment value of the aquifer to be determined (Table 19.7). The decay of pathogens in groundwater during MAR is influenced by a range of factors such as the activity of indigenous groundwater microorganisms, temperature, redox status, oxygen concentrations and organic carbon concentrations (Gordon & Toze, 2003; Toze *et al.* 2004). Research has shown that bacteria tend to survive for much shorter times in aquifers than enteric viruses and protozoa (Toze *et al.* 2004) but the relative times can be aquifer-dependent. Virus decay rate is not always linear. The decay of some pathogens, in particular the more resistant viruses have been observed to have a decline in decay rate with time. Thus, in these cases a broken stick model of decay with different rates of decay may be more appropriate than a single rate of decay. The most appropriate decay rates to use will need to be verified in future risk assessments.

19.5.2 Case study sites human health risk assessment

The results in DALYs of the risk assessment are reported in Table 19.8. In the case studies where the recovered water is used for crops irrigation, green space irrigation or streets cleaning, the scenario considered is the accidental aerosols inhalation by growers and/or irrigators, as it is the worst case among the scenarios tested in Ayuso-Gabella *et al.* (submitted). The other scenarios tested were accidental aerosols inhalation by local communities and crops consumption. All these scenarios are recommended to be evaluated in the WHO Guidelines for Wastewater Reuse in irrigation (WHO, 2006).

Table 19.8. Mean, Median and 95th percentile residual risk assessment in DALYs.

Pathogen		**Atlantis**	**Nardò**	**Parafield**	**Sabadell**	**Shafdan**	**Torreele/St-André**
Campylobacter	Mean	$<1.0 \times 10^{-10}$	$\mathbf{4.2 \times 10^{-5}}$	$<1.0 \times 10^{-10}$	2.9×10^{-10}	$<1.0 \times 10^{-10}$	$<1.0 \times 10^{-10}$
	Median	$<1.0 \times 10^{-10}$	$\mathbf{9.7 \times 10^{-6}}$	$<1.0 \times 10^{-10}$	$<1.0 \times 10^{-10}$	$<1.0 \times 10^{-10}$	$<1.0 \times 10^{-10}$
	95th	$<1.0 \times 10^{-10}$	$\mathbf{1.9 \times 10^{-4}}$	$<1.0 \times 10^{-10}$	5.1×10^{-10}	$<1.0 \times 10^{-10}$	$<1.0 \times 10^{-10}$
Cryptosporidium	Mean	$\mathbf{7.0 \times 10^{-6}}$	8.9×10^{-7}	7.7×10^{-9}	5.3×10^{-8}	7.6×10^{-7}	$<1.0 \times 10^{-10}$
	Median	5.3×10^{-9}	2.1×10^{-7}	2.0×10^{-10}	2.0×10^{-8}	8.6×10^{-8}	$<1.0 \times 10^{-10}$
	95th	$\mathbf{1.2 \times 10^{-5}}$	$\mathbf{3.8 \times 10^{-6}}$	1.8×10^{-8}	2.1×10^{-7}	$\mathbf{3.0 \times 10^{-6}}$	$<1.0 \times 10^{-10}$
Rotavirus	Mean	$\mathbf{2.3 \times 10^{-4}}$	$\mathbf{1.7 \times 10^{-6}}$	8.5×10^{-7}	9.1×10^{-8}	1.4×10^{-10}	$<1.0 \times 10^{-10}$
	Median	$\mathbf{4.9 \times 10^{-5}}$	2.5×10^{-8}	5.0×10^{-8}	1.6×10^{-10}	$<1.0 \times 10^{-10}$	$<1.0 \times 10^{-10}$
	95th	$\mathbf{8.3 \times 10^{-4}}$	$\mathbf{5.7 \times 10^{-6}}$	$\mathbf{3.1 \times 10^{-6}}$	1.1×10^{-7}	5.2×10^{-10}	$<1.0 \times 10^{-10}$

Bold value indicates value exceeds guideline of 1×10^{-6} DALYs; limit of calculation is 1×10^{-10} DALYs.

Atlantis had acceptable risk for *Campylobacter*, but higher risk for *Cryptosporidium* and very high risk for rotavirus. Parafield and Shafdan had low risks for each of the pathogens. Torreele/St-André had a very low risk for each pathogen due to the large pre- and post-recovery treatment trains. Sabadell had also a very low risk for each pathogen, due to the large post-recovery treatment trains. Nardò had medium risks for *Campylobacter* and rotavirus, even though the dose considered is low as the water is used for irrigation purposes. At Atlantis, Parafield and Torrele/St. André case studies the recovered water will be used for drinking water and potable purposes, assuming a daily ingestion of 2 L. In Sabadell, the recovered water is not intended for drinking water purposes, but for urban parks irrigation and streets cleaning. In Shafdan and Nardò the recovered water is used for crops irrigation, but at Nardò the water is recovered from a different well than the one considered for the calculations. The dose considered is according to an accidental aerosol ingestion by a grower or irrigator, and it ranges from 0.0001 to 0.001 mL, which is a dose between 3–4 orders of magnitude lower than the drinking water dose.

While the mean gives an assessment of the average risk, and the median its central tendency, the 95th percentile gives an estimate of the variability and reasonable maximum of the risk. Where the 95th percentile was below the acceptable risk threshold, the risk assessment was considered to be robust. As such the risk assessment from rotavirus for Parafield is not as robust and further work is required to reduce the uncertainty of this risk estimate or further treatment is required to reduce the risk. Something similar is observed for *Cryptosporidium* in Shafdan and Nardò. For the other case study sites, the median and 95th percentile are similar indicating that the assessments are generally robust.

To evaluate the risk from enteric pathogens during MAR the potential presence of these pathogens and their numbers need to be determined. The major source of all enteric pathogens is faecal contamination, particularly from human faecal material. The largest number of enteric pathogens can be expected to be detected in untreated wastewater (Tables 19.5 and 19.6) with numbers reducing through the non-aquifer and aquifer treatment processes (Table 19.7). The potential presence of enteric pathogens in the recharge water is directly linked to the potential of human faecal matter contaminating the water. Thus, in this study, the pathogen risk for Torreele/St-André was assessed to be very low due to the high level of treatment prior to MAR. On the other hand, the level of treatment of the wastewater prior to MAR is much lower in Sabadell, Shafdan and Nardò, thus highly increasing the risk of pathogen presence in the recharged water. The Atlantis scheme has less opportunity for the presence of microbial pathogens due to the blending of treated wastewater and stormwater, while the risk in the Parafield system is limited to the potential for sewer pump-station overflows and contamination from animal faeces.

A number of limitations to the QMRA approach with MAR systems have been previously identified by Toze *et al.* (2010), including factors such as variability in pathogen decay rates. An accurate risk assessment also requires the input of accurate pathogen numbers. The initial pathogen numbers in the recharge water are influenced by a range of factors such as disease burden of the local population and the level of treatment for the recharge water. The numbers of some pathogens are also less accurate due to the difficulties in detection. For example, the detection of *Cryptosporidium* oocysts and rotavirus is difficult due to the lack of suitable culture methods and the low numbers (≤100 units) usually present in large volumes of water (>1 L). Numbers in river, canal and recreation water for *Cryptosporidium* oocysts have been quoted as between 5 and 240 oocysts per 10 litres (Schets *et al.* 2008; Plutzer *et al.* 2008; Mons *et al.* 2009). In comparison rotavirus numbers in similar water types have been reported to be between 2 and 200 detectable units per litre (Mehnert *et al.* 1993; Lodder *et al.* 2005).

In general, the risks evaluated for each of the MAR sites (Table 19.8) were in the order Atlantis > Nardò > Parafield and Shafdan > Sabadell > Torreele/St-André for *Cryptosporidium* and rotavirus, and all of the sites but Nardò had low risks for the bacterial pathogen, *Campylobacter*. Only Atlantis and Nardò did not meet the tolerable risk value for all the reference pathogens (Table 19.8).

19.5.3 Valuing the aquifer barrier in MAR schemes

A sensitivity analysis was performed for each barrier in the treatment train for each case study site and the factor sensitivity (FS) calculated. The FS calculation standardises the comparison between each of the water treatment barriers and the aquifer and thereby aids in valuing the aquifer as part of the larger treatment train. A value of 1.0 indicates a ten-fold increase in risk. Table 19.9 gives a comparison of the FS values for each of the treatment barriers across the MAR systems.

For Atlantis and Shafdan the FS analysis indicated that the aquifer was the single most important barrier in determining risk from all pathogens. For Atlantis, >6 orders of magnitude increase in risk would result if the aquifer was removed from the treatment train for *Campylobacter*. For Shafdan, 6 orders of magnitude increase in risk would result if the aquifer was removed from the treatment train for *Campylobacter* and rotavirus. If the aquifer barrier is in place at Atlantis the other barriers have little influence in determining the residual risk from *Campylobacter*. For *Cryptosporidium*, the secondary wastewater treatment plant had a slightly lower capacity to reduce the residual risk than the aquifer. In the case of Shafdan, if the aquifer barrier is in place the other barriers have little to no influence in determining the residual risk for all of the pathogens.

Table 19.9. Factor Sensitivity ratio – relative importance of barriers.

	Atlantis	Nardò	Parafield	Sabadell	Shafdan	Torreele/St-André
Campylobacter						
Constructed wetland	0.00*	–	0.00*	–	–	–
Secondary treatment	0.00*	1.3	–	1.8	2.1	0.00*
Ultra filtration	–	–	–	–	–	0.00*
Reverse osmosis	–	–	–	–	–	0.00*
UV disinfection	–	–	–	–	–	0.00*
Channel/River mixture	–	0.6	–	0.1	–	–
Aquifer	7.57	0.9	1.29	1.3	6.0	0.00*
Rapid sand filtration	–	–	–	–	–	0.00*
UV disinfection	–	–	0.00*	2.7	–	0.00*
Chlorination	0.00*	–	0.00*	2.9	2.9	–
Rapid sand filtration	–	–	–	0.2	–	–
Cryptosporidium						
Constructed wetland	0.78	–	0.61	–	–	–
Secondary treatment	1.65	0.9	–	0.9	0.9	1.24
Ultra Filtration	–	–	–	–	–	3.48
Reverse Osmosis	–	–	–	–	–	3.48
UV disinfection	–	–	–	–	–	2.57
Channel/River mixture	–	0.8	–	0.6	–	–
Aquifer	1.93	1.5	2.03	0.1	2.3	0.00*
Rapid sand filtration	–	–	–	–	–	1.92
UV disinfection	–	–	2.78	2.7	–	2.57
Chlorination	0.05	–	0.14	0.1	0.2	–
Rapid sand filtration	–	–	–	0.2	–	–
Rotavirus						
Constructed wetland	0.00	–	0.00	–	–	–
Secondary treatment	0.35	1.0	–	1.0	1.1	1.14
Ultrafiltration	–	–	–	–	–	4.51
Reverse osmosis	–	–	–	–	–	3.49
UV disinfection	–	–	–	–	–	2.23
Channel/River mixture	–	0.5	–	0.1	–	–
Aquifer	0.55	2.2	0.94	1.2	5.7	2.23
Rapid sand filtration	–	–	–	–	–	0.92
UV disinfection	–	–	1.94	1.6	–	2.23
Chlorination	0.43	–	1.66	1.5	1.7	–
Rapid sand filtration	–	–	–	0.1	–	–

*FS score could not be calculated as the resultant risk was equal to the residual risk.

For Parafield the aquifer barrier again dominated the risk from *Campylobacter*, resulting in over ten fold increase in risk if it were not present. The aquifer was the third most important barrier with respect to rotavirus and second for *Cryptosporidium* risk (UV slightly higher), but post-recovery UV and chlorine disinfection was each superior to the aquifer in reducing risk for rotavirus.

For Torreele/St-André the aquifer only played a measurable role in reducing residual risk for rotavirus. The most important barriers were ultrafiltration and reverse osmosis for each of the reference pathogens. The FS value of the aquifer could not be calculated for *Cryptosporidium* and *Campylobacter* as the revised risk in removing the barrier was equal to the initially calculated residual risk, $<1.0 \times 10^{-10}$ DALYs.

For Sabadell the aquifer was not the most important barrier to reduce the risk, achieving a similar effect to the secondary treatment for *Campylobacter* and rotavirus. The key barriers to reduce the risk for all the pathogens were the UV disinfection and chlorination.

At Nardò the results were rather different, being the aquifer the most important barrier to reduce *Cryptosporidium* and rotavirus, and the second most important barrier to reduce the risk for *Campylobacter*.

From the FS analysis of Table 19.8, the subsurface treatment steps were identified as being highly variable in the treatment train in reducing the calculated residual risk. The initial pathogen numbers in the water to be recharged for each MAR site is a function of the pre-treatment barriers. Torreele/St-André with its large pre-treatment train (average $\log_{10}$ removals of 14.7, 10.8, 12.4 for rotavirus, *Cryptosporidium* and *Campylobacter* respectively) begin with very low numbers of pathogens in the recharge water. Atlantis and Parafield have lower numbers than Torreele/St-André but Parafield has much lower numbers of pathogens than Atlantis as its recharge water was solely urban stormwater as opposed to reclaimed effluent with a minor component of stormwater. The pathogen numbers for each site steadily decreased as a function of the decay rate and the residence time in the aquifer reported in Table 19.5 and Table 19.6.

19.5.4 Integrating aquifer treatment with engineered treatments

To date aquifer treatment is being slowly integrated into an engineered water treatment train due to the difficulty in measuring a quantifiable reduction in risk. This is in part due to the adoption of risk-based management systems, such as the Hazard Analysis and Critical Control Point (HACCP) approach. HACCP concepts have been adopted by the water industry and promoted as a more proactive approach to managing drinking water supplies (WHO, 2004; NRMMC-EPHC-NHMRC, 2008), as well as recycled water systems (NRMMC-EPHC-AHMC, 2006) and even MAR systems (NRMMC-EPHC-NHMRC, 2009b). Yet, aquifer treatment remains difficult to integrate as there are no easily identifiable critical limits and control points such as for the more common water treatment technologies, like chlorination which uses contact time, UV disinfection which uses UV-transmittance and membrane treatments which use pressure and electrical conductivity.

It is proposed that an extension of the FS sensitivity analysis could also be used to provide a means of generating evidence-based critical limits to manage critical control points. While there are no health-based targets for pathogen numbers QMRA can be used to address the issue of setting critical limits. This is done by treating the DALYs estimates as representing acceptable estimates of "absolute" risk and comparing them to the agreed international human health risk benchmarks, 1.0×10^{-6} DALYs (WHO 2004). In this instance, the comparison of the Parafield risk estimate indicated that the residual risk was acceptable for *Campylobacter* when compared to this benchmark and this conclusion was robust as indicated by the 95th percentile being less than the benchmark value. However, for rotavirus the assessment was less robust and the required aquifer residence time was just great enough for the scheme to support so additional post-recovery treatment could be required. An illustrative example for setting of critical limits for mean aquifer residence time comes from the *Cryptosporidium* for the Atlantis site, where the mean residence time needs to exceed ~550 days to achieve tolerable levels of risk. Again, this assumes that the pathogen decay rates of Sidhu *et al.* (2010) are linear and are representative of the processes occurring in the subsurface of this site. Use of the residence time critical limit could also be used to design infiltration and extraction pumping regimes to ensure the mean residence time in the aquifer is achieved. Where aquifer residence time is not accurately known, such as in the Sabadell case study site, it can be determined by use of suitable groundwater tracers. This can include both applied tracers, substances injected into the groundwater intentionally and thereby in controlled doses, time intervals and locations (such as SF_6) or natural tracers (such as the recharge water electrical conductivity if this has marked variation from the ambient groundwater). Knowledge of the residence time in the aquifer coupled with pathogen decay rates could then be used to fully appreciate the water treatment function of the subsurface and integrate the aquifer barrier with the engineered treatments in the provision of safe drinking water.

19.6 CONCLUSIONS

MAR case studies were evaluated for chemical (organic compounds) and microbiological risks.

For the organic compounds, the RQ methodology was used, and the results showed no risk for the human health in the case studies evaluated.

To assess the microbiological risks, the QMRA methodology was used. QMRA provides a means of quantifying the combined effects of aquifers and engineered treatments for reference pathogens in terms of $\log_{10}$ removal characteristics. The use of QMRA was found to be a useful tool in establishing the value of the aquifer within the treatment train and allowed the assessment of human health risk from pathogens in terms of DALYs. In general, the risks were below the benchmark value of 1.0×10^{-6} DALYs for all the reference pathogens, except for some exceptions. A sensitivity analysis was used to assess which of the treatment barriers was most important in each of the MAR systems. In this case, some sites relied completely in the aquifer to reduce the risks, whereas in others the aquifer had nearly no effect in reducing the human health risks. Nevertheless, the QMRA approach allows the integration of the aquifer treatment characteristics into the larger engineered treatment train and could be used in the future to quantitatively assess the reduction of human health risk for MAR systems more generally.

REFERENCES

Aharoni A., Ayuso Gabella M. N., Cauwenberghs J., Chavez A., Cikurel H., Genthe B., Gibson R., Jeffrey P., Jimenez B., Le Corre K., Masciopinto C., Page D., Regel R., Rinck-Pfeiffer S., Salgot M., Steyn M., van Houtte E., Tredoux G., Xuzhou C. and Zhao X. (2008). Deliverable 4.2: Technical water reclamation and aquifer recharge case studies. RECLAIM WATER project.

Ayuso-Gabella M. N., Barry K., Bixio D., Dillon P., Genthe B., Jefferson B., Jeffrey P., Kopač I., Page D., Pavelic P., Purdie M., Regel R., Rinck-Pfeiffer S., Salgot M., Van Houtte E. and Wintgens T. (2007). *Deliverable D6.1 Report on case-study specific risk assessment*, RECLAIM WATER.

Ayuso-Gabella N., Page D., Masciopinto C., Aharoni A., Salgot M. and Wintgens T. (in press). Quantifying the effect of Managed Aquifer Recharge on the microbiological human health risks of irrigating crops with recycled water, *Agricultural Water Management*.

Böckelmann U., Dörries H.-H., Ayuso-Gabella M. N., Salgot de Marçay M., Tandoi V., Levantesi C., Masciopinto C., Van Houtte E., Szewzyk U., Wintgens T. and Grohmann E. (2009). Quantitative PCR monitoring of antibiotic resistance genes and bacterial pathogens in three european artificial groundwater recharge system. *Applied and Environmental Microbiology*, **75**(1), 154–163.

Cleary S. A. (2005). Sustainable drinking water treatment for small communities using multistage slow sand filtration. PhD thesis, University of Waterloo, tOntario, Canada.

Dillon P. (2005). Future management of aquifer recharge, *Hydrogeology Journal*, **13**, 313–316.

Dillon P. and Toze S. (2005). Water quality improvements during aquifer storage and recovery, Vol 2. Compilation of Information from Ten Sites. *AWWARF Project 2618*, Final Report, p. 203.

Dillon P., Pavelic P., Page D., Beringen H. and Ward J. (2009). 'Managed aquifer recharge: an introduction', *National Water Commission Waterlines Report Series* No 13, February 2009.

Ernst M., Asmin J., Benner J., Fink G., Amy G., Hollender J., Hein A., Krauss M., McArdell C., Reemtsma T., Sharma S. and Ternes T. (2009). *Milestone 3.2: Monitoring of developed and adapted parameters*, RECLAIM WATER.

Gordon C. and Toze S. (2003). Influence of groundwater characteristics on the survival of enteric viruses. *Journal of Applied Microbiology,* **95**(3), 536–544.

Jiménez B. (2004). Chapter 12.3 El Mezquital, Mexico: The biggest irrigation district that uses wastewater. In: Water Reuse for Irrigation: Agriculture, Landscape and Turf Grass. Lazarova and Bahri Editors. CRC Press, 535–562.

Jiménez B. and Chávez A. (2004). Quality assessment of an aquifer recharged with wastewater for its potential use as drinking source: "El Mezquital Valley" case. *Water Science and Technology*, **50**(2), 269–273.

Jiménez B. and Asano T. (2008). Water reuse around the World. In: Water Reuse: An International Survey of Current Practice, Issues and Needs. Jiménez and Asano (eds), IWA Publishing, Inc. London, 628, ISBN 156 670 6491.

John D. E. and Rose J. B. (2005). A Review of factors affecting microbial survival in ground water. *Environmental Science and Technology*, **39**(19), 7345–7356.

Kocwa-Haluch R. and Zalewska B. (2002). Presence of rotavirus hominis in sewage and water. *Polish Journal of Environmental Studies*, **11**, 751–755.

Kopač I., Ayuso-Gabella M. N. and Salgot M. (2007). Integrating disability adjusted life-years (DALYs) as a tool for human health risk assessment in the RECLAIM WATER. *Proceedings of the 6th Conference on Wastewater Reclamation and Reuse for Sustainability*. October 9-12, Antwerp, Belgium.

Kremer S., Pavelic P., Dillon P. and Barry K. (2008). Flow and solute transport observations and modelling from the first phase of flushing operations at the salisbury ASTR site. CSIRO: Water for a Healthy Country National Research Flagship, www.csiro.au/org/HealthyCountry.html

La Mantia R., Masciopinto C., Levantesi C. and Tandoi V. (2008). Fate and transport of faecal contamination microbial indicators, pathogenic protozoa and Campylobacter in the artificially recharged fractured aquifer of Salento, Italy. *Water Science and Technology*, **57**(6), 849–856.

Le Corre K., de Heyder B., Masciopinto C., Aharoni A., Cikurel H., Zhao X., Salgot M., Ayuso Gabella M. N., Saperas N., Cartmell E., Jefferson B. and Jeffrey P. (2007). Preliminary results of managed aquifer recharge with reclaimed wastewater in five operational case studies from around the world. *Proceedings of the 6th Conference on Wastewater Reclamation and Reuse for Sustainability*. October 9-12, Antwerp, Belgium.

Le Corre K., Aharoni A., Ayuso Gabella M. N., Cauwenberghs J., Chavez A., Cikurel H., Genthe B., Gibson R., Jeffrey P., Jimenez B., Masciopinto C., Page D., Regel R., Rinck-Pfeiffer S., Salgot M., Steyn M., van Houtte E., Tredoux G., Xuzhou C. and Zhao X. (2009). *Deliveralbe 4.2: Integrated report from each test site on technology performance and contaminant monitoring results and a cross case analysis*, RECLAIM WATER.

Levantesi C., La Mantia R., Masciopinto C., Böckelmann U., Ayuso-Gabella M. N., Salgot M., Tandoi V., Van Houtte E., Wintgens T. and Grohmann E. (2010). Quantification of pathogenic microorganisms and microbial indicators in three wastewater reclamation and managed aquifer recharge facilities in Europe. *Science of the Total Environment*, **408**, 4923–4930.

Lodder W. J. and Husman A. M. D. (2005). Presence of noroviruses and other enteric viruses in sewage and surface waters in The Netherlands. *Applied and Environmental Microbiology*, **71**(3), 1453–1461.

Mehnert D. U. and Stewien K. E. (1993). Detection and distribution of rotavirus in raw sewage and creeks in Sao Paulo, Brazil. *Applied and Environmental Microbiology*, **59**(1), 140–143.

Mons C., Dumetre A., Gosselin S., Galliot C. and Moulin L. (2009). Monitoring of *Cryptosporidium* and *Giardia* river contamination in Paris area, *Water Research* **43**(1), 211–217.

Montemayor M., Valero F., Cofre J. and Lucena F. (2005). Occurrence of *Cryptosporidium* spp. oocysts in raw and treated sewage and river water in north-eastern Spain. *Journal of Applied Microbiology*, **99**, 1455–1462.

Nasser A. M. and Oman S. D. (1999). Quantitative assessment of the inactivation of pathogenic and indicator viruses in natural water sources. *Water Research,* **33**, 1748–1752.

NRMMC-EPHC–NHMRC (Natural Resource Management Ministerial Council, Environment Protection, Heritage Council, Australian Health Ministers Conference). (2006). Australian Guidelines for Water Recycling: Managing Health and Environmental Risks (Phase1). (http://www.ephc.gov.au/sites/default/files/WQ_AGWR_GL__Managing_Health_Environmental_Risks_Phase1_Final_200611.pdf)

NRMMC-EPHC–NHMRC (Natural Resource Management Ministerial Council, Environment Protection, Heritage Council, Australian Health Ministers Conference). (2008). *Australian Guidelines for Water Recycling: Managing Health and Environmental Risks. Phase 2A: Augmentation of Drinking Water Supplies*, Natural Resource Management Ministerial Council, Environment Protection and Heritage Council, and National Health and Medical Research Council, www.ephc.gov.au/taxonomy/term/39.

NRMMC-EPHC–NHMRC (Natural Resource Management Ministerial Council, Environment Protection, Heritage Council, Australian Health Ministers Conference). (2009a). *Australian Guidelines for Water Recycling: Managing Health and Environmental Risks. Phase 2B: Stormwater Harvesting and Reuse*, Natural Resource Management Ministerial Council, Environment Protection and Heritage Council and National Health and Medical Research Council, www.ephc.gov.au/taxonomy/term/39.

NRMMC-EPHC–NHMRC (Natural Resource Management Ministerial Council, Environment Protection, Heritage Council, Australian Health Ministers Conference). (2009b). *Australian Guidelines for Water Recycling: Managing Health and Environmental Risks. Phase 2C: Managed Aquifer Recharge*, Natural Resource Management Ministerial Council, Environment Protection and Heritage Council and National Health and Medical Research Council, www.ephc.gov.au/taxonomy/term/39.

Page D., Barry K., Pavelic P., Dillon P. and Chassagne A. (2008). Preliminary quantitative risk assessment for the Salisbury stormwater ASTR project. Water for a Healthy Country National Research Flagship report CSIRO, http://www.clw.csiro.au/publications/waterforahealthycountry/2008/wfhc-Salisbury-ASTR-HACCP.pdf.

Page D. W, Vanderzalm J., Barry K., Levett K., Kremer S., Ayuso-Gabella M. N., Dillon P., Toze S., Sidhu J., Shackleton M., Purdie M. and Regel R. (2009). Operational residual risk assessment for the Salisbury stormwater ASTR project. CSIRO: Water for a Healthy Country National Research Flagship, http://www.csiro.au/files/files/psgp.pdf.

Page D., Dillon P., Vanderzalm J., Bekele E., Barry K., Miotlinski K. and Levett K. (2010a). Managed aquifer recharge case study risk assessments. *Water for a Healthy Country Flagship Report series*, ISSN: 1835-095X.

Page D. W., Dillon P., Toze S. and Sidhu J. P. S. (2010b). Characterising aquifer treatment for pathogens in managed aquifer recharge, *Water Science and Technology*, **62**(9), 2009–2015.

Page D., Dillon P., Toze S., Bixio D, Genthe B., Jiménez Cisneros B. and Wintgens T. (2010c). Valuing the subsurface pathogen treatment barrier in water recycling via aquifers for drinking supplies, *Water Research*, **44**, 1841–1852.

Page D., Vanderzalm J., Barry K., Levett K., Sidhu J., Toze S., Kremer S. and Dillon P. (2010d). Managed Aquifer Recharge (MAR) staged risk assessment framework applied to the Parafield Aquifer Storage Transfer Recovery system to assess human and environmental risks. *Journal of Environmental Quality*, **39**(6), 2029–2039.

Pavelic P., Dillon P. and Robinson N. (2004). Groundwater modelling to optimise well-field design and operation for the ASTR trial at the Greenfield Railway Station site, Salisbury, South Australia, *CSIRO Land and Water Technical Report No. 27/04*.

Pedley S., Yates M., Schijven J. F., West J., Howard G. and Barrett M. (2006). Pathogens: Health relevance, transport and attenuation. In: Protecting Groundwater for Health. Section 1: Managing the Quality of Drinking-water Sources. O. Schmoll, G. Howard, J. Chilton and I. Chorus (eds), WHO, IWA publishing, London, pp. 49–76.

Plutzer J., Karanis P., Domokos K., Torokne A. and Marialigeti K. (2008). Detection and characterisation of *Giardia* and *Cryptosporidium* in Hungarian raw, surface and sewage water samples by IFT, PCR and sequence analysis of the SSUrRNA and GDH genes. *International Journal of Hygiene and Environmental Health*, **211**(5–6), 524–533.

Robertson L. J., Hermansen L. and Gjerde B. K. (2006). Occurrence of *cryptosporidium* oocysts and *Giardia* cysts in sewage in Norway. *Applied and Environmental Microbiology*, **72**(8), 5297–5303.

Schets F. A., van Wijnen J. H., Schijven J. F., Schoon A. and de Husman A. M. (2008). Monitoring of waterborne pathogens in surface waters in Amsterdam, The Netherlands, and the potential health risk associated with exposure to Cryptosporidium and Giardia in these waters. *Applied and Environmental Microbiology* **74**(7), 2069–2078.

Schriks M., Heringa M. B., van der Kooi M. M. E., de Voogt P. and van Wezel A. P. (2009). Toxicological relevance of emerging contaminants for drinking water quality. *Water Research*, **44**(2), 461–476.

Sedmak G., Bina D., MacDonald J. and Couillard L. (2005). Nine-year study of the occurrence of culturable viruses in source water for two drinking water treatment plants and the influent and effluent of a wastewater treatment plant in Milwaukee, Wisconsin (August 1994 through July 2003). *Applied and Environmental Microbiology*, **71**(2), 1042–1050.

Sidhu J. P. S., Toze S., Hodgers L., Shackelton M., Barry K., Page D. and Dillon P. (2010). Pathogen inactivation during passage of stormwater through a constructed reedbed and aquifer transfer, storage and recovery. *Water Science and Technology*, **62**(5), 1190–1197.

Smeets P., Rietveld L., Hijnem W., Medema G. and Stenström T.-A. (2006). Efficacy of water treatment processes, MICRORISK Research report, http://www.microrisk.com/uploads/microrisk_efficacy_of_water_treatment_processes.pdf.

Swierc J., Page D., Van Leeuwen J. and Dillon P. (2005). Preliminary Hazard Analysis and Critical Control Points Plan (HACCP) – Salisbury Stormwater to Drinking Water Aquifer Storage Transfer and Recovery (ASTR) Project. *CSIRO Land and Water Technical Report No. 20/05*.

Toze S., Hanna J., Smith T., Edmonds L. and McCrow A. (2004). Determination of water quality improvements due to the artificial recharge of treated effluent. *Wastewater Reuse and Groundwater Quality*, IAHS Publication, **285**, 53–60.

Toze S., Bekele E, Page D., Sidhu J. and Shackleton M. (2010). Use of static quantitative microbial risk assessment to determine pathogen risks in an unconfined carbonate aquifer used for managed aquifer recharge. *Water Research*, **44**, 1038–1049.

US EPA (United States Environmental protection Agency) (1987). Risk Assessment Guidelines of 1986. EPA/600/8-87/045. Office of health and Environmental Assessment. Washington, DC.

US EPA (United States Environmental Protection Agency) (1998). Human Health Risk Assessment Protocol for Hazardous Waste Combustion facilities. Volume 1, EPA530/D/98/001A.

US EPA (United States Environmental Protection Agency) (2002). A review of the reference dose and reference concentration processes. EPA/630/P-02/002F. *Risk Assessment Forum*, US EPA, Washington, DC.

Van Houtte E. and Verbauwhede J. (2005). Artificial recharge of treated wastewater effluent enables sustainable groundwater management of a dune aquifer in Flanders, Belgium. *5th International Symposium on Management of Aquifer Recharge*, Berlin, Germany, 11–16th June 2005.

Van Houtte E. and Verbauwhede J. (2008). Operational experiences with indirect potable reuse at the Flemish coast, *Desalination*, **218**, 198–207.

Westrell T. (2004). Microbial risk assessment and its implications for risk management in urban water systems. PhD dissertation.

WHO (World Health Organisation). (2004). Guidelines for Drinking-water Quality. Third Edition. Volume 1. Recommendations. World Health Organisation, Geneva.

WHO (World Health Organisation). (2005). Water Safety Plans – Managing drinking-water quality from catchment to consumer. World Health Organisation, Geneva.

WHO (World Health Organization). (2006). Guidelines for the safe use of wastewater, excreta and greywater. Volume 2: Wastewater use in agriculture. World Health Organisation, Geneva

Yates M. V., Sterzenbach L. D., Gerba C. P. and Sinclair N. A. (1990). The effect of indigenous bacteria on virus survival in ground water. *Journal of Environmental Science and Health*, **25**, 81–100.

Ying G-G., Toze S., Hanna J., Yu X-Y., Dillon P. J. and Kookana R. S. (2008). Decay of endocrine-disrupting chemicals in aerobic and anoxic groundwater. *Water Research*, **42**(4–5), 1133–1141.

Zwietering M. H. and van Gerwen S. J. C. (2000). Sensitivity analysis in quantitative microbial risk assessment. *International Journal of Food Microbiology*, **58**, 213–221.

Chapter 20

Risk perception and communication for managed aquifer recharge

Kristell Le Corre, Susan Baggett, Melanie Muro and Paul Jeffrey

20.1 INTRODUCTION

Understanding the risks associated by different actors with a water reuse project, the specifics of any perceived threats and their consequences are key steps in building trust with stakeholders. Reconciling stakeholders' concerns with the potential benefits that could be achieved through water reuse schemes may be challenging without clear communication between all parties. Low levels of stakeholder awareness and support for a proposed MAR scheme can have a significant impact on the feasibility, costs, and sustainability of the project. This chapter provides general guidance on risk communication for water reuse schemes which includes key principles adapted from and relevant to many other types of capital project. It has been written with reference to current literature and using insights provided by several existing water reuse projects. The objective is to provide guidance for water managers to communicate the technical aspects of their plans and get more out of the stakeholder engagement process. The content focuses on the importance of communication in situations where specific strategies and actions need to be informed by local circumstances. In delivering this, we illustrate the main principles of incorporating evidence, clarity, wide relevance and transparency into the communication process. We also argue that risk communication should be regarded as an opportunity for an exchange of views and opinions in order to prevent misconceptions and address the concerns of those who perceive potential hazards. The agenda for dialogue and debate should therefore be driven by those impacted by a MAR scheme – not those implementing it.

20.2 REASONS FOR AND OBJECTIVES OF RISK COMMUNICATION

The risks of water reuse should be as clearly communicated as the benefits in order to inform consultation and engagement processes. Proactive risk communication demonstrates transparency and acts as a counterweight to uninformed speculation. Risk is an inherent component of all water resource schemes. Responsibility for shaping and managing the debate needs to be explicitly taken by a trusted body. There are no direct obligatory requirements for risk communication. However, in many jurisdictions, the general public and other interested parties have the right of access to information that impacts their environment (e.g. through the Århus Convention, 1998).

The risks identified by scientists are often very different from the risks perceived by the community (Daughton, 2004). In the past, the main goals of risk communication have been orientated towards the 'alignment of a community's risk perception with that of the risk experts' and to 'reduce the fear and community resistance of risk-related technology' (Khan & Gerrard, 2006). However, a more rewarding and iterative means of risk communication can be achieved by including the objectives established by Renn & Levine (1991) presented in Table 20.1. The goal of risk communication for a MAR scheme should not be to achieve water reuse without complaints from the stakeholders, but to support the emergence of an informed and involved public who can play a constructive role in project design and implementation. Communication of risks relies on a receptive audience. Communication should broadly seek to reduce uncertainty, ensure a perception of equity, and build trust.

Table 20.1 Communication objectives (Renn & Levine, 1991).

Role	Explanation
Enlightenment	Aiming to improve risk understanding among target groups
Right-to-know	Designed to disclose information about hazards to those who may be exposed
Attitude modification	To legitimise risk-related decisions, to improve the acceptance of a specific risk source, or to challenge such decisions and reject specific risk sources
Legitimacy	To explain and justify risk management routines to enhance the trust in the competence and fairness of the management process
Risk reduction	To enhance public protection through information about individual risk reduction measures
Behavioural change	To encourage protective behaviour or supportive actions towards the communicating agency
Emergency readiness	To provide guidelines or behavioural advice for emergency situations
Public involvement	Aiming at educating decision-makers about public concerns and perceptions
Participation	To assist in reconciling conflicts about risk related controversies

Understanding risk perceptions

While experts tend to define risk in terms of the high value placed by the public on health issues, the public viewpoint on what risk is and the value they attribute to their definition of risk is more complex (Bierle & Cayford, 2002). Perceptions of risk are unique to an individual, and are shaped by their values, education, experiences, their interest in the issue at stake and a multitude of other factors, such as voluntariness, trust and familiarity. For example, people can base judgements regarding actions taken or technologies used not only on how they think about it but also on how they feel towards it. If they favour an activity then they tend to judge the risks as low and the benefits as high; conversely, if they feel unfavourable towards it then they will judge it as high risk and of low benefit. Also, if the risk is seen to be under one's own direct control then it seems to be more acceptable than if the risk is controlled by others. MAR schemes will typically be perceived as the latter type of activity.

These risk features, often referred to as "outrage factors" have been shown to greatly influence the level of concern, or outrage, that people feel regarding a real or potential hazard (Covello & Sandman 2001). Table 20.2 outlines some water reuse examples of the 'outrage factors' that may influence stakeholder reactions to MAR schemes.

Table 20.2 Potential outrage factors for MAR schemes (adapted from Covello & Sandman, 2001).

Outrage factor	Relevant question for MAR scheme
Voluntariness	Can people choose whether they use recycled water?
Familiarity	How familiar are people with the concept of MAR. Do they know anyone else who lives with a MAR scheme?
Fairness	Do people feel that the selection process by which some receive recycled water whilst others don't has been fair?
Benefits	Are the benefits of water reuse obvious and transparent? Are there positive experiences available to examine?
Catastrophic potential	Is the risk of pollution and contamination understood and appropriately managed?
Understanding	Are the workings of, and risks associated with, the MAR scheme explained in layman's terms?
Delayed effects	What long term effects might there be from the scheme?
Effects on children	What additional risks might there be to young people who have prolonged exposure to using recycled water?
Trust	Which institutions and organisations are trusted by stakeholders to set standards for and monitor the scheme?
Accident history	Is there a history of public health incidents for MAR schemes of similar design?
Personal stake	What forms and intensity of personal exposure to the scheme and its impacts are associated with the various stakeholders?
Ethical/moral nature	Is there an ethical or moral dimension to stakeholder concerns about the scheme?
Human *vs.* natural origin	To what extent can the proposed scheme be seen as 'natural'?

The outrage factors listed can be broadly categorised into those relating to uncertainty, hazard potential, equity, and trust, illustrating that risk perceptions are not just shaped by the characteristics and impacts of a scheme but also a community's attitudes to and relationships with the institutions in charge. Indeed, how believable the public view any information conveyed to them about risk from any institution is closely related to the level of credibility and trust they attribute to those institutions. This multidimensional social trust, or confidence, can be seriously diminished if national agencies or institutions are judged to be incompetent, biased or compromised, or seem to arrive at decisions without regard for public concerns. Erosion of the social dimensions of trust can seriously impair the ability of institutions to communicate with risk bearers.

Although we acknowledge that outrage factors will vary depending on the specific case at hand (particularly the quality of existing relationships between communities and the authorities and previous engagement initiatives), several factors are bound to shape risk perceptions and subsequently public outrage in the context of water reuse. A very basic guide might be that the greater the potential for human contact with the recycled water, the greater the perceived risk. Hartley (2006) lists some other influential factors in addition to the degree of human contact affecting the acceptability of water reuse.

They suggest that community acceptance of reused water is higher when:

- It is clear that public health is ensured
- Promotion of water conservation or environmental protection is a clear benefit
- Cost of treatment and distribution technologies and systems is reasonable
- Perception of wastewater as the source of reused water is minimal
- Awareness of water supply problems in the community is high
- Role of water reuse in the overall water supply scheme is clear
- Perception of the quality of reused water is high
- Confidence in local management of public utilities and technologies is high

An outraged public is extremely difficult to engage in any constructive manner. Therefore, a conscious consideration of the outrage factors affecting a particular situation can help policy-makers and practitioners to address those which are more likely to shape the interaction with an affected community.

20.3 PRINCIPLES OF RISK COMMUNICATION

Against this background, it is crucial to extend the assessment and management of risk beyond the circle of 'experts' and engage in a meaningful and substantive dialogue with affected communities to explore and understand different perceptions of risk. There are a number of useful texts that provide guidance on risk communication (e.g. Morgan *et al.* 2002; Powell & Leiss, 1997). Valuable insights are also available from other industries, such as food safety, flood management, crisis management and more. Common themes run through the literature; one of the most important being that risk communication and dialogue is an enduring two way process and it is how questions are dealt with that defines the success of an engagement process. The provision of opportunities for easy communication between parties involved in a MAR scheme ensures that concerns are acknowledged and clarification is sought and received. However, the success of any engagement effort and its subsequent impact on decision outcomes and project delivery is not only dependent on the quality of the involvement opportunities offered to the affected public, but is equally shaped by the quality of relationships, trust and credibility between the parties involved. This section will highlight key challenges to consider when structuring a risk dialogue before presenting selected risk communication and engagement techniques.

20.3.1 Building and keeping trust

Successful communication and engagement is not simply a function of using the right methods and terminology. In the assessment of risks there are inherent uncertainties. A major aim of any communication strategy should therefore be to ensure that there is sufficient trust for those taking part to accept that uncertainty or lack of information is legitimate and reasonable. Establishing stakeholders' trust in the management and objectives of a MAR project is equally important. For effective communication, stakeholders should be able to trust the parameters of the communication process. Khan & Gerrard (2006) give a list of conditions to be met to build trust with the public in a water reuse project. In addition to the earlier points raised, the focus is more about communicating with the public, not simply to the public. The conditions they give are that:

- Dialogue is sustained
- The community has independent sources of information, not linked to the sponsoring agency
- The community can ask questions
- The community is involved early
- Information is available to everyone
- Behaviour is non coercive. It is considered a reasoned and fair way to make a decision
- Everyone's opinion matters, and there is a willingness to listen to all views and expand the discussion if necessary
- Citizens have some level of control in the process (such as by contributing to the agenda or ground rules)

Within this context, one should remember that social and cultural differences can act as barriers to communication, both in the sense that they may restrict access to the process, and that they might limit understanding, even if access is secured. Variation in individual or even community socio-cultural characteristics (e.g. ethnicity, language, religion) can make it difficult to design communication processes as the detail of what constitutes a legitimate process may be questioned. The credibility of different scientific inputs and frameworks for dialogue (e.g. the format of meetings) can also be influenced by cultural predispositions of valuing some forms of knowledge and process over others. Often, credibility may be enhanced by conveying the message through locally trusted parties, and not by an international consensus system or state based authority. Indeed, intermediaries with moral authority in the local cultural setting are often essential if dialogue is to happen at all.

20.3.2 Framing and managing communication

Developing a relationship based upon transparency and honesty is crucial for good risk communication. This does not mean that all available information should be actively communicated to all stakeholders, but thought should be given to their technical knowledge and how they may relate to it. Such information should have relevance for the stakeholder and it should be accessible. This means that the writing style should be informative and concise, and technical information and statistics should be presented clearly with an obvious reason for presentation of data. Information should not be too complex for a stakeholder, but expressing issues too simply can generate suspicion and be perceived to be deliberately vague. Care should be taken to highlight the research behind the facts presented. It is important to emphasise the professionalism of the project. Making it look too simple might be misleading and it might also cause a decline of trust in the project.

Complex risk information might require some explanation and this can be achieved in many different ways. A common approach for explaining risk information is by making a comparison with other projects or situations of a similar nature or, where unfamiliarity is a concern, a thematic example that might be more familiar. Simply quoting the probability of an event might not convey useful information to the stakeholder without some familiar context to identify it with. Yet it is important to be careful with comparisons. They should not be used to imply acceptability or to trivialise the concerns of the stakeholder. Therefore the comparison should be relevant for the specific subject. Comparisons may also be made in order to explore alternative scenarios, including the risks involved with doing nothing. Such a comparison should be able to convey to people what the actual consequences for them will be.

20.3.3 Some comments on bias and transparency

The influence of the media on public perceptions of risk is now well documented. The print and broadcast media have been criticised for the sensationalist tone of some of their reporting. Such coverage can amplify public concern about minimal risks, obfuscate the nature of significant risks, or hamper efforts to respond to a disaster. Risk assessment scientists and public interest activists have frequently noted their frustration in getting their messages to the general public through traditional media outlets. Where messages about risks and benefits have become distorted by over-simplistic or incompetent reporting, public trust in both science and governance regimes has been challenged.

But the media is not the only sector to have attracted criticism in recent years for its role. Indeed, the very nature of science's interfaces with other constituencies is rapidly changing, moving from work which is defined in relation to the cognitive and social norms that govern academic science, to knowledge produced in the context of application. Whilst the need for improvements in the way science interacts with the social and governance spheres is widely accepted, this realisation has not been matched by the linguistic and conceptual revisions needed to make the next step.

Important principles to guide the use of language and the design of discourses for public understanding of risk include the need for information to be jargon free or "jargon explained", as much as possible, for transcripts to use value explicit and undeterministic wording, and for ignorance to be acknowledged rather than being hidden by ambiguity. Issues of uncertainty and quality of information are always present.

20.4 COMMUNICATING RISK

What information should be available and when should action be taken to engage stakeholders? Should communication be pre-emptive, responsive or ongoing? Starting the communication process early can help to guide the process and should enable more constructive feedback to the benefit of the project. On the other hand at such an early stage much may still be unknown; particularly if a risk assessment has yet to be conducted or a risk management process formalised.

Effective communication at the beginning of a project can save time and pre-empt problems that might delay progress at a later stage. But issues of concern may emerge later on in the project. Traditionally risk management has been depicted as a linear process, a single step in a long management chain leading to the completed project. It is now more often described as a cyclic process with risk communication as an integral component as shown in Figure 20.1. Every risk management activity requires some kind of risk communication. The type of stakeholder, their level of involvement and interest defines who should be the main target for communication during each stage of the cycle. Therefore a different version of the cycle might be necessary for the different types of stakeholder. This cyclic view of risk management also ensures that risk communication is an ongoing process throughout a project.

Figure 20.1 The risk management cycle (Adapted from Lang *et al.* 2001)

The stages throughout a project at which risk communication is necessary may vary a great deal depending on the nature of the work being undertaken. When there is already a lot of attention given to the project at a very early stage, it might not be possible to choose the timing of risk communication. If for example the public media request information about general or specific risks, information should be given that holds relevance to that stage of the project. Repeating the same information package time after time will be noticed and may be perceived to show either a lack of progress or that no more consideration has been given since the original report was made.

Table 20.3 lists some common communication methods together with comments on their strengths and weaknesses in the context of a MAR scheme. Readers should note, however, that even the timeliest and best managed process may face opposition and may be unable to achieve positive outcomes on deeply decisive issues.

20.5 RECONCILING CONFLICTING VIEWS

Finally, it should be acknowledged that even the best-managed process may be unable to achieve consensus on deeply divisive issues. If a project faces opposition it is important first to identify the nature of the opposition. With information available from so many sources, there is potential for wasting time and resources responding to unfounded and unjustified complaints, but severe repercussions may follow if valid concerns are ignored. A clear criteria needs to be established to sort the complaints by validity in a consistent and transparent way. It is vital to ensure that legitimate contentions are seen to be given consideration.

Opposition to a project generally occurs when the risk perception of stakeholders differs from the risk perception of the project. This difference in perception can be caused by a disagreement about the analysis of the risks, but is often also highly influenced by the outrage factors outlined in Table 20.2.

20.5.1 Objections over hazard risk

Just like normal risk communication it is important to take stakeholders seriously when responding to opposition. Listen to what opposition has to say and do not respond simply by denial. It is important to avoid a confrontation. Lang *et al.* (2001) give a standard way of responding to opposition emerging based upon concerns regarding a specific perceived technical risk or hazard. The key point of the response is an effective explanation of the methods used to mitigate or nullify the risk, or the way in which it may have been established that a risk has been established to be negligible. Three guiding principles given by Lang *et al.* (2001) (pp. 329) are:

- Acknowledge the initial plausibility of the link, might they be correct?
- Explain what evidence would be expected if such a link existed
- Show that serious, well-conducted investigation has not found such evidence

Table 20.3 Risk communication methods for a MAR scheme.

Technique & description	Positive features	Negative features
Open/public meetings Widely advertised and free access event lasting perhaps a couple of hours. Various formats possible but would provide opportunities for questions to be asked by attendees.	• Provides opportunities for people to comment on matters that affect them close to home. • Requires no special training to implement (although professional facilitators are often used). • Provides a convenient and transparent way to organise public consultation and build relationships. • Effective as a means for encouraging debate and flows of information between stakeholders.	• Can be difficult to accurately record proceedings. • People who attend may not be representative of the local population. • Attendees' ability to contribute can be limited by a lack of knowledge and possible lack of interest. • Contributions can be dominated by local, topical or personal concerns.
Focus groups Focus groups are similar to citizens' juries in that they bring together people to discuss a specific issue. However, focus groups need not be representative of the general population and they respond to a pre-determined set of issues selected by the focus group organisers.	• A well facilitated group setting often allows people to generate ideas collectively. • Easy to use a variety of presentation media (visual, audio, text, even field visits). • Useful way of exploring why interviewees hold certain beliefs and attitudes. • Holding several sessions can provide feedback on how perceptions and attitudes change over time and in response to new information.	• Requires experienced facilitator. • Group dynamics can result in a few individuals dominating the discussion or others being wary of making a contribution (good facilitation can help avoid this). • Current awareness and understanding of issues may vary greatly between participants. • Easy to rely on focus groups as the sole point of contact with a community – their opinions will not necessarily be representative.
Written consultation exercises This support technique involves an invitation to interested parties to comment on a specific proposal. Documents and perhaps other forms information are made available for examination and stakeholders are encouraged to make a written submission with their comments, suggestions, opinions and so on.	• Provides an opportunity to distribute detailed, comprehensive information. • Respondents can work together to formulate a response. • Responses are likely to be well thought through, comprehensive, and measured. • Although considered somewhat old fashioned as a support tool for participation, this approach remains popular with institutions as it often fits in well with existing procedures.	• Low response rates are to be expected from some response groups (e.g. those who's first language is not the same as that of the consultation call) • The costs of printing and distributing documents can be significant. • Timescales are significantly longer than for some other methods of consultation. • Analysis of responses can be prolonged and resource-intensive.

Open days/roadshows/exhibitions Event which is open to all and where visitors are invited to either speak with staff about a particular planning issue, view displays or perhaps get involved in a small discussion group.	• Allows a variety of presentational and communication media to be used. • Provides an informal and non-stressful environment within which people can contribute • Flexible in terms of when and where participation takes place. • Allows a participation opportunity to be taken to geographically remote or difficult to access communities.	• Attendees will not necessarily be representative of specific groups. • Preparation of exhibited material can be expensive. • May require additional promotion (e.g. adverts in newspapers) to encourage attendance. • Responses to exhibited material may be difficult to capture.
Internet based interaction Use of web-sites to provide information, support discussion/debate, or elicit opinions.	• Access to an engagement opportunity is available 24 hours a day • Revisions, additions and updates to the provided information can be made quickly and by all participating parties. • Allows links to supporting or complementary information and initiatives to be provided • Can be accessed from home.	• Limits participation to the computer literate and to those who have access to a computer. • If site supports open discussion then this may need to be moderated. • Requires technical skills to implement. • Analysis of information elicited through a web-site can be difficult to analyse and interpret.

20.5.2 Objections over outrage factors

If the problem is outrage and not the actual hazard, the response should be built upon outrage mitigation. Outrage factors may be project specific and responses should be tailored accordingly, but there are some broad guidelines, influenced by Sandman (1999), such as:

- Moderate the statements given. Making claims of '100% no risk' may be very difficult to prove or justify.
- Acknowledge prior failings if they are relevant and the topic of the conflict. But place them in context and detail what has been learnt and why the same is avoidable now. The prerogative is with the stakeholder to decide.
- Ensure transparency. Identify where there are gaps and attempt to explain. Where there are uncertainties, these should be explained.
- Where project parameters are outside the immediate control of the project, or where the processes are regulated or controlled by external bodies, an effort should be made to make a clear distinction between volitional and externally imposed actions. These can be used to allow assessment of compliance and reliability to be made.

If the problem is outrage it can be useful to evaluate your previous communication efforts. Is there a way outrage could have been prevented? This may be too late for replying to the current opposition, but it might help to prevent further problems. In both cases it is important to communicate directly with those expressing concern. People should know that the goal is the production of safe water, not to impose risks.

If a project is facing a lot of opposition and a significant group of citizens prefer other options than suggested, it might be useful to think about the goals of water recycling and for whom it is intended. Even after good risk communication people might still find the risks too high or may just not want it. Although this outcome might be very unsatisfactory, if the negative public reaction is so strong, it might be worth considering if the long term viability of the project will hold.

20.6 CONCLUSIONS

As noted at the start of this document, there can be no universal shopping list of actions for effective risk communication – context and circumstances are hugely influential factors in the selection of appropriate messages and forms of dialogue. What works in Rome may not work in Lisbon and what has been effective for an indirect potable reuse scheme will likely not be useful for an environmental flow enhancement scheme. However, we have highlighted primary considerations for the design of risk communication processes and indicated what risk elements should be addressed at various stages of a reuse scheme.

In concluding we would emphasise the following point. Whilst project managers may well feel that risk communication is something for their marketing or public relations departments to deal with, it is only through better understanding of the science and engineering behind water reuse schemes that the public will become more informed and trusting partners.

REFERENCES

Århus Convention (1998). *Convention on Access to Information, Public Participation in Decision-Making and Access to Justice in Environmental Matters*, Done at Århus Convention, Denmark, on 25 June 1998. United Nations Economic Commission for Europe (UNECE), Geneva, Switzerland.

Bierle T. C. and Cayford J. (2002). Democracy in Practice: Public Participation in Environmental Decisions. RFF Press Book, Washington, DC.

Daughton C. G. (2004). Ground water recharge and chemical contaminants: Challenges in communicating the connections and collisions of two disparate worlds. *Ground Water Monitoring and Remediation*, **24**(2), 127–138.

Hartley T. W. (2006). Public perception and participation in water reuse. *Desalination*, **187**, 115–126.

Khan S. J. and Gerrard L. E. (2006). Stakeholder communications for successful water reuse operations. *Desalination*, **18**, 191–202.

Lang S., Fewtrell L. and Bartram J. (2001). Risk Communication. In: Water Quality: Guidelines, Standards and Health, L. Fewtrell and J. Bartram (eds), Water Quality: Guidelines, Standards and Health. IWA Publishing, London, pp. 317–332.

Morgan M. G., Fischhoff B., Bostrom A. and Atman C. J. (2002). Risk communication: A Mental Models Approach. Cambridge University Press, Cambridge, UK.

Powell D. and Leiss W. (1997). Mad cows and mother's milk: The Perils of Poor Risk Communication. McGill-Queens University Press, Ontario, CA.

Renn O. and Levine D. (1991). Credibility and trust in risk communication. In: Communicating Risk to the Public, R. Kasperson and P. Stallen (eds), Kluwer Academic Press, Boston, pp. 175–218.

Sandman P. M. (1999). Risk = Hazard + Outrage: Coping with Controversy About Utility Risks, Engineering News Record, October 4/1999, pp. A19–A23.

Chapter 21

Decision support for MAR planning in the context of Integrated Water Resources Management: The Gabardine DSS

Bernd Rusteberg, Mohammad Azizur Rahman, Jacob Bear, Jacob Bensabat, Radu Constantin Gogu, Serge Brouyère, Piotr Wojda, Muath Abu Sadah, João Paulo Lobo-Ferreira and Martin Sauter

21.1 MAR AND INTEGRATED WATER RESOURCES MANAGEMENT

Among the challenges faced in the next century are an ever increasing population, unsustainable agricultural and industrial development, and water quality degradation. The stress on water resources in many regions of the world is potentially great enough to spark conflict. Without mitigation, conflicts and environmental degradation may be imminent. Minimizing water scarcity is thus the best mitigation and prevention strategy for possible future "water" conflicts that may seem to be unavoidable, but which can be hindered with innovative water resources management response strategies and technologies. These strategies and technologies can be best lumped together as belonging to the more encompassing view of Integrated Water Resources Management (IWRM). At the core of the IWRM concept is the integrated management of all available water resources as based on participative planning and decision making. Due to considerable water losses by evaporation from open water storage, groundwater management is gaining a special importance. IWRM considers all potential water resources when being implemented as the mitigation management strategy. IWRM must also help to ensure that demand of water can be met and the goal is to achieve water resources sustainability. Non-conventional water resources, such as treated wastewater, saline water, water from flash floods, and imported water are all considered as potential sources under the IWRM strategy.

Managed Aquifer Recharge (MAR) permits the managed conjunctive use of surface- and ground-water resources and, therefore, is an important component of IWRM. The implementation of a MAR system requires careful planning in terms of achieving efficient integration into the water resources system and the overall water resources management objectives (Rusteberg *et al.* 2008a). Environmental and socio-economic impacts of MAR planning options will have to be investigated in this context. The roles that MAR may play within the framework of a water resources system are: short and long term storage for later recovery during dry seasons, recovery of groundwater level of over-exploited aquifers, provision of barriers to seawater intrusion in coastal areas, improvement of water quality, use of the aquifer as a water distribution system for individual users, and flood-prevention by deviating peak-flows.

To recharge the water, different infiltration and injection technologies are available. The most common methods are: infiltration basins (spreading basins), sink-pits and canals, induced recharge by bank filtration, and injection wells. Use of these methods depends upon basic planning parameters, such as the type of water or effluent, the geological conditions, prevailing groundwater quality, water reuse purpose, and land availability. In case of water infiltration through ponds or ditches, the quality of the recharge water can be significantly improved due to filtration and other natural attenuation processes. MAR projects using surface waters are known to exist in many countries in Europe, USA, Latin America, former USSR, Mediterranean countries and the Middle East, South Africa, Australia and Southeast Asia, attending to different water management objectives.

MAR may be applied for low-cost water reclamation and reuse in irrigated agriculture with low water quality requirements, especially if reused for restricted irrigation. Israel may be considered one of the pioneers of Soil-Aquifer-Treatment, a MAR technique which has been practiced over decades in Israel at various locations. Treated effluent of 124 million m^3 per year of the Dan region is recharged into the coastal aquifer by means of infiltration ponds, and is pumped after a residence time of at least six months. As based on the achieved water quality standard, the water is used for unrestricted irrigation in the Negev desert.

Drinking water standards are mostly required of MAR process design if the groundwater reservoir is intended to be served as a drinking water source. According to the site specific conditions, the most economic and beneficial uses are identified based upon a comprehensive feasibility study encompassing technical, socio-economic, administrative and legal aspects. Public acceptance is an important factor and must also be addressed, as the difficulties in the introduction of indirect potable water reuse in Singapore and Australia has shown. Public concern issues are best addressed in a site and case specific manner to ensure greater support from the stakeholders and outside funding. The willingness to pay for recycled water, which is often regarded as a low value good but in fact is often of higher quality than conventional drinking water, is also a top issue. The set of technologies providing improved water quality vary widely from natural systems such as wetlands, (maturation) ponds, river bank filtration plants, and soil aquifer treatment to more engineered system such as sand filtration, disinfection, membrane filtration, and oxidation processes (to retain pathogens, particulates, bulk and trace organics, inorganic toxicants, and salinity).

21.2 DECISION SUPPORT FOR MAR PLANNING

Often times MAR plant operation is not successful and even satisfactory results may not be achieved. Maliva and Missimer (2010) stated, "It is not an overstatement that the single most important process for the successful implementation of an Aquifer Storage and Recovery (ASR) project is planning". In spite of this fact, the today available decision support to MAR planning is restricted to some regulatory frameworks, literature on specific planning or MAR technology issues (e.g. Pyne, 2005; Dillon & Molley, 2006; Dillon *et al.* 2007), recommendations with regards to the planning process (e.g. Brown, 2005; ASCE, 2001) and some science-based procedures for specific planning aspects (e.g. Kallali, 2007; Ghayoumian *et al.* 2007). Dillon *et al.* (2007) emphasised the need for an improved governance of MAR addressing water allocation and water quality protection. The author discussed the science and technology base necessary to support good guidelines especially for water reuse.

Regulatory frameworks have been developed in a number of countries in order to support the MAR implementation process, focusing on groundwater and public health protection. Commonly addressed issues refer to application of MAR structures, authorization to recharge and store water in the underground, requirements concerning water quality, or environmental control (NRC, 2008). An analysis of different regulatory approaches from Australia, USA, and the United Kingdom, which focuses on Aquifer Storage and Recovery (ASR) systems, leads to the conclusion that each approach shares a common goal which is to protect the water quality in the storage zone or aquifer from contamination (Maliva & Missimer, 2010).

The French MAR regulatory framework has been summarized by Ward and Dillon (2009). French water policy relies on regulatory and planning instruments combined with various degrees of negotiated agreement, which is different from the Australian and UK framework. In the French framework, water management is subject to compliance with three levels of water policy: the European Union, the National level, and the level of the hydrographic basin.

Australian water recycling guidelines, consisting of several documents, have solely advisory character and not prescriptive (Hochstrat *et al.* 2010). Instead of setting water quality parameters, the documents provide principles and a framework for safe implementation of a MAR scheme using a multi barrier approach as a key concept (Hochstrat *et al.* 2010). An overview of the MAR regulatory issues in the United Kingdom is provided by Gale *et al.* (2001) and summarized by Maliva and Missimer (2010). The authorization or licensing of the source water extraction, the quality of the recharged water and potential environmental impacts are addressed in the regulatory documents.

In the United States, the extraction of water from surface water bodies and aquifers is regulated by the States themselves or by local regulatory agencies. Therefore, the regulation for MAR in the USA is fragmented. A variety of different Federal, State, and local regulatory agencies have been working on this issue (Maliva & Missimer, 2010). A number of codes of regulations regarding wastewater reuse have been issued by the California Department of Public health (e.g. CDPH, 2008) and these are designed for recharging aquifers designated as a source for drinking water supply (Hochstrat *et al.* 2010). The Australian approach to MAR is probably the more science-based, commonsense permitting process, which matches the current and future uses of the storage zone with the appropriate water quality standards (Maliva & Missimer, 2010).

There are a number of further initiatives to develop guiding procedures with regards to the required planning process for the implementation of MAR projects. For brackish-water storage zone ASR systems, Brown (2005) developed a 12-step "ASR Planning Decision Framework". The main focus of the framework is desktop investigation, evaluation of project alternatives, feasibility checking and pilot plant experiment. NRC (2008) suggested the following five-step processes: Phase I: Feasibility evaluation; Phase II: Field investigation and experiment using pilot plants; Phase III: Project design; Phase IV: ASR system construction; Phase V: Project review and adaptive management. Another example is the "Standard Guidelines for Artificial Recharge of Groundwater" developed by the American Society of Civil Engineers (ASCE, 2001). The planning process is divided into five main phases, referring to (I) preliminary activities, such as data collection and resources evaluation, (II) field investigation and test program, (III) design phase, (IV) construction of the MAR plant and, (V) project operation, maintenance and review. All planning steps are addressed in detail, including aspects such as project economics and potential problems related to MAR implementation.

The actual operation of MAR projects is an extremely important aspect for MAR planning, but which has not been sufficiently addressed. Dillon (2009) summarizes some operational issues that are mentioned in the Australian guidelines for wastewater recycling, which are: (1) clogging, (2) recovery efficiency, (3) interactions with other groundwater users/stakeholders, (4) salinity intrusion, (5) operations designed to protect groundwater dependent ecosystems (GDEs), (6) management of recharge facilities. These issues are general for most MAR projects. Data acquisition systems and monitoring networks are important tools to improve system operation. Good operational schemes for MAR facilities result in risk minimization.

Up to now, research has been focused on the achievement of in-depth knowledge of MAR technologies (e.g. UNESCO-IHP, 2005; Bouwer, 2002) and related processes, focusing on few aspects of MAR planning. Specific research has been conducted on the performance of soils and aquifers for water quality improvement (e.g. Idelovitch *et al.* 2003; Fox *et al.* 2001), natural attenuation of organic matter (e.g. Amy *et al.* 2007; Quanrud *et al.* 2003), fate of pharmaceuticals (e.g. Cordy *et al.* 2004; Drewes *et al.* 2003), transformation and removal of inorganic matter (e.g. Fryar *et al.* 2000, Johnson *et al.* 1999), clogging effect of infiltration ponds (e.g. Pavelic *et al.* 2006), and human and environmental health risk (e.g. Asano and Cotruvo, 2004; Brissaud *et al.* 2003), as well as water quality aspects with regards to Soil-Aquifer-Treatment (SAT) and wastewater reuse (WHO, 2006; CDPH, 2008; NRMMC-EPHC-NHMRC, 2009). Some effort has been invested in applying GIS and operations research procedures to specific planning aspects, for example, to analyze the suitability of areas for MAR implementation (e.g. Anbazhagan *et al.* 2005; Calijuri *et al.* 2004; Gomes & Lins, 2002). Existing GIS based analysis methods are poor in dealing with uncertainty, risks, and potential conflicts. Complex regional characteristics, heterogeneities in surface and/or subsurface characteristics, and variable groundwater qualities make site selection for MAR difficult (Anbazhagan *et al.* 2005). Political and social factors are also extremely important in the decision-making process. National and international water policies, natural conservation regulations, environmental impact assessments, and socio-economic considerations are all needed for the site selection. Further developments on flexible Spatial Multi-Criteria-Analysis (SMCA) procedures to support MAR site selection are required.

Few studies have been performed on Multi-Criteria-Analysis (MCA) for MAR impact assessment (e.g. Escalante *et al.* 2005). MAR impact assessment is an essential MAR planning step. Environmental and health impacts can be assessed using mathematical modelling, which is not yet well explored in the field of MAR. In the case of SAT, water flow and reactive transport models may support pond operation. Tang *et al.* (1996) and Li *et al.* (2000) developed an optimization methodology for operation and design of SAT systems. These studies developed a SALQR (Successive Approximation Linear Quadratic Regulator) algorithm to solve the SAT operation problem. Tang and Mays (1998) developed a combinational scaling method with a special evaluation procedure to improve the performance of the genetic algorithm. Efforts may be undertaken to couple optimization procedures with advanced water flow and reactive transport codes such as MODFLOW-SURFACT for analysis of contaminant transport using vadose zone flow and transport equations (Panday & Huyakorn, 2008). Under SAT conditions, the fate and transport of organic and nitrogen species has been studied in limited scale using MODFLOW SURFACT (Kim *et al.* 2004). The authors concluded that the model successfully described fate and transport of the key constituents during the wet/dry operation periods in both unsaturated and saturated subsurface environments.

In the area of IWRM, numerous Decision Support Systems (DSS) have been developed in order to study and compare different water resources planning and management options in a structured way, leading to transparent results with regards to the most promising measures to achieve the given water management goals. The DSS Mulino and Water Strategy Plan, developed in the context of European research projects, may be cited as an example (Giupponi *et al.* 2008; WSM, 2004).

These often GIS-based systems are linked to surface and groundwater hydrology tools in order to study the environmental impacts of alternative measures, related to specific water management problems or tasks, such as flooding, water allocation, or pollution. The DDS SMS (EMRL, 2004), WMS (EMRL, 2004), DELFT TOOLS (Delft Hydraulics, 2004); MIKE BASIN (DHI, 2008), MODULUS (Oxley *et al.* 2004) and WISDOM (BMBF, 2010a) may be mentioned in this context.

Meanwhile, initiatives are missing with regards to the development of an expert system which could provide comprehensive decision support towards the implementation of MAR projects. There are a number of reasons why the development of such a MAR planning support tool is a complex task and a real challenge:

(1) The planning of a MAR system is a non-straightforward decision making process. Numerous, mostly interdependent decisions will have to be taken, for example, with regards to the selection of a source of water to be recharged, the identification of suitable recharge sites and adequate recharge technologies, the pre-treatment of water before recharge, the volume of water to be recharged and, the subsequent recovery of the recharged water. An intelligent, interactive approach is required to be able to follow the suggested planning procedure or step back to earlier planning levels in order to restudy and update decisions.

(2) The implementation of a MAR project is not only a local or site specific task, which is restricted to the underground storage of water or to water quality improvement due to recharge, but should also be part of a regional IWRM approach on the basin level. MAR permits the joint management of surface water, groundwater, and wastewater resources and is therefore an important integrating element of IWRM. MAR implementation should contribute to the achievement of the overall water management goals of the river basin. The most promising MAR options will have to be studied in terms of environmental impact, socio-economic efficiency, as well as their contribution to the prevailing or future water resources problems as seen on the basin level. In the context of IWRM, MAR has to be understood as an optional measure, competing with other water resources planning and management solutions. A decision support tool for MAR planning should take this into consideration.

(3) There are several relevant research gaps, for example, with regards to the process-understanding of the fate of and transport of pharmaceuticals and other emerging pollutants. Furthermore, the complexity of the underground system hampers the validation of mathematical models for water flow and natural attenuation processes, which should support environmental impact assessments as well as the sustainable operation of the water recharge and recovery facilities. Quality thresholds for the water to be recharged as a function of the natural attenuation capacity of the underground which strongly depend on the local conditions are difficult to determine.

(4) There is a lack of common international legal standards and regulatory frameworks with regards to MAR implementation.

21.3 THE GABARDINE DECISION SUPPORT SYSTEM

For comprehensive support of MAR project planning under water scarce conditions, an innovative geospatial decision support system (DSS) has been developed under the scope of the European Research Project GABARDINE (Rusteberg, 2006). The system, developed on the ArcGIS 9.2 environment with Visual Studio.NET and ArcObjects library, offers a highly interactive planning approach, taking the non-straightforward decision structure of MAR planning challenges into account. The Gabardine DSS (G-DSS) has to be understood as an expert system to be applied by planners and managers of surface and ground water resources on the hydrological basin level. The system supports the planning process of a MAR plant via a science-based approach, starting with the problem analysis, water budgets, identification of potential sources of water, suitability mapping and ranking of potential MAR sites, selection of MAR technology, definition of alternative MAR planning and management options as well as the evaluation and ranking of MAR options, based on their impacts and performance towards the mitigation of the water-deficit problem. Alternative MAR projects and management strategies may be evaluated and ranked based on MCA, considering environmental, socio-economic, and technical decision criteria. Meanwhile, the quantification of the selected decision criteria is still an external process. The system does not substitute the expertise of a water resources planner, but guides planners through a challenging decision making process within a complex water resources system and natural environment (Rusteberg *et al.* 2008b).

The following section gives an introduction to the MAR planning process, as considered by the G-DSS, the system structure, and its Graphical User Interface (GUI), referring briefly to the MAR planning module as the most relevant planning tool.

21.3.1 MAR planning process

Figure 21.1 presents the most important steps of the MAR planning process as considered in the G-DSS in the form of a simplified flowchart. The process starts with river basin characterization, followed by an analysis of the water resources system and the prevailing water resources problems. The Driver-Pressure-Impact-State-Response (DPSIR) approach may be applied in order to answer the question if MAR is a potential response to the prevailing water resources problems. For specific attention to the water scarcity problem, a water budget study should be undertaken to provide information about the mean monthly water deficits to be mitigated by MAR.

Following this pre-analysis, the planner (DSS user) will work towards the development of a set of alternative MAR planning and management options. Decisions have to be taken with regards to the sources of water to be applied for aquifer recharge, the most appropriate MAR technology, best locations to infiltrate or to inject the water, required water treatment before recharge and, roughly, the design of the MAR plant, indicating for example. the required area for the installation of infiltration ponds. For later definition of the system operation, the total annual volume of water to be recharged per recharge site should be specified. An analysis of the interdependencies between these decision steps reveals the non-straightforward nature of the MAR decision making process. The selection of a water source for recharge, MAR technology, location, pre-treatment, and operation refer to decisions at the same level within the overall decision making process, which are intimately linked to each other and, therefore, are presented in one decision box (Figure 21.1).

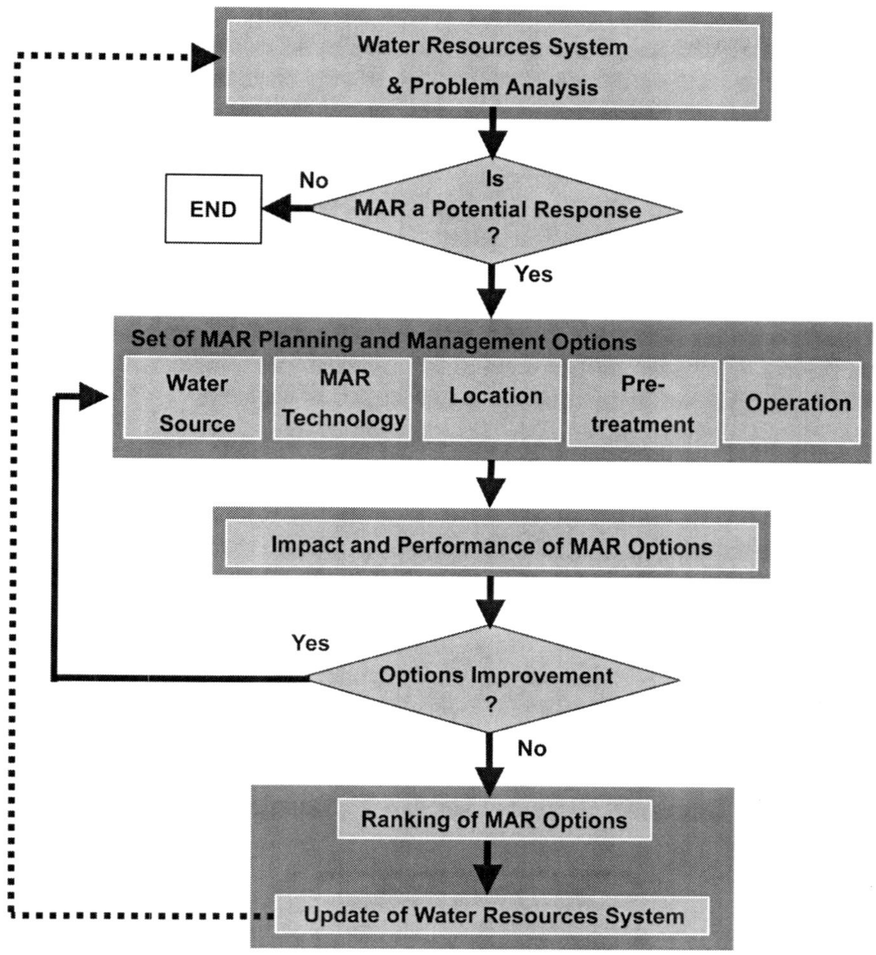

Figure 21.1 Simplified MAR Planning Flow-Chart

While developing a MAR option, for example, to store water in the underground and cover deficits in irrigation, the planner may be obliged to review earlier decisions in order to create a feasible MAR response measure. Imagine the following situation: The land use situation or presence of impermeable layers between the soil surface and a phreatic aquifer does not permit the installation of large infiltration ponds near to the irrigated areas. Therefore, potential in-situ water quality improvement, for example, by SAT, is not possible. Eventually, the planner will have to consider other more distant sources of water or an adequate pre-treatment of the first source and then implement injection wells as the alternative recharge technology. Alternatively, the planner could infiltrate the water in upstream areas of the river basin where water infiltration would be possible, using the phreatic aquifer as a water transfer system with water recovery by wells downstream, near to the irrigated areas. The latter option may lead to a significant water quality improvement due to natural attenuation processes. In this way, the planner, supported by the DSS, will develop a set of alternative MAR planning and management options.

The next planning step consists of the analysis of socio-economic and environmental impacts as well as an evaluation of MAR options' performance toward the main water resources management objectives. In order to evaluate both, the planner will select a number of the most representative decision criteria, which will have to be quantified for each option. While criteria selection is supported by the DSS, the quantification of the selected decision criteria is still an external process. The selection of representative decision criteria should be handled as a participative process, involving relevant stakeholders, since their choice, similar to criteria weighting (discussed later), affects the evaluation of alternative MAR options.

Environmental and water related decision criteria, such as groundwater level rise or mean nitrate concentration, are determined by means of mathematical simulation models for underground flow and transport, applied over an adequate planning horizon, which is commonly at least 20 years. The system does not offer a full integration to a groundwater model. MODFLOW applications for Gabardine case studies have been handled as external processes. Economic viability may be judged on the basis of a Cost-Benefit-Analysis (CBA) while social aspects, such as social acceptance or willingness to pay, are quantified by means of questionnaires as supplied by the DSS example applications (Sadah *et al.* 2009b). The evaluation of typical performance criteria or project benefits depend on the degree of water deficit reduction, the increase in irrigated land, or the agricultural productivity in case of a water scarcity problem.

The environmental analysis may show that certain MAR options are actually non-feasible, for example, due to legal environmental constraints and regulations. These options will have to be excluded from further analysis. In general terms, the DSS user may try to improve the set of options by slightly changing decision variables related to MAR management (system operation), water recovery, location and others. Decision rules to support the optimization of MAR planning and management options are not yet included in the DSS.

The MAR options are ranked by MCA methods together with weighting procedures as presented in the DSS example application from the Northern Gaza-Strip, Palestine (see case study). Decision criteria weights are best when achieved through an involved and considerable participative process with stakeholder participation. The MCA permits an evaluation of the MAR planning options in terms of socio-economic and environmental impacts, system performance, and comparisons of these. The final planning step is an update of the overall water resources system. Close cooperation with water resources planners on the river basin level is required, since the MAR project will impact the overall water resources system management. Abstracted surface water to be recharged for example, will reduce the water availability downstream and restrict water resources development in the lower part of the basin.

21.3.2 DSS structure and GUI

The following DSS modules have been developed and integrated into the GIS platform, being accessed by the main menu of the system in the top of the screen: (a) Geo DATA-base management module, (b) DPSIR module, (c) MAR PLANNING module, and (d) a MCA module for MAR site selection, as well as MAR option comparison and ranking. The so-called tool-box contains a number of software tools and models which have been developed in the context of the Gabardine project, such as a time series generator, a hydro budget tool, and a groundwater simulation software, based on the mesh-less approach (Figure 21.2). These tools are not yet integrated into the GIS platform, which is an important objective for further system development. Therefore, the quantification of decision criteria is still an external process. Figure 21.3 presents the Graphical User Interface of the Gabardine DSS:

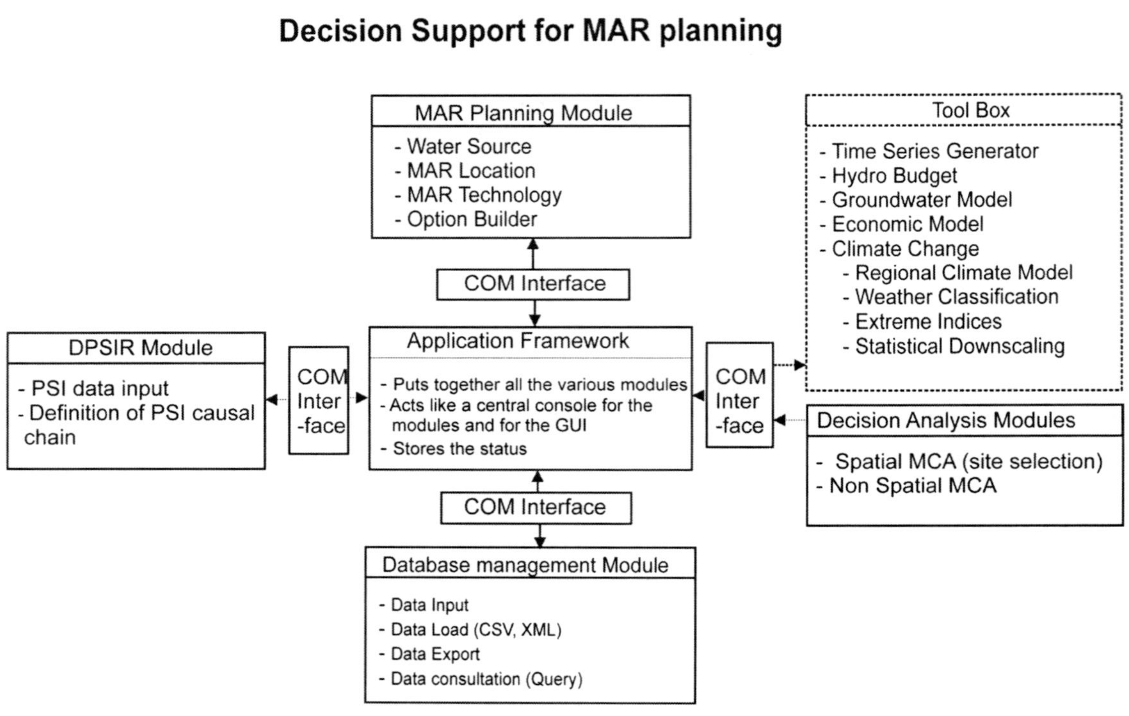

Figure 21.2 Gabardine DSS modular system structure

The left-hand menu of the DSS may be chosen to show the common display of the geographical information (DISPLAY), the work modus for analysis and presentation of potential water sources for MAR (SOURCE), the procedure for MAR suitability mapping and site selection (SELECTION) (see Figure 21.7 and Figure 21.8), and the general MAR Planning module (PLANNING). In this manner, all relevant information is geographically linked and easy to update at any time of the planning process, jumping between different decision steps, if required, and strengthening the interactive system operation.

The geo-database management module is a core element of the Gabardine Decision Support System (G-DSS) on which is built the entire system structure. Each module presents a number of functionalities. The MAR planning module is the main decision support tool of the system, accompanying the entire planning process, as roughly specified above. More details are given in the following sections. Figure 21.3 presents the modular structure of the G-DSS, controlled by an application framework.

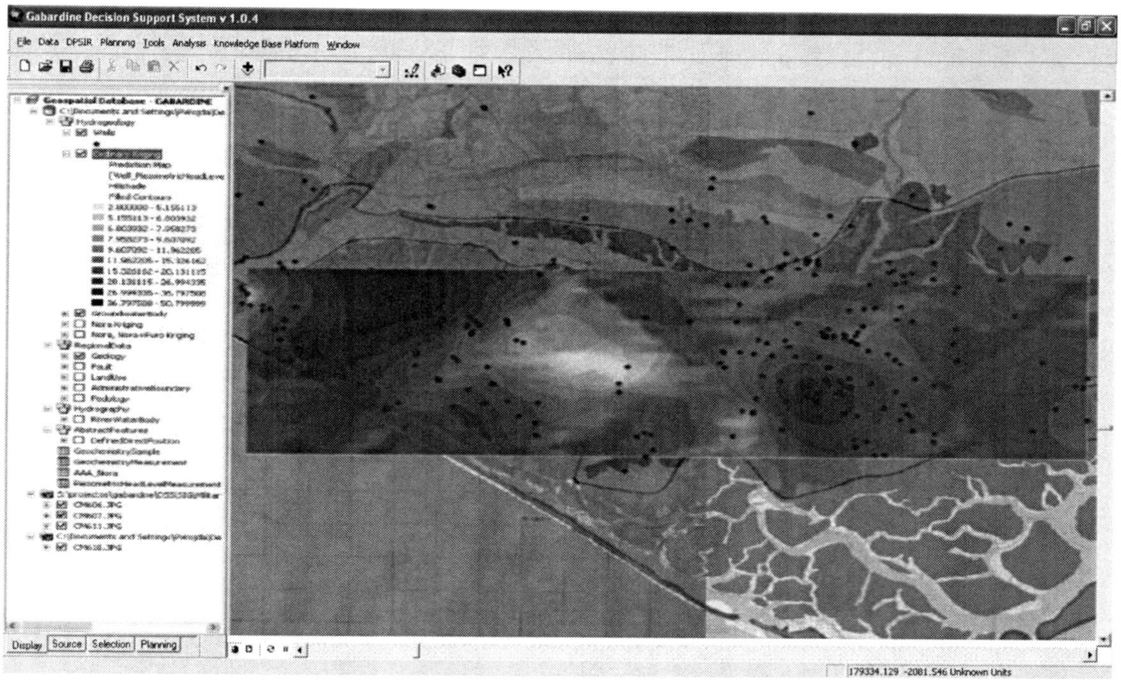

Figure 21.3 Graphical User Interface and Main Menu of the G-DSS

21.3.3 Spatial database and GIS platform

The database module of the G-DSS allows for spatial and temporal data management. A number of crucial functions for data presentation and analysis have been implemented. It supports performing queries and visualization of the spatial data and the time series. This module together with its geospatial database offers the possibility of integrating data coming from different sources and of various kinds (e.g. spatial, non-spatial and temporal).

As the basis of the geospatial database, an object-oriented data model has been built upon the project-based approach developed in the HydroCube relational database (Wojda *et al.* 2010a) currently in use in the Walloon Region of Belgium by the groundwater authorities and university research centres. In the GABARDINE context the latter has evolved in order to use the advantages of object-oriented modeling, to follow the international standards for transfer of geospatial information (ISO/TC211 and OGC) and to be compliant with the recommendations from the European Geospatial Information Working Group (Vogt, 2002) and the specific objectives of the project related to artificial recharge.

The GABARDINE data model allows for a precise outlook on data and more efficient data management (Wojda, 2009). A holistic project-based approach covers a full set of concepts and features, allowing for effective GABARDINE project management. Project data such as locality, hydrogeological equipment, and related observations and measurements can be stored. It also focuses on specialized hydrogeological field experiments such as hydraulic and tracer tests. Figure 21.3 presents an example for geographical data presentation and analysis.

The architecture of the object-oriented data model follows international standards concerning geospatial data encoding and transfer. This is reflected in its object-oriented approach supported by the Open Geospatial Consortium (OGC) and the International Organisation for Standardisation (ISO), which enables, for instance, an easy transposition and translation between real-world and informatics objects. The conceptual model of data for the Geospatial Database has been developed and described using an actual business and industrial standard of the Unified Modelling Language (UML) (Wojda *et al.* 2006).

The object-oriented data model can be implemented in different spatial database environments. In order to test the versatility and flexibility of the object-oriented hydrogeological data model presented, it has also been implemented in the Web2GIS Open Source free system developed in the Geomatics Unit of the University of Liege. Web2GIS is a web-based spatial database conception environment, which is supported and maintained on a server using exclusively Open Source software such as Apache, PHP, PostgresSQL/PostGIS and PhpMapScript/MapServer (Laplanche, 2006, 2008). This modular solution promotes the use of international standards coming from ISO/TC211 and OGC.

Clear structure and unambiguous description, following the geomatic principles and platform-independent data models, allow for easy data integration and data exchange by programmed interfaces and eventually assimilation within broader, international spatial data infrastructures (Wojda *et al.* 2010b).

21.3.4 The G-DSS planning module

The planning module of the G-DSS explicitly supports the decision steps that are required to define a set of alternative MAR planning and management options as a substantial part of the overall planning process as described in the before chapter. The decision steps, referring to water source identification, water quality check, MAR site selection, choice of MAR technology, and definition of MAR options are all considered in the left-hand menu of the planning module as presented in Figure 21.4.

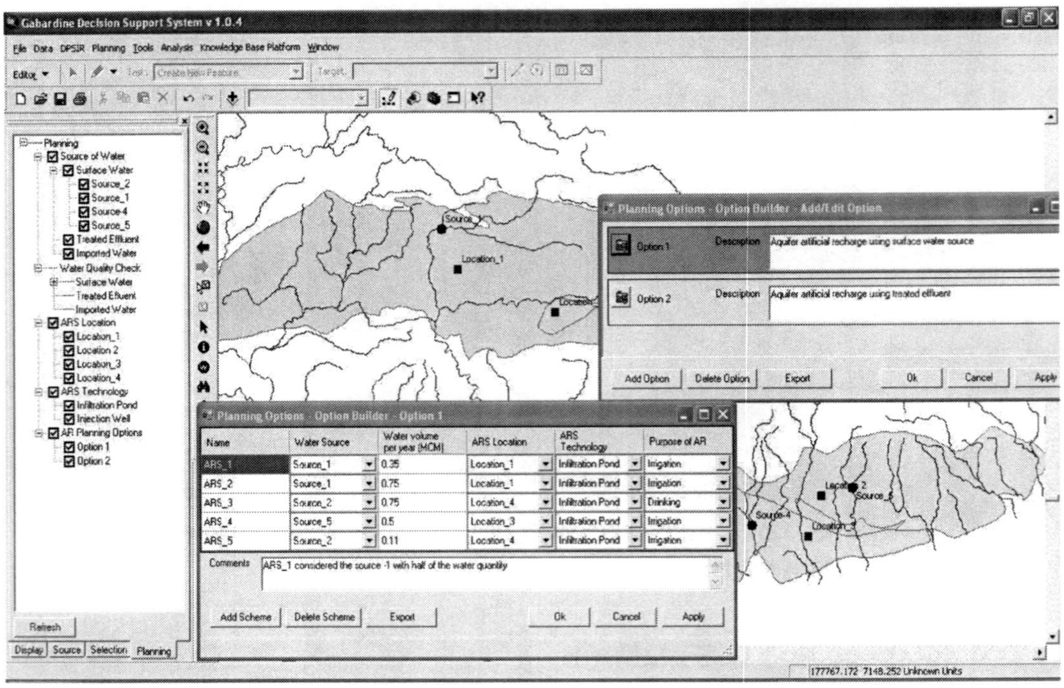

Figure 21.4 The MAR planning module with its corresponding left-hand menu and the Option Builder

For efficient system application, the geo-database of the DSS should have been supplied with the most relevant information about basin characteristics, such as soil maps, topography, land use, geological maps, water pollution sources, waste water treatment plants, river discharge series and water quality data.

For many regions in the world, GIS information systems, or at least an adequate database with land and water resources information, are already available and should be transferred to the DSS geo-database before starting the MAR planning module. The system supports the characterization of the water resources system on the basin level, providing a wide range of functions for graphical data representation of the relevant temporal and spatial data.

The system may consider any type of potential source of water for managed aquifer recharge. Potential sources of water may be classified in surface water (river discharge surplus, flash-floods, stored water in reservoirs), treated effluents, and imported water from neighbouring basins or other sources. Current monthly water availability and water quality of different sources may be reviewed by the planner and DSS user. The system contains statistical tools for the analysis of monitoring data (Figure 21.5). Strategic locations to abstract water for MAR are defined interactively on the map.

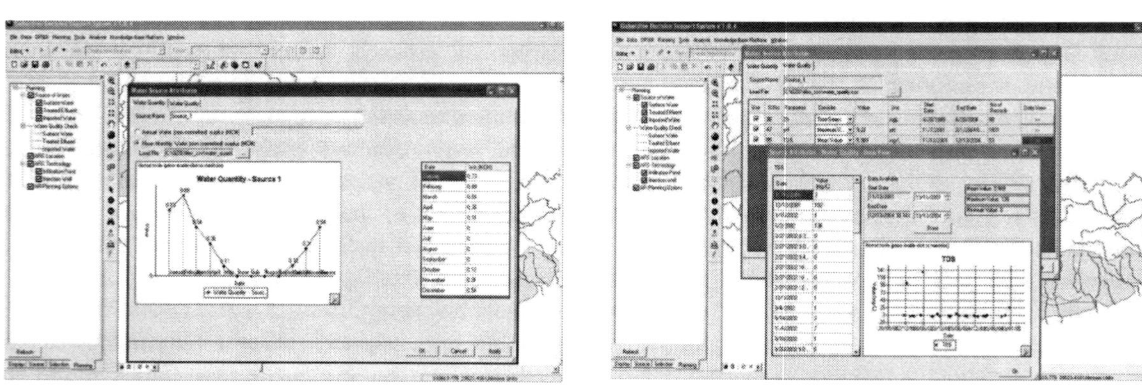

Figure 21.5 Water availability and water quality analysis

The DSS offers a water quality check where the water quality of a specific source is compared with local and international standards for irrigation, drinking water supply, and aquifer recharge (WHO, 2006).

Depending on the results, the analysed source will have to be restricted to infiltration and agricultural use or will need to be further treated to improve the water quality before recharge. The water may even be excluded from becoming potential source. The recommended MAR technology will be selected accordingly. The user may select between different infiltration and injection technologies for MAR.

MAR site selection

A specific procedure to support the selection of MAR sites based on Spatial Multi-Criteria-Decision-Analysis (SMCDA) has been developed and incorporated in the G-DSS (Rahman *et al.* 2009; Rahman *et al.* 2012). The overall methodology of the new site selection model is presented in Figure 21.6. The flowchart shows the main decision steps for spatial analysis. In general, the entire process involves the following main steps: (a) constraint mapping, (b) suitability mapping, (c) sensitivity analysis, and (d) site ranking. Figure 21.7 presents the left-hand menu of the system for MAR site selection (switch to SELECTION – Figure 21.4) which facilitates an interactive working.

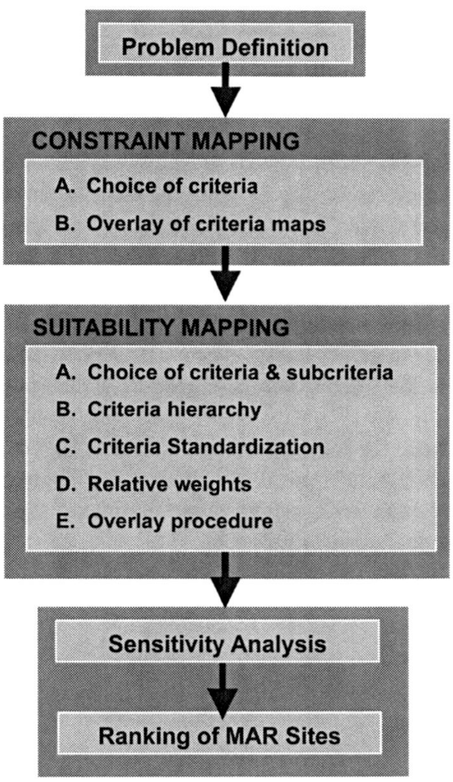

Figure 21.6 MAR site selection procedure (modified after Rahman *et al.* 2012)

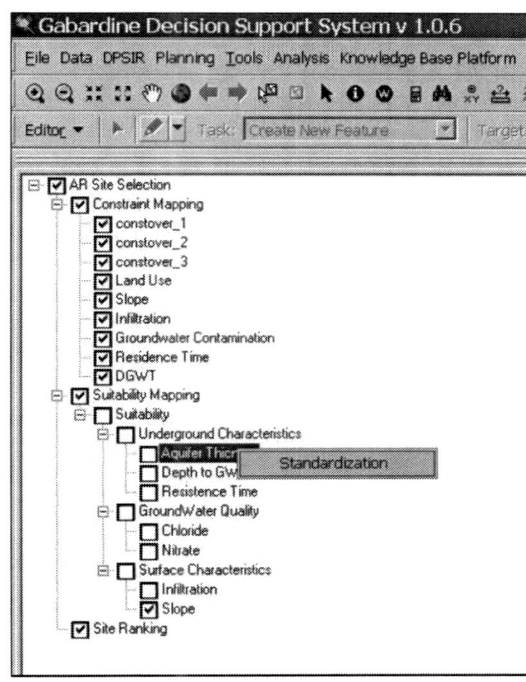

Figure 21.7 Left-hand menu for site selection

Before starting the procedure, the planner (G-DSS user) should have a clear definition of the water resources problem to be mitigated and the water management objectives to be achieved by MAR. In case of water scarcity mitigation, water storage and recovery by MAR as well as water quality improvement by SAT are important objectives. Based on a DPSIR analysis (OECD, 1993; OECD, 2003) offered by the system, the most relevant decision criteria will be selected.

The main objective of constraint mapping is to screen out the non-feasible solutions; this simplifies the further analysis and avoids conflicts in decision making. The sites, which are of prime interest to other planning projects, or simply not available or completely non-feasible for MAR implementation, are screened out in this step. A conjunctive screening approach was chosen for constraint mapping. Under conjunctive screening, an alternative is accepted if it meets specified thresholds for all evaluation criteria. Non-feasibility of MAR may be given due to land use, hydrogeological conditions, groundwater pollution, or topographical conditions. The developed constraint map serves as a mask for suitability mapping. Figure 21.8 presents the system interface for constraint criteria selection and assignment of threshold values.

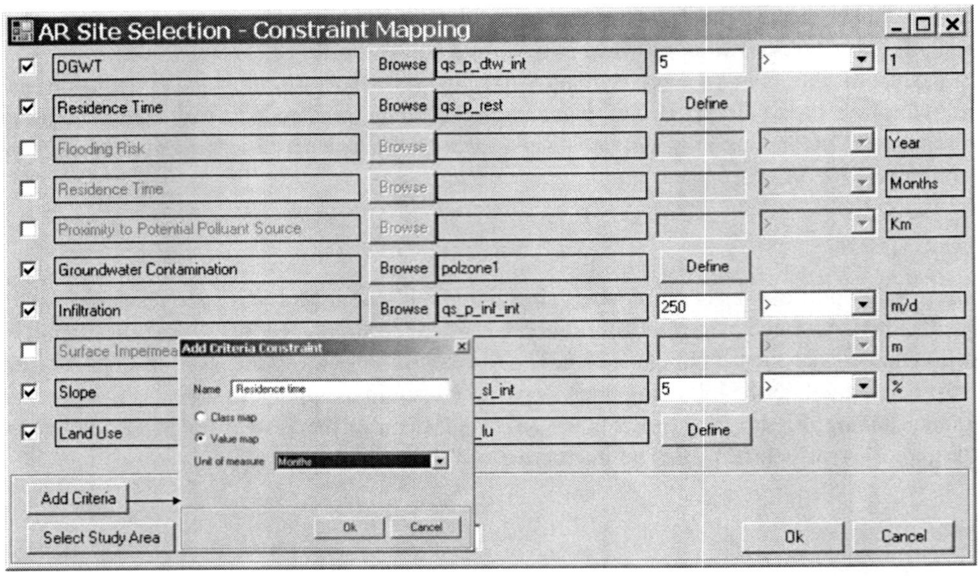

Figure 21.8 DSS interface for constraint criteria selection and assignment of threshold values

For suitability mapping, all relevant decision criteria that impact site suitability, with regards to MAR implementation, are selected. The system offers the possibility to create a hierarchical structure according to Saaty (1980), as shown in Figure 21.14. Surface, subsurface, or other characteristics may be considered as main criteria. Corresponding sub-criteria are grouped under the main criteria. Each sub-criterion in the criteria tree is represented by a map of different types such as a classified map (e.g. land use) or a value map (e.g. slope, infiltration). For decision analysis, the values and classes of all the maps should be standardized to reduce the dimensionality (Sharifi and Retsios, 2004). Different standardization methods can be applied to different maps. The system offers linear, piece-wise linear, and step functions for standardization. The outcome of the function is always a value between 0 and 1. Figure 21.9 illustrates an example of the system interface for criteria standardization.

The next step in the site selection procedure is assigning values of importance for each criteria and sub-criteria, which is done by assigning a weight to each criterion. Different weighting methods are available. Pair-wise comparison (Figure 21.10) and direct weighting are used here. The sub-criteria under each main criterion are compared amongst themselves and a weight is assigned to each one. The main criteria are also evaluated in this way.

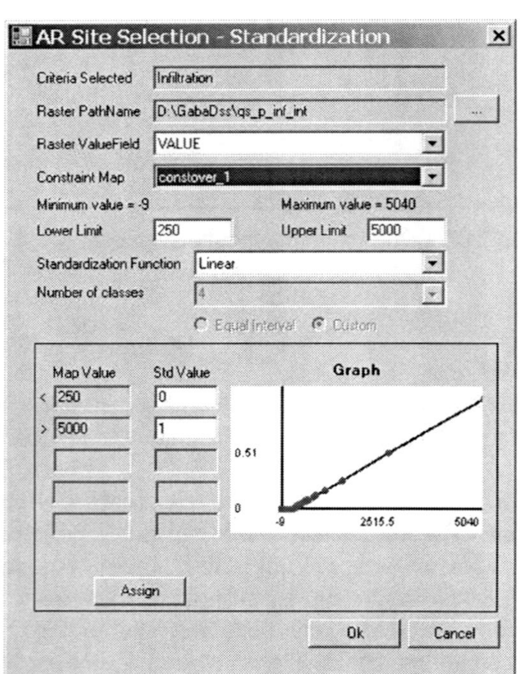

Figure 21.9 Interface for criteria standardization

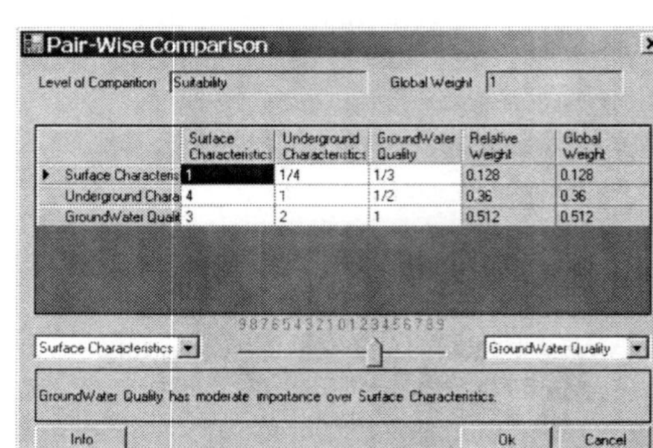

Figure 21.10 Criteria weighting by pair-wise comparison

The final "suitability map" is then produced by the overlay of the resulted criteria maps, determining the overall suitability index of each alternative for each cell of the map. Weighted Linear Combination (WLC) and Ordered Weighted Average (OWA) with fuzzy linguistic quantifiers are overlay methods supported by the DSS. WLC is the simplest and most commonly used aggregation method in spatial analysis (Eastman *et al.* 1993), while OWA is a class of multi-criteria combination operators, involving two sets of criteria weights.

Based on the OWA concept and the fuzzy linguistic quantifier, represented by the α-value, a sensitivity analysis is possible by varying the α-value (Rahman *et al.* 2009). The DSS also permits the study of robustness of identified MAR sites with regards to different weighting schemes and standardization. These analyses are useful when uncertainty exists in the construction of hierarchy and in the assignment of relative importance (Store & Kangas, 2001). For the final site ranking, a water source related approach is applied, taking the distance to potential water sources and economic aspects into account.

The definition of a set of alternative MAR planning options is supported by the instrument called "MAR Options Builder." Figure 21.4 presents the corresponding graphical user interface of the builder. Viable options of potential water sources and available amount of water to be recharged, MAR technologies, recharge location are linked to options for water recovery and reuse. Pre-treatment requirements with regards to water quality improvement may be introduced as comments. MAR schemes may be defined, considering several recharge locations for one water source or groundwater recharge of different water sources at one or more locations. Once viable sets of different MAR schemes have been defined, supported by the DSS planning module, further MAR planning is developed, as described in the chapter on the "MAR planning process."

21.4 G-DSS APPLICATIONS

Two example applications of the G-DSS have been selected, presenting a MAR site suitability analysis of the aquifer system of Querença-Silves, Algarve, Portugal and the analysis of alternative MAR response strategies in the Gaza-Strip, Palestine.

21.4.1 The Querença-Silves case study, Portugal

The developed SMCA procedure for MAR site identification, suitability analysis and ranking, integrated in the G-DSS, has been applied to the Querença-Silves case study in the Algarve region in South-Portugal. The main steps in applying the G-DSS and some of the results are presented and discussed.

Introduction

The Algarve region in South-Portugal has been affected by droughts during the last decades. The geographical location of the Algarve contributes to a very high risk regarding droughts. The hydrological year of 2004/2005 was extremely dry in the entire Portuguese mainland and especially in the Algarve region. The drought caused severe problems, considering the availability of water resources. Surface water reservoirs reached volumes that were below acceptable levels, and the Querença-Silves aquifer system was over-exploited. The Querença -Silves aquifer system was a major source of drinking water to the urban areas within the Algarve region, during the 90′s. Today the Funcho Dam is considered to be the most important drinking water source in the western part of the Algarve. Another dam, the Arade Dam is located downstream of Funcho Dam in the Arade river. More than 50 hm^3 of river water was lost to the sea during the wet hydrological year of 2000/2001, and in dry hydrological year of 2004/2005 there was an equivalent shortage of water resources (54 hm^3) (Lobo-Ferreira & Oliveira, 2007).

Case study area

The Querença-Silves Aquifer System is a 318 km^2 aquifer system, the largest of the Algarve. It is located in the municipalities of Silves, Loulé, Lagoa, and Albufeira (Central Algarve). The aquifer is mainly composed by karstified Lower Jurassic (Lias-Dogger) dolomite structures. The south-western part of the aquifer is mainly unconfined. The general groundwater flow direction is from northeast to southwest. According to the characterization of the Querença-Silves aquifer system (Almeida *et al.* 2000), the hydraulic parameters are heterogeneously distributed and the aquifer productivity values are high. The hydraulic properties of the aquifer system are rather heterogeneous, with transmissivity values ranging between 83 and 30,000 m^2/day and storage coefficients between $5 * 10^{-3}$ to $3 * 10^{-2}$. INAG (2001) presents the recharge value as being 220 ± 54 mm/year. This represents a percentage of precipitation of around $40 \pm 10\%$. Monteiro (2005) obtained an average recharge of 292.5 mm/year. These estimations have incorporated the average precipitation values in the area. With reduced precipitation, as was the case during the hydrological year of 2004/2005 due to precipitation around 40% of the average, recharge is much lower.

For the present study, only the southwestern part of the Querença -Silves Aquifer is taken into account to put in place the below stated water management strategy. The groundwater catchment area is about 114 km^2. Based on earlier modeling studies, the area has been divided into four zones (Figure 21.11), according to the residence time of groundwater in the aquifer; namely Zone I (residence time less than 6 months), Zone II (residence time 6 months to 1 year), Zone III (residence time 1 year to 3 years) and Zone IV (residence time greater than 3 years).

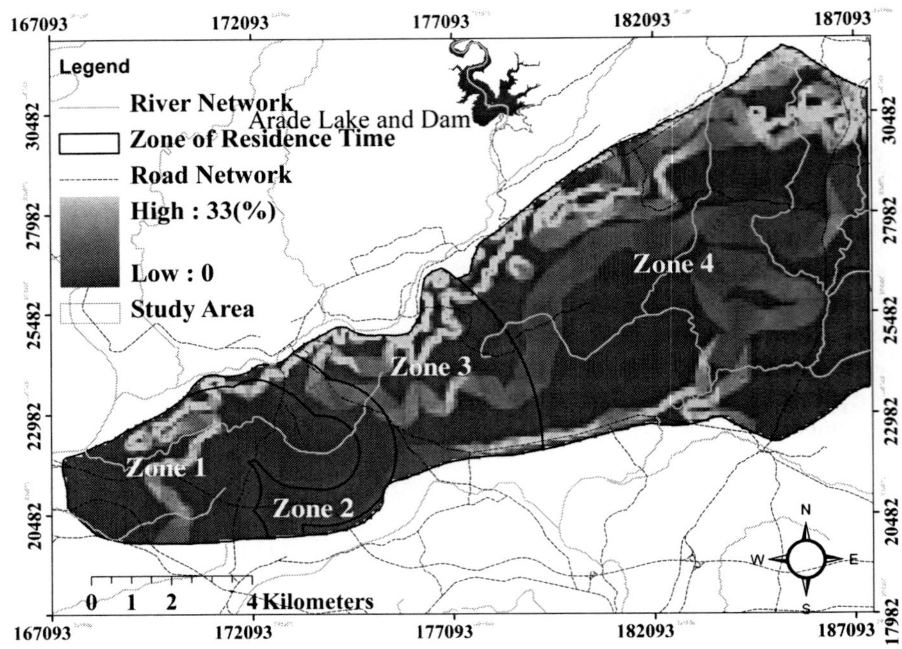

Figure 21.11 Case study area Querença-Silva, Algarve, South-Portugal showing the slope (modified after Rahman *et al.* 2012)

Water resources management challenges

The recent water supply crises caused by recent droughts within the Algarve region show the necessity to implement new measures for improved water resources management in the region. The implemented surface water reservoirs in the northern areas suffer from significant water losses due to high evaporation rates and are not able to cover the water demand under drought conditions. During the rainy season huge spillway losses have been observed at these reservoirs. The river water flows to the sea. Water availability during the dry period is nowadays the main constraint for the development of the Algarve region. Further exploitation of the already over-exploited Querença-Silves aquifer system is no sustainable response strategy to the water supply problem.

MAR response strategy

The enormous losses of surface water resources in connection to the over-exploitation of the local aquifer systems indicate the requirement of the conjunctive management of surface water and groundwater resources in order to guarantee water supply during droughts, to mitigate their impacts and provide conditions for sustainable development of the Querença-Silves region.

Therefore, MAR is a potential response measure to the prevailing water management problems of that region, transferring surface water of the nearby Arade Dam to the upper part of the study area and recharging this water to the Querença-Silves aquifer. The proposed water management response strategy uses the aquifer as a water transfer and storage system. The high residence times of the infiltrated water in the aquifer system would guarantee long-term storage of the recharged water in the underground with water recovery by wells in the lower part of the aquifer system located at strategic places. The application of infiltration techniques would permit the recharge of large volumes of water during the time of surface water surplus. The present study analyzes the viability to implement infiltration ponds in the upper parts of the aquifer system by studying the suitability of areas for MAR implementation. The water recovery by pumping well groups in the lower aquifer system requires further groundwater modeling efforts which are subject to further studies in the region.

Decision criteria for spatial analysis

Under the prevailing site characteristics and study objectives, the following spatial decision criteria have been selected in order to proceed with a) constraint and b) suitability mapping. Some important criteria were selected for both steps after analysing their importance and relevance.

Constraint mapping

According to Table 21.1, the decision criteria (land use, slope, infiltration rate, surface impermeable layer thickness, groundwater depth, groundwater contamination and residence time) are considered in order to exclude areas from the further evaluation process, which are unfeasible for MAR implementation by infiltration ponds. With regards to the decision criterion, land use, unfeasible areas for MAR are agricultural systems, agricultural areas outside irrigation perimeters, irrigated areas, quarries, marshy places, salt-pits, and isolated urban areas. Generally feasible areas for MAR are areas of natural vegetation, grasslands, poor pasture, and orchards. Rivers are not excluded, since MAR facilities may be constructed directly in the river. The minimum required infiltration rate is 25 cm per day. For the groundwater depth, a threshold value of 5 m has been defined. The maximum acceptable surface slope is 5%, while the residence of water in the aquifer should be at least 6 months in order to avoid drainage losses already within few months after recharge (Figure 21.11). With regards to contamination, sources that are at a minimum radial distance of 500 m away are considered to be feasible and sites in the upstream flow path of pollution sources are excluded from further evaluation. Figure 21.12 shows the thematic map of the decision criterion "landuse" and its conversion to the corresponding constraint map.

Table 21.1 Spatial decision criteria for constraint and suitability mapping (modified after Rahman *et al.* 2012).

Criteria	Mapping	Description
Land use	Constraint mapping	Land availability for MAR, for example, urban zones are non-feasible areas for infiltration pond implementation.
Slope (topography)	Constraint & suitability	Steeper slopes do not allow the implementation of infiltration basins. Higher priority is given to gentle slopes.
Infiltration rate (soil)	Constraint & suitability	Water penetration into the underground. The higher the infiltration capacity, the more suitable the site.
Surface impermeable layer thickness	Constraint & suitability	Impermeable layers at the surface may be excavated. The lower the thickness, the more suitable the site.
Groundwater depth	Constraint & suitability	Improved natural attenuation and storage capacities of the site due to higher unsaturated zone thickness.
Groundwater contamination	Constraint mapping	Sufficient distance from groundwater pollution sources and no contamination in the pathway of groundwater transfer.
Aquifer thickness	Suitability mapping	Suitable sites should have high thickness values. Transmissivity and aquifer storage volume depends on the aquifer thickness.
Groundwater quality (chloride & nitrate)	Suitability mapping	Adequate groundwater quality is required at the place of recharge. The better the groundwater quality, the more suitable the site.
Residence time	Constraint & suitability	Sufficient residence time of the recharged water in the aquifer are required to use the aquifer as water storage and transfer system.

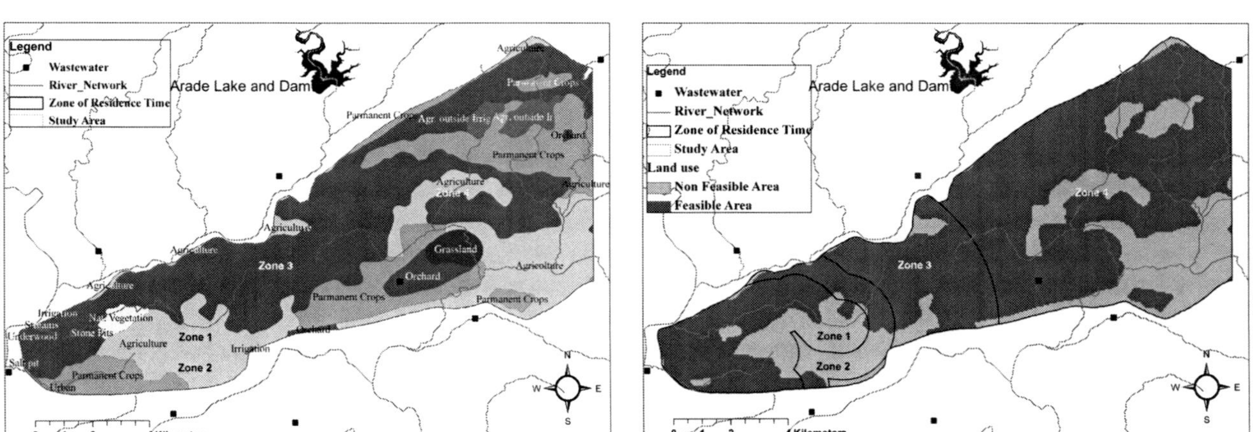

Figure 21.12 Thematic map "landuse" and its conversion into the corresponding constraint map

Figure 21.13 illustrates the final constraint map as a Boolean overlay operation result of the criteria constraint maps. The results show that just 11% of the total area is viable with regards to MAR implementation by means of infiltration ponds. Most of the area is located in Zone 4, which is characterized by higher residence times.

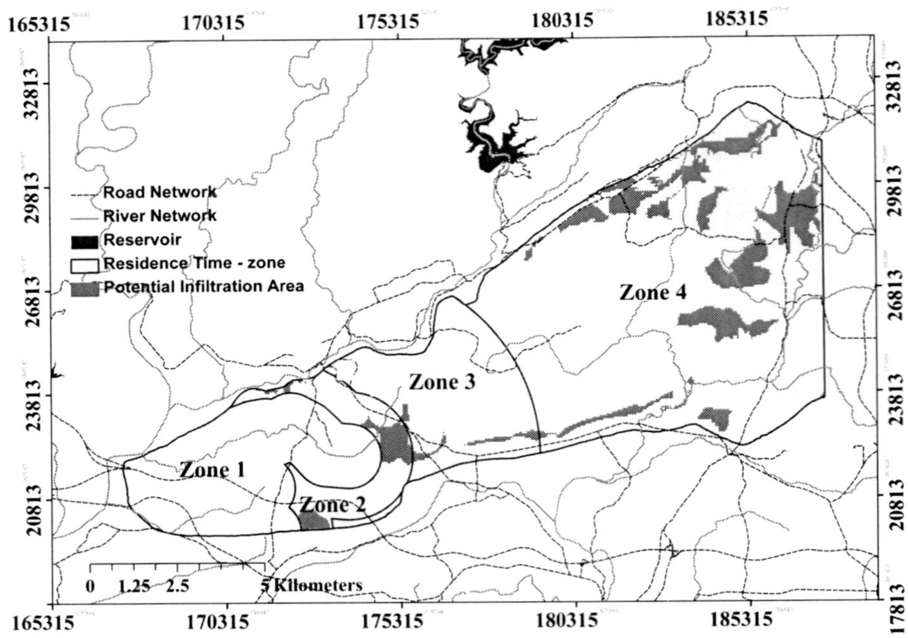

Figure 21.13 The final constraint map (modified after Rahman *et al.* 2012)

Suitability mapping

Figure 21.14 presents the selected decision criteria for suitability mapping in a hierarchical structure. The numbers indicate the local (left) and global (right) weights of the sub-criteria as well as the weights of the main criteria groups.

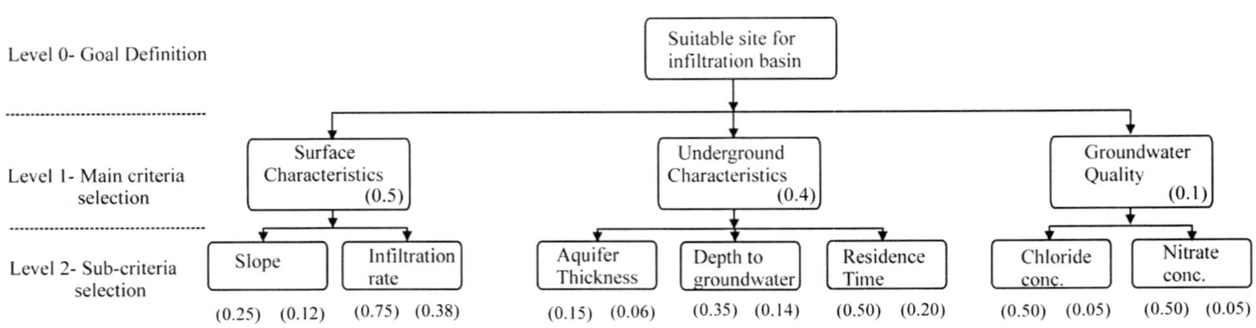

Figure 21.14 Hierarchy of the spatial decision criteria and weights (modified after Rahman *et al.* 2012)

The importance of each sub-criterion has been calculated using pair-wise comparison method (Saaty, 1980). Infiltration rate of the soil, residence time of groundwater, and depth to groundwater were given higher priority in the analysis. As the groundwater quality of the area is good and does not vary spatially too much, it was considered as lower priority criterion. The thematic layers of the sub-criteria were standardized by linear, piece-wise linear, and step-wise linear value functions as presented in Figure 21.15. The range indicates the existing limit of the criteria values in the study area.

Figure 21.16 presents the final suitability map with regards to MAR implementation by infiltration ponds based on WLC as an overlay option. The most suitable sites are located in the northeastern part of the study area. Suitable and medium suitable sites are located at different places in the region.

According to the overall suitability score (Figure 21.15), 1% of the total aquifer is very good (suitability score 80–100, see circle), 3.2 % is good (suitability score 60–80), 6.4 % is moderate (suitability score 40–60), and 0.6% is poor for MAR (see details in Rahman *et al.* 2012). The rest of 88.8% of the aquifer surface is not suitable due to the constraints of MAR implementation. The most suitable areas shows soil high infiltration capacity and do not require additional excavation efforts. The results of the sensitivity study based on Ordered Weighted Average (OWA) are presented by Rahman *et al.* (2009, 2010, 2012).

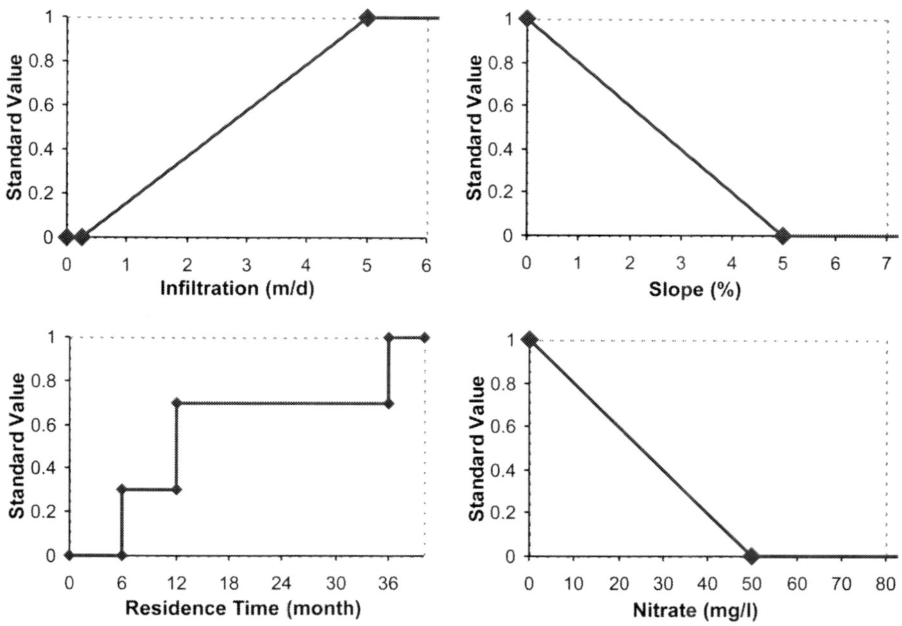

Figure 21.15 Procedure for criteria standardization used in this study

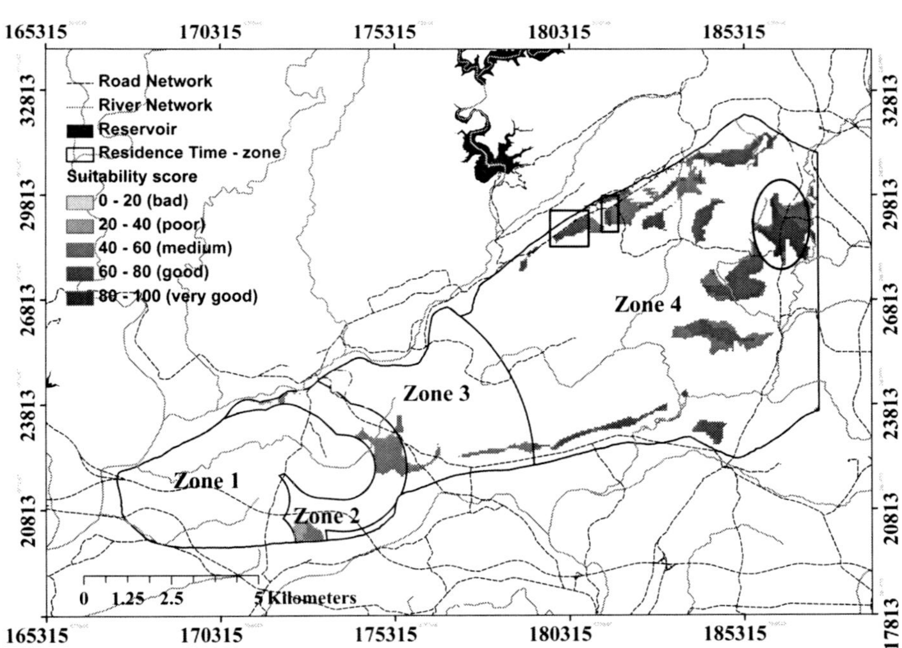

Figure 21.16 Final suitability map according to spatial multi-criteria-analysis procedure (G-DSS) (modified after Rahman *et al.* 2012)

MAR site selection

The highest ranked areas with regards to MAR implementation are located in the northeastern part of the Querença-Silves aquifer system (see circle in Figure 21.16), at about 8 km distance from the Arade Dam. Pumping and pipeline construction would result in elevated costs for water transfer from the Arade Dam to these areas.

The most promising sites with regards to the water source are located near the northern border of the aquifer system (rectangles in Figure 21.16). To minimize the cost, the natural riverbed downstream to the Arade Dam could be used for water transfer. The water would need to be pumped from riverbed to the selected MAR feasible areas. The computed residence time of the recharged water in the aquifer resulted more than 3 years. Considering this parameter, the most promising MAR sites are located in Zone 4. This would permit a long-term storage of water in the underground considering also the water quality improvement by natural attenuation processes. Some other *in-situ* parameters, such as soil salinity, organic carbon content, and sediment chemistry are still being studied in order to achieve more information about the water cleaning potential. A similar site selection study may be conducted to study the potential of MAR injection technologies.

Further MAR planning process

The studies at the Querença-Silves aquifer system related to MAR implementation should be further developed according to the planning flow-chart presented in Figure 21.1. Based on the results of the MAR site selection study, a set of most promising MAR planning and management options may be defined, supported by the Option Builder. The surface water surpluses at the Arade Reservoir should be used for groundwater recharge. The impact of each option on groundwater availability and quality is to be studied by mathematical modeling, defining optimal locations of the well systems for water recovery, considering the actual flow path of the infiltrated water, the spatial water demand structure and required water distribution into consideration. Once these MAR schemes are defined, the economic viability of each option needs to be analyzed. The evaluation of environmental project viability and impacts will rely again on groundwater modeling. Social benefits and impacts related to water scarcity mitigation should be quantified, too. The final evaluation, comparison, and ranking of MAR options will be based on a MCA as shown in the following G-DSS application at the Gaza-strip, Palestine. The most relevant socio-economic and environmental decision criteria will be taken into account. A participatory process, involving the most important stakeholders, governmental decision makers, and water resources planners is required. MAR implementation will change the water resources system significantly. The regional plan for IWRM implementation has to be updated accordingly.

Conclusions

This example demonstrates the function of the new GIS-based SMCA tool for MAR site ranking as implemented in the G-DSS (Rusteberg *et al.* 2008b). Site selection analysis involves a number of criteria, alternatives, and decision factors, resulting in a manageable decision environment. The decision steps are explicitly given to the user according to the overall analysis procedure in order to tackle an unstructured problem. Standard criteria and decision rules are offered to the user in order to reduce the analysis efforts.

The case study application on MAR implementation at the Querença-Silves aquifer system in South-Portugal shows that the aquifer could be used for water transfer and long-term storage purposes, contributing to water scarcity mitigation especially under drought conditions. There are promising locations to infiltrate surface water of the Arade Dam in the upper part of the aquifer system by means of ponds. The development of MAR planning and management options, their evaluation and ranking are subject to further studies in the region. Close cooperation with the stakeholder community, especially governmental water resources planners is required.

The new spatial multicriteria tool, due to its non-site specific, adaptive and comprehensive concept may serve as complementary element for any GIS based Water Resources Management support system. By modifying the input criteria and using the relevant dataset and decision rules, this spatial multi-criteria analysis tool can be applied to a wide range of disciplines, such as groundwater vulnerability assessment, land use planning, site selection for waste disposal, and so on. A detailed version of the site selection tool and case study has been prepared and is described by Rahman *et al.* (2012).

21.4.2 The Gaza-Strip case study, Palestine

The second G-DSS application focuses on the evaluation, comparison, and ranking of different strategies for MAR implementation in Northern Gaza Strip, Palestine. MAR has been identified as a potential response measure to the significant water resources problems of the region.

Introduction – MAR strategy evaluation and ranking

The Gaza Strip, located on the eastern coast of the Mediterranean Sea, is a region facing severe water resources problems. The climate here provides for approximately 200–400 mm/yr of rainfall, with an average of 300 mm/yr and high temperatures throughout most of the year (Al Agha, 1997). The population density and demographic development is very high, contributing to water pollution. This calls for efficient measures for water resources protection and development, based on sustainable long-term strategies for water resources planning and management. Due to the hot and dry climate, small amount of surface water is available. The water supply consists mostly of groundwater resources of the Coastal Aquifer of Gaza. This aquifer provides 97% of total water to the people of the Gaza Strip (Sánchez-Vila &

Barbieri, 2007). The current rate of abstraction far exceeds natural recharge rates, which has lead to a increased drawdown of the groundwater level over time. Concerns involving saltwater intrusion into the aquifer have risen due to the increased levels of chloride found in the abstracted water. In Northern Gaza Strip, is the Beit Lahiya wastewater treatment plant (BLWWTP), which is dysfunctional and has created severe public health and environmental problems. The BLWWTP is located on the edge of the town Beit Lahiya (Figure 21.17).

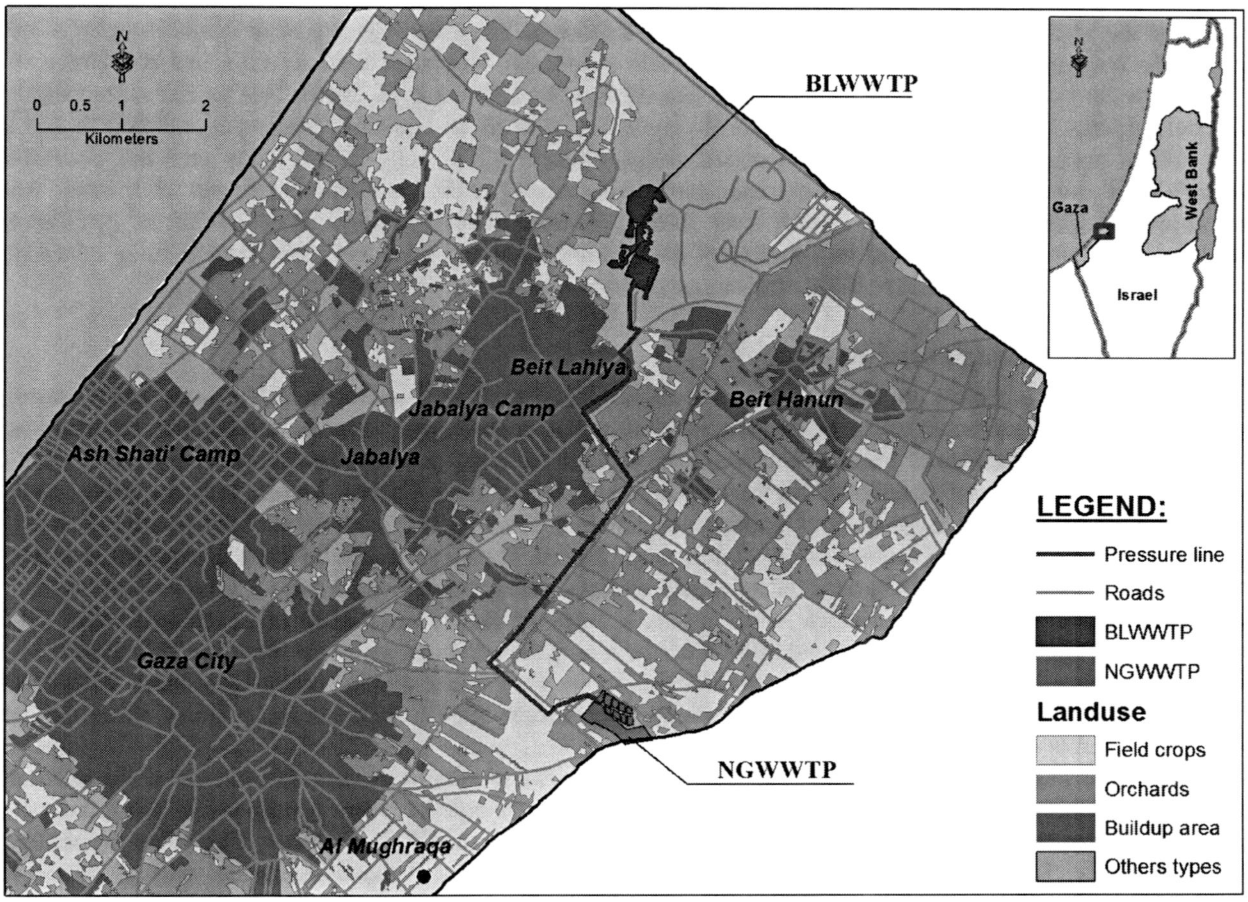

Figure 21.17 Case study area – Northern Gaza Strip

The BLWWTP was designed to be a system of aerobic and anaerobic lagoons, achieving a secondary level of treatment. The effluent of the final polishing lagoon was to be discharged directly on the sand outside of the BLWWTP compound, owing to the fact that no previously existing surface water bodies were present in the area. Due to the increasing number of population and the overloading of the wastewater treatment plant, the effluent slowly began overflowing on the sand. This was due to the creation of a slowly permeable to impermeable layer created by non-decaying organic matter which accumulated on top of the sand. The resulting pond has become a large water body containing over 1.5 million m^3 of partially treated effluent. A precarious situation has developed where the water level of the lake has reached the same level of the polishing bed, resulting in hydraulic problems where treatment of the wastewater coming through the lagoons in the BLWWTP has drastically been reduced (Sadah *et al.* 2009a).

A three-phase 20-year project involving the construction of a new WWTP and a pipeline transferring the water from the old WWTP to the new one is in progress. The new WWTP will involve MAR of effluent and a pipeline connecting the effluent lake to the new WWTP. The Palestinian Hydrology Group (PHG) and the Palestinian Water Authority (PWA) along with international support have decided to use practical, already established MAR technologies such as infiltration ponds with SAT to replenish the coastal aquifer in order to meet the continually rising demand of water for domestic, industrial, and agricultural use in this water-parched region. MAR is a promising solution not only for groundwater supply, but also to the effluent lake problem. By using SAT, the wastewater is naturally filtered by the sediments in the unsaturated zone. After sufficient residence time in the subsurface and water improvement by natural attenuation, the water then can be reused to meet domestic and agricultural demand. Monitoring should be performed adequately by the use of observation wells.

The following DSS case study application involves the analysis of several strategies for the implementation and management of MAR in northern Gaza Strip based on their potential impacts on agriculture, environment, society, and the economy. All strategies have been compared to each other and ranked according to their capability to promote sustainable development in Northern Gaza Strip. The following study also identifies the best water resources management strategy for the sustainable development of Northern Gaza strip.

Study area

With an area of 365 km^2 and a population of roughly 1.26 million, Gaza Strip is located on the southwestern part of Palestine at the Mediterranean Coast on the edge of Sinai Peninsula. The climate is dry, with rainfall roughly at 400 mm/yr. The soil types in the area are mainly Loessial arid brown soil, Regosols, Sand regosols, and arid brown soil. 97% of the water used in Northern Gaza is from the Coastal Aquifer (Shomar *et al.* 2005). This aquifer is composed of Pleistocene marine sand, sandstone, and intercalated clay layers. 30% of the aquifer is freshwater, which is tapped by 3,800 private agricultural wells and 90 municipal wells. Around 1,059 wells in the study area are abstracting 91.5 million m^3/year. 1,015 agricultural wells correspond to a pumping of 49.2 million m^3/year and 44 domestic wells abstract around 42.3 million m^3/year (Sadah *et al.* 2009a). Average natural recharge is 47 million m^3/yr. The net groundwater exploitation has resulted in a decline of groundwater level and more and more brine is being mixed into the upper aquifer, causing increased chloride concentration with time.

Water resources management challenges

The water resource problems at Gaza Strip are immense. 1.26 million inhabitants suffer from water scarcity, overexploitation of groundwater resources, and water pollution due to inadequate wastewater collection, treatment, and disposal. The overexploitation of groundwater resources in Gaza Strip refers to the unsustainable extraction rates of groundwater from the Coastal Aquifer of Northern Gaza Strip. Currently, the Coastal Aquifer is facing severe quality and quantity problems, which are a result of excessive exploitation, resulting in a water deficit between natural recharge and extraction of about 40 to 50 million m^3/yr (Sadah *et al.* 2009a). The water deficit for 2020 is expected to increase to 100 million m^3/yr, if the "Business as usual approach" to water management in Gaza Strip is followed until then.

Currently, Gaza's main industry, which is agriculture, is suffering due to salinization of agricultural wells. Citrus produce, tomatoes, potatoes, cucumbers, and other fruits such as strawberries are grown in the Gaza Strip (Sadah *et al.* 2009a). Methaemoglobinemia is common in infants due to the poor water quality, which is a result of not only salinization but also contamination by wastewater. The poverty level in Gaza is high and many cannot afford the costs of water treatment or desalinization (Sadah *et al.* 2009a).

Gaza Strip faces water quality and quantity problems from the point of view of the unsustainable use of the sole water resource, which is the coastal aquifer. Increased salinity is reason enough for many people not to drink the water. Additionally, Gaza suffers health problems due to increasing water pollution by wastewater. This refers especially to infants. Currently, Gaza Strip gives major attention to combat wastewater contamination as well as decreasing groundwater levels. The contamination by wastewater of North Gaza Strip can be directly linked to the current wastewater treatment system which exists in Northern Gaza Strip. Sewage is collected by a sewer system and piped to the Beit Lahiya (old) wastewater treatment plant (BLWWTP) (Figure 21.17). Wastewater has been disposed of at the BLWWTP since the late 1970's when the plant was constructed. The initial design of the BLWWTP was for the residents in the municipality of Jabalya (Sadah *et al.* 2009a). 50,000 people live in this municipality and the maximum wastewater generation estimated was to be 5,000 m^3/day (World Bank, 2008). This volume was the value incorporated into the design of the BLWWTP. Due to the fact that the BLWWTP is now connected to several other municipalities and one refugee camp, the BLWWTP now receives 12,000 m^3 of wastewater per day, which is over twice the volume of the designed capacity of the system. 180,000 people live in these municipalities which now have their wastewater being treated at the lagoon treatment system, the BLWWTP. These municipalities include Jabalya, Beit Hanoun, Beit Lahiya, Umm Al Nasr, and the Jabalya Refugee Camp (Enshassi, 2000).

The overloading of the BLWWTP has lead to a complex picture which has complicated the initial problem of groundwater exploitation. The poor management of the BLWWTP and the incomplete treatment and improper disposal of its effluent has caused serious environmental, socio-economic, political, and agricultural impacts for the people of Gaza.

Due to the permanent overloading of the BLWWTP, Northern Gaza currently suffers from the presence of a 1.5 million m^3 wastewater lake which has slowly grown to this size since the BLWWTP's inception in 1976. Initially, the effluent from the BLWWTP was intended to be reused for agricultural purposes. Since this has not been realized, the effluent discharged from the BLWWTP has been deposited just on the sand dunes outside of the plant (Sadah *et al.* 2009a). Overtime ponding began and since then the ponds have accreted and the volume has accumulated to a 1.5 million m^3 lake, contained just by sand dams. The Palestinian Water Authority (PWA), responsible for environmental and health protection has taken other emergency measures in order to avoid inundation of communities within Beit Lahiya (Sadah *et al.* 2009). Now, the BLWWTP is blocked due to the height of the lake and treatment is at an all-time minimum. In March 2007, 5,000 people were affected and six people died due to a dam break at the BLWWTP (Sadah *et al.* 2009a).

Further use of the BLWWTP is likely to contribute to catastrophes such as the one in March 2007, together with further water quality degradation by faecal matter. Odor emissions from the lake also disturb those who live in the area. Archaeological sites from the Roman-Byzantine Empire exist near the effluent lake and are also endangered by a potential dam collapse (Sadah *et al.* 2009a).

MAR response strategies

Based on water resources problem analysis, a number of MAR response strategies were formulated. Considering the water management plans for the timeframe 2005–2025, the following seven water resources management scenarios were established. Table 21.2 shows the MAR management strategies formulated in this study. Causal chain analysis techniques, applied to the prevailing water resources problems, confirmed the choice of Managed Aquifer Recharge as a potential response action to the main pressures on Gaza groundwater resources.

Table 21.2 MAR strategies towards sustainable development of water resources at northern Gaza.

Strategy no	MAR response strategy for WR development	Waste water reuse option
1	'Business as usual' approach	No MAR – no reuse
2	Phase 1: Construction of infiltration ponds and pipeline	No reuse – just storage and groundwater level recovery
3		Water reuse for agriculture
4	Phase 2: Construction of the New Wastewater Treatment Plant (NGWWTP)	No reuse – just storage and groundwater level recovery
5		Water reuse for agriculture
6	Phase 3: Extension of the New Wastewater Treatment Plant (NGWWTP)	No reuse – just storage and groundwater level recovery
7		Water reuse for agriculture

Due to specifications of the Israeli Authorities, there are no alternatives with regards to the location of the infiltration ponds at the eastern border as shown in Figure 21.17. Therefore, the above listed strategies differ with regards to wastewater treatment, water quality of the recharged water, recharged volumes, investment cost and water reuse.

The water management strategies as presented in Table 21.2 consider three phases in terms of water resources development. The first phase considers the construction of a pipeline (Figure 21.18a) to conduct the preliminary treated effluent of BLWWTP to the new wastewater treatment plant and the infiltration ponds (Figure 21.18b), both located at the Israeli border (Figure 21.17).

Figure 21.18 Pipeline contstruction between old and new WWTP (a) and infiltration ponds (b)

The second phase includes the construction of the NGWWTP, while the third phase considers the possibility to extend the new plant up to a treatment capacity of around 24 million m^3 per year in 2025.

A further classification is done between a) maintaining the current water withdrawal for agriculture and annual production, and b) sustainable development of agriculture according to the demand. Including the so-called "Business as usual approach," a total number of seven water resources planning and management strategies have been defined.

Decision criteria selection and hierarchy

A total of 21 criteria were selected and quantified in close cooperation with Palestinian researchers and authorities as well as further relevant stakeholders. A wide range of indicators was considered for the selection of criteria. The criteria were grouped into four categories, namely environmental, social, health and economic. Figure 21.19 shows the hierarchical structure of the criteria and sub criteria.

This analysis encompasses three multi-criteria analysis techniques: Analytical Hierarchy Process (AHP), Weighted Linear Combination (WLC) and Composite Programming (CP). The procedures are discussed below briefly:

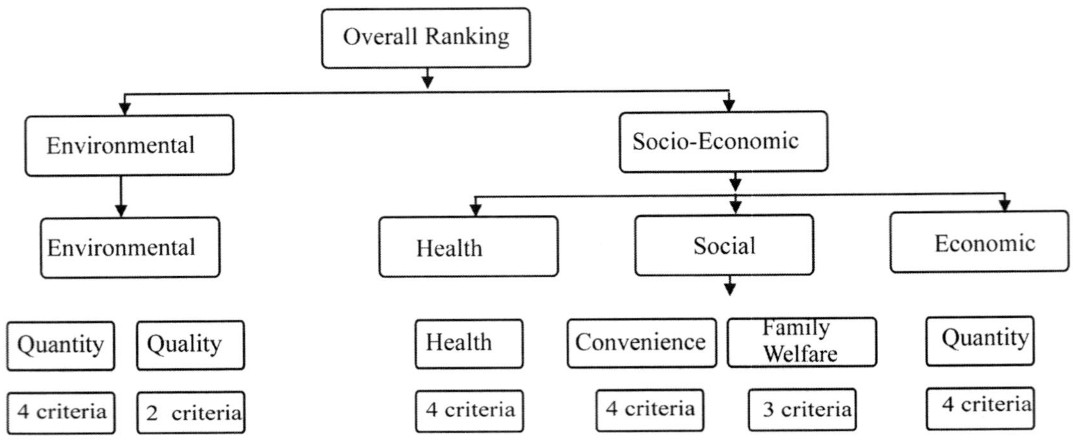

Figure 21.19 Criteria groups and hierarchical structure

Criteria Quantification

The selected criteria are quantified using different state-of-art analysis techniques. The environmental and health related criteria refer to groundwater quality and were quantified based on groundwater modeling. Visual Modflow was used to simulate the groundwater flow and solute transport for each of the planning and management scenarios. The quantification of social and economic criteria has been supported by field surveys, stakeholder meetings and questionnaires. The quantified criteria have been passed through a process of normalization.

Figure 21.20 shows the water level development in the Northern Gaza strip under the different water resources planning and management scenarios. This confirms that the so-called "Business as usual approach" results in continuously decreasing groundwater levels in the project area with a total dropping of 3 m during the planning period until 2025.

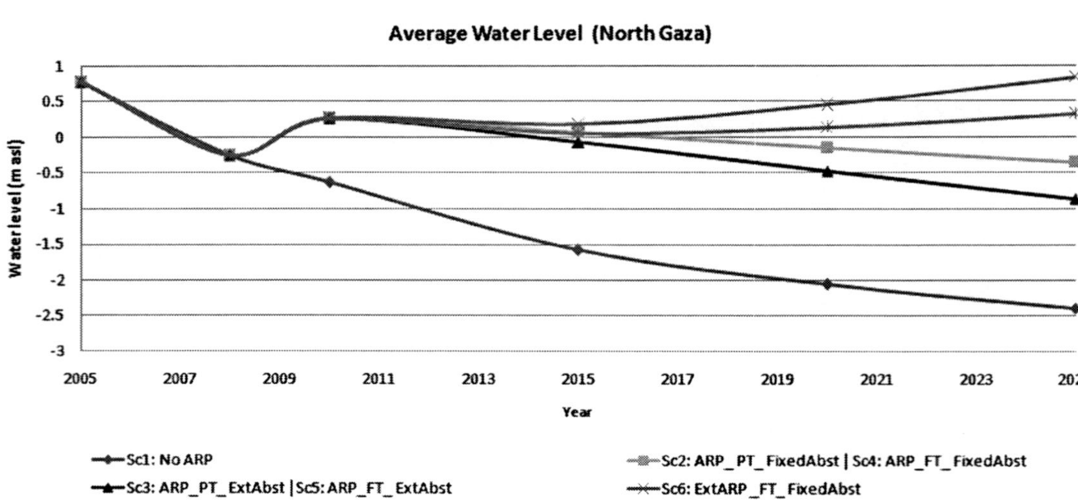

Figure 21.20 Groundwater level depending on the MAR response strategy

Best performance, in terms of groundwater level recovery, is projected to be with Strategies 6 and 7 (Table 21.2), which consider the extension of the new wastewater treatment plant and treatment and infiltration to the aquifer of secondary treated effluent via the basins. Strategies 6 and 7 also allow possible operation up to maximum production rates. Under these strategies, the groundwater level may rise. With respect to sustainable agricultural development, Strategy 7 does not perform as well as Strategy 6. The ground water level in Strategy 7 will not recover as significantly as under Strategy 6, due to the higher extraction rates by agricultural wells.

MAR strategy ranking

After quantification of all 21 decision criteria, the normalized matrix has been prepared for further analysis. Figure 21.21 presents the ranking of strategy alternatives based on MCA. It shows the ranking of alternative strategies using the AHP together with WLC analysis techniques. MAR Strategy 7 is the best ranked option. More comprehensive studies on ranking robustness have proven that neither the MCA procedure nor the criteria and preference structure represented by the weights significantly change the ranking order (Rusteberg *et al.* 2010). Strategy 7 is always ranked first and can be considered as a stable solution. The "Business as usual approach" is not an acceptable option towards the sustainable development of the Gaza strip.

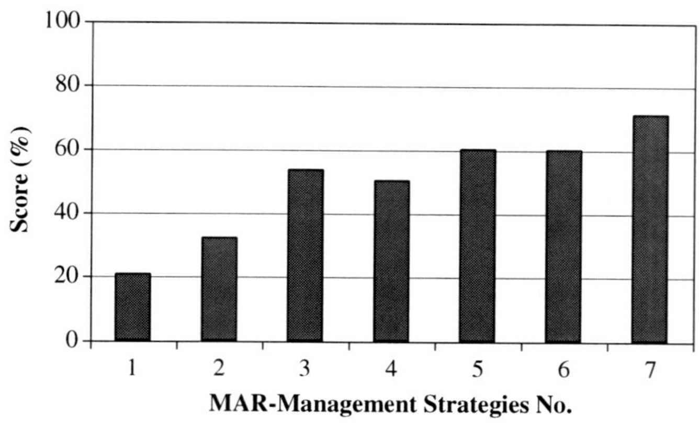

Figure 21.21 Ranking of MAR strategies based on AHP and weighted linear combination

Conclusions

The case study application shows that appropriate water resources planning and management measures are urgently required to combat water scarcity, water pollution, and health problems in Northern Gaza-Strip. The so-called "Business as usual approach" may not be considered as an option for Gaza, contributing to further groundwater level decline and groundwater quality deterioration while increasing the health risks for the Palestinian population. The performance analysis of the developed water resources planning and management strategies shows clearly that MAR by means of infiltration ponds as SAT with secondary treated effluent is a viable response to the increasing water resources problems of the region. Further investments should be undertaken to better maintain and further extend the existing wastewater collection network as well as the capacity of the new wastewater treatment plant (NGWWTP) at the Israeli border, accompanying the rapidly increasing wastewater production. The infiltration of these effluents, adequately treated and monitored, contributes to groundwater recharge, recovery of groundwater levels, and water quality protection of deeper groundwater resources which still meet international drinking water standards. The study showed that MAR might ensure sustainable agricultural development in Northern Gaza Strip. The comparison of water management options revealed that increasing investments in wastewater collection, treatment, and later MAR results in increasing water management strategy performance with regards to the considered environmental, social, and health criteria. Obvious drawbacks are the infrastructural investments and their impact on economic feasibility that are to be more properly discussed as based on comprehensive cost-benefit studies which aim toward cost minimization and maximization of related environmental and health benefits. Those benefits are vital for the sustainable development of the Gaza-Strip.

21.5 SUMMARY AND OUTLOOK

This section discusses the role of Managed Aquifer Recharge (MAR) in the context of Integrated Water Resources Management (IWRM) as well as the need for comprehensive decision support with regards to the planning of MAR systems. Intended to be made very clear was that the implementation of MAR is not only a local or site specific task, restricted to aquifer storage and water quality attenuation via recharge, but rather is part of a regional basin scale IWRM approach. MAR permits the joint management of surface water, groundwater, and wastewater resources and is therefore an important integrating element of IWRM. The planning of a MAR system is complex and requires a specialized decision making process. Numerous interdependent decisions always have to be made in the planning of a MAR system.

The innovative Gabardine Decision Support System (G-DSS) for MAR planning, developed in the framework of the FP6-project Gabardine, was briefly introduced. The planning process, system structure, main menus, and the spatial geo-database were also briefly presented. Some examples of the graphical user interface of the G-DSS were given. The system offers a highly interactive scientifically based planning approach, taking the non-straightforward decision structure of MAR planning into account. The G-DSS should be understood as an expert system to be applied by planners and managers of surface and ground water resources on the hydrological basin level. It guides one through the planning process of a MAR facility, starting with the problem analysis, water budgets, identification of potential sources of water, suitability mapping and ranking of potential MAR sites, selection of MAR technology, definition of alternative MAR planning and management options as well as the evaluation and ranking of MAR options, based on their impacts and performance towards the mitigation of the water deficit problem. Alternative MAR projects and management strategies may be evaluated and ranked based on Multi-Criteria-Analysis (MCA), considering environmental, socio-economic, and technical decision criteria.

Two applications of the DSS have been shortly presented. The Algarve region in South-Portugal suffers from water scarcity, especially under droughts. During drought years the water supply problem has not been solved by the construction of surface water reservoirs. Managed Aquifer Recharge is a potential response measure for water scarcity mitigation. Surface water surpluses at the nearby Arade dam will be transferred to the northern part of the Querença-Silva aquifer system and recharged by means of infiltration ponds. This application focuses on the identification of suitable sites for MAR implementation as well as their ranking, based on an innovative decision support tool, integrated in the Gabardine DSS. The definitive selection of a specific location for MAR implementation will depend on the evaluation of the overall MAR planning scheme, as demonstrated in the DSS example application at the northern Gaza-Strip.

Based on the Palestinian water policy (period 2005–2025) on wastewater reclamation and reuse, a set of seven MAR management schemes were developed and studied in close cooperation with the local stakeholder community. The study shows that MAR, by means of infiltration ponds as SAT with secondary treated effluent, is a viable response to the increasing water resources problems of the region. The comparison of water management options showed that increasing investments in wastewater collection, treatment and subsequent MAR results in increasing water management strategy performance with regards to the considered environmental, social, and health criteria.

The quantification of decision criteria in the MAR planning process is still an external process. Future development of the G-DSS should aim toward the full integration of the software tools and models required in the decision making process and developed within the context of the Gabardine project. Special attention with regards to further system development may be given to the simulation of natural attenuation processes under infiltration conditions, risk assessment, regulatory aspects, design and operation of MAR facilities, as well as decision rules for the optimization of system performance.

The transfer of MAR technologies, considering unconventional sources of water, and their large scale implementation in semiarid regions requires further research and technology adaptation, for example, with regards to an adequate water treatment system prior to MAR.

ACKNOWLEDGEMENTS

The authors acknowledge the funding of the present research by the European Commission in the context of the FP6 GABARDINE project (contract number: 518118) and would like to thank the Palestinian Hydrology Group (PHG) and the Palestinian Water Authority (PWA) for the highly productive cooperation towards peaceful and sustainable development of the Gaza-Strip. Many thanks to the Portuguese National Laboratory of Civil Engineering of Portugal (LNEC) for their support in the Algarve case study region.

REFERENCES

Al-Agha M. R. (1997). Environmental management in the gaza strip. *Environ Impact Asses Rev.*, **17**, 65–76.

Almeida C., Mendonça J. L., Jesus M. R. and Gomes A. J. (2000). Sistemas Aquíferos De Portugal Continental, Relatório (Aquifer Systems in Portugal, Report). INAG, Instituto da Água. Lisboa. Electr. Doc., CD-ROM.

Anbazhagan S., Ramsamy S. M. and Gupta S. D. (2005). Remote sensing and GIS for artificial recharge study, runoff estimation and planning in Ayyar Basin, Tamil nadu, India. *Environmental Geology*, **48**, 158–170.

Asano T. and Cotruvo J. (2006). Groundwater recharge with recycled municipal wastewater: health and regulatory considerations, WHO guidelines for the safe use of wastewater. *Excreta and Greywater*, **4**(6), 9.

ASCE (American Society of Civil Engineers) (2001). Standard guidelines for artificial recharge of ground water. *ASCE Standard*, EWRI/ASCE34-01. 106.

BMBF (2010). WISDOM: Wasser-Information System für das Mekong Delta. Website: http://www.wasserressourcen-management.de/de/174.php, (March 2010).

Bouwer H. (2002). Artificial recharge of groundwater: hydrogeology and engineering. *Hydrogeol. J.*, **10**, 121–142.

Brissaud F., Xu P. and Aiset M. (2003). Extensive reclamation technologies, assets for the development of water reuse in the Mediterranean. *Water Supply*, **3**(4), 209–216.

Brown C. J., Itani S. and Zhang M. (2005). Aquifer Storage & Recovery (ASR) regional study: A scientific evaluation of potential pressure induced constraints and changes in the Florida aquifer system and the Hawthron Group (Final Draft, July 2005), U.S. Army Crops of Engineers, Jacksonville District, 70.

Calijury M. L., Marques E. T., Lorentez J. F., Azevedo R. F. and Carvalho C. B. (2004). Multi-criteria analysis for the identification of waste disposal areas. *Geotechnical and Geological Engineering*, **22**, 299–312.

CDPH (California Department of Public Health) (2008). Groundwater Recharge Reuse, Draft Regulation. Website: http://www.cdph.ca.gov/certlic/drinkingwater/Documents/Recharge/DraftRegulations.pdf (February 2010).

Cordy G. E., Duran N. L., Bouwer H., Rice R. C., Flurlong E. T., Zaugg S. D., Meyer M. T., Barber L. B. and Kolpin D. W. (2004). Do pharmaceuticals, pathogens, and other organic water compounds persist when waste water is used for recharge? *Ground Water Monitoring and Remediation*, **24**(2), 58–69.

Delft Hydraulics (2004). http://www.wldelft.nl/soft/tools/index.html (Accessed on March 2008).

DHI (Danish Hydraulics Institute) (2008). MIKE SHE- Integrated River Basin Planning. Website: http://www.mikebydhi.com/Products/WaterResources/MIKEBASIN.aspx (Jan 2008).

Dillon P. and Molloy R. (2006). Developing Aquifer Storage & Recovery (ASR) Opportunities in Melbourne, Technical Guidance for ASR: CSIRO Land Water Client Report Prepared for the Smart Water Fund. 24.

Dillon P., Ward J. and Cunliffe D. (2007). Innovation in governance in managed aquifer recharge. In: *ISMAR6 Proceedings*, Phoenix, USA.

Dillon P. (2009). Water recycling via managed aquifer recharge in Australia. *Boletín Geológico y Minero*, **120**(2), 121–130.

Drewes J. E., Heberer T. and Reddersen K. (2003). Fate of pharmaceuticals during groundwater recharge. *Ground water Monit. & Remedit.*, **23**(3), 64–72.

Eastman J. R., Kyem P. A. K., Toledano J. and Jin W. (1993). GIS and decision making. united nations institute for training and research (UNITAR). *Explorations in Geographic Information Systems Technology*, **4**, 112.

EMRL (Environmental Modeling Research laboratory) (2004). Birgham Young University, Provo, Utah. Website: http://www.emrl.byu.edu/gms.htm (Accessed March 2008).

Enshassi A. (2000). Environmental concerns for construction growth in gaza strip. *Building and Environment*, **35**, 273–279.

Escalante E. F., Rodríguez M. G. and Gil F. V. (2005). A Methodology for Wetland Classification Attending to the Capacity for Artificial Aquifer Recharge. In: *ISMAR 5 Proceedings*, Berlin, Germany, pp. 789–798.

Fox P., Narayanaswamy K., Genz A. and Drewes J. E. (2001). Water quality transformations during soil aquifer treatment at the Mesa Northwest Water Reclamation Plant, USA. *Water Science and Technology*, **43**(10), 343–350.

Fryar A. E., Macko S. A., Mullican W. F., Romanak K. D. and Bennett P. C. (2000). Nitrate reduction during ground-water recharge, southern high plains. *Texas. J. Contam. Hydrol.*, **40**, 335–363.

Gale I. N., Williams A. T., Gaus I. and Jones H. K. (2001). ASR – UK: Elucidating the Hydrogeological Issues Associated with Aquifer Storage and Recovery in the UK, BGS Report No. CR/02/156/N. 54.

Giupponi C., Mysiak J., Fassio A. and Cogan V. (2008). MULINO-DSS: a computer tool for sustainable use of water resources at the catchment scale. *Mathematics and Computers in Simulation*, **64**, 13–24.

Gomes E. G. and Lins M. P. E. (2002). Integrating geographical information systems and multi-criteria methods: a case study. *Annals of Operations Research*, **116**, 243–269.

Hochstrat R., Wintgens T., Kazner C., Jeffrey P., Jefferson B. and Melin T. (2010). Managed aquifer recharge with reclaimed water: approaches to a European guidance framework. *Water Science and Technology*, **62**(6), 1265–1273.

Idelovitch E., Icekson-Tal N., Abraham O. and Michail M. (2003). The long-term performance of soil aquifer treatment (SAT) for effluent reuse. *Water Science and Technology: Water Supply*, **3**(4), 239–246.

Johnson J. S., Baker L. A. and Fox P. (1999). Geochemical transformations during artificial groundwater recharge: soil-water interactions of inorganic constituents. *Wat. Res.*, **33**(1), 196–206.

Kim J. W., Choi H. and Schwartz F. W. (2004). Modeling the fate and transport of organic and nitrogen species in soil aquifer treatment process. *Water Science and Technology*, **50**(2), 255–261.

Laplanche F. (2006). Environnement de Conception De Bases De Donnees Spatiales Sur Internet. PhD thesis, Geomatics Unit, University of Liege.

Laplanche F. (2008). Web2GIS: A Spatial Database Conception Environment on Internet. Manuscript in preparation, Geomatics Unit, University of Liege.

Li G., Tang Z., Mays L. W. and Fox P. (2000). New methodology for optimal operation of soil aquifer treatment systems. *Water Resources Management*, **14**, 13–33.

Lobo Ferreira J. P. and Oliveira L. (2007). Aquifer storage and recovery and applicability to algarve (Portugal). In: L. Ribeiro, A. Chambel and M. T. Condesso de Melo (eds), *Proceedings XXXV IAH Congress, Groundwater and Ecosystems*, Lisboa, pp. 10.

Maliva R. G. and Missimer T. M. (2010). Aquifer Storage and Recovery and Managed Aquifer Recharge Using Wells: Planning, Hydrogeology, Design, and Operation, Methods in Water Resources Evaluation – Series No. 2, Schlumberger Water Services, Texas, USA, 578.

Monteiro J. P. (2005). A Importacia Do Dimentionamento de Captacoes de Agus Subterraneas No Algarve – Passado, Presente e Futuro, Manual Técnico de Engenharia, Águas Subterrâneas e Efluentes Tratados da GRUNDFOS, Lisboa, Expresso Gráfico.

NRC (National Research Council) (2008). Prospects of Managed Underground Storage of Recoverable Water. National Academy Press, Washington D.C., 337.

NRMMC, EPHC, NHMRC (2009). Managed Aquifer Recharge Guidelines. Australian Guidelines for Water Recycling. Natural Resources Management Ministerial Council, Environment Protection and Heritage Council, and National Health and Medical Research Council. http://www.ephc.gov.au/taxonomy/term/39.

OECD (Organisation for Economic Cooperation and Development) (1993). OECD Core set of Indicators for Environmental Performance Reviews. *Environment Monographs*, **83**, Paris.

OECD (Organisation for Economic Cooperation and Development) (2003). OECD Environmental Indicators, Development, Measurement and Use. Reference Paper, Website: http://www.oecd.org/dataoecd/7/47/24993546.pdf (Jan 2010).

Oxley T., McIntosh B., Winder N., Mulligan M. and Engelen G. (2004). Integrated modeling and decision-support tools: a mediterranean example. *Environmental Modeling & Software*, **19811**, 999–1010.

Panday S. and Huyakorn P. (2008). Modflow surfact: a state-of-art use of vadose zone flow and transport equations and numerical techniques for environmental evaluations. *Vadose Zone J.*, **7**, 610–631.

Pavelic P., Dillon P. J. and Nicholson B. C. (2006). Comparative evaluation of the fate of disinfections by products at eight aquifer storage and recovery. *Environ. Sci. Technol.*, **40**, 501–508.

Pyne R. D. G. (2005). Aquifer Storage Recovery, a Guide to Groundwater Recharge Through Wells. ASR Systems, Gainsville, FL, 606.

Rahman M. A., Rusteberg R., Gogu R. C., Ferreira J. P. L. and Sauter M. (2009). Innovative Tool for Selection of Potential Site for Managed Aquifer Recharge, Gabardine Deliverable D55, 27.

Rahman M. A., Rusteberg B., Gogu R. C, Ferreira J. P. L. and Sauter M. (2010). Spatial multi-criteria decision support for site suitability mapping with regards to aquifer artificial recharge projects. In: 7th International Symposium on Managed Aquifer Recharge (ISMAR7), October, 2010, Abu Dhabi, UAE.

Rahman M. A., Rusteberg B., Gogu R. C, Ferreira J. P. L. and Sauter M. (2012). A new spatial multi-criteria decision support tool for site selection for implementation of managed aquifer recharge. *Journal of Environmental Management*, doi: 10.1016/j.jenvman.2012.01.003.

Rusteberg B. (2006). Groundwater artificial recharge by alternative sources of water to guarantee water supply in the future: The European Research Project Gabardine. Conference AQUA 2006, Water Science and Technology: Integrated Management of Water Resources, Athens, Hellas.

Rusteberg B., Bear J., Bensabat J., Brouyère S., Wojda P. and Gogu R. C. (2008a). Decision Support to the Planning and Management of Groundwater Artificial Recharge Systems, First joint workshop ReclaimWater-Gabardine. Sabadell, Barcelona, Spain, pp. 21–23, March.

Rusteberg B., Rahman M. A., Bear J., Bensabat J., Brouyère S., Wojda P., Gogu R. C. and Power H. (2008b). The gabardine decision support system for artificial recharge planning and management. In: Conference on Water Reclamation and Aquifer Recharge, Maribor, Slovenia.

Rusteberg B., Rahman M. A., Sadah M. A., Rabi A. and Sauter M. (2010). Water Management Strategies Analysis Using Multiple Criteria Decision Techniques Towards Sustainable Development Of Northern Gaza Strip. Water Observation and Information for Decision Support, Macedonia.

Saaty T. L. (1980). The Analytical Hierarchy Process. Mcgraw-Hill, New York.

Sadah M. A., Rabi A., Ghannam S. and Tamimi A. R. (2009a). Impact of the "Do-nothing" Approach at the Gaza Strip – Palestine, GABARDINE Deliverables – D 71, 21.

Sadah M. A., Rabi A., Ghannam S. and Tamimi A. R. (2009b). Economic Analysis and Feasibility of MAR at Gaza-Strip, Palestine, GABARDINE Deliverables – D 72, 16.

Sánchez-Vila X. and Barbieri M. (2007). Current Necessities of Artificial Recharge in the Test Sites, GABARDINE Project- Deliverable 24, 17.

Sharifi M. A. and Retsios V. (2004). Site selection for waste disposal through spatial multiple criteria decision analysis. *Journal of Telecommunications and Information Technology*, **3**, 1–11.

Shomar B. H., Müller G. and Yahya A. (2005). Geochemical features of topsoils in the gaza strip: natural occurrence and anthropogenic inputs. *Environmental Research*, **98**, 372–382.

Store R. and Kangas J. (2001). Integrating spatial multi-criteria evaluation and expert knowledge for GIS-based habitat suitability modeling. *Landscape and Urban Planning*, **55**, 79–93.

Tang Z., Li G., Mays L. W. and Fox P. (1996). Development of methodology for the optimal operation of soil aquifer treatment systems. *Wat. Sci. Tech.*, **33**(11), 433–442.

UNESCO-IHP (2005). Strategies for managed aquifer recharge (MAR). In: Semi-arid Areas, Ian Gale (eds), pp. 65.

Vogt J. (2002). Guidance Document on Implementing the GIS Elements of the Water Framework Directive.

Ward J. and Dillon P. (2009). Robust Design of Managed Aquifer Recharge policy in Australia. CSIRO Water for a Healthy Country Flagship Report to National Water Commission for Raising National Water Standards Project: Facilitating Recycling of Stormwater and Reclaimed Water via Aquifers in Australia - Milestone Report 3.1, April 2009.

WHO (World Health Organization) (2006). Guidelines fro drinking water quality, 3rd edn. 595.

World Bank (2008). North gaza emergency sewage treatment additional financing, project information document, 7, Website: http://www-wds.worldbank.org/external/default/WDSContentServer/WDSP/IB/2008/04/08/000076092_20080411104258/Original/Final0PID.doc (January, 2010).

Wojda P. (2009). Hydrogeological Data Modelling in Groundwater Studies. PhD thesis, University of Liège, Faculty of Applied Sciences, Belgium.

Wojda P., Brouyère S., Derouane J. and Dassargues A. (2010a). HydroCube: an entity-relationship hydrogeological data model. *Hydrogeology Journal*, doi: 10.1007/s10040-010-0653-6.

Wojda P., Brouyère S. and Dassargues A. (2010b). Geospatial information in hydrogeological studies. In: Encyclopedia of Hydrological Sciences, M. G. Angerson (eds), Wiley, Chichester, UK. doi: 10.1002/0470848944.hsa304.

Wojda P., Gadin N., Brouyère S. and Gogu R. (2006). Report on computerized database system, GABARDINE Report- DL 63, 80. Web-link: www.Gabardine-fp6.org

WSM (Water Strategy Man) (2004). The WaterStrategyMan DSS - A Comprehensive Decision Support System for the Development of Sustainable Water Management Strategies. In: S. R. L. Pro GEA (ed.), Deliverable 21.4 submitted to EC (FP-5 program). Website: http://environ.chemeng.ntua.gr/wsm/Uploads/Deliverables/ThirdYear/Deliverable_21_4.pdf (Jan 2009).

Part E

Promoting MAR Systems for Water Recycling

Chapter 22

Managed Aquifer Recharge as a component of sustainable water strategies – a brief guidance for EU policies

Thomas Wintgens, Rita Hochstrat, Christian Kazner, Paul Jeffrey, Bruce Jefferson and Thomas Melin

22.1 INTRODUCTION

22.1.1 Water reuse as a water stress mitigation option

The benefits of water reclamation and reuse within an integrated water management framework are well documented in the literature (Anderson, 2003; Asano *et al.* 2007; EC, 2006). Water recycling provides additional drought-proof water supply, favours a more local sourcing of water and avoids the use of drinking water quality sources where such high quality is not needed. Reclaimed wastewater is a potential source for all the various purposes that freshwater is used for, appropriate pre-treatment presupposed. Types of water reuse include agricultural irrigation, urban and domestic applications, industrial uses, such as cooling water and boiler feed make-up, environmental application aimed at flow augmentation or wetland restoration as well as groundwater recharge. The anthropogenic water cycle as illustrated in Figure 22.1, involves regular though unintended reuse of wastewater. Once discharged to the aquatic environment, wastewater treatment plant effluents are either withdrawn for various purposes or may contribute to environmental flows or the recharge of aquifers. This fact has to be acknowledged in evaluating any direct and planned reuse activities.

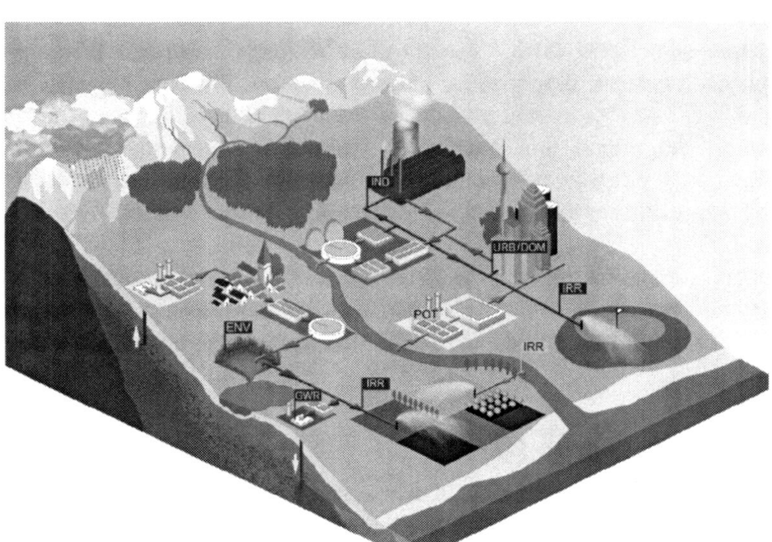

Figure 22.1 The anthropogenic water cycle with direct and indirect water reuse (modified from Veolia Water, Durham 2005, where ▒: indirect reuse ■: direct reuse, GWR: groundwater recharge, IRR: irrigation, POT: potable reuse, IND: industrial reuse, URB/DOM: urban & domestic reuse, ENV: environmental enhancement)

The status of water reuse in Europe has been comprehensively investigated in the AQUAREC project and by the MED-EUWI Wastewater Reuse Working Group (Bixio *et al.* 2006, Hochstrat *et al.* 2006, EC, 2006, MED WWR-WG, 2007) where water scarcity has been identified as a main driver for developing wastewater reuse practice. Climate change is expected to aggravate the availability of water, particularly in the Mediterranean and adaptation measures like more efficient use of water and water reuse are increasingly being considered as mitigation options (EEA, 2007).

Among the different water reuse options Managed Aquifer Recharge (MAR) with reclaimed water plays a particular role, due to both the large potential benefits achieved in terms of sustainable water resources management as well as the complexity of the planning and implementation of such schemes (Bouwer, 2002; Dillon et al. 2009; Steenbergen and Tuinhof 2009). Within the European Union, strategies for groundwater augmentation, including recharge with conventional and non-conventional water are proposed as key features of drought management plans (EC, 2007a). For the water reuse case studies described in RECLAIM WATER the motivations for managed aquifer recharge fall into various categories and represent a wide range of drivers known to promote water recycling.

Mitigation of water scarcity and drought is a major driver for implementing managed aquifer recharge. Such drivers are related to the deterioration of water resources, for example, the intrusion of saltwater into coastal aquifers, or simply to the lack of sufficient water of appropriate quality to meet regional demand. In this context the objectives of managed aquifer recharge can be defined as:

- Augmentation of limited drinking water resources,
- Restoration of groundwater resources impaired by seawater intrusion or overabstraction,
- Provision of treatment and storage capacity,
- Establishing a barrier against seawater intrusion.

As a complementary fresh water resource, especially for coastal areas where discharges into the sea are irreversible, MAR also helps to secure water supply and to reconcile the demand for water from competing sectoral users. MAR allows;

- Supply of irrigation water to agriculture without further exploiting drinking water resources,
- Utilisation of stormwater for industrial purposes,
- Balancing of seasonal peak demand caused by tourist activities.

Environmental aspects also play a role in the decision to implement recharge projects as such schemes may be intended to maintain the ecological quality of for example, hydraulically connected wetlands or to temper environmental objections to groundwater abstraction. Reuse may also be driven by environmental concerns associated with discharge and disposal of treated wastewater. From an environmental perspective the positive impacts of water recycling are;

- Reduced diversion and withdrawal from natural water systems and
- Reduced discharge of treated wastewater and limited pollution due to discharge of stormwater or treated wastewater, particularly under low flow regimes.

Economic considerations sometimes favour recycling or recharge schemes. Water recycling can have cost advantages over other options for water supply or wastewater disposal. Through resource replacement it contributes to water conservation and thus may postpone, reduce or even avoid investment in new infrastructure. Reuse is often a viable alternative to the import and transfer of freshwater over long distances. It may also be more economically attractive than deep groundwater extraction or seawater desalination. In other cases, the high cost of constructing and operating a suitable marine outfall for discharging treated wastewater and stormwater to the ocean might be avoided.

Using an aquifer as a natural storage circumvents the construction of surface reservoirs or tanks to store reclaimed water. Underground storage is also not prone to recontamination and does not experience losses through evapotranspiration. The natural attenuation capacity of the sub-soil, particularly of the vadose zone can be utilised in this context.

22.1.2 Managed aquifer recharge (MAR)

Aquifers have been recharged artificially for centuries. In most cases surface water has been used to replenish groundwater reservoirs to allow more intensive use/abstraction. But long-term and sustainable operation strongly depends on effective management of water quality and clogging issues. Figure 22.2 presents schematics of the most common forms of MAR. Among the most widely applied methods are: Aquifer Storage and Recovery (ASR), Aquifer Storage Transfer and Recovery (ASTR), river bank filtration and Soil Aquifer Treatment (SAT), which can be performed for example, with effluents from sewage treatment plants (STP).

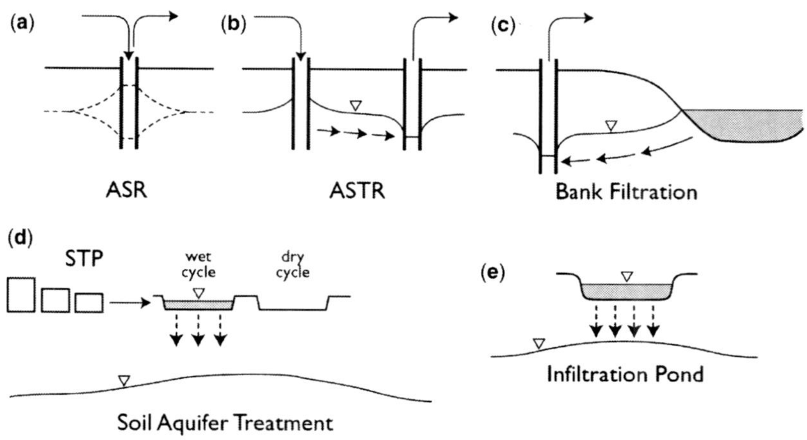

Figure 22.2 Schematic types of managed aquifer recharge, where ASR: aquifer storage and recovery, ASTR: aquifer storage, transfer and recovery (Dillon, 2005)

In principle, all considerations applicable for recharge involving conventional water sources have to be taken into account when recharging with reclaimed water. Additional challenges are provided by the presence of a higher pollutant load and the implications of pollution retention for aquifer degradation. Wastewater contains problematic compounds in higher concentrations than are present in surface water. Hydrogeological settings are paramount for the physical success of recharge schemes. They include permeability and transmissivity of the overlying soil and the hydraulics of the aquifer, such as storage capacity and drainage behaviour (UNESCO-IHP, 2005; NRC, 2008). There are many aspects to be considered when planning, implementing and operating a MAR scheme, particularly if alternative water sources such as reclaimed water are to be utilised. Among the main aspects to be considered are:

- Conditions of the local aquifer (quality and quantity aspects)
- Geological profile and hydro-geological characteristics of the aquifer
- Quality and quantity of water available for recharge
- Hydrological (e.g. transmissivity) and hydrogeochemical (e.g. impact on redox conditions) aspects of water infiltration
- Pre-treatment and infiltration systems
- Sustainability of the infiltration method (e.g. evolution of the infiltration capacity)
- Residence time underground
- Observation and recovery methods
- Post-treatment and use (e.g. for potable or non-potable purposes)
- Water quality monitoring
- Risk assessment and management

A concise overview of the practical aspects to be addressed when planning a managed aquifer recharge scheme is provided by Dillon et al. (2009). A range of these aspects have been addressed in the RECLAIM WATER project. Some of the issues are addressed in regulations and guidelines relevant for MAR, which are briefly discussed later in the chapter. The legal aspects considered most relevant for MAR in the European Union are reported in the following section.

22.2 LEGAL FRAMEWORK

Given the various benefits of MAR schemes it is important to ensure that MAR with reclaimed water does not compromises any of the protection aims related to water bodies and water uses as laid down in European or national legislation. In the absence of dedicated legislation concerning MAR with reclaimed water in Europe, the practice must at least not contravene existing regulations but should rather support their protection aims. This particularly refers to impacts on groundwater quality and human health and applies to water treatment processes, recharge techniques and final uses.

Figure 22.3 gives a schematic overview of a recharge scheme, indicating possible points of compliance (A: reclaimed water to be infiltrated, B: infiltrated water in vadose and saturated zone, C: recovered water intended for use). The most relevant European directives are included. The scope of these directives and their possible connection with MAR are outlined in the following section. Emphasis is given to the conceptual approaches used to achieve the defined protection objectives.

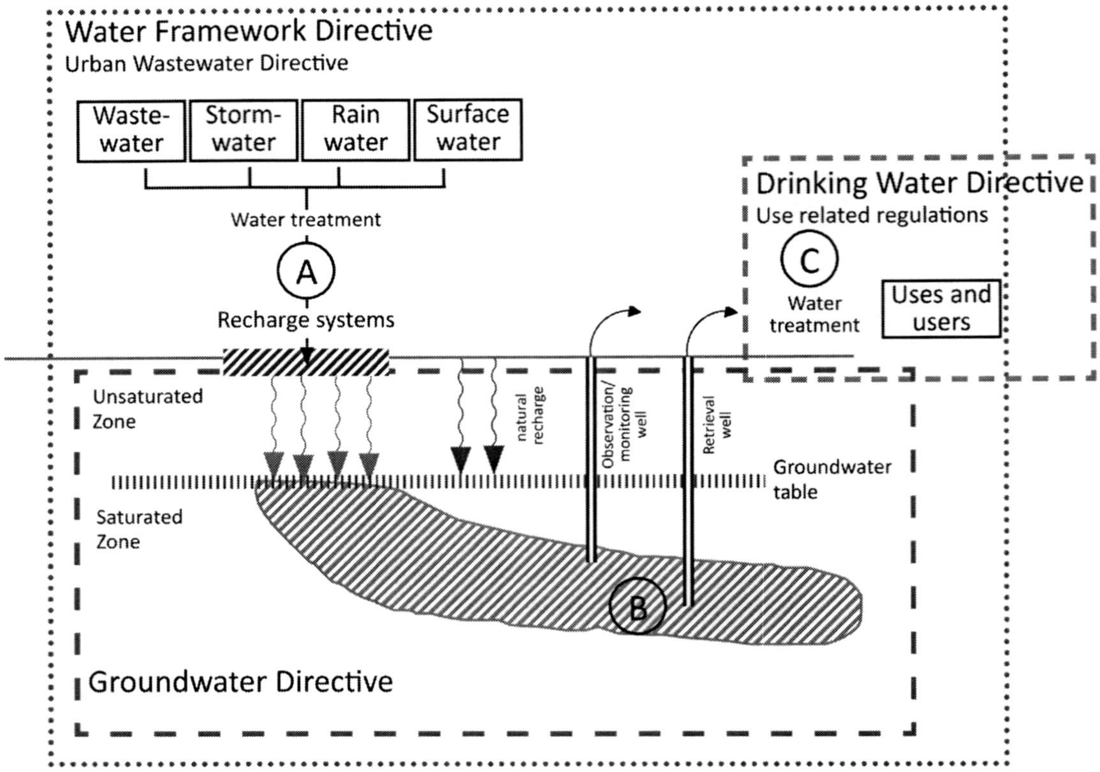

Figure 22.3 Managed aquifer recharge and relevant European legislation, with A-C representing possible points of compliance for which quality definitions can be defined

22.2.1 European legislation relevant to Managed Aquifer Recharge

Water Framework Directive (2000/60/EC)

Currently, policy making in the water sector is occurring within the context of the implementation of the Water Framework Directive (2000/60/EC). The directive streamlines several existing pieces of legislation which are now either repealed or revised. Moreover, the new framework for water management prompts the establishment of new structures and types of consultation with stakeholders and the public. Its primary objectives are to promote the sustainable use of water, reduce water pollution, particular by 'priority' and 'priority hazardous' substances, lessen the effects of floods and droughts, while introducing a co-ordinated approach to water management based on the river basin as the appropriate scale for planning.

The directive has a broad agenda, setting the protection of aquatic ecosystems as a priority and promoting public participation and information dissemination. Surface waters are required to meet 'good ecological and chemical status' and groundwater bodies to meet 'good chemical and quantitative status' by 2015. Groundwater bodies are defined as distinct volumes of groundwater within an aquifer or aquifers. Significantly, the WFD also requires that no deterioration in water status may take place and that protected area objectives are met.

A programme of measures is supposed to pave the way to meeting these objectives (Article 11 WFD). As referred to under Article 11(4), artificial recharge of aquifers is mentioned as a possible supplementary measure in part B of Annex VI. However, 'controls, including a requirement for prior authorisation of artificial recharge or augmentation of groundwater bodies are mandatory. The water used may be derived from any surface water or groundwater, provided that the use of the source does not compromise the achievement of the environmental objectives established for the source or the recharged or augmented groundwater body. These controls shall be periodically reviewed and, where necessary, updated [Article 11(3)(f)].

Recharge and reuse measures should not compromise the environmental objectives for groundwater quality as specified in the Groundwater Directive. In the process of evaluating the risk of failing to achieve the objectives (see Figure 22.4) the characterisation of groundwater bodies with respect to their chemical status (conductivity, concentrations of pollutants) is required and has to be related to defined quality standards or threshold values. This characterisation and evaluation procedure should therefore account for the potential impacts arising from recharge and withdrawal. Further details on how groundwater should be protected are provided in the Groundwater Directive. Some key aspects of the Water Framework Directive as relevant for managed aquifer recharge activities are summarised in Box 22.1.

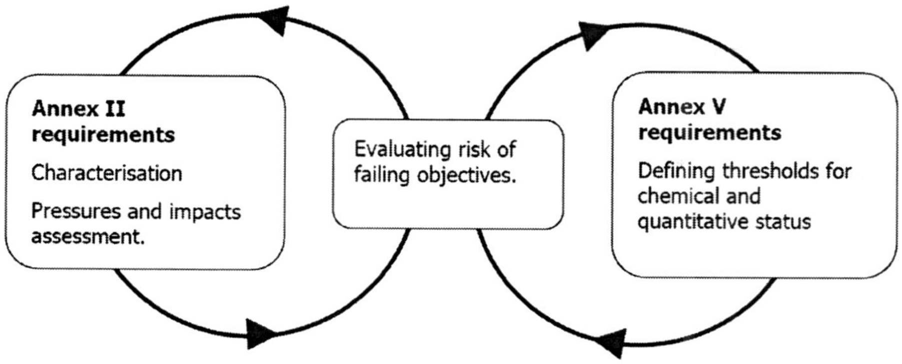

Figure 22.4 Evaluation process of groundwater status [adapted from EC (2007)]

Box 22.1 WFD key aspects related to MAR

- Managed aquifer recharge is considered a possible measure to achieve the 'good status' objectives
- WFD does not specify any particular methodologies to implement measures such as MAR

Groundwater Directive (2006/118/EC)

The directive on the protection of groundwater against pollution and deterioration is a daughter directive of the WFD. As provided for in Article 17 of the WFD, strategies to prevent and control pollution of groundwater should be adopted. The GWD seeks to establish specific measures to prevent and control groundwater pollution. In order to protect groundwater from deterioration and pollution, criteria to assess the chemical status of the resource are needed. According to Article 3(1) (a) and (b) of the GWD, these are defined as;

- Groundwater quality standards (for nitrate and pesticides) and
- Threshold values for pollutants and indicators of pollution

Member States are expected to define threshold values on a national, river basin or other appropriate level having regard to regional or even local conditions. Substances to be considered should be evaluated with regards to 'the interference with actual or potential legitimate uses or functions of groundwater'. The European Commission has recently published a Guidance Document on how to proceed in identifying relevant compounds and how to derive threshold values (EC, 2009). Article 6 of the GWD demands establishment of measures to prevent and control input of pollutants into groundwater, where *preventing input* refers to **hazardous substances**. The input of any other non-hazardous substance 'considered to present an existing or potential risk of pollution' is to be *limited* so as to ensure that the status of groundwater does not deteriorate or cause an upward trend in concentration of pollutants (see Figure 22.5). Measures to limit input so that maximum allowable concentrations (MAC) or quality standards are not exceeded shall at least be based on established best practices and best available techniques [Article 6(1)(b) GWD].

Following the interpretation provided in the Common Implementation Strategy (CIS) Guidance Document (EC 2007) 'preventing the input into groundwater' means 'taking all measures deemed necessary and reasonable to avoid the entry of hazardous substances' into groundwater. 'Reasonable' means technically feasible without involving disproportionate costs. It is acknowledged that it is often not technically or economically feasible to stop all input of hazardous substances, therefore exemptions are defined.

The natural attenuation capacity of the unsaturated zone may be taken into account when defining measures for both the *preventing* and *limiting* objective. For *limiting* even processes taking place in the saturated zone may be considered.

In summary, *'artificial recharge of aquifers'* is mentioned in Article 11(3)(f) of the WFD and article 6(3)(d) of the GWD. However, in these two directives, aquifer recharge is not defined precisely. Only the source of water is specified: *'the water used may be derived from any surface water or groundwater, provided that the use of the source does not compromise the achievement of the environmental objectives established for the source or the recharged or augmented body of groundwater.'*

In principle, managed aquifer recharge with treated wastewater is not explicitly excluded by either the Water Framework Directive or the Groundwater Directive (GWD). Therefore aquifer recharge may be implemented as long as Member States take the following measures: permit or authorisation and control and monitoring (WWR-WG, 2007). Nevertheless a more explicit commitment to the legitimacy of unconventional water sources for groundwater recharge is desirable so as to avoid potentially controversial decisions. The requirements for managed aquifer recharge arising from the Groundwater Directive are highlighted in Box 22.2.

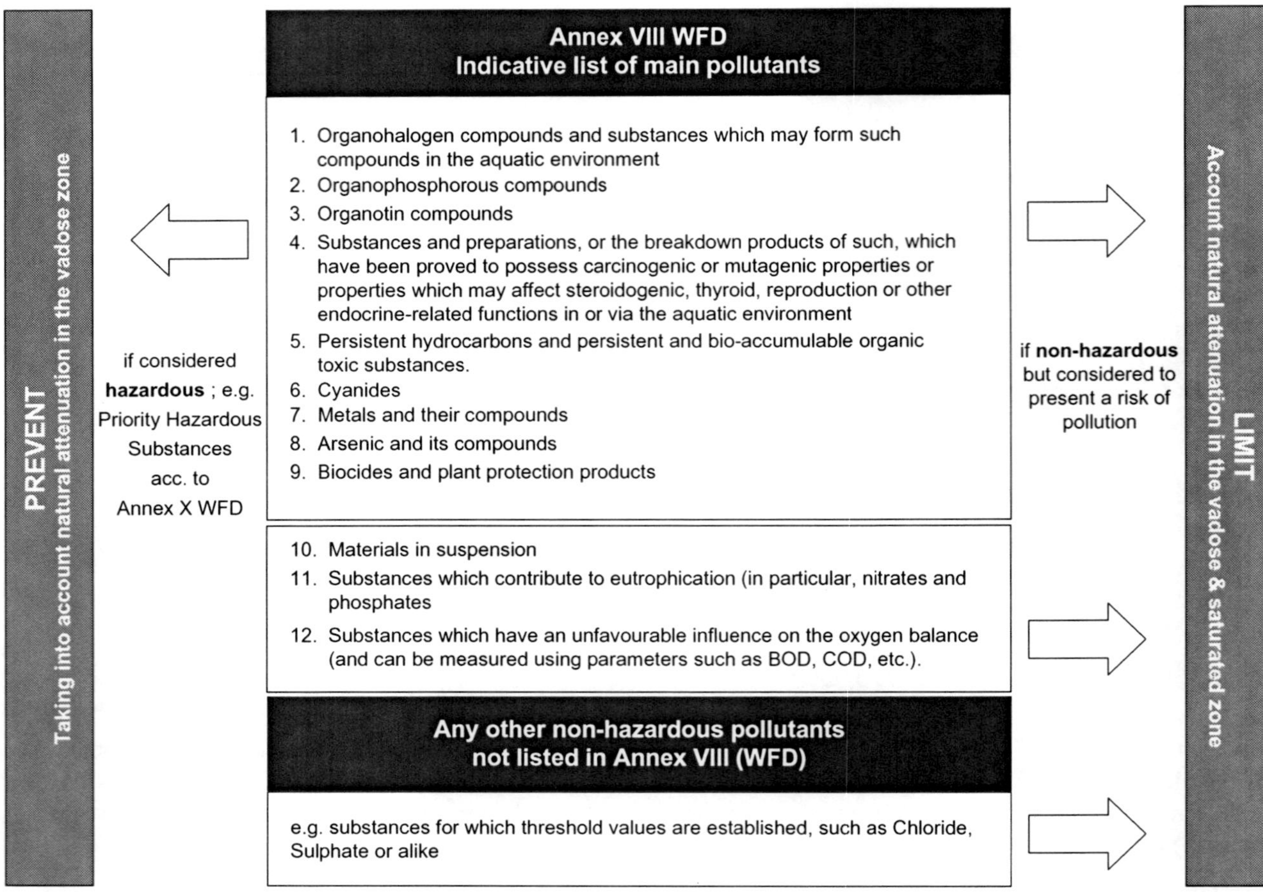

Figure 22.5 Differentiation of 'preventing' and 'limiting' input of pollutants into groundwater as demanded by Article 6 of the GWD

Box 22.2 GWD key aspects related to MAR

- Managed aquifer recharge with reclaimed water must not introduce any hazardous substances into the groundwater
- Any other input of substances shall be limited so the groundwater quality does not deteriorate (anti/non-degradation policy)
- Processes in the vadose zone that result in attenuation or degradation of substances may be taken into account when considering measures to prevent or limit input into groundwater
- Option to exempt authorised groundwater recharge from measures required to prevent and limit inputs, provided appropriate **monitoring** of the groundwater is being carried out [Article 6(3)(d)]

Urban Wastewater Treatment Directive (91/271/EEC)

The Council Directive 91/271/EEC of 21 May 1991 concerning urban wastewater treatment (Urban Wastewater Treatment Directive - UWWTD) aims to protect the environment from any adverse effect due to the discharge of wastewater which may cause eutrophication and damage aquatic life in the long term. It defines which wastewaters have to be collected and treated thus determining the available amount of potentially reusable flows. Secondly, it stipulates the minimum treatment level thus giving a first rough estimate of the quality of wastewater treatment plant effluents. At the same time it explicitly mandates for water reuse when stating "Treated waste water shall be reused whenever appropriate" (Article 12 UWWTD) but the term "appropriateness" remains legally undefined.

Water use related regulations

A third optional point of compliance for water quality is the point of use. Recovered water from recharge schemes has to be fit for purpose. Depending on the intended use, different requirements may have to be met. On a European level, directives specifying water quality for particular uses such as supporting fish life or drinking water production (Freshwater Fish, Shellfish, and Surface Water Directive) have been superseded by the Water Framework Directive. Its objective of achieving 'good status' of all water bodies encompasses the suitability of the water bodies for the above mentioned uses. Apart from this general approach, few national regulations or guidelines exist regarding the water quality for other uses, for example, irrigation water. The quality required for industrial applications is usually determined by technical requirements and defined by industry itself. Authorities focus their activities on regulating water uses that may pose a potential danger to human health which is particularly true for the production of drinking water.

Drinking Water Directive (98/83/EC)

The objective of the Directive on the Quality of Water Intended for Human Consumption (DWD) is to 'protect human health from the adverse effects of any contamination of water intended for human consumption by ensuring that it is wholesome and clean' (Article 1). In its current version, the directive pursues a product oriented approach in defining drinking water quality. The emphasis is on **meeting mandatory or indicator values for water quality parameters**. It covers both microbial and chemical constituents as well as some physical characteristics. The DWD does not explicitly define requirements for the source of water used to produce potable water, it simply requires member states to 'ensure that the measures taken to implement the Directive in no circumstances have the effect of allowing, directly or indirectly, (...) any increase in the pollution of waters used for the production of drinking water' (Article 4 DWD).

Next to the 48 microbial and chemical-physical parameters limited by the directive, the competent Health Authorities of Member States may define additional limit values for other compounds. The point of compliance is defined as the point of use, that is, the tap. In that respect, the relevance of the DWD for managed aquifer recharge includes the aspects listed in Box 22.3. Table 22.1 provides an overview on the legal aspects to be considered in the context of MAR as they can be derived from the European environmental legislation in force.

Box 22.3 DWD key aspects related to MAR

- Focus on quality control of the end-product (drinking water)
- Additional water quality parameters may be set with specific compliance values
- More stringent parametric values may be defined on a Member State basis
- Possible additional monitoring of substances or microorganisms for which no parametric value has been set on a case-by-case basis

Table 22.1 Summary of aspects to be addressed based on legislation in force.

Directive	Aspect/principle	Relevance for MARs
WFD/GWD	No limit values for admissible compounds in reclaimed water at the point of injection or infiltration,	Definition of point of compliance Specification of water quality criteria for either the point of recharge or of withdrawal to be clarified with the authorisation
UWWTD	Definition of wastewater treatment requirements and definition of minimum effluent quality; no specification of when reuse is appropriate	Legal mandate to reuse treated waste water, Provision of legitimacy of water reuse in general
GWD	Prevent input of any hazardous substances	specification of pollutants the recharge water must not contain
GWD	Limit input of non-hazardous substances	Degradation and attenuation processes taking place during soil passage shall be taken into account
GWD	Definition of threshold values on member states level taking into account regional aspects	Scheme and site specific requirements, option to define certain parts of an aquifer as groundwater body
DWD/GWD	Monitoring of unregulated substances – future concepts?	Adoption of risk assessment framework

Future prospects in water safety concepts

Whereas this policy brief summarises the current state of affairs, reflecting the legislation in force, various developments are ongoing which may result in revision of the relevant documents. The drinking water sector is increasingly adopting the

Water Safety Plan (WSP) concept promoted by the World Health Organisation (WHO, 2005). The primary objectives of a WSP in ensuring good drinking-water supply practice are the **minimization of contamination of source waters**, the reduction or **removal of contamination** through treatment processes and the **prevention of recontamination** during storage, distribution and handling of drinking-water.

The WSP helps to assess the hazards, to characterise the risks, and anticipates the use of monitoring and control measures. It suggests a procedural set of tools for safe performance of drinking water supply through risk assessment and preventive management approaches in a framework extending from water resource to consumer, managing the whole water supply chain (Figure 22.6).

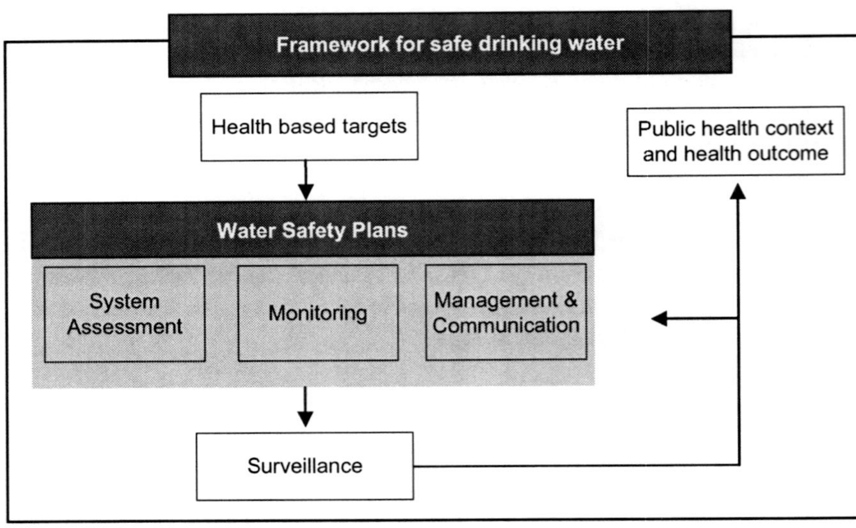

Figure 22.6 Safe drinking water framework (adopted from WHO, 2007)

A revision of the EU Drinking Water Directive would give regulatory impetus for implementing these principles. Nevertheless a considerable number of larger water utilities already follow objectives to improve aspects of bacteriological contamination, chemical substances including construction products in contact with drinking water, small water supplies as well as risk assessment and risk management (WHO, 2007). Such a change from end-point parametric compliance monitoring to confirmation of risk management throughout the production chain is also an appropriate approach for recharge schemes especially when sensitive end-uses like indirect potable reuse is foreseen, as demonstrated by Bixio *et al.* (2008) and as recently adopted in Australia (NRMMC, EPHC, NHMRC 2009).

22.2.2 Aquifer recharge specific regulations and guidelines

The above described legislation can only give an indication of what decision and management issues to consider when designing aquifer recharge schemes. However, they do not provide a dedicated framework or guidance along which aquifer recharge may be implemented. According to Dillon *et al.* (2006) there is a need for rules and science-based guidelines for recharge. They suggest 'simplistic approaches to water quality management based on rules of thumb are not an adequate basis for health and environmental protection in MAR operations. Simply treating water to a standard where problems are believed to be avoided, as adopted in some guidelines does not guarantee that environmental and human health protection will be met'.

In the following sections a selection of national or regional approaches towards regulating groundwater recharge are described and discussed. They represent different concepts for achieving the paramount objective of protecting human health against transmission of infectious disease by microbial pathogens or other harmful constituents. A crucial issue is the point of compliance with water quality standards; either at the point of recharge or the point of use. This raises the question of whether it is preferable to require a high degree of treatment before injection/infiltration or after water is withdrawn.

Legal situation in Spain

In December 2007 the Spanish Government adopted the Royal Decree 1620/2007, of 7 December establishing a legal regulation for the reuse of treated wastewater (R.D. 1620/2007, de 7 de diciembre, por el que se establece el régimen jurídico de la reutilización de las aguas depuradas). The decree defines the conditions for water reuse authorisation. It also addresses issues of establishing plans and programmes to most efficiently integrate water reuse in water management, taking into account economic and financial analyses and environmental effects.

The regulation specifies various types of reuse, excludes certain uses and particularly sets water quality criteria for the different kinds of application. This Spanish legislation explicitly prohibits the use of recycled water for human consumption and classifies groundwater recharge by either percolation or injection as environmental use. Any specified water quality requirements have to be met at the point of handing over the reclaimed water to the user.

Regulations in California (USA)

The California Department of Public Health (CDPH) has issued a number of codes of regulation concerning wastewater reuse. Title 22 of these regulations has become a synonym for highly treated wastewater and stringent quality requirements. It encompasses a variety of aspects related to reuse, such as treatment technology, sampling procedures, catchment management and water quality specifications. The section dealing with groundwater recharge (Article 5.1. §60320) currently just states that:

(a) Reclaimed water used for groundwater recharge of domestic water supply aquifers by surface spreading shall be at all times of a quality that fully protects public health. The State Department of Health Services' recommendations to the Regional Water Quality Control Boards for proposed groundwater recharge projects and for expansion of existing projects will be made on an individual case basis where the use of reclaimed water involves a potential risk to public health.

(b) The State Department of Health Services' recommendations will be based on all relevant aspects of each project, including the following factors: treatment provided; effluent quality and quantity; spreading area operations; soil characteristics; hydrogeology; residence time; and distance to withdrawal.

(c) The State Department of Health Services will hold a public hearing prior to making the final determination regarding the public health aspects of each groundwater recharge project. Final recommendations will be submitted to the Regional Water Quality Control Board in an expeditious manner (CCR, 2009).

A recent proposal for amendment of this section was drafted in August 2008 (CDPH, 2008). To ensure safe projects multiple protection barriers along the entire treatment process are prerequisite (Johnson, 2009). For groundwater recharge and reuse projects augmenting sources of domestic water supply it suggests comprehensive monitoring and control of the **recycled water prior to recharge** encompassing;

- Pathogenic microorganisms
- Regulated chemicals and physical characteristics, which must meet maximum contaminant levels (MCLs)
- Additional constituent monitoring for priority toxic pollutants and chemicals with state notification levels
- Additional constituent monitoring regarding pharmaceuticals, endocrine disruptors, and other wastewater indicator chemicals
- Nitrogen compounds
- Total organic carbon

A minimum residence time of six months is also specified 'prior to extraction for use as a drinking water supply, although such an arbitrary value does not necessarily properly reflect site specific conditions (NRC, 2008, chapter 5). The regulations promote sophisticated treatment before recharge and on close tracking of further evolution of (ground) water quality during underground passage at monitoring wells. Regulatory efforts are ongoing, as the California Water Code (Section 13560) demands to adopt uniform water recycling criteria for indirect potable reuse for groundwater recharge by December 2013. The ambition for water recycling stated therein even goes further with the commitment to investigate the feasibility of developing the uniform water recycling criteria even for direct potable reuse, in this case surface water augmentation.

Guidelines in Australia

Within its National Water Quality Management Strategy, the Council of Australian Governments, comprising Commonwealth and all State and Territory governments, has issued a suite of documents that make up Australia's water recycling guidelines. These target 'anyone involved in the supply, use and regulation of recycled water schemes, including government and local government agencies, regulatory agencies, health and environment agencies, operators of water and wastewater schemes, water suppliers, consultants, industry, private developers, body corporates and property managers'.

The Guidelines consist of four documents. Phase I (NRMMC–EPHC–AHMC 2006) provides the foundations that provide a scientific basis to assess and manage health and environmental risks. The three Phase II documents cover the specialised requirements for augmentation of drinking water supplies (NRMMC–EPHC–NHMRC 2008), stormwater harvesting and reuse (NRMMC–EPHC–NHMRC 2009a) and managed aquifer recharge (NRMMC–EPHC–NHMRC,

2009b). These guidelines refer to earlier NWQMS guidelines for scientifically defensible acceptable values of water quality parameters for various uses, However this is extended in Phase I which provides comprehensive information for irrigation water quality targets and the Augmentation of drinking water supply guidelines show methods for calculating acceptable values for parameters (such as organic chemicals) where these were undefined in earlier guidelines and demonstrate quantitative microbial risk assessment methodology. The elements of the framework for managing water quality and use are summarised in Figure 22.7 which illustrates that water quality management involves many factors. Safe water will be the result of proper system analysis, risk assessment and risk management that incorporates multiple barriers.

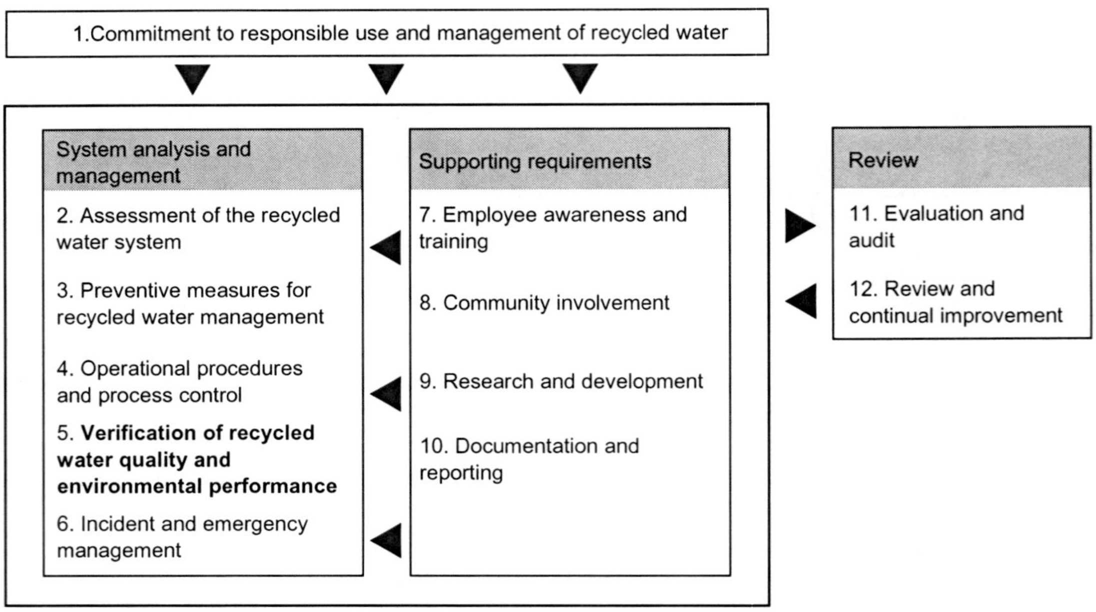

Figure 22.7 Elements of the framework for managing water quality and use (NRMMC–EPHC–AHMC 2006)

The risk management approach for managed aquifer recharge is based on scientific findings on fate of contaminants and pathogens in aquifers and takes into account interactions between recharged water and the aquifer material. Twelve hazard types are considered, seven of which focus on water quality. These include hazards that are demonstrably sustainably attenuated during residence time in the aquifer (e.g. inactivation of pathogens and biodegradation of some organic chemicals). Hazards ineffectively removed in the aquifer must be removed sufficiently prior to recharge by natural or engineered treatments. Examples of such hazards include salinity, excessive nutrients, non-biodegradable compounds. This is required to ensure that water quality beyond a defined attenuation zone are permanently less than their allowable values (i.e. the aquifer is protected for its defined environmental values), the recovered water achieves its required quality, and the treatment processes within the aquifer are sustained. The MAR Guidelines also require evaluation of the potential for hazards to arise as a result of geochemical interactions between recharge water and the aquifer materials. Treatment of recharge water provides a way to avoid such adverse consequences (e.g. metal mobilization, hydrogen sulfide) (Dillon, 2009).

The Australian MAR Guidelines allow reliance on demonstrated sustainable aquifer treatment processes and this requires a staged process for project establishment. Examples of risk assessments that follow the guidelines were presented for nine MAR demonstration projects by Page *et al.* (2010).

Guideline Summary

Table 22.2 summarises those parameters controlled under the regulations or guidelines in Spain. California and Australia discussed above. The information provided refers to the three different possible points of compliance as illustrated in Figure 22.3 and gives an overview on the relevant water quality parameters as well as the recommended or prescribed treatment and monitoring criteria. Whereas the Spanish and Californian regulations have legal status, the Australian Guidelines are themselves advisory. However as the States are signatories to these Guidelines they are being adopted by state regulators so that compliance with these guidelines is deemed to meet the legal requirements of jurisdictions. The guidelines also include non-mandatory best practice suggestions for MAR operations. Whilst they fully address all relevant hazards, they do not specify *a priori* the water quality for recharge, because they take account of the growing body of knowledge of aquifer treatment including at the specific MAR site. Compliance is required on the boundary of the attenuation zone and at the point of recovery, and a critical control point normally precedes recharge.

Table 22.2 Regulations and guidelines for groundwater recharge – regulated parameters and point of compliance.

A **Quality of recharged water**	B **Quality of receiving water (groundwater)**	C **Quality of withdrawn water or water put to use**
REAL DECRETO 1620/2007, of 7 December establishing a legal regulation for the reuse of treated wastewater		
Maximum admissible value for – Nematode eggs – E. coli – Total suspended solids – Turbidity – N_{tot} – Nitrate Provisions for high priority pollutants (list I) and of other pollutants (list II) of former GWD 80/68/EEC Authorities may fix additional values for contaminants or parameters, or set stricter limit values	Not specified in this decree, apart from reference to directive 80/68/EEC	Regulations for sectoral uses may be applied when setting additional values (see column A)
US, Title 22, California Code of Regulations (Draft by CDPH – August 2008)		
(1) Control of parameters for meeting set **Maximum Contaminant Levels (MCL)** – Pathogenic micro-organisms (compliance with specifications for filtered wastewater[1] or disinfected tertiary recycled water[2]) – Average and maximum turbidity (depending on the type of filtration) • Total coliforms • Virus removal – Nitrogen compounds – Regulated chemicals and physical characteristics • Various organic and inorganic chemicals • Disinfection by-products • Lead and copper • TOC limit (depending on recycled water contribution, i.e. a degree of dilution) (2) Monitoring of additional constituents in recycled water – Priority Toxic Pollutants – Chemicals with state notification level – Constituents indicating the presence of municipal wastewater (e.g. EDCs, pharmaceuticals)	*Prior to operation*: Check of background concentrations *During operation*: Monitoring of downgradient wells to be analysed each quarter for – Total coliforms – TOC – N_{tot}, Nitrate, Nitrite – Organic contaminants – Other specified constituents	Retention time of 6 months No further specifications in this Code
Australian Guidelines for Water Recycling: Managing Health and Environmental Risks		
Water quality of recharge water is required so as to meet recovered water quality and aquifer protection for existing environmental values accounting for an attenuation zone in the aquifer. Validation of aquifer treatment processes required before or during commissioning. Monitoring is required for operation and verification.		

(*Continued*)

Table 22.2 Regulations and guidelines for groundwater recharge – regulated parameters and point of compliance (*Continued*).

A **Quality of recharged water**	B **Quality of receiving water (groundwater)**	C **Quality of withdrawn water or water put to use**
Water quality hazards 1. Pathogens 2. Inorganic chemicals 3. Salinity and sodicity 4. Nutrients 5. Organic chemicals 6. Turbidity/particulates 7. Radionuclides *Verification monitoring*: – at point of recovery – at obs wells on bdy of attenuation zone.	*Other hazards to be addressed*: 8. Pressure, flow rates, volumes and levels 9. Contaminant migration in fractured rock and karstic aquifers 10. Aquifer dissolution and aquitard and well stability 11. Impacts on groundwater dependent ecosystems 12. Greenhouse gas emissions Clogging and recovery efficiency advisory.	*Water quality targets* Set by NWQMS documents accounting for health and environment protection for all intended uses and for existing environmental values of aquifer, Phase 2 Guideline- Augmentation of drinking water supplies – provides guidance on QMRA and methods to establish GL values for organic compounds without existing GLs

[1]"**Filtered wastewater**" means an oxidized wastewater that meets the criteria in subsection (a) or (b):
(a) Has been coagulated and passed through **natural undisturbed soils or a bed of filter media** pursuant to the following:
(1) At a rate that does not exceed 5 gallons per minute per square foot of surface area in mono, dual or mixed media gravity, upflow or pressure filtration systems, or does not exceed 2 gallons per minute per square foot of surface area in travelling bridge automatic backwash filters; and
(2) So that the turbidity of the filtered wastewater does not exceed any of the following: (A) An average of 2 NTU within a 24-hour period; (B) 5 NTU more than 5 percent of the time within a 24-hour period; and (C) 10 NTU at any time.
(b) Has been passed through a **microfiltration, ultrafiltration, nanofiltration, or reverse osmosis membrane** so that the turbidity of the filtered wastewater does not exceed any of the following: (1) 0.2 NTU more than 5 percent of the time within a 24-hour period; and (2) 0.5 NTU at any time
[2]**Disinfected tertiary recycled water.**
"Disinfected tertiary recycled water" means a filtered and subsequently disinfected wastewater that meets the following criteria:
(a) The filtered wastewater has been disinfected by either: (1) A chlorine disinfection process following filtration that provides a CT (the product of total chlorine residual and modal contact time measured at the same point) value of not less than 450 milligram-minutes per litre at all times with a modal contact time of at least 90 minutes, based on peak dry weather design flow; or (2) A disinfection process that, when combined with the filtration process, has been demonstrated to inactivate and/or remove 99.999 percent of the plaqueforming units of F-specific bacteriophage MS2, or polio virus in the wastewater. A virus that is at least as resistant to disinfection as polio virus may be used for purposes of the demonstration
(b) The **median concentration of total coliform bacteria** measured in the disinfected effluent does not exceed **an MPN of 2.2 per 100 mL** utilizing the bacteriological results of the last seven days for which analyses have been completed and the number of total coliform bacteria does not exceed an MPN of 24 per 100 millilitres in more than one sample in any 30 day period. No sample shall exceed an MPN of 240 total coliform bacteria per 100 mL

Regulations with enforceable law status put much more focus on setting verifiable limit values for selected parameters to assure safe reuse. The Spanish approach relies on defining the quality of the recharge water. The Californian regulation additionally demands regular monitoring of the underground water quality. The Californian guidelines are designed for recharging aquifers designated as a source for drinking water supply, hence careful monitoring and definition of new drinking water guidelines are justified. The Australian MAR Guidelines address all sources of water for recharge (including reclaimed water, stormwater, drinking water, rivers, lakes and groundwater from other aquifers), all types of recharge methods and all types of end uses of recovered water and aquifer environmental values. However, these guidelines do not apply to recharge of aquifers where no recovery or environmental benefit is intended.

22.3 RISK ASSESSMENT AND MANAGEMENT APPROACHES

Managed aquifer recharge schemes using reclaimed water obviously need to be designed, monitored and managed to assure that potential risks are consciously controlled (Asano & Cotruvo, 2004). An appropriate strategy would be to adopt existing risk management tools already in use for other activities. The concept of Hazard Assessment and Critical Control Points (HACCP) is such a well established practice in the food and other industries. The WHO's Water Safety Plans approach has fitted it to the water production chain. Within the RECLAIM project it has been applied to several test sites.

22.3.1 Microbial risks

A major health risk in water reuse applications is the transmission of water borne diseases through pathogenic microorganisms, such as bacteria, viruses, protozoa or helminth eggs. In view of the variety of potential microbial hazards, (WHO, 2003) the question of appropriate and relevant indicator organisms is repeatedly under discussion as pathogens may have a different survival potential and fate in reuse systems than that of the indicator.

Existing guidelines on water reuse mostly rely on regular monitoring of a few, easy to detect, indicator organisms to represent microbial contamination. In order to overcome the intrinsic drawback of routine monitoring which is often not real-time but ex-post, time consuming and rather expensive, multi-barrier concepts are implemented. WHO and Australian Guidelines require preventive measures to achieve 10^{-6} DALYs per person per year.

22.3.2 Chemical risks – new substances of concern

Human health risks from organic chemicals are often associated with long-term or chronic effects rather than acute toxicity. This is particularly true for emerging trace contaminants, which are only sparsely tested for their effects and whose hazard is consequently difficult to assess. Despite this lack of information and knowledge, different strategies may be pursued to consider these substances when establishing water reuse guidelines.

Establishing limit values for these substances

There are various options for setting limits or guide values for water quality at a point of compliance, as has been shown in the section on Legal Frameworks. For priority substances Environmental Quality Standards (EQS) have been defined (EC, 2001; EC, 2008). Those may also be deemed the limit for any source water for drinking water production, without prejudice, that this purpose may require even more stringent standards. Currently the list does not contain substances increasingly under discussion, such as personal care product residues, pharmaceutically active compounds and endocrine disrupting compounds. Nevertheless, in the process of adopting this directive, a Proposition of the European Parliament (2007) to amend and extend Annex X of the WFD by a number of trace organics has been raised (EP, 2008).

Likewise, the International Association of Waterworks in the Rhine Catchment Area (IAWR, 2008), has put forward such more detailed water quality targets for raw water intended for the production of drinking water. Next to 'conventional' parameters they also address organic and emerging pollutants. The target values are specified in Table 22.3. These values have been derived aiming at production of drinking water by means of exclusively natural purification methods. Therefore, values sometimes correspond to drinking water standards whereas unregulated substances were limited having regard to the precautionary principle.

Table 22.3 Suggestion for target values for raw water for drinking water production (IAWR, 2008).

Anthropogenic non-natural substances with known biological effects	**Target value [μg/L] maximum permissible values**
Pesticides and their metabolites per individual substance	0.1*[,a]
Endocrine active substances per individual substance	0.1*
Pharmaceuticals (incl. antibiotics) per individual substance	0.1*
Biocides per individual substance	0.1*
Other organic halogen compounds per individual substance	0.1*
Evaluated anthropogenic non-natural substances with no known effects	
Substances with low biodegradability per individual substance μg/L 1.0	1.0[b]
Synthetic complexing agents per individual substance μg/L 5.0	5.0

*Unless toxicological information necessitates a lower value
[a]equal to the drinking water standard
[b]if other non-natural organic substances have passed proper toxicological screening and are regarded as harmless, a target value of 1 μg/L is justified, similar to other official precaution targets. Only for complexing agents is a temporary value of 5 μg/L acceptable for technical reasons

Occasionally, different institutions arrive at very different limit values for the same compound. For the carcinogenic disinfection by-product N-Nitroso-dimethylamine (NDMA) for example, the public health goal in the US is 3 ng/L NDMA with a notification level of 10 ng/L for all nitrosamines in drinking water (CDPH, 2007; EPA, 2008) while the WHO set the guideline value for the same contaminant at 100 ng/L (WHO, 2008).

The Australian guidelines apply a specific procedure for setting guidelines for chemicals in drinking water augmented with recycled water (NRMMC-EPHC-NHMRC, 2008). It basically adopts the WHO approach used to derive drinking water guideline values. The approach is slightly varied depending on the nature of compounds under consideration, distinguishing;

(a) Substances with health or toxicological effects: threshold chemicals and non-threshold chemicals (NOEL, TDI)
(b) Substances without sufficient toxicological information: threshold of toxicological concern, (genotoxicity) structural alerts on toxicity, neurotoxicity
(c) Pharmaceuticals (based on lowest daily therapeutic dose)

The suitability and robustness of data is taken into account by establishing safety or uncertainty factors when calculating the guideline value.

Application of best available technology and techniques

In the absence of a detailed investigation of the relevance of specific compounds and related mitigation options, the prescription of a specific treatment level with well investigated purification capacity might be an option, while no single technology should be prescribed in detail, the level of best available technologies for this area should be documented and updated periodically and used as a reference point.

22.4 TECHNOLOGY OPTIONS

Wastewater is generally more polluted than most drinking water sources (e.g. river or reservoir waters) and hence requires more extensive treatment prior to aquifer recharge, particularly where the reabstracted water is to be used for indirect potable use. The treatment technology for these ambitious goals is generally available and implemented in full scale (e.g. in Orange County/California, in the Singaporean NEWater plants and in Wulpen/Belgium). Examples exist where the aquifer is recharged with secondary treated effluent only and the soil-aquifer system is used as treatment and storage, as it is the case in economically developed countries such as Israel, USA, and Australia (International Water & Irrigation Review, 1999) as well as in countries with transition economies such as Mexico, South Africa, Thailand and Peru (Dillon & Jiminez, 2007). The natural attenuation processes occurring in the soil and sub-soil, particularly in the vadose zone, have been shown to be quite effective with respect to trace organic removal (Ternes *et al.* 2007). On the other hand, in the last decade the common practice in industrialised countries is rather to apply tertiary (e.g. coagulation, filtration, disinfection), or quaternary treatment (various types of double membrane systems, as it is the case in Belgium, California, Singapore and Australia) before infiltration and potable use (Bixio *et al.* 2005).

The effectiveness of a range of intensive water reclamation technologies such as conventional and membrane coupled (MBR) activated sludge treatment, membrane effluent filtration with porous and dense membrane processes, activated carbon adsorption as well as different oxidation processes (Ozone, UV+Ozone, UV+Peroxide) have been investigated in a number of studies (Asano *et al.* 2007; Ternes & Joss, 2006; Snyder *et al.* 2007). It seems clear from the result of these investigations that an almost complete retention of a wide range of emerging contaminants can be achieved with multi-barrier treatment processes. Particularly difficult is the removal of NDMA. Advanced oxidation processes with UV and H_2O_2 have been implemented in indirect potable reuse schemes in the US and Australia to cope with this problem (Asano *et al.* 2007; Walker *et al.* 2007).

The technical case studies investigated in RECLAIM WATER give an overview on the applicable treatment technologies prior to aquifer recharge (Table 22.4). Those case studies provide examples for a range of different managed aquifer recharge options utilising different

- Feed sources (e.g. reclaimed waste and storm water),
- Quantity of reclaimed water,
- Pre-treatment options (from no treatment to double membrane),
- Infiltration systems (infiltration ponds, periodically filled recharge basins, dug wells, river bed etc.),
- Geological formations (sandy to carstic aquifers),
- Residences times,
- Types of uses of extracted water (e.g. indirect potable use and irrigation).

Water reclamation technologies reflect the socio-economic circumstances, risk perceptions and final uses envisaged. This compilation shows the case-sensitivity of the technology selection. Despite the benefits of each individual scheme some of the pre-treatment applied (e.g. no pre-treatment at all in Mexico and secondary treatment in Italy) show a need for improvement from a pre-cautionary perspective. More details about the performance of the different treatment trains investigated can be obtained from Le-Corre *et al.* (2007).

Table 22.4 Water reclamation and aquifer recharge schemes investigated in RECLAIM WATER.

Site location & capacity	Scheme description
Sabadell 30 km from Barcelona, Spain Capacity: 25,000 m^3 d^{-1}	Secondary treated wastewater effluent discharged into a river bed where it infiltrates and is recovered. The water is then disinfected (UV) and distributed for parks irrigation.
Nardò Salento Region, south of Bari, Italy Capacity: 12,000 m^3 d^{-1}	Secondary treated municipal effluent is transported to aquifer injection. Recharge acts as a salt intrusion barrier and resource is also used as drinking water source.
Shafdan Greater Tel Aviv, Israel Capacity: 342,000 m^3 d^{-1}	Secondary wastewater from the Tel-Aviv area is recharged to an aquifer via a soil aquifer treatment (SAT) system. Recovered water is primarily used for irrigation but has accidental drinking water quality. Pilot dug-well system after ultrafiltration with short residence time was tested.
Gaobeidian Beijing, China Capacity: 650 m^3 d^{-1}	Tertiary effluent is used for aquifer recharge. Treatment is provided by coagulation, filtration and ozonation prior to infiltration and recharge.
Parafield/Salisbury near Adelaide, South Australia Capacity: 1100 m^3 d^{-1}	Wetland treated urban stormwater injected into a brackish aquifer. Water recovered via separate recovery wells. Recovered water intended for drinking supplies, and until proven will be used for irrigation.
Torrele/Wulpen, Belgium Capacity: 8640 m^3 d^{-1} Infiltration: 2.5 million m^3/year	Tertiary treated municipal effluent is upgraded by microfiltration and reverse osmosis, and then infiltrated via an infiltration pond to recharge an aquifer used for drinking water production.
Mezquital Valley State of Hildalgo, Mexico Capacity: 630 million m^3/year	Wastewater mixed with stormwater and surface water is discharged to an irrigated area of more than 76,000 ha. About 40 % of the irrigation water infiltrates into the aquifer. The water is recovered via separate wells and springs. Recovered water is chlorinated and distributed for drinking water supply, industrial use, irrigation and other purposes (bathing, swimming, washing).
Atlantis near Cape Town, South Africa Capacity: 15 000 m^3/d Infiltration: 2.7 million m^3/year	Urban stormwater run-off, is collected via a series of detention basins and blended with secondary treated domestic wastewater and recharged up-gradient of the production well field for augmenting the water supply. The blend of natural groundwater and recharged water abstracted from the well field is used as potable water supply for the town of Atlantis.

22.5 RECOMMENDATIONS ON SCHEME AUTHORISATION AND FUTURE DEVELOPMENTS

22.5.1 Which investigations shall be carried out during MAR planning?

A rigorous assessment of the hydrogeological characteristics and geochemical behaviour of the aquifer to be recharged is paramount (Bouwer, 2002). In-depth knowledge of the site properties is essential to support assessment of the best technical option for a planned recharge scheme. First investigations can draw on existing information such as soil maps and hydrogeological reports to estimate soil permeability and hydraulic conductivity of the aquifer. But field investigations are essential to determine infiltration rates and for designing hydraulic loads. These may encompass pumping test, flow metering, soil sampling and setting up of a conceptual hydrological model.

The US National Research Council (NRC, 2008) recommends that analyses using groundwater flow and solute transport modelling should become a routine part of planning for, designing, and adaptively operating MAR systems.3D modelling as a supportive tool and tracer testing for verification of residence time or travel distance is already partly requested by legal entities (e.g. in the California Code of Regulation). Tracer studies can also be successfully applied to assess the mixture of infiltrated water with surrounding native groundwater (Kloppmann *et al.* 2008).

Next to hydrogeological parameters, constituents that lead to aquifer clogging or dissolution, or other reactions that improve or degrade water quality are to be evaluated. The geochemical compatibility of groundwater and recharge water has to be assured to avoid unwanted reactions such as carbonate precipitation, iron hydroxide formation, mobilisation of minerals and so on, especially in case of injection wells (Bouwer, 2002; NRC, 2008). In Australia pathogen inactivation rate needs to be validated in the soil or aquifer in order to claim $\log_{10}$ removal credits for this barrier (NRMMC-EPHC-NHMRC, 2009b).

22.5.2 What can be controlled and how?

The appropriate extent of water quality control and monitoring has often been the subject of intense debate (Salgot, 2001). Requirements are typically related to the intended use and increase with the assumed potential human health risk of an application. Based on the schemes investigated during the RECLAIM WATER project control should cover a wide range of aspects, including both scheme operation and water quality. The operation of recharge schemes (with an emphasis on clogging control) should encompass:

- Monitoring of pre-treatment performance (reduction of suspended solids, removal of nutrients and organic carbon, pH)
- Monitoring of aquifer characteristics (levels, temperature, quality parameters, changes in transmissivity),
- Identification of relevant water quality parameters (microbiological and chemical constituents);
- Use of the HACCP concept to identify critical control points defined for processes and the produced water as well as points of attention.
- Distinction between controls aimed at verifying compliance with existing quality standards and regular monitoring of additional, unregulated constituents;
- Structured approach to monitoring and sampling frequency as a function of delivered water end-uses and the time period over which the scheme proves to operate properly. Appropriate reduction of sampling and analysis effort are provided for in the Spanish and Californian guidelines.

22.5.3 What role will MAR with reclaimed water play in the future?

In European debates about water scarcity, drought, and climate change adaptation, water reuse does not yet play a very prominent role. In the European Commission's recent communication addressing the challenge of water scarcity and droughts, alternative options like desalination or wastewater re-use are considered as potential solutions across Europe. However, the communication goes on to state that 'any definitive Commission position on these options will have to be based on further work on risk and impact assessment, taking into account the specific bio-geographical circumstances of Member States and regions' (EC, 2007b).

On the other hand, the MED-EUWI Working Group on Water Scarcity and Droughts identified treated wastewater reuse as a way of addressing long term imbalances between demand and water supply, and the MED-EUWI Working Group on Groundwater has recognised the importance of its potential impact on the quality of groundwater. This Working Group has contributed a Mediterranean Wastewater Reuse Report (MED WWR WG, 2007) which seeks to identify the main objectives of a treated wastewater reuse policy and the existing barriers and constraints.

In addition, there are also activities in member states to further explore MAR. A systematic investigation of zones in Spain potentially suitable for artificial recharge within the framework of sustainable development has been carried out in the DINA-MAR project (2008). This project focuses on recharging with natural resources (or captured rainwater) but also considers treated wastewater as a source for groundwater augmentation. A GIS-based analysis of geological formations suitable for artificial recharge has been performed. In a second phase, pilot areas will be established to study and design specific mechanisms to achieve high infiltration rates, also considering environmental aspects (DINA-MAR, 2008).

It is clear that MAR with reclaimed water can play a valuable role in securing water supplies for a range of uses if there are suitable sources, aquifers and appropriate investment in preventive measures commensurate with risk. The examples in this report give evidence of the effectiveness of MAR where it has been implemented and the extent of investment required to achieve intended outcomes.

22.5.4 What should be done to promote the strategic adoption of MAR with reclaimed water as a water stress mitigation option?

Firstly, there is no need to promote a particular technology, unless there are policies inhibiting its uptake and the economics, environmental impacts and social acceptance are superior to existing technologies that are unencumbered. Alternatively, if the above statement is anticipated but there are inadequate data to verify its truth, investment in research and demonstration may be highly efficient given the future massive scale of investment required in augmenting public water supplies in Europe and around the world.

There are number of actions which could be taken at the European level to promote more wide-spread implementation of managed aquifer recharge where appropriate and in a safe as well as sustainable manner. The RECLAIM WATER project has made advances that will help with the achievement of these goals, however further research will accelerate and extend the benefits of appropriate confident adoption of MAR. Among the potential actions are:

Policy and management

- Further development of a policy supporting utilisation of reclaimed water in Europe and the Mediterranean region as a continuation of the EU-MED Joint Process (MED WWR WG, 2007);
- Development of policies to consider all options to augment water supplies, acknowledging climate change, so that options including MAR are accounted for, with their inherent uncertainties. Investigations costs to prove MAR may in some cases be considerably smaller than the incremental cost to the next cheapest established method for water supply;
- Further clarification of objectives and scientific constraints in using reclaimed water for MAR, accounting for current state of information and practice and for projected priorities for future water supplies and associated pre-requisite knowledge;
- Better clarification and guidance how the "limit and prevent principles" in the GWD should be adopted to the authorisation of managed aquifer recharge schemes utilising reclaimed water;
- Further development of policies to assign water allocations and trading rights to compliant rechargers of aquifers;
- Surveying the potential contribution MAR could make to mitigate the non-compliance risk of groundwater bodies in the river basins with the objectives of the WFD and GWD;
- Mapping of areas geologically suitable where MAR is likely to be in greatest demand in Europe.

Integration of results of demonstration projects

- Further demonstration of MAR with reclaimed water as a water management option over a wider range of aquifers and pre-treatments, with economics, environmental externalities and social acceptance quantified in relation to alternatives;
- Addressing open questions remaining with respect to the relevance of many priority pollutants which are now regarded as a potential barrier to further implementation of water reuse or may lead to over-engineered systems;
- More research on the toxicological evaluation of these compounds and fast-track testing options;
- More research on the identification of the attenuation capacity of the saturated and unsaturated zone for different Geo-hydrological conditions which could not be covered in this particular project;
- Establishment of a best available technologies reference document for MAR with reclaimed water would support decision makers and authorities, which could be developed in the context of the Common Implementation Strategy (WFD) including clarifications on suitable methodologies for selection and implementation of measures such as aquifer recharge.
- Economic evaluation and comparison of different MAR methods and treatment trains for various end uses of the recovered water.

Data base of international projects.

- Integration of research results from national and international projects and closely link with stakeholders. This may include information on geochemical reactions, fate of pathogens and organics, with a view to building models to better predict quality of recovered water related to source water quality, groundwater quality, soil and aquifer mineralogy and environmental conditions (temperature and redox state);
- Compilation of a data base with successful MAR implementations and addressing classes of project that present most opportunity and future challenges.

These recommendations strongly reinforce those of the Water Supply and Sanitation Technology Platform Task Force on Managed Aquifer Recharge (2009) in its report for the European Water Industry on the State-of-the-Art and Research Needs.

REFERENCES

Asano T. and Cotruvo J. A. (2004). Groundwater recharge with reclaimed municipal wastewater: health and regulatory considerations. *Water Research*, **38**(8), 1941–1951.

Asano T., Burton F. L., Leverenz H. L., Tsuchihashi R. and Tschobanoglous G. (2007). Water reuse: issues, technologies and applications. McGraw Hill, New York.

Anderson J. (2003). The environmental benefits of water recycling and reuse. *Water Supply*, **3**(4), 1–10.

Bixio D., de Heyder B., Cikurel H., Muston M., Miska V., Joksimovic D., Schäfer A. I., Ravazzini A., Aharoni A., Savic D. and Thoeye C. (2005). Wat.Sci.Tech. *Water Supply*, **5**(1), 77–85.

Bixio D., Thoeye C., De Koning J., Joksimovic D., Savic D., Wintgens T. and Melin T. (2006). Wastewater reuse in Europe. *Desalination*, **187**, 89–101.

Bixio D., Van Houtte E., Thoeye C., Van De Steene B., Verbauwhede J., Wintgens T. and Melin T. (2008). Water Safety Plan of an Artificial Groundwater Recharge Site with Reclaimed Water for Drinking Water Production 6th IWA World Water Congress and Exhibition, 06–12 September 2008, Vienna, Austria.

Bouwer H. (2002). Artificial recharge of groundwater. *Hydrogeology Journal*, (10), 121–142.

California Water Code, Division 7, Chapter 7.3. Direct and indirect potable reuse, http://www.leginfo.ca.gov/cgi-bin/calawquery?codesection=wat&codebody=&hits=20 (last accessed May 17, 2011).

CCR (2009). California Code of Regulation, Title 22, Title 22. Social Security, Division 4. Environmental Health, Chapter 3. Water Recycling Criteria, Article 5.1. Groundwater Recharge, § 60320. Groundwater Recharge. Available at: http://weblinks.westlaw.com/result/default.aspx?cite=22CAADCS60320&db=1000937&findtype=L&fn=%5Ftop&pbc=DA010192&rlt=CLID%5FFQRLT29121161251 75&rp=%2FSearch%2Fdefault%2Ewl&rs=WEBL11%2E04&service=Find&spa=CCR%2D1000&sr=TC&vr=2%2E0 (accessed May 17, 2011).

CDPH (2007). Drinking Water Notification Levels and Response Levels: An Overview California Department of Public Health.

CDPH (2008). Draft Groundwater Recharge Reuse Regulation, August 5, 2008; Available at http://www.cdph.ca.gov/HealthInfo/environhealth/water/Pages/Waterrecycling.aspx (accessed May 2011).

Dillon P., Ward J. and Cunliffe D. (2007). Innovation in governance of managed aquifer recharge. *Proceedings of the 6th Symposium on managed artificial recharge of groundwater*, ISMAR6, 28 October – 2 November 2007, Phoenix, Arizona, USA.

Dillon P. J. and Jiménez B. (2008). Water reuse via aquifer recharge, intentional and unintentional practices. In: Water Reuse: an International survey of current practice, issues and needs by B. Jiménez and Takashi Asano. ISBN 1843390892. IWA Publishing, London, UK.

Dillon P., Pavelic P., Toze P., Rinck-Pfeiffer S., Martin R., Knapton A. and Pidsley D. (2006). Role of aquifer storage in water reuse. *Desalination*, **188**(1–3), 123–134. *Proc. Intl. Conf. Integrated Concepts in Water Recycling*, Wollongong, 14–17 Feb 2005.

Dillon P. (2009). Ground water replenishment with recycled water. *An Australian Perspective*, **47**(4), 492–495, doi: 10.1111/j.1745-6584.2009.00587.x, GROUND WATER–July-August 2009.

Dillon P., Pavelic P., Page D., Beringen H. and Ward J. (2009). Managed Aquifer Recharge: An Introduction Waterlines Report Series No. 13, February 2009. The National Water Commission.

DINA-MAR (2008). Project Material Available at: http://www.dina-mar.es/Default.aspx?s=2 (project flyer and presentation at Expo Zaragossa).

EEA (2007). Climate Change and Water Adaptation Issues, Technical Report No 2/2007. European Environment Agency, Copenhagen.

EC (1998). Council Directive 98/83/EC of 3 November 1998 on the Quality of Water Intended for Human Consumption.

EC (2006). Water Reuse System Management Manual – AQUAREC; Davide Bixio and Thomas Wintgens (eds); Luxembourg: Office for Official Publications of the European Communities, ISBN 92-79-01934-1.

EC (2004). Groundwater Risk Assessment – Technical Report on Groundwater Risk Assessment Issues as Discussed at the Workshop of 28th January 2004.

EC (2007). Guidance Document n°17 Guidance on the Application of the Term 'Direct and Indirect Inputs' in the Context of the Groundwater Directive 2006/118/EC, Technical Report 2007-012, European Communities. See Website: http://circa.europa.eu/Public/irc/env/wfd/library?l=/framework_directive/guidance_documents/guidance_document/_EN_1.0_&a=d

EC (2007a). Drought Management Plan Report: Including Agricultural, Drought Indicators and Climate Change Aspects Water Scarcity and Droughts Expert Network. European Communities.

EC (2007b). Communication from the Commission to the European Parliament and Teh Council: Addressing the Challenge of Water Scarcity and Droughts in the European Union, COM(2007) 414 final, Commission of the European Communities.

EC (2008). Directive 2008/105/EC of the European Parliament and of the Council of 16 December 2008 on Environmental Quality Standards in the Field of Water Policy, Amending and Subsequently Repealing Council Directives 82/176/EEC, 83/513/EEC, 84/156/EEC, 84/491/EEC, 86/280/EEC and Amending Directive 2000/60/EC of the European Parliament and of the Council.

EC (2009). Guidance on Groundwater Status and Trend Assessment, CIS Working Group on Groundwater Guidance Document No. 18, Technical Report - 2009 - 026.

EPA (2008). Emerging Contaminant - N-Nitroso-dimethylamine (NDMA) - Fact Sheet, April 2008 United States Environmental Protection Agency.

Hochstrat R., Wintgens T., Melin T. and Jeffrey P. (2006). Assessing the European wastewater reclamation and reuse potential — a scenario analysis. *Desalination*, **188**, 1–8.

Johnson T. A. (2009). Ground water recharge using recycled municipal waste water in Los Angeles country and the California department of public health's draft regulations on aquifer retention time. **47**(4), 496–499, GROUND WATER–July-August 2009.

IAWR (2008). Danube, Meuse and Rhine MEMORANDUM 2008, International Association of Waterworks in the Rhine Catchment Area.

International Water & Irrigation Review (1999). **19**(1), ISSN 0334–5807.

Le Corre K., Aharoni A., Ayuso Gabella M. N., Cauwenberghs J., Chavez A., Cikurel H., Genthe B., Jimenez B., Masciopinto C., Rinck-Pfeiffer S., Regel R., Salgot M., Tredoux G., Zhao X., Xuzhou C., Cartmell E., Jefferson B. and Jeffrey P. (2007). Managed Aquifer Recharge with Reclaimed Wastewater: a Comparative Study of Eight Operational Sites. *Proceedings of IWA Specialist Conference on Wastewater Reclamation and Reuse for Sustainability*, 9–12 October 2007, Antwerp, Belgium.

MED WWR WG (2007). Mediterranean Wastewater Reuse Report, Produced by the Mediterranean Wastewater Reuse Working Group. Available Online at http://www.emwis.net/topics/WaterReuse

NRC (2008). Prospects for Managed Underground Storage of Recoverable Water, Committee on Sustainable Underground Storage of Recoverable Water, National Research Council of the National Academies. The National Academies Press, Washington, D.C. Available Online at: http://www.nap.edu/catalog.php?record_id=12057

NRMMC–EPHC–AHMC (2006). Australian Guidelines for Water Recycling (Phase 1): Managing Health and Environmental Risks. Natural Resources Management Ministerial Council, Environment Protection and Heritage Council and Australian Health Ministers' Conference, Canberra. www.ephc.gov.au/taxonomy/term/39

NRMMC–EPHC–NHMRC (2008). Australian Guidelines for Water Recycling (Phase 2): Augmentation of Drinking Water Supplies. Natural Resource Management Ministerial Council, Environment Protection and Heritage Council and National Health and Medical Research Council, Canberra. www.ephc.gov.au/taxonomy/term/39

NRMMC–EPHC–NHMRC (2009a). Australian Guidelines for Water Recycling (Phase 2): Stormwater Harvesting and Reuse. Natural Resource Management Ministerial Council, Environment Protection and Heritage Council and National Health and Medical Research Council, Canberra. www.ephc.gov.au/taxonomy/term/39

NRMMC–EPHC–NHMRC (2009b). Australian Guidelines for Water Recycling (Phase 2): Managed Aquifer Recharge. Natural Resource Management Ministerial Council, Environment Protection and Heritage Council and National Health and Medical Research Council, Canberra. www.ephc.gov.au/taxonomy/term/39

Page D., Dillon P., Vanderzalm J., Bekele E., Barry K., Miotlinski K. and Levett K. (2010). Managed Aquifer Recharge Case Study Risk Assessments. CSIRO Water for a Healthy Country Flagship Report, 144. Dec 2010, http://www.clw.csiro.au/publications/waterforahealthycountry/2010/wfhc-MAR-case-study-risk-assessments.pdf

Salgot M., Campos C., Galofré B. and Tapias J. C. (2001). Biological control tools for wastewater reclamation and reuse, a critical review. *Water Science and Technology*, **43**(10), 195–201.

Snyder S. A., Adham S., Redding A. M., Cannon F. S., DeCarolis J., Oppenheimer J., Wert C. and Yoon Y. (2007). Role of membranes and activated carbon in the removal of endocrine disruptors and pharmaceuticals. *Desalination*, **202**, 156–181.

Steenbergen F., van and Tuinhof A. (2009). Managing the Water Buffer for Development and Climate Change Adaptation: Groundwater Recharge, Retention, Reuse and Rainwater Storage. MetaMeta Communications, Wageningen, The Netherlands.

Ternes T. A. and Joss A. (2006). Human Pharmaceuticals, Hormones and Fragances – The Challenge of Micropollutants in Urban Water Management. ISBN 1843390930, IWA Publishing, London.

Ternes T. A., Bonerz M., Herrmann N., Teiser B. and Andersen H. R. (2007). Irrigation of treated wastewater in Braunschweig, Germany: an option to remove pharmaceuticals and musk fragrances. *Chemosphere*, **66**(2007), 894–904.

UNESCO IHP (2005). Strategies for Managed Aquifer Recharge (MAR). In: semi-arid areas (ed.), Ian Gale.

Walker T., Roux A. and Owens E. (2007) Western corridor recycled water project - the largest recycled water scheme in the southern hemisphere. In: Water Reuse and Recycling, S. J. Khan, R. M. Stuetz and J. M. Anderson (eds), UNSW Publishing, Sydney, pp. 498–511.

Water Supply and Sanitation Technology Platform Task Force "Managed Aquifer Recharge" (2009). State-of-the-Art and Research Needs. November 2009.

WHO (2003). State of the art report - health risks. In: Aquifer Recharge Using Reclaimed Water, R. Aertgeerts and A. Angelakis (eds), Water, Sanitation and Health Protection and the Human Environment World Health Organization, Geneva.

WHO (2005). Water Safety Plans: Managing drinking-water quality from catchment to consumer; Water, Sanitation and Health Protection and the Human Environment, World Health Organization, Geneva.

WHO (2007). Support of the development of a framework for the Implementation of Water Safety Plans in the European Union.

WHO (2008). Guidelines for Drinking-Water Quality. 3rd edn, Including 1st and 2nd Addenda, World Health Organization, Geneva.

Lightning Source UK Ltd.
Milton Keynes UK
UKOW021915261112

202813UK00002B/10/P

9 781843 393443